D0706907

HEAT PUMPS

HEAT PUMPS
Theory and Service

Lee Miles

I(T)P™

NOTICE TO THE READER

Publisher does not warrant or guarantee any of the products described herein or perform any independent analysis in connection with any of the product information contained herein. Publisher does not assume, and expressly disclaims, any obligation to obtain and include information other than that provided to it by the manufacturer.

The reader is expressly warned to consider and adopt all safety precautions that might be indicated by the activities described herein and to avoid all potential hazards. By following the instructions contained herein, the reader willingly assumes all risks in connection with such instructions.

The publisher makes no representations or warranties of any kind, including but not limited to, the warranties of fitness for particular purpose or merchantability, nor are any such representations implied with respect to the material set forth herein, and the publisher takes no responsibility with respect to such material. The publisher shall not be liable for any special, consequential or exemplary damages resulting, in whole or in part, from the readers' use of, or reliance upon, this material.

Cover Photo: Inter-city Products Corporation (USA)

Delmar Staff
Senior Administrative Editor: Vernon Anthony
Editorial Assistant: Pat Konczeski
Project Editor: Eleanor Isenhart
Production Coordinator: Dianne Jensis
Design Coordinator: Cheri Plasse
Art Coordinator: Megan DeSantis

For information, address Delmar Publishers Inc.,
3 Columbia Circle Drive, Box 15-015
Albany, New York 12212-5015

Printed in the United States of America
Published simultaneously in Canada
by Nelson Canada
A Division of The Thomson Corporation

10 9 8 7 6 5 4 3 2 1 XXX 99 98 97 96 95 94

Library of Congress Cataloging-in-Publication Data

Miles, Lee.
Heat pumps: theory and service/Lee Miles.
p. cm.
Includes index.
ISBN 0-8273-4956-4 (textbook)
1. Heat pumps. I. Title
TH7638.M55 1994
621.402'5--dc20 92-42119
CIP

BRIEF CONTENTS

CONTENTS

PREFACE

The heat pump has developed in the last 25 years into a sophisticated system designed to do a specific task. The first heat pumps were intended for human comfort conditioning primarily in the residential field. These used the outside air as a source of energy for the winter heating cycle and as a heat sink for summer air conditioning. Since the introduction of the designed heat pump in the late 1960s, a variety of equipment has been designed and marketed for specific applications.

With the rapid rise in the design, manufacture, and sale of heat pumps of all varieties, a need has developed for the gathering of information into book form. The aim of this book is to cover the four types of heat pump systems in a general manner, outlining that information that is common to the various manufacturers' products in each category.

The aim of this book is to provide background information to assist the reader in understanding the operating principles and language of the heat pump industry. This will assist the reader in the understanding of performance as well as possible problem diagnosis in specific equipment.

The reader should be aware of the fact that not all manufacturers' equipment can be discussed in a textbook. An attempt is made, however, to present as much information of a general nature as possible in each system category. Therefore, it is recommended that specific product information and installation and service material be obtained from the manufacturer of the product involved.

The information in this book is not possible without the generous contribution of the companies that have developed and marketed the products discussed. My appreciation goes to the following who generously provided illustrations and printed material:

Addison Products Co., Addison, MI, 49220
Air Conditioning Contractors of America (ACCA), Washington, D. C., 20036
Amana Refrigeration, Inc., Amana, IA, 52204
Bacharach Instrument Co., Pittsburgh, PA, 15238
Bard Manufacturing Co., Bryan, OH, 43506
Blue-White Industries, Inc., Westminster, CA, 92683
Borg-Warner Central Environmental Systems, Inc., York, PA, 17405
Climate-Master, Oklahoma City, OK, 73125
The Coleman Company, Inc., Wichita, KS, 67219
Command-Aire Corp., Waco, TX, 76714
D.E.C. International, Therma-Stor Products Group, Madison, WI, 53708
Drake Industries, Inc., Port St. Lucie, FL, 34986
Dwyer Instruments, Inc., Michigan City, ID, 46360
E-TECH Inc., Atlanta, GA, 30341
EXTECH Instruments Corp., Waltham, MA, 02154
Goulds Pumps Inc., Seneca Falls, NY 13148
Ray Leonard Well Drilling, Battle Creek, MI, 49017
Mammoth, A Nortek Company, Minneapolis, MN, 55441

March Manufacturing Company, Glenview, IL, 60025
Marshalltown Instruments Inc., DESCO, Marion, OH 43302
New England Valve Co., Middleton, CT 06457
Orangeburg Industries, Inc., Ashville, NC 28804
Refrigeration Components Group, Parker-Hannefin Corp., Broadview, IL, 60153
P S G Industries, Inc. Perkasie, PA, 18944
Ranco Controls, Ranco North America, Plains City, OH, 43206
Robinaire Div., S P X Corporation, Montpelier, OH, 43543
Shortridge Instruments Inc., Scottsdale, AZ 85260
Sporlan Valve Co., St. Louis, MO, 63143
Taco Inc., Cranston, RI, 02920
Tecumseh Products Co., Tecumseh, MI, 49286
Vanguard Plastics, Inc., McPherson, KS, 67460
The Whalen Co., Easton, MD, 21601
White Rodgers Div., Emerson Electric Co., St. Louis, MO, 63123

Special appreciation also goes to the following people who not only provided and approved the use of their personally published material, but also reviewed copy for accuracy and updating:

Addison Products Co.: James Kirkland, Training Director; Ray Orfait, National Sales Manager
Bard Manufacturing Co.: John Briggs, Director of Purchasing and Production Scheduling;
Fred Paepke, Training Director
Blue-White Industries, Inc.: Robert E. Gledhill, Marketing Manger
D E C International, Therma-Stor Products Group: Ken Gehring, Vice-President
Ray Leonard Well Drilling: Ray Leonard, Owner
Vanguard Plastics, Inc.: Allen Skouby, Market Specialist

Finally, thanks to the following reviewers for the final review of the manuscript.

James Bussey
Griffin Technical Institute
Griffin, GA

Patrick Dennis
Northern Virginia Community College
Woodbridge, VA

Clyde Perry
Gateway Community College
Phoenix, AZ

Henry Puzio
Lincoln Technical Institute
Union, NJ

James Roberts
Tarrant County Jr. College
Ft. Worth, TX

Greg Skudlarek
Minneapolis Technical College
Minneapolis, MN

Darius Spence
Northern Virginia Community College
Woodbridge, VA

Bill Wyatt
Morris County Vocational School
Denville, NJ

Section ONE
General

BASIC PRINCIPLES 1

1–1 GENERAL

The heat pump, which is the same as an air conditioning system, uses the change in state of liquid (refrigerant) to absorb heat from one place (the heat source) and put it in another place (the heat sink). This is done by causing a liquid to boil at a temperature far enough below the temperature of the heat source to obtain the desired heat quantity transfer rate. At the same time, the heat energy picked up by the system, as well as the energy required to do the work, has to be disposed of at a higher temperature than the place the heat is intended for (the heat sink).

The basic principle, as in the refrigeration or air conditioning system, is the control of the boiling point and condensing temperature of the refrigerant. This is accomplished by use of pressure-reducing devices for feeding the refrigerant into an evaporator coil and compressors for raising the condensing temperatures above the temperature of the heat sink used for disposing of the heat. The only change in thinking required to understand the heat pump is that the system attempts to air condition or cool a separate source of heat and disposes of the heat into the occupied or conditioned area.

Basically, the heat pump in the heating cycle is an air conditioning unit operating in reverse, taking heat from an outside source and putting it into the occupied area. With this type of operation, the heat energy that is produced by the electrical energy supplied to the unit is converted into mechanical energy and finally into heat energy. The heat energy is added to the heat picked up from the outside source and the entire amount is transferred to the occupied area. In the heating mode, the entire heat quantity is used to condition the occupied area. When the heat pump is in the cooling mode (air conditioning), the heat picked up from the occupied area and the electrical energy are ejected to the outside heat sink.

The heat pump is rated, in the heat mode, by the total amount of heat produced compared to the heat energy equivalent of the electrical energy used to perform the operation. This ratio of heat output to electrical input in btuh is called *Coefficience of Performance* (COP).

The heat source for the heat pump system in the heating mode may be of many sources—outside air, water supply, ground or earth, or the sources of heat such as material dumps. The only requirement is a constant source of heat energy at a rate sufficient to satisfy the requirements of the system. The use of various sources of heat has produced two categories of systems—Air Source and Water Source.

Heat pumps are also used to control the temperature of liquids for process work. Thus, a separate category of heat pumps has been developed for application in Hydronic systems.

A third factor that has been introduced in this field is the single-usage system. This is designed to operate as a heating-only system. This compares to an air conditioning system with the condensing unit in the occupied area and the evaporator/air handling system outside.

The general categories of equipment marketed today fall into the following categories:

1. Air-to-air
2. Liquid-to-air
3. Air-to-liquid
4. Liquid-to-liquid

Specific information on operating characteristics will be given in the sections covering each category of equipment.

1–2 REFRIGERATION CYCLE—AIR HEAT SOURCE

As previously stated, a refrigeration or air conditioning system transfers heat from one place to another by the change in state of a liquid. Liquid refrigerant, at a pressure high enough to keep the boiling point higher than the temperature surrounding the line carrying the refrigerant, is forced through a pressure-reducing device and then into a heat exchanger coil. Two coils are used in the unit—an outdoor coil and an indoor coil. Each coil acts as an evaporator or condenser, depending upon the mode of operation of the system. As the refrigerant passes through the pressure-reducing device, the pressure and corresponding boiling point of the liquid is reduced. The desired pressure and boiling point are far enough below the temperature of the product to be cooled in the heat exchanger to get the desired heat transfer rate from the product to the boiling refrigerant. When the boiling refrigerant absorbs heat, it becomes vapor. It then passes out of the heat exchanger to begin the reclamation process.

The first step in this process is raising the pressure on the vapor and the corresponding condensing temperature. This is done in the compressor or refrigerant pump. The high-pressure vapor is fed from the compressor into another heat exchanger. Here the hot high-pressure vapor is cooled below its condensing temperature and condenses back to the original high-pressure liquid. The process is a continuous one as long as the compressor operates to maintain the pressure difference between the high and low sides of the system.

Figure 1–1 shows the refrigerant flow through the pressure-reducing device into the heat exchanger where heat is absorbed (the evaporator). From the evaporator, the vapor passes through the suction line to the compressor. Here the pressure increases or the compression process occurs. The high-pressure vapor then passes through the discharge line to the heat exchanger where heat is given off (the condenser).

In the condenser, the refrigerant changes to a liquid by being cooled below the condensing temperature of the high-pressure vapor. The liquid refrigerant is cooled below the condensing temperature of the refrigerant or is *subcooled.* This is desired to ensure the refrigerant will remain in a liquid state as it flows through the liquid line to the pressure-reducing device.

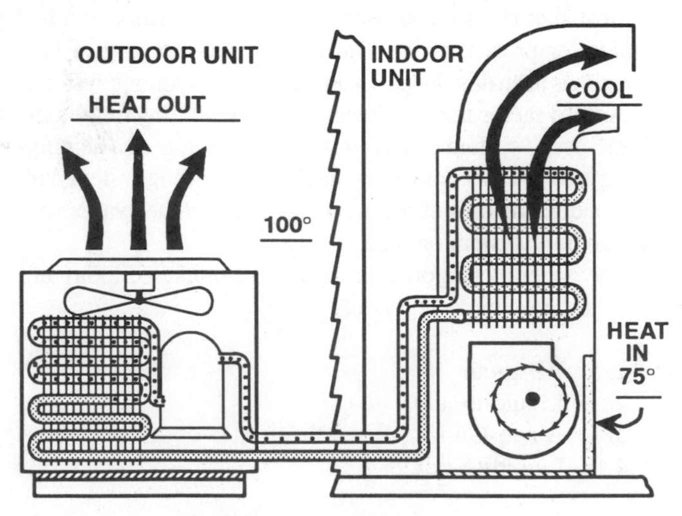

Figure 1–1 Standard air conditioning circuit (Courtesy Addison Products Co.)

The system in Figure 1–1 is capable of moving heat in only one direction, from the evaporator to the condenser. This circuit would be used in an air-to-air, heating-only-type heat pump system. When both heating and cooling are desired from one system, a means of reversing the refrigerant flow has to be provided. It is not possible to reverse the action of the compressor. Therefore, it is necessary to change the connections to the suction and discharge ports of the compressor to direct the refrigerant flow to and from the desired heat exchangers. This is done by adding a four-way reversing valve into the system.

Figure 1–2 shows the basic refrigeration circuit with the four-way reversing valve added and operating in the cooling mode. The indoor coil is the evaporator and the outdoor coil is the condenser. The heat that is picked up in the indoor coil (the evaporator) plus the heat energy of the electrical energy needed to drive the compressor flows to the outdoor coil (the condenser) where the total heat quantity is extracted to the outdoor air. Vapor from the indoor coil flows into the suction side of the compressor and vapor from the discharge side of the compressor flows to the outdoor coil.

Figure 1–3 shows the same system with the reversing valve in the heating position. The vapor is from the outdoor coil (the evaporator) to the suction side of the compressor and from the discharge side of the compressor to the indoor coil (the condenser). The heat into the conditioned area now consists of the heat picked up from the outdoor air plus the heat energy equal to the electrical energy used to drive the compressor. The major advantage of this operation is that the heat pump will pick up from 2 1/2 to 3 1/2 times as much heat from the outdoor air as compared to the amount of heat produced by the electrical energy used. A heat pump will produce 2 1/2 to 3 1/2 times the heat per kilowatt used as compared to resistance heat. This fact is called the Coefficience of Performance (COP).

Due to the fact that the system operates at different temperature conditions in the heating and cooling modes, different operating pressures must be produced. In Figures 1–2 and 1–3, two pressure-reducing devices are in the circuit. Each coil has its matching pressure-reducing device to be used when the coil is the evaporator.

Check Valves

To remove the pressure-reducing device out of the circuit when the coil is used as a condenser, a check valve is installed around the pressure-reducing device. With the system in the heating mode, the indoor heat exchanger is the condenser and its matching pressure-reducing device is not used. The liquid refrigerant flows through the parallel check valve circuit around the pressure-reducing device into the liquid line. When the liquid refrigerant reaches the outdoor coil (the

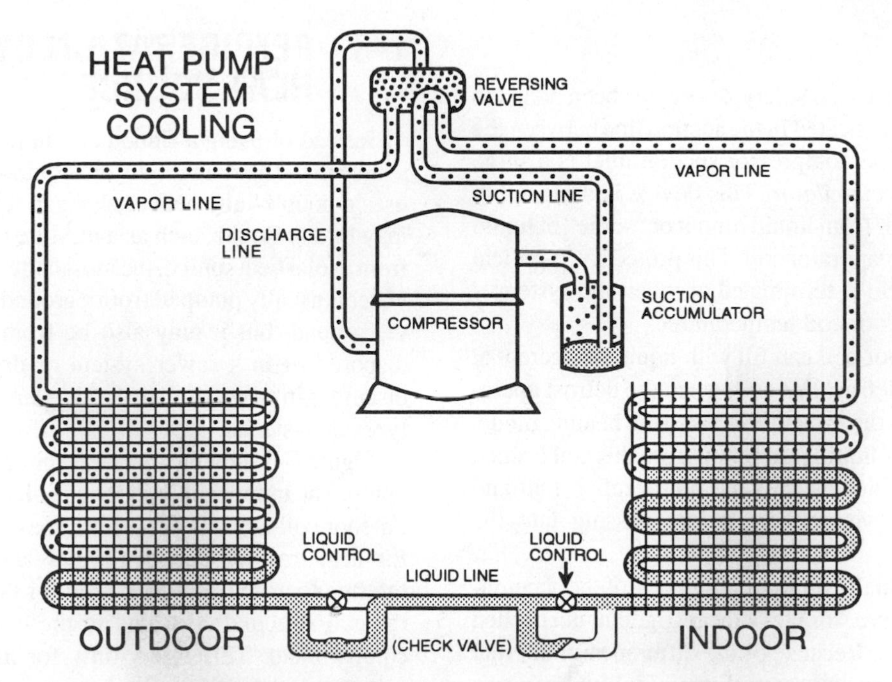

Figure 1–2 Reversing valve added—cooling mode (Courtesy Addison Products Co.)

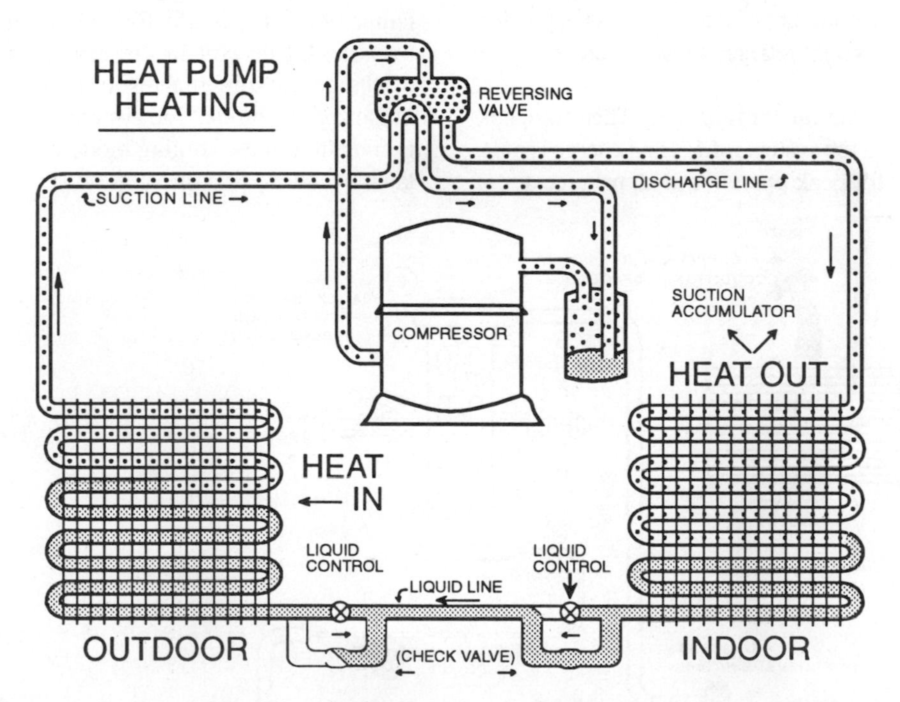

Figure 1–3 Reversing valve added—heating mode (Courtesy Addison Products Co.)

evaporator), the check valve closes and forces the liquid refrigerant through the pressure-reducing device. The desired reduction in pressure and boiling point is accomplished.

In the cooling mode, the outdoor coil is the condenser. Therefore, its matching pressure-reducing device is not needed. The liquid refrigerant flows through the check valve, around the pressure-reducing device, through the liquid line to the indoor coil's pressure-reducing device. Here the check valve is closed, forcing the liquid refrigerant to flow through the pressure-reducing device. The desired reduction in pressure and boiling point of the liquid refrigerant is accomplished, heat is absorbed and the air through the coil (the evaporator) is cooled. From this you can see that the check valves are a vital part of the circuit to get the desired results.

Accumulator

In Figures 1–2 and 1–3, a safety device has been added to the refrigerant circuit. Located in the suction line between the reversing valve and the compressor suction inlet is a surge chamber called an *accumulator*. This device is designed to protect the compressor from liquid runout or "surge" of liquid refrigerant from the evaporator coil. This protection is critical when the defrost cycle is terminated or when the system is operating at low outdoor coil temperatures.

Because the outdoor coil can fill with liquid refrigerant at high temperature to defrost this coil, when the defrost operation is completed and the system returns to the heating mode, an extensive runout of liquid refrigerant from this coil occurs. The suction accumulator catches the liquid refrigerant and ensures that it is in vapor form before passing into the compressor.

The second and equally important function of the accumulator is to act as a reserve storage of the refrigerant not needed in the operating mode. Because of the difference in internal volume of each heat exchanger coil, the required operating quantity of refrigerant charge changes with the operating mode as well as the outside ambient. During the period of operation when the lesser amount of refrigerant is needed for proper operation, the excess refrigerant will gather in the accumulator.

The capacity of the accumulator is limited. Therefore, it is extremely important that the system refrigerant charge is only the amount necessary for peak operation with no excess.

1–3 REFRIGERATION CYCLE—LIQUID HEAT SOURCE

Instead of using a finned coil in the outdoor air as a heat source in the heating mode, the liquid-heat-source heat pump uses a double-tube heat exchanger. Though the heat source may be any liquid, such as a mixture of antifreeze and water from a solar heat source, the most common liquid is water. The water is usually pumped from a ground source and returned to the ground, but it may also be from a central supply and disposed of in a sewer system or dry well. These will be discussed in more detail in the chapters covering the particular types of systems.

Figure 1–4 shows a refrigerant circuit of a liquid-to-air heat pump. The indoor coil is a finned coil for treatment of air. The outdoor coil is a tube-in-a-tube coaxial-type heat exchanger for heat transfer between water and refrigerant. Though referred to as the "outdoor" coil for the sake of explanation, the coil is actually located in the conditioned area or unit compartment. This is not only for the convenience of the manufacturer but also to reduce the possibility of freeze up.

The refrigerant circuit is less complicated in the liquid-to-air system than in the air-to-air system. Using supply water temperatures in the 45° F to 75° F range and 70° F DB air over the inside coil in the heating mode, the suction and discharge pressures that result are very close to those obtained in the cooling mode. The system is a water-cooled air conditioning unit in the cooling mode with condenser water regulated by a head pressure control.

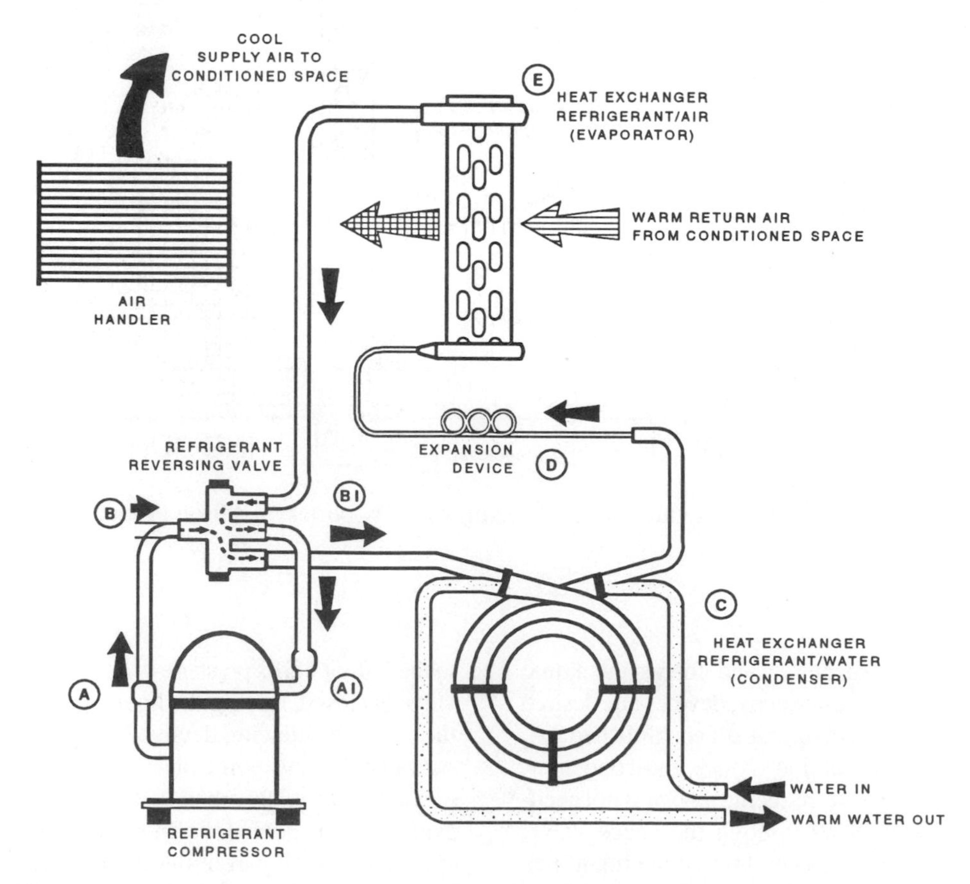

Figure 1–4 Liquid-to-air circuit—cooling mode (Courtesy Climate Master, Friedrich)

With operating suction and discharge pressures practically the same for both the heating and cooling modes, the same pressure-reducing device is often used in both the heating and cooling modes. No check valves are required for the control of the refrigerant flow. A four-way reversing valve is used to control the direction of refrigerant flow to produce the desired heating or cooling effect.

Figure 1–4 shows the system in the cooling mode. Heat is picked up in the inside coil (the evaporator) producing refrigerant vapor. The vapor travels to the reversing valve (B1) through the valve to the compressor suction intakc (A1). Compressed by the action of the compressor, the hot high-pressure vapor leaves the compressor (A) and travels to the high-pressure connection (B) of the four-way reversing valve. Continuing through the valve and out the lower port, the hot high-pressure vapor travels to the outer tube connection of the tube-in-a-tube heat exchanger. As it flows through the heat exchanger, heat is given up to the liquid flowing through the inner tube. For higher heat transfer efficiency, the refrigerant flow and the liquid flow are in opposite directions.

As the heat is removed from the hot refrigerant vapor, it is condensed to a liquid and subcooled before leaving the heat exchanger (condenser). The subcooled liquid refrigerant then flows to and through the pressure-reducing device. The pressure and boiling point of the refrigerant is lowered far enough below the temperature of the air to produce the desired rate of heat transfer. The liquid refrigerant absorbs the heat, vaporizes, and the process is repeated. Actually, the process is continuous as long as the compressor operates to maintain the difference in operating pressures.

Figure 1–5 shows the liquid-to-air unit in the heating mode. The four-way reversing valve has operated to shift the refrigerant vapor flow. The tube-in-a-tube heat exchanger is now the evaporator, absorbing heat from water in the tube. The refrigerant vapor flows to the compressor, through the lower and middle connections of the four-way reversing valve, from the middle connection to the compressor, through the compression system, and out the high-pressure connection of the four-way reversing valve. The hot high-pressure refrigerant vapor reaches the refrigerant-to-air heat exchanger. This coil, now the condenser, transfers the heat to air passing over the fins and on to the occupied area.

With the heat removed, the refrigerant vapor is condensed to a liquid and subcooled. It then flows out to the pressure-reducing device. Now flowing in the reverse direction through the pressure-reducing device, the pressure and boiling point of the liquid refrigerant is lowered far enough below the temperature of the water supply to pick up heat at the desired rate.

The air-to-air heat pump has a fixed amount of air flowing through the outdoor coil. The capacity of the unit depends upon the outdoor air temperature. The liquid-to-air unit has a regulating valve on the water supply that operates to maintain the desired refrigerant boiling point in the evaporator. This control, operating to open on pressure fall, regulates the water during the heating mode.

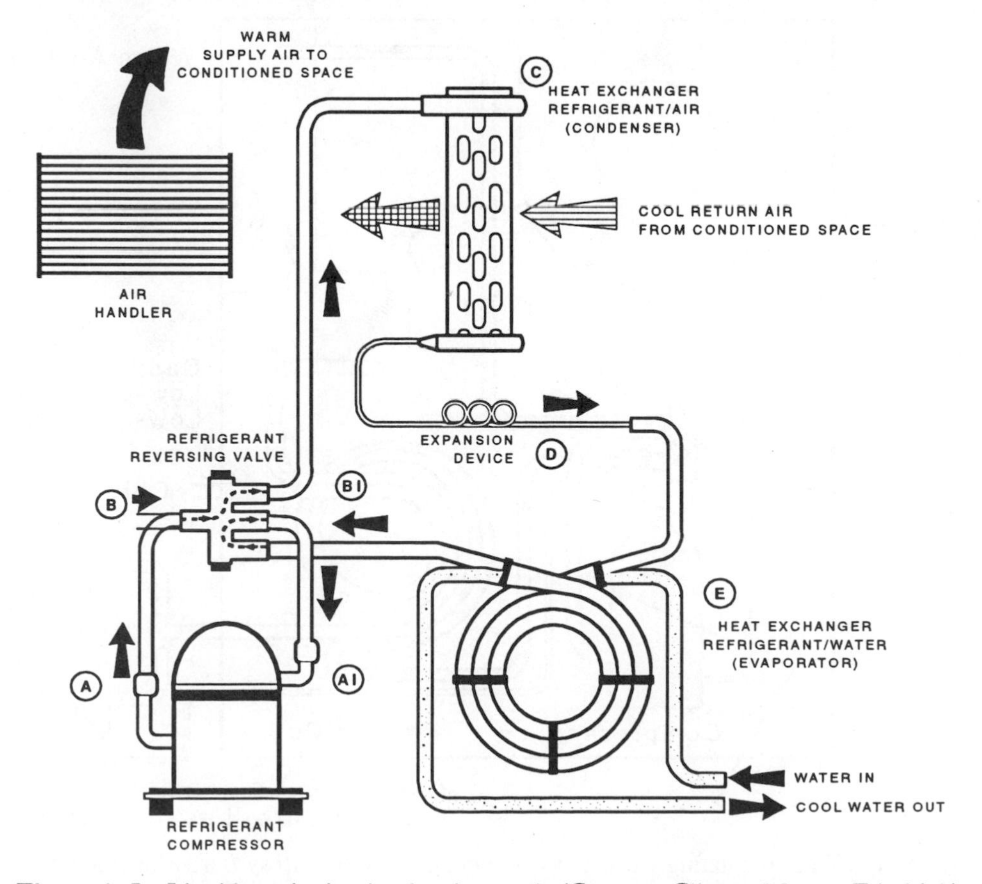

Figure 1–5 Liquid-to-air circuit—heating mode (Courtesy Climate Master, Friedrich)

A separate water flow control valve is connected in parallel with the heating flow control valve. This water flow control valve operates to control the condensing pressure and temperature during the cooling mode. This is the same as any other water-cooled air conditioning unit. The valve is set to maintain 105° F condensing temperature in the cooling mode. In the heating cycle, practically all liquid-to-air systems operate with a 45° F boiling point of the refrigerant in the evaporator.

Figure 1–6 shows a liquid-to-air heat pump unit with the water-regulating valves attached to the water supply system. Notice that the valves are attached to the outlet or downstream side of the heat exchanger. This is to prevent a drop in pressure in the water in the coil, which could cause air release from the water and a reduction in the heat transfer efficiency of the coil. Shown also are the pressure connections of the valves into the outlet of the coil refrigerant circuit ahead of the four-way reversing valve.

For the sake of clarity of artwork, the connections are shown at the bottom of the refrigerant vapor tube. The connections to a pressure control *MUST ALWAYS COME OFF THE TOP OF THE VAPOR LINE*. This reduces the possibility of oil being trapped in the line to the valve and causing delayed operation of the valve.

Figure 1–7 shows the refrigerant and water flow through the circuits of the liquid-to-air unit in the heating mode. One of the water-regulating valves opens on pressure fall to maintain an evaporator pressure of 45° F minimum in the cooling mode. The other valve operates to maintain a maximum operating pressure of 105° F condensing temperature and opens on pressure rise.

Because of the wide range of operating pressure requirements and the fact that one operates on pressure fall while the other operates on pressure rise, it is not possible for both valves to be operating at the same time. The switch over of valve operation is automatic with a no-flow period in between.

1–4 ELECTRICAL CYCLE

General

The electrical power and control systems for the air source heat pump are the most complicated. The basic control and power systems for the compressor and outdoor fan are the same as for a standard air conditioning unit. Both items are under the control of the compressor contractor, which, in turn, is controlled by the thermostat in the condition area. Additional controls are added into the heat pump circuit to provide changeover from heating to cooling and reverse as well as defrost operation of the outdoor coil when required.

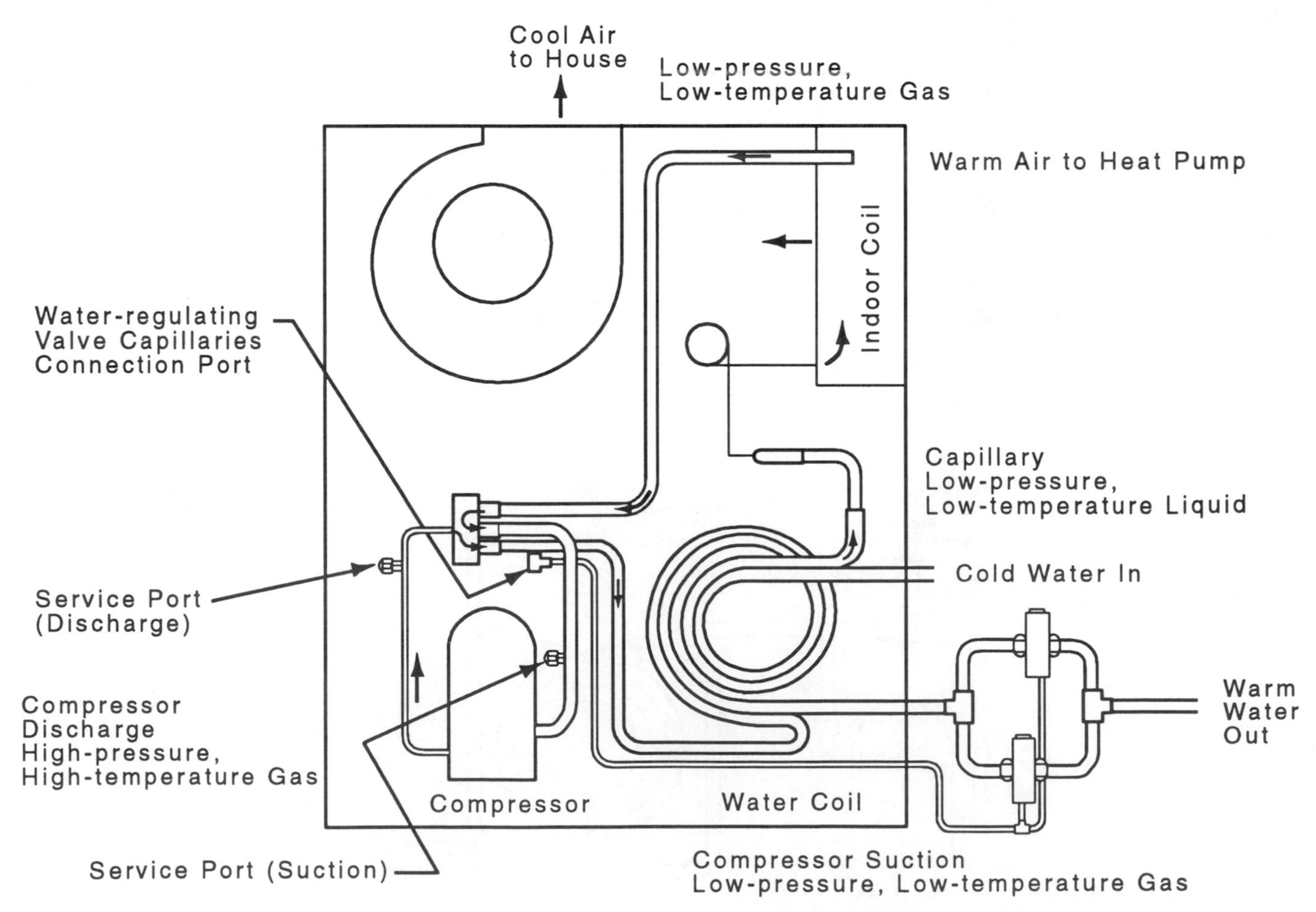

Figure 1–6 Water regulating valve added—cooling mode (Courtesy Bard Manufacturing Co.)

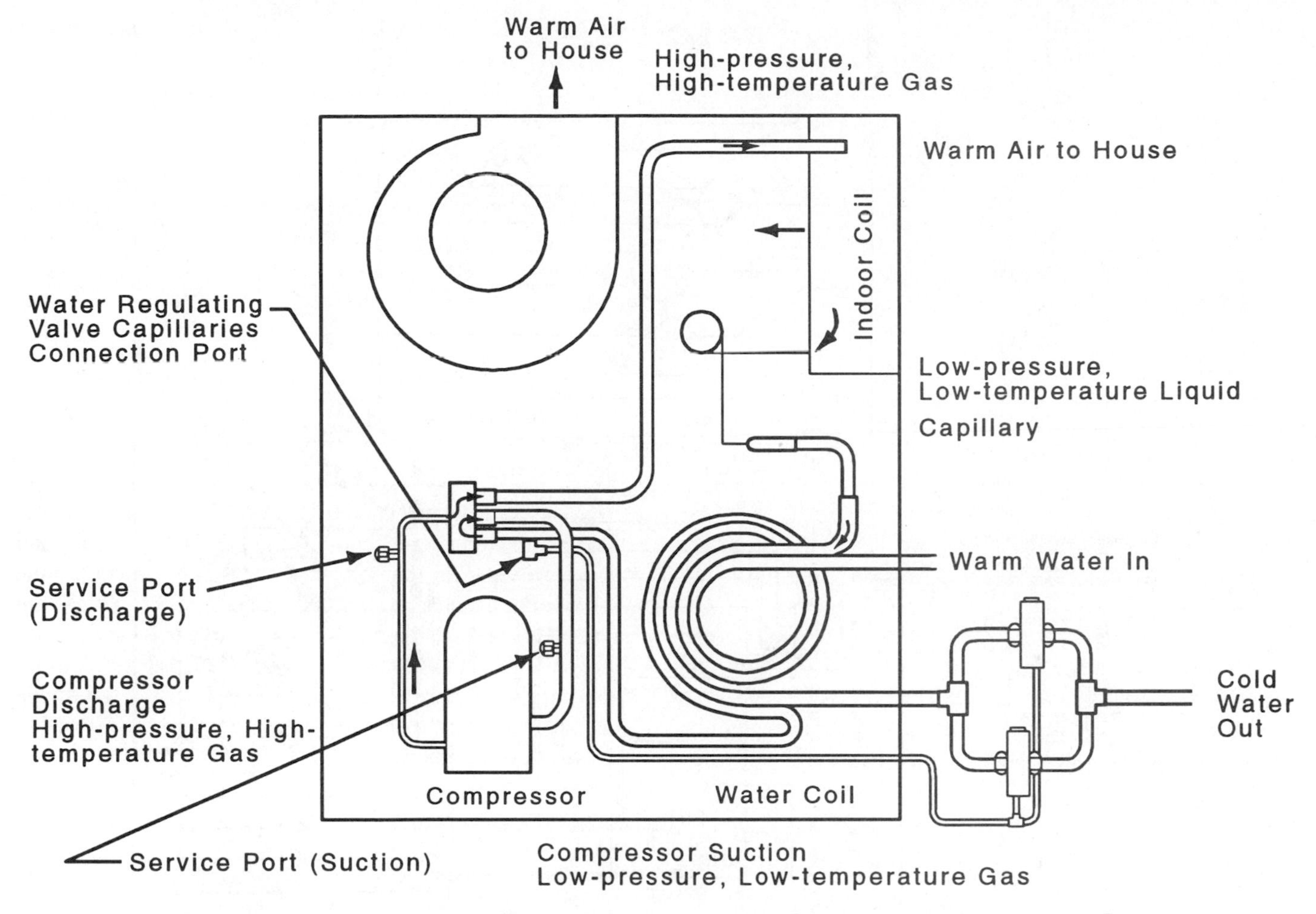

Figure 1–7 Water regulating valve added—heating mode (Courtesy Bard Manufacturing Co.)

Control Circuits

Figure 1–8 is the control circuit of an air-to-air heat pump with connected room thermostat and auxiliary heat. The basic compressor/condenser fan control circuit has been emphasized. As stated before, this could be an air conditioning unit that provides only cooling. To provide both heating and cooling, a four-way reversing valve is added.

Figure 1–9 shows the reversing valve circuit emphasized. This particular system uses a 240-volt coil in the four-way reversing valve under the control of the cooling relay contact (CR1). The cooling relay is controlled by the first stage thermostat through the cooling relay coil (CR). By this means, a choice between heating and cooling can be made by the thermostat either automatically or by the area occupant if the thermostat does not have an automatic changeover feature. This portion of the control system is also used by liquid-to-air units for switchover between heating and cooling.

An additional control system is required in air-to-air heat pumps because under weather conditions below 45° F and with high relative humidity, the outside coil can become cold enough to cause moisture to condense from the air and freeze on the coil. Since this situation cannot be prevented, it is necessary to provide some type of defrost means when the amount of frost reaches a predetermined level. The many types of defrost systems will be discussed in detail in the section in air-to-air units.

For the sake of illustration of the amount of control circuitry that is added, Figure 1–10 has the defrost control circuit emphasized. This system uses a pressure switch (D20) that closes a circuit when frost buildup on the outside coil reaches the amount to cause an air-flow resistance pressure drop sufficient to actuate the control. This action operates a relay (ADR), which operates a timer motor (TM), which in turn operates a defrost relay (DR). The defrost relay operates the reversing valve and cuts off the outdoor fan motor. This puts the system into a hot gas defrost mode. When the frost has been removed, the system is returned to the heating mode.

All defrost systems, regardless of the type of controls used, accomplish the reversal of the system to remove the frost when required. These systems will be discussed in detail in the section on air-to-air systems. The purpose of discussion here is to point out the difference in the amount of control work needed between air-to-air and liquid-to-air systems.

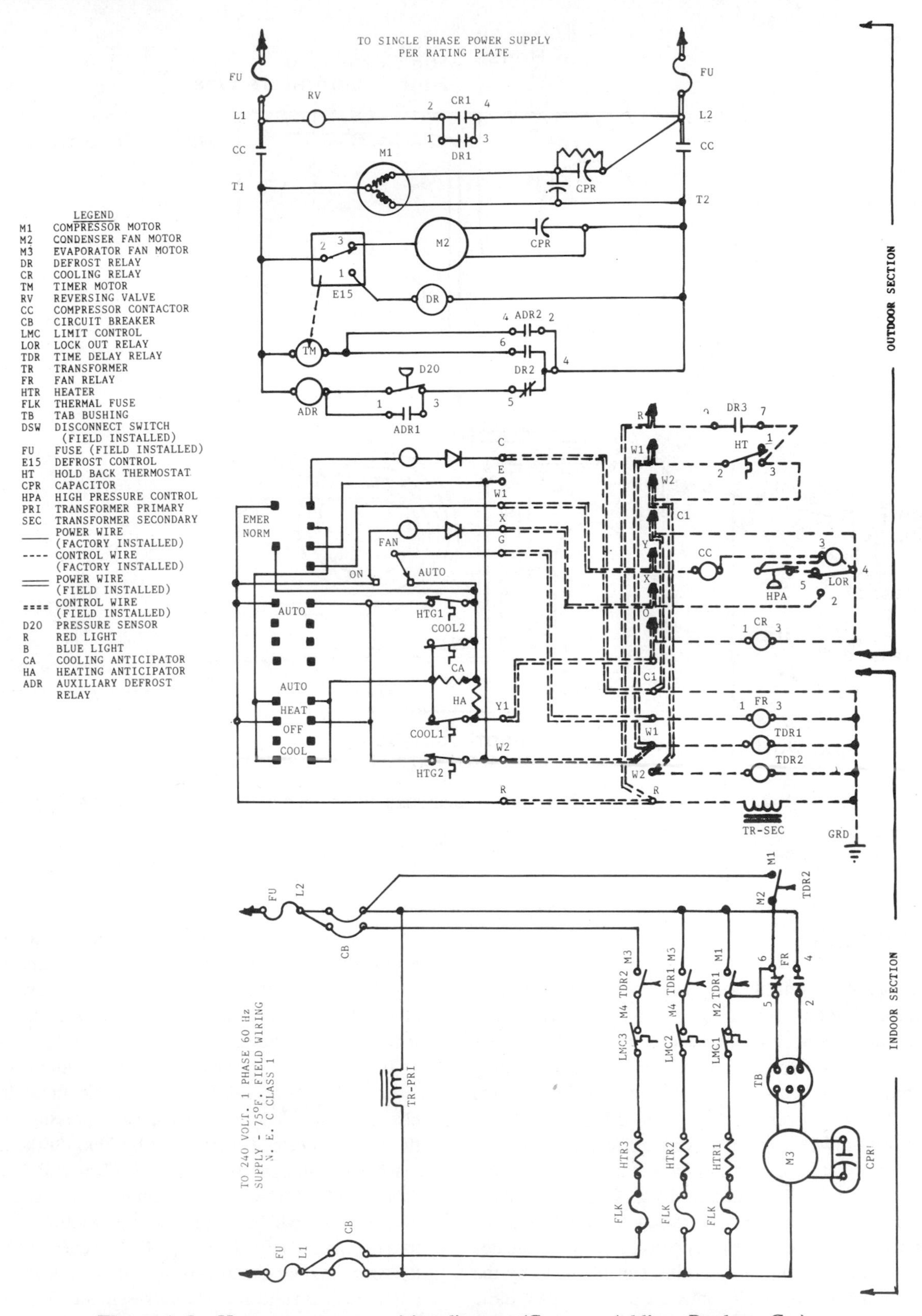

Figure 1–8 Heat pump system wiring diagram (Courtesy Addison Products Co.)

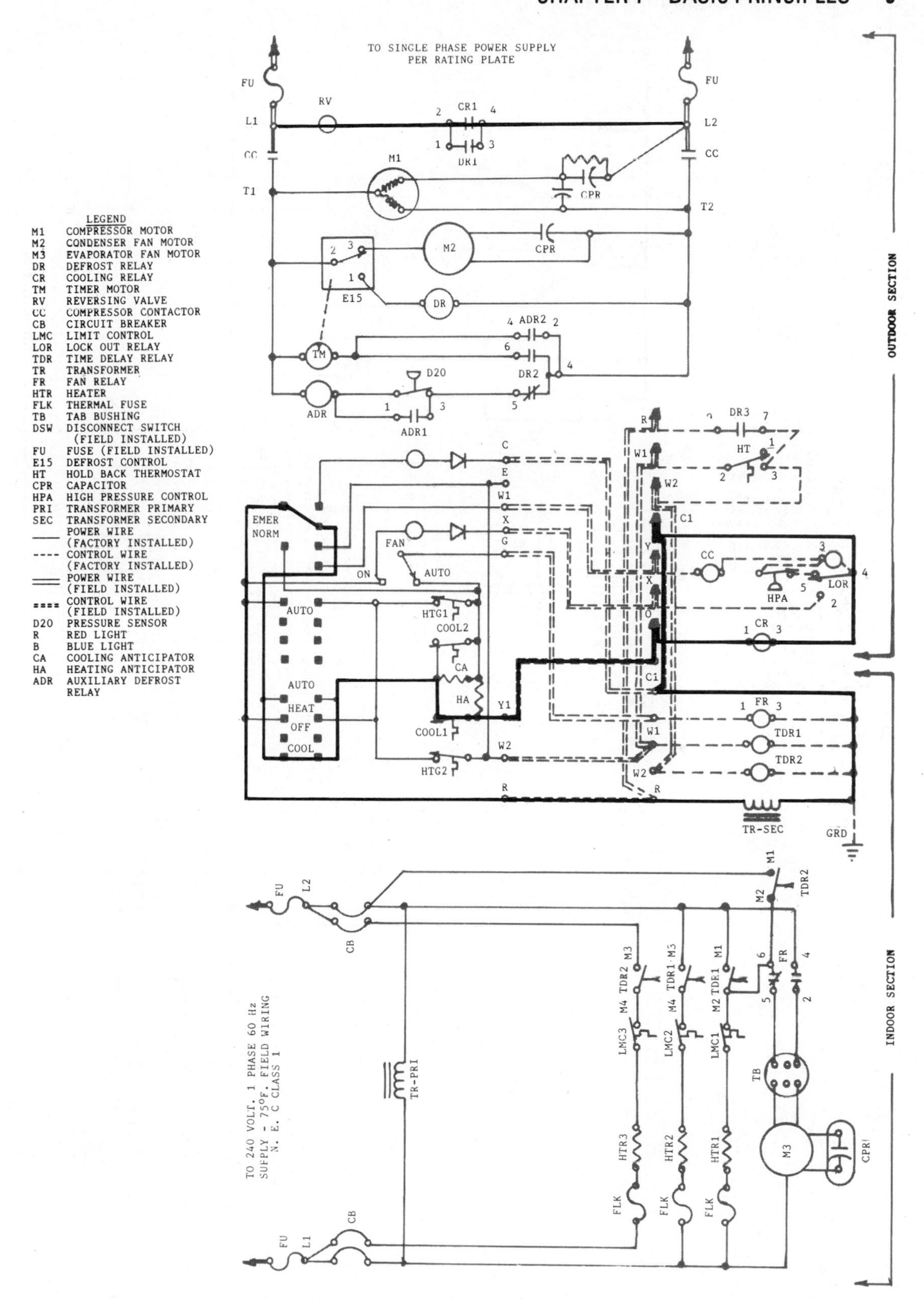

Figure 1–9 Reversing valve circuit (Courtesy Addison Products Co.)

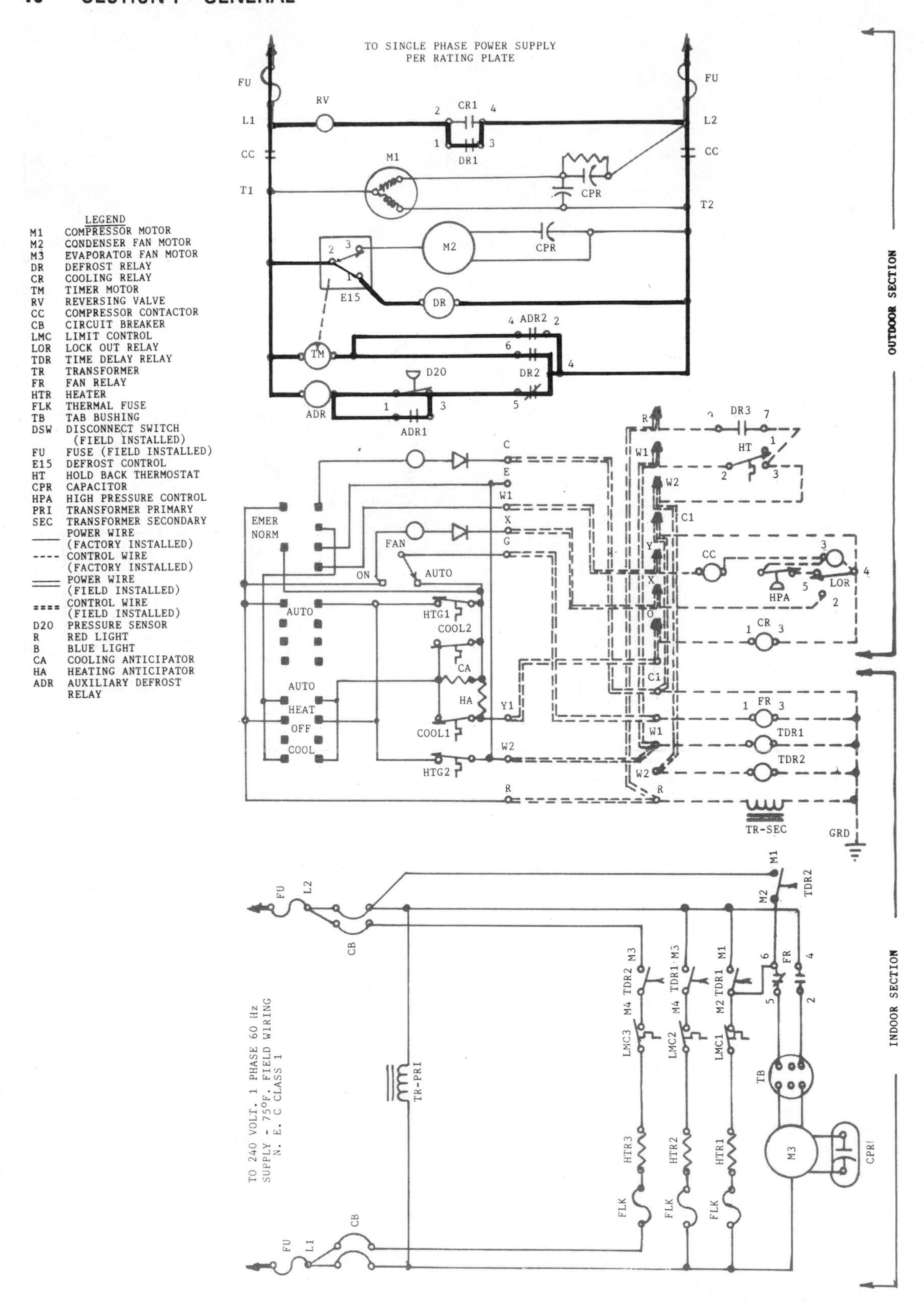

Figure 1–10 Defrost control circuit (Courtesy Addison Products Co.)

REVIEW QUESTIONS

1. The heat pump operates on a different principle than an air conditioning unit to obtain the heat needed to control the temperature of the conditioned area. True or False?
2. In an air conditioning system, the basic principle is the control of the boiling point of the refrigerant in the evaporator. In a heat pump, the basic principle is the control of the condensing temperature of the refrigerant in the inside coil (condenser). True or False?
3. A heat pump in the heating mode cools the outside atmosphere. True or False?
4. The heat picked up from the outside atmosphere represents the total heating capacity of the unit. True or False?
5. The COP of a heat pump, in the heating mode, is the total amount of heat picked up from the outside atmosphere as compared to the electrical energy in kwh used. True or False?
6. In what general categories are heat pumps produced and marketed?
7. The term *heat pump* means a unit that is capable of both heating and cooling an occupied area. True or False?
8. Subcooling of the liquid refrigerant off the condenser of an air conditioning unit is beneficial but detrimental to the heat pump system. True or False?
9. To reverse the action of the heat pump system, it is only necessary to reverse the rotation of the compressor. True or False?
10. What is the purpose of the check valve in the refrigerant circuit?
11. An accumulator is added to the system to reduce the need for a critical refrigerant charge quantity. True or False?
12. Heat pump units using liquid as the heat source are designed to use water only as the heat source. True or False?
13. Air-source heat pumps are rated at what inside and outside temperatures in the heating mode?
14. Heat pumps are rated at what inside and outside temperatures in the cooling mode?
15. The reversing valve is called a four-way valve because it has four pipe connections. True or False?
16. All liquid-to-air heat pumps are capable of operating at liquid temperatures down to 5° F. True or False?

2 TEST INSTRUMENTS

2–1 GENERAL

Many different types of test instruments are required to service heat pumps. To provide careful, accurate measurements, the instruments must be of high quality with high accuracy ratings. Since the introduction of solid state circuits into the instrument field, readings within ±0.1° F or °C are available in solid state thermometers. Direct-reading humidity meters are available that eliminate the need for the psychrometric chart. Direct-reading pH meters for water quality testing are marketed. The service technician should have the necessary test instruments, know how and when to use them, and appreciate their sensitivity with good care and maintenance.

2–2 TEMPERATURE MEASUREMENT

With the small temperature differences that develop in heat pump systems, especially in the liquid side of the liquid-to-air units, digital-type thermometers that register in the 0.1° F accuracy should be used for greatest accuracy in testing. Digital-type thermometers are available in both pocket-type and portable models.

Pocket Thermometers

An excellent pocket thermometer is the dial-type thermometer. Equipped with a carrying case and clip, it is convenient and practical for measuring temperatures in air streams or insertion in P/T plugs in the liquid system. Figure 2–1 shows a typical dial-type thermometer.

For increased accuracy, especially when measuring small temperature changes in the liquid system, the digital pocket thermometer is recommended (see Figure 2–2). This instrument is capable of measuring to 0.2° F with ±2% accuracy.

Figure 2–1 Pocket dial thermometer (Courtesy Robinaire Division, Sealed Power Corp.)

Figure 2–2 Pocket digital thermometer (Courtesy PSG Industries, Inc.)

Both instruments are available in various ranges. The most common range for use in heat pump service is the 0° F to 160° F range.

Electronic Thermometers

Where several individual thermometers are needed for complete testing of the system, the number of instruments needed can be reduced by using multicouple digital thermometers. Figure 2–3 shows a digital temperature test using four permanently attached test leads. Two 15 ft. and two 8 ft. leads are included. Operating off self-contained batteries, the instrument can be used in any location, regardless of power availability.

Figure 2–4 shows the same type of instrument using leads that are detachable. This permits the use of various types of test leads such as air sampling, surface contact, insertion-type, and so on.

Psychrometer

The most popular instrument for measuring the dry bulb and wet bulb temperatures of air is the sling psychrometer. Show in Figure 2–5 is the Weksler sling psychrometer using two glass thermometers with one carrying a wet sack for measuring the wet bulb (WB) temperature and the other plain for measuring the dry bulb (DB) temperature. This instrument requires the use of a psychrometric chart.

Another type of sling pyschrometer that has the slide rule as part of the case is pictured in Figure 2–6.

The electronic digital system has also been used to produce a direct-reading humidity/temperature meter. Shown in Figure 2–7 is a humidity/temperature meter that will check the

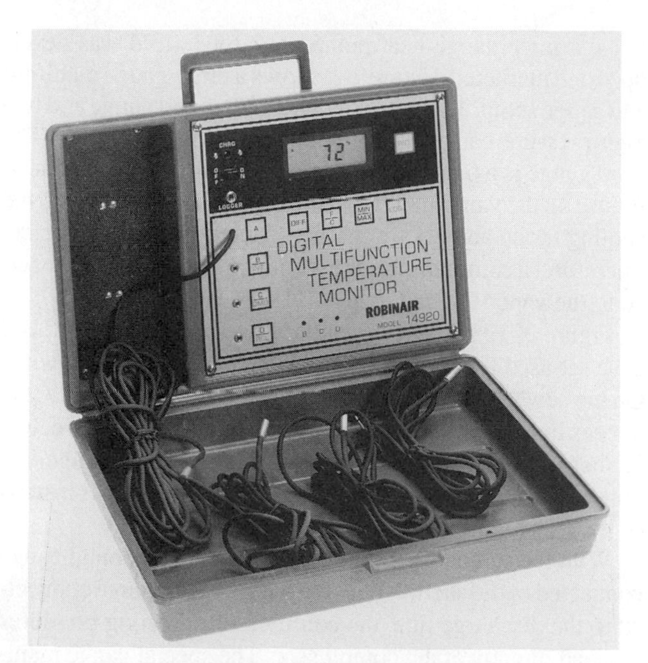

Figure 2–3 Digital temperature tester with permanent leads (Courtesy Robinaire Division, Sealed Power Corp.)

Figure 2–4 Digital temperature tester with portable leads (Courtesy Robinaire Division, Sealed Power Corp.)

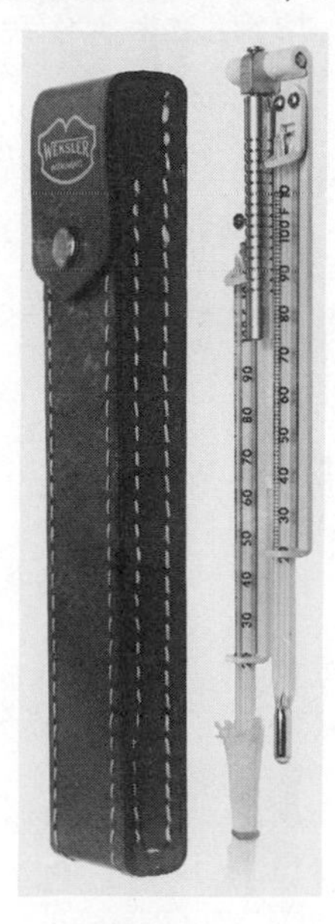

Figure 2–5 Wekesler sling psychrometer (Courtesy Robinaire Division, Sealed Power Corp.)

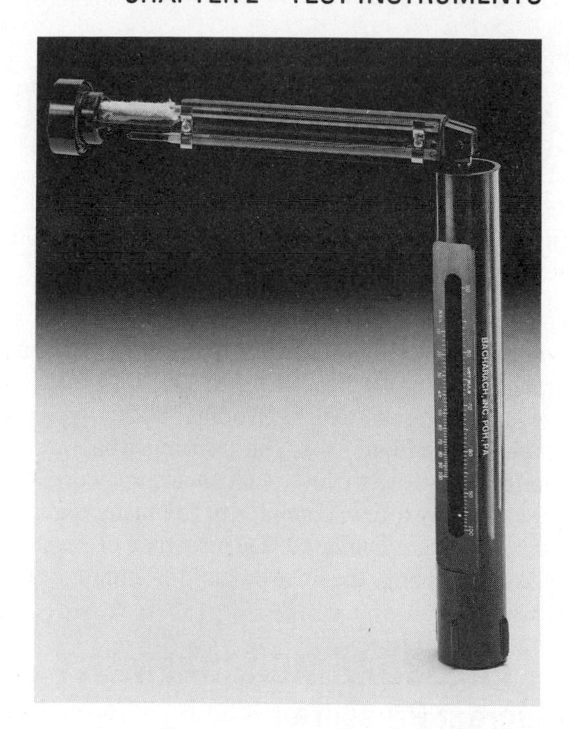

Figure 2–6 Sling psychrometer in case (Courtesy Bachrach Inc.)

Figure 2–7 Digital direct reading humidity meter (Courtesy EXTECH Instruments Corp.)

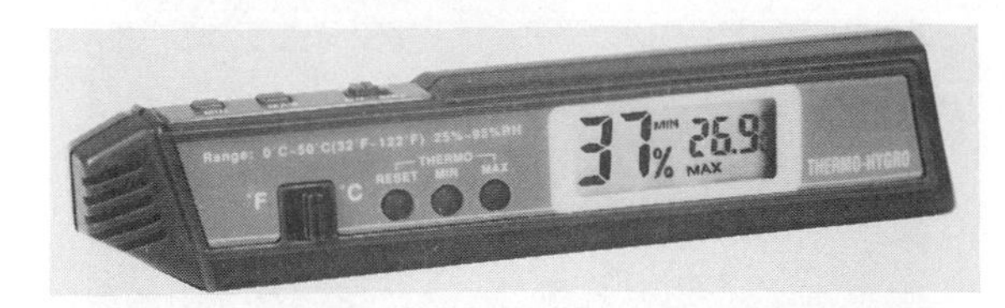

Figure 2–8 Pocket-type digital direct reading humidity meter (Courtesy EXTECH Instruments Corp.)

percentage of relative humidity from 5% to 95%. The temperature range is 0° C to 60° C or –40° F to 200 ° F. For measuring relative humidity only, a direct-reading, pocket-type instrument is pictured in Figure 2–8. This instrument has a measurement range from 10% to 90%, with an accuracy of ±3%.

These are only a few examples of the many digital-type instruments that are marketed. The accuracy of such instruments is well worth the investment for improvement of service quality.

2–3 PRESSURE MEASUREMENT

Refrigerant Pressure

On older model heat pumps, three pressure taps were used. These were the liquid line, the vapor line, and the suction line between the reversing valve and the compressor. For these units, a three-gauge heat pump gauge manifold was developed and marketed. Figure 2–9 shows a three-gauge manifold and gauge setup. This set contains a compound gauge and two high-pressure gauges along with the connecting hoses. The third gauge is a 0 to 500 psig standard gauge that is intended for use on the vapor line. This line carries suction vapor in the cooling mode and hot discharge vapor in the heating mode. Therefore, a compound gauge would be destroyed if connected to the vapor line in the heating mode.

Figure 2–10 shows the standard two-gauge manifold and hose set used for refrigeration and air conditioning systems. On the newer heat pump units with the fourth gauge tap in the hot gas line between the compressor and reversing valve, two of these gauge sets could be used. The second gauge set, however, would have to have two 0 to 500 psig gauges instead of the compound and standard gauge setup.

With the standard manifold set with the compound gauge connected to the suction line and the standard gauge connected to the discharge line, the compressor operating pressures are read directly at the compressor. The special gauge manifold is connected to the liquid line and vapor line. This reads the pressures between the coil pressure-reducing devices as well as the pressure on the coil side of the reversing valve. System operating problems can then be read by determining pressure differences in various parts of the system.

Figure 2–9 Heat pump gauge manifold (Courtesy Robinaire Division, Sealed Power Corp.)

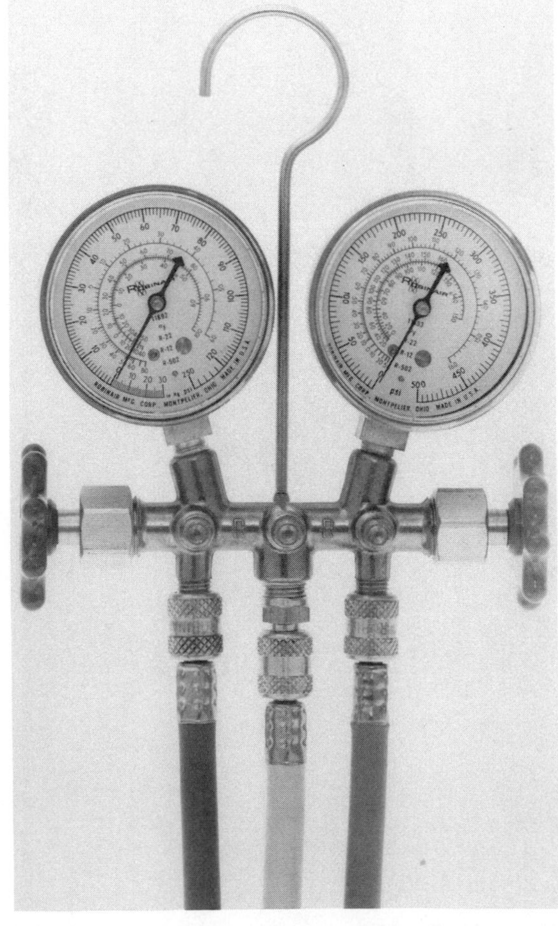

Figure 2–10 Standard gauge manifold (Courtesy Robinaire Division, Sealed Power Corp.)

Water Pressure

The pressure drop across the liquid coil in liquid-to-air heat pumps must be checked. For this purpose, a standard pressure gauge with a 0 to 50 psig range equipped with a tap needle is used. When the system has been provided with P/T plugs in the liquid circuit, liquid pressures and temperatures can be measured very easily.

2–4 ELECTRICAL

The instrument used most in the electrical testing process is the clamp-type ammeter/voltmeter. The digital type has been developed to provide a wider range of tests than the meter dial type. Figure 2–11 shows a clamp-type meter that has a transformer jaw clamp that measures up to 400 amperes AC. It will measure current surges from motor startup and recall the highest point of surge. The instrument will also measure low amperage or volts in circuit, and it has an accurate resistance meter.

For instrument protection, the meter should be set on the highest range when the electrical load is applied. The instrument range is then lowered for more accurate readings. Motors should never be started without a meter measuring the inrush and operating current to make sure the motor starts. Such instruments are usually marketed with adapters that will widen the usage range. Adapters are also available to measure temperature, humidity, air velocity, and outputs in the milliampere as well as 0 to 100 DC range. When selecting a test instrument, the most reliable and usable over the widest range is usually the best investment.

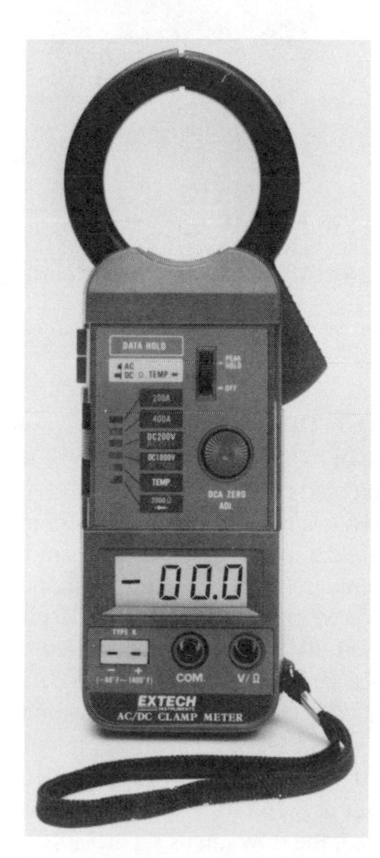

Figure 2–11 Clamp-type meter (Courtesy EXTECH Instruments Corp.)

Figure 2–12 Air volume meter (Courtesy Shortridge Instruments Inc.)

2–5 AIR HANDLING

In the past, a Pitot tube and inclined manometer have been used to determine the pressure in duct work to calculate the velocity and CFM. Direct-reading instruments are available that eliminate the calculations measuring only the CFM. Velocity and static pressure must be measured by other instruments.

Direct Reading Instruments

Figure 2–12 shows an instrument called a flow hood. Using a solid state digital read-out-type meter, the instrument gives direct readings in CFM. Measuring supply, return, or exhaust CFM, the instrument is excellent for balancing systems as well as determining the total CFM supplied by the unit.

Pressure Reading

When checking the air pressures in duct work in the pressure range from 0.0 to 1.5" WC, the pressure gauge or "inclined manometer" as shown in Figure 2–13 is used. The

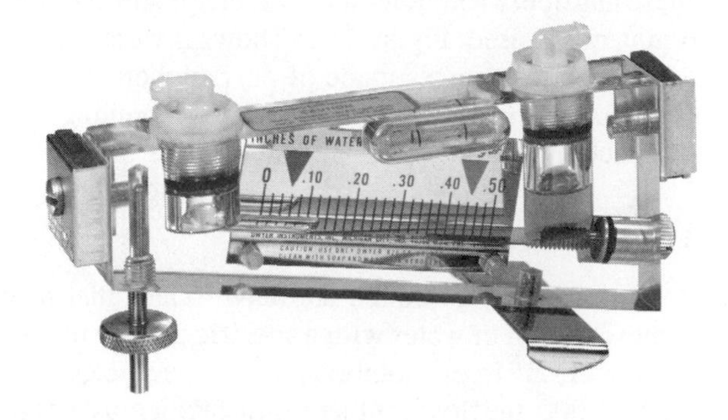

Figure 2–13 Inclined manometer (Courtesy Dwyer Instruments Inc.)

reason for the incline of the pressure-indicating fluid is to spread out the scale into larger spaces for more accurate readings. The instrument pictured has a range of 0.0 to 1" WC extended over an 8" length. If extreme accuracy is needed, a 12" scale length is used.

2–6 LIQUID HANDLING

In liquid-to-air as well as liquid-to-liquid systems, the liquid circuits must be checked for pressure as well as flow rate. These systems may use water from a constant source or they may have a closed recirculation system. In closed systems, the liquid may be purified water or antifreeze solution. If antifreeze solution is used, the antifreeze material usually is chloride salts, glycols, or alcohol. Therefore, the instruments needed include pressure measurement, specific gravity, flow rate, and alkalinity/acid ratio (pH factor).

Specific Gravity

The specific gravity of a liquid as compared to water (which has a specific gravity of 1.000) is important from a heat transfer standpoint as well as the flow rate required. This subject will be discussed in depth in Section 3, Liquid-to-Air Heat Pump Systems. A float-type hydrometer is used for measuring the specific gravity of liquids. This instrument has a range of 0.985 to 1.190, which is within the range of chloride and glycol antifreeze solutions down to the minimum freezing point of 0° F. This instrument is used in the process of making wine and is easily obtainable from any wine supply outlet.

Flow Rate

A flow meter is a must in liquid-to-air heat pump installations. This provides a check on the flow rate and quality of the liquid. Flow meters used in purified water recirculation systems or in well supply systems can be made of acrylic resins. The flow meter in Figure 2–14 is an example of such an instrument. This meter operates in the 2 GPM to 10 GPM range; this type of meter can be obtained in a wide variety of ranges.

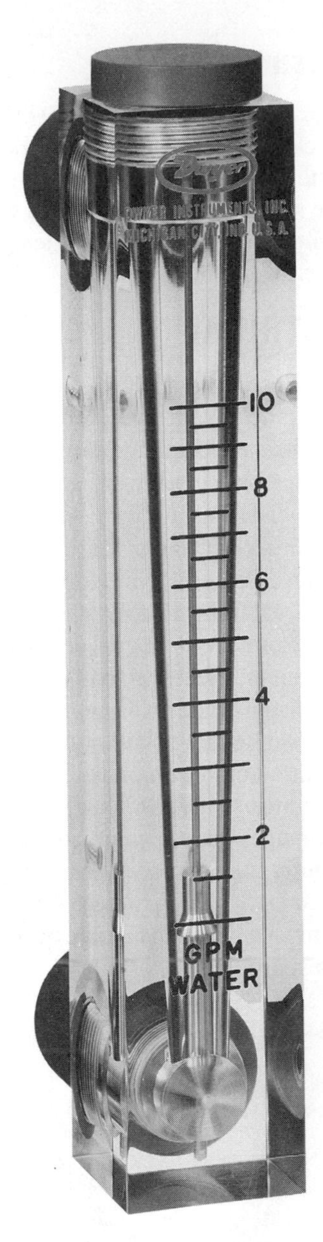

Figure 2–14 Acrylic-flow meter (Courtesy Dwyer Instruments Inc.)

The glycols, chlorides, and alcohols used in antifreeze solutions will attack acrylic materials and conduct a slow process of cracking or crazing the contacted surface. Therefore, it is not advisable to use acrylic meters on these products. For these antifreeze solutions, a flow meter made of polysulphon material is used. Figure 2–15 shows a variety of flow meters of different ranges made of polysulphon. These are also available in a variety of styles as well as mounting means and connections.

Antifreeze Compensation

Flow meters are calibrated and have scales that record flow rates in GPM of water with a specific gravity of 1.000. When used on antifreeze solutions with a specific gravity of other than 1.000, the flow compensation factor must be taken into consideration and the flow meter reading adjusted accordingly. For example, if the flow rate for the unit is established at 6 GPM for water and a propylene glycol concentration of a 0° F freeze point is to be used, what flow rate should be read on the flow meter?

Figure 2–16 is a table of the multipliers for the two most popular antifreeze materials used at the most predominate solution percentage—20%. These multipliers mean that if the unit uses 6 GPM flow rate of water, if propylene glycol solution is used, the flow rate is increased by 36% to 8.16 GPM to get the same heat transfer rate ($6 \text{ GPM} \times 1.36 = 8.16$ GPM). For calcium chloride solution, the multiplier is 1.23 ($6 \text{ GPM} \times 1.23 = 7.38$ GPM). Therefore, to obtain the 6 GPM flow rate for the system using propylene glycol solution, the flow rate is set at 8.16 GPM on the flow meter. If calcium chloride is used, the flow rate is set at 7.38 GPM on the flow meter. These adjustment factors must always be kept in mind when using antifreeze solutions.

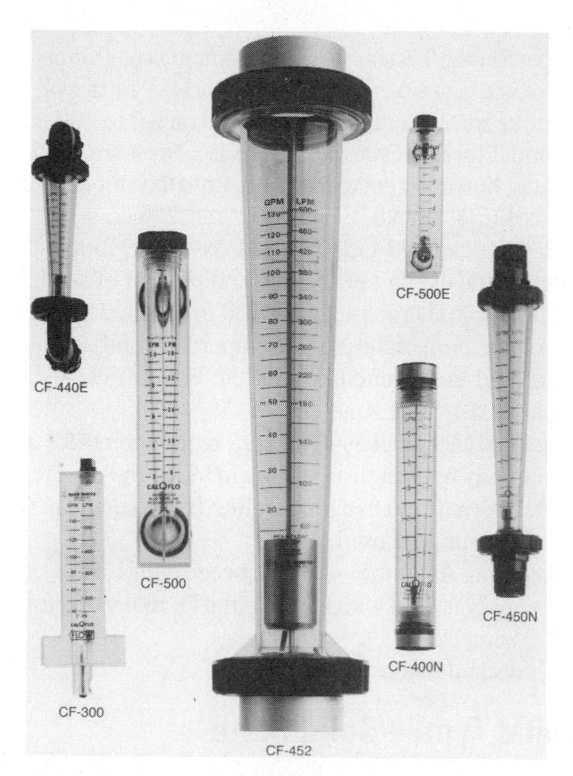

Figure 2–15 Polysulphon flow meters (Courtesy Blue White Industries Inc.)

Fluid	Multiplier
20% Propylene Glycol	1.36
20% Calcium Chloride	1.23
20% Methanol Alcohol	1.25

Figure 2–16 Flow compensation factor for antifreeze solutions (Courtesy Bard Manufacturing Co.)

After the flow meter rate has been established, the meter should be marked at the reading. Red nail polish is excellent for this purpose. If an indicating point is included on the meter, the pointer should be glued in position to prevent tampering.

Alkalinity/Acid Ratio—pH Factor

The alkalinity/acid ratio factor (pH factor) of the solution in closed loop systems is very important. The neutral alkalinity/acid ratio of a liquid is established at 7 on a range from 1 to 14. Below 7 the material is acid and above 7 the material is alkaline. If the solution is acid, it will attack all metal parts in the system including the liquid-to-refrigerant coil. The lower the number, the stronger the acid solution. If the pH factor is above 7, the solution is alkaline. If the alkaline content is too high, the alkaline materials will tend to precipitate out and cause fouling of the liquid system. The desired pH rating is between 7.5 and 8.5 for longest equipment life.

To measure the pH factor of a sample of the liquid in the system, a pH meter is used. There are several varieties on the market from permanently mounted to desk type to hand-held type. A hand-held type such as the one shown in Figure 2–17 is recommended for the service technician. Accuracy to within 0.01 pH as well as compensation for the temperature of the liquid tested are desirable. A meter for pH testing is a must to check the solution quality.

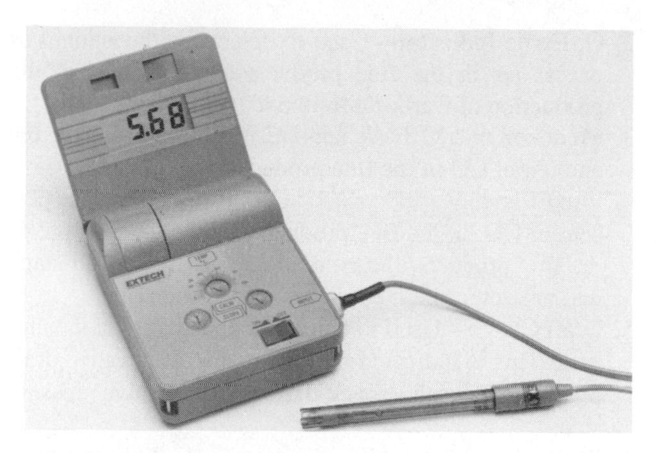

Figure 2–17 Hand-held pH meter (Courtesy EXTECH Instruments Corp.)

2–7 EFFICIENCY TESTING

To determine the capacity of an air-to-air heat pump when using fossil-fuel-fired units as the auxiliary heat, the efficiency of the fossil-fuel unit must be determined. The step-by-step process is described in Section 2, Air-to-Air Heat Pump Systems. The instruments used may be individual units as shown in Figure 2–18 for gas-fired units or Figure 2–19 for oil-fired units.

Test Kit for Gas-Fired Units

Figure 2-18 shows a typical set of instruments for efficiency testing a gas-fired unit. The kit consists of the following items:

1. CO_2 Fyrite Indicator—Used to determine the percentage of CO_2 in the flue products. The flue gas CO_2 along with the flue temperature is used to determine the unit efficiency on the Fire Efficiency Slide Rule.

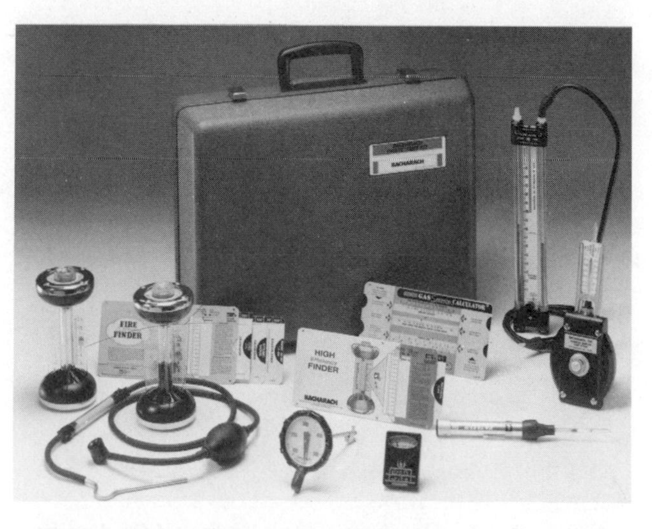

Figure 2–18 Combustion test kit for gas-fired units (Courtesy Bachrach Inc.)

2. O_2 Fyrite Indicator—Used to determine the amount of excess air in the flue products to guard against the production of Carbon Monoxide (CO).
3. Monoxor or CO Indicator—Used to read directly the amount of CO in the flue products.
4. Fire Efficiency Finder/Stack Loss Rule—Using the percent of CO_2 in the flue products and the temperature of the flue products, the combustion efficiency and stack loss are determined by the use of the slide rule.
5. Draft Gauge—Used to determine the quality of the unit vent. This is indicated by the amount of negative draft pressure or possible back draft on the positive pressure side.
6. Stack Temperature Thermometer—A dial-type thermometer in the 200° F to 1000° F range for measuring flue gas temperatures.
7. Gas Service Calculator—Used to determine the orifice size needed for the type of gas burned (natural, mfg., propane, butane, or mixture) at the specific gravity and BTU content of the gas as well as the unit design input and manifold pressure.
8. Gas Pressure Manifold Gauge—Used to measure the manifold pressure in the unit to determine input and/or orifice size.
9. Monoxor-Carbon Monoxide Detector—Used for testing for CO concentrations in areas other than direct flue product tests.
10. All controls in a carrying case.

Test Kit for Oil-Fired Units

Figure 2–19 shows a typical set of instruments for efficiency testing oil-fired units. This set consists of:

1. CO_2 Indicator—Used to determine the amount of CO_2 in the unit flue products.
2. O_2 Indicator—Used to determine the amount of excess air to determine the amount of CO produced in the flue products.
3. Spot Smoke Tester—This instrument uses a filter paper to produce a smoke spot for comparison to the oil burner smoke scale. Comparison is made on a 0 to 10 scale. The normal for domestic oil burners is a No. 1 smoke. In some areas, however, code does not allow the smoke content to exceed "0" smoke.
4. Fire Efficiency Finder/Stack Loss Slide Rule—Used to determine the unit efficiency and percent of stack loss.
5. Dial Stack Thermometer—Used to measure the stack flue products temperature. This temperature and the stack CO_2 are used in conjunction with the Fire Efficiency Finder/Stack Loss Slide Rule.
6. Inclined Manometer—The draft requirements for oil-fired units may be as high as 1.5"wc in industrial oil-fired units. Therefore, a draft gauge or inclined manometer in the 0"wc to 2"wc range is used.
7. Sampling Assembly—This squeeze pump, filter, and hose assembly is used with the CO_2 and O_2 indicators for taking flue samples.
8. All packed in a carrying case.

Digital Type—Solid State

As in all instruments, the solid state or digital principle is incorporated into an efficiency meter. Figure 2–20 shows a direct-reading, hand-held instrument that will directly read the combustion efficiency, CO_2 percentage, excess air per-

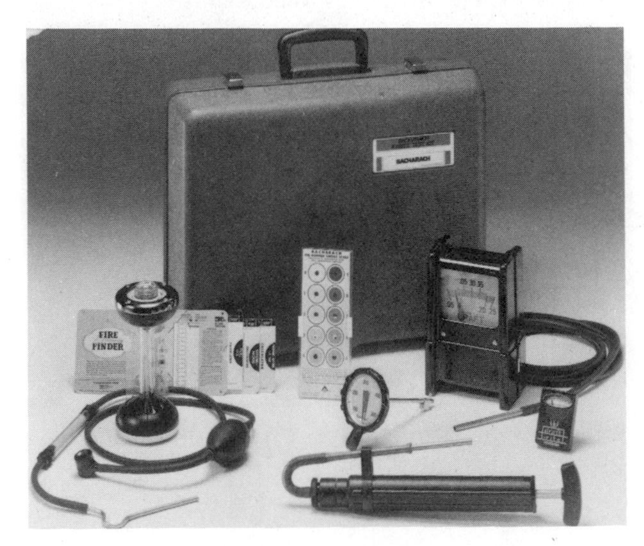

Figure 2–19 Combustion test kit for oil fired units (Courtesy Bachrach Inc.)

Figure 2–20 Hand-held direct reading efficiency meter (Courtesy Bachrach Inc.)

centage, and percentage of stack loss for the six most commonly used fuels. It will not measure draft so it is not a total replacement for the efficiency kit, but it represents a considerable savings in time. The use of the instruments in this chapter will be discussed more extensively in the individual sections on unit types.

REVIEW QUESTIONS

1. Digital-type instruments are preferred for testing heat pump systems. True or False?
2. A sling psychrometer measures what temperature of air?
3. To determine the relative humidity in the air from the sling psychrometer readings, a ___________ chart is used.
4. To measure the pressure in a heat pump system, three gauges are used. Name the three lines that these gauges are connected to.
5. Gauges are made in two different style ranges. List them.
6. The three gauges used in questions 2-4 are two ___________ type and one ___________ type.
7. Sometimes a fourth pressure tap is provided in the refrigeration system. If so, a ___________ type is used.
8. A clamp-type ammeter will work equally well on both AC and DC current. True or False?
9. When testing for amperage draw, the dial-type ammeter should be set on the highest range and then lowered to get an accurate reading. True or false?
10. A flow-hood meter is used to measure liquid flow through the liquid circuit of a liquid-to air heat pump. True or False?
11. A pressure gauge used to measure air pressure in duct work in the 0.0" to 1.5" wc range is called an ___________.
12. The instrument used to measure specific gravity is called a ___________.
13. A flow meter is a necessary part of every liquid source heat pump. True or False?
14. Flow meters are calibrated based on the specific gravity of water, which is ___________.
15. When using a flow meter, the GPM flow rate through the unit is standard regardless of the antifreeze and water solution used in the system. True or False?
16. The alkalinity/acid condition of a liquid is measured by what type of meter?
17. A solution in a neutral condition of alkalinity/acid has a meter reading of:
    a. 3
    b. 5
    c. 7
    d. 9
    e. 0
18. The instrument in column 1 is used to measure what item in column 2?

Column 1:

1. Monoxor
2. Flow hood
3. Pitot tube
4. *U* tube monometer
5. Spot smoke tester
6. CO_2 indicator
7. Clamp-type ammeter
8. Compound 30", 250 psig
9. Flow meter
10. pH meter
11. O_2 indicator
12. Inclined manometer
13. Standard 0 to 500 psig
14. Stack temperature thermometer
15. Hydrometer liquid
16. Draft gauge
17. Standard 0 to 50 psig

Column 2:

a. Amperage flow in a wire
b. Negative pressure of a vent
c. Pressure in the liquid heat source system
d. Compressor discharge pressure
e. Alkalinity/acid ratio of liquid
f. CO_2 content of flue products
g. Compressor suction pressure
h. Flue product temperature
i. Air pressure in duct system
j. Excess oxygen in flue products
k. Vapor line pressure
l. Specific gravity of a liquid
m. Carbon dioxide content of flue products
n. Air flow from registers and grills
o. Carbon monoxide level
p. Flow rate of a liquid
q. Liquid line pressure
r. Gas unit manifold pressure

3 COMPONENT PARTS

3–1 REFRIGERATION CYCLE

The vapor compression cycle of a heat pump uses a heat exchanger to pick up heat from a heat source, a heat exchanger to get rid of the heat to a heat sink, a compressor to raise the pressure and condensing temperature, and a pressure-reducing device to lower the pressure and boiling point. In those systems that provide both heating and cooling, a four-way reversing valve is used in addition to the check valves. Where a liquid surge is likely, such as in air-to-air units, an accumulator is also incorporated into the system.

3–2 HEAT EXCHANGERS

Two different types of heat exchangers are used—air-to-refrigerant and liquid-to-refrigerant. In both types, heat transfer may be in either direction depending upon the location of the heat exchanger in the system and the operating mode.

Air-to-Refrigerant Type

Air-to-refrigerant-type heat exchangers are of fin and tube construction for the highest level of heat transfer rate. They may be of a flat or *U* shape or *A* or *H* shape. The coil located in the portion of the system that conditions the air for the occupied area is usually of the *A* type if the system is a split type. Smaller tonnage units will use a coil of one-half the *A* type called a slab type.

Figure 3–1 shows an air-handling unit that is the indoor section of the heat pump system. The heat exchanger located in the bottom of the cabinet is an *A*-type fin coil. This particular unit is designed to be used in an up-flow configuration as shown, but it may also be used in a down flow as well as horizontal flow. To be used in a down-flow configuration, an extra drain pan is located at the bottom or *V* end of the coil.

When used in the horizontal configuration, a drain pan is added. This lies in the cabinet under the coil extending from the large coil end to a couple inches beyond the small end or *V* end of the coil. A pan insert of perforated metal is also included to catch the condensate off the small end of the coil and prevent splash and carry out of the water out of the pan. This coil may also be supplied as the coil only for addition to an existing heating system where the heat pump will be the primary heat source and the present heating system will supply the auxiliary heat. In the unit shown in Figure 3–1, the electric elements above the blower are the auxiliary heat source.

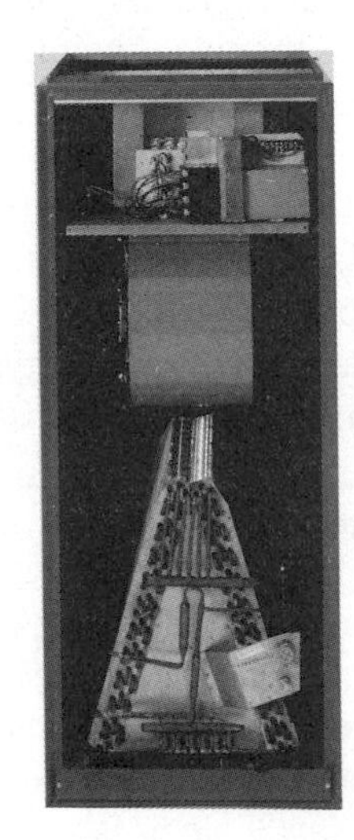

Figure 3–1 Air handling unit (Courtesy Addison Products Co.)

Figure 3–2 is a view of the outdoor section of a remote air-to-air system with the top removed. A flat-type heat exchanger is incorporated in a wraparound configuration. The *U*-shaped heat exchanger presents considerably more heat transfer surface to the air passing over it with lower air flow resistance. This unit uses a horizontal fan discharge with the fan and motor assembly located in the end of the unit cabinet. Other makes and models locate the fan and motor assembly in the cover for vertical up discharge. Both types of configuration have their advantages and disadvantages.

A. Horizontal Discharge
Advantage—The compressor is located in the air stream for cooling by air flow as well as refrigerant vapor. Servicing the unit is easy by removal of the top panel.
Disadvantage—Compressor operating sound is higher due to sound passing out through the heat exchanger.

B. Vertical Discharge
Advantage—Fan sound level is discharged upward with less side travel. Unit can be smaller in physical dimensions.
Disadvantage—The unit is more difficult to service because the fan assembly must be removed to obtain access to the motor/compressor assembly.

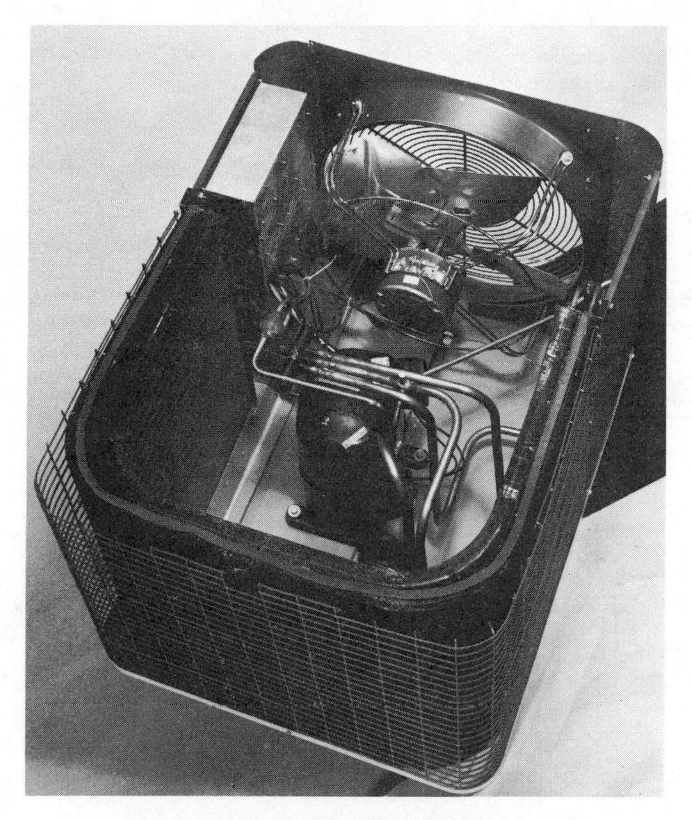

Figure 3–2 Outdoor section (Courtesy Amana Refrigeration Inc.)

In each case, the advantages and disadvantages have little effect upon the performance of a properly installed and adjusted system.

Figure 3–3 shows a heat pump system in one package. The indoor coil is a single flat coil in a vertical position. Air is drawn through the coil by the blower and discharged out the top over the coil. Electric auxiliary heating elements are located in the blower discharge opening.

On the opposite side of the unit is the outdoor coil. This also is a fin-type heat exchanger of a single flat configuration in a vertical position. A propeller-type fan is used to move the required amount of air through the heat exchanger. Because a propeller fan is used, no duct resistance can be added to the air circuit without seriously affecting the performance of the unit.

De-ice Subcooler Coil

The problem of outdoor tube collapse due to incomplete defrosting of the coil is constantly referred to in this text. This is a problem that has convinced one manufacturer to incorporate a "de-ice" section in the outdoor coil. By supplying heat to the lower section of the coil, the melting of frost and/or ice from these tubes is accelerated (see Figure 3–4).

The de-ice subcooler coil uses the cold outdoor coil to remove heat from the liquid refrigerant and keep ice from forming on the outdoor coil at the unit base in the heating mode. The subcooler coil is integrated into the bottom of the outdoor coil where ice could form and push upward, due to expansion of freezing, and crush the tubes in the coil. The liquid refrigerant coming from the warm indoor coil still has heat available even though it is what is termed *low grade heat.* The temperature of the liquid is approximately 90° F and has a heat content of approximately 36 BTUs per pound of refrigerant circulated. The change of phase from hot gas to liquid has been made and very little of the indoor coil has been used to subcool the liquid in the low-grade-heat mode. The major portion of the indoor coil is used in the change of state process of the refrigerant at approximately 70 BTUs per pound circulated, rather than in low-grade-heat liquid at 5 BTUs per pound of circulated refrigerant. In effect, the subcooler has increased the working size of the indoor coil over a conventional capillary tube nonsubcooler system.

The low-grade-heat liquid has a temperature of approximately 90° F. Piped into the subcooler loop in the bottom of

Figure 3–3 Package heat pump (Courtesy Amana Refrigeration Inc.)

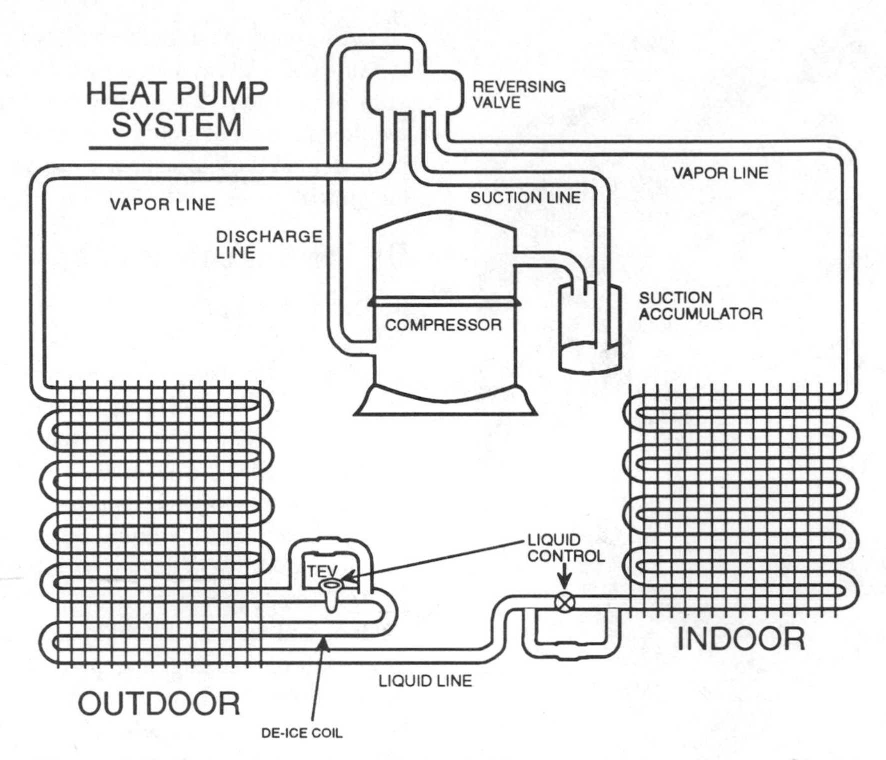

Figure 3–4 De-ice subcooler coil circuit (Courtesy Addison Products Co.)

the outdoor coil, heat is given up to melt any frost and/or ice from the low-temperature outdoor coil. The air temperature to refrigerant temperature outdoors is much greater than it would be indoors: 90° F – 45° F outdoors = Δ55° F versus 90° F – 80° F = Δ10° F indoors. With the large temperature spread outdoors, the liquid refrigerant can be reduced approximately 30° F across the subcooler loop. The heat content of the liquid at 90° F is approximately 36 BTUs per pound of refrigerant circulated. The heat content of the liquid at 60° F is approximately 27 BTUs per pound or a heat loss of 9 BTUs for each pound circulated. A 3-ton system, containing R-22, circulates approximately 360 pounds per hour. Therefore, 350 pounds per hour × 9 BTUs per pound = 3240 BTUs heat removed from the liquid refrigerant and transferred into the ice and metal at the bottom of the outdoor coil.

The refrigerant side capacity of the outdoor coil is equal to the heat content of the liquid refrigerant in minus the heat content of the refrigerant vapor out times the pounds of refrigerant circulated per hour. The refrigerant vapor leaving the outdoor coil has a heat content of approximately 109 BTUs per pound. 109 BTUs per pound vapor minus 27 BTUs per pound liquid means that each pound of refrigerant picks up 82 BTUs in the vaporizing process. This 82 BTUs per pound × 360 pounds of refrigerant circulated = 29520 BTUs absorbed from the air. The compressor and fan motor watts × 3.413 BTUs per watt = the additional heat to bring the capacity up to 35820 BTUh heating rate.

If the subcooler coil were not in the circuit, the absorption rate would be 29520 minus 3150 or 26370 BTUs per hour absorbed by the outdoor coil. The inclusion of the subcooler coil will result in an increase in the outdoor unit performance of approximately 12%.

A secondary benefit is that the extra cost of the T. X. valve is more than offset by the decrease in coil price. We can now control the refrigerant flow from –20° F up to 80° F outdoor temperature and never allow liquid refrigerant to return to the compressor even in heavy frost conditions. The T. X. valve will not lose the liquid seal as the subcooler coil will condense any refrigerant vapor due to the large temperature difference that it has to work with. This allows 100% of the outdoor coil to be used to condense the hot refrigerant vapor to liquid rather than wasting a portion of the coil to hold liquid refrigerant for subcooling.

The system in the cooling mode is essentially the same as a standard cooling system except that all of the liquid refrigerant is routed through the subcooler coil to the indoor pressure-reducing devices (capillary tubes). The subcooler coil helps to assure a liquid refrigerant seal on the indoor pressure-reducing device as it acts as a liquid collector. We are using the full outdoor coil as a condenser due to the circuitry involved.

The wide range capability of the T. X. valve during extreme frost and load conditions will give the compressor a much better range to operate in during the heat mode. We would expect longer life and less lubrication problems than with a capillary-tube-type pressure-reducing device on the outdoor coil. The indoor coil can still use a capillary-tube-type pressure-reducing device as the operating load is much narrower.

Figure 3–5 Package-type unit, liquid-to-air (Courtesy Friedrich Air Conditioning and Refrigeration Co., Climate Master Division)

Figure 3–6 Package-type unit with coaxial in horizontal position (Courtesy TempMaster Enterprises Inc.)

Liquid-to-Refrigerant Type

Figure 3–5 shows a typical package-type liquid-to-air heat pump unit. The inside coil (air type) is located in the upper compartment. A centrifugal blower forces air through the coil and air distribution system to the conditioned area. This heat exchanger is the same as used in air-to-air units. The liquid refrigerant is fed through the bottom of the unit in the cooling mode and the hot vapor is fed through the top in the heating mode.

For minimum refrigerant flow resistance, the coil is divided into several circuits in a stacked configuration. Therefore, a manifold is used on the vapor end to connect to each circuit. Capillary tubes connect the refrigerant distributor to the individual circuits of the coil.

The "outside" coil is a tube-in-a-tube coil called a coaxial coil. Usually made of cupro-nickel for greatest resistance to corrosion, the double tube carries liquid through the outside tube and refrigerant through the inner tube. These coils are also circuited for counterflow operation—the liquid and refrigerant flow in opposite directions. This provides a more even temperature difference between the liquid and the refrigerant for highest heat transfer efficiency.

This type of coil will operate efficiently in either a vertical or horizontal position. Figure 3–5 shows the coil in a vertical position. The unit shown in Figure 3–6 is designed with vertical coil. Either position is strictly the design engineer's choice for best fit in the cabinet configuration desired.

Not all units are the coaxial tube-in-a-tube heat exchanger. Figure 3–7 shows a shell-and-tube heat exchanger. This type uses a coil carrying the liquid and a shell containing the refrigerant vapor. During the cooling cycle, the hot refrigerant vapor enters the top of the shell and liquid refrigerant leaves the bottom of the shell. The cooling liquid flows through the tubing coils, entering at the bottom of the coil and leaving at the top of the coil. This gives the counterflow heat exchange desired for highest efficiency.

In the heating mode, liquid refrigerant enters the bottom of the shell and surrounds the liquid coils. Heat from the liquid coils vaporizes the liquid refrigerant. This vapor then flows from the top of the shell to the compressor.

The pressure-reducing device used with this type heat exchanger is the capillary tube. The refrigerant charge is critical and limited to the quantity needed to flow the major portion of the heat exchanger. The actual charge, however, would be determined by the operation of the system in the cooling mode where the fin coil is used as the evaporator.

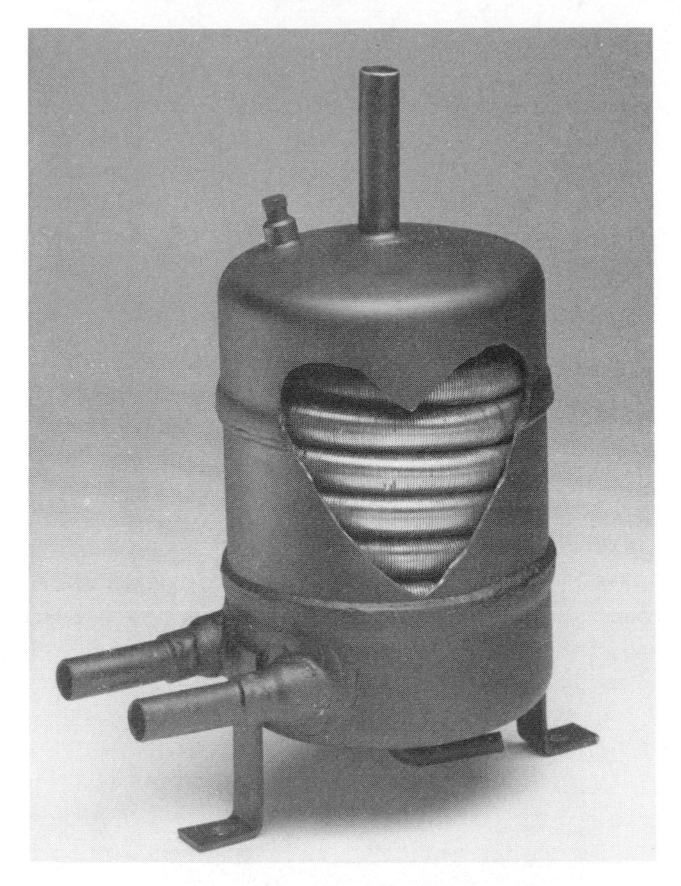

Figure 3–7 Shell and tube heat exchanger (Courtesy Command Aire Corp.)

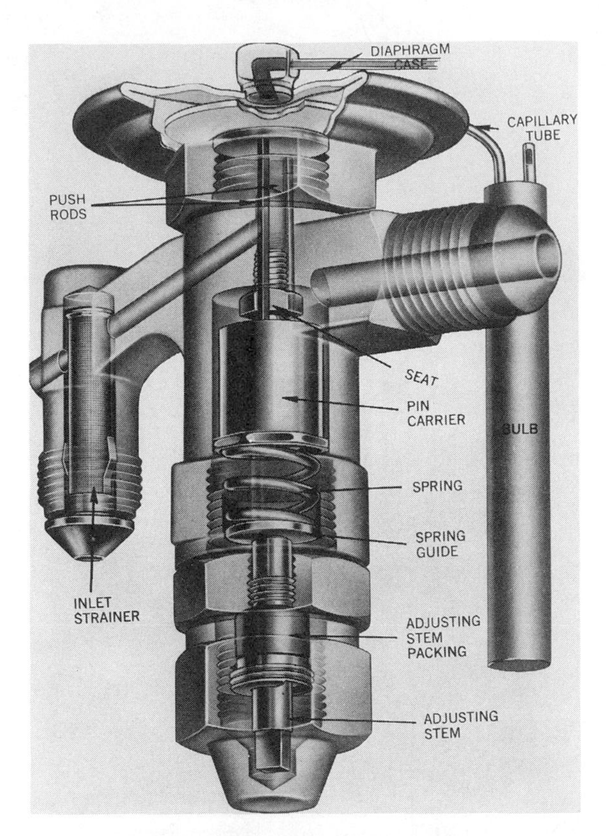

Figure 3–8 Thermostatic expansion valve (Courtesy Sporlan Valve Co.)

3–3 PRESSURE-REDUCING DEVICES

Two types of pressure-reducing devices, the thermostatic expansion valve and the capillary tube or restrictor, are used to produce the desired operating pressure and boiling point in the evaporator.

Thermostatic Expansion Valve

Figure 3–8 shows a typical thermostatic expansion valve used in heat pump systems. Basically, this valve is the same construction as those used in refrigeration and air conditioning systems. Consisting of a valve body, pressure spring, and power element, the valve operates to maintain a desirable superheat in the evaporator. With the pressure in the power element, which is determined by the temperature of the feeler bulb, counteracting the valve body pressure plus the spring pressure, the proper superheat range is maintained. Usually this is 5° F to 7° F, the same as for air conditioning systems.

Though air conditioning T.X. valves and heat pump T.X. valves look alike, *THEY MAY NOT BE INTERCHANGEABLE.* The T.X. valve used in heat pumps has a special power element. The power charge that this element contains has a pressure-limiting ability to limit the pressure that can be produced above the element diaphram. The reason for this special feature is that the feeler bulb of this valve is fastened to the vapor line off the coil to which it controls the refrigerant flow. When the coil is used as the evaporator, the valve operates at normal temperatures and pressures. However, when the coil is used as a condenser, the refrigerant flow is bypassed around the valve in the reverse direction and the valve body is up to condensing pressure. The tube the feeler bulb is connected to is now the hot gas line. The bulb rapidly reaches the hot gas temperature of the compressor discharge gas. If the power element of the T.X. valve did not have the pressure-limiting ability, the pressure buildup would warp the diaphram and destroy the valve.

Some manufacturers have T.X. valves in their product lines that are interchangeable between heat pumps and air conditioning units. These may not, however, be satisfactory replacements for the valve the unit manufacturer specifies.

Capillary Tubes or Restrictors

The predominant pressure-reducing device used in heat pumps is the capillary tube. This small-diameter liquid line has the flow resistance to drop the liquid refrigerant pressure and boiling point into the range required for proper operation.

Each coil circuit, air or liquid, has its own set of capillary tubes sized to perform at the conditions encountered during the particular mode of operation.

In the cooling mode, the inside-coil capillary tubes provide the pressure drop and resulting boiling point for heat absorption in the coil. Unit design conditions are, for an air-to-air heat pump, 95°F DB outdoor air with 80°F DB and 50% RH inside and 45° F evaporator boiling point. For a liquid-to-air heat pump, the condensing pressure would be the equal of 105° F condensing temperature along with the 80°F DB and 50% RH air and 45°F boiling point in the coil.

In the heating mode, in the air-to-air unit, the outside coil is the evaporator with 45°F air over it. The inside coil is the condenser with 70°F DB air over it. The capillary tubes on the outside coil are designed for these conditions. The liquid-to-air unit, in the heating mode, will have the liquid flow rate controlled to produce a 45°F evaporator boiling point along with the 70°F DB air over the condenser (inside coil).

From these various conditions, you can see that different capillary tube flow resistances are required to get the proper results. Capillary tubes are selected very carefully for size and length to obtain the desired results. They should never be altered or changed in length or inside diameter. Any time they are changed, the new tube must be of the same diameter and length as the original tubes.

3–4 COMPRESSORS

The motor-compressor assemblies used in residential and small commercial heat pump systems are predominately the totally hermetic type. The compressor and motor are inside a welded shell. The only special feature of the heat pump compressor is that it is the high efficiency version of the standard assembly. Most manufacturers use the same compressor assemblies in their high-efficiency air conditioning units and their heat pumps.

3–5 REVERSING VALVES

To provide the heating and cooling functions of a dual-function heat pump, it is necessary to be able to reverse the refrigerant flow between the two heat exchangers. Thus, each heat exchanger is either an evaporator or a condenser, depending upon the mode the unit is operating in. In the first heat pumps, this was accomplished by means of four solenoid valves, two open when the power was off (normally open) and two closed when the power was off (normally closed). The erratic operation that this provided plus the maintenance problems encountered required the development of a more certain way of accomplishing this function.

This led to the development of a double-acting solenoid valve with a slide valve selector under the control of a direct-acting solenoid. This type valve had its inherent problems due to the distance the slide had to travel as well as the high-pressure differences between discharge and suction pressures the valve encountered. Erratic valve action was the result. To reduce the load on the solenoid valve and reduce the solenoid piston travel, a pilot-operated valve was developed. This valve also uses the difference in suction and discharge pressure to operate the main slide valve.

Figure 3–9 shows the exterior of a typical reversing valve. There are three connection tubes on one side of the valve and one on the other side. These tubes connect the valve to the suction and discharge of the compressor as well as the vapor line connections of the heat exchangers.

Figure 3–9 Typical reversing valve (Courtesy Ranco Inc.)

The tube on the single connection side of the valve always connects to the compressor discharge. This puts high pressure vapor into the main body of the valve between the two piston heads of the valve slide assembly.

The center tube of the three-tube side of the valve always connects to the suction side of the compressor. The compressor receives the suction vapor from whichever heat exchanger coil is acting as the evaporator. This selection is determined by the position of the slide valve and piston assembly covering the combination of center port and outside tube port.

When looking at the valve with the three-tube side down and the single tube on the top, with the valve de-energized, the position of the slide valve will be to carry the suction gas from the left-hand tube to the center tube and compressor discharge gas from the single tube on the top to the right-hand tube of the three-tube side. Where this action uses the outdoor coil as the evaporator and the indoor coil as the condenser in this position depends upon the design selection for the particular unit.

This selection usually depends upon the major usage factor for the equipment design. This also will depend upon the particular market the manufacturer is targeting. Each manufacturer trys to design the unit to operate with a de-energized valve solenoid coil during the season of greatest demand. This not only tends to reduce the yearly consumption of electricity but also improves the valve solenoid life by reducing the annual number of operating hours. In Phoenix, Arizona, for example, the design requirements would be a de-energized valve during the cooling season. The demand for cooling is a greater number of hours than the demand for heating. In this case, the left-hand tube would be connected to the indoor coil (the evaporator) in the cooling mode. If the unit is designed for northern climates, the valve is de-energized during the heating season and the left-hand connection is connected to the outdoor coil (the evaporator) in the heating mode.

The problem is that units are not designed and built for each type of weather condition and location. Therefore, the service technician must be careful to note where connections are made to obtain proper operation if the valve should ever be changed.

Valve Operation

The valve is operated under the control of a pilot valve, but the actual changeover is accomplished by using the pressure difference between the discharge and suction pressures of the operating compressor. To see how this is accomplished, see Figure 3–10, which shows the valve connected to a compressor as well as an indoor heat exchanger and an outdoor heat exchanger.

The pilot valve is de-energized with the right port closed and the left port open. The pressure tube (No. 5) has been drained to the suction side of the valve and the slide has moved to the left, closing off the port to the left capillary (No. 5). In this position, the hot gas from the compressor flows to the outdoor heat exchanger and the cold vapor from the inside heat exchanger flows through the slide ports to the suction side of the compressor. This action shows that the unit is in the cooling mode with the valve de-energized.

To cause the valve to change position, a pressure difference of 75 psig or more must be exerted between the suction and discharge sides of the valve. Therefore, the valve will not operate with the unit idle. The compressor must have been operating long enough to build up to the minimum pressure or higher.

When the pilot valve is energized, the valve piston is raised off the right or coil end port. At the same time, pressure is taken off the left piston and the left piston spring pushes the piston onto the seat—closing the left-hand port (see Figure 3–11).

This action of the pilot piston causes the high pressure in the right capillary to drain into the common capillary and into the suction side of the compressor. The vapor in the right capillary drains through the bleed hole in the main piston. Therefore, the pressure in the chamber between the piston end and the cylinder drops. This creates a difference in pressure across the right end of the slide and the slide starts to move. The sliding action does several things in sequence:

1. The slide needle in the left end of the slide is removed from the seat and high pressure builds up in the left capillary tube and the end of the main valve body.
2. As the slide progresses, it starts to close the left tube port and open the right tube port. When the slide reaches the mid-position, high-pressure vapor is allowed to pass into the suction side of the compressor. This is a benefit to the compressor action as it raises the suction pressure and lowers the discharge pressure to reduce the pressure change shock on the compressor. The compressor will end up with a suction pressure higher than the discharge pressure when the slide action is completed. This momentary bypass of high-pressure vapor helps to reduce this changeover load.
3. When the slide reaches the right-hand position, the left port is open to the high-pressure vapor in the valve body, the

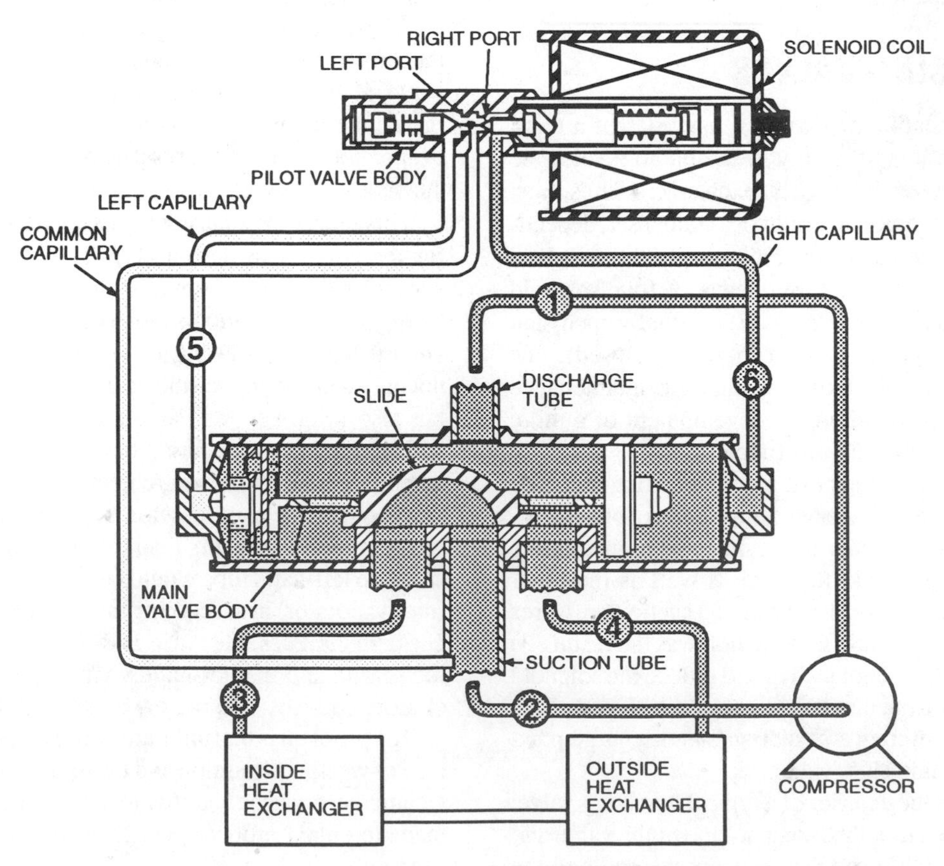

Figure 3–10 Reversing valve circuit de-energized (Courtesy Ranco Inc.)

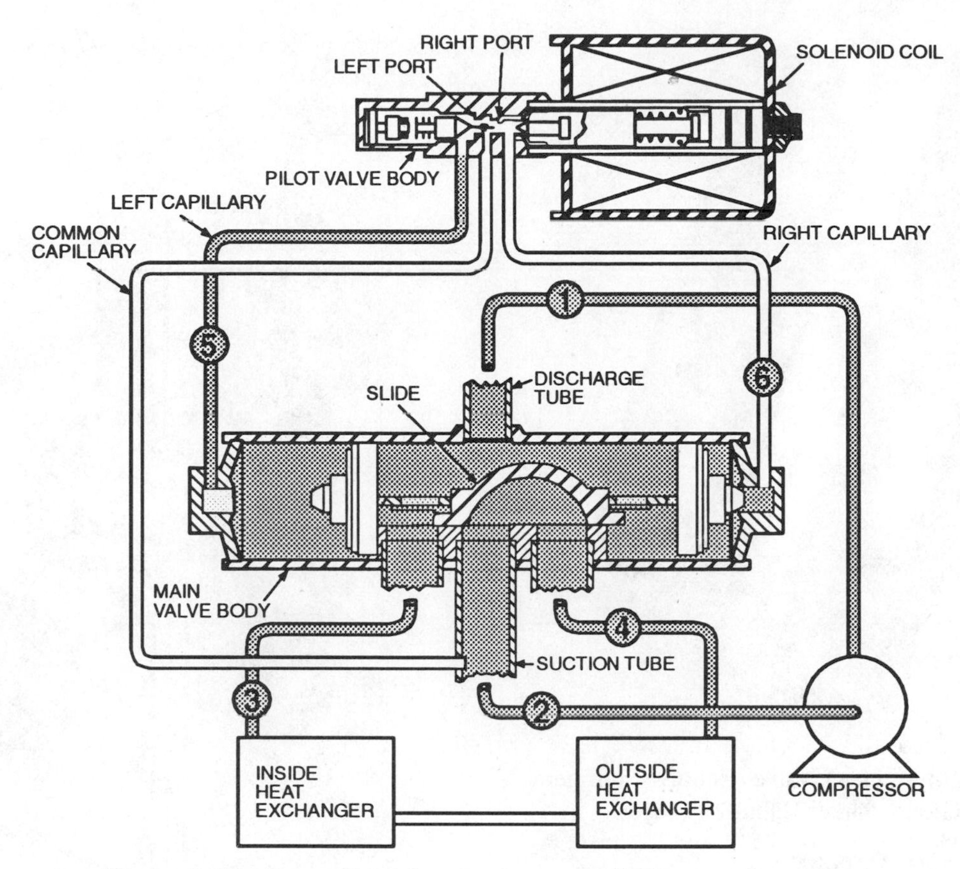

Figure 3–11 Reversing valve circuit energized (Courtesy Ranco Inc.)

right port is connected to the compressor suction, and the slide needle is inserted in the cut-off seat. Pressure in the right-hand capillary tube drops to suction pressure and the pressure on the slide needle builds to the discharge pressure. This pressure difference keeps the needle tight on the seat to prevent leakage and to keep the slide in the proper position. The slide stays in this position as long as the pilot solenoid valve is energized.

When the pilot valve solenoid coil is de-energized, the reverse action takes place:

1. The pilot valve right port is closed and the left port opened.
2. Pressure in the left capillary tube and left end of the slide valve chamber drops.
3. Pressure difference across the left end of the slide causes it to move.
4. The right port closes and the left port opens with the momentary pressure equalization when the slide is in the mid-position.
5. The slide reaches the left end and closes off the slide needle and seat. The unit is now in the original mode of operation.

Testing

Problems in reversing valves can be classified as either electrical or mechanical. This subject is covered in Section 2, Air-To-Air Type Systems, Chapter 7, Troubleshooting.

3–6 CHECK VALVES

As outlined in Chapter 1, check valves are used in parallel with the pressure-reducing devices on each heat exchanger coil to either force the refrigerant through the pressure-reducing device when the coil is the evaporator or bypass the pressure-reducing device when the coil is the condenser. The first check valves used in heat pumps were a flat disc type which occasionally caused a problem. The high-pressure shocks that can occur during a system changeover sometimes forced the disc into a position where it could not seat and the valve stuck open.

This problem was corrected by the introduction of the ball-and-cage check valve (see Figure 3–12). Examination of the valve shows that there is no spring to return the ball to the seat when no refrigerant flow occurs. It is, therefore, necessary to install the valve in a vertical position with the seat at the bottom. The ball returns to the seat by gravity when the refrigerant flow stops and assured shut off occurs with the reverse or downflow of refrigerant.

Upward flow of the refrigerant easily lifts the ball from the seat. This type of check valve is practically trouble free. The only possible potential problem is the ball sticking from cage warping due to excessive body heat when the valve is soldered in place. A Thermo-mastic material must be used on the body of the valve to keep it cool when the installation is made. Troubleshooting check valves will be discussed in the chapters covering each category of heat pumps.

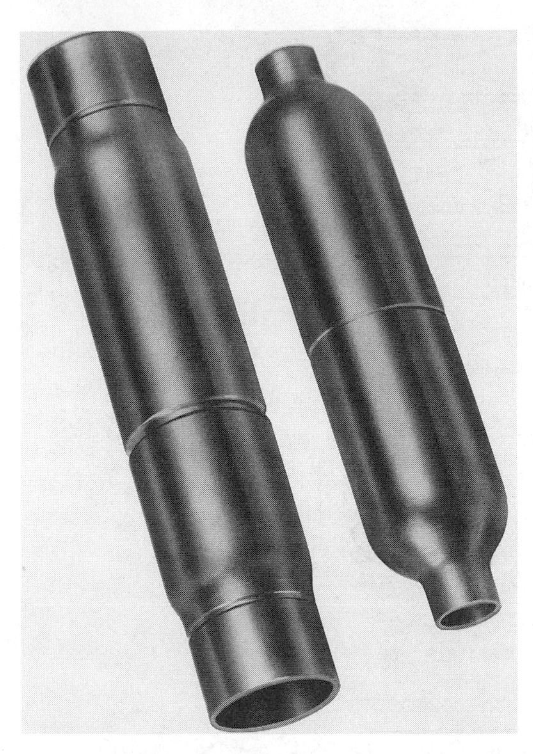

Figure 3–12 Ball-type check valve (Courtesy Refrigeration Components Group, Parker Hannifen Corp.)

3–7 ACCUMULATORS

During certain conditions of weather on the outside coil, in the heating mode, the humidity in the air is high and the coil operates below 32° F boiling point. Under these conditions, heavy frost buildup on the outside coil will occur. When the amount of frost buildup reaches the point where operating efficiency is seriously affected, a system of defrost control puts the heat pump in the defrost mode. This mode is the cooling mode with the outdoor fan off.

In the defrost mode, heat is taken from the inside air and forced into the outdoor coil by the evaporation and condensation of refrigerant. When the hot high-pressure vapor is forced into the outside coil, it gives up heat and condenses. In so doing, it heats the coil and melts the frost off the coil. The process continues until the coil liquid outlet temperature reaches the required temperature to ensure complete removal of the frost. This means the coil can be more than 50% full of hot liquid refrigerant.

At the defrost termination point, the defrost period terminates and the system reverts back to the heating mode. With this changeover, the outdoor coil is again the evaporator, containing hot liquid refrigerant, and connected to the suction side of the compressor. The suction pressure falls very rapidly and heavy boiling action of the liquid refrigerant takes place in the coil. This boiling causes liquid refrigerant to be carried out of the coil through the reversing valve and into the compressor unless a protection system is provided. Protection against this action is provided by the use of a device called an accumulator.

Figure 3–13 shows a cut-away of a typical accumulator. The flow of refrigerant vapor from whichever coil is the evaporator enters the accumulator through the side connection. Any liquid refrigerant and oil falls to the bottom. The liquid refrigerant vaporizes and, with the vapor from the coil, enters the *U* tube at the top of the chamber. From here, it passes to the compressor. The purpose of this action is to ensure that the refrigerant to the compressor is in vapor form. As the oil accumulates in the accumulator, it will build up until the oil-metering orifice is covered. When it reaches this level, oil bleeds into the tube and is carried by the flowing vapor to the compressor. This prevents excessive oil accumulation in the accumulator.

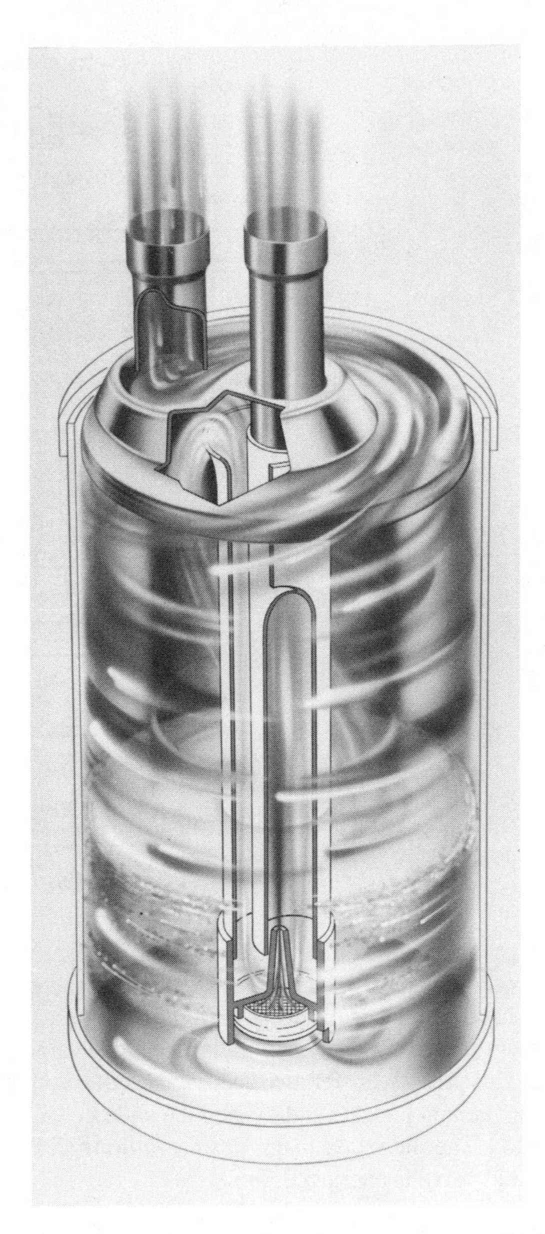

Figure 3–13 Cut-away of typical accumulator (Courtesy Tecumseh Products Co.)

This gathering of the liquid refrigerant to prevent the flow of liquid to the compressor is very important in heat pumps that require a defrost mode operation. An accumulator is a necessity in air-to-air heat pumps. It is connected in the system between the center port of the reversing valve and the compressor suction inlet (see Figure 3–14). This ensures accumulator protection regardless of the position of the slide in the reversing valve.

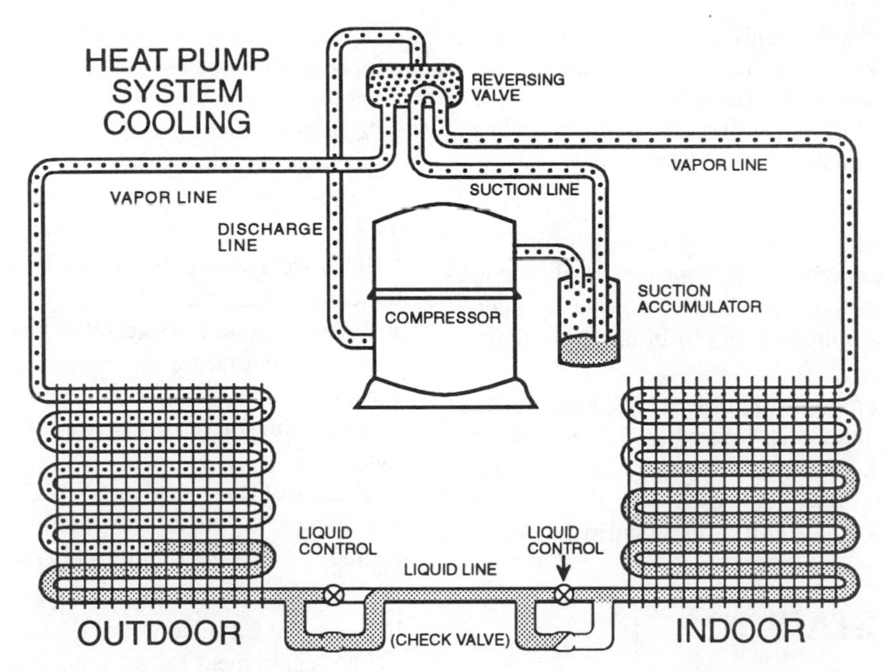

Figure 3–14 Accumulator added in refrigerant circuit (Courtesy Addison Products Co.)

3–8 ELECTRICAL SYSTEM

The electrical components in the text are purposely limited to those used in heat pump systems.

Contactors

Contactors used in heat pumps are the same in size and capacity as those used in other refrigeration and air conditioning systems.

Relays

In addition to standard fan, lockout, and isolation relays, a special relay is required in air-to-air heat pumps. Called the *defrost relay,* it is used in conjunction with the defrost control system to accomplish the defrost function. To effectively accomplish the defrost function in a minimum of time, four functions must take place:

1. The reversing valve must be operated to put the system into the defrost (cooling) mode.
2. The outdoor fan must be cut off to promote rapid buildup of heat in the outdoor coil.
3. The system must remain in the defrost mode until the outdoor coil has reached the required cut-off temperature.
4. Auxiliary heat must be supplied to reduce the supply air chill during the defrost cycle.

Figure 3–15 shows a typical defrost relay. This example has three sets of single-pole double-throw (SPDT) contacts to accomplish the necessary functions. Not all contacts provided are used. Rather than a special-made item, a standard item is used to accomplish the purpose. The contact sets are used as follows:

Contact Set No. 1. The normally open contacts between terminals 1 and 3 are used to control the action of the reversing valve. This action means that the system is designed to have a de-energized valve solenoid during the heating season. If the cooling season were the greater requirement and the unit were designed to have the reversing valve solenoid energized during the heating season, the normally closed contacts between terminals 1 and 2 would be used for controlling the reversing valve.

Contact Set No. 2. This set of contacts is usually used to hold the defrost circuit in by bypassing the defrost initiation means. This holding action is to prevent the defrost mode from terminating before the outside coil termination thermostat can respond from the proper coil temperature produced when the coil reaches a completely defrosted condition. In this set, the normally open contact set, terminals 4 and 6, is used. In addition, the outdoor fan motor is supplied power through the

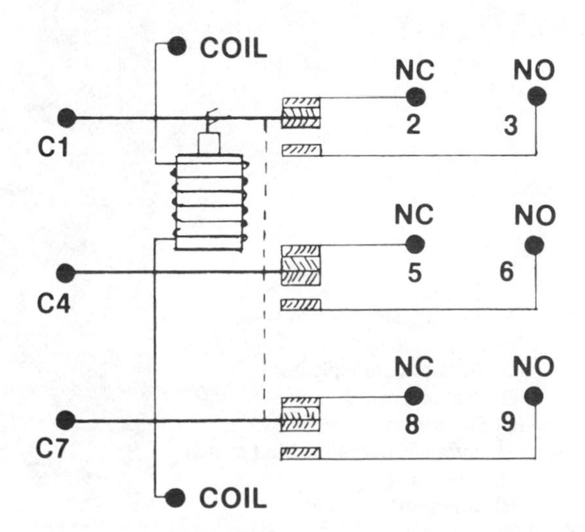

Figure 3–15 Typical defrost relay contact arrangement (Courtesy Addison Products Co.)

normally closed contacts, terminals 4 and 5. Thus, when the relay is pulled in to make the holding circuit, the circuit to the outdoor fan is opened and the fan stops. This combination of contacts is used to ensure the removal of the outdoor fan when the reversing valve is actuated for the defrost mode.

Contact Set No. 3. The third set of contacts is used to operate the auxiliary heat supply to reduce supply air chill when the system is removing heat from the supply air to accomplish the defrost function. This is a normally open contact requirement (terminals 7 and 9) in the low voltage circuit to the thermostat.

In older model heat pumps, multiple relays were used to accomplish the same functions. Regardless of the type or number of relays used, the four functions must be accomplished. Various arrangements are used with the different types of defrost systems. These will be discussed in the section on air-to-air heat pumps.

3–9 THERMOSTATS

Area Thermostats

All types of heat pump systems require the use of some type of area temperature control or thermostat. It may be a simple mechanical type with a single setting for the heating function and another for the cooling function with the selection of either function left to the area occupant, or it may be a totally automatic control (see Figure 3–16). Other features may be automatic set back for both the heating and cooling functions on a daily basis or be sophisticated to the degree of multiple set backs per day with each day of the week selected individually.

Regardless of the degree of sophistication, there are several basic functions the control must perform:

1. System totally off. A master switch is provided to cut off the entire system at the discretion of the area occupant.
2. System on cooling only. The thermostat must be able to operate the system—indoor blower as well as outdoor motor-compressor and outdoor fan in the cooling mode.
3. System on heating only. The thermostat must be able to operate the system—indoor blower as well as the outdoor motor-compressor and outdoor fan in the heating mode.

Here we see that functions 2 and 3 operate the same devices for both the heating and cooling modes. To accomplish the separate functions, the thermostat must provide a means of controlling the reversing valve to produce the desired operating mode. This is done by a contact in the season selector switch that controls a relay that controls the reversing valve.

These are the required functions. Most thermostats used on heat pumps, however, have additional functions that are desired and in some localities are required:

1. Room Air Circulation Only. This function is not an absolute requirement for the control of the area temperature. However, it has been a feature for so many years that it is accepted as standard. The inside blower operation is controlled by a fan "Auto" or "On" switch on the thermostat sub-base. This control, working in conjunction with the fan relay, provides the control function.
2. Two Stages of Heat. Heat pumps installed in areas where auxiliary heat is required to handle the heating design conditions must have separate stages of thermostat action for each individual heat supply. The thermostat, therefore, must have two separate switching elements or "two-stage heat."
3. Automatic Changeover. This is an optional feature supplied in some thermostats. This gives the area occupant the choice of heating only, cooling only, or the ability to have the system maintain a minimum and maximum temperature in the conditioned area.

(A) Heating temperature selector
(B) Cooling temperature selector
(C) Temperature indicator
(D) Systems switch
(E) Fan switch
(F) Normal/Emergency heat switch
(G) Check-Lite
(H) Emergency heat light

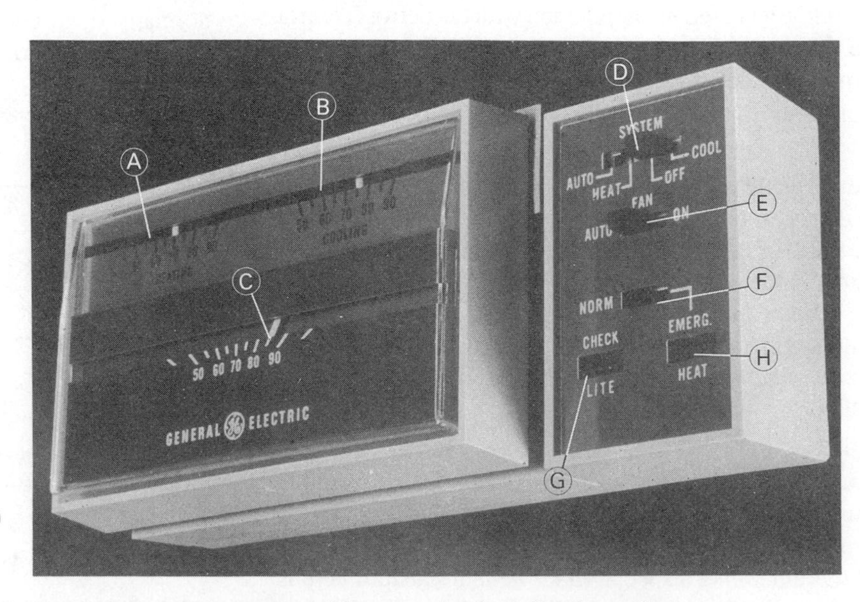

Figure 3–16 Typical heat pump thermostat (Courtesy Addison Products Co.)

4. Emergency Heat. A desirable feature is the ability to change the thermostat over to operate only the auxiliary heat in the case of malfunction of the heat pump system. When this feature is incorporated, a separate switch is provided for choice between "Emerg" and "Normal." A pilot light is usually incorporated with the switch to remind the area occupant that the more costly type of heat energy is being used.
5. Trouble Indication. To meet UL requirements, heat pumps are required to have manual-reset-type, high-pressure-cutout control operation. This is accomplished by either a manual reset control or through the use of an automatic reset control and a lockout relay combination. When the lockout relay combination is incorporated into the system, the thermostat usually has an indicating pilot light to show the lockout condition exists.

The actual tie in and operation of the thermostat in conjunction with each type of system will be covered in the sections on each type of system.

Defrost Termination Thermostats

In air-to-air heat pump systems, the outdoor coil is subject to frost formation and requires some type of defrost control system to take care of this problem. A necessary part of each control system is a temperature-sensitive device to terminate the defrost function when the process is completed. In some systems, the termination function is incorporated into the basic control. In other types of controls, a separate thermostat is incorporated into the system.

Figure 3–17 shows a defrost termination thermostat spring clip fastened to a piece of tubing. These thermostats are selected to provided defrost mode termination at a given temperature of the liquid refrigerant in the outdoor coil. The thermostat may be mounted on the liquid line leaving the coil or a portion of the coil itself. This position is determined by the manufacturer of the particular equipment. Any replacement of the thermostat should find the new part in the same position as the original part.

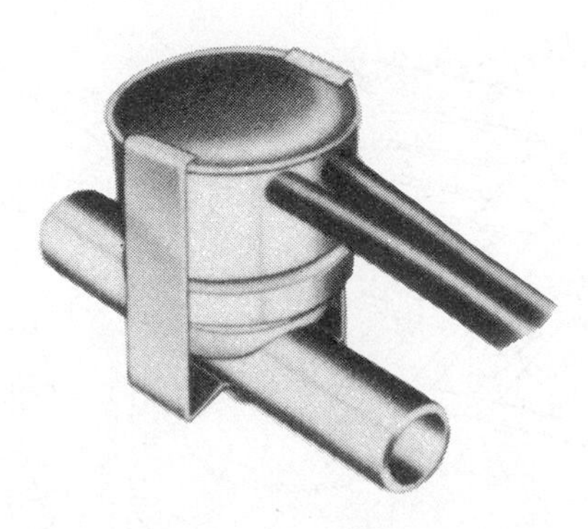

Figure 3–17 Defrost termination thermostat (Courtesy Addison Products Co.)

These thermostats are available in a range of cutout temperatures from 45° F to 90° F, usually in 5° increments. Therefore, care is necessary to select the correct cutout temperature as well when selecting a replacement part.

3–10 DEFROST CONTROLS

Applicable only to the type of unit using outside air as a source of heat energy, defrost controls are required to ensure operation of the system in cold humid climates. The greatest frost problems occur in the outdoor ambient range from 30° F to 45° F when the relative humidity is 50% or higher. With the outdoor coil, the evaporator in the heating mode, operating at 15° F to 20° F colder than the air entering the coil, moisture is condensed out of the air and freezes on the coil.

When frost occurs on the coil, it does two things:

1. It increases the resistance to air flow through the coil fins by blocking off the spaces between the fins.
2. It acts as an insulator. This requires the coil to operate at a colder temperature than normal to pick up heat from the outside air. The effect of frost accumulation is the same as reducing the size of the coil. The smaller coil operates at a higher split (the difference in temperature between the temperature of the air entering the coil and the coil operating temperature).

When the coil heat transfer rate is reduced, the coil temperature drops, suction pressure drops, and the system capacity drops. Therefore, when the capacity is reduced enough to affect the heating ability of the system, a defrosting action becomes a necessity.

Defrost Methods

Many methods of defrosting the outside coil have been tried. These include water wash, electric heating elements, supply air bypass, and so on. The most successful method is operating the system in the cooling mode to provide hot gas defrost using the heat picked up in the inside coil and the heat of compression from the motor-compressor assembly. This method, using various types of controls, is used by all heat pump manufacturers.

To accomplish the hot gas defrost method of clearing the outside coil, several things must occur at the same time:

1. The system is reversed and put into the cooling mode. Heat picked up in the inside coil is added to the motor-compressor heat and this gross heating capacity is used to warm up and clear the outside coil.
2. The outdoor fan is cycled off. To defrost the outdoor coil as rapidly as possible, the outdoor fan is turned off. This reduces air movement over the coil.
3. There must be a means of initiating the defrost action when sufficient amount of frost has accumulated on the outside coil.
4. A means of keeping the system in the defrost mode until the outdoor coil is clean of frost and ice must be supplied.

5. There must be a means of heating the air off the inside coil during the defrost function (cooling mode) to reduce cold air drafts in the conditioned area.

All of the functions are accomplished with various types of defrost control systems developed over the years.

3–11 TYPES OF DEFROST SYSTEMS

Defrost systems can be classified in the following categories according to the factors used for initiation and termination:

1. Temperature initiation/temperature termination.
2. Time initiation/temperature termination.
3. Static pressure initiation/temperature termination.
4. Static pressure-time initiation/temperature termination-time termination.

Practically all defrost control systems are a variation of these four categories.

Temperature Termination

Before discussing each type of defrost control system, there is one common denominator of all four categories—the temperature of the outdoor coil is used to determine when the defrost action is terminated. Testing is done by each manufacturer to determine where the last bit of frost is melted and the coil reaches a temperature high enough above freezing temperatures to ensure this action.

This location is usually the liquid off the bottom of the outside coil where the temperature of the liquid refrigerant leaving the condenser is measured. A thermostat located at this point will open and interrupt the defrost function when the liquid temperature rises to its opening set point. Some manufacturers use a location two or three passes above the bottom of the coil fin area. Regardless of the location used, the service technician should replace controls using the same characteristics as the original control and temperature-sensing locations the same as the original ones.

Temperature Initiation/Temperature Termination

Mechanical Type. The first defrost controls were the mechanical temperature initiation/temperature termination type. Produced and marketed by Ranco Inc., the control used two temperature-sensing bulbs. As seen in Figure 3–18, the temperature-sensing bulb on the left is a straight-tube type. This is mounted in the air stream on the air inlet side of the outdoor fan. This measures the temperature of the air before any work is done on the air by the fan or coil. The temperature-sensing bulb on the right is mounted against the tube in the coil. This measures the operating temperature of the coil (see Figure 3–19). The manufacturer determined that when this point in the coil reaches the desired temperature, the entire coil is free of frost and ice and is in a condition to resume the heating mode.

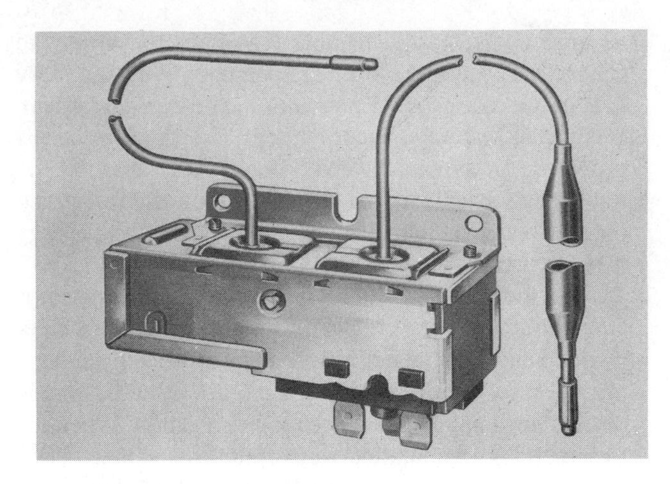

Figure 3–18 Two-bulb defrost control (Courtesy Ranco Inc.)

Under normal clear conditions, the outdoor coil (evaporator in the heating mode) operates with a design temperature difference between the temperature of the air entering the unit (outside ambient temperature) and the coil operating temperature of 15° F to 25° F. On most of the units using this defrost control, the temperature difference was about 15° F at 40° F outside ambient temperature down to 10° F at 10° F outside ambient temperature. This was well within the range of expected frost formation. Frost formation is practically nonexistent above 45° F to 48° F. To prevent the possibility of false defrost functions, the temperature-sensing bulb in the air stream was purposely charged to prevent the possibility of control operation above 45° F.

The control operated on the "split" of the coil—the difference in temperature between the ambient air temperature and the coil-operating temperature. When the split remained constant, with a clean coil, the control remained in the heating mode. When frost began to form on the coil, the frost acted as an insulation and caused the coil temperature to lower in order to absorb heat. This increased the split of the coil. When the split increased to 20° F to 22° F, the control contacts closed. This energized a defrost relay. The relay closed and operated various sets of contacts to perform the four functions of the defrost mode.

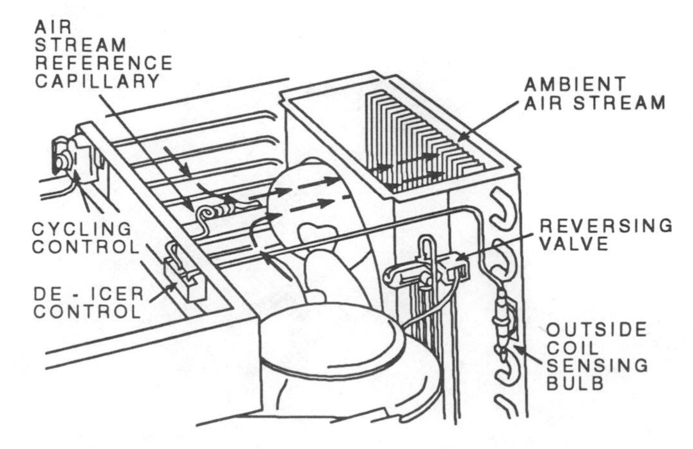

Figure 3–19 Defrost control installation (Courtesy Ranco Inc.)

When the contacts in the control closed, a locking device tripped and anchored the contacts in a closed position. This position was maintained until the temperature-sensing bulb on the coil reached the required temperature. This was usually 65° F. The locking device eliminated the effect of the ambient air temperature-sensing bulb to ensure complete defrost regardless of the outside temperature.

When the temperature of the coil temperature-sensing bulb reached the clean coil temperature, the locking device was released. The contacts opened, the relay was deactivated, and the system returned to the heating mode.

This system worked as intended when the refrigeration system was properly charged with refrigerant and the *AIR QUANTITY THROUGH THE INSIDE COIL WAS CORRECT*. The unfortunate part was that the control was used on improperly designed units, applied to improperly designed air distribution systems by an industry with little experience with heat pumps. The control depends upon the correct amount of air over the inside coil (condenser) in the heating mode to keep the condensing pressure of the system in the normal range. This allows the unit to operate with the correct operating temperatures.

Unfortunately, the consumer was not ready to accept "cool air" heating. This caused the reduction of air through the inside coil to raise the temperature of the supply air. The result was an increase in the condensing pressure in the system. The increase in pressure raised the flow rate of the pressure-reducing devices and increased the amount of liquid refrigerant to the outside coil. This raised the coil operating temperature.

Although the rise in coil operating temperature was only in the 3° F to 5° F range, it forced the coil to build up more frost before the coil operating temperature became low enough to raise the coil split to the initiation temperature. In some cases, the frost buildup was so great that frost back to the compressor caused high compressor failure rates. As a result, repeated control replacements failed to correct the problem and the use of this control was discontinued. The control was an excellent control, but the industry was not ready for it.

Electronic Type. The development of the solid state (electronic) control is one of the improvements made in the heat pump industry. Along with the knowledge of the proper methods of installing and adjusting heat pump systems, the industry designed and built true heat pump equipment. This reduced the possibility of air problems as well as adverse operating conditions. The consuming public became better acquainted with how a heat pump operates. This naturally lead to the development of better defrost control methods. This lead to the development of the solid state defrost control.

Figure 3–20 is the board assembly of a solid state defrost control. Using two temperature-sensing thermistors, the control measures the outdoor ambient temperature and coil operating temperature and compares the two to determine the coil split. When the split is in the normal operating range, the unit functions in the heating mode.

The wiring diagram in Figure 3–21 shows a solid state defrost control using two thermistors connected to the control body and connections to control a defrost relay. The thermistor on the left reduces resistance on temperature rise and measures outdoor ambient temperature. The one on the right reduces resistance on temperature fall and measures coil operating temperature.

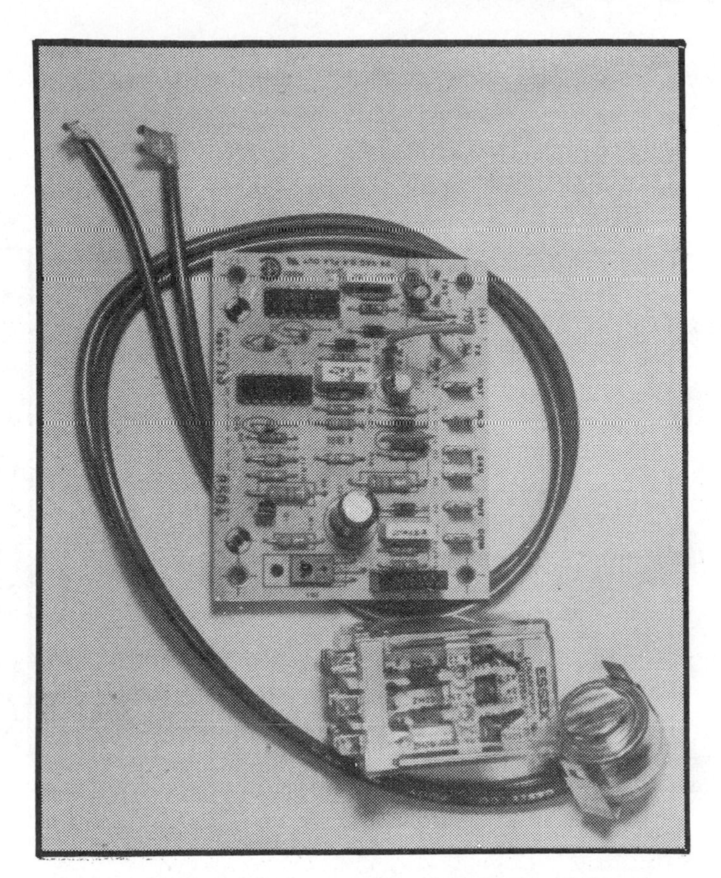

Figure 3–20 Solid state defrost control board (Courtesy Rheem Manufacturing Co.)

When the difference in temperature between the two thermistors increases, the change in the resistances creates an unbalance in the board electrical circuit and the relay control circuit closes. This supplies power to the defrost relay and energizes the defrost relay coil. By means of opening and closing contacts in the defrost relay, all four required functions of the defrost mode occur. When the outside coil temperature reaches the required temperature, the circuit through the coil temperature-sensing thermistor reaches the override point and the control de-energizes the defrost relay. This puts the system back in the heating mode.

This system also has the advantage that the defrost control, through the action of the defrost relay contact (terminals 4 and 6) bypasses the control of the area thermostat. This means that the defrost function is completed even if the area thermostat is satisfied and opens the HTG-1 contact in the thermostat.

Time Initiation/Temperature Termination

When the industry encountered air problems with heat pumps causing erratic defrost control operation, another means of defrost control was developed. This involved the use of a time clock to determine when the defrost function should start. A coil-mounted thermostat was used to terminate the

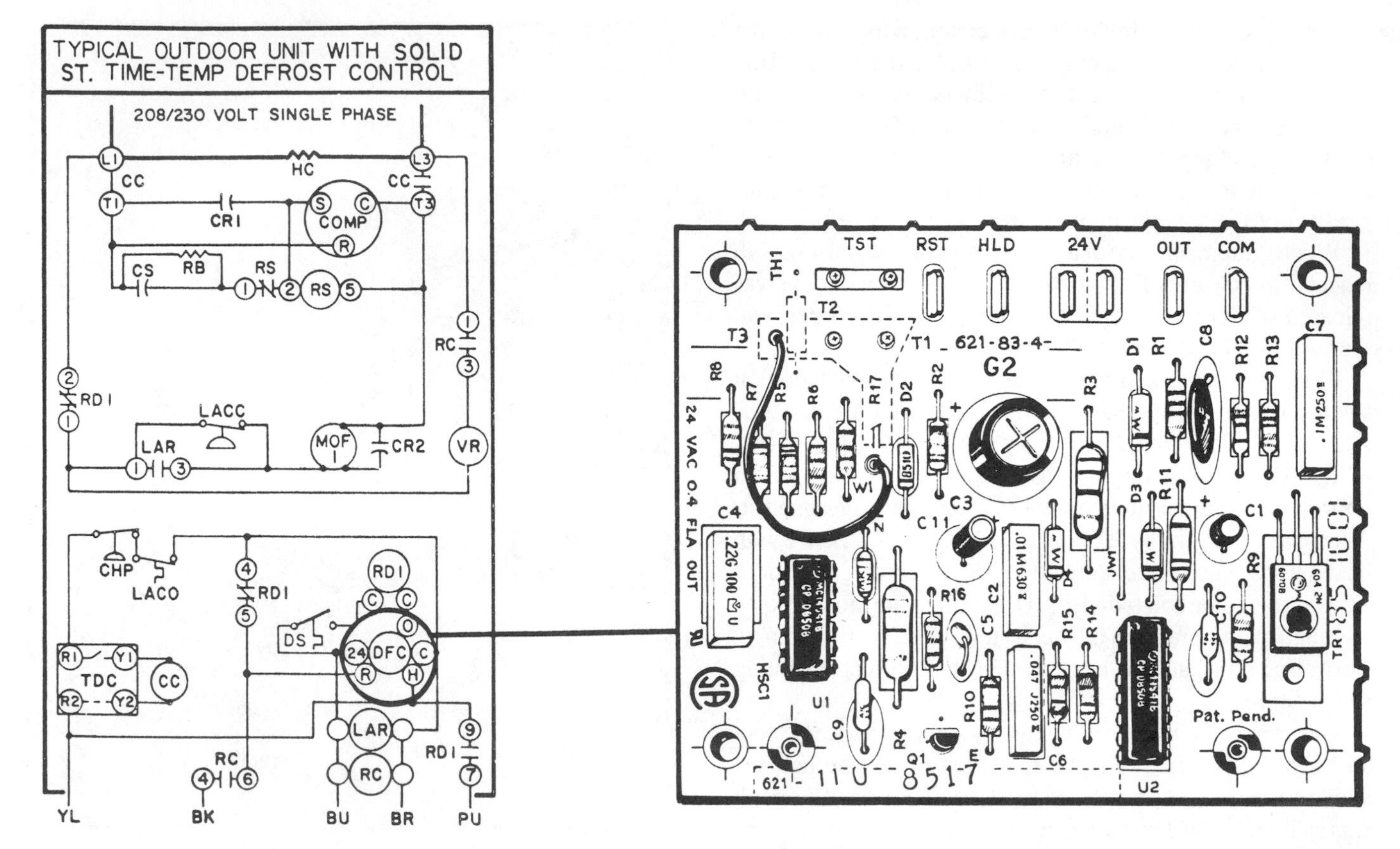

Figure 3–21 Solid state defrost control circuit (Courtesy Rheem Manufacturing Co.)

defrost function. The thermostat pictured in Figure 3–17, mounted on the outside coil was used for two purposes:

1. To allow the defrost timer to operate if the coil temperature dropped below 26° F.
2. To open the defrost relay control circuit when the coil temperature reached the design temperature that produced a clean coil. This could be any temperature between 45° F and 65° F, depending upon the design of the particular unit. Most units were designed to use a termination thermostat with an opening set point of 55° F.

A typical installation of a time initiated/temperature terminated defrost control system is shown in Figure 3–22. The defrost timer is located in the upper portion of the control panel and controls the defrost relay. The termination thermostat, which is mounted on the coil behind the left panel, is not shown.

Power to operate the timer motor is usually taken off the load side of the compressor contactor. This means the defrost timer can only operate if the compressor motor is operating. This means that under potential frosting conditions, the system will go into defrost mode every 30 to 90 minutes of compressor operation. To prevent false defrost periods, the termination thermostat, in the timer motor circuit, keeps the circuit open and the timer off if the coil temperature is too warm to gather frost. This limiting temperature is usually 26° F.

Therefore, for every 30 to 90 minutes that the system is operating with an outside coil temperature below 26° F, a defrost function will be initiated automatically. In areas of mild possibilities, the 90-minutes cam setup is used. In areas of high humidity around freezing temperatures (30° F to 40° F), the 30-minute time period is required. A double set of cams is usually supplied with the timer so time selection can be done in the field.

Figure 3–22 Typical heat pump electrical box (Courtesy Addison Products Co.)

The timer is used to start the defrost function and the terminating thermostat is used to stop the defrost function. Both work in conjunction with a holding contact in the defrost relay. A typical time clock defrost control circuit is shown in Figure 3–23.

As shown, power is taken off the load side of the CC contact at 3 on the left of the diagram. This is carried through the defrost terminating thermostat (DFTH), through the timer motor (TM) to the compressor contactor contact (T2-CCO). With the compressor motor operating (CC contacts closed),

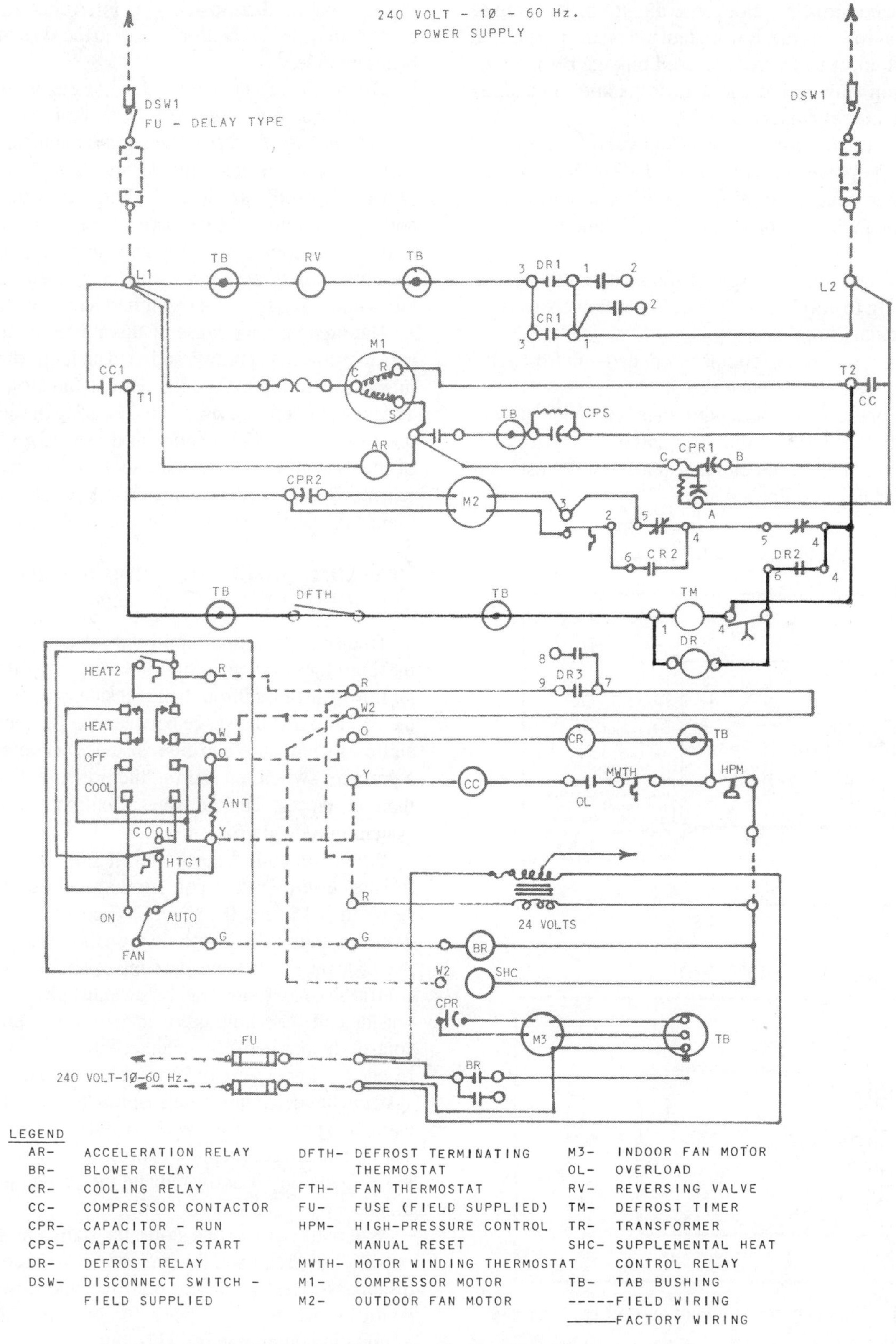

Figure 3–23 Time initiated/temperature terminated defrost control circuit

when the temperature of the defrost terminating thermostat closes, the timer motor is actuated.

The timer motor shown in Figure 3–24 has a set of contacts controlled by a cam. At the end of the operating period, either 30 or 90 minutes depending on the cam arrangement used, the bottom leaf of the switch arrangement drops into the cam groove. The cam turns in a clockwise direction. The drop of the bottom leaf relieves the upward tension on the top portion of the contact imposed by the insulated linkage between the bottom and top leaves. The top leaf drops and the making contact of the circuit closes.

This completes the circuit from the defrost relay coil to the other side of the power supply (terminal T2). The relay coil magnetic action causes the relay to pull in and perform the necessary functions to complete the defrost function:

1. The reversing valve is energized and switches the system from heating to cooling mode—defrost relay contacts set DR1, terminals 1 and 3.
2. The outdoor fan motor circuit is opened—defrost relay contact set DR2, terminals 4 and 5.
3. Reheating of the air off the inside coil is activated—defrost relay contact set DR3, terminals 7 and 9.
4. A holding contact in the circuit is closed—defrost relay contact set DR2, terminals 4 and 6.

90 MIN. CAM AS SUPPLIED

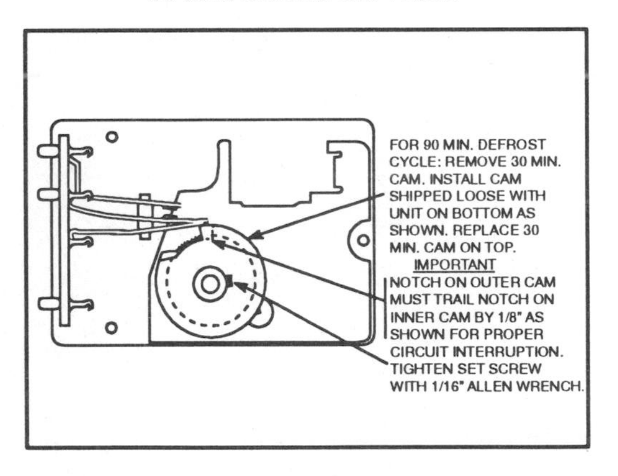

30 MIN. CAM

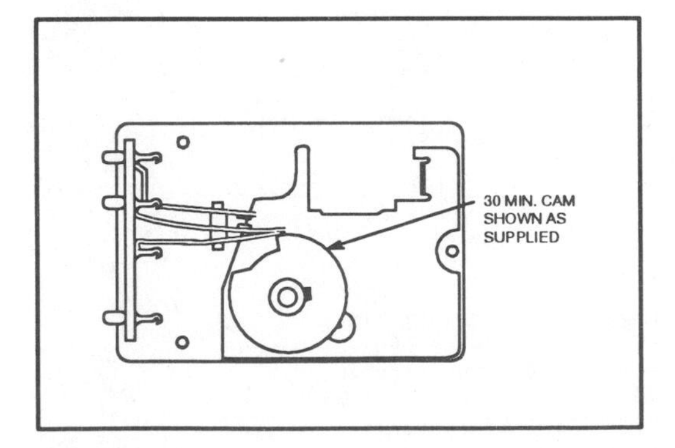

Figure 3–24 Timer motor and cam assembly (Courtesy Addison Products Co.)

As the time clock continues to turn, the cam movement causes the middle leaf of the contact set to drop. This opens the making contact of the defrost relay circuit. The relay stays in, however, because it is energized through the holding contact in the defrost relay itself. The system stays in the defrost mode until the coil temperature reaches the opening temperature of the terminating thermostat. This interrupts the power to the defrost relay, the relay drops out, and the system returns to the heating mode.

The major problem with this system is that *any power interruption to the compressor motor contactor will automatically terminate the defrost function.* This statement is emphasized because this premature termination of the defrost cycle will leave quantities of ice at the bottom portion of the outside coil. The melting of the ice has not been completed. As a result of this ice accumulation and the thawing and freezing of the ice immediately adjacent to the coil tubes, collapse of the tubes under the fins will occur and leaks develop.

The predominant cause of this power interruption is the area thermostat becoming satisfied and interrupting the compressor operation before the defrost function is completed. The area temperature rises rapidly during the defrost function due to too much reheat being used. An ice ring at the bottom of the outside coil is an indication of the use of too much reheat. The amount of reheat used should not exceed the "sensible" heat capacity of the unit in the cooling mode.

Pressure Initiation/Temperature Termination

To eliminate the possibility of false defrost periods with the time clock system, a control was developed to measure the air flow resistance through the outside coil. As frost forms on the coil and fills the space between the fins, the resistance to air flow increases. When this resistance becomes high enough, a pressure switch, measuring the pressure difference across the coil, closes. This activates a defrost relay and puts the system into the defrost mode.

When the coil is clean and at full air flow, the design resistance through the coil (static pressure drop) is usually between 0.15 and 0.25" wc. When the frost on the coil develops enough resistance to raise the static pressure drop to the 0.5 to 0.65" wc range, the contact in the control closes. The control also has a temperature-sensing bulb that attaches to the outside coil. This bulb is connected to an override arm in the control that prevents the contact from closing until the bulb reaches a temperature of 26° F or below. This means that the system cannot go into a defrost mode unless the coil is cold enough to produce frost. With the bulb cold enough, when the static pressure reaches the control initiation pressure point, the defrost relay is activated and the system is in the defrost mode.

When the outdoor fan stops, the static pressure across the outside coil drops and the pressure difference in the control disappears. The termination bulb override, however, locks the control contacts closed until the bulb reaches the desired termination temperature. This releases the contact, which opens and interrupts the power to the defrost relay. The system returns to the heating mode.

The major problem with the control is the wind. If the wind blows against the discharge opening of the unit, it prevents the development of the necessary static pressure to start the defrost function until excessive frost formation occurs. This affects the heating of the occupied area. It may even prevent the defrost function from functioning.

If the control is adjusted for the wind effect, the control becomes too sensitive and false defrost functions will occur.

Pressure-Time/Temperature Termination

To overcome the false defrost functions possible with the time clock method of initiation and the effect of wind pressure on the pressure initiation system, a combination of the two was developed. Using a clock-driven cam to control the defrost relay and the outdoor fan motor, under the control of a static pressure switch, the static pressure drop has to be at the initiation point for a period of time long enough for the control to act.

The wiring diagram in Figure 3–25 shows the wiring of a Ranco E15/D20 defrost system. In normal operation, the system is powered when the compressor contactor is activated. The system is connected across the same contactor terminals as the motor-compressor circuit. In the normal operating position, the D20 pressure switch is open, the clock in the E15 is not operating and the contact between 2 and 3 in clock assembly is closed. The outdoor fan is operating along with the motor-compressor assembly.

When conditions exist that frost formation can accumulate and the coil temperature drops below 26° F, the action temperature of the outdoor coil sensing bulb, any frost formation will raise the static pressure drop across the coil. When the static pressure drop reaches the closing point of the D20 pressure control contacts, a circuit is made and the timer motor is energized.

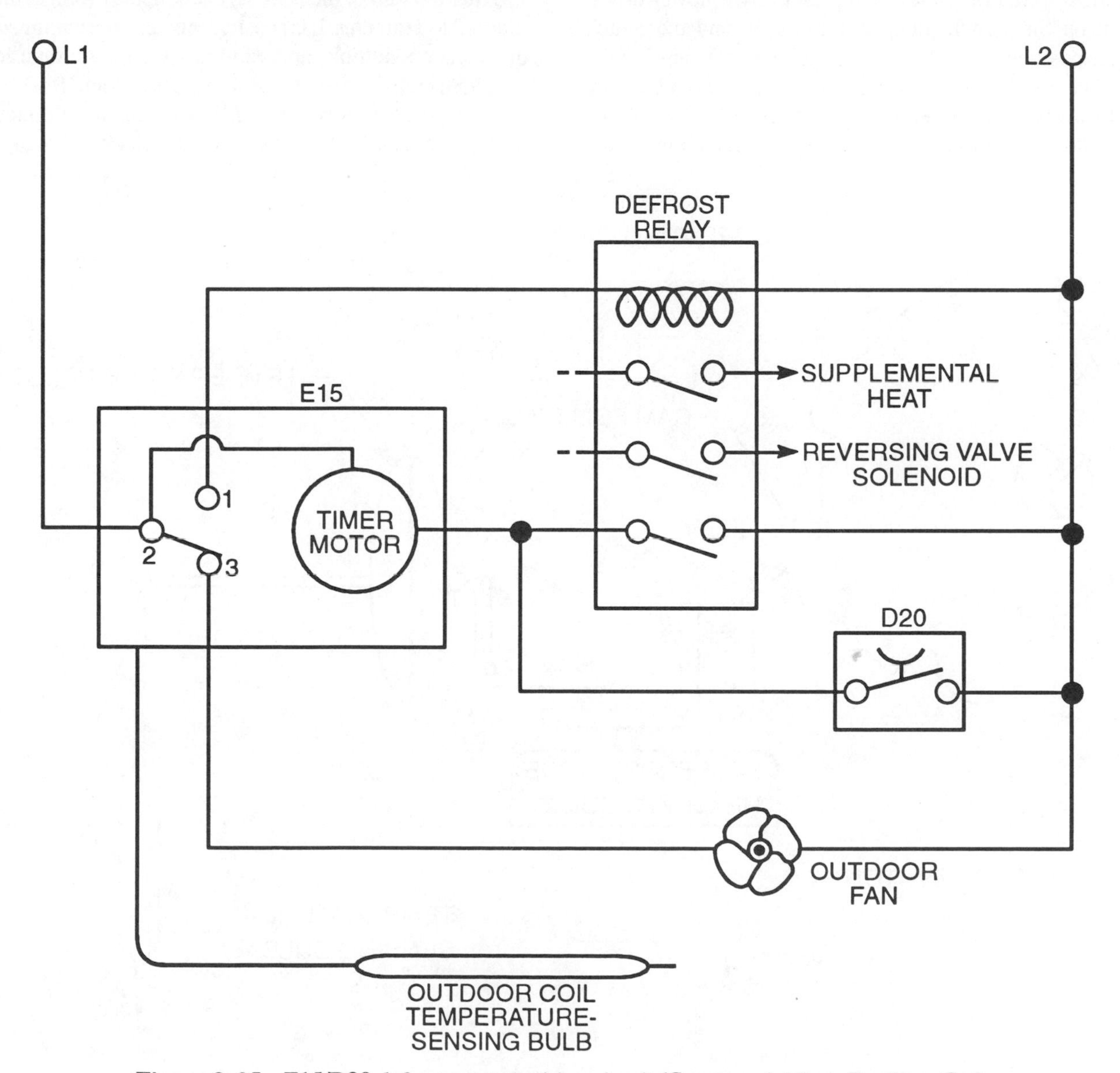

Figure 3–25 E15/D20 defrost control wiring circuit (Courtesy Addison Products Co.)

The timer motor drives a cam that is divided into four segments of 15 minutes. Figure 3–26 shows the arrangement of the cam with its follower and the bellows of the temperature-sensing bulb. These two devices are linked via the crank arms to the SPDT switch that controls the outside fan motor and the defrost relay. In this illustration, we see that the temperature bellows and crank arm prevent any switch action when the bellows pressure is above the 26° F pressure equivalent. Regardless of any action by the cam and cam follower, the control cannot operate the contacts. When the temperature-sensing bulb is below 26° F, the bellows pressure is down and its crank arm is released. The action of the timer cam can actuate the switch contacts through the primary switch crank arm.

The illustration shows the cam follower on the top elevation, which terminated the previous defrost mode. When the next demand for the defrost function occurs, the cam turns in a counterclockwise rotation. When the cam follower reaches the defrost initiation valley in the cam, the follower drops, tension on the switch plunger is released, and the switch changes position. Contacts 2 and 3 open and contacts 2 and 1 close. The opening of contacts 2 and 3 stops the outdoor fan. The closing of contacts 2 and 1 energizes the defrost relay. The relay action performs the remaining functions to initiate the defrost function, including a holding contact around the D20 pressure switch contacts.

The system is now in the defrost mode. When the outdoor coil is defrosted and the temperature-sensing bulb reaches the correct temperature, the pressure of the bellows, through the action of the crank arm, forces the switch to reverse its action by raising the primary crank arm away from the follower. This deactivates the defrost relay, stops the timer, and reactivates the outside fan motor.

The minimum time between defrost functions is regulated by the time necessary to complete each defrost function. Defrost functions are on a 15-minute interval. As illustrated in Figure 3–27, the time between defrost initiation and the next initiation is 15 minutes. The time from when the follower reaches level 1 on the cam and the defrost function is completed until the follower reaches level 3 to initiate the next defrost function is 5 minutes. This means that the minimum time between defrost functions that can occur is 5 minutes.

After the start of the defrost function, the average function time is 2 to 3 minutes. It is possible, under some circumstances of weather conditions and wind to increase the time for the bulb temperature to rise to the termination point. It is possible in extreme conditions for the unit to fail to terminate the defrost function. With this system, the clock cam will termi-

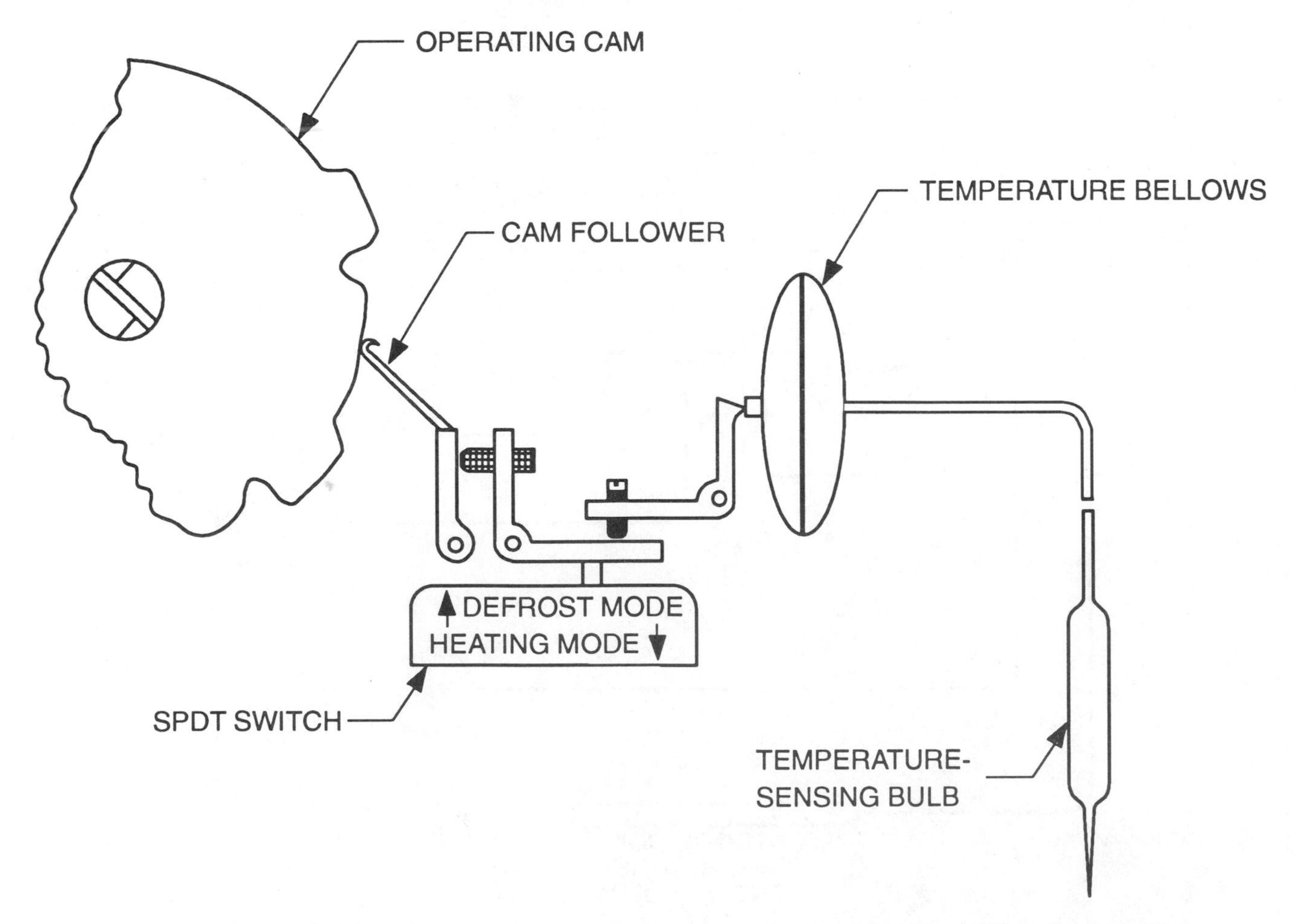

Figure 3–26 E15/D20 time/temperature control operating mechanism (Courtesy Addison Products Co.)

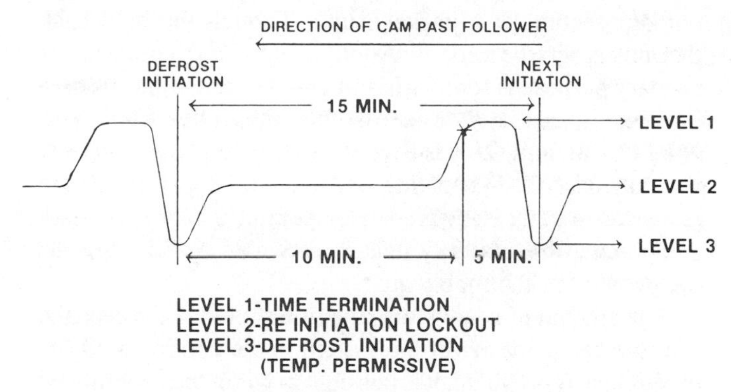

Figure 3–27 Operating cam configuration and sequence (Courtesy Addison Products Co.)

nate the defrost function after 10 minutes regardless of the termination bulb temperature.

Figure 3–28 shows the possible defrost function time for three different weather conditions. Keeping in mind the fact that the total time period on the cam is 15 minutes, the time of the defrost function and the time between defrost functions will vary but not exceed the 15-minute time period.

In the first example, with no wind to affect the buildup of heat in the outside coil and a relatively warm outside temperature, the defrost function termination is strictly by the rise in bulb temperature. If this is a 2-minute time period, the next defrost function will require 13 minutes of the bulb below 26° F and the pressure switch closed.

If wind effect causes a delay in the coil temperature buildup to, for example, 8 minutes before the defrost period is terminated by the bulb reaching termination temperature, only 7 minutes of pressure switch closure and bulb at 26° F is required.

In those rare occasions where the wind is very strong and the outdoor temperature is low enough to prevent the coil temperature from reaching the termination setting, the clock will automatically terminate the defrost function after 10 minutes. Only a 5-minute period of pressure contact closure and termination bulb temperature below 26° F is required to initiate the next defrost function. It is possible under extreme conditions for the defrost function to repeat in 15-minute increments—5 minutes off and 10 minutes on until the coil is completely clear.

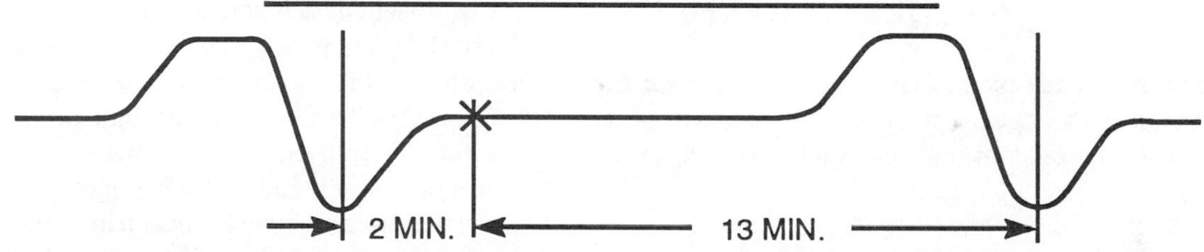

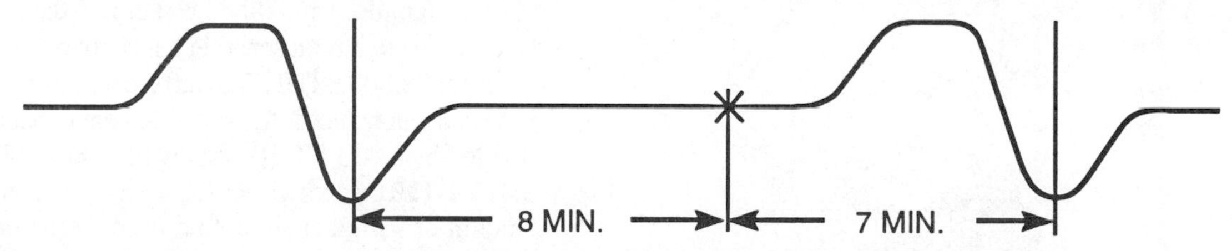

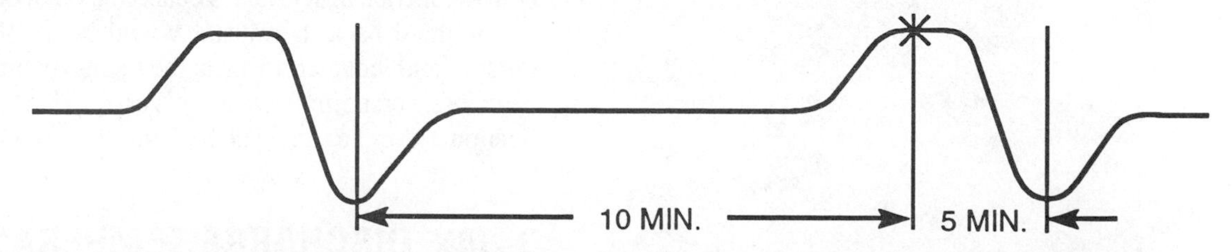

Figure 3–28 Examples of typical defrost cycles (Courtesy Addison Products Co.)

Many varieties of defrost controls have been specified by various manufacturers of heat pumps. Service technicians should gather and retain such information.

3–12 OUTDOOR AMBIENT THERMOSTATS (OATs)

When auxiliary electric heat is required in the heat pump installation, the auxiliary heat operation is under the control of the second stage of the conditioned area thermostat. If the auxiliary electric package is more than 10kwh, electric utilities may require that the elements in the package be staged under the control of an outdoor ambient thermostat (OAT). Figure 3–29 shows a remote bulb thermostat with a set of single-pole double-throw (SPDT) contacts. In an enclosure case, it can be mounted outdoors under a protective overhang to measure the outdoor temperature. This would act as an outdoor ambient thermostat (OAT) to control the second stage of the auxiliary heat.

When three or more stages of auxiliary heat are involved, multi-stage thermostats are available or duplicates of the one in Figure 3–29 can be used. Available in a range from –30° F to +90° F, they can cover practically all applications. Specific application of outdoor ambient thermostats will be covered in Section 2, Air-to-Air Type Systems, Chapter 4, Application.

3–13 HOLD BACK THERMOSTATS

The term *hold back* thermostat is another name for the outdoor ambient thermostat. It is usually supplied as an integral part of the outdoor unit in a split system or in the outdoor section of a package unit. Though the hold back thermostat acts the same as an outdoor ambient thermostat, its primary purpose is for a different type of protection. When a heat pump goes into the defrost function, it becomes an air conditioning unit. This is necessary to obtain heat to defrost the outside coil. When this occurs, the supply air to the conditioned space is lowered in temperature. To prevent draft complaints, the auxiliary heat is activated to bring up the temperature of the supply air.

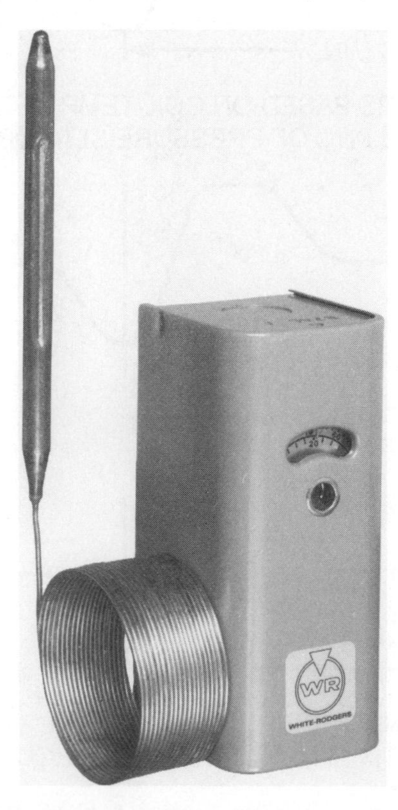

Figure 3–29 Remote bulb-type thermostat (Courtesy White Rodgers, a Division of Emerson Electric Co.)

The amount of heat in this situation should not exceed the sensible heat capacity of the system in the cooling mode. If excess heat is provided, it is possible to warm the conditioned area to the temperature setting of the area thermostat before the defrost function is completed and all the ice has been removed from the outside coil. If this occurs, the room thermostat will cut off the power to the compressor and the defrost control system. This is the same as terminating the defrost function, but the actual defrost function is not completed. Water is held between the ice coat and the coil on the outdoor coil.

When the system starts again, the water refreezes and forces water between the fin collar and the tube it is wrapped around. Repeated cycles of this action will eventually cause the refrigerant tube to collapse in the coil and a leak to develop. The only prevention of this damage is a complete defrost function each time the defrost function is demanded.

The problem is easy to detect as it consists of an ice layer at the bottom of the outdoor coil. The predominant cause of the "poor defrost" ice layer is too much reheat in the defrost function. The prevention, therefore, is to limit the amount of reheat produced in the first stage of the auxiliary heat to the sensible heat capacity in the heat pump system in the cooling mode. If the sensible heat capacity is not stated in the manufacturer's literature, a safe assumption is 80% of the total capacity of the unit in the cooling mode.

For example, a 36,000 BTUH unit in the cooling mode can be assumed to have a sensible heat capacity of 36,000 BTUH × 0.80 or 28,800 BTUH. To find the maximum kwh of electric heat that can be used, the sensible heat capacity of the unit is divided by 3,413 BTUH per kwh (28,800 BTUH divided by 3,413 BTUH per kwh = 8.4 kwh). This is the maximum amount of kwh that should be used. If the unit has 4.8 kwh elements, only one element should be used. It is better to sacrifice a small amount of air temperature change during the defrost function than to take a chance on outdoor coil breakage.

The hold back thermostat would be set the same as an outdoor ambient thermostat at the second balance point of the heat loss versus unit capacity curve. This is explained in Section 2, Air-To-Air Type Systems, Chapter 4, Application.

3–14 DISCHARGE TEMPERATURE THERMOSTATS

When heat pumps are added to fossil fuel units, the fossil fuel unit is used as the tempering means to prevent the cold air complaints during the defrost function. If excessive tempera-

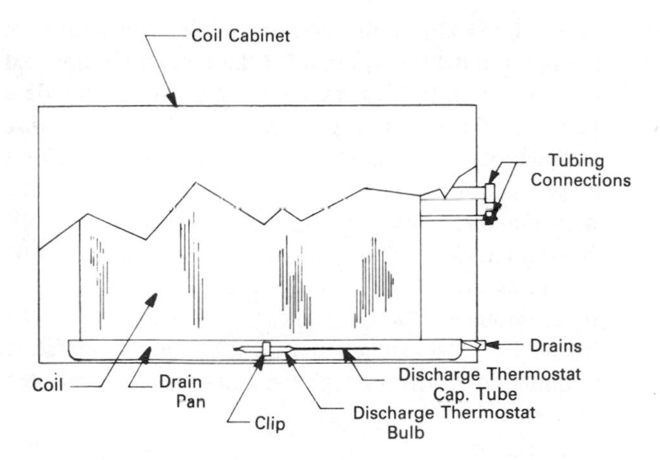

Figure 3–30 Discharge thermostat bulb installation (Courtesy Addison Products Co.)

tures are supplied to the indoor coil, which is the evaporator during the defrost function, damaging temperatures and pressures could result in the refrigeration system.

Figure 3–30 shows a remote-type thermostat temperature-sensing bulb located adjacent to the inside coil and used to cycle the fossil fuel supply to provide a maximum air temperature of 95° F to the coil. This thermostat would be the same as shown in Figure 3–29, using the normally closed set of contacts adjusted to open at 95° F.

REVIEW QUESTIONS

1. What two types of heat exchangers are used in heat pumps?
2. Heat pump heat exchangers are designed to transfer heat in one direction only. True or False?
3. Air-to-refrigerant type coils are made in four different configurations. Name them.
4. When a propeller fan is used to move air through the outdoor coil, duct work to carry the air is allowed up to 10' distance. True or False?
5. The main purpose of a de-ice subcooler coil is to increase the amount of subcooling of the liquid refrigerant. True or False?
6. The use of a de-ice subcooler coil has no effect on the performance of the unit. True or False?
7. What is the advantage of using a TX valve as the pressure-reducing device on the outdoor coil of a heat pump?
8. Air-to-refrigerant coils used in heat pumps will work equally well with bottom or top feed of the liquid refrigerant into the coil. True or False?
9. To get the highest heat transfer rate in a coaxial-type heat exchanger, the liquid and refrigerant flow in the same direction for constant temperature difference. True or False?
10. A different type coaxial coil is used for horizontal applications than for vertical applications. True or False?
11. It is only necessary for a service technician to carry one type of TX valve when servicing A/C units and heat pumps. True or False?
12. Standard size capillary tubes are used on each size of heat pumps in the United States. True or False?
13. Reversing valves will provide trouble-free operation regardless of the position in which they are mounted. True or False?
14. On a four-way reversing valve, the compressor discharge is connected to what connection on the valve?
15. The compressor suction is always connected to what connection of the four-way reversing valve?
16. What is the minimum pressure differential across the reversing valve to ensure proper shifting of the valve slide?
17. If a reversing valve will not operate, the first thing to check is:
    a. The unit operating suction and discharge pressures.
    b. The supply voltage to the unit.
    c. The condition of the contacts in the defrost relay.
    d. The reversing-valve solenoid coil.
18. The maximum temperature the valve body will tolerate when soldering tube connections is:
    a. 200° F.
    b. 250° F.
    c. 300° F.
    d. 350° F.
19. The best operating position for a check valve is:
    a. Horizontal refrigerant liquid flow.
    b. Vertical down refrigerant liquid flow.
    c. Vertical up refrigerant liquid flow.
20. The accumulator in a heat pump system is always connected:
    a. Between the reversing valve and the outdoor coil.
    b. Between the reversing valve and the indoor coil.
    c. Between the reversing valve and the compressor.
21. What four functions must the defrost control system perform to produce satisfactory defrosting of the outdoor coil in an air-to-air heat pump system?
22. What four functions must the defrost control system perform to produce satisfactory defrosting of the liquid-to-refrigerant coil in a liquid-to-air heat pump?
23. In the defrost relay, the contact controlling the outdoor fan is a normally open type. True or False?
24. In the defrost relay, the contact controlling the reversing valve is always a normally open type. True or False?
25. Heat pump thermostats and air conditioning thermostats are interchangeable. True or False?
26. The *Y* terminal of the heat pump thermostat controls the compressor contactor. True or False?
27. In UL approved heat pumps, automatic reset of the safety controls is required. True or False?
28. Defrost termination thermostats control the defrost function by measuring the:
    a. Temperature of the air leaving the outdoor coil.
    b. Temperature of the air leaving the indoor coil.
    c. Temperature of the refrigerant entering the outdoor coil.
    d. Temperature of the refrigerant liquid leaving the outdoor coil.

e. Temperature of the refrigerant vapor leaving the indoor coil.
f. Temperature of the refrigerant vapor entering the indoor coil.

29. The most popular method of defrosting the outdoor coil is by:
    a. Electric heat.
    b. Hot gas.
    c. Supply air.
    d. Water spray.
30. List the four types of defrost systems.
31. The primary cause of failure of the defrost control system to properly initiate the defrost mode is:
    a. Improper voltage at the control.
    b. Defective control.
    c. Insufficient air over the indoor coil.
    d. Insufficient air over the outdoor coil.
32. A change in temperature of the thermistor causes a change in:
    a. The size of the thermistor.
    b. The thickness of the thermistor.
    c. The electrical insulating value.
    d. The electrical resistance value.
33. Power to operate the defrost control system is taken off the line side of the compressor contactor. True or False?
34. If a power outage occurs during the defrost mode, the defrost mode will continue when power returns. True or False?
35. The most common cause of incomplete defrosting of the outdoor coil is:
    a. Interruption of the main power supply.
    b. Premature cut-off of the defrost mode by the defrost control.
    c. Excessive amount of reheat.
    d. Insufficient amount of reheat.
    e. Sticking defrost relay contacts.
36. An ice ring at the bottom of the outdoor coil is always an indication of:
    a. Stuck contacts in the defrost relay.
    b. Failure of the defrost timer.
    c. Loose termination thermostat.
    d. Excessive amount of reheat.
37. An outdoor coil plugged with dirt and debris will always cause the unit to go into the defrost mode. True or False?
38. When the unit goes into the defrost mode, the defrost mode is held until the outdoor coil reaches the termination temperature by:
    a. Action of the fan operation maintaining the defrost termination switch closed.
    b. Current drawn by the compressor.
    c. Action by the reversing valve changing the direction of the refrigerant flow.
    d. Action of the holding contact in the defrost relay.
    e. The contacts in the timer motor.
39. In the E15/D20 static pressure initiation/temperature termination defrost system, list the two conditions that must occur for the system to go into the defrost mode.
40. In the E15/D20 static pressure initiation/temperature termination defrost system, the maximum length of a defrost period is:
    a. 5 minutes.
    b. 10 minutes
    c. 15 minutes.
    d. 20 minutes.
41. In the E15/D20 static pressure initiation/temperature termination defrost system, the minimum time between defrost periods is:
    a. 5 minutes.
    b. 10 minutes.
    c. 15 minutes.
    d. 20 minutes.
42. The maximum kwh of electrical energy that electric utilities will allow to be brought on at the same time is:
    a. 5 kwh.
    b. 10 kwh.
    c. 15 kwh.
    d. 20 kwh.
43. The thermostat that controls the auxiliary electric heat according to the outdoor temperature is called the:
    a. Auxiliary temperature control.
    b. Outdoor auxiliary thermostat.
    c. Outdoor ambient thermostat.
    d. Auxiliary heat control thermostat.
    e. Hold back thermostat.
44. The total amount of reheat used in the defrost mode should not exceed 80% of the sensible heat capacity of the unit in the cooling mode. True or False?
45. In an add-on system, the discharge temperature thermostat controls the temperature of the flue products when the fossil fuel unit is operating during the defrost mode. True or False?

Section TWO
Air-to-Air Type Systems

APPLICATION 4

In order to successfully promote heat pumps, several rules were developed for successful operation of the system and customer satisfaction:

1. The equipment must be sized according to the cooling load.
2. Enough auxiliary heat must be installed to be able to handle the entire heating requirement in case the heat pump malfunctions.
3. The air distribution system must be draft free in spite of handling supply air with temperatures in the 80° F to 105° F range.
4. Installation, evacuation, and charging the system must be done in accordance with the best industry practices.
5. Maintenance programs are a must, especially regarding air filter changes.

Each step in the sizing, selection, installation, and adjustment of the equipment must be done properly to help ensure satisfactory operation.

4–1 SIZING

It is an absolute must that the size of the equipment selected for the application be within the required range to provide the most economical, yet satisfactory, conditions. This means that rules 1 and 2 definitely apply at this point:

1. The unit must be sized for the cooling load.
2. There must be enough auxiliary heat to handle the entire heating load.

The first step, therefore, is to accurately determine the heating and cooling requirements of the conditioned area. At this point, there is no need to "reinvent the wheel." Heat gain and heat loss calculation methods have been developed and are constantly being updated by the Air Conditioning Contractors of America (ACCA). Their Environmental Systems Library contains manuals on the sizing and application of heating and air conditioning systems as well as heat pumps. Also included are manuals on air distribution and adjustment of the systems. These manuals are highly recommended for use in covering the load calculation and air distribution design. For this reason, these subjects are not covered in this text.

The house plan shown in Figure 4–1 has been calculated according to the procedure outlined in Manual J of the ACCA Environmental Systems Library for weather conditions in Omaha, Nebraska and Phoenix, Arizona. The Omaha, Nebraska application is for a home with a full basement. The design weather conditions are –5° F in winter with 70° F inside, and summer weather conditions of 95° F with 75° F WB at 40% RH and 75° F inside temperature at 65% RH. The daily range of temperatures is medium.

The Phoenix, Arizona application is for a house on a concrete slab with no edge insulation (no footings are used). The design weather conditions are +34° F with 70° F inside in the winter, and summer weather conditions of 105° F DB – 71° F WB at 50% RH and 80° F DB – 40% RH inside. The daily range of temperatures is high. These calculations resulted in the following heating and cooling loads:

	Omaha, Nebraska	Phoenix, Arizona
Winter Heating	53,782 BTUH	19,996 BTUH
Summer Cooling	18,833 BTUH	27,265 BTUH

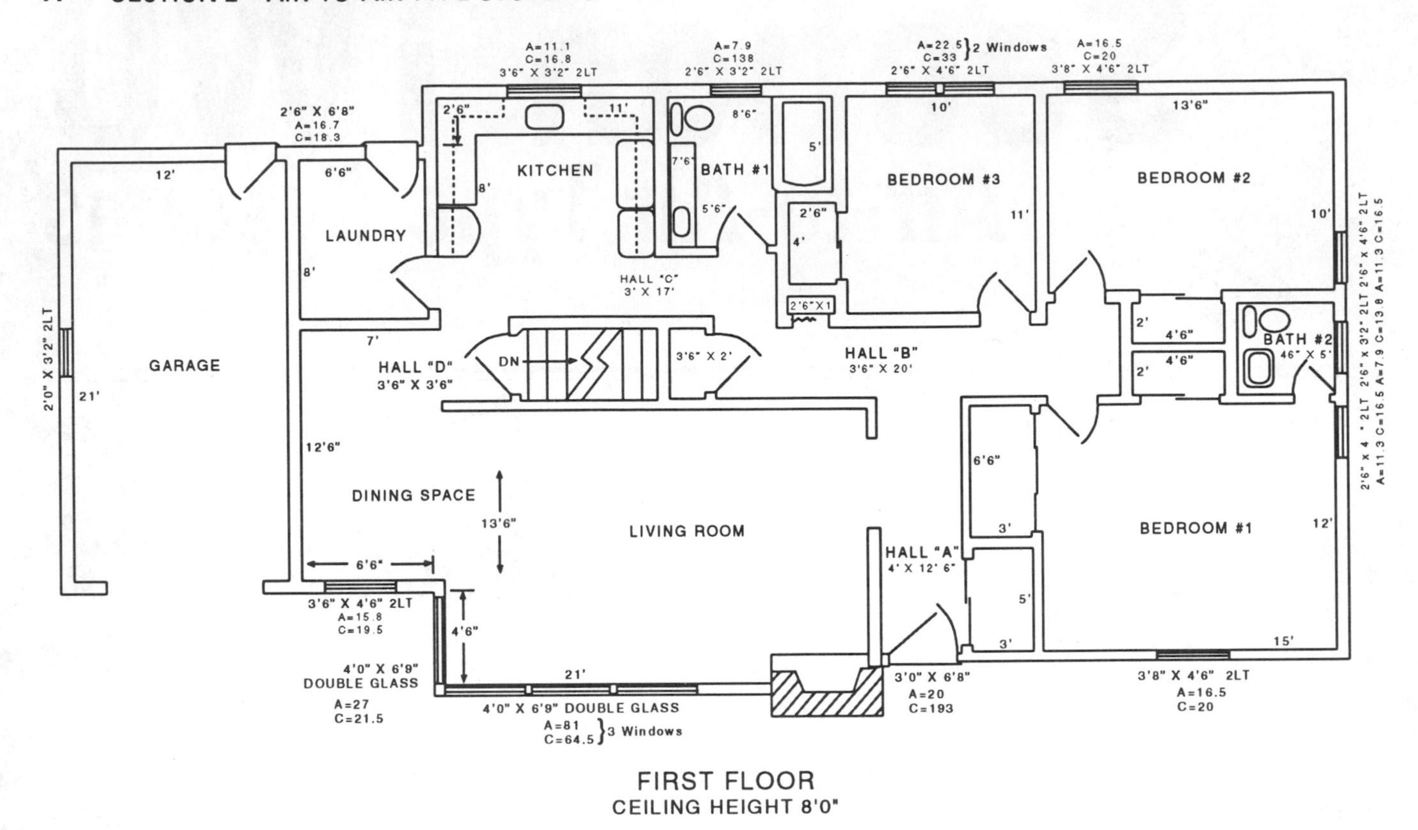

Figure 4–1 Basic house plan

Each installation will require a different combination of heat pump unit and auxiliary heat because of the different heating-to-cooling load ratios.

Using these two load ratios, each installation, will end up with different size equipment, even though the same steps for application will be used. Therefore, each application has to be reviewed according to its individual requirements.

Another item that must be taken into account is the change in the heating and cooling capacity of the air-to-air heat pump with the change in outdoor temperatures. In the heating mode, the capacity of the unit decreases as the outdoor temperature drops. In the cooling mode, the cooling capacity decreases with the rise in outdoor temperature. The change in capacity in the heating mode per degree temperature change is more radical than in the cooling mode. In this section on application, we will discuss both the split system and the package system.

4–2 SPLIT SYSTEM UNITS

In Figure 4–2, the two sections of a split system heat pump are shown. The section on the right is the indoor section. This contains the indoor heat exchanger with its matching pressure-reducing device and check valve assembly, room air circulating blower, auxiliary electric heating elements, and controls. The outdoor section on the left contains the outdoor heat exchanger with its matching pressure-reducing device and check valve assembly as well as the compressor, fan, and motor assembly and controls. These two sections are connected by field-installed refrigerant lines to complete the system. Electrical controls and power are field connected to each to provide the necessary control operation.

The heating capacity range of an air-to-air split system heat pump is given in Figure 4–3. Two separate items are given for each size of heat pump in 10° F increments from –10° F to +70° F DB. These items are the BTU per hour (BTUH) heating capacity and the Coefficient Of Performance. (COP). The heating output of a heat pump, called the *gross heating capacity,* is made up of the heat absorbed into the outside evaporator coil from the outside heat source (air through the outdoor coil) and the electrical energy used by the compressor motor.

For example, for the model QH836/AH68HF 3 HP unit at 40° F outdoor temperature, if the output is 33,000 BTUH and the unit has a COP of 2.52, the total output is 2.52 times the energy put in by the compressor motor. The motor heat would be 33,000 BTUH divided by 2.52 or 13,095 BTUH. The watts the motor is using to produce this result would be 13,095 BTUH divided by 3.413 BTU/watt or 3,837 watts per hour (3.837 KWH).

At –10° F, this same unit will produce a gross capacity of 10,800 BTUH with a COP of 1.27. This means the compressor motor is producing 8,504 BTUH and using 2.392 KWH in the process.

Inspection of the entire chart shows that gross heating capacity and COP decrease as the outdoor temperature drops. Not shown on this particular chart, all units reach a point where the gross heating capacity of the unit equals the heat produced by the compressor motor and the COP is 1.0. At this

Figure 4–2 Remote type unit (Courtesy Addison Products Co.)

	OUTDOOR AMBIENT (°F DB)																	
	-10°		0°		10°		20°		30°		40°		50°		60°		70°	
MODEL	BTUH	COP	BTUH	COP	BTUH	COP	BTUH	COP	BTUH	COP	BTUH	COP	BTUH	COP	BTUH	COP	BTUH	COP
QH818/AH65HC	4300	1.22	6200	1.42	10000	1.63	12700	1.87	15200	2.12	17800	2.35	20400	2.58	22900	2.80	25300	2.97
QH824/AH65HD	6800	1.32	9800	1.55	12900	1.75	15900	2.00	19100	2.27	22200	2.55	25200	2.75	28100	2.95	30900	3.10
QH830/AH65HE	7700	1.32	11700	1.55	15800	1.75	19900	2.00	24000	2.25	27900	2.50	31700	2.72	35400	2.92	39000	3.10
QH836/AH68HF	10800	1.27	14800	1.45	18800	1.70	23600	1.92	28400	2.20	33000	2.52	37500	2.73	41700	2.93	45300	3.10
QH842/AH68HG	14500	1.45	18500	1.65	23000	1.85	28000	2.10	33000	2.30	38700	2.50	44300	2.70	45500	2.85	50000	3.00
QH848/AH68HH	18800	1.70	24000	1.85	29400	2.10	35100	2.35	40500	2.50	47100	2.75	53600	2.95	59900	3.10	65200	3.25
QH860/AH68HK	22000	1.30	26500	1.50	32500	1.75	38500	2.00	44300	2.20	53700	2.45	63300	2.70	70600	2.85	74200	3.05

(1) Above at standard rating airflow with 70° entering air temperature.

Figure 4–3 Heating application ratings (Courtesy Addison Products Co.)

	OUTDOOR AMBIENT (°F DB)				
	75°	80°	85°	90°	95°
MODEL	BTUH	BTUH	BTUH	BTUH	BTUH
QH818/AH65HC	19000	18700	18500	17900	17300
QH824/AH65HD	23900	23600	23300	22600	21800
QH830/AH65HE	31600	31200	30900	30300	29000
QH836/AH68HF	36400	36000	35500	34600	33400
QH842/AH68HG	43000	42500	42000	41300	40000
QH848/AH68HH	50600	50000	49350	48300	47000
QH860/AH68HK	57400	56900	55900	54800	53500

(1) Above at standard rating airflow with 80°F DB, 67°FWB entering air.

Figure 4–4 Cooling application ratings (Courtesy Addison Products Co.)

point, the unit is operating with the same heat output ratio as the electric resistance heat.

In years past, it was a practice to turn the refrigeration portion of the system off when the outdoor temperature dropped to a point where the COP was reduced to 1.5 or below. With the compressor off at these low temperatures, the problem of refrigerant migration from the warm inside coil to the colder outdoor compressor was accelerated and a high percentage of compressor failures resulted. Though crankcase heaters were used on the compressor assembly, the heat produced was not enough to protect the compressor. Today's practice is to *keep the compressor running.*

This reduces the possibility of liquid refrigerant forming in the compressor assembly. As a result of this practice, compressor failures from this situation have practically been eliminated.

The cooling capacity rating of the same unit is given in Figure 4–4. This rating is the net capacity rating of the total amount of heat in BTUH (both sensible and latent) picked up by the inside coil (the evaporator) in the cooling mode.

ARI STANDARD 240 RATING CONDITIONS FOR HEAT PUMPS	Air Entering Temp (°F)			
	Indoor Coil		Outdoor Coil	
	DB	WB	DB	WB
Cooling	80	67	95	—
Heating (High Temp.)	70	60[1]	47	43
Heating (Low Temp.)	70	60[1]	17	15

[1]Maximum — May be lower temperature

Figure 4–5 ARI standards rating (Courtesy Addison Products Co.)

In this situation, the reverse occurs. As the outdoor temperature increases, the net capacity of the system decreases. The units are rated up to 95° F outdoor temperature, which is the ARI rating standard. For areas such as Phoenix, where the design temperature is above 95° F, the capacity will drop approximately 4% for each 5° F rise in outdoor temperature above the 95° F.

Figure 4–5 shows a table of the ARI Standard 240 Rating Conditions for heat pumps using dry bulb (DB) and wet bulb (WB) temperatures of the air over the inside coil for heating and cooling as well as the air through the outdoor coil in the heating mode. The amount of moisture in the outdoor air in the cooling mode has little effect on the operation of the system. It is ignored and a wet bulb (WB) temperature is not specified.

4–3 EQUIPMENT SELECTION

The first step in the application process is the selection of the equipment size to handle the calculated heating and cooling loads. The basic rule is "the heat pump is selected to handle the cooling load."

In some instances, the installer has tried to compromise between the heating and cooling requirements when selecting the unit size. This invariably results in oversized cooling capacity, insufficient running time, and high humidity problems in the conditioned area. An air conditioning unit is primarily a dehumidifier and it must operate to do its job. The more off time in the operating cycle, the greater the humidity problems.

This problem inspired the Lennox Corp. to develop the two-speed compressor unit. By reducing the capacity of the compressor (by reducing the speed), the unit operating time is increased and better humidity levels are maintained.

Omaha, Nebraska

The cooling load is 18,833 BTUH at 95° F and the heating load is 53,782 BTUH at –5° F. Using the table in Figure 4–4, we would select the QH824/AH65HD unit with a capacity of 21,800 BTUH at 95° F. Only 18,833 BTUH is required but the QH18/AH65HC unit with a rating of 17,300 BTUH at 95° F does not have sufficient capacity.

Unit ratings and cooling loads will not always match. Therefore, a tolerance should be considered. For best results, you should select a unit that has a range of –0% to +10% of the cooling requirements.

The equipment selected is a split-type system using an outdoor section (QH824) and an indoor air handler (AH65HD) with the option of electric auxiliary heat. This takes care of the cooling requirements.

The heating requirement in Omaha, Nebraska is 53,783 BTUH at –5° F. The amount of auxiliary heat needed is the difference between the calculated heat load and the heating capacity of the unit at the –5° F outdoor temperature. This would be 53,782 BTUH minus 9,700 BTUH (the capacity midway between the capacity at –10° F and at 0° F) or 44,082 BTUH.

The selection of the auxiliary heat capacity is not based on this difference. The total amount of auxiliary heat capacity selected must be equal to or exceed the total heating load at design conditions. This selection method is a requirement with most electric utilities to handle the heat load in the event of system malfunction.

In this case, we need electric heating elements that will provide the 53,782 BTUH. By dividing the 53,782 BTUH by 3,413 BTUH per KWH, we find the requirement to be 15.76 KWH. As in the case of selecting the cooling unit size, the tolerance is –0% so the next larger unit is required. If the unit used 4.8 KWH elements, a 19.2 KWH unit would be used.

In Figure 4–6, the various BTUH capacity ratings are given for different applied voltages. In our example, we will assume that a standard voltage of 240 volts is supplied by the electric utility. Using the air handler model number AH65HE and the 20 KWH requirement, we would select the AH65HE420C unit. This will give a heating capacity of 68,200 BTUH, which will handle our heating requirement of design conditions and below.

The unit delivering 68,200 BTUH is based on an applied voltage of 240 volts when the full heater load is activated. If the applied voltage is to be other than the 240 volts, an adjustment in the rated capacity has to be made to find out if the heaters will supply the required 53,782 BTUH heating load.

To do this, it is necessary to find the actual resistance of the heater deck. In this case, the resistance is made up of four resistances in parallel across the applied voltage. To find the resistance, Ohm's Law, "amperage (I) is equal to voltage (V) divided by resistance (R)" is used.

The first step is to find the number of watts required to produce the 68,200 BTUH at 240 volts. The watts are found by dividing the heater output (68,200 BTUH) by the BTU per watt (3.413), which results in a figure of 19,982 watts.

The next step is to find the amperes required to produce this wattage. This is found by dividing the watts (19,982) by the applied voltage (240 volts), which equals 83.25 amperes.

Using the applied voltage of 240 volts and the amperage of 83.26 amperes, Ohm's Law is revised to solve for the resistance (R) instead of the amperage (I). The law now becomes "The resistance (R) equals the voltage (V) divided by the amperage (I)." The voltage (240 volts) divided by the amperage (83.26 amps) results in a resistance of 2.88 ohms. This reveals that the total resistance of the heater bank is 2.88 ohms.

MODEL	MINIMUM FIELD WIRE (AWG) (1)	MINIMUM FUSE (DELAY) (1)	ELECTRIC RESISTANCE HEATING CAPACITY							
			KW		STAGING		BTUH (2)			
			240V	208V	1	2	240V	230V	220V	208V
AH65HDB405B	10	30A	5.0	3.8	5	-	17100	15700	14400	12800
410B	6	60A	10.0	7.5	10	-	34100	31300	28700	25600
415C	2	90A	15.0	11.3	10	5	51200	47000	43000	38400
AH65HEB405B	8	35A	5.0	3.8	5	-	17100	15700	14400	12800
410B	4	60A	10.0	7.5	10	-	34100	31300	28700	25600
415C	2	90A	15.0	11.3	10	5	51200	47000	43000	38400
420C	1	110A	20.0	15.0	10	10	68200	62700	57400	51200
AH68HFB410B	4	60A	10.0	7.5	10	-	34100	31300	28700	25600
415C	2	90A	15.0	11.3	10	5	51200	47000	43000	38400
420C	1	110A	20.0	15.0	10	10	68200	62700	57400	51200
425B	2/0	150A	25.0	18.8	10	15	85300	78400	71700	64000
AH68HGB410B	4	60A	10.0	7.5	10	-	34100	31300	28700	25600
415C	2	90A	15.0	11.3	10	5	51200	47000	43000	38400
420C	1/0	125A	20.0	15.0	10	10	68200	62700	57400	51200
425B	2/0	150A	25.0	18.8	10	15	85300	78400	71700	64000
430B	3/0	175A	30.0	22.5	15	15	102300	93900	85900	76800
AH68HHB415C	2	90A	15.0	11.3	10	5	51200	47000	43000	38400
420C	1/0	125A	20.0	15.0	10	10	68200	62700	57400	51200
425B	2/0	150A	25.0	18.8	10	15	85300	78400	71700	64000
430B	3/0	175A	30.0	22.5	15	15	102300	93900	85900	76800
435B	4/0	200A	35.0	26.3	15	20	119400	109600	100300	89600
AH68HKB415C	2	90A	15.0	11.3	10	5	51200	47000	43000	38400
420C	1/0	125A	20.0	15.0	10	10	68200	62700	57400	51200
425B	2/0	150A	25.0	18.8	10	15	85300	78400	71700	64000
430B	3/0	175A	30.0	22.5	15	15	102300	93900	85900	76800
435B	4/0	200A	35.0	26.3	15	20	119400	109600	100300	89600

(1) Values shown for 240V. (2) Add capacity to APPLICATION RATINGS - HEATING to obtain total capacity available at given outdoor temperature.

Figure 4–6 Indoor unit specifications (Courtesy Addison Products Co.)

By reversing the process, we can use the resistance figure to determine the heat output of the heater bank at voltages other than the 240 volts. For example: suppose the applied voltage were 230 volts instead of the 240 volts. The following steps would be taken to find the amount of heat that would be produced:

1. Ohm's Law $I = \dfrac{V}{R}$

$$I = \frac{230 \text{ volts}}{2.88 \text{ ohms}}$$

$$I = 79.86 \text{ amperes}$$

2. Watts $= \text{volts} \times \text{amperes}$

$$= 230 \text{ volts} \times 79.86 \text{ amperes}$$

$$= 18{,}368 \text{ watts}$$

3. BTUH output $= \text{watts} \times 3.413 \text{ BTUH per watt}$

$$= 18{,}368 \times 3.413 \text{ BTU per watt}$$

$$= 62{,}690 \text{ BTUH}$$

This output is more than required to handle the design load of 53,782 BTUH. Therefore, the unit selected will perform satisfactorily with an applied voltage of 230 volts.

Power supplies for electric heater banks are sometimes taken off a 208-volt, three-phase system with the heaters operating across each single phase of the three-phase supply. With 208 volts supplied to the heater bank, will we get enough heat from the 20-KWH assembly?

We already have the basic resistance of the heater bank—2.88 ohms. To find the amperage at 208 volts:

1. Ohm's $I = \dfrac{V}{R}$

$$I = \frac{208 \text{ volts}}{2.88 \text{ ohms}}$$

$$I = 72.22 \text{ amperes}$$

2. Watts $= I \times V$

$$= 208 \text{ volts} \times 72.22 \text{ amperes}$$

$$= 15{,}022 \text{ watts}$$

3. BTUH $= \text{watts} \times 3.413 \text{ BTU per watt}$

$$= 51{,}270 \text{ BTUH}$$

With an applied voltage of 208 volts, the 51,270 BTUH of the heater deck selected will not satisfy the load of 53,782 BTUH. The system will be 2,512 BTUH short. To meet utility requirements of satisfying the heat loss at design conditions, the next larger size unit (25 KWH) will have to be used.

Phoenix, Arizona

In Phoenix, Arizona, the conditions are reversed. The cooling load of 27,265 BTUH is greater than the heating load of 18,833 BTUH, so it is obvious that the unit should be selected to handle the cooling load. From the table in Figure 4–4, we select a unit that will produce 27,265 BTUH at 105° F.

The first selection is the QH830/AH65HE rated at 29,000 BTUH at 95° F. Because we can expect a capacity reduction of around 4 to 5% for each increase of 5° F in outdoor temperature above the 95° F, we can expect a capacity of 26,172 BTUH at 105° F design conditions. This size unit will not handle the cooling load at design conditions and will definitely be too small when the temperature reaches the 110° F to 112° F mark. Therefore, the next larger unit (QH36/AH68HF) is selected.

What capacity will the unit have when it is handling the heating load? The Phoenix, Arizona winter design temperature is +34° F. At design temperature, the unit selected has a heating capacity of 28,400 BTUH with a COP of 2.2. The calculated heating load at design temperature is 19,996 BTUH. Therefore, this unit will handle the heating requirements.

Should auxiliary heaters be installed? Yes, for reheating during the occasional defrost cycle and for emergency heat. This means that the 19,996 BTUH heating load divided by 3,413 BTUH per KWH, or 5.86 of auxiliary heat is required. It may seem reasonable to assume that a 5 KWH element is close enough. However, design temperatures are for a temperature reached 2 ½% of the winter heating season. When the temperature drops below design, the unit will not be able to handle the load. 10 KWH of electric heat in two stages must be included in the air handler in this application. An AH68FB410B air handler should be selected for this installation.

4–4 PACKAGE SYSTEMS

In some applications, a package-type unit (see Figure 4–7) is more appropriate than a split-type unit. This type of application is prevalent in smaller homes and in row house construction.

The placing of the package unit on the roof puts it into position for close coupling to an overhead air supply system. It also eliminates potential damage from traffic around the unit. Where snow depths are sufficient to affect the operation of the unit, overhead installation eliminates this problem. To select the proper package-type unit for each application, information such as that given in Figure 4–8 is used.

Omaha, Nebraska

With a cooling load of 18,833 BTUH at 91° F design temperature, selection of the Model PH018R-1 will be too close to provide adequate capacity when the outdoor temperature exceeds the design temperature. For example, if the outdoor temperature were to rise to 95° F, the cooling load would be 23,541 BTUH. Therefore, the model PH025R-1, which has a cooling capacity of 25,600 BTUH at 95° F, will have to be used.

In the heating mode, the unit will not provide sufficient heating capacity to handle the heating load down to the design temperature of –5° F. Therefore, an auxiliary electric heat package must be added to the unit. With package-type units, the auxiliary heat may be an integral part of the unit or a duct heater may be added into the supply duct system.

Figure 4–7 Package unit (Courtesy Addison Products Co.)

Model Number	Cooling Capacity							Heating Capacity(70°F. Indoor Air)						DOE Region IV
								Outside Air 47°F.D.B./43°F.W.B. DOE High Temperature			Outside Air 17°F.D.B./15°F.W.B. DOE Low Temperature			
	ARI Std.† Cap BTU/Hr.	Net.Sens. Cooling Cap.	Net. Lat. BTU/Hr.	Single Phase SEER	Single Phase Total Watts	ARI Sound Rating	Approx. CFM	BTU/Hr.	C.O.P.	Power Input Watts	BTU/Hr.	C.O.P.	Power Input Watts	HSPF BTU/Watt-Hr.
PHO18R-1	19,000	13,800	5,200	8.75	2260	7.6	675	18,800	2.80	1973	10,800	1.90	1680	6.50
PHO25R-1	24,000	18,100	5,900	9.00	2705	7.8	900	22,600	2.80	2385	11,000	1.58	2065	6.55
PHO30R-1	30,800	22,800	8,000	8.60	3580	7.8	1075	30,600	2.80	3180	15,600	1.70	2680	6.35
PHO38R-1	36,400	27,800	8,600	8.85	4430	8.2	1350	36,600	2.80	3842	19,600	1.70	3364	6.40
PHO43R-1	41,500	30,100	11,400	8.05	5548	8.2	1450	43,500	2.74	4644	25,000	1.90	3870	6.65
PHO48R-1	49,000	37,100	11,900	8.15	6527	8.4	1750	50,000	2.74	5293	29,000	1.94	4315	6.75
PHO60R-1	58,000	43,000	15,000	8.10	7542	8.6	2000	60,000	2.74	6340	35,800	2.04	5114	6.75

†Design Conditions: ARI Rating Temperatures of 80°F.D.B.,67°F.W.B.indoor, return air 95°F.D.B. outdoor

Figure 4–8 Package unit performance data (Courtesy Addison Products Co.)

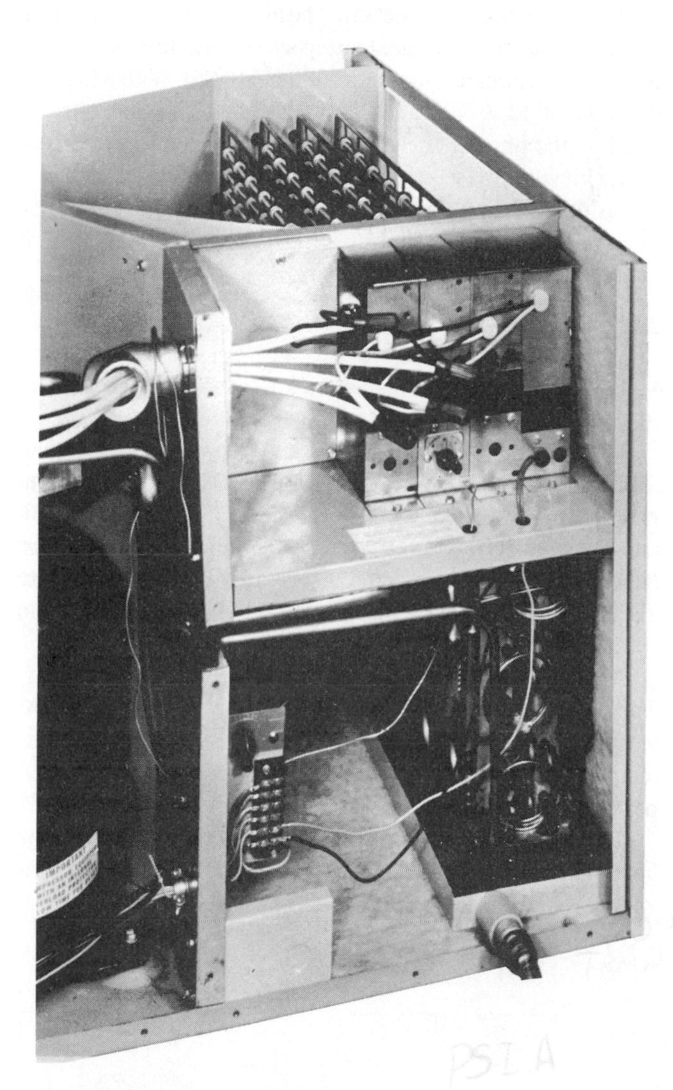

Figure 4–9 Typical field installation of heater elements (Courtesy Addison Products Co.)

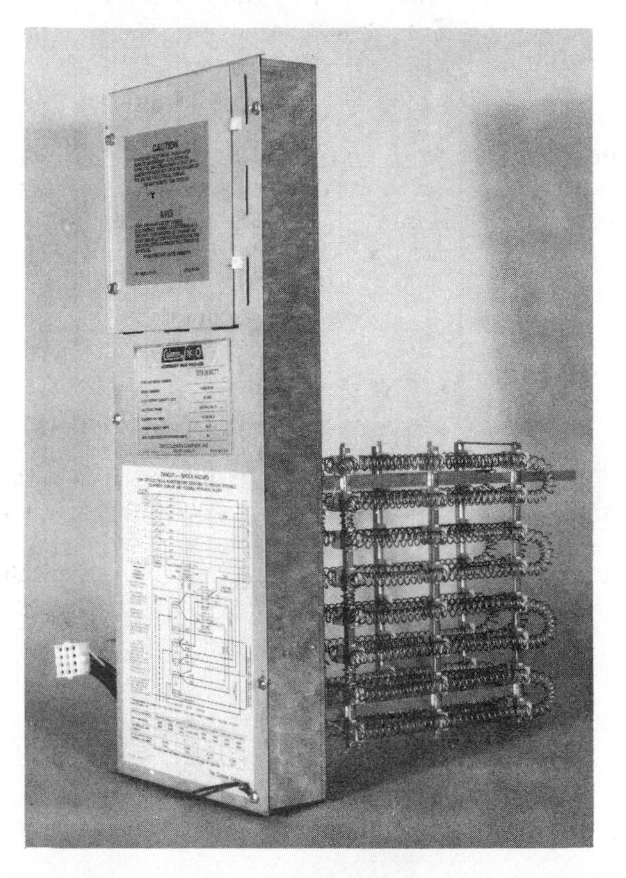

Figure 4–10 Heater element package (Courtesy The Coleman Co.)

Figure 4–9 shows a package-type unit with slide-in heater assemblies. Each heater consists of the heater element and frame plus a limit control, fusible link, and sequence relay. This illustration shows the method of controlling the auxiliary heat capacity by the number of element assemblies used.

Figure 4–10 shows a different concept of auxiliary heat application. This manufacturer markets a heater package in one assembly. Sizes from 0 KWH (0 BTUH) to 21.6 KWH (67,600 BTUH) in five assemblies are physically sized to fit and fill the opening provided in the cabinet. Each assembly is factory wired to simplify the installation and connection of the power and control wiring.

The heat pump and heater package are connected together via a male/female connector assembly. Control wiring to the heater package is then connected to the heat pump via the connector. The heat pump will not operate unless a heater package has been installed. This type of design ensures that matching equipment will be used in the application.

The equipment selected may not use built-in type auxiliary heaters. In this case, a duct-type heater assembly (Figure 4–11) would be used. These assemblies are designed to be installed in the duct system down stream from the unit blower.

Figure 4–11 Duct heater assembly (Courtesy Addison Products Co.)

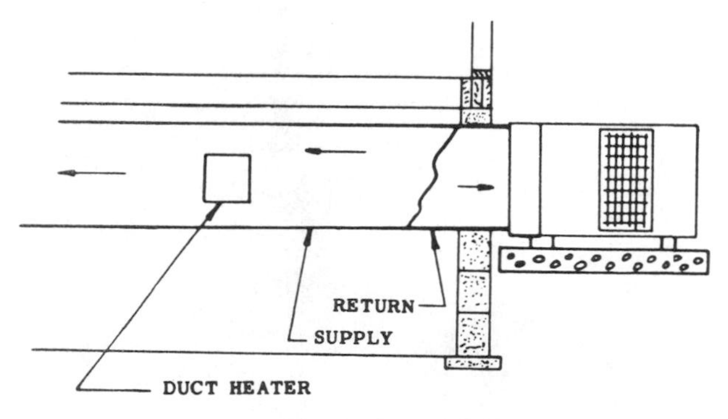

Figure 4–12 Crawl space installation package unit (Courtesy Addison Products Co.)

Figure 4–12 shows the duct heater installed in the supply duct in a crawl space installation. Manufactured in different sizes, they are usually a complete assembly of elements, safety devices, and controls. Wiring is by separate power supply and low-voltage wiring to the control thermostat.

The major precaution in installing duct heater assemblies is that they must be installed with not less than 4' (48") of metal duct between the outlet of the duct heater and ductwork made of material other than metal. This is a fire safety regulation in case of blower failure with the heater elements active.

Figure 4–13 shows a typical table of characteristics for auxiliary heater kits. Information is given covering BTUH output at different voltages as well as suggested unit sizes each kit can be connected to.

The kit we need to select must be able to handle the heating requirements in case of heat pump system malfunction. From this equipment chart, the 8604-070 heater kit, with a heating capacity of 19.2 KWH (65,530 BTUH) with 240 volts applied, is the one for this application. This is a four-element heater kit with two sequence steps for close control of capacity according to the outdoor temperature. This is done by the use of an outdoor ambient thermostat. Discussion on this will be covered under "Balance Point."

Phoenix, Arizona

For the Phoenix, Arizona area application with a cooling load of 27,265 BTUH and a winter heating load of 19,996 BTUH, the PH038R-1 package heat pump would be selected. This unit is rated at 38,500 BTUH at 95° F. At 105° F, it can be expected to have a capacity of 35,482 BTUH. Both split and package refrigeration systems can be expected to lose approximately 4% for each 5° F rise in outdoor temperature above 95° F. This unit will still give satisfactory results up to 111.5° F.

At this temperature, the unit will operate continuously. This is the system cooling load/capacity balance point. At temperatures above this balance point, the conditioned area temperature will rise a couple of degrees during the short period of these extreme temperatures.

The heating capacity of the selected unit will handle the heating load of the building down to +24° F when the heat loss will be 24,650 BTUH.

DUCT HEATERS

PART NO.	PH	VOLTS	KW	MINIMUM AMPACITY	WIRE SIZE* Cu	WIRE SIZE* A1	MAX. FUSE	DIMENSIONS H&A	W	B	C
8604-067	1	240	4.8	25	#10	#8	25	8	12	17	6
8604-068	1	240	9.6	50	#6	#4	50	8	16	17	6
8604-069**	1	240	15.0	79	#3	#1	80	12	18	27	12
8604-070**	1	240	19.2	100	#1	#0	100	12	18	27	12

* Use wire suitable for at least 90°C. **Fused units (over 48 amperes).
NOTE: All duct heaters are supplied with backup protection and internal fusing as required by NEC.

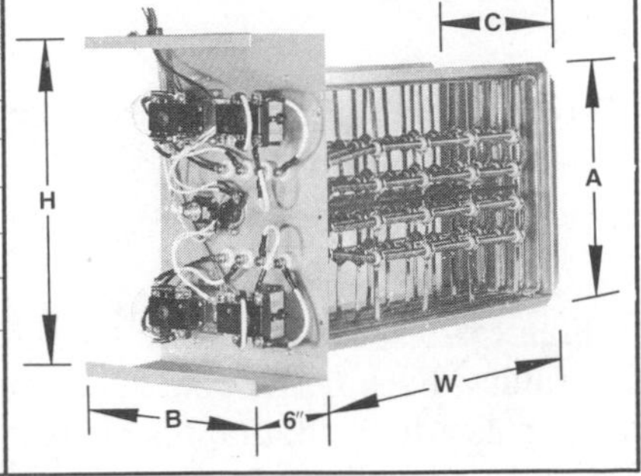

Figure 4–13 Duct heater specifications (Courtesy Addison Products Co.)

From these performance figures, it seems that no auxiliary heat is needed. This may not be the case. The local electric utilities may require the standby safety factor of sufficient heat to handle the heating load in the event of heat pump system malfunction.

4–5 ADD-ON SYSTEMS

We have been discussing the application of heat pumps using electric auxiliary heat. There are applications where it is desirable to operate a heat pump in conjunction with fossil fuel heating equipment—natural gas, propane, or fuel oil. Such a choice depends upon the cost of such fossil fuels as compared to the cost of electricity to operate the heat pump. When comparing operating costs, it must be remembered that the heat pump and the electric heat operate at the same time. The advantage of the heat pump is that it is used down to the 1.0 COP level. When adding a heat pump to a fossil fuel heat source, the fossil fuel heat source is used *instead of* the heat pump. It is a choice of either heat source, *not both.*

Because of this requirement of heating system choice, the fossil fuel system has to take over when the heat pump cannot handle the heat load. The changeover thermostat that accomplishes this is set at the initial balance point.

The reason for this requirement of either system, but not both, is twofold:

1. The indoor coil of the heat pump must be located downstream from the fossil fuel heating unit. This is required to preserve the life of the heat exchanger in the fossil fuel unit. If the coil were located ahead or upstream of the heating unit, the heat exchanger would sustain heavy condensate formation when in the cooling mode. Heavy rusting and short heat exchanger life would result.
2. When the fossil fuel unit operates, the proper temperature rise of the air through the heating unit is 80° F. This puts 150° F air to the inside coil in the heating mode. The coil is now acting as a condenser. The extreme temperatures and condensing pressures under these circumstances would destroy the compressor.

To accomplish the selection of heat pump or fossil fuel for the heat source, the outside ambient thermostat (OAT) has to be a single-pole, double-throw (SPDT) type in the control circuit of the thermostat that normally controls the operation of the motor-compressor assembly.

This will be covered in more detail under "Installation—Controls."

Equipment

The fact that a heat pump system is being added to an existing forced warm air fossil fuel unit eliminates the need for any additional air-handling equipment. The installation is the same as adding an air conditioning unit to an existing heating system.

The equipment for this installation is the outdoor unit, a heat-pump type indoor coil, and a heat pump thermostat control center for addition to the heating unit (see Figure 4–14.)

Outdoor Section. This section contains the outdoor heat exchanger with its matching pressure-reducing device and bypass check valve assembly, compressor, fan and motor assembly, and controls. The controls in this section consist of operating controls for the motor-compressor assembly and fan motor as well as the reversing valve and relay and the defrost controls.

Some manufacturers include an outdoor ambient thermostat (OAT) as part of the control system. Others require a separate control be used.

Figure 4–14 Add-on system (Courtesy Addison Products Co.)

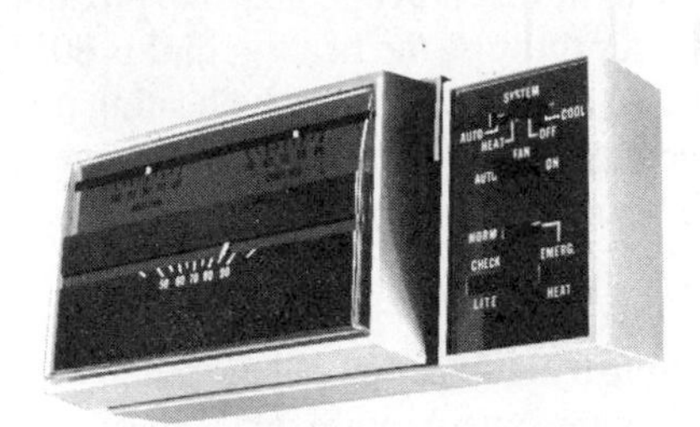

Figure 4–15 Control center (Courtesy Addison Products Co.)

Indoor Section. The indoor section of an add-on system is the heat exchanger coil with its matching pressure-reducing device and check valve assembly. The inside heat pump coil is the condenser in the heating mode and is larger in physical size than an air conditioning coil. The reason for this is the need for more heat exchanger surface to transfer the gross heating capacity of the unit to the inside air system. This physical difference between the heat pump coil and an air conditioning coil is usually in the height of the assembly. Both coils, being designed to be applied to the same fossil fuel unit, will have the same base pan dimensions. Therefore, any additional heat exchanger surface is reflected in the height of the coil.

There is no standard of design on this subject. Therefore, the installer must determine ahead of time if there is sufficient room for insertion of the coil in the system.

Control Center. The heat pump area thermostat part of the control center is shown in Figure 4–15. It looks the same as the thermostat used in a standard air conditioning system but its function and control circuitry are different. The difference will be discussed under "Control Wiring."

In addition to the area thermostat, this manufacturer requires a relay assembly for isolation of the operation of the heat pump and the fossil fuel unit to prevent both systems from operating at the same time. The application and wiring of this control is covered under "Control Wiring."

4–6 BALANCE POINT

When selecting the heat pump capacity to be used in a particular application, rule number one states that the unit must always be selected according to the cooling capacity required. Most publications on this subject usually follow with a statement to the effect that if the unit is selected according to the heating load requirements, the cooling capacity will be too great. The author has deliberately picked two different locations to illustrate the point that this in not always true.

In Omaha, Nebraska, yes, the cooling unit would be considerably oversized. A 5-hp unit with 55,000 BTUH capacity at 95° F handling a cooling load of 18,833 BTUH would produce very poor results. The statement applies to this application.

However, in Phoenix, Arizona, we would have a heat pump with a heating capacity of 20,650 BTUH at 34° F trying to handle a cooling load of 27,265 BTUH at 105° F. The cooling capacity of this unit at 105° F would be 20,091 BTUH—approximately 73.6% of the capacity needed to obtain satisfactory results.

Again the rule is emphasized—the unit must be selected according to the cooling load requirements.

Omaha, Nebraska

In the application in Omaha, Nebraska, where the heating load is considerably higher than the cooling load, the heat pump will not handle the heating load at design conditions. Therefore, some form of auxiliary heat is needed.

To determine the amount of auxiliary heat required, it is necessary to find at what outdoor temperature the unit will run continuously to satisfy the heating requirements. This point is called the *initial balance point* of the unit. This point is the result of the increase in the heat loss of the conditioned area versus the decrease in unit capacity as the outside temperature falls.

The procedure for determining the initial balance point is very simple. A plot is made of both the building heat loss and the unit capacity. Where the two plot lines cross is the initial balance point. To plot the initial balance point, ten square graph paper is required.

The graph shown in Figure 4–16 is a straight line graph showing the heat loss of a building between 70° F and the heating design temperature of –5° F for the Omaha, Nebraska area. The line between these two points is the basic heat loss line (dashed line).

Experience, however, shows that a building does not require heat until the outdoor temperature drops to 65° F. The reason for this is that enough heat is generated from lighting, appliances, human activities, and so on to take care of the building heat loss down to 65° F outside.

Therefore, to promote accuracy in plotting the expected results, a plot line is drawn from the 65° F point on the ambient temperature scale to the heat loss point on the building heat loss scale at the design temperature vertical solid line. For our Omaha, Nebraska application, the design temperature is -5° F. By extending the solid line to the left of the design temperature versus heat loss point, an estimation of the heat loss requirements below the design temperatures can be determined. For example, the home in Omaha, NB can be expected to have a heat loss of 64,000 BTUH at -20° F and at +20° F the heat loss is only 34,000 BTUH.

4–7 SECOND BALANCE POINT

To find the capacity of the heat pump in the heating mode at various outdoor temperatures, the capacity is plotted. In Figure 4–17, the heating capacity of the unit selected for the Omaha, NB installation has been plotted. This is done by locating the point of the heating capacity of the unit selected for the application at various outdoor temperatures. The manufacturer of the unit being discussed has provided capacities in 10° F increments from –10° F to +70° F.

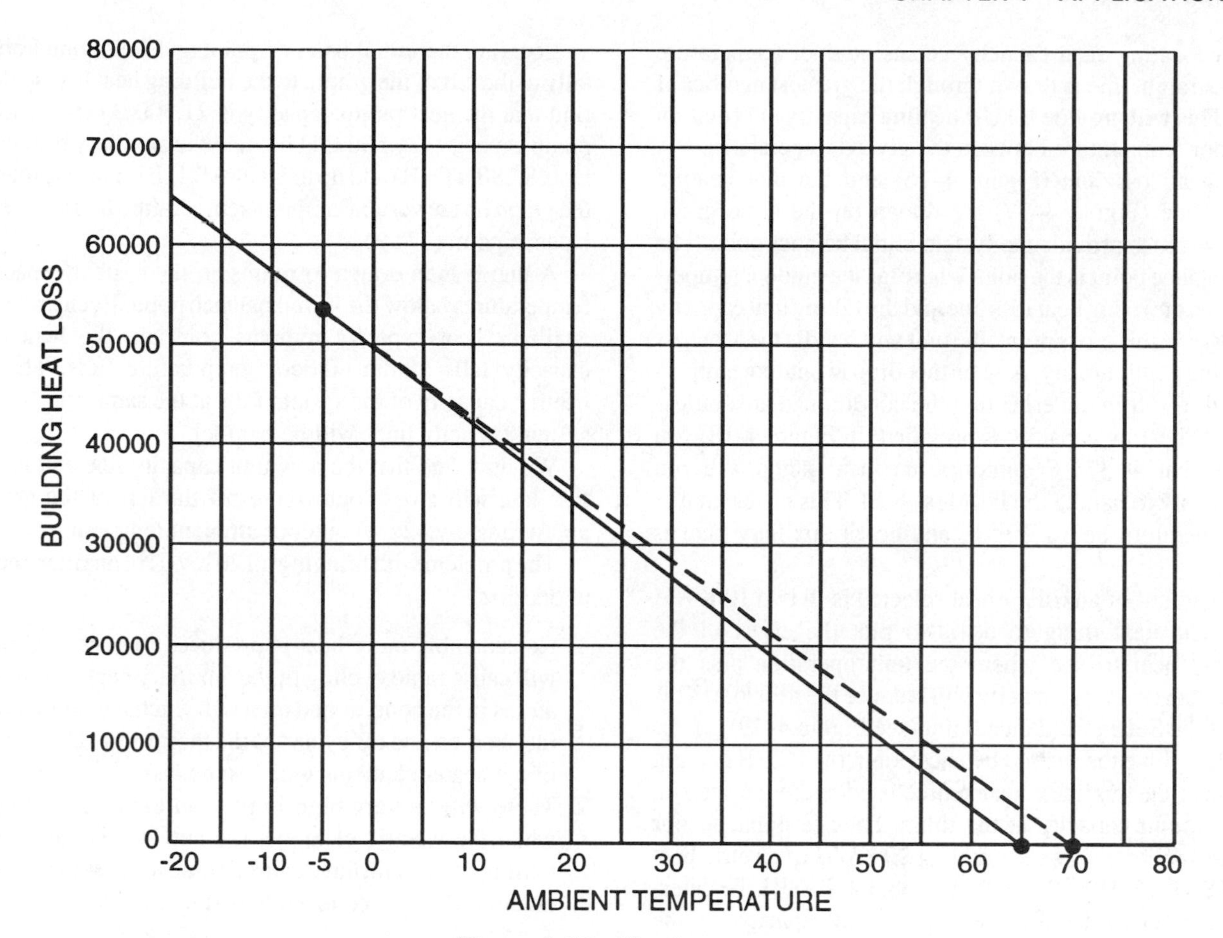

Figure 4–16 Heat loss curve

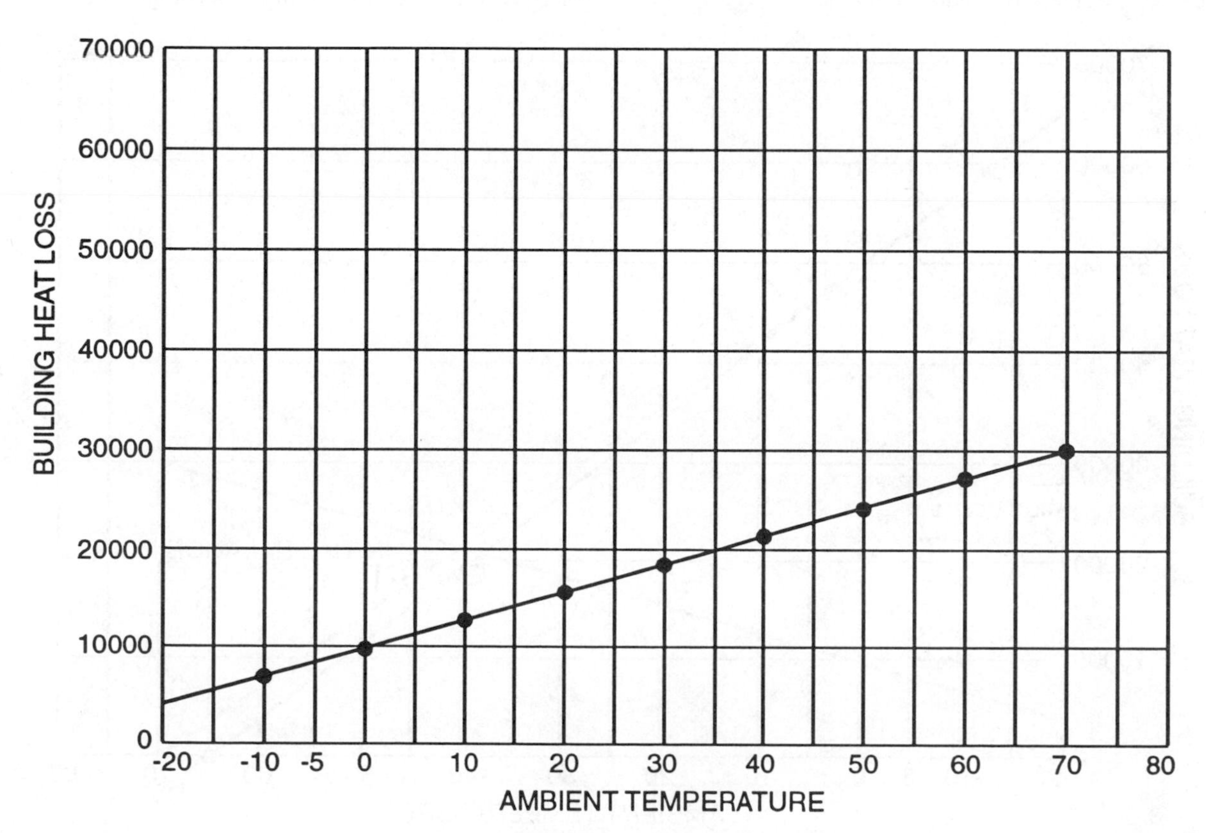

Figure 4–17 Unit capacity curve

After locating each capacity versus outdoor temperature point, a straight line is drawn through the greatest number of points. This will provide BTUH heating capacity information at outdoor temperature between the given temperatures.

The heat loss line (Figure 4–16) and the unit heating capacity line (Figure 4–17) are drawn on the same graph (Figure 4–18) to provide the system initial balance point. The initial balance point is the point where, as the outdoor temperature falls, the rising heat loss line and the falling unit capacity line cross. At this cross point, the unit will handle the heat loss by running continuously. Any further drop in outdoor temperature will result in underheating the conditioned area unless additional heating capacity is provided. In Figure 4–18, you will see that at 37° F outdoors, the heat pump will run continuously to handle the heat loss load. This means that at any temperature below 37° F, additional auxiliary heat is needed.

The amount of auxiliary heat selected is in two 10 KWH stages. The next thing to do is to plot the effect of the additional heat to see where the unit operation plus the auxiliary heat system capacity will balance the heat loss. This is called the Second Balance Point (see Figure 4–19).

To determine the second balance point, the BTUH heating capacity of the auxiliary heat source is added to the system, above the unit capacity at the initial balance point. In our example application, we are adding 20 KWH of electric heat or 20 KWH × 3,413 BTU/KWH of 68,280 BTUH. To determine the point on the graph where the unit capacity and the auxiliary heat will fall, we need to determine the total system heating capacity at the initial balance point.

Locating the initial balance point and following horizontally to the left of the graph, to the building heat loss scale, we find that the heat pump capacity is 21,500 BTUH. The heat pump capacity (21,500 BTUH) and the auxiliary heat capacity (68,280 BTUH) add up to 89,780 BTUH. This is plotted on the graph on the vertical ambient temperature line at the initial balance point (37° F).

A line is then drawn to represent the system capacity at temperatures below the initial balance point. Even though the auxiliary heat capacity remains constant, the heat pump capacity falls as the outdoor temperature falls. The total heating capacity of the system falls at the same rate. The two system capacity lines will be parallel.

We now find that the new unit capacity line and the heat loss line will cross somewhere off the left of the graph at approximately –26° F outdoor ambient temperature.

The problem with bringing all 20 KWH of auxiliary heat on at once is:

1. Considerably more heat is produced than needed, which will cause rapid cycling of the auxiliary heat. The temperatures in the conditioned area will fluctuate rapidly due to the short on and off cycles of the thermostat. Expected life of the heater elements will also be less.
2. There will be very high impact on the electrical supply when the electric elements are cut in. The high surge currents that result have caused some utilities to restrict the amount of single connected load to 10 KWH.

The unit selected has been designed to provide two stages of 10 KWH each. The staging is accomplished by the use of

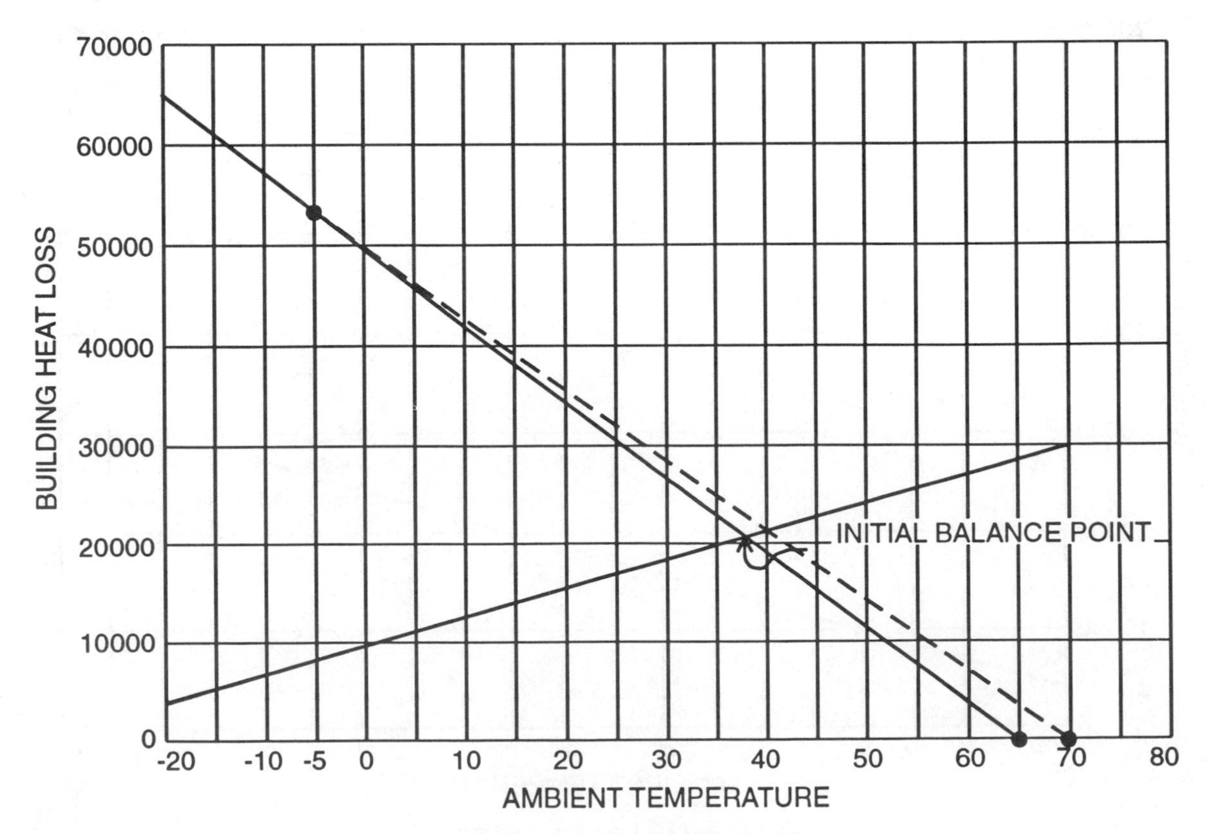

Figure 4–18 Initial balance point plotted

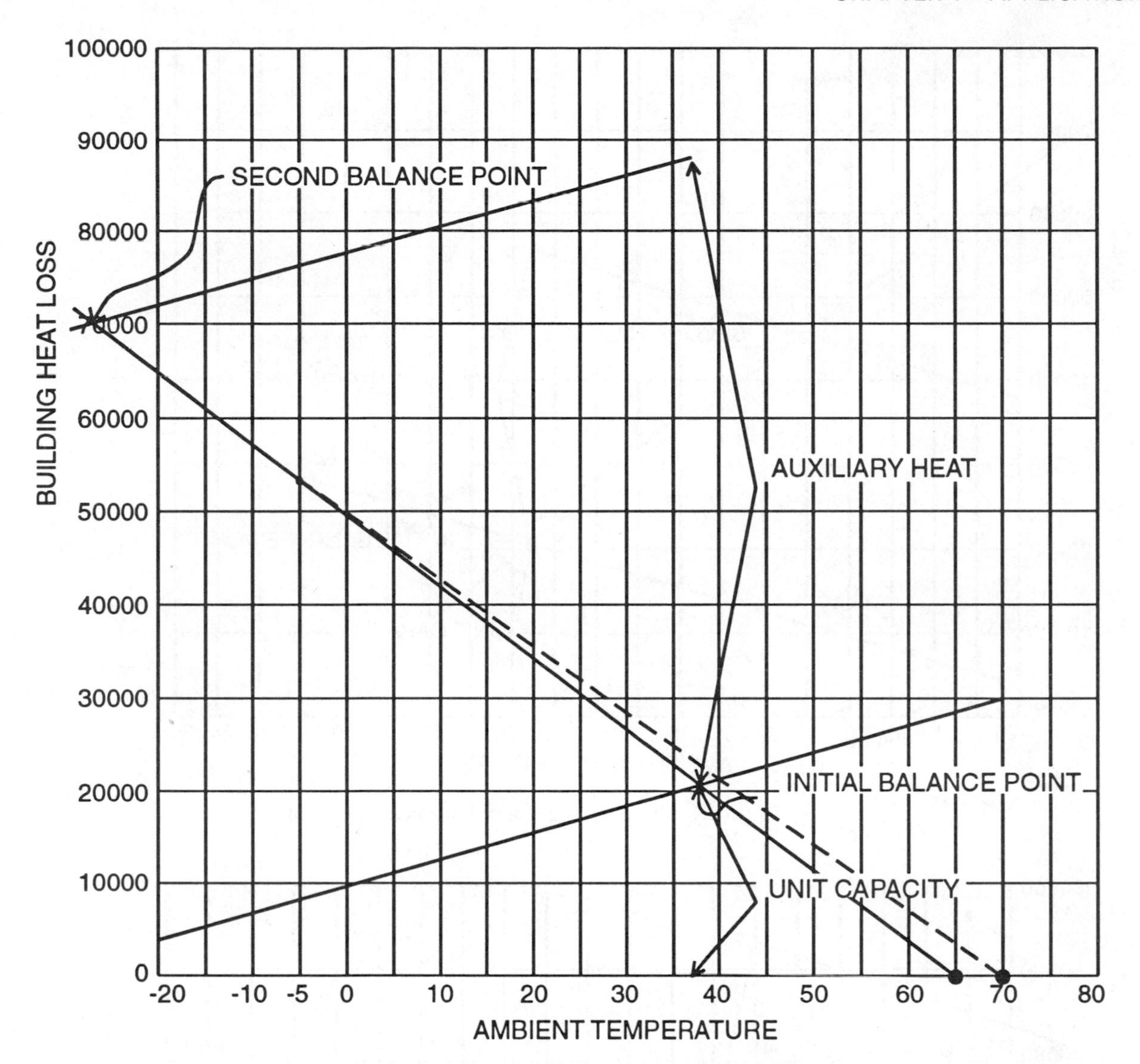

Figure 4–19 Second balance point plotted

an outdoor ambient thermostat (OAT) set to keep the second stage of electric heat off until the outdoor ambient temperature drops to a point where the heat loss is enough to require it. This temperature setting is the setting point of the outdoor ambient thermostat (OAT). To determine this point, the auxiliary heat capacity is plotted on the graph in two steps, 10 KWH per step.

In Figure 4–20, the second balance point is the heating capacity of the heat pump (21,500 BTUH) plus the amount of heat from the first 10 kwh stage of the auxiliary heat (34,130). This system capacity of 55,630 BTUH is plotted on the vertical +37° F line. A line representing the heating capacity of the two heat sources is drawn parallel to the unit capacity line to represent the system capacity at the falling outdoor ambient temperatures (see Figure 4–21.)

We now find the second balance point, the combination of auxiliary heat BTUH and unit BTUH, is +6° F. Down to this point, only one 10-KWH stage is needed to handle the load. The OAT would be set at +6° F. Actually, the OAT would be set at the second balance point plus 3 to 5° F. This additional temperature is a safety factor to allow for wind effect on the heat loss. The original heat loss is figured with 15-mph wind. At wind speeds higher than this, at the second balance point, uncomfortable conditions would exist in the conditioned area. Therefore, the second stage auxiliary heat would be brought in at a temperature above the plotted second balance point. In our example, we will use 4° F and set the OAT at +10° F.

By splitting the auxiliary heat into two stages, the longer unit cycles will result in a more even comfort in the conditioned area at lower operating costs.

Phoenix, Arizona

In the Phoenix, Arizona application, a different picture results when the load requirements are graphed (see Figure 4–22). When this application is plotted, we find the balance point of the heat pump versus the heat loss is +20° F. This is lower than recorded temperatures in the Phoenix area. Therefore, the auxiliary heat source would not be required with the exception of unit malfunction and during the defrost mode for supply air tempering.

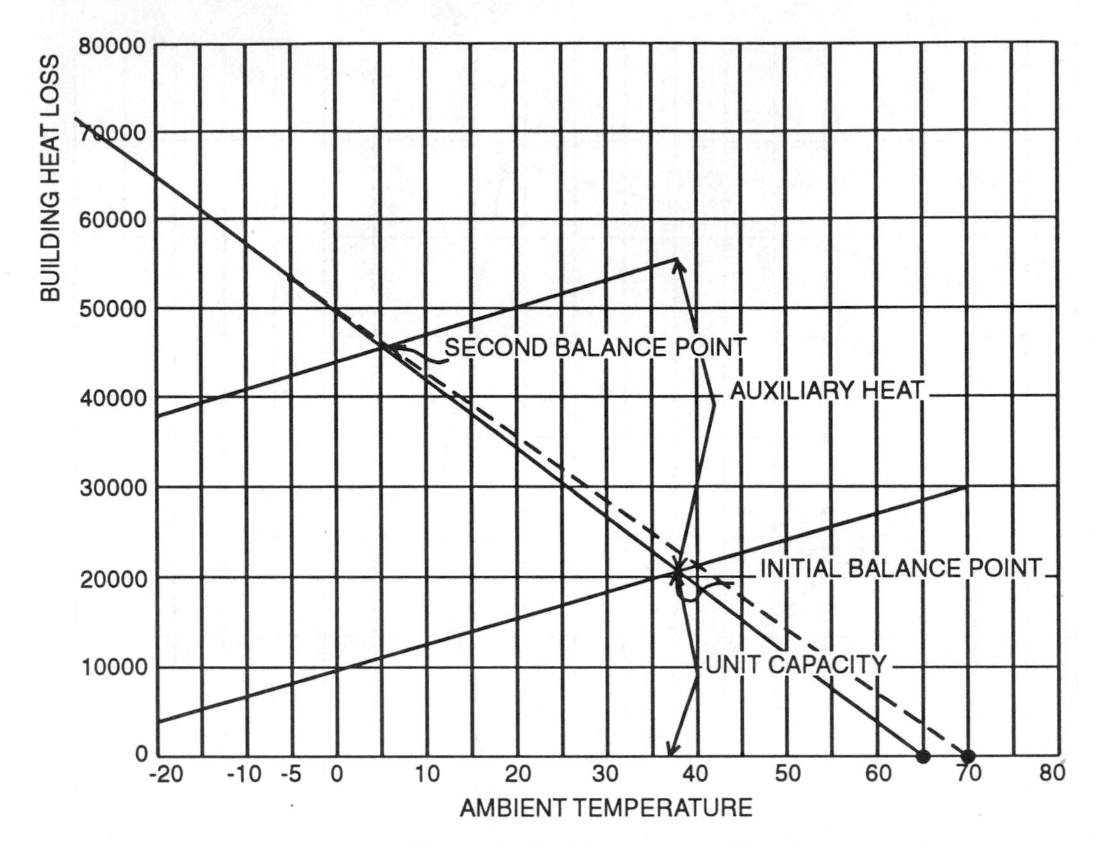

Figure 4–20 First stage auxiliary heat

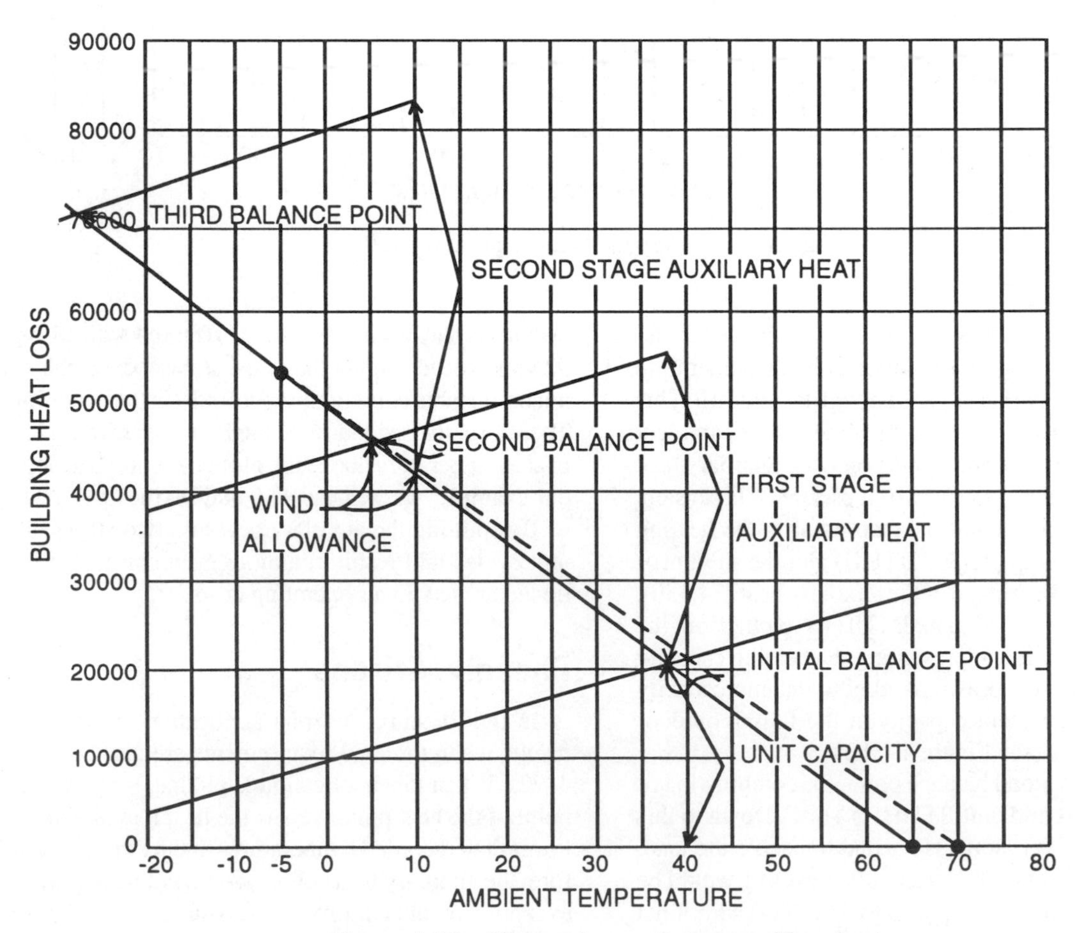

Figure 4–21 First and second stage auxiliary heat

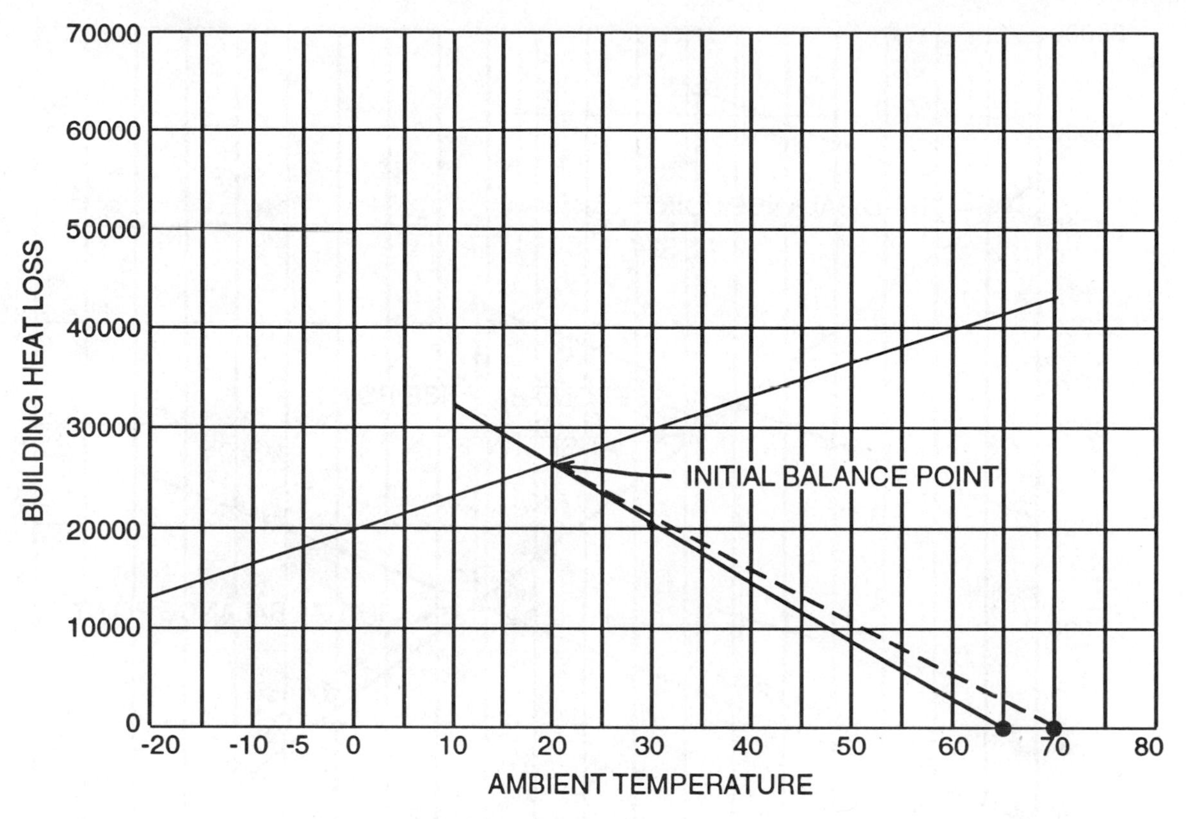

Figure 4–22 Initial balance point—north package

Omaha, Nebraska (Package-Type System)

To illustrate the application of the package-type system in Omaha, NB, the graph in Figure 4–23 has been drawn to show the following plotting steps completed:

1. Building heat loss line from 65° F outdoor ambient temperature through the heat loss at design temperature point of –5° F.
2. Unit capacity line. For this particular unit, the manufacturer has provided only two points of unit heating capacity, at 45° F and at 17° F (ARI Standards). These are used to draw the unit heating capacity line. Development of these two lines reveals a cross point or initial balance point of 36° F. The auxiliary heat selected has two stages of heat output. This consists of four 4.8 KWH elements in two stages of 9.6 KWH each.
3. Stage 1 brings in 9.6 KWH or 30,154. The unit capacity at 36° F (22,800 BTUH) plus the 30,154 BTUH auxiliary heat produces a system capacity of 52,954 BTUH. This has been plotted to show the second balance point.

   A line through the second system balance point, parallel with the original system capacity line crosses the heat loss line at slightly under +10° F. This temperature plus 4° F wind allowance means the OAT controlling the second stage of auxiliary heat will close at +14° F.
4. Stage 2 will bring in another 9.8 KWH (30,154 BTUH) at +14° F when the second OAT closes. This will raise the system heating capacity to 60,308 BTUH (two stages of auxiliary heat) plus the unit capacity of 14,000 BTUH at the +14° F for a total of 74,308 BTUH. This will handle the heat loss of the home down to –26° F.

Both systems used in the Omaha, NB application produced an initial balance point of approximately 36° F. The average winter temperature in Omaha, is 35.6° F. The heat pump should handle the heating load for 50% of the winter heating season.

Phoenix, Arizona (Package-Type System)

To illustrate the application of a heat pump in the Phoenix, AZ area, the graph in Figure 4–24 has been drawn to show all the plotting steps completed.

1. Building heat loss line from 65° F outdoors to the outdoor design temperature point of 34° F.
2. Unit capacity line from 40,000 BTUH at +45° F to 25,000 BTUH at +17° F. This produces an initial balance point of 22.5° F. This is lower than recorded winter temperatures in the Phoenix area. The heat pump will handle the heating load 100% of the winter heating season. Auxiliary heat would only be required for any malfunction or air tempering during defrost.

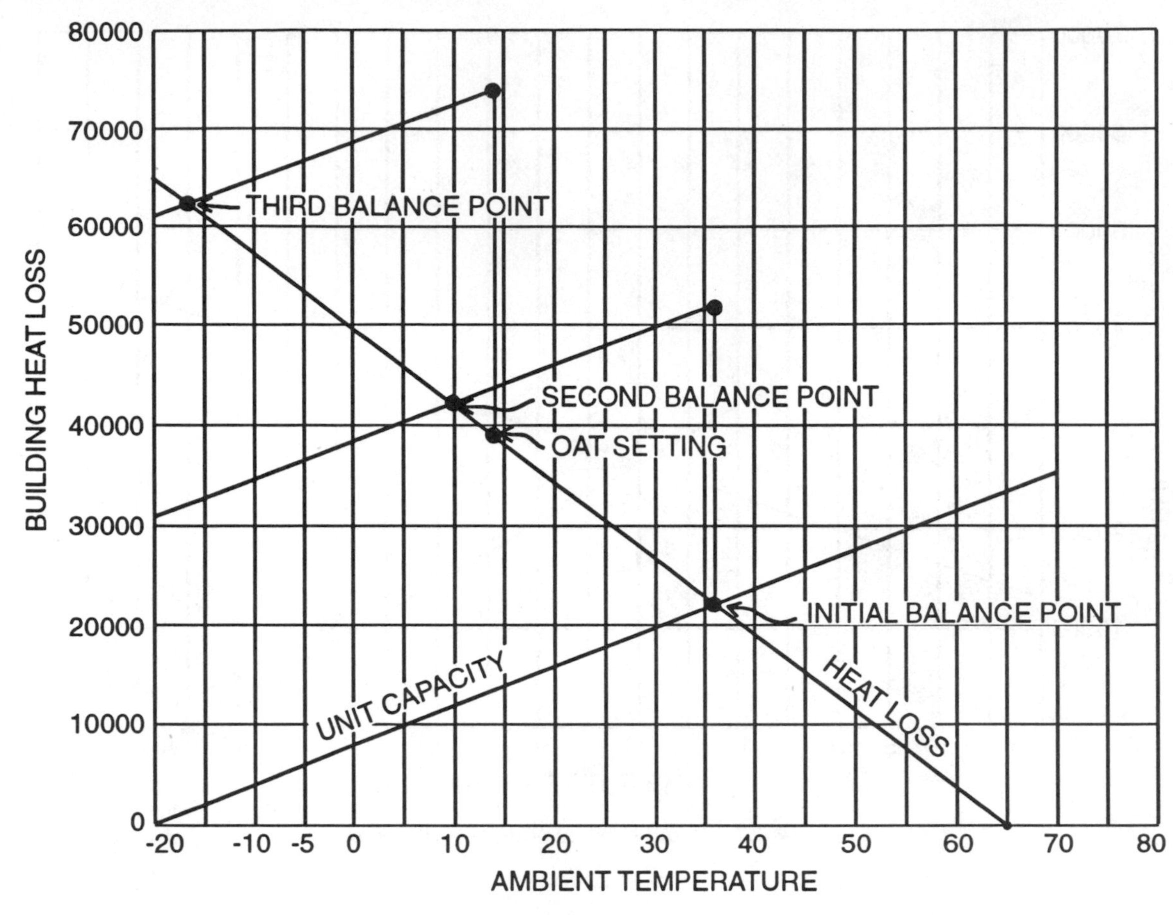

Figure 4–23 First and second stage auxiliary heat—north package

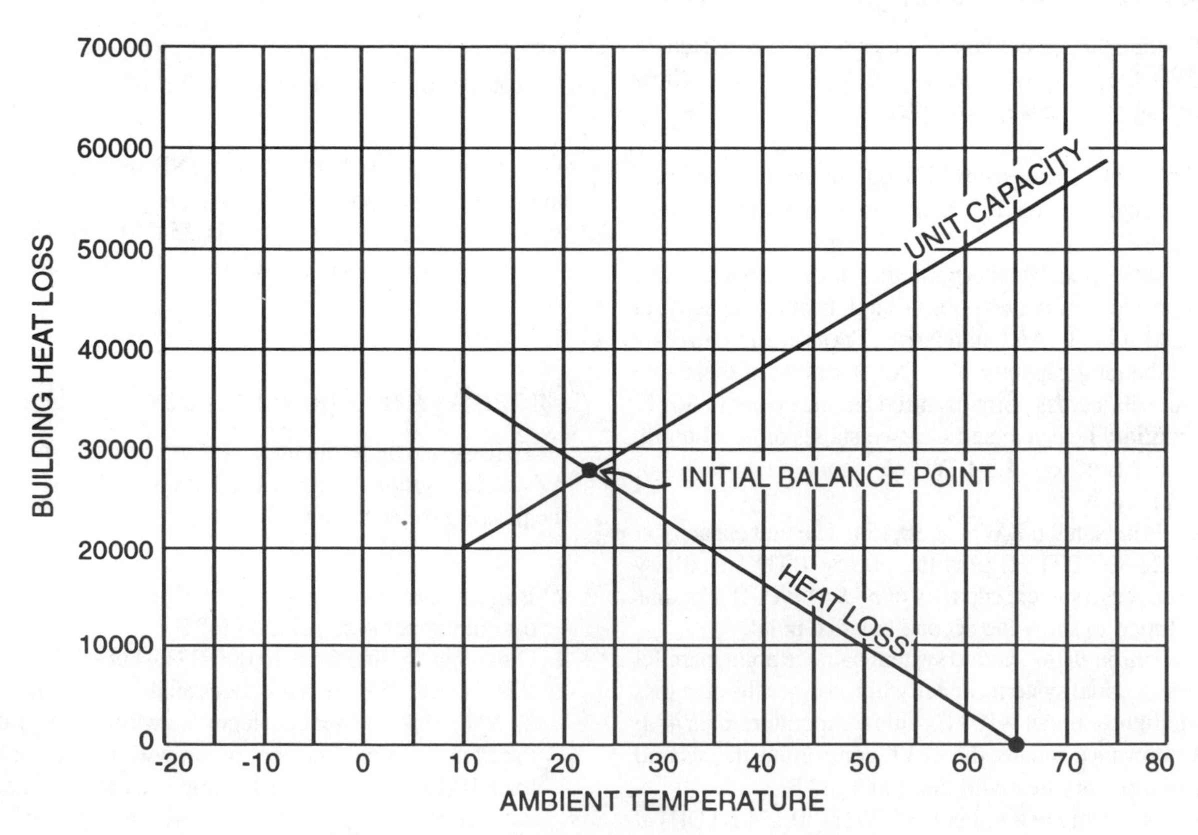

Figure 4–24 Initial balance point—south package

REVIEW QUESTIONS

1. The heat pump is sized to handle the heating load because its prime function is heating. True or False?
2. The amount of auxiliary heat capacity is equal to the heating load minus the unit heating capacity. True or False?
3. Heat pumps, like fossil fuel units, deliver supply air in the 130° F to 150° F range. True or False?
4. Maintenance of a heat pump is more critical than maintenance of a gas-fired unit. True or False?
5. The first step in sizing a heat pump is to determine the:
   a. Supply voltage available.
   b. Space available.
   c. Heat loss of the conditioned area.
   d. Heat gain of the conditioned area.
   e. Amount of money the buyer has.
6. Determining the heat gain and heat loss of the conditioned area by the cubic content method is sufficiently accurate for heat pump applications. True or False?
7. In the heating mode, the capacity of an air-to-air heat pump increases or decreases with a decrease in outdoor temperature?
8. In the cooling mode, the capacity of an air-to-air heat pump increases or decreases with a decrease in outdoor temperature?
9. Match the component parts with the section of the split-type system in which the part is located. Use an *I* for indoor section and *O* for outdoor section.
   a. _____Indoor heat exchanger
   b. _____Compressor
   c. _____Fan and motor assembly
   d. _____Pressure-reducing device
   e. _____Circulating blower
   f. _____Check valve
   g. _____Auxiliary electric heating elements
   h. _____Controls
   i. _____Outdoor heat exchanger
10. The heating output of a heat pump is called the _____ _____.
11. The heating output of a heat pump system is made up of the heat picked up from the outside air plus the electrical energy of the compressor. True or False?
12. A heat pump system operating at +40° F, producing 45,182 BTUH, uses 4,814 KWH. What is the COP rating?
13. For best unit performance and minimum maintenance cost, the heat pump should be cut off at the outdoor temperature too low to produce a COP of 1.5. True or False?
14. The cooling capacity rating of a heat pump is the amount of sensible heat picked up by the inside coil. True or False?
15. When selecting a unit capacity as compared to the calculated cooling load, the tolerance range of capacity versus the cooling load is:
   a. –10% to +10%.
   b. –5% to +10%.
   c. –10% to +5%.
   d. –10% to -0%.
   e. –0% to +10%.
16. The calculated heating requirement is 32,765 BTUH. How many 5 KWH electric elements are required to handle the heating load?
17. If the 5 KWH elements required in Problem 16 are connected to a 208-volt power supply, how many elements will be required?
18. What is the cooling capacity of a 42,000 BTUH heat pump when operating at an outdoor ambient temperature of 110° F?

5 INSTALLATION

The methods used and the precautions followed when installing an air-to-air heat pump system can be the difference between satisfactory performance at reasonable operating cost with little maintenance or the opposite—poor performance, high operating cost, and constant problems. An air conditioning system and an air-to-air heat pump need the proper amount of air over both heat exchangers to operate properly. Heat pumps also have motors and a compressor that can produce vibrations. These vibrations can cause objectionable noises in the occupied area if the equipment is not properly located and supported. Adequate power to operate the equipment is a must. These are general statements which will be discussed from a more specific angle with each type of system.

5–1 SPLIT SYSTEM INSTALLATION

When installing a split-type heat pump system, there are rules that apply to the outdoor unit as well as the indoor unit. Each section of the system has its own special requirements.

Outdoor Section

Location. The main consideration in the location of the outdoor section is the length of the interconnecting refrigerant lines. The length of the liquid line is not as critical as the length of the vapor line. The liquid line should be as short as possible to keep the liquid flow resistance within the proper range and the refrigerant charge to a minimum.

The vapor line is not called the "suction line" because it is a dual-purpose line. It is the "suction line" in the summer cooling mode and the "hot gas line" in the winter heating mode. When it is the "hot gas line," the length becomes critical.

When the compressor operates, hot refrigerant vapor enters the hot gas or vapor line to be carried to the inside condenser coil and give up heat to the air over the coil to be supplied to the conditioned area. Before any amount of heat enters the condenser coil, the vapor line has to rise in temperature to reduce the heat loss through the line. The longer the line, the longer it takes to get heat to the condenser and the longer the compressor must operate before the system is up to full heating capacity. Therefore, the length of the line not only affects the system heating performance but also has a decided effect on the system operating cost.

Airflow. The outdoor unit must be located to provide maximum airflow through all sides of the condenser supply air openings. The air openings of a unit, either entering or leaving air, must not be located any closer to a surface parallel to the coil face than 1½ times the greater dimension of the opening. For example, if the opening in which the air is received is 20" × 30", the minimum distance to a parallel flat surface would be 45". The resistance to air flow increases very rapidly as this dimension is reduced.

If the unit uses a propeller fan for air movement, the fan motor can be overloaded, the cfm of air reduced, and the unit can have operating problems because of poor location too close to a flat surface.

Serviceability. The unit has to be serviced. Do not locate the service access side of the unit any closer than 30" to a vertical flat surface.

Air Recirculation. The unit should not be located under an overhang or under a horizontal surface such as a porch floor. When the discharge air from the unit is trapped and forced to recirculate through the unit, winter temperatures drop and unit heating capacity suffers. During the summer, entering temperature will rise, the unit will work harder, capacity will drop, and operating costs will rise.

Figure 5–1 shows the clearance requirements for a vertical-discharge outdoor unit that has a condenser air opening on

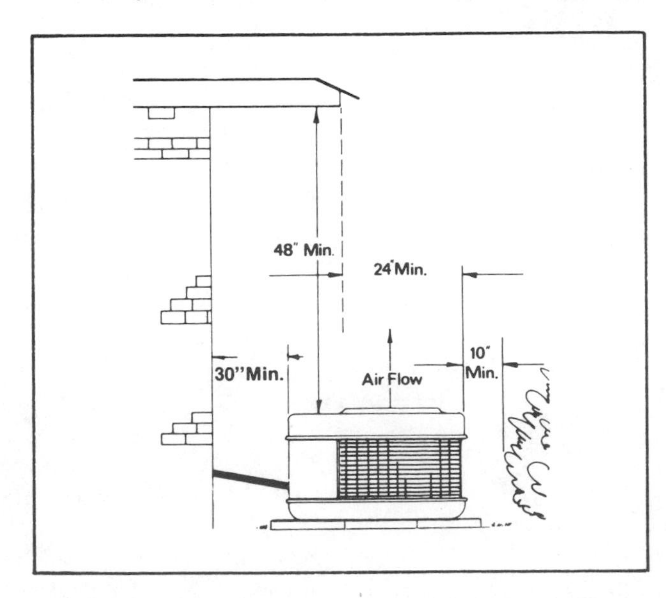

Figure 5–1 Outdoor unit location (Courtesy Addison Products Co.)

three sides. The fourth side (in this illustration) contains the motor-compressor and controls. Illustrated are the minimum placement dimensions for air supply to the condenser, reduction of possible recirculation, and proper service accessibility.

Foundation. The unit must be on a sturdy foundation. It must be high enough above the ground level to reduce the effect of ice buildup during the defrost function. The height must be sufficient to allow for snow accumulation. There are several types of unit foundations marketed. Regardless of the type used, it is advisable to use a bed of gravel, well tamped to a firm foundation, for the unit base. The unit foundation selected for the installation should be located so as not to have any physical connection to the building. Unit operating vibrations can be transmitted into the building if there is physical contact.

The secret to proper installation techniques is to arrange for adequate air flow through the unit, serviceability, and location for minimal noise complaints.

Power Wiring. The installation of electrical power circuits is closely regulated by local, state, and national codes. Most areas require that such installations be installed by licensed electricians. A universal requirement from a safety angle is the installation of an outdoor disconnect switch in the service distribution box. It should not be mounted directly on the unit but a short distance away to promote safety to the operator of the unit.

Figure 5–2 shows a typical power service arrangement. The outdoor disconnect switch is located on the wall adjacent to the unit but far enough away to protect the operator in the event of an electrical flash from the unit.

All outdoor connections must be of weatherproof cable or conduit and fittings of the type approved by local code. In any case, the exposed cable or conduit must be protected from mechanical damage. The size of the wire used will depend upon the amperage draw of the unit supplied with power and the length of the power circuit.

All manufacturers supply the electrical information that applies to their equipment. Supplied in table form, Figure 5–3 is a typical listing in the manufacturer's installation instructions. This table supplies the locked rotor amperage (LRA), the full load amperage (FLA) or rated load amperage (RLA), the minimum copper wire size, the maximum length of wire that will permit use of the minimum wire size, and the recommended fuse size.

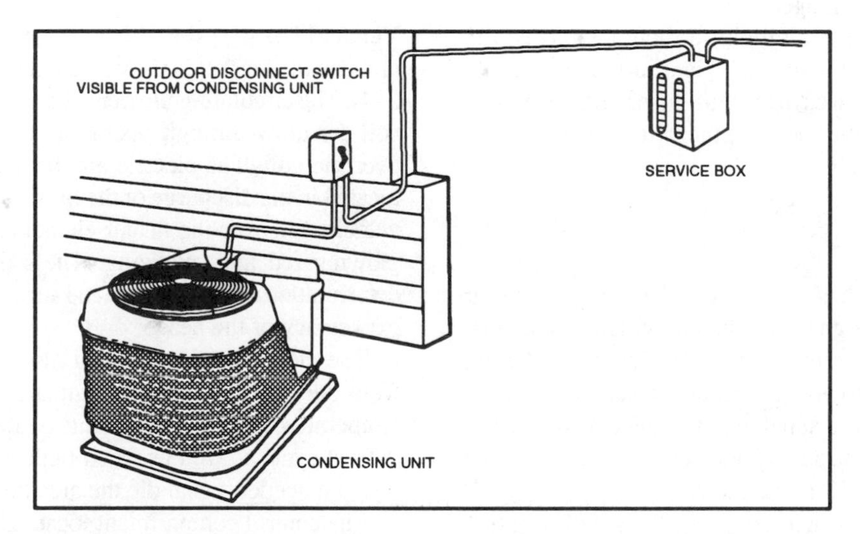

Figure 5–2 Electrical service to outdoor unit (Courtesy Addison Products Co.)

Model	Volts	Phase	Current at Rating Conditions Amps	Total Locked Rotor Amps	Min. Wire Size	Max. Run For Min. Wire Size	Recommended Delay Fuse Size	Max. Fuse Size
QH824	230/208	1	11.6/13	61.4	12	40	20	25
QH830	230/208	1	14.7/16	69.4	12	35	20	30
QH836	230/208	1	18.2/20	82.4	10	45	25	40
QH842	230	1	23.5	104.4	10	37	40	40
QH848	230	1	24.6	111.4	8	56	45	45
QH860	230/208	1	33.9/35.4	175	6/6	57	60	60
QH848	230/208	3	15.5/16.8	79.4	10	53	25	30
QH860	230/208	3	21.8/23.2	136	8	49	35	40

Figure 5–3 Electrical characteristics—split heat pump (Courtesy Addison Products Co.)

Two things must be noted here:

1. Copper wire is specified. Aluminum wire is not approved for power circuits because of the high LRA current involved.
2. The fuse size is specified as "time-delay type." When a motor such as a compressor motor starts, the in-rush current (LRA) is many times higher than the running current (FLA). The example: at 230 volts, the QH830 unit selected for the Phoenix, AZ application has an LRA of 69.4 amperes and a FLA (RLA) of 14.7 amperes. This unit would require No. 12 wire with a maximum length of 35' before using the next larger wire. The recommended time-delay type fuse would be 20 amperes with a supply voltage of 230 volts. If the supply voltage were 208 volts, the 30-ampere time-delay fuse would be used.

Control Wiring. Residential and small commercial heat pump systems are controlled by low voltage controls in practically all cases. 18 Ga. thermostat wire cable is used to supply the control voltage to the contactors and relays in the outdoor unit. The recommended cable would be eight-wire type, each wire of a different colored insulation. Attempting to use multiple cable of like color only leads to confusion and sometimes equipment damage.

The use of telephone cable is highly discouraged. This is 22 Ga. wire and is too small to carry sufficient current to properly operate the electrical devices in the unit. To be on the safe side and give the best operation and equipment life, use thermostat cable in all control circuits.

Indoor Section

The indoor section has to be located to supply the air distribution system designed for the particular application. Practically all manufacturers of standard residential and small commercial heat pumps use the industry standards of air distribution design. These standards designate a static pressure (Ps) of 0.2" WC in the duct system, using 0.15" WC in the supply duct and 0.05" WC in the return duct.

It is very important, when designing the air distribution system for a heat pump, to remember that a heat pump must move 2 to 2.5 times as much air in the heating mode as would be required by a fossil fuel unit. The heat pump works with a maximum of 35°F temperature rise of the air through the indoor coil when it is 70°F inside and 70°F outside the conditioned area. This low temperature rise is necessary to keep the compressor discharge pressure and amperage draw within the proper operating range.

The recommended temperature rise through a fossil fuel unit, according to the ACCA manuals, is 80°F with continuous air circulation (CAC). This calculates out to 2.28 times as much air for the heat pump. Using the recommended static pressure of 0.15" WC in the supply and 0.05" WC in the return, the heat pump requires a larger duct system with more supply outlets.

Figure 5–4 Air handler—Vertical up airflow (Courtesy Addison Products Co.)

Flexible Duct Connections. All heat pump units develop vibrations because of the motors and blowers involved. To reduce the possibility of these vibrations being transmitted into the duct system and causing noise problems, flexible duct connections should be used between the unit and the duct system.

Vertical Up Airflow. The basic design of practically all units on the market is the vertical up airflow type (see Figure 5–4). The circulating blower is on the downstream side of the coil for draw-through operation and better air distribution over the coil. The electric auxiliary heat resistor banks are located in the discharge of the blower. This is to provide high velocity air over the heater elements to prevent them from glowing red and oxidizing. The oxidizing process causes deterioration of the heater metal surfaces and shortens the life expectancy of the heater.

The electric elements must also be located downstream from the inside coil. This eliminates the effect of the high-temperature air off the elements on the inside coil during the defrost cycle as well as when both heat pump and auxiliary heat are needed to handle the area heat loss.

The control compartment located immediately in front of the electric elements usually contains the 24-volt power source for the low-voltage control system. Also included are the time-delay controls for the electric elements. These are slow-acting heat-motor-operated contactors called *sequencers.*

The purpose of using sequencers instead of relays is to provide a time delay when bringing in the heater loads. This limits the inrush current to one element at a time. Figure 5–5 shows an example of a dual switch assembly, with one switch controlling a single element and one switch controlling two elements in sequence. Shown also is the contact schematic that this particular manufacturer produces in this type sequencer.

The heat motors are 24 volt with the contacts designed to control high voltage loads. When used in heat pump air handlers, the blower motor is supplied power through the first

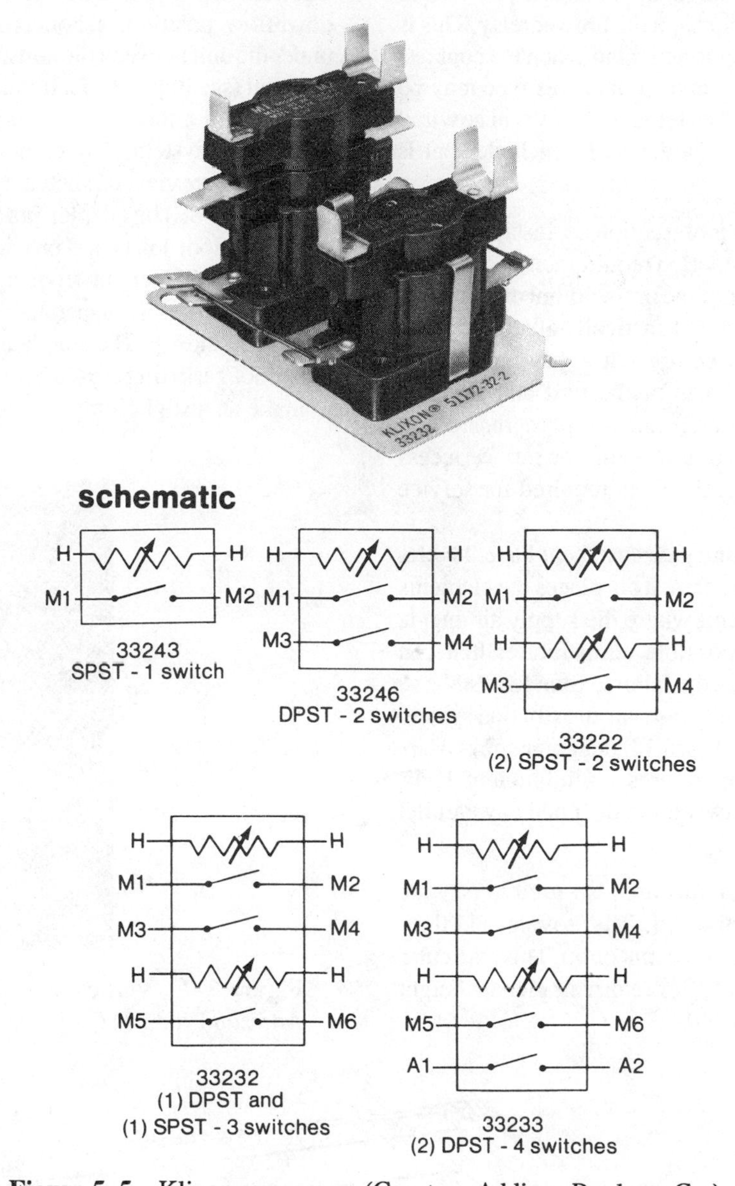

Figure 5–5 Klixon sequencers (Courtesy Addison Products Co.)

contact to close. This contact controls the first stage element as well as the blower motor. The purpose of this action is to ensure blower operation when the elements are active.

The safety action requires a blower relay in the unit that has a SPDT set of contacts controlling the blower motor. The blower motor power supply lead is connected to the common terminal of the contact set. Power directly from the high-voltage supply circuit is connected to the normally open contact. Power from the first stage element, controlled by the sequencer, is connected to the normally closed contact. The purpose of this arrangement is to start the blower immediately upon a call for heat if the second stage of the thermostat is satisfied, until all the elements are off.

Figure 5–6 shows a different type of arrangement. This device has two stages of high-voltage control to the heater

Figure 5–6 Robertshaw sequencers (Courtesy Robertshaw Controls Co.)

elements. In addition, a separate set of contacts is provided to control the blower operation through the blower relay. This is to ensure that the blower relay is activated when the contacts are calling for heat. Several sequencers of this type may be used for random operation of the elements. They will be wired to provide blower operation regardless of which element is activated.

Foundation. When the indoor section is installed in an upright position, the only foundation requirement is sufficient strength to support the weight of the unit without distortion or excessive vibration transmission. Practically all manufacturers design and construct their equipment to have zero clearance between the sides and rear of the unit and adjacent combustible material—wallboard, paneling, plasterboard, and so on. Because the front of the unit contains the service access panels, a minimum of 20" clearance is required for service accessibility.

Indoor sections with auxiliary electric heat have the elements in the blower discharge area. This places the elements on the discharge side of the unit where the supply air duct is connected. Because of the possible temperatures that can develop due to blower operation failure, four feet (48") of metal duct must be used in the supply air distribution system before any change to other types of duct material. Also, because of such possible temperatures, a minimum of 1" air space must be maintained between the duct and any parallel combustible material.

Return Air. Return air plenums are also used to provide return air duct connections to the unit. This provides additional height to the assembly for drain connection. This particular manufacturer provides 10" to 13" of return air plenum height in its return air plenum assemblies.

Vertical Down Airflow. If the unit is installed in a vertical downflow position, a noncombustible base must be used under the unit between the unit and any flooring (combustible) material (see Figure 5–7). If the unit is set on a supply plenum formed in a concrete slab, such as used in a perimeter air distribution system, a noncombustible adapter is not required.

A cutaway view of such a noncombustible base is shown in Figure 5–8. The adapter fits into the frame opening of the floor and floor joists and provides a raised portion to isolate the supply air plenum from the combustible material. The supply plenum is suspended inside the adapter with the flanges on the top. The unit then sits upon the plenum flanges. Fireproof gasketing is used between the adapter and the flange to make an airtight joint.

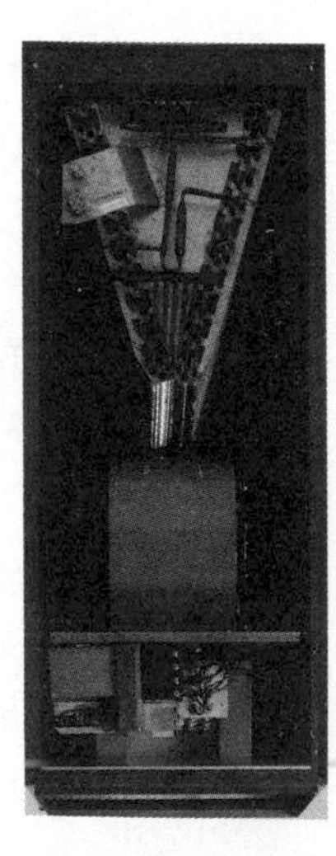

Figure 5–7 Vertical downflow air handler (Courtesy Addison Products Co.)

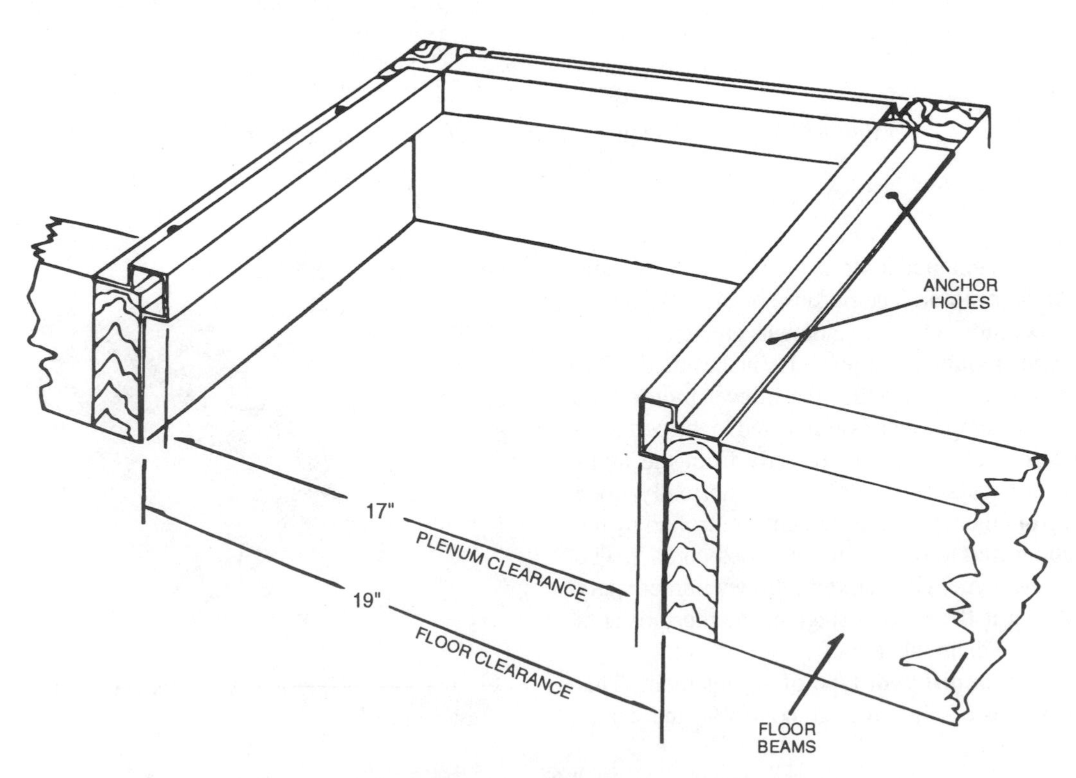

Figure 5–8 Noncombustible floor base (Courtesy Addison Products Co.)

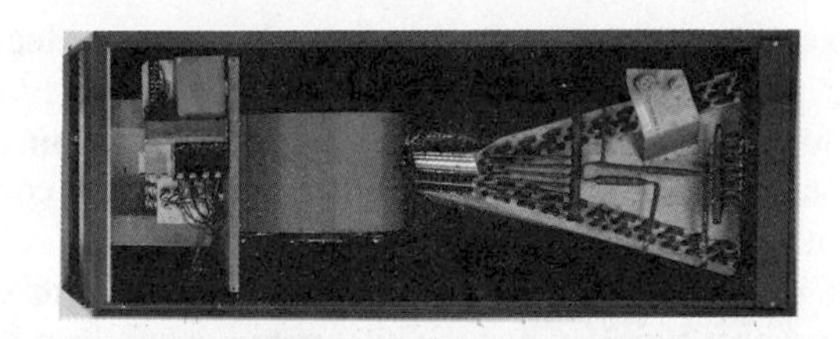

Figure 5–9 Air handler—horizontal air flow (Courtesy Addison Products Co.)

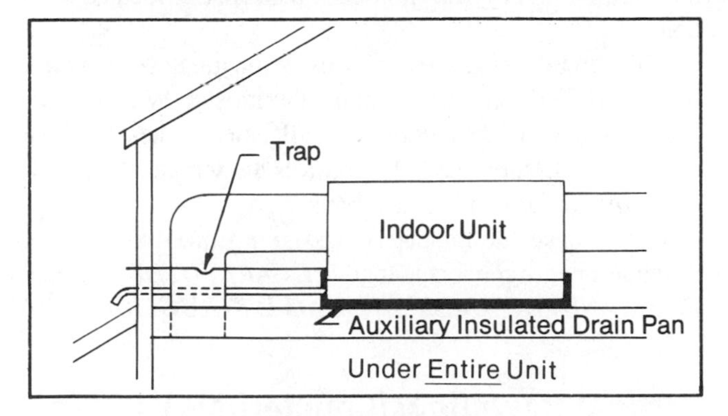

Figure 5–10 Auxiliary drain pan (Courtesy Addison Products Co.)

Horizontal Airflow. When the indoor unit is used in a horizontal position (see Figure 5–9), a complete foundation must be used to support the entire bottom surface of the unit. This is to eliminate cabinet strain.

Auxiliary Drain Pan. If the unit is to be placed in any location where drain overflow can cause damage, an auxiliary drain pan must be installed under the entire unit. The auxiliary drain pan is shown in Figure 5–10. The pan must be insulated on the outer surface and provided with a separate complete drain system. No trap is needed.

When the unit is located in the attic, the drain from the insulated auxiliary drain pan is usually run down in one corner of the bath tube or shower area on the first floor. This serves as a warning that the main drain is clogged if water comes out of the auxiliary drain line.

Coil Drain. In Figure 5–4, the coil is located in the bottom of the cabinet with the drain pan as part of the coil assembly. The air filter compartment below the coil assembly provides some elevation to the drain pan for connection of a drain line and trap. Access means to the air filter must be allowed when installing the drain line. Usually a rubber drain trap with a slip-fitting connection is used for easy removal when filter service is required.

Condensate Drain Trap. A condensate drain trap such as shown in Figure 5–11 must be used to prevent backflow of air through the drain line and blockage of the condensate flow out of the drain line. All air handling units that operate with negative pressure in the coil area require a condensate drain trap.

Power Wiring. The power supply to the inside air handler section has to be able to furnish full voltage at the full amperage draw of the unit. This means not only do the blower motor and control transformer constitute part of the electrical load, the electric heater bank is also included.

Each manufacturer includes the electrical characteristics and requirements in its printed instructions. These should be followed. Figure 5–12 is an electrical characteristic listing for

INSTALL CONDENSATE DRAIN TRAP SHOWN BELOW. USE DRAIN CONNECTION SIZE OR LARGER. DO NOT OPERATE UNIT WITHOUT TRAP. UNIT MUST BE LEVEL OR SLIGHTLY INCLINED TOWARD DRAIN.

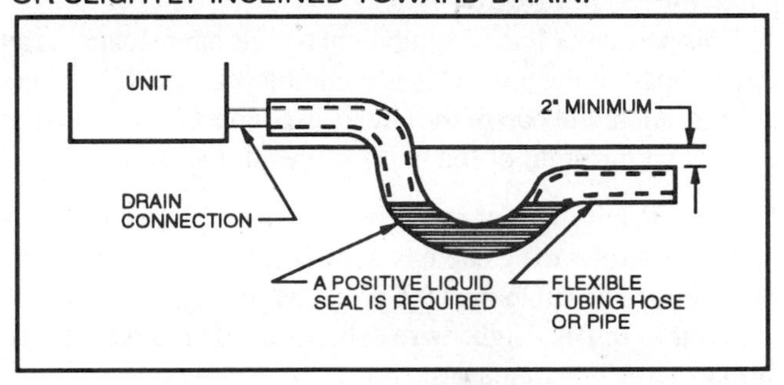

Figure 5–11 Condensate drain trap installation (Courtesy Addison Products Co.)

B/M	Phase	KW 208V	KW 240V	FLA 240V	BTUH 208V	BTUH 240V	Min. Wire Size 208V	Min. Wire Size 240V	Max. Fuse Size 208V	Max. Fuse Size 240V
65HDB405B	1	3.8	5	22.8	12,800	17,100	10	10	25	30
65HDB410B	1	7.5	10	43.7	25,600	34,100	6	6	50	60
65HDB415C	1	11.3	15	64.5	38,420	51,200	4	2	70	90
65HEB405B	1	3.8	5	24.7	12,800	17,100	10	8	30	35
65HEB410B	1	7.5	10	45.6	25,600	34,100	6	4	50	60
65HEB415C	1	11.3	15	66.4	38,400	51,200	3	2	80	90
65HEB420C	1	15	20	87.2	51,200	68,200	1	1	100	110
68HFB410B	1	7.5	10	45.6	25,600	34,100	6	4	50	60
68HFB415C	1	11.3	15	66.4	38,400	51,200	3	2	80	90
68HFB420C	1	15	20	87.2	51,200	68,200	1	1	100	110
68HFB425B	1	18.8	25	108.1	64,000	85,300	1/0	2/0	125	150
68HGB410B	1	7.5	10	47.7	25,600	34,100	6	4	60	60
68HGB415C	1	11.3	15	68.5	38,400	51,200	3	2	80	90
68HGB420C	1	15	20	89.3	51,200	68,200	1	1/0	100	125
68HGB425B	1	18.8	25	110.2	64,000	85,300	1/0	2/0	125	150
68HGB430B	1	22.5	30	131.0	76,800	102,300	2/0	3/0	150	175
68HHB415C	1	11.3	15	69.0	38,400	51,200	3	2	80	90
68HHB420C	1	15	20	89.8	51,200	68,200	1	1/0	100	125
68HHB425B	1	18.8	25	110.7	64,000	85,300	1/0	2/0	125	150
68HHB430B	1	22.5	30	131.5	76,800	102,300	2/0	3/0	150	175
68HHB435B	1	26.3	35	152.3	89,600	119,400	4/0	4/0	175	200
68HKB415C	1	11.3	15	69.0	38,400	51,200	3	2	80	90
68HKB420C	1	15	20	89.8	51,200	68,200	1	1/0	100	125
68HKB425B	1	18.8	25	110.7	64,000	85,300	1/0	2/0	125	150
68HKB430B	1	22.5	30	131.5	76,800	102,300	2/0	3/0	150	175
68HKB435B	1	26.3	35	152.3	89,600	119,400	4/0	4/0	175	200

Figure 5–12 Electrical characteristics of an air handler (Courtesy Addison Products Co.)

the manufacturer whose equipment we have been using in the application examples.

The six different models of inside air handlers with the available sizes of electrical heat capacities have the amperage draw at two voltages (208 V and 240 V) as well as the BTUH capacities at the two voltages. Of prime interest at this point are the wire and fuse sizes. You will notice that no maximum length per wire size is given. This table is based on the National Electrical Code requirements, which give wire sizes based on 100 ft. of wire.

This particular line of equipment has circuit breakers as an integral part of the unit. This is common practice. If breakers are not an integral part of the unit, a disconnect switch must be located within sight of the unit for operator safety.

Control Wiring. For inside wiring installations, the thermostat control wiring cable is usually run without enclosure in conduit. The cable must be protected from damage where exposed to traffic. Eight-wire cable is needed to operate the system's various functions.

Depending upon the type of control system used and connection locations, the number of wires in the cable will vary. Most manufacturers use the indoor handler as the main connection terminal location. The control cables then radiate from this unit to the thermostat and to the outdoor section.

A typical wiring diagram showing the cable between the wall thermostat and the indoor unit as well as the cable between the indoor unit and the outdoor unit is shown in Figure 5–13. Each of the three locations use industry standard terminal designations. The problem this can create is that terminal designations do not always mean the same function.

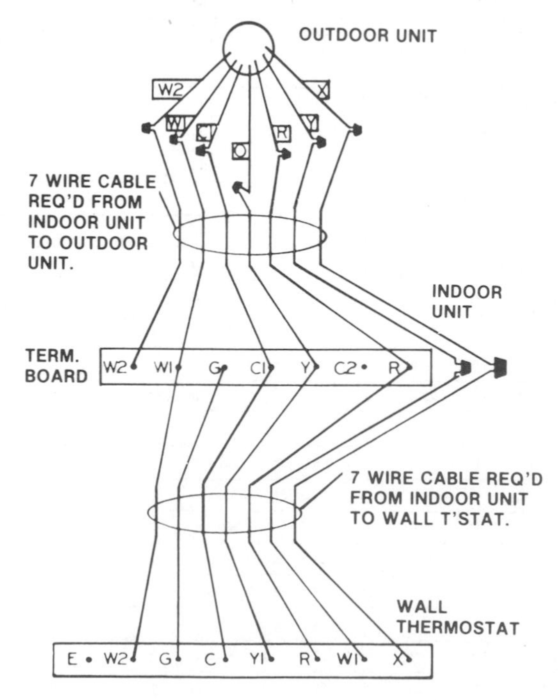

Figure 5–13 Low-voltage field wiring (Courtesy Addison Products Co.)

For example, the Y1 terminal on the thermostat does not control the cooling operation. It controls the system reversing valve to change the operation mode of the unit. Terminal W1 (first-stage heat or first-stage cool) on the thermostat controls the motor-compressor operation.

The second-stage heat function (W2) actually controls the auxiliary heat supply when the heat pump cannot handle the load. How much auxiliary heat is brought in will depend upon the outdoor ambient thermostat and the setting of the OAT. In this unit, the OAT is the "holdback thermostat" located in the unit.

This circuit has given the industry the greatest amount of confusion. The second stage of the thermostat (W2) connects to the first stage of the indoor air handler electric heat bank and to the OAT. In turn, the OAT controls the second stage (W2) of the indoor air handler heat bank.

Don't make the mistake of assuming that like terminal designations are connected together. *CAREFULLY FOLLOW THE MANUFACTURER'S WIRING INSTRUCTIONS.*

Vertical Downflow (Counterflow)

A popular application, particularly in slab-home construction with perimeter duct distribution systems or manufactured homes with under-the-floor distribution systems, uses the indoor air handler in an "upside-down" or "counterflow" position.

The performance is the same as the vertical upflow. However, safety factors have to be considered for an approved installation. When inverting the assembly, the coil is inverted and the drain pan portion of the coil can no longer be used. This means that the manufacturer must provide a means of catching the condensate off the coil at the V point of the coil.

In addition, for the unit to be approved for operation in this position, an antifrost thermostat must be installed on the coil. This thermostat prevents the buildup of ice on the coil. If any ice accumulation occurs, the melting of the ice could allow quantities of ice and water to bypass the drain pan and fall into the heater element deck at the bottom of the unit. Thus, the antifrost thermostat requirement.

Usually, these items are supplied in kit form called a downflow adapter kit.

Horizontal Airflow

When the heat pump inside air handler is used in a horizontal position, the major change is the addition of a drain pan assembly located under the coil (see Figure 5–9). The pan must be longer than the coil, must extend beyond the coil and it must contain some type of antisplash means to keep condensate off the coil from leaving the drain pan and damaging the cabinet.

The only position for the cabinet is on either the left or right side. The access panels must be on the front to provide accessibility for service. The cabinet cannot lay on the back as this will prevent proper drainage of the coil.

5–2 ADD-ON SYSTEMS

The outdoor section of an add-on type system is the same as the outdoor section used in the split-type system using electric heat for auxiliary heating capacity. All rules that apply to the installing of the split-type outdoor section apply to the add-on outdoor section.

Control Wiring

When adding a heat pump to a fossil fuel heating unit, considerably more control circuits and control wiring are involved. In addition, the fossil fuel heating unit is used *instead* of the heat pump when the system is operating below the initial balance point. The control system selects which equipment will supply heat on demand from the area thermostat—heat pump above the initial balance point or fossil fuel below the initial balance point.

There are two ways to control the changeover at the initial balance point—an outdoor ambient thermostat (OAT) set at the initial balance point plus wind allowance or the second stage of the area thermostat through the use of a changeover relay.

The first item that should be discussed is the change in the thermostat circuits in an add-on type thermostat as compared to a standard system thermostat. This difference is the way the power is supplied from the control power source transformer. Basically, the standard thermostat uses power from one power terminal, the R terminal. The add-on thermostat uses power through two different circuits, each feeding a separate thermostat contact circuit.

Standard Thermostat

The thermostat used in the standard heat pump with electric auxiliary heat has a single power supply terminal, R. This circuit is connected directly to the R terminal of the control power transformer and is hot at all times (see Figure 5–14).

In the circuits of the thermostat subbase, with the selector switches in the "heat" position, power is supplied through the season switch to a jumper circuit connected to the HTG-1 and the HTG-2 control circuit contacts. These contacts are connected together mechanically to maintain 2° F above HTG-2 as the area temperature falls. This action cycles the heat pump as long as the heat loss is less than the heating capacity of the heat pump.

When the initial balance point is reached and the heat pump operates continuously to handle the heating load, any further increase in the heating load will cause the conditioned area temperature to drop below the first stage thermostat setting. When the conditioned area temperature falls 2° F below the first stage setting, the second stage thermostat contact (HTG-2) closes and brings on the auxiliary electric heat. Both heat sources operate simultaneously.

Add-on Thermostat

The add-on type area thermostat uses two separate power sources to operate either the heat pump or the fossil fuel heating unit. The control circuit in Figure 5–15 shows a heat pump operating in conjunction with a gas-fired heating unit. This thermostat does not have a single R terminal for power. Instead, it uses terminal W3 to supply power through the first stage thermostat contact (HTG-1) to terminal Y to control the motor-compressor assembly. Terminal W2 is used to supply power through the second stage thermostat contact (HTG-2) to terminal W to control the gas valve circuit in the heating unit.

Outdoor Thermostat

The selection of the heating action is determined by the action of the outdoor ambient thermostat (OAT) located in the outdoor section of the system. This particular manufacturer has labeled this control the "holdback" thermostat (HT). This is a single-pole, double-throw (SPDT) thermostat, set at the corrected initial balance point of the system operating curves. Above the corrected initial balance point, the circuit is made between terminal 2 (common) and terminal 1 (normally closed) to supply power to thermostat terminal W3. Below the corrected initial balance point, the circuit between terminal 2 (common) and terminal 3 (normally open) is made and power is supplied to terminal W2 of the thermostat subbase. This places the selection of the heat method strictly under the control of the OAT.

The control system in Figure 5–16 does not use an OAT to select the system to supply the necessary heating capacity. This system uses a changeover relay to select the source of heating capacity. Both the W2 and W3 terminals of the thermostat are supplied power directly from the R terminal of the control power source transformer by means of a jumper wire between terminals W2 and W3.

The heat pump compressor and the gas-fired heating unit are controlled through a twin relay assembly. One relay is used to control the furnace heat light. This is an indicator light on the thermostat that indicates when the fossil fuel heating unit is operating.

The second relay uses two SPST sets of contacts, one set normally open (R2-1 and R2-3) and one set normally closed (R2-4 and R2-5). The normally closed set of contacts (terminals R2-4 and R2-5) is used to complete the circuit from the first stage contact in the thermostat (HTG-1) through the Y terminal on the thermostat to the Y terminal and compressor contactor in the outdoor unit.

The normally open set of contacts (R2-1 and R2-3) is relay 2 is in the power circuit to the gas valve circuit (R to W) in the heating unit. The relay is wired to have the outdoor section on

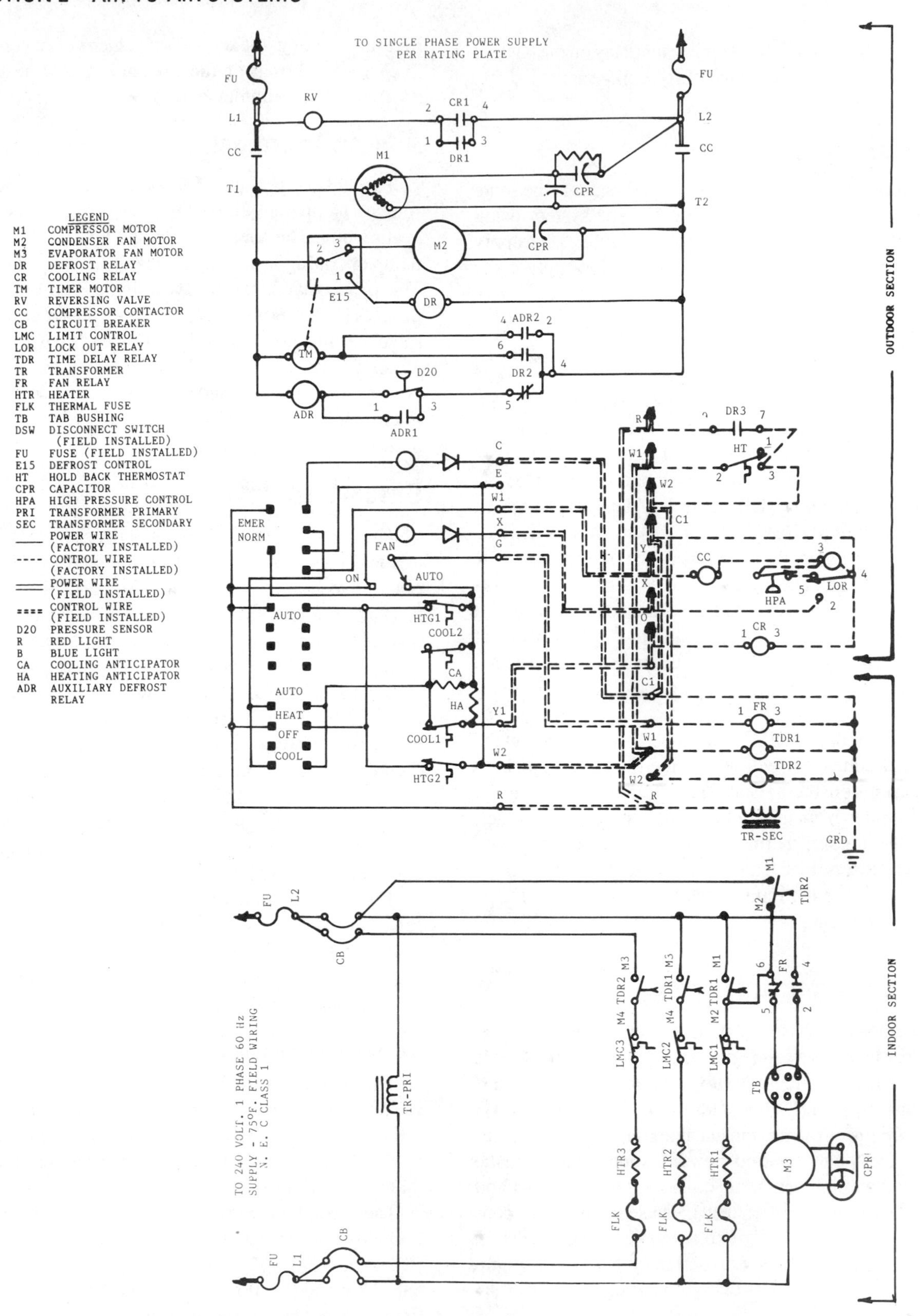

Figure 5–14 Heat pump system wiring diagram (Courtesy Addison Products Co.)

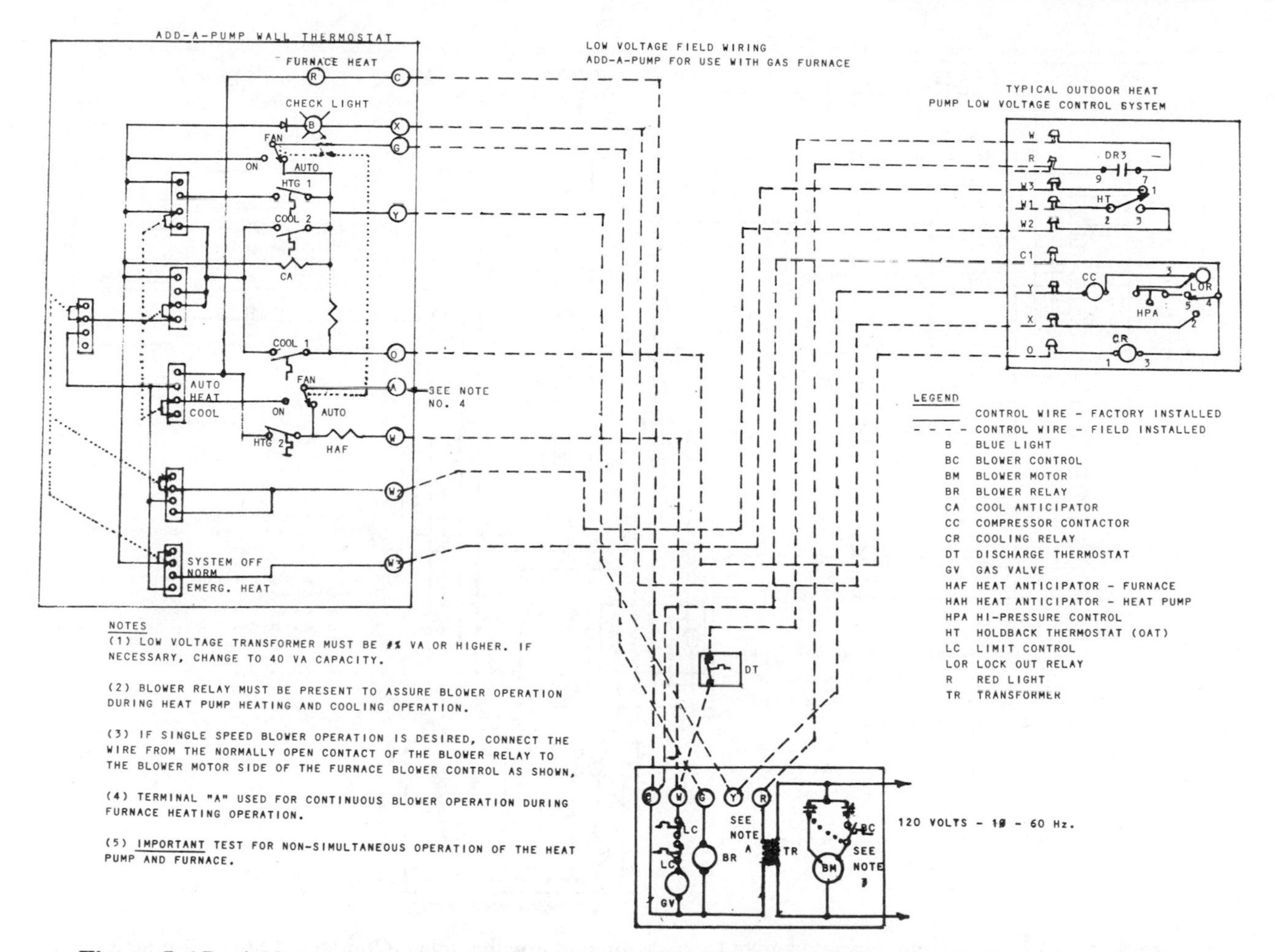

Figure 5–15 Add-on system wiring diagram for a gas-fired heating unit (Courtesy Addison Products Co.)

the heat pump operate under the control of the first stage of the area thermostat through the normally closed contacts (R2-4 and R2-5).

When the heat loss is more than the heat pump can provide for, the area temperature drops the 2°F to close the second stage contact of the thermostat (HTG-2). This energizes the twin relay assembly. Relay number 1 (R-1) closes contact R1-4 to R1-6 to light the furnace heat light in the thermostat. Relay 2 pulls in and reverses the contact positions. The normally closed contact (R2-4 and R2-5) opens and stops the outdoor section of the heat pump. The normally open contact (R2-1 and R2-3) closes and activates the gas valve circuit of the heating unit.

The heating unit is sized to handle the heat loss of the structure. The heating unit will produce more heat than is required and the system will cycle on the second stage of the thermostat.

In the system using the OAT for the heat source selection (Figure 5–15), either the heat pump or the heating unit is cycled by the action of the indoor thermostat. This means that regardless of the heat source used, the heat source is either on or off according to the demand of the indoor thermostat.

In the system using the changeover relay (Figure 5–16), the second stage of the indoor thermostat is selecting either heat pump operation if the heat pump will handle the heat loss or the fossil fuel heating unit if the heat pump cannot handle the heat loss. Because either unit is operating by the selection of the indoor thermostat the heating system is supplying two-stage heating capacity rather than on or off.

This results in longer cycles and a more even heat level in the conditioned area. This system also increases the usage time of the heat pump and promotes the more economical operation. The heat pump operating in the off cycle of the second stage of the thermostat will lengthen the off-cycle period much longer than the straight on-off cycle operation when the OAT is used. The motor-compressor assembly is not subject to long off periods in cold weather with the possible damage from refrigerant migration.

When an add-on type heat pump is added to an oil-fired heating unit, control circuitry such as that illustrated in

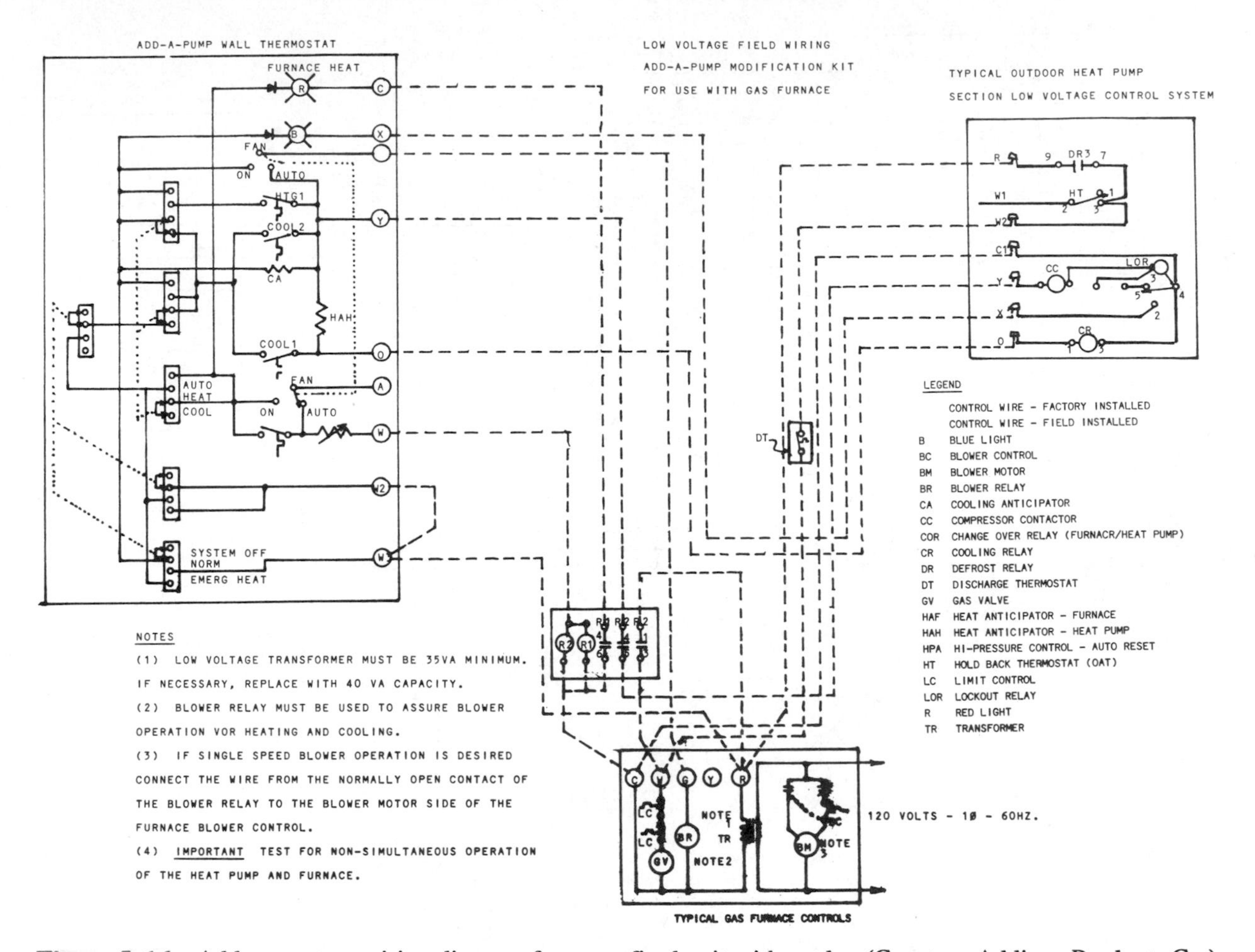

Figure 5–16 Add-on system wiring diagram for a gas-fired unit with a relay (Courtesy Addison Products Co.)

Figure 5–17 is used. This system is the same action as shown in Figure 5–16 for the gas-fired heating unit application.

The area thermostat operates the outdoor section of the heat pump system through the first stage thermostat contact (HTG-1), the Y terminal of the thermostat, through the normally closed contact of the dual-relay assembly (contacts R2-4 and R2-5) to the Y terminal of the outdoor section of the heat pump, and to the compressor contactor coil (CC).

When the heat load is more than the heat pump capacity, the area temperature drops and the second stage of heat in the thermostat is activated (HTG-2). This energizes the twin-relay assembly. Relay 1 energizes the furnace heat light in the thermostat subbase.

Relay 2 switches the action from the heat pump to the oil-fired heating unit. In this application, another relay is required. The control system has a power source and the oil burner protecto-relay has a power source. These two power sources cannot be connected into the same control system. An isolating relay is required.

This relay has a SPST normally open set of contacts that are connected to the T-T terminals of the oil-burner protecto-relay. When the isolating relay is activated by contacts in the twin-relay assembly, the contacts close and activate the thermostat circuit in the oil-burner protecto-relay. Power circuit isolation is achieved and no cross currents or voltages will occur.

When adding an add-on heat pump to an electric heating unit, the cooling coil should be in the air section ahead of the blower and electric heating elements. This allows the use of a standard heat pump thermostat and heat pump operation with the electric heat supplement.

In those cases where it is necessary to put the coil after the electric heater elements so that the air passes over the electric elements before it passes through the coil, this application must be treated the same as a fossil fuel application. This means that the either/or type of control must be used. Either the heat pump or the electric elements can be operated, not both at the same time.

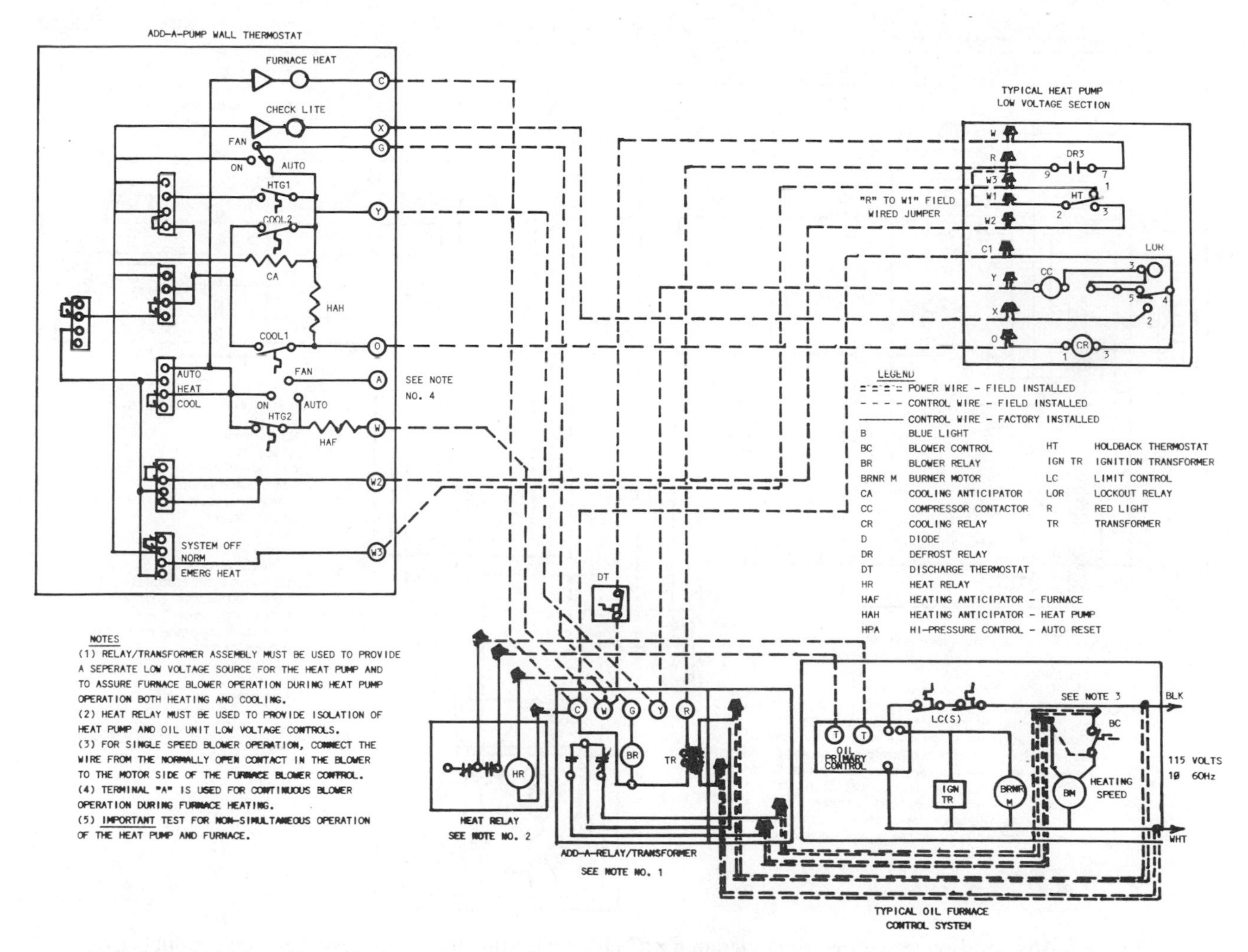

Figure 5–17 Add-on system wiring diagram for an oil-fired heating unit (Courtesy Addison Products Co.)

The control wiring diagram for this application is shown in Figure 5–18. Here again, the changeover relay is used to extend the heat pump operating time when below the corrected initial balance point.

Discharge Thermostat

Air-to-air heat pumps have the problem of frost accumulation on the outside coil (evaporator in the heating mode). Under certain weather conditions, this necessitates the operation of the unit in the defrost mode (cooling mode) in order to clean the frost and ice from the outside coil. In this situation, it is necessary to reheat the air off the inside coil (evaporator during the defrost mode) to reduce the possibility of discomfort in the conditioned area.

The air is reheated by operating the gas- or oil-fired unit or the electric heat bank. All these applications have considerably more heating capacity than is required. In addition, the heating units are located upstream from the inside coil. This means that the inside coil can be overheated very quickly unless a limited control is used.

All the control circuits in Figures 5–14, 5–15, 5–16, 5–17, and 5–18 have a limiting thermostat in the circuit from the contact in the defrost relay that controls the reheat function. This thermostat is used to prevent the alternate heat source from supplying air at excessive temperatures. Usually this thermostat is set to limit the air temperature to a maximum of 95°F.

This thermostat is a remote bulb type with a temperature-sensing bulb in the air entering the coil. If the coil is located in an upflow position with air entering the bottom of the coil, the bulb must be located under the coil fin area. This can be accomplished by a hole in the front close-off plate of the coil assembly. With a rubber grommet in the hole to protect the heat-sensing bulb, the bulb can be inserted into the area beneath the coil. See Figure 5–19.

If the coil is located under a downflow-type heating unit, the bulb can be located on the edge of the drain pan. This

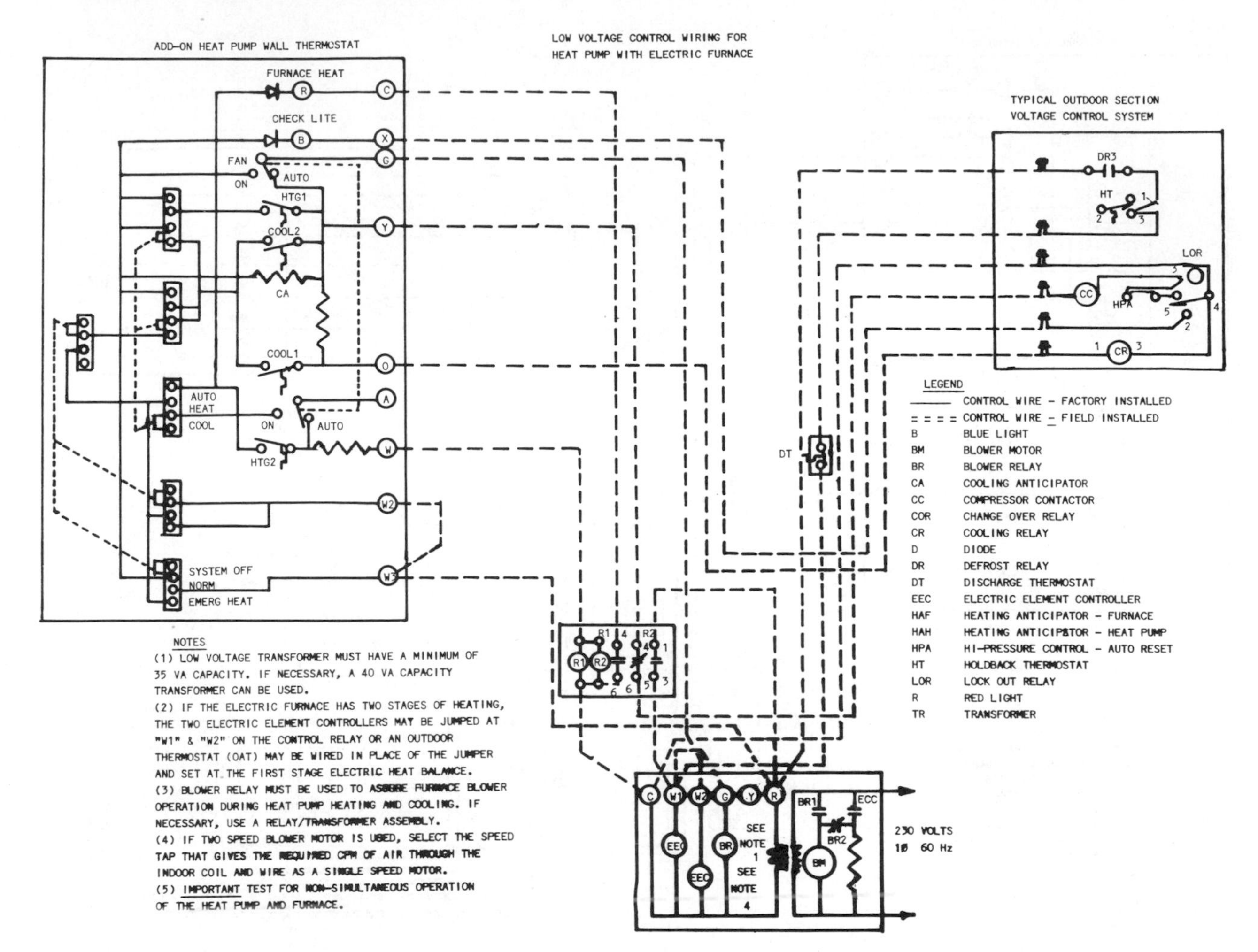

Figure 5–18 Add-on system wiring diagram for an electric heating unit (Courtesy Addison Products Co.)

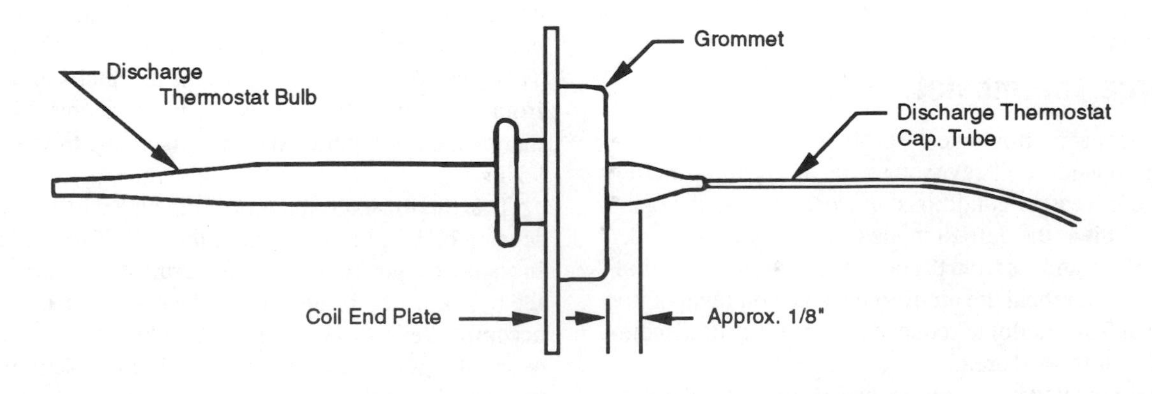

CAUTION:
Do not use a tool to push the bulb into the grommet as a dent in the bulb will offset the calibration.

Figure 5–19 Changeover control bulb installation—upflow (Courtesy Addison Products Co.)

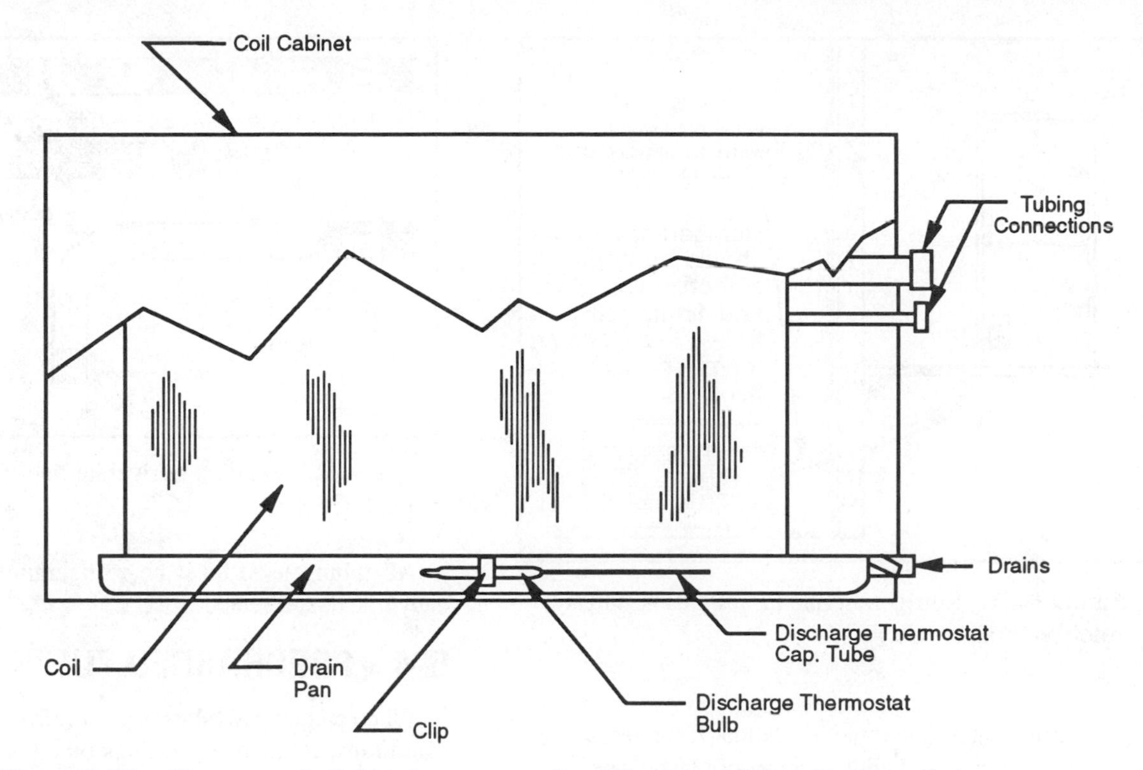

Figure 5–20 Changeover bulb location—counterflow (Courtesy Addison Products Co.)

location would be on the entering side of the coil. See Figure 5–20. A slip-type clamp or pipe strap with bolt and nut can be used to fasten the bulb in this location.

Be careful when locating the hole for the bolt. It should be above the bulb and as close to the top of the pan as possible and still get fastening strength in the pan metal. Holes drilled too low can promote water leaks.

5–3 REFRIGERANT LINES

Refrigerant lines were discussed under "Split-type Unit—Outdoor Section—Location" and "Add-on Split-type System—Outdoor Section—Location."

Rules for Installation

There are other do's and don'ts that must be covered. It is also necessary to protect copper lines as well as keep heat loss or gain to a minimum. Therefore, several rules on line installation must be followed:

Length. The length of the lines should be kept as short as possible, both from a pressure drop standpoint and from a heat loss or heat gain standpoint. The pressure drop standard is a maximum of 3 psig per 100 equivalent feet of line length for minimum capacity loss. The minimum pressure loss per 100' equivalent feet is 1½ psig. This results in a gas velocity through the line of 1,000 to 1,200 feet per minute for best oil return.

Insulation. Insulation is an absolute requirement on the vapor line. In the cooling mode, insulation helps to minimize the temperature rise of the cold vapor from the inside evaporator as it travels from the coil outlet through the reversing valve, accumulator, and to the compressor, as well as minimize condensation on the tubing. The motor cooling by the cold vapor has a direct effect upon the motor performance and operating life. Any excessive temperature gain in the vapor line will shorten motor life.

In the heating mode, the insulation is of greater importance. The vapor line carries the hot refrigerant vapor from the compressor to the inside condenser. In order for the condenser to receive the vapor with as high a temperature as possible, the vapor line must be insulated. The heavier the insulation, the less time it will take for the unit to reach the desirable temperature balance point to perform properly.

The first part of each operating cycle is used to heat up the vapor line. The faster this is accomplished, the higher the system operating efficiency and the seasonal efficiency ratio (SEER) and the lower the operating cost.

Installation. When handling rolled tubing, the line should be straightened as it is rolling off a spool. Pulling the tubing off the side of the coil will cause twisting of the tube, waves in the tubing run, and a possibility of kinking or flattening of the tube.

Do not route the tubing across access openings or access hole covers. When passing the insulated tubing through a

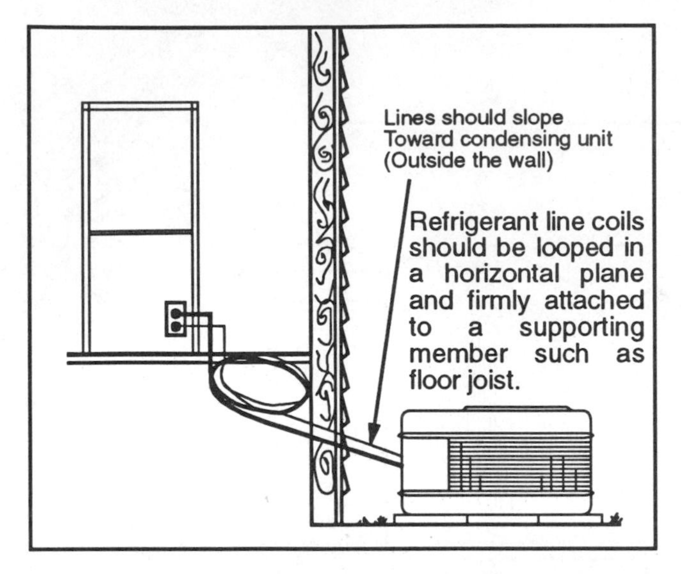

Figure 5–21 Refrigerant line installation (Courtesy Addison Products Co.)

wall, be careful not to tear the insulation. If the insulation is torn, it must be repaired with waterproof plastic tape.

If the tubing is to go through a masonry wall or through masonry foundation, *it must be protected by PVC pipe or orangeburg pipe.* Do not allow the bare copper tube to touch any concrete or masonry material. If contact is made, and moisture is present, electrolysis action will occur and pin holes will develop in the tube. These are very difficult to find and repair.

The vapor line must slope downward in the direction of vapor flow in the cooling mode to ensure proper oil return to the compressor. The proper positioning of the refrigerant lines is illustrated in Figure 5–21. The loops of excess tubing must be in a horizontal position. Vapor must flow from the top loop downward. If the loops are in a vertical position, each loop will act as an oil trap. This will increase the tubing flow resistance and seriously affect the operation of the system.

The tubing must be securely fastened in place with a minimum of bends. Each bend adds to the resistance to vapor flow or the pressure drop through the tubing. Long sweeping bends in the vapor line should be made wherever possible.

Where short radius bends are required, less then 12" radius, a tubing bender is required. Short bends done by hand will flatten the tube and greatly increase flow resistance.

When it is necessary to use a bender on the insulated vapor line, cut the insulation around its circumference at a distance far enough beyond the point of the bend so as to clear the tubing bender. Slip the insulation out of the way, make the bend, and slip the insulation back to its original position (see Figure 5–22). Be sure to seal the cut joint of the insulation with vaporproof plastic tape.

Solder Joints. If hard drawn tube and solder fittings are used in the installation of the refrigerant lines, nonlead solders must be used to prevent joint leaks. Inert gas such as nitrogen must be used in the lines when soldering to reduce the formation of copper oxides in the solder joint areas.

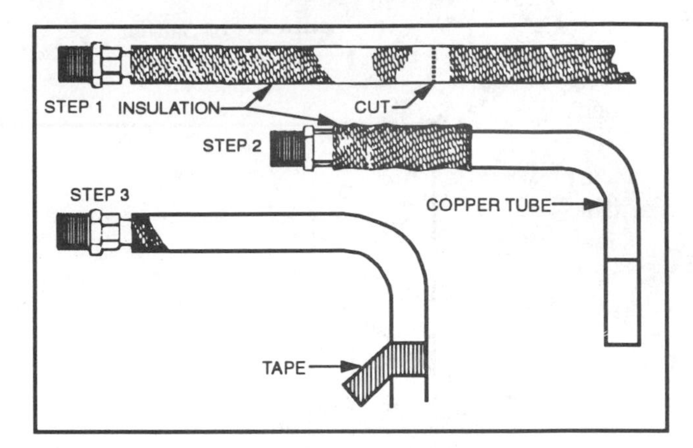

Figure 5–22 Bending refrigerant lines (Courtesy Addison Products Co.)

All tubing used must be refrigeration grade—cleaned, dehydrated, and sealed.

5–4 PRECHARGED TUBING

When precharged tubing sets of refrigerant lines are used, careful inspection of the fittings on the lines and the indoor and outdoor sections must be done to determine that the two matching halves of each coupling are compatible. Different types with different inside diameters of the fastening nut as well as the number of threads per inch are used in the industry.

Most line sets are produced with a gauge port on the fitting on one end. The fittings are then arranged to have the gauge port at the outside section. The prime reason for this is service accessibility.

The two sections of the fitting are illustrated in Figure 5–23. The female section, with the gauge port, is on the end of the line and the male section is on the outside unit. On the inside coil, the female section, without a gauge port, is part of the coil and the male section is on the refrigerant line. To assemble a fitting of this type, careful procedure is required:

1. After removing the protective cap and plug from the two parts of the fitting to be joined, lightly coat the interior parts of the female section of the joint with refrigerant oil.
2. Line up couplings to be joined to have the gauge port in a position for easy connection of the gauge lines.

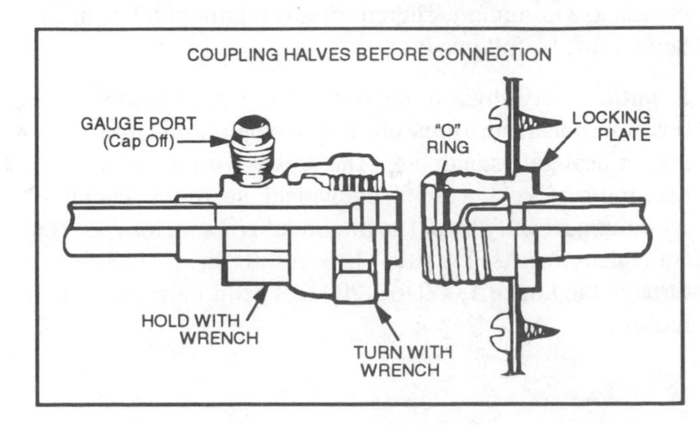

Figure 5–23 Quick couple fittings for a precharged system (Courtesy Addison Products Co.)

3. Start the connection by hand. *Be sure the threads are not crossthreaded.* Engage at least two threads before using wrenches.

Open end wrenches are the preferred type to use. An adjustable wrench, *in good condition,* may be used. *Do not use electrician's pliers or slip joint pliers on these fittings.* Leaks in these fittings are caused by squeezing the nut to an egg shape. Once this is done, the joint cannot be straightened and repaired. Replacement is the only solution.

WARNING: USE OF ANY TOOL THAT REQUIRES A SQUEEZING PRESSURE ON THE NUT WILL DISTORT THE NUT AND DESTROY THE FITTING.

Using the open end wrench on the hex nut of the female section of the fitting, tighten the coupling unit it "bottoms out." A hissing sound may occur. Do not stop the tightening process. Continue until the bottoming out occurs. After the bottoming out occurs, tighten the nut an additional quarter turn to close the metal seal ring.

CAUTION: If more than two threads are engaged, do not back the nut off or take the fitting apart. Loss of refrigerant will occur.

If, in the future, the lines must be disconnected, the fittings can be undone and remade. The system refrigerant charge must be recovered before the fittings are taken apart. The fittings will again make a leak tight joint if assembled using the proper procedure and tools.

5–5 LEAK TESTING

A leak test should be done using an electronic leak detector, Halide torch, or soap bubble solution. Children's "Blow Bubble" solution is an excellent leak detector. When using a Halide torch, be sure you are in a well-ventilated area. When chlorinated hydrocarbon refrigerants are in the air, the byproducts of combustion from a Halide torch are hydrochloric acid (HCL) and Phosgene gas. Both byproducts are very injurious to human health.

The valve stem in the gauge port is *not* leak tight. Gauge port caps, preferably bronze type, must be used as a seal and must be wrench tight. Fingertight, as dust caps only, will greatly increase the chance of refrigerant loss from the system.

5–6 PACKAGE UNIT

The package unit contains the indoor section that must be connected to the air distribution system as well as the outdoor section. Therefore, the factors that affect the entire split system will affect the package system.

When selecting the location for the package unit heat pump, there are several factors that must be considered. Not all of the factors can be given equal consideration. Some affect system performance more than others.

Supply Duct System

It is important to keep the air distribution system as short as possible, both the supply and the return portions, to hold air flow resistance to the design resistance used. The supply and return distribution systems should be designed according to the method given in Manual D of the ACCA Environmental Systems Library. This will provide the proper supply and return air flow resistance for the desired performance of the unit. The duct connections between the unit and the air distribution system should be as direct as possible to keep flow resistance down. Flexible connectors should be used to keep mechanical vibration from being transmitted to the building.

To prevent possible fire hazard, when auxiliary electric heat is included in the package unit, the first four feet (48") of the supply duct from the unit must be of sheet metal and insulated. A minimum of one inch (1") clearance must be maintained between the duct and combustible material. These requirements are dictated by UL standards and the NFPA code.

Air Recirculation

The minimum duct length is important but not to the extent of having the unit located in a confined area where air recirculation becomes a problem. Locating the unit in the corner of two walls will trap the discharge air from the unit and force recirculation of the air into the intake openings. This will affect the heat pump in both operating modes. In the heating mode, the entering temperature is lowered, reducing the heating capacity. In the cooling mode, the entering air temperature is raised. This raises the discharge pressure and the condensing temperature as well as the amperage draw. The result is reduction of the cooling capacity while the operating cost is increased.

The unit should be located along a single flat surface for best air distribution.

Service

The unit has to be serviced both initially and periodically. Therefore, the recommendation is a minimum of 30" between the service access panels and any vertical flat surface. Most package units have service access panels on both sides of the unit. The common practice is to place the supply and return openings toward the building with the unit perpendicular to the flat surface.

Noise

There are operating sounds to any mechanical piece of equipment. The classification of sound or noise depends upon the effect of the sound on the occupants of the conditioned area. Location outside windows of sleeping areas should be avoided. In this case, it is better to add some air distribution duct resistance to locate the unit away from the window.

Snow Load

In climates where snow becomes a problem, the unit should not be located where a natural snow drift occurs. This piling up of snow around the unit will seriously affect the air circulation through the unit and the heating capacity. The average winter snow depth must be determined. The mounting means must take this into consideration when determining the height of the unit above grade.

Foundation

The selection of the type of foundation to be used must include the consideration of snow depth, noise transmission into the conditioned area, serviceability, defrost drainage, and condensate drainage.

Naturally, the length of electrical lines as well as drain lines is important. However, the difference between long and short lines for power and drain are minor compared to operating efficiency and noise problems.

Several different foundation means are used.

Slab Mounted. The majority of package units are slab mounted. This method of mounting is ideal for slab floor construction, areas over crawl spaces or even basement-type homes where the first floor level is far enough above grade level to allow straight-through installation of the supply and return air ducts.

Figure 5–24 shows the application of the unit to a distribution system using the furred return in the bottom of the closet. The supply air duct is to the overhead distribution system. The illustration shows a space between the foundation slab and the building to prevent noise transmission into the conditioned area.

Where the first floor of a basement-type building is below grade level, a package unit is not recommended. Any excavation outside the building to install the duct system will only produce serious water trapping problems.

Roof Mounted. The second most common method of mounting a package unit is "roof mounted." When mounting on the roof, the unit support must be designed to distribute the weight over as wide an area as possible. The roof structure must be strong enough to hold the weight of the unit without excessive distortion (see Figure 5–25).

On flat roofs, the unit must be located high enough above the roof to allow full drainage of defrost water off the outside coil during the defrost function. This could cause ice buildup where defrost requirements are excessive. This usually requires a height of 12" or more.

On a hip roof, an angle support must be provided to have the unit in a level position.

Figure 5–25 shows a cross section of a roof/attic combination type of construction. The unit is mounted on an angle platform. The supply air distribution system is a multi-outlet duct system in the attic area. The return air system is a single return grill which contains the air filter. This location of the air filter is preferred over a filter in the unit or return air duct. The return air filter grill assembly is more accessible and will receive better service attention than air filters located outside the building on the roof area.

Figure 5–26 shows a suggested method of connecting to a furred down ceiling supply system in a hallway in the conditioned area. A return air filter grill assembly should also be used in this type of installation.

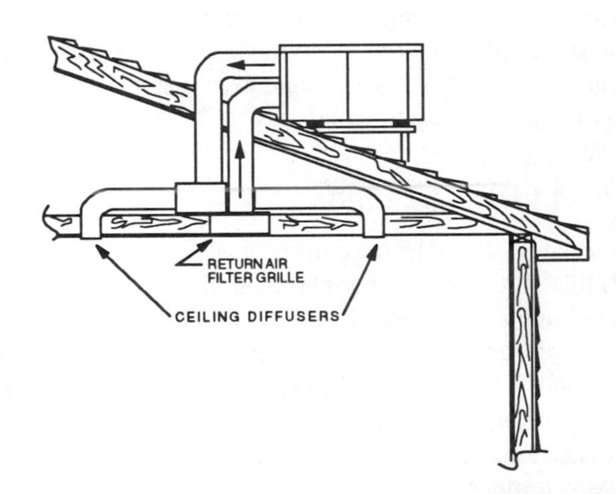

Figure 5–25 Unit installed on roof with attic supply duct (Courtesy Addison Products Co.)

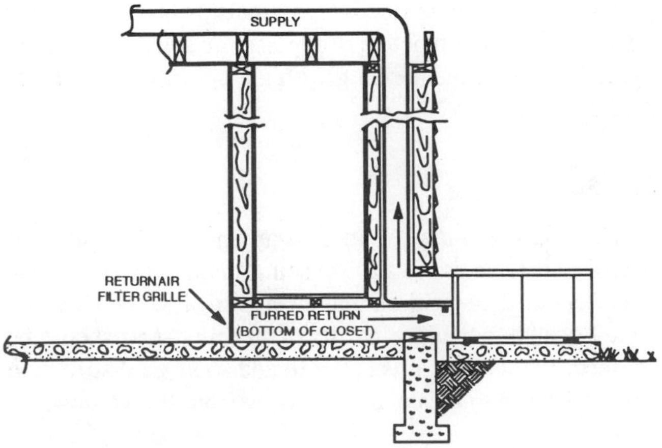

Figure 5–24 Unit mounted on slab at ground level (Courtesy Addison Products Co.)

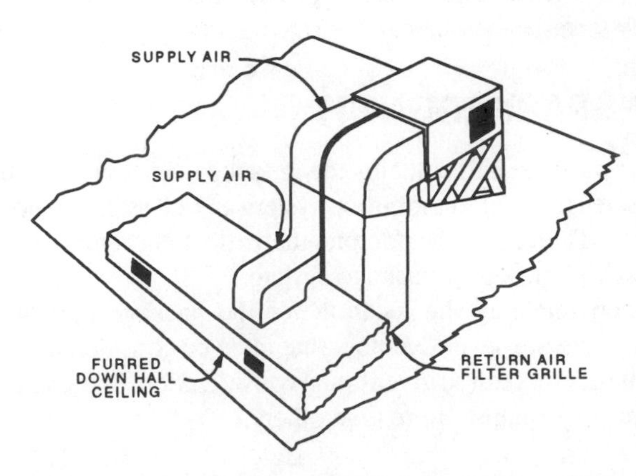

Figure 5–26 Unit mounted on roof with drop ceiling supply duct (Courtesy Addison Products Co.)

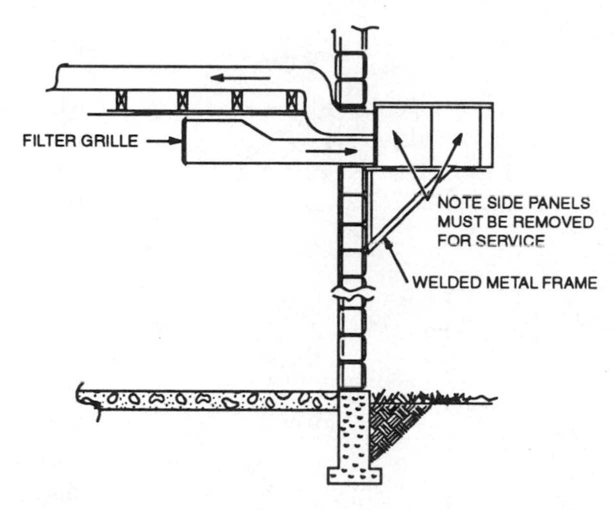

Figure 5–27 Unit cantilevered from wall (Courtesy Addison Products Co.)

Cantilevered from the Wall. Another type of installation is the "cantilevered from the wall." This is shown in Figure 5–27. This type of installation is not recommended if another choice exists.

1. The wall must be of sufficient strength to support the twisting action of the unitary weight.
2. The wall must be of sufficient density to absorb the operating vibration of the unit.
3. The outdoor section must be high enough so as not to interfere with traffic under it. However, the height must not be so great as to limit accessibility.

This type of application is usually limited to smaller package units.

Duct Connections

1. All duct work passing through unconditioned area must be insulated according to local requirements. If there are no local requirements, 2" of insulation is recommended for duct distribution systems, both supply and return, in attics and outdoors.

   In crawl spaces, usually 1" of insulation is sufficient.
2. The insulation used must contain a vapor barrier.
3. Outdoor ducts must also have weather protection from the sun, rain, and so on.
4. In heat pump installations, outdoor air should never be brought into the building through the return air system. Mixing cold air with the return air will lower the temperature to the inside coil. On the heating cycle, when the coil is the condenser, the lower temperature air will reduce the condensing pressure in the system and reduce the system capacity. Heat pumps are not designed to operate in the heating mode with entering air to the condenser below 65°F. Air temperatures below this can cause sufficient capacity loss to prevent adequate supply of heat to the conditioned area.
5. To help prevent noise transmission from the unit into the duct system, flexible duct connections must be used to connect the duct system to the unit. This must be done in both the supply and the return. A shield method should be used to protect the flexible duct collars from the weather without destroying the flexibility of the connectors.

Drain Systems

The condensate drain line from the outside coil is usually located on the side of the unit. To ensure proper flow of condensate, the drain line must have a continuous down slope of at least 1" in 10' to prevent the water from becoming trapped and causing the drain system to fail.

Drain Trap

The side location of the drain connection enables the easy installation of a drain trap. *A trap is necessary.*

The coil compartment of practically all units is located on the upstream side of the blower. This results in a negative pressure in the coil compartment. To prevent air flow into the coil compartment and disruption of the coil drainage flow, a condensate drain trap (see Figure 5–28) must be installed. To promote good drainage from the coil and condensate pan, the unit must be level or inclined slightly toward the drain.

On units where the coil is located on the downflow side of the blower and a positive pressure is developed in the coil compartment, a drain trap is also recommended. The drain trap prevents a conditioned air loss from the blower compartment.

Power Wiring

The installation of electrical power circuits is closely regulated by local, state, and national codes. Most areas require that such installations be done by a licensed electrician.

A universal requirement from a safety angle is the installation of an outdoor disconnect switch in the power supply along with the circuit breaker in the service distribution panel. This disconnect switch must be installed so as to be visible from the unit (see Figure 5–29). It should not be mounted on

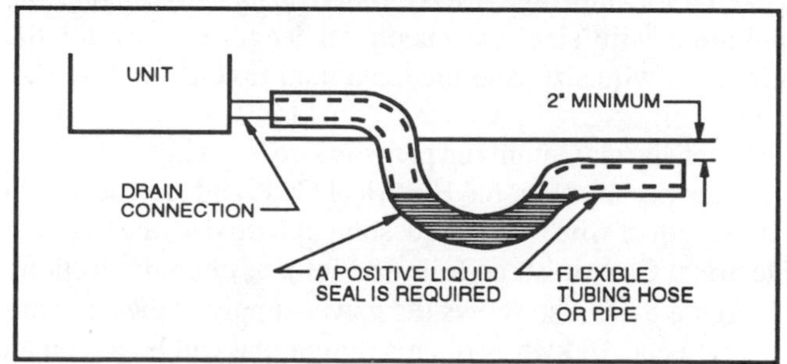

Figure 5–28 Condensate drain trap installation (Courtesy Addison Products Co.)

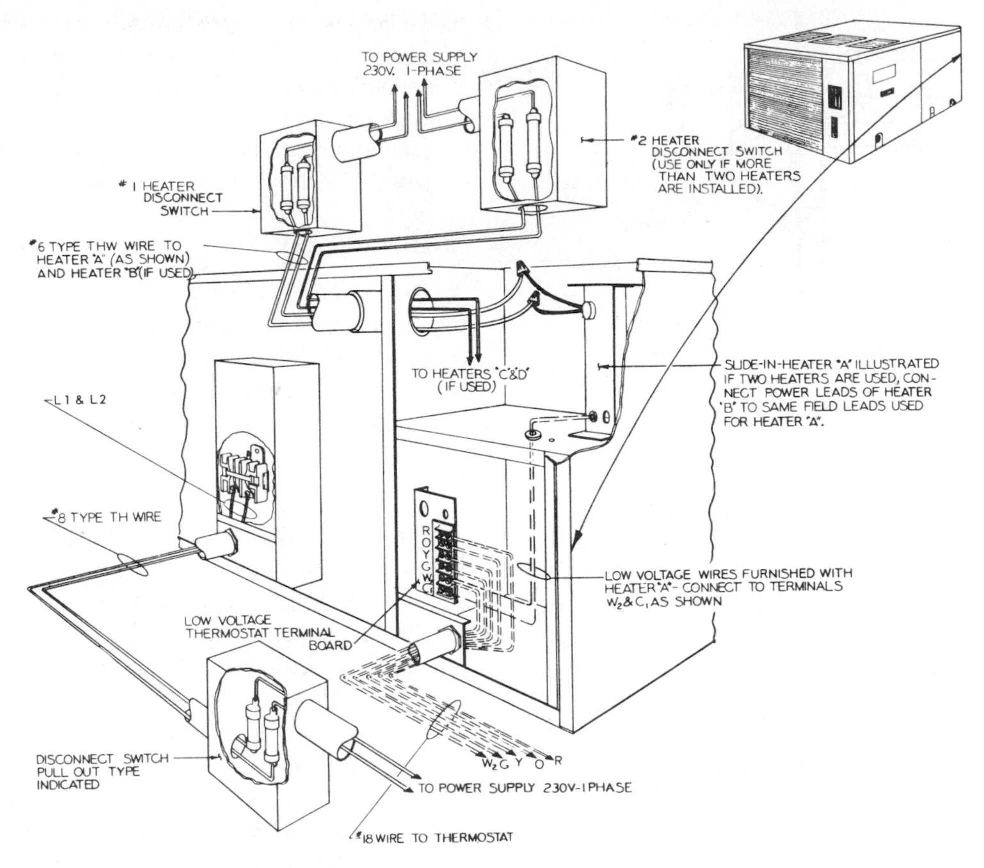

Figure 5–29 Typical pictorial field wiring diagram (Courtesy Addison Products Co.)

the unit. The preferred mounting means is on the wall or on a stake at the same level and within 50', with no walls in between, to promote safety for the operator of the unit.

Included in Figure 5–29 is the wiring of the pullout disconnect switch in the power supply. Examples of wire sizes are shown for illustration purposes only. The actual wire size required will depend upon the unit size as well as the length of the wire required. Each manufacturer supplies the electrical requirements for the equipment in the installation instructions as well as on the model and serial number plate on the unit.

Figure 5–30 is the electrical wiring information supplied by one manufacturer. Listed by unit model number are the electrical supply characteristics (voltage and amperage), the full load amperage (FLA) or rated load amperage (RLA), and locked rotor amperage (LRA). Based on this information, the minimum wire size, the maximum length of wire for the minimum wire size, and the maximum time-delay fuse size are given.

When the maximum run per wire size is not given, the wire sizes are per the National Electrical Code and are based on a 100' length of wire. On any question of wire size, the National Electrical Code must be followed for maximum protection.

Figure 5–29 also shows the power supply to the auxiliary electric heat. 10 kwh is the maximum that can be put on an individual branch circuit (60 amperes). The National Electrical Code states that the current draw of 41.67 amperes must be protected with not over a 60-ampere fuse. If more than 10 kwh are required, additional branch circuits and fuses are required for each 10-kwh stage.

Control Wiring

As pictured in Figure 5–29, the control wiring of a package heat pump is simplified in that it is concentrated in the unit. Usually the manufacturer will mark the connection terminals to correspond with the terminal markings of the thermostat that should be used with the unit.

In all cases, the manufacturer's wiring diagram supplied with the unit should be used.

REVIEW QUESTIONS

1. Package units are roof mounted to make them harder to maintain. True or False?
2. When adding a heat pump to a fossil fuel system, the total capacity of the system is determined by the sum of the fossil fuel unit and heat pump capacities. True or False?
3. The outdoor temperature at which the heat pump will handle the heat loss with continuous operation is called the:
   a. Outdoor ambient temperature
   b. Heat loss/unit capacity balance point
   c. Initial balance point
   d. Indoor capacity balance point
   e. Second balance point

Model	Volts	Phase	Current @ Rating Conditions Amps	Total Locked Rotor Amps	Min. Wire Size	Max. Run * For Min. Wire Size	Recommended Max. Delay Fuse Size
PH24-1D	230	1	Cool—18.5 Heat—16.0	57.0	10	54 Ft.	25 Amp.
PH36-1D	230	1	Cool—27.6 Heat—24.0	115.6	8	59 Ft.	40 Amp.
PH36-3D	208/240	3	Cool—23.0/22.0 Heat—17.0/16.0	72.0	10	54 Ft.	30 Amp.
PH42-1E	230	1	Cool—28.5 Heat—24.0	108.0	8	51 Ft.	45 Amp.
PH60-1E	230	1	Cool—39.5 Heat—31.4	175.0	6	64 Ft.	60 Amp.
PH60-3D	208/240	3	Cool—26.5 Heat—21.5	132.0	8	59 Ft.	45 Amp.

***Note:** For runs which are up to 60% longer, use next larger, even gauge size, wires. Minimum wire sizes shown are based on National Electric Code requirements. For this size, normal voltage drop is less than 1%; and inrush voltage drop is 3½% for maximum run specified.

Figure 5–30 Electrical wiring information for a package unit (Courtesy Addison Products Co.)

4. When plotting the heat loss line, the outdoor temperature used is:
   a. 65°F
   b. 68°F
   c. 70°F
   d. 72°F
5. The outdoor temperature at which the unit and auxiliary heat handle the heat loss with continuous operation is called the:
   a. Outdoor ambient temperature
   b. Heat loss/unit capacity point
   c. Initial balance point
   d. Indoor capacity balance point
   e. Second balance point
6. The heat pump has three stages of electric auxiliary heat controlled by three OAT controls. How many balance points would be plotted on the heat loss/unit capacity chart?
7. When installing a split-type heat pump system, the following factors must be considered. Name the factor that has the highest priority.
   a. Height of the indoor section above the outdoor section
   b. Maximum air flow to the outdoor coil
   c. Possible air recirculation
   d. Height of the outdoor unit above the grade
   e. Length of the connecting lines
   f. Type of outdoor unit foundation
8. Aluminum wire should be used in power supplies to heat pumps because it is lighter and more economical. True or False?
9. The low voltage control cable for the 24-volt circuits should be _____-gauge wire.
10. Multicable telephone cable is an excellent substitute for thermostat cable. True or False?
11. When designing heat pump air distribution systems, the supply and return air duct pressures are the same as for fossil fuel systems. Therefore, the duct sizes will be the same. True or False?
12. Electric auxiliary heat resistance banks are located in the blower discharge because they are easier to mount in this location. True or False?
13. Downflow heat pump air handlers do not require a noncombustible base such as a gas- or oil-fired unit would require. True or False?
14. When a horizontal air handler is installed in an attic or upper area, an auxiliary drain pan is required. However, one drain from both the coil and auxiliary drain pan is approved as long as the drain lien is of sufficient size to handle both requirements. True or False?
15. Name two ways to control the changeover from the heat pump to the fossil fuel unit.
16. In two-stage-heating thermostats, the control point temperature difference between the stages is _____°F.
17. When adding a heat pump system to an oil-fired fossil fuel unit, two changeover relays are needed where only one is needed for a gas-fired unit. Why?
18. The purpose of the discharge thermostat in an add-on system is to make sure the air supply to the conditioned area is high enough to prevent cold drafts. True or False?

19. Insulation is an absolute requirement on which of the following refrigerant lines?
    a. Suction line
    b. Liquid line
    c. Vapor line
    d. Discharge line
    e. Liquid return line
20. Plastic pipe used as a shield is required when running copper lines through concrete or masonry. True or False?
21. The refrigerant velocity through the vapor line is always high enough to keep the line clear regardless of the line position. True or False?
22. When working in a confined area, which of the following leak detectors should be used?
    a. Soap bubbles
    b. Halide torch
    c. Electronic leak detector
23. Introduction of fresh air into the return air system so as to filter it before introduction into the conditioned area is recommended. True or False?
24. The low temperature limit of return air when operating the heat pump in the heating mode is 65°F. True or False?
25. The minimum slope of the drain line in the drain system is:
    a. ¼" in 10'
    b. ½" in 10'
    c. 1" in 20'
    d. 1" in 10'
26. On practically all air handlers, a drain trap is required. True or False?
27. In the cooling mode, the maximum outdoor temperature that will allow satisfactory operation is:
    a. 105°F
    b. 110°F
    c. 115°F
    d. 120°F
28. In the heating mode, the maximum outdoor temperature that will allow satisfactory operation is:
    a. 80°F
    b. 75°F
    c. 70°F
    d. 65°F
    e. 60°F
29. Before connecting power to a unit, all electrical connections should be tightened, including those that are factory made. True or False?

STARTUP, CHECKOUT, AND ADJUSTMENT 6

6–1 GENERAL

The benefits of installing a heat pump system are greatly reduced if the system is not properly set up and adjusted after the installation is completed. The largest percentage of owner complaints on the performance and operating cost of their new heat pump system can be traced to the lack of final tune-up of the system. In the automobile field, the saying is that "The automobile is no better than the dealer you buy it from." The automobile that is inspected and adjusted by a trained mechanic promotes the greatest customer satisfaction. A heat pump is a piece of mechanical equipment designed and built to do a specific task. If the final adjustments are not made on the system, the system will "get by" but with decreased owner satisfaction.

The heat pump is a heat transfer device. In the air-to-air system, the correct amount of air through the outside coil, the correct amount of air through the inside coil, and the correct amount of refrigerant in the system all add up to the system doing what it is designed to do. Therefore, the final adjustments of the system are as important as proper installation.

To perform the task of setup, checkout, and adjustment, the technician performing the operation must have specific information. To maintain proper records on the performance and service history, as well as any warranty period records, certain information must be recorded. This is especially important in the event of a disagreement over warranty questions.

The Job Record

The following items should be part of the job record of the installation.

Owner's Name, Address, and Telephone Number. With most manufacturers, the address is the important item. Units are under warranty at a certain address. The owner may change but the address does not. Practically all manufacturers will honor a change in ownership but a change in address will cancel the warranty agreement.

Unit Identification. A record should be kept of the unit identification information:

- Outside unit model No. _____
- Outside unit serial No. _____
- Inside unit model No. _____
- Inside unit serial No. _____
- Auxiliary heat unit model No. _____
- Auxiliary heat unit serial No. _____

If the heat pump has been installed in conjunction with a fossil fuel unit, the inside coil information should also be recorded:

- Inside coil model No. _____
- Inside coil serial No. _____

Pressure-reducing Devices. The type of pressure-reducing devices used on the inside and outside coils should be recorded. The operating characteristics of the systems using T.X. valves are different from those using capillary tubes or restrictors. Therefore, this information is necessary for proper checkout and adjustment.

- Outdoor coil
  - ___ T.X. valve
  - ___ Capillary tube or restrictor
- Indoor coil
  - ___ T.X. valve
  - ___ Capillary tube or restrictor

Fuse Protection. The size and type of fuses installed in the original installation must be recorded. Amateur electricians tend to change fuse sizes with no thought to consequences. Installed circuit breakers will reduce this possibility as compared to the cartridge-type fuses.

Power Supply Wire Size. The size of the wire as well as the length of the wire run should be recorded for the power supply to both the outdoor and indoor sections of the system. With a package-type unit, the power supply to the unit proper as well as to any auxiliary heat unit should be recorded.

Inside Air Filter. A check should be made to confirm that clean air filters have been installed before the unit is operated. This is especially important in new construction installations. Operation of the system while inside finishing is done must be done with air filters in place. Removal of saw dust and plaster board debris from the coil fins is very difficult and totally unnecessary when air filters are in place.

6–2 STARTUP

When startup is performed on an air-to-air dual-operation heat pump, the operating mode at startup will depend upon the season of the year and the end results required. The system has limits for each operating mode that cannot be exceeded if accurate checkout is to be performed. Therefore, the system cannot be operated and checked out in both operating modes unless the outdoor and indoor temperatures are in the narrow overlapping range of temperatures for each mode of operation.

Cooling Mode

The heat pump will operate in the cooling mode down to 65°F outdoor temperature. Below this temperature, condensing pressures and temperatures drop too low to provide sufficient refrigerant flow for acceptable operating results. The system will not operate satisfactorily in outdoor temperatures above 115°F. The operating capacity falls very drastically at temperatures exceeding 115°F.

The temperature of the air entering the inside coil should also be at least 70°F and not over 95°F to be able to begin to adjust the system. The final checkout requires the inside air temperature to be in the 70°F to 85°F range.

Heating Mode

In the heating mode, the unit will operate with outdoor temperatures up to 65°F. Some units will operate to 70°F. Operation at outdoor temperatures above 70°F will overload the outdoor evaporator, raise compressor pressures, and cause cutout. Cutout will normally be on the head pressure safety control system if one is included in the unit. If not, the compressor internal overload will actuate. If this happens, the compressor can take several hours to cool to the point where the overload will reset. Compressors have been condemned as open and replaced because of this compressor safety device action.

The system will not operate satisfactorily in the heating mode with air temperatures entering the inside coil below 65°F. The condensing temperatures and pressures are too low to force enough refrigerant into the outside evaporator to establish any type of heat balance. The unit will operate with very little heat from the inside coil.

When starting the system, three sets of information are needed, one before startup, one immediately after startup, and one after several minutes of operating time.

Prestarting Checklist

Several checks should be made before power is actually applied to the unit and the unit put into operation:

1. Has the outdoor unit been installed at the proper height to compensate for snow depth?
2. Is the outdoor unit in a level position, allowing for coil defrost water disposal?
3. Is the outdoor unit located for free air travel into and away from the unit?
4. Are all mechanical parts operating properly? Check the condenser fan to make sure it turns freely without striking the coil or fan shroud. Is the fan secure on the motor shaft? Is the motor and fan assembly properly aligned in the fan shroud? On package units, check the inside blower assembly for security and alignment.
5. Are the refrigerant lines in the split system properly installed, securely fastened, and properly insulated?
6. Are evaporator drain pans properly sloped toward the drain outlet for proper condensate removal?
7. Are condensate drain lines properly trapped, are they the correct size, and do they have proper slope to an "open drain"?
8. Is the system wired according to the wiring diagram that applies to this particular unit?
9. Are the different sections of the system properly fused (type and size) in accordance with the information supplied by the manufacturer?
10. Are all electrical connections tight, including those in the sections? Check and tighten *all* connections, including those done by the manufacturer.
11. Is the area thermostat level and correctly wired? Is it in a good location—away from drafts and heat sources such as lamps, TV sets, and so on?
12. Are the clean air filters in place?
13. Is the duct work correctly run and well-insulated where necessary, with vapor barriers where necessary and taped joints?
14. Are refrigerant couplings or joints tight and leakproof?

Some of these items may seem unnecessary, but they are items over which litigation has resulted.

6–3 STARTUP PROCEDURE

The startup procedure begins with the main power to both outside and inside sections off.

1. The voltage at the service distribution panel should be measured with the system off. This gives the no-load voltage.
2. Two electrical checks should be made when starting each section of the system, the inside blower motor, the outside unit, and the auxiliary heat. Each in turn will be checked in the order given.

To make each check, a voltmeter should be connected across the high-voltage supply to each section on the line side of the controlling contactor or relay. Place a clamp-type ammeter (preferably a digital type with a high-amperage lock) around one of the hot leads to the contactor or relay.

The purpose of these applied voltage and amperage draw tests is to determine if the electrical supply system is of sufficient size to carry the load. When the contactor or relay closes, bringing on the electrical load, *the voltage at the contactor or relay should not drop more than 5 volts*. If it does, one or more of the following situations could be the problem.

1. The wire size could be too small for the length installed. Manufacturers are required to give the wire size in ampacity (amperage capacity) per wire length for each unit. If a specific length is not given, the National Electrical Code requires a size based on 100' wire length.
2. The building distribution service could be too small. The starting test should be repeated measuring the voltage at the circuit breaker in the distribution panel. If the starting voltage drops at the distribution panel less than 5 volts when the drop at the unit is more than 5 volts, the branch circuit wires to the unit are too small. If the drop in starting voltage at the distribution panel is more than 5 volts, either the service drop to the building is too small or the distribution power transformer is too small. In this case, the local power utility must be contacted.

At the same time that the voltage check is made, the amperage draw of the unit under test must be observed. Both the starting and running amperage of the unit must be checked. Any time an induction load (a motor) is started, a high starting current occurs. This starts the motor and the motor should reach full running speed within 1 second. The amperage draw should then drop to the full load amperage (FLA) or run load amperage (RLA). The amperage draw must be observed during the startup to make sure the amperage draw drops to the full load amperes. If the amperage draw remains high, near the LRA, the unit must be disconnected immediately.

After the electrical problem is found and corrected, the test is repeated to recheck the operating results.

6–4 PERFORMANCE CHECK

To check the performance of the system, temperature and pressure readings have to be taken. The temperature and pressure readings will depend upon whether the unit is in the cooling or heating mode. Since heat pumps are sized to handle the cooling load, the most accurate testing and adjusting is done in the cooling mode. If the unit is started in the heating season, an approximate adjustment can be made, but the system must be rechecked and set during the next cooling season.

Air System Check

Because the heat pump system is selected to handle the cooling load, the amount of air over the inside coil (the evaporator in the cooling mode), must be correct for the cooling load. With the correct amount of air through the inside coil, the unit will produce the desired temperature drop of the air. The inside design conditions are usually 75°F but ARI test conditions are 80°F DB and 50% RH. The objective of air adjustment through the coil is to obtain the combination of temperature and relative humidity that will satisfy the greatest number of the area occupants.

The first step is to measure the return air DB and WB temperatures at the return air grill. By measuring the temperatures at this point, the return air duct can be checked for air leaks.

By using the psychrometric chart, the conditioned area relative humidity is obtained. Using this information, the proper temperature drop across the evaporator coil can be determined from the load ratio chart in Figure 6–1. Figure 6–1 has been plotted for the various conditions of air within the operating range of the unit. A range of dry bulb (DB) temperatures from 56°F to 92°F are given.

Using the DB temperature and the %RH found in the air to the return air grill, we can determine the temperature drop of the air through the evaporator coil that is needed to give the desired conditions over the widest range of outside conditions (the cooling load).

For example, assuming that the conditioned area has a measured dry bulb (DB) temperature of 80°F and a wet bulb (WB) temperature of 66.5°F, the relative humidity (RH) would be 50%. The psychrometric chart in Figure 6–2 was used to determine this RH.

Using the 80°F DB and 50% RH figures, referring to the load ratio chart (Figure 6–1) we find that the cross point of 80°F horizontal line and the 50% RH slant line is on the vertical line of 20°F temperature drop of air through the evaporator coil. These conditions of 80°F DB, 50% RH, and ΔT 20°F of the air through the coil are considered ideal from the ARI test standards. However, when other than these conditions of DB and RH are found, the ΔT 20°F would not give the desired results.

The heat pump system, in the cooling mode, adjusts to the load conditions by varying the temperature drop of the air through the evaporator coil according to the "total load" in the air to be conditioned.

Also, the heaviest load factor in this dimension is the amount of moisture in the air. If we assume a constant air temperature of, for example, 80°F DB and vary the relative humidity (RH) from 30% to 70%, we find that the temperature drop produced across the evaporator coil will drop from 26°F to 15.5°F. On the other hand, if we keep the relative humidity (RH) constant, at 50%, and vary the dry bulb (DB) temperature from 88°F to 68°F, we find a temperature drop only from 21°F to 18°F.

Because both the temperature and humidity in the conditioned area will vary, the correct temperature drop across the evaporator coil will result regardless of the change in either or both conditions. To obtain the correct temperature drop, the CFM of air through the evaporator coil is increased or decreased depending upon the direction of change needed. Increasing the CFM will reduce the temperature drop. Decreasing the CFM will increase the temperature drop.

Evaporator Coil ΔT °F

To obtain a starting point for air adjustment, the DB temperatures of the air entering and leaving the evaporator coil are obtained. This can be done by measuring the DB temperature in the supply and return ducts connected to the unit. Figure 6–3 shows the location for measuring the temperatures in the supply and return ducts on a vertical up air handler installation. Subtracting the supply air temperature reading from the return air temperature reading will produce

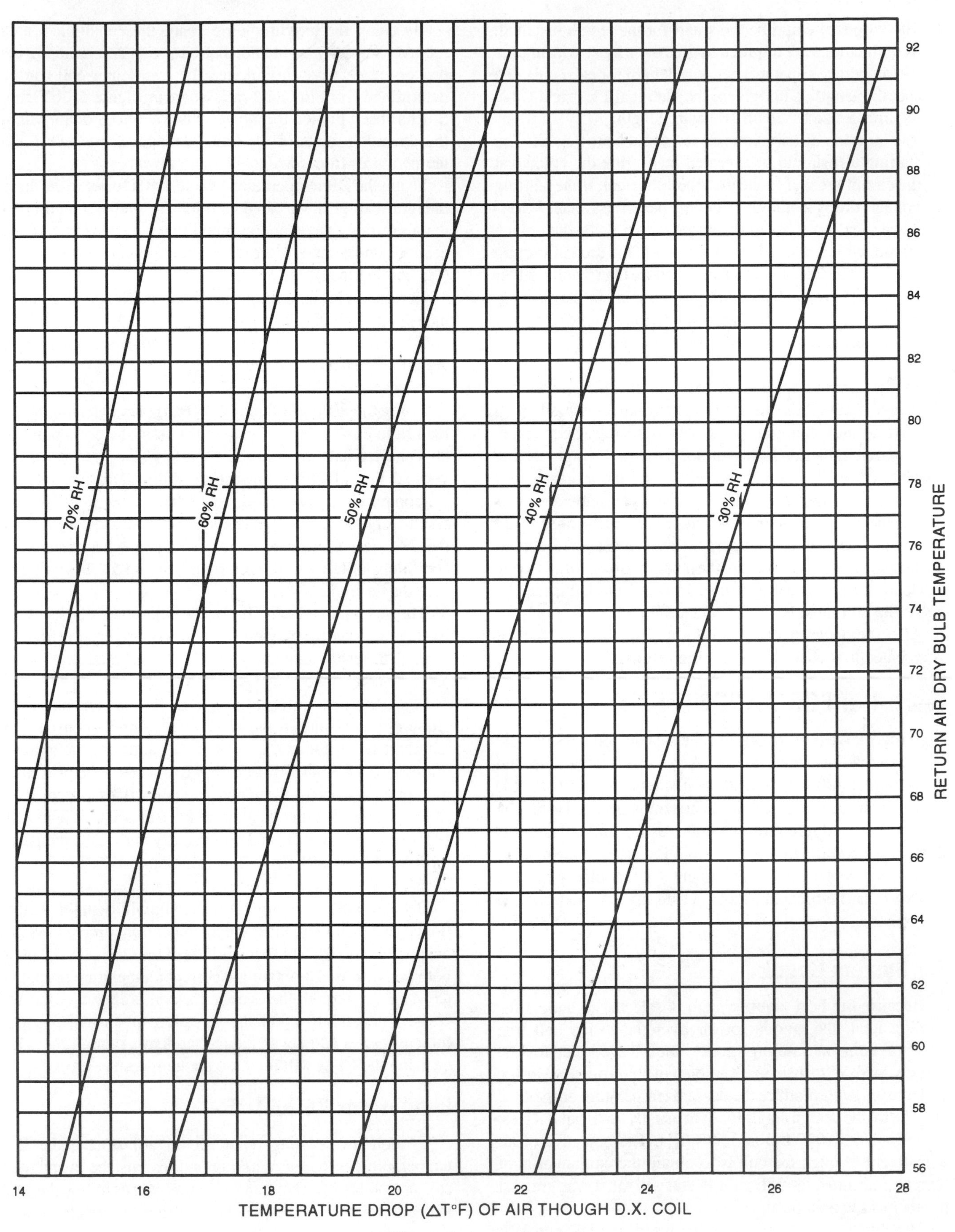

Figure 6–1 D. X. coil

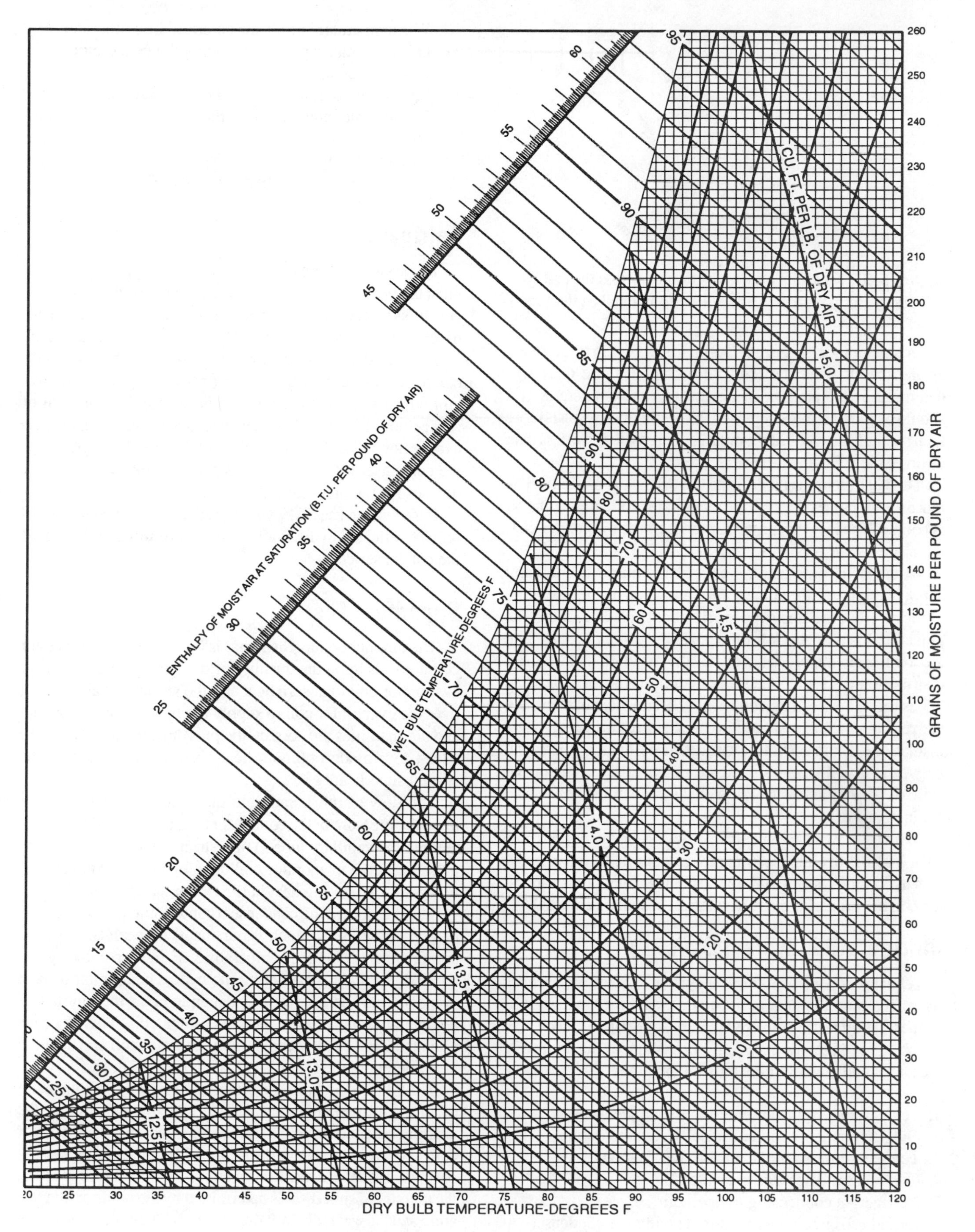

Figure 6–2 Psychrometric chart

Figure 6–3 Measuring air temperatures across the coil (Courtesy Borg-Warner, Central Environmental System Inc.)

the actual temperature drop. Comparison of this result to the desired result from the load ratio chart (Figure 6–1) will determine if the amount of air through the evaporator coil should be increased or decreased. If the obtained result is more than desired, the CFM of air must be increased. If it is less than desired, the CFM must be decreased.

The heat pump has a definite heat removal capacity. The more the air passes through the evaporator coil, the less the air is cooled. The less the air passes through the evaporator coil, the more the air is cooled.

At the same time that the area temperatures and the unit supply and return air temperatures are taken, the temperature of the air from the area supply grill should also be taken. The object of this is to check for the possibility of air leakage into the return air supply as well as the insulation on the air distribution duct system. If a temperature rise of more than 2°F is found between the return air grill and the unit return air, the return air system should be checked for air leakage and insulation. If a temperature rise of more than 3°F is found between the unit supply air and the air from the area supply air grill, the supply air distribution duct system must be checked for insulation.

With a 20°F temperature drop through the evaporator coil, each 1°F loss in either section of the duct system means a 5% loss in cooling capacity. With the CFM through the evaporator coil set properly, the refrigeration system is checked.

Refrigeration System Check

A heat pump is a heat transfer system. Therefore, to check the refrigeration system performance, the heat input must be proper. This has been covered under "Air System Check." To check the refrigeration system performance, the following pressures and temperatures must be obtained:

1. Temperature of the air entering the condenser coil
2. Outside condenser coil air discharge temperature
3. Temperature rise of the air through the condenser
4. Compressor discharge pressure
5. Refrigerant condensing temperature
6. Liquid refrigerant temperature leaving the condenser
7. Refrigerant liquid subcooling
8. CFM through the condenser coil
9. DB temperature of the air entering the evaporator
10. WB temperature of the air entering the evaporator
11. DB temperature of the air leaving the evaporator
12. Compressor suction pressure
13. Refrigerant boiling point at the evaporator coil
14. Suction line temperature at the outlet of the evaporator coil
15. Evaporator superheat
16. Quantity of system refrigerant charge

Refrigerant Charge

To produce the desired heat transfer capacity, not only must the proper amount of air pass through the heat exchangers, the system refrigerant charge must be correct. Each factor of refrigerant charge versus CFM affects the system capacity.

After the desired temperature drop through the evaporator coil has been obtained, the system pressures and temperatures will adjust to reflect the refrigerant charge. The amount of refrigerant charge will be revealed by the superheat of the evaporator coil and the subcooling of the liquid refrigerant off the condenser if the system uses capillary tubes or restrictor devices for the pressure-reducing devices.

If the system uses T.X. valves, the refrigerant quantity will only be reflected in the amount of subcooling of the liquid off the condenser coil.

Superheat Test

The amount of superheat that is built up in the evaporator coil depends upon the quantity of refrigerant forced into the evaporator coil to absorb heat and the amount of air flowing over the evaporator coil to supply the heat. If the refrigerant and air quantities are correct, the pressures in the system will adjust to provide optimum heat absorption, producing the correct superheat.

The flow of refrigerant into the evaporator coil depends upon the pressure in the condenser forcing the liquid refrigerant into the capillary tube(s) or restrictor and opposed by the pressure in the evaporator coil developed by the expanding refrigerant absorbing heat in the evaporator coil.

The pressure in the condenser (condensing temperature) is determined by how much vapor (heat) is received from the compressor versus the temperature of the air entering the condenser coil and extracting the heat. The amount of heat (vapor) from the compressor depends upon the amount of refrigerant vaporized in the evaporator coil. This depends upon the CFM of air through the evaporator coil and the total heat content both sensible and latent heat, of the air.

The CFM of air has been established by proper adjustment of air quantity to obtain the desired temperature drop for the existing air conditions. Therefore, the load on the evaporator coil and the outdoor air temperature will determine the amount of superheat developed in the evaporator. To be more specific, the amount of superheat is determined by the outdoor temperature and the wet bulb (WB) temperature of the air entering the inside coil.

To measure the superheat, the following items are determined:

Compressor Suction Pressure. The suction pressure will give the boiling point of the refrigerant at the point where the pressure is measured.

Refrigerant Boiling Point at the Evaporator. When the suction pressure is measured at the outdoor section of the split system or at the compressor line tap in a package system, the flow resistance of the vapor through the line must be considered. Suction lines should be sized to have a 3 psig pressure drop per 100 equivalent feet of line. A fairly safe assumption is 2 psig for the average size installation. By adding 2 psig to the suction pressure gauge reading, the operating pressure in the evaporator coil is found. This pressure, when converted to temperature on the pressure/temperature chart for the particular refrigerant, will give the boiling point.

For example, suppose the system uses R-22 refrigerant and the measured suction pressure is 82 psig. Adding the 2 psig pressure loss in the suction line means an 84 psig operating pressure in the evaporator coil. Using the temperature-pressure chart in Figure 6–4, the saturation temperature (boiling point) of the R-22 refrigerant is 50°F.

Suction Line Temperature at the Evaporator Coil Outlet. After vaporization of the liquid refrigerant is completed, the vapor will continue to rise in temperature depending upon how long it has to travel from the point of vaporization to the coil outlet. The physical temperature of the refrigerant at this point will reflect the heat gain.

Evaporator Superheat. The amount of superheat is the difference in temperature between the physical temperature and the evaporating temperature (boiling point).

Physical temperature – Boiling point = Superheat

In the example, if the physical temperature of the vapor off the evaporator coil is 61°F with a 50°F refrigerant boiling point, the superheat will be 61°F less 50°F or 11°F. To determine if the superheat is correct, other factors must be determined.

Temperature of the Air Entering the Condenser Coil. The flow rate of the refrigerant depends upon the heat pressure. The head pressure is affected by the outdoor ambient temperature (condensing air inlet temperature). The outdoor air temperature must be determined at the same time the other readings are taken. Because this temperature can change in short time periods, a reading should be taken for each test.

Return Air Wet Bulb Temperature. The wet bulb temperature of the return air supply was determined when adjusting the amount of air through the evaporator coil. This test must be repeated at this time to obtain accurate results. The effect of air temperature changes must be eliminated. Therefore, the wet bulb temperature of the return air must be determined at the same time as the outdoor ambient temperature, coil operating pressure, and the refrigerant vapor temperature.

Vacuum - Inches of Mercury - Italic Figures

SPORLAN **TEMPERATURE PRESSURE CHART**

Pressure-Pounds Per Square Inch Bold Figures

TEMPERATURE °F.	12-F	22-V	500-D	502-R	717-A	TEMPERATURE °F.	12-F	22-V	500-D	502-R	717-A	TEMPERATURE °F.	12-F	22-V	500-D	502-R	717-A
-60	*19.0*	*12.0*	*17.0*	*7.2*	*18.6*	12	15.8	34.7	21.2	43.2	25.6	42	38.8	71.5	48.2	83.8	61.6
-55	*17.3*	9.2	*15.0*	*3.9*	*16.6*	13	16.5	35.7	21.9	44.3	26.5	43	39.8	73.0	49.4	85.4	63.1
-50	*15.4*	6.2	*12.8*	*0.2*	*14.3*	14	17.1	36.7	22.6	45.4	27.5	44	40.7	74.5	50.5	87.0	64.7
-45	*13.3*	2.7	*10.4*	1.9	*11.7*	15	17.7	37.7	23.4	46.5	28.4	45	41.7	76.0	51.6	88.7	66.3
-40	*11.0*	0.5	*7.6*	4.1	*8.7*	16	18.4	38.7	24.1	47.7	29.4	46	42.7	77.6	52.8	90.4	67.9
-35	*8.4*	2.6	*4.6*	6.5	*5.4*	17	19.0	39.8	24.9	48.9	30.4	47	43.6	79.2	54.0	92.1	69.5
-30	*5.5*	4.9	*1.2*	9.2	*1.6*	18	19.7	40.9	25.7	50.0	31.4	48	44.7	80.8	55.1	93.9	71.1
-25	*2.3*	7.4	1.2	12.1	1.3	19	20.4	41.9	26.5	51.2	32.5	49	45.7	82.4	56.3	95.6	72.8
-20	0.6	10.1	3.2	15.3	3.6	20	21.0	43.0	27.3	52.5	33.5	50	46.7	84.0	57.6	97.4	74.5
-18	1.3	11.3	4.1	16.7	4.6	21	21.7	44.1	28.1	53.7	34.6	55	52.0	92.6	63.9	106.6	83.4
-16	2.1	12.5	5.0	18.1	5.6	22	22.4	45.3	28.9	54.9	35.7	60	57.7	101.6	70.6	116.4	92.9
-14	2.8	13.8	5.9	19.5	6.7	23	23.2	46.4	29.8	56.2	36.8	65	63.8	111.2	77.8	126.7	103.1
-12	3.7	15.1	6.8	21.0	7.9	24	23.9	47.6	30.6	57.5	37.9	70	70.2	121.4	85.4	137.6	114.1
-10	4.5	16.5	7.8	22.6	9.0	25	24.6	48.8	31.5	58.8	39.0	75	77.0	132.2	93.5	149.1	125.8
- 8	5.4	17.9	8.8	24.2	10.3	26	25.4	49.9	32.4	60.1	40.2	80	84.2	143.6	102.0	161.2	138.3
- 6	6.3	19.3	9.9	25.8	11.6	27	26.1	51.2	33.3	61.5	41.4	85	91.8	155.7	111.0	174.0	151.7
- 4	7.2	20.8	11.0	27.5	12.9	28	26.9	52.4	34.2	62.8	42.6	90	99.8	168.4	120.6	187.4	165.9
- 2	8.2	22.4	12.1	29.3	14.3	29	27.7	53.6	35.1	64.2	43.8	95	108.3	181.8	130.6	201.4	181.1
0	9.2	24.0	13.3	31.1	15.7	30	28.5	54.9	36.0	65.6	45.0	100	117.2	195.9	141.2	216.2	197.2
1	9.7	24.8	13.9	32.0	16.5	31	29.3	56.2	36.9	67.0	46.3	105	126.6	210.8	152.4	231.7	214.2
2	10.2	25.6	14.5	32.9	17.2	32	30.1	57.5	37.9	68.4	47.6	110	136.4	226.4	164.1	247.9	232.3
3	10.7	26.5	15.1	33.9	18.0	33	30.9	58.8	38.9	69.9	48.9	115	146.8	242.7	176.5	264.9	251.5
4	11.2	27.3	15.7	34.9	18.8	34	31.7	60.1	39.9	71.3	50.2	120	157.7	259.9	189.4	282.7	271.7
5	11.8	28.2	16.4	35.9	19.6	35	32.6	61.5	40.9	72.8	51.6	125	169.1	277.9	203.0	301.4	293.1
6	12.3	29.1	17.0	36.9	20.4	36	33.4	62.8	41.9	74.3	52.9	130	181.0	296.8	217.2	320.8	—
7	12.9	30.0	17.7	37.9	21.2	37	34.3	64.2	42.9	75.9	54.3	135	193.5	316.6	232.1	341.2	—
8	13.5	30.9	18.4	38.9	22.1	38	35.2	65.6	43.9	77.4	55.7	140	206.6	337.3	247.7	362.6	—
9	14.1	31.8	19.0	39.9	22.9	39	36.1	67.1	45.0	79.0	57.2	145	220.3	358.9	264.0	385.0	—
10	14.6	32.8	19.7	41.0	23.8	40	37.0	68.5	46.1	80.5	58.6	150	234.6	381.5	281.1	408.4	—
11	15.2	33.7	20.4	42.1	24.7	41	37.9	70.0	47.1	82.1	60.1	155	249.5	405.1	298.9	432.9	—

Figure 6–4 Temperature pressure chart (Courtesy The Sporlan Valve Co.)

CONDENSER AIR INLET TEMP.-°F-D.B.	EVAPORATOR AIR INLET TEMP.-°F-W.B.											
	54	56	58	60	62	64	66	68	70	72	74	76
60	13	17	18	20	24	26	28	30				
65	11	13	17	17	18	22	25	28	30			
70	8	11	12	14	16	18	22	25	28	30		
75		7	10	12	14	16	18	23	26	28	30	
80			6	8	12	14	16	18	23	27	28	30
85				6	8	12	14	17	20	25	27	28
90					6	9	12	15	18	22	25	28
95						7	11	13	16	20	23	27
100							8	11	14	18	20	25
105							6	8	12	15	19	24
110								7	11	14	18	23
115									8	13	16	21

Figure 6–5 Superheat table (Courtesy Addison Products Co.)

With the inside air WB temperature and the outdoor ambient DB temperature, the table in Figure 6–5 is used to determine the correct superheat for a capillary tube or restrictor-type system. If, for example, the return air WB temperature is 66°F and the outdoor ambient temperature is 95°F, the superheat should be 11°F. The horizontal line at 95°F DB and the vertical down line at 66°F WB cross at 11°F. If the inside air humidity content were higher, for example 70°F WB, at the 95°F outdoor ambient temperature, the superheat should be 16°F. If the inside air humidity content were lower (64°F WB), at the same outside ambient temperature (95°F DB), the superheat should be 7°F.

This shows the effect of the amount of total heat in the air over the evaporator coil. The greater the total heat content of the air, the faster the refrigerant will boil in the evaporator coil, producing a higher superheat. After determining the actual superheat that the system is producing and the theoretical superheat desired at the conditioned area WB and outdoor air DB temperatures, a comparison of the two is made. Using average service gauges and thermometers, if the two results are within ±1°F, the refrigerant charge is correct. The use of digital thermometers is recommended for greater accuracy and a charge tolerance of ±0.5°F.

If the actual superheat is more than the theoretical superheat, the system is short of refrigerant charge. To correct the situation, refrigerant must be added. The addition should be in 4 oz. increments, allowing time for the system to reach a thermal balance with the increased refrigerant charge. This will usually occur within a 5- to 15-minute period. This procedure should be repeated until the superheat is lowered to the correct amount.

If the actual superheat is less than the theoretical, the system is overcharged. Removing the excess refrigerant into a DDT-approved cylinder is required to reduce atmospheric pollution. The refrigerant should be recovered from the liquid line in 4 oz. increments, allowing sufficient running time between recovery to allow the system to reach thermal balance. Charging or recovering as necessary should be continued until the proper superheat is obtained. Remember to repeat the pressure and temperature tests when the thermal balance has been reached. Changes will occur with the change in refrigerant quantity as well as temperatures.

Subcooling Test

A second method of checking the refrigerant charge is by the amount of liquid refrigerant subcooling generated in the condenser coil. This applies to systems using T.X. valves, capillary tubes, or restrictor-type flow valves as long as the system does not have a receiver. The subcooling is the drop in temperature of the liquid refrigerant below the condensing temperature. The condenser coil

1. Lowers the temperature of the superheated vapor from the compressor.
2. Condenses the vapor to a liquid.
3. Cools the liquid below the condensing temperature.

The condenser has a fixed amount of heat exchanger surface to transfer heat from the vapor to the outdoor air over the condenser coil surface. The objective is to use each portion for the maximum heat extraction in each of the three phases of heat transfer.

To realize the importance of subcooling, the performance of the entire system has to be considered. Liquid refrigerant is forced through the pressure-reducing device to lower the pressure and boiling point of the refrigerant far enough below the temperature of the air to be cooled to get the desired heat transfer rate.

In our example, the boiling point of the refrigerant was found to be 50°F. This was produced with an evaporator coil operating pressure of 84 psig. The vapor is then compressed by the action of the motor-compressor assembly to a pressure and corresponding condensing temperature high enough above the outside ambient temperature to eject the heat to the outside air at the desired rate. After the vapor is de-superheated and condensed, it is at the high condensing temperature. When this liquid leaves the condenser coil through the liquid line, it is at a high temperature and contains considerable heat.

When it passes through the pressure-reducing device, the liquid pressure and corresponding boiling point are reduced. The physical temperature of the liquid cannot remain higher than the corresponding boiling point. Some of the liquid refrigerant is vaporized to absorb the heat from the liquid and lower its temperature to the boiling point temperature in the evaporator coil. The vapor produced in this process is called *flash gas* and represents a reduction in system efficiency. The compressor must handle this vapor along with the vapor produced in the evaporator coil from absorbing the heat from the air.

To reduce the amount of flash gas formed, the liquid refrigerant temperature is lowered before it leaves the condenser. This is done by having enough liquid refrigerant in the system to have liquid refrigerant in the last passes of the condenser coil when the system is operating.

There are limits on the amount of subcooling that the system can tolerate. Most heat pumps, in the cooling mode, are designed to operate with 18°F to 20°F of subcooling. This represents the best balance between the head pressure the compressor must operate against and the amount of flash gas formed in the coil.

If no subcooling were to take place in the condenser coil, the entire coil surface would be for de-superheating and condensing the vapor to a liquid. This would produce the heat transfer with a low temperature difference between the air and the refrigerant vapor. This temperature difference is called *split.*

On the other hand, the higher temperature of the liquid would result in abnormally high quantities of flash gas. This would reduce the system capacity by causing the compressor to handle less vapor formed in the evaporator coil.

Using this information, reasoning indicates the greater the amount of subcooling, the higher the system efficiency. This is true to a point. As refrigerant is added to the system, more condenser coil surface is used for subcooling, which reduces the amount of surface for the vapor-condensing function.

As this amount of surface is reduced, the remaining surface must have a greater temperature difference (split) between vapor and air to get the heat transfer needed. To accomplish this, the condensing temperature is raised. This means the compressor discharge pressure is raised to raise the condensing temperature. As the discharge pressure is raised, the efficiency of the compressor decreases.

Here we see that the reduction in liquid temperature increases the system efficiency while the increase in compressor discharge pressure decreases the system efficiency. It is now a matter of how much a change in either one or both of the factors affects the final results. As subcooling starts to occur, the decrease in liquid temperature (decrease in the amount of flash gas) has a greater influence on raising the efficiency than the detrimental effect of the increase in compressor discharge pressure.

As the liquid temperature is reduced, the amount of subcooling is increased. However, the beneficial effect on the efficiency is a diminishing factor. At the same time, as the compressor discharge pressure is increased, the detrimental effect of this change is an increasing factor. The point where each effect balances the other is the desirable refrigerant charge. At this point, the net cooling capacity of the system is at peak.

As stated previously, most systems are designed to operate with 18°F to 20°F of liquid subcooling. The amount of refrigerant in the system is the dominating factor in determining the amount of subcooling of the liquid off the condenser coil. However, the system must be operating within the outdoor temperature range limits of 115°F down to 65°F and the inside DB temperatures and RH% must be within the ranges of the load ratio chart (see Figure 6–1).

To determine the amount of subcooling, only two test readings are needed:

Compressor Discharge Pressure. The compressor discharge can be measured at the discharge outlet of the compressor, if such a means is provided. This will give the pressure which the compressor is producing. This may not be the actual operating pressure (condensing temperature) in the condenser coil.

On package units, as well as outdoor condensing units, the hot gas line between the compressor and the condenser coil (in the cooling mode) is short enough that any pressure drop in the line is ignored.

If the condensing pressure is measured at the liquid line out of the condenser coil, the condenser pressure drop of 1.5 psig is added to the liquid line pressure gauge reading to determine the condensing temperature of the refrigerant in the condenser coil.

Condensing Temperature. Using the temperature-pressure chart (Figure 6–4), the condensing temperature of the refrigerant vapor is determined by finding the temperature equivalent of the pressure for the particular refrigerant in the system. For example, the system using R-22 with a gauge reading of 241.2 psig would have a condensing temperature of 115°F. 241.2 psig gauge reading + 1.5 psig condenser coil pressure loss = 242.7 psig. For R-22, the boiling point or condensing temperature at 242.7 psig is 115°F.

Refrigerant Temperature Leaving the Condenser. The physical temperature of the liquid refrigerant leaving the condenser should be measured within 6" of the condenser coil outlet manifold. This is to reduce, as much as possible, the line temperature loss that will affect the test results. A digital thermometer is recommended.

Refrigerant Liquid Subcooling.

The condensing temperature – the physical temperature of the liquid = the amount of liquid subcooling

Quantity of Refrigerant Charge. The previous methods of determining the correct charge are for a system that is operating. The check is to determine if the system is operating at peak capacity and efficiency. To remove all the refrigerant charge is an unnecessary expense and waste of refrigerant. However, if the system is so short of refrigerant that there is no refrigerant in liquid form, the preferred method is to recover the remaining refrigerant, find and repair the leak, evacuate the system, and charge by weight.

All manufacturers are required by U.L. standards to list the type and amount of refrigerant on the rating plate of the unit or portion of the system. In the case of a package unit, the entire system charge is given. In the split system, the amount of normal operating charge is usually given for the sections marketed by the particular manufacturer. To determine the total charge, the amount of refrigerant in the liquid and vapor lines is calculated. Depending upon the type of refrigerant, each line will contain a certain weight per foot.

For example, from the table in Figure 6–6, a ¼" liquid line in a system using R-22 will contain 0.39 oz. per foot. If the line were 25' long, the refrigerant charge for the liquid line would be 25' × 0.39 oz. per foot or 9.75 oz. for the entire line. If the vapor line were ½" OD size and 25' long, the refrigerant charge for the vapor line would be 25' × 0.03 oz. per foot or

REFRIGERANT LINE SIZE	CONTENTS PER LIQUID LINE	FOOT IN OUNCES	SUCTION LINE	
IN. O.D.	R-22	R-12	R-22	R-12
1/4"	.39	.54	.01	.01
3/8"	.48	.64	.02	.01
1/2"	1.12	1.28	.03	.02
5/8"	1.76	2.08	.05	.03
3/4"	2.08	3.04	.08	.06
7/8"	3.84	4.16	.10	.08

Figure 6–6 Copper tube refrigerant volume (Courtesy Addison Products Co.)

0.75 oz. The total charge would be 9.75 oz. in the liquid line plus 0.75 oz. in the vapor line for a total of 10.5 oz. by measured weight.

This line refrigerant quantity, added to the refrigerant charge for the indoor and outdoor sections, is the quantity of refrigerant needed in the system for proper operation.

System Capacity

The first step after starting the heat pump system was the establishment of the correct temperature drop of the air through the evaporator coil for the DB temperature and % RH of the return air. The second step was obtaining the correct refrigerant charge by any of the following methods:

1. Adjusting the refrigerant charge to the correct superheat
2. Adjusting the refrigerant charge to the correct subcooling
3. Completely evacuating the system and weighing in the calculated charge

Regardless of the method used to correct the refrigerant charge, the drop of the air through the evaporator coil must now be rechecked and adjusted accordingly. Usually, correcting the refrigerant charge will increase the operating capacity of the system, resulting in an increase in the temperature drop of the air through the evaporator coil. Usually, the CFM of air through the evaporator coil needs to be increased to reduce the temperature drop to the proper amount. After the adjustment has been made, the system cooling capacity can be checked against the manufacturer's rating to determine if the system capacity is correct.

Figure 6–7 shows a table of cooling capacities of different size outdoor and indoor section combinations at different outdoor ambient temperatures. These capacities are the net cooling capacities of the system. The heat rejected by the condenser coil is made up of the heat (both sensible and latent) picked up by the evaporator coil plus the motor heat put in by the condenser fan motor and the motor-compressor assembly.

The net cooling capacity of the system can be determined by either of two methods:

MODEL		COOLING		HEATING @ 47° F (1)		HEATING @ 17 ° F (1)		CONNECTIONS	
OUTDOOR	INDOOR	Capacity	EER	Capacity	COP	Capacity	COP	SUC.	LIQ.
QH824	AH65HD	24000	8.1	25000	2.7	12500	1.8	3/4	1/4
QH830	AH65HE	30000	7.9	33000	2.8	19500	2.0	3/4	1/4
QH836	AH68HF	35000	8.0	39000	2.8	24000	2.0	3/4	1/4
QH842	AH68HG	40000	7.7	44000	2.8	27000	2.0	7/8	3/8
QH848	AH68HH	46000	7.5	48000	2.7	30000	1.9	7/8	3/8
QH860	AH68HK	59000	7.3	69000	2.8	38000	1.9	7/8	3/8

(1) Per ARI Standard 240-77 Rating Conditions

Figure 6–7 Heat pump specifications (Courtesy Addison Products Co.)

1. Determine the gross cooling capacity and subtract the motor heat. If the system does not have any form of auxiliary heat that can be used to determine the CFM through the inside evaporator coil and the air distribution system is not easily accessible, the gross cooling capacity method is used.
2. If a means of accurately measuring the CFM through the evaporator coil is available, the net cooling capacity method is used.

Gross Cooling Capacity Method

To determine the gross cooling capacity of the system, the amount of heat ejected by the condenser coil, two factors must be determined—the CFM of air through the outdoor section and the temperature rise of that air.

Air Quantity. In their specification literature, practically all manufacturers list the CFM of air through the outdoor section. Figure 6–8 is a table of outdoor section specifications where the CFM of air through the unit is listed under "Condenser Fan CFM." The QH836–1D outdoor section that was specified for the example in the Phoenix, AZ area would have 3,680 CFM of air through the unit. The QH824-1D unit selected for the Omaha, NB area would have 2,670 CFM through the unit.

Air Temperature Rise. To measure the air temperature rise, both entering and leaving air temperatures are recorded.

Temperature of Air Entering the Condenser. The entering air temperature is the ambient temperature surrounding the unit. Using a sling psychrometer, the air temperature should be taken on all sides of the unit that have air openings. The readings should be taken from a distance of 12" or more from the condenser surface. This is to reduce the radiant heat effect of the hot condenser surface.

Outside Coil Air Discharge Temperature. The temperature of the leaving air needs to be an average temperature of readings taken over the area of the air outlet grill. On vertical discharge units, such as shown in Figure 6–9, the thermometers are located on the top of the fan discharge grill. A reading should be taken above each condenser inlet face. For the unit shown in Figure 6–9, three readings are taken and an average temperature is calculated. If the unit is a complete wraparound type, such as in Figure 6–10, four readings are taken and averaged.

MODEL	Electrical Characteristics	Operating Current (Amps) (1)	Minimum Fieldwire (AWG) (2)	Maximum Wire Run (FT.) (3)	Minimum Fuse (Delay)	Compressor L.R. Current	Condenser Face Area (SQ. FT.)	Condenser Fan CFM	Fan Motor	Weight	
										Net	Ship
QH818-1D	1-60-230/208	11.1/11.8	14	33	20A	42	9.1	2670	PSC-¼	166	192
QH824-1D	1-60-230/208	13.9/13.9	12	42	20A	53	9.1	2670	PSC-¼	166	192
QH830-1D	1-60-230/208	17.2/17.2	10	48	25A	75	9.1	2670	PSC-¼	189	215
QH836-1D	1-60-230/208	19.5/21.0	10	43	30A	81	9.1	3680	PSC-⅓	197	228
QH842-1D	1-60-230/208	23.0/23.0	10	35	35A	97	13.0	3620	PSC-⅓	242	297
QH848-1D	1-60-230/208	26.5/26.5	8	50	40A	118	13.0	3920	PSC-⅓	267	322
QH848-3D	3-60-230/208	19.0/19.0	10	40	30A	90	13.0	3950	PSC-⅓	267	322
QH860-1D	1-60-230/208	32/9/32.9	6	66	50A	139	13.0	4450	PSC-½	293	348
QH860-3D	3-60-230/208	20.4/20.4	10	34	30A	104	13.0	4450	PSC-½	293	348

(1) Per ARI Standard 240 Rating Conditions. (2) Copper wire only.
(3) For longer run, increase wire gauge one size.

Figure 6–8 Outdoor section specifications (Courtesy Addison Products Co.)

Figure 6–9 Draw-through vertical discharge unit (Courtesy Addison Products Co.)

Figure 6–10 Outdoor section vertical air flow (Courtesy Addison Products Co.)

In both units, the condenser fan is a draw-through type with a fair amount of mix of the air leaving the condenser face. If the unit has a blow-through type air discharge design, the average discharge air temperature is more difficult to arrive at because of the wide variety of air quantities off each section of a propeller fan blade.

Figure 6–11 Outdoor section horizontal air flow (Courtesy Addison Products Co.)

Figure 6–11 shows a horizontal discharge blow-through type unit. In this case, more readings must be taken to arrive at the air temperature average. The face of the discharge grill is divided into equal size segments and the air temperature from each segment is measured. In this case, the condenser outlet air face is divided into 20 different segments for the greatest accuracy in determining the average air temperature.

Temperature Rise of the Air Through the Condenser. After determining the average leaving air temperature, the temperature rise of the air through the condenser is calculated.

The average leaving air temperature minus the ambient air temperature equals the temperature rise of the air through the condenser.

The heat rejected from the unit is determined by using the standard heat content formula:

$$\text{ft.}^3/\text{min.} = \frac{\text{BTU/hr.}}{\Delta T°F \times 1.08}$$

where:

ft.³/ min. = amount of air traveling per minute through the coil from which sensible heat is being extracted from or ejected into.

BTU/hr. = amount of sensible heat added or extracted in 1 hour.

$\Delta T°F$ = temperature change of the air being treated.

1.08 = a constant used to convert from cubic feet per pound and cubic feet per minute to pounds per hour. It also contains the quantity of heat needed to change the temperature of each pound of air. This figure is derived from the following formula:

$$1.08 = 60 \text{ min./hr.} \times 0.075 \text{ lb./ft.}^3 \times 0.24 \text{ BTU/lb./°F}$$

To convert the BTU/hr. to BTU/min., we divide the hour figure by 60. We cannot use the cubic foot of air to determine heat content because air expands or contracts with temperature change. However, a pound of air is the same regardless of how much space it occupies.

A standard has been established to use this formula. Air at 70°F and at a barometric pressure of 29.92 inches of mercury is considered to be standard air. At these conditions, air occupies 0.075 ft.³ per pound and has a specific heat of 0.24 BTU per pound per 1°F change in temperature. Therefore, 60 minutes per hour × 0.075 pounds per cubic foot × 0.24 BTU per pound per 1°F temperature change = 1.08.

To determine the heat rejected from the condenser coil in the outside section, the basic air formula is changed to solve for the BTUH (total heat).

$$\text{BTUH} = \text{ft.}^3/\text{min.} \times \Delta T°F \times 1.08$$

CFM of Air Through the Outside Coil. Using the CFM of air through the unit from the manufacturer's literature, and the temperature rise of the air through the unit, the figures are multiplied together along with the constant of 1.08. This gives the gross capacity of the system.

Motor Heat

The amount of heat added to the vapor as it passes through the compressor operation is practically 100% of the electrical energy needed to operate the motor-compressor assembly. The small amount of heat loss from the compressor shell by conduction is ignored.

The formula used to calculate the motor heat is:

$$\text{Motor BTUH} = \text{volts} \times \text{amperes} \times \text{PF} \times 3.413 \text{ BTU/watt}$$

where:

Volts = voltage at the load side of the compressor contactor with the unit operating. Regardless of what the unit is rated at, the actual applied voltage must be used. If this is a three-phase unit, the voltage across each of the phases should be the same as the others. If a voltage difference of more than 3% is found, the voltage unbalance will cause amperage unbalance, increase motor winding temperature, and shorten motor life. The utility supplying the power should be contacted regarding this problem.

Amperage = the total amperage the unit is drawing. This includes the amperage draw of both the motor-compressor assembly and the condenser fan motor.

PF = The power factor of the entire outdoor section. The power factor of condensing units will vary between 86% and 94%, depending upon the applied voltage and the load on the unit. An average of 90% (0.9) power factor is used in the formula. For more accurate calculations, a "true watt" meter should be used.

BTU/Watt = Heat energy of electrical energy – 3.413 BTU per each watt of electrical energy

Multiplying the applied voltage × the total amperage × 0.9 power factor × 3.413 BTU per watt = motor heat input.

Net Capacity

Subtracting the motor heat input from the gross capacity will leave the system net capacity. Comparing these results with the capacities listed in the manufacturer's literature at the outside ambient used for the test will show the system actual net capacity as compared to the printed capacity. Because of the inaccuracy of field instrumentation a tolerance of comparison of ±10% must be allowed.

6–5 NET COOLING CAPACITY METHOD

To use the net cooling capacity method, the CFM through the evaporator coil and the entering and leaving wet bulb (WB) temperatures of the air through the evaporator coil must be measured. The method for determining the CFM through the evaporator coil will be covered in Air System Check.

Referring to the psychrometric chart in Figure 6–2, the scale in the open to the left of the wet bulb temperature curve indicates the total heat content (enthalpy) of a pound of air at its saturation temperature. For example, if the wet bulb temperature of the air entering the coil is 65°F WB, each pound of the air holds 30.05 BTU.

If, when the air passes through the evaporator coil, the wet bulb temperature of the air is reduced to 55°F WB, each pound of that air contains 23.4 BTU. This means that 6.65 BTU were removed from each pound of that air. 30.03 BTU per pound of air at 65°F WB minus the 23.4 BTU per pound of air at 55°F WB equals 6.65 BTU removed from each pound of air.

The next step is to determine the pounds of air that are traveling through the evaporator coil. This is done by converting the cubic feet of air per minute (CFM) through the evaporator coil to pounds of air per minute by dividing the cubic feet per minute (CFM) by the cubic feet per pound as found on the psychrometric chart. The methods of determining the cubic feet per minute (CFM) are explained in the section Air System Check.

On the psychrometric chart, the oblique lines that slant upward to the left and are marked 12.5, 13.0, 13.5, 14.0, 14.5, and 15.0 are the number of cubic feet of volume a pound of air will occupy. This is the specific volume of air (V_s). As the air increases in temperature, it expands and each pound of air occupies a larger volume. The volume changes with a change in temperature, but the weight in pounds remains the same.

If the return air in our example is at 80°F DB and 65°F WB, the cross point (where the vertical dry bulb line and the slant wet bulb line cross) is between the 13.5 and the 14.0 cubic feet per pound lines. If the cross point were midway between the two cubic feet per pound lines, the cubic feet per pound at the cross point would be 13.75 cubic feet per pound. It is, however, closer to the 14.0 line than the 13.5 line.

By spacing off the distance on the 3/16" range on a drafting scale, we find that the distance between the 13.5 and 14.0 lines is 7 increments on the $^3/_{16}$" scale. We also find that the cross point is 2.5 increments from the 14.0 line. By dividing the 2.5 increments by the 7 increments, we find that the cross point is 35.7% of the distance between the lines. The difference between the two cubic feet per pound lines is 0.5 cubic feet per pound. Therefore, if the cross point is 35.7% (0.357) of the 0.5 difference, the cross point is 0.2 below the 14.0 line or 13.8 cubic feet per pound.

The equation to calculate this is:

$$\frac{NUS \times 0.5}{TUS} = ANUS$$

where:

NUS = number of scale units that the cross point of the dry bulb and wet bulb lines is below the larger cubic foot per pound line.
TUS = the number of scale units between the cubic foot per pound lines involved.
0.5 = the difference between the cubic foot per pound lines.
ANUS = the difference between the cubic feet per pound at the cross point and the larger cubic feet per pound line.

Subtracting the difference (ANUS) from the larger cubic feet per pound line results in the value in cubic feet per pound at the cross point.

The pounds of air per minute is then determined by dividing the CFM through the evaporator coil by the determined cubic feet per pound (V_s) to get the pounds of air through the evaporator coil per minute. For example, we will assume the unit to be moving 1,200 CFM of 80°F DB – 65°F WB (50% RH) air through the evaporator coil. We have calculated that the air at these conditions has a specific volume of 13.8 cubic feet per pound. By dividing the CFM by the cubic feet per pound (specific volume—V_s), we determine the pounds of air per minute through the evaporator coil.

$$\frac{\text{CFM}}{\text{cu. ft./pound}} = \text{Pounds per minute}$$

$$\frac{\text{1,200 CFM}}{\text{13.8 cu. ft./pound}} = \text{86.96 pounds per minute}$$

Therefore, 86.96 pounds of air per minute times 6.65 BTU per pound equals 598.17 BTU per minute being picked up in the evaporator coil. This is total heat, both sensible and latent.

Units are rated in BTU per hour (BTUH) so 598.17 BTU per minute times 60 minutes = 35,890.05 BTUH, the net cooling capacity of the unit.

Allowing for inaccuracy of the instruments used, the calculated capacity should be within ±10% of the manufacturer's rating.

6–6 AIR SYSTEM CHECK

The CFM through the inside coil can be determined by any of the following methods:

1. Duct velocity
2. Static pressure drop external to the inside air handling unit
3. Temperature rise of the air through the inside air handler unit
4. Total CFM supply from the supply outlet grills
5. Auxiliary heat

Each of the methods of determining the CFM of air through the inside condenser coil requires test instruments. These will be discussed under the test methods.

Duct Velocity. To determine the total CFM through an air duct, the area of the duct in sq. ft. and the velocity of the air through the duct in ft./min. are needed. The area of the duct in sq. ft. times the velocity of the air through the duct in ft./min. will produce the cubic feet of air per minute (CFM) flowing through the duct.

$$\text{CFM} = \text{Duct area in sq. ft.} \times \text{air velocity in ft./min.}$$

Duct work is sized in inches in 2 in. increments for production standards. With all dimensions in 2 in. increments, it is possible to use conversion charts to convert from the duct area in sq. in. to sq. ft. Figure 6–12 is a conversion table for converting rectangular duct with inch dimensions to sq. ft. area, as well as the diameter equivalent in round duct.

Side Rect. Duct	4.0	4.5	5.0	5.5	6.0	6.5	7.0	7.5	8	9	10	11	12	13	14	15	16	Side Rect. Duct
3.0	3.8	4.0	4.2	4.4	4.6	4.8	4.9	5.1	5.2	5.5	5.7	6.0	6.2	6.4	6.6	6.8	7.0	3.0
3.5	4.1	4.3	4.6	4.8	5.0	5.2	5.3	5.5	5.7	6.0	6.3	6.5	6.8	7.0	7.2	7.4	7.6	3.5
4.0	4.4	4.6	4.9	5.1	5.3	5.5	5.7	5.9	6.1	6.4	6.8	7.1	7.3	7.6	7.8	8.1	8.3	4.0
4.5	4.6	4.9	5.2	5.4	5.6	5.9	6.1	6.3	6.5	6.9	7.2	7.5	7.8	8.1	8.4	8.6	8.9	4.5
5.0	4.9	5.2	5.5	5.7	6.0	6.2	6.4	6.7	6.9	7.3	7.6	8.0	8.3	8.6	8.6	9.1	9.4	5.0
5.5	5.1	5.4	5.7	6.0	6.3	6.5	6.8	7.0	7.2	7.6	8.0	8.4	8.7	9.0	9.4	9.6	9.8	5.5

Side	6	8	10	12	14	16	18	20	22	24	26	28	30	Side
6	6.6													6
8	7.5	8.8												8
10	8.4	9.8	10.9											10
12	9.1	10.7	11.9	13.1										12
14	9.8	11.5	12.9	14.2	15.3									14
16	10.4	12.2	13.7	15.1	16.3	17.5								16
18	11.0	12.9	14.5	16.0	17.3	18.5	19.7							18
20	11.5	13.5	15.2	16.8	18.2	19.5	20.7	21.9						20
22	12.0	14.1	15.9	17.6	19.1	20.4	21.7	22.9	24.1					22
24	12.4	14.6	16.6	18.3	19.8	21.3	22.6	23.9	25.1	26.2				24
26	12.8	15.2	17.2	19.0	20.6	22.1	23.5	24.8	26.1	27.2	28.4			26
28	13.2	15.6	17.7	19.6	21.3	22.9	24.4	25.7	27.1	28.2	29.5	30.6		28
30	13.6	16.1	18.3	20.2	22.0	23.7	25.2	26.7	28.0	29.3	30.5	31.6	32.8	30
32	14.0	16.5	18.8	20.8	22.7	24.4	26.0	27.5	28.9	30.1	31.4	32.6	33.8	32
34	14.4	17.0	19.3	21.4	23.3	25.1	26.7	28.3	29.7	31.0	32.3	33.6	34.8	34
36	14.7	17.4	19.8	21.9	23.9	25.8	27.4	29.0	30.5	32.0	33.0	34.6	35.8	36
38	15.0	17.8	20.3	22.5	24.5	26.4	28.1	29.8	31.4	32.8	34.2	35.5	36.7	38
40	15.3	18.2	20.7	23.0	25.1	27.0	28.8	30.5	32.1	33.6	35.1	36.4	37.6	40
42	15.6	18.5	21.1	23.4	25.6	27.6	29.4	31.2	32.8	34.4	35.9	37.3	38.6	42
44	15.9	18.9	21.5	23.9	26.1	28.2	30.0	31.9	33.5	35.2	36.7	38.1	39.5	44
46	16.2	19.2	21.9	24.3	26.7	28.7	30.6	32.5	34.2	35.9	37.4	38.9	40.3	46
48	16.5	19.6	22.3	24.8	27.2	29.2	31.2	33.1	34.9	36.6	38.2	39.7	41.2	48
50	16.8	19.9	22.7	25.2	27.6	29.8	31.8	33.7	35.5	37.3	38.9	40.4	42.0	50
52	17.0	20.2	23.1	25.6	28.1	30.3	32.4	34.3	36.2	38.0	39.6	41.2	42.8	52
54	17.3	20.5	23.4	26.1	28.5	30.8	32.9	34.9	36.8	38.7	40.3	42.0	43.6	54
56	17.6	20.9	23.8	26.5	28.9	31.2	33.4	35.5	37.4	39.3	41.0	42.7	44.3	56
58	17.8	21.1	24.2	26.9	29.3	31.7	33.9	36.0	38.0	39.8	41.7	43.4	45.0	58
60	18.1	21.4	24.5	27.3	29.8	32.2	34.5	36.5	38.6	40.4	42.3	44.0	45.8	60

Side	32	34	36	38	40	42	44	46	48	50	52	56	60	Side
32	35.0													32
34	36.0	37.2												34
36	37.0	38.2	39.4											36
38	38.0	39.2	40.4	41.6										38
40	39.0	40.2	41.4	42.6	43.8									40
42	39.9	41.1	42.4	43.6	44.8	45.9								42
44	40.8	42.0	43.4	44.6	45.8	46.9	48.1							44
46	41.7	43.0	44.3	45.6	46.8	47.9	49.1	50.3						46
48	42.6	43.9	45.2	46.5	47.8	48.9	50.2	51.3	52.6					48
50	43.5	44.8	46.1	47.4	48.8	49.8	51.2	52.3	53.6	54.7				50
52	44.3	45.7	47.1	48.3	49.7	50.8	52.2	53.3	54.6	55.8	56.9			52
54	45.0	46.5	48.0	49.2	50.6	51.8	53.2	54.3	55.6	56.8	57.9			54
56	45.8	47.3	48.8	50.1	51.5	52.7	54.1	55.3	56.5	57.8	58.9	61.3		56
58	46.6	48.1	49.6	51.0	52.4	53.7	55.0	56.2	57.5	58.8	60.0	62.3		58
60	47.3	48.9	50.4	51.8	53.3	54.6	55.9	57.1	58.5	59.8	61.0	63.3	65.7	60

Figure 6–12 Duct dimension table (Courtesy Air Conditioning Contractors of America)

The sheet metal industry has generally standardized on 8" high man ducts for residential use and 10" high main ducts for commercial use. This allows the use of standard collars, connectors, duct takeoffs, and so on in standard installations. In large commercial and industrial use, however, both height and width of the duct will vary.

To use the conversion table in Figure 6–12, the height and width of the duct are measured. If, for example, we find the height of the duct is 8" and the width is 22", what is the sq. ft. area of the duct? The vertical column under 8 and the horizontal line at 22 cross and show a sq. ft. area of 1.08 and the equivalent round duct of 14.1" diameter. This sq. ft. area × the velocity of the air through the duct will give the CFM of air through the duct.

To measure the velocity of air through the duct, a Pitot tube and manometer are needed. Figure 6–13 shows the setup of a Pitot tube and manometer measuring the total pressure and static pressure as they are applied to the inclined manometer.

The total pressure in a duct is equal to the velocity pressure of the air moving through the duct plus the pressure of the air volume in the duct. Therefore, the total pressure minus the static pressure equals the velocity pressure. The total pressure is forcing the liquid down the inclined tube of the manometer while the static pressure is forcing the liquid up the inclined tube. The actual reading is the difference in the two pressures.

Two different scales are shown on the inclined manometer in Figure 6–14. The top scale is in inch water column (″WC) and the lower one is in ft./min. velocity. The lower scale would be used for this test.

After determining the velocity in ft./min. the duct area in ft.^{2} × the velocity will produce the CFM of air through the particular duct. If there is only one duct, this is the CFM of air through the coil. If there is more than one main duct, all the ducts must be tested and all CFM quantities added together to get the total CFM through the coil.

Static Pressure Drop Through the Indoor Air Handler Section. By measuring the static pressure in the supply and return air duct systems, the total ESP (external static pressure) the indoor handler is working against can be determined if the manufacturer's literature gives this information. The table on indoor section specifications shown in Figure 6–15 gives the CFM of air through the air handler for the different capacity air handlers and at ESP from 0.10" WC to 0.5" WC. If, for example, the AH65HE indoor section that was selected for the Omaha, NB application is operating at 0.20" WC ESP, the air through the unit would be 1,220 CFM. Remember, the positive pressure in the supply duct and the negative pressure in the return duct are added together as though both were positive pressures to determine the total ESP. A positive pressure of +0.15" WC in the supply duct plus a negative pressure of –0.05" WC in the return duct add to 0.20" WC ESP.

SUPPLY DUCT
POSITIVE PRES. + .15
SUPPLY +.15
RETURN (-)-.05
TOTAL ESP +.20
MOD. AH65HD
810 CFM AT .20 ESP
RETURN DUCT
.20 INS. W.C.
MANOMETER
NEGATIVE PRES. - .05

Figure 6–13 Pitot tube and manometer assembly (Courtesy Addison Products Co.)

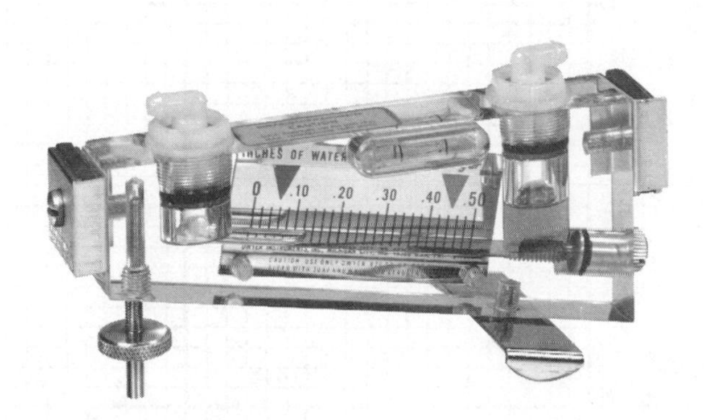

Figure 6–14 Inclined manometer (Courtesy Dwyer Instruments Inc.)

MODEL	BLOWER MOTOR	EVAPORATING COIL CFM (1) @ EXTERNAL STATIC PRESSURES (INS. OF WATER)							WEIGHT	
		.10	.15	.20	.25	.30	.40	.50	NET	SHIP
AH65HC	¼	700	670	645	620	600	550	500	110	125
AH65HD	¼	870	840	810	780	750	690	620	113	128
AH65HE	½	1260	1240	1220	1195	1170	1100	1025	121	136
AH68HF	½	1410	1380	1350	1310	1290	1220	1150	158	173
AH68HG	¾	1760	1740	1720	1685	1650	1575	1450	155	170
AH68HH	¾	1870	1845	1830	1790	1750	1690	1590	160	175
AH68HK	¾	1870	1845	1830	1790	1750	1690	1590	170	186

(1) Since filter furnished with unit, total E.S.P. available to ducts and grills.

Figure 6–15 Indoor section specifications (Courtesy Addison Products Co.)

If we assume that the system is performing as designed, a check of the air temperature rise through the inside condenser coil will allow an approximate CFM setting of the blower.

The amount of heat the system will transfer will depend upon the temperature of the air entering the indoor condenser coil and the outdoor evaporator coil. The outdoor air CFM is factory set by unit design. The CFM of air through the inside condenser coil is affected by the external static pressure in the duct system as well as the blower RPM. This can be field adjusted within the speed adjustment range of the blower motor and/or changes in the air distribution duct system.

Figure 6–16 shows the temperature rise that would result from the various outdoor temperatures from a range of –10°F to +70°F and indoor return air temperatures from +60°F to +80°F.

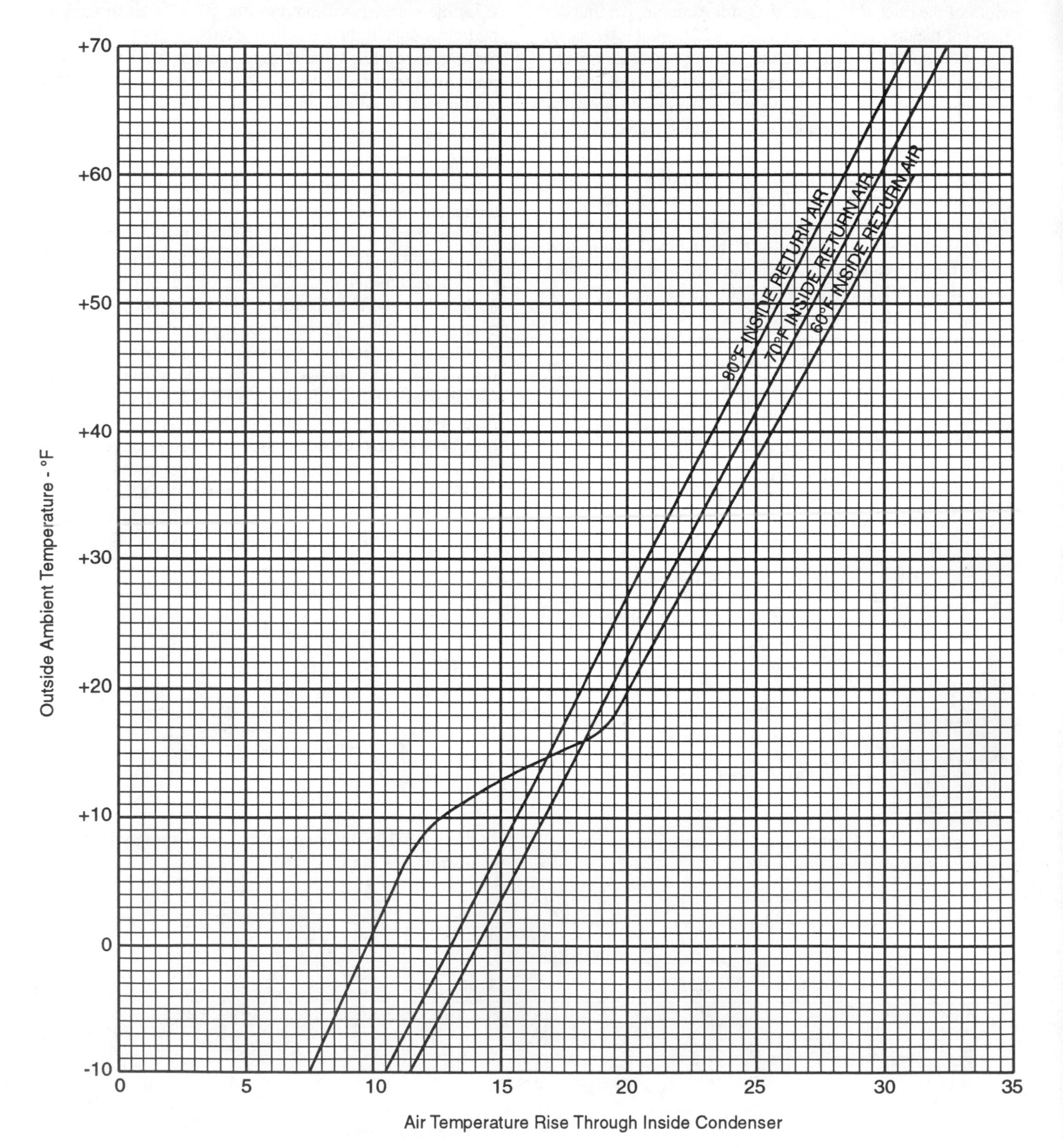

Figure 6–16 Unit capacity curves

Figure 6–17 Air volume meter (Courtesy The Shortridge Instrument Co.)

These curves represent an average of ±2°F at each condition point. For example, with a return air temperature of +70°F and an outdoor temperature of +50°F, the temperature rise through the inside condenser coil should be 27°F, ±2°F. If the actual temperature rise is between +25°F and +29°F, the system is delivering the design capacity.

Total CFM from Supply Outlet Grills. The easiest way to measure the total CFM of air through the inside coil is to measure the CFM of each of the supply outlet grills in the supply duct distribution system. The easiest and most accurate way to measure these air quantities is by the use of a balometer (balance meter), Figure 6–17.

The use of an instrument of this type is recommended to reduce test time and promote accuracy.

6–7 AUXILIARY HEAT METHOD

To determine if the system is delivering rated capacity, the gross heating capacity of the system is determined. This is the same test used to determine the gross cooling capacity of the outdoor section in the cooling mode. The CFM of air through the condenser coil times the temperature rise of the air through the condenser coil times 1.08 equals the gross heat transfer capacity.

The temperature rise of the air through the inside evaporator coil is easily measured. With the thermometers in the supply and return air ducts (see Figure 6–3), the difference between the two temperatures will be the temperature rise. This factor is taken into consideration in the curves of Figure 6–1.

Unitary CFM

The CFM of air through the inside coil can be determined by the use of the auxiliary heat that is used in conjunction with the heat pump. When electric auxiliary heat is used, the procedure is less complicated than when fossil fuels are used. In all calculations, the standard air formula is used:

$$\text{CFM} = \frac{\text{BTUH}}{\Delta\text{T°F} \times 1.08}$$

The difference in the procedures between electric and fossil fuel auxiliary heat is finding the BTUH output of the auxiliary heat source. The output of any heat source is equal to the heat input multiplied by the unit efficiency. The efficiency of electric heat elements is 100%. All the electric energy in comes out in the form of heat. In the case of fossil fuels, however, the efficiency of these units is less than 100%. An efficiency cannot be assumed; it must be measured and calculated.

Electric Heat

The efficiency of an electric element heat source is 100%. To calculate the heat output, it is only necessary to determine the BTUH input. The unit is 100% efficient – input = output.

The formula used is the same as the one used to calculate motor heat input:

$$\text{BTUH} = \text{volts} \times \text{amperes} \times \text{PF} \times 3.413 \text{ BTUH/watt}$$

where:

BTUH = the total electrical energy into the electric resistance banks. This also represents the output.
volts = the measured voltage across the electrical supply when all the elements in the resistance element bank are active and the blower motor is supplying the air.
amperes = the measured amperage drawn by the elements in the resistance element bank plus the amperage drawn by the blower motor supplying the air.
PF = the power factor. The PF of resistance-type electrical loads is 100%. Therefore, the PF number is 1.
3.413 BTUH/watt = the heat equivalent of 1 watt of electrical energy is 3.413 BTU. For each watt of electrical energy used in 1 hour, 3.413 BTUH will be produced.

Therefore:

$$\text{Watts} = \text{volts} \times \text{amperes} \times \text{PF}$$
$$\text{BTUH} = \text{watts/hr.} \times 3.413 \text{ BTU/watt}$$

For example, we will assume that the heater bank in the AH65HE420C unit selected for the Omaha, NB application is drawing 80.5 amperes at 232 volts. What would be the heat output of the unit?

BTUH = 232 volts × 80.5 amperes × PF × 3.413 BTUH/watt × 100% efficiency
BTUH = 63,741.2 BTUH

Assuming that the unit operation, on auxiliary electric heat, with the heat pump off, is supplying 118°F air with 70°F return air, what is the CFM through the unit?

$$CFM = \frac{BTUH}{\Delta T°F \times 1.08}$$

where:

CFM = the amount of air through the unit.
BTUH = the heat output of the heater bank. In this example, 63,741.2 BTUH.
ΔT°F = the difference in temperature between the return air and supply air. In this example, the air temperature rise is 48°F. (The maximum recommended supply air temperature off the electrical heater elements is 120°F.)
1.08 = standard air factor

Therefore:

$$CFM = \frac{63,741.2 \text{ BTUH}}{48°F \times 1.08}$$
$$CFM = 1,229.6$$

Fossil Fuel Auxiliary Heat Unit

To determine the CFM through a fossil fuel unit, the efficiency of the unit must be determined using the air temperature rise method. This procedure is more complicated than electric heat and may discourage the procedure. The procedure must be learned, however, as it may be necessary to perform the process in the case of pending litigation.

The instruments needed for an efficiency test on a fossil fuel heating unit are: two air temperature thermometers, a high-temperature stack thermometer, a CO_2 analyzer, an oil pressure gauge, a stop watch, and a combustion efficiency and stack loss calculator. These were discussed in Chapter 4–Testing Instruments and Equipment.

To determine the CFM, the steps are:

1. Obtain the BTU/ft.^{3} content and specific gravity of the fuel supplied to the unit. In the case of natural or mixed gas, contact the local gas utility. In the case of propane, butane, or propane/butane mixture, contact the L.P. gas supplier. The heat content of a gallon of No. 2 fuel oil is assumed to be 140,000 BTU/gallon.
2. Make sure the input to the unit is correct.

Natural Gas. Time and record the gas supply meter operating time. By using the BTU/ft.^{3} and the meter dial rate, the input can be determined. It can be calculated on a meter rate chart such as shown in Figure 6–18. ***CAUTION: ALL OTHER CASE LOADS EXCEPT PILOT LIGHTS MUST BE SHUT OFF.***

By determining the time of one revolution of the meter test dial, and comparing to the time column in the chart, the cubic feet per hour of gas into the unit can be determined. For example, if the test dial of the gas meter takes 42 seconds per revolution and the meter has a ¼ ft.^{3} test dial, the input is 21 ft.^{3}/hr. For a ½ ft.^{3} test dial, the input is 45 ft.^{3}/hr. For a 1 ft.^{3} test dial, the input is 90 ft.^{3}/hr. and so on.

Using the BTU/ft.^{3} of the gas and the ft.^{3}/hr. of gas burned, the BTUH input is determined:

$$\text{Input BTUH} = \text{BTU/ft.}^3 \times \text{ft.}^3\text{/hr.}$$

If, for example, the natural gas has 1,050 BTU/ft.^{3} and 45 ft.^{3} is used each hour, the input is 47,250 BTUH. If the design input of the unit is 48,000 BTUH, the input is considered correct. The input can be within +0% and –10% of design. If the design input of the unit is beyond the recommended input range of the calculated load, the input must be corrected to as close to the design input as possible.

LP Gas. The input of an LP gas heating unit is set by the manifold pressure. The manifold pressure must be 11" WC +0% and –10%. To determine the actual input, the main burner orifice sizes are measured by means of a number drill set.

Figure 6–19 shows the input using propane, butane, or propane/butane mixture for orifice hole sizes from No. 18 drill down to 0.008" drill. These are all based on a manifold pressure of 11" WC.

Oil-fired Heating Unit. The input to an oil-fired unit will depend upon the capacity of the nozzle(s) and the operating pressure. The nozzle size is stamped on one of the wrench flats of the nozzle in portions of a gallon per hour, e.g., 75 GPH, 0.85 GPH, 1.00 GPH, and so on. Physical removal of the oil burner firing head is necessary to determine the nozzle size.

The capacity of a standard oil burner nozzle is based on 100 psig operating pressure. Therefore, when running an efficiency test on an oil-fired unit, the operating pressure on the nozzle must be 100 psig.

The BTUH input of the unit is based on 140,000 BTU/gal. of No. 2 fuel oil. If the nozzle used is, for example, a 0.85 GPH, the BTUH input will be 140,000 BTU per gallon at 100 psig nozzle pressure × 0.85 GPH or 119,000 BTUH.

CO_2 Test

After making sure the input to the heating unit is within the correct range and the unit is operated a sufficient period of time for the air temperature thermometers to stabilize their readings, a CO_2 test must be conducted on the unit.

Gas Rate, Cubic Feet per Hour

Seconds for one Revolution	SIZE OF TEST DIAL 1/4 cu. ft.	1/2 cu. ft.	1 cu. ft.	2 cu. ft.	5 cu. ft.	Seconds for one Revolution	SIZE OF TEST DIAL 1/4 cu. ft.	1/2 cu. ft.	1 cu. ft.	2 cu. ft.	5 cu. ft.
10	90	180	360	720	1800	50	18	36	72	144	360
11	82	164	327	655	1636	51	••	••	••	141	355
12	75	150	300	600	1500	52	••	••	69	138	346
13	69	138	277	555	1385	53	17	34	••	136	340
14	64	129	257	514	1286	54	••	••	67	133	333
15	60	120	240	480	1200	55	••	••	••	131	327
16	56	113	225	450	1125	56	16	32	64	129	321
17	53	106	212	424	1059	57	••	••	••	126	316
18	50	100	200	400	1000	58	••	31	62	124	310
19	47	95	189	379	947	59	••	••	••	122	305
20	45	90	180	360	900	60	15	30	60	120	300
21	43	86	171	343	857	62	••	••	••	116	290
22	41	82	164	327	818	64	••	••	••	112	281
23	39	78	157	313	783	66	••	••	••	109	273
24	37	75	150	300	750	68	••	••	••	106	265
25	36	72	144	288	720	70	••	••	••	103	257
26	34	69	138	277	692	72	12	25	50	100	250
27	33	67	133	267	667	74	••	••	••	97	243
28	32	64	129	257	643	76	••	••	••	95	237
29	31	62	124	248	621	78	••	••	••	92	231
30	30	60	120	240	600	80	••	••	••	90	225
31	••	••	116	232	581	82	••	••	••	88	220
32	28	56	113	225	563	84	••	••	••	86	214
33	••	••	109	218	545	86	••	••	••	84	209
34	26	53	106	212	529	88	••	••	••	82	205
35	••	••	103	206	514	90	10	20	40	80	200
36	25	50	100	200	500	92	••	••	••	78	196
37	••	••	97	195	486	94	••	••	••	••	192
38	23	47	95	189	474	96	••	••	••	75	188
39	••	••	92	185	462	98	••	••	••	••	184
40	22	45	90	180	450	100	••	••	••	72	180
41	••	••	••	176	439	102	••	••	••	••	178
42	21	43	86	172	429	104	9	17	35	69	173
43	••	••	••	167	419	106	••	••	••	••	170
44	••	41	82	164	409	108	••	••	••	67	167
45	20	40	80	160	400	110	••	••	••	••	164
46	••	••	78	157	391	112	••	••	••	64	161
47	19	38	76	153	383	116	••	••	••	62	155
48	••	••	75	150	375	120	7	15	30	60	150
49	••	••	••	147	367						

Figure 6–18 Test dial gas rate table (Courtesy Addison Products Co.)

Gas-fired Units. If the unit has an open diverter, the CO_2 sampling tube must be extended far enough into the heat exchanger outlet to eliminate the possibility of outside air mixing with the flue gas under test. Figure 6–20 shows the use of a CO_2 analyzer on a gas-fired unit. The metal probe, in this case, is a piece of ¼" copper tube 14" long in order to reach beyond the flue restrictors into the upper portion of the heat exchanger section.

Oil-fired Units. On oil-fired units, the flue sample must be taken at the heat exchanger outlet ahead of the barometric draft control.

LP Gases	*Propane*	*Butane*
BTU per Cubic Foot =	2,500	3,175
Specfic Gravity =	1.53	2.00
Pressure at Orifice, Inches Water Column =	11	11
Orifice Coefficient =	0.9	0.9

Drill Size (Decimal or DMS)	Gas Input, BTU Per Hour For: Propane	Butane or Butane-Propane Mixture
0.008	500	554
0.009	641	709
0.010	791	875
0.011	951	1,053
0.012	1,130	1,250
80	1,430	1,590
79	1,655	1,830
78	2,015	2,230
77	2,545	2,815
76	3,140	3,480
75	3,465	3,840
74	3,985	4,410
73	4,525	5,010
72	4,920	5,450
71	5,320	5,900
70	6,180	6,830
69	6,710	7,430
68	7,560	8,370
67	8,040	8,910
66	8,550	9,470
65	9,630	10,670
64	10,200	11,300
63	10,800	11,900
62	11,360	12,530
61	11,930	13,280
60	12,570	13,840
59	13,220	14,630
58	13,840	15,300
57	14,550	16,090
56	16,990	18,790
55	21,200	23,510
54	23,850	26,300
53	27,790	30,830
52	31,730	35,100
51	35,330	39,400
50	38,500	42,800
49	41,850	45,350
48	45,450	50,300
47	48,400	53,550
46	51,500	57,000
45	52,900	58,500

Figure 6–19 LP gas orifice size table (Courtesy American Gas Association)

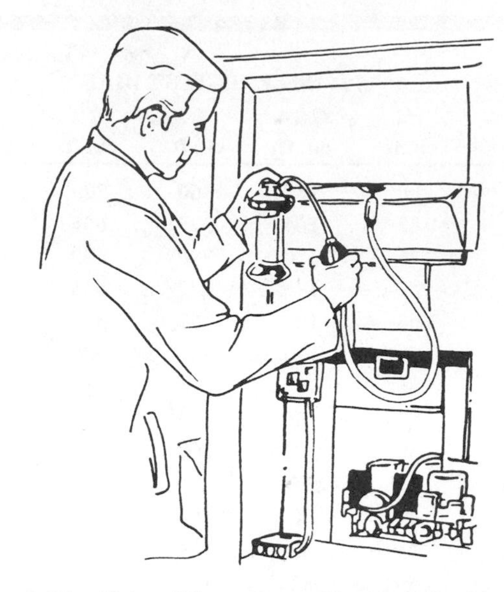

Figure 6–20 Using CO_2 analyzer (Courtesy Bacharach Inc.)

Stack Temperature

The third item needed for the efficiency test is the flue gas temperature in the vent outlet of the heat exchanger. These are taken in the same place as the CO_2 test.

Efficiency Determination

Using a combination efficiency and stack loss calculator such as shown in Figure 6–21, the CO_2 reading and flue gas temperature, within the nearest 50°F increment on the chart, the operating efficiency of the fossil fuel heating unit is determined. This efficiency is added to the basic CFM formula:

$$CFM = \frac{\text{BTUH output (BTUH input} \times \text{efficiency)}}{\Delta T°F \times 1.08}$$

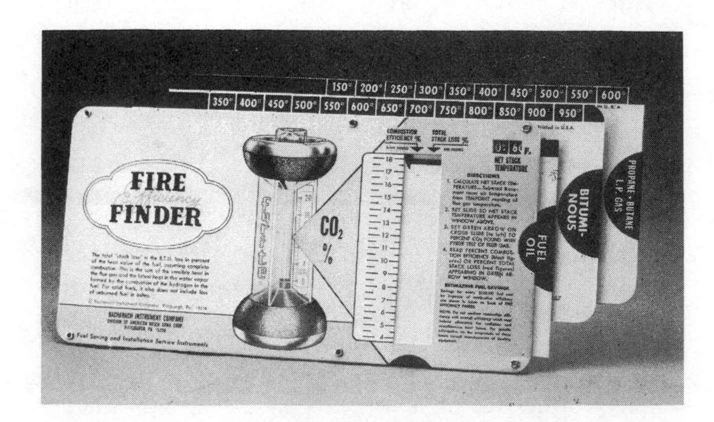

Figure 6–21 Combustion efficiency and stack loss calculator (Courtesy Bacharach Inc.)

where:

CFM = the total quantity of air through the heating unit.

BTUH output = the BTUH input to the unit multiplied by the % efficiency. If, for example, the unit is 80% efficient, the output is only 0.8 times the input.

$\Delta T°F$ = the difference in temperature between the supply and return air through the unit—the temperature rise. (On gas-fired units, the AGA recommendation is 80°F.)

6–8 HEAT PUMP GROSS CAPACITY

After the CFM through the unit has been established, the heat pump is operated with the auxiliary heat off. A new temperature rise will be established with the heat output of the heat pump system. When the supply and return air temperature readings reach stability, the difference in the readings is the new temperature rise. Using the standard air formula and solving for the gross BTUH output, the formula is:

$$\text{Gross BTUH output} = \text{CFM} \times \Delta T°F \times 1.08$$

where:

Gross BTUH output = total amount of heat ejected into the conditioned area air supply by the heat pump system.

CFM = amount of air through the unit established by the unit efficiency test.

$\Delta T°F$ = temperature rise of the air through the inside condenser coil.

1.08 = standard air conversion factor.

After determining the gross BTUH output of the heat pump system, this output must be compared to the manufacturer's Application Ratings—Heating found in the specification literature covering the particular unit.

The output of a heat pump depends upon the outdoor ambient DB temperature. This has to be taken into consideration in the comparison.

Figure 6–22 shows the Application Rating—Heating table published by one manufacturer. The outdoor range of temperatures is from −10°F to +70°F. If the gross heating capacity of the heat pump system is found to be within ±10% of the published rating, the unit is considered to be operating correctly. This is within the normal accuracy range of field test instruments. The use of electronic instruments can reduce the range tolerance by more accurate temperature and pressure readings.

Capacity Above Rated Output of the System

If the calculated output of the system is more than the +10% of the manufacturer's rating, the test should be repeated with close observance of instrument results.

Capacity Below Rated Output of the System

If the output of the system is below the allowable range of the rated output of the system, further testing is required to determine the reason. The two major factors that affect the system performance are the CFM of air over the inside coil and the refrigerant charge in the system.

The CFM of air through the inside coil was determined from the input efficiency test. If the CFM found during the test is between 360 and 400 CFM per each 12,000 BTUH of the unit cooling capacity at 95°F, this eliminates the possibility of air quantity affecting the system performance. If the CFM quantity is above or below the 360 to 420 CFM range, the indoor blower speed should be adjusted accordingly.

After the CFM through the inside coil is set within the correct range, the refrigerant charge of the unit is checked. If the refrigerant charge is low enough to not have any refrigerant in liquid form in the system, the remaining refrigerant vapor must be reclaimed, leaks found and repaired, and the system evacuated and recharged with the correct charge. See Performance Check—Cooling Mode.

	OUTDOOR AMBIENT (°F DB)																	
	-10°		0°		10°		20°		30°		40°		50°		60°		70°	
MODEL	BTUH	COP	BTUH	COP	BTUH	COP	BTUH	COP	BTUH	COP	BTUH	COP	BTUH	COP	BTUH	COP	BTUH	COP
QH824/QHA24	7200	1.27	9100	1.46	11200	1.68	13500	1.92	17000	2.2	21800	2.44	25900	2.73	29200	2.82	30900	2.98
QH830/QHA30	9700	1.32	11700	1.53	15600	1.82	19400	2.16	23500	2.42	29000	2.73	33500	2.98	36400	3.19	38000	3.29
QH836/QHA36	15200	1.38	17700	1.66	20400	1.93	24000	2.18	28500	2.47	34000	2.80	39000	2.99	42500	3.15	45000	3.25
QH842/QHA42	17000	1.32	19500	1.59	23000	1.86	27000	2.13	32000	2.45	38000	2.90	44500	3.03	49500	3.18	52000	3.34
QH848/QHA48	17500	1.34	21000	1.63	25000	1.86	29500	2.14	35000	2.42	41500	2.58	48000	2.97	54000	3.18	58000	3.28
QH860/QHA60	20000	1.22	25000	1.43	31000	1.65	38000	2.0	48000	2.3	58500	2.67	69500	2.98	77000	3.19	81.500	3.29

(1) Above at Standard rating airflow with 70° entering air temperature.

Figure 6–22 Unit application ratings—Heating (Courtesy Addison Products Co.)

If the refrigerant charge is sufficient to have some heating operation but not sufficient for full operating capacity, the operating pressures can be measured and compared to the manufacturer's service information on the particular unit.

Figure 6–23 shows the heating performance table for the QH836–1E unit that was selected for the Omaha, NB application. The first column is the inside coil inlet air DB temperature in the ranges of 60°F, 70°F, and 80°F. Column No. 2 is the outside coil inlet air DB temperatures from +70°F down to −10°F for the 80°F and 70°F inside coil air temperature range. The maximum recommended outside coil operating temperature is 60°F when the inside temperature is 60°F or below.

Below 60°F inside coil temperature, the compressor discharge pressure is too low to produce sufficient vapor pressure at low enough superheat to protect the compressor and raise the temperature in the conditioned area. The temperature should be brought up to at least 65°F via the use of the auxiliary heat source. The heat pump will then produce sufficient heat to have an effect on the conditioned area.

Column No. 3 is the range of air temperature rise through the inside coil at various indoor and outdoor temperatures. Column No. 4 is the expected amperage draw of the condenser fan motor and motor-compressor assembly at the various indoor and outdoor temperatures. Column No. 5 is the low side (suction pressure) range for the various indoor and outdoor temperatures. Column No. 6 is the high side (compressor discharge pressure) range for the various indoor and outdoor temperatures.

Using this table, a comparison can be made to determine if the QH836 unit is operating correctly. By this method, the refrigerant charge is checked. If the operating pressures are low, with the correct CFM of air through the inside coil, an assumption can be made that the system is low on refrigerant charge.

Add refrigerant in ¼# increments, allowing time for the system to reach a thermal balance before taking new pressure readings. When the low side and high side pressures have been raised to the low point of the pressure range, the charging process should end. The final check and adjustment of charge should be done in the cooling mode when outdoor temperatures permit.

Because the operating characteristics of heat pumps vary widely with different manufacturers as well as different style units, standard operating pressures cannot be used. The range of pressures would be too great to be of any value to any particular unit. Therefore, if the manufacturer's service information is not available, the evacuation and recharge method should be used to obtain the correct refrigerant charge.

INSIDE COIL INLET AIR DRY BULB °F	OUTSIDE COIL INLET AIR DRY BULB °F	INSIDE COIL INLET-OUTLET AIR TEMP. DIFFERENTIAL °F		TOTAL AMPS OF UNIT AT 230V INPUT* SINGLE PHASE		LO SIDE PRESSURE PSIG		HIGH SIDE PRESSURE PSIG	
		MIN	MAX	MIN	MAX	MIN	MAX	MIN	MAX
80	70	29	33	16.3	20.3	88	92	312	332
	60	27	31	15.5	19.5	75	79	295	315
	50	25	29	15.0	19.0	63	67	280	300
	40	21	25	13.6	17.6	53	57	267	287
	30	17	21	12.2	16.2	44	48	255	275
	20	15	19	11.3	15.3	35	39	243	263
	10	12	16	10.8	14.8	27	31	232	252
	0	10	14	10.4	14.4	18	22	221	241
	-10	8	12	10.3	14.3	10	14	210	230
70	70	30	34	15.8	19.8	83	87	286	306
	60	28	32	15.4	19.4	70	74	270	290
	50	26	30	14.5	18.5	59	63	256	276
	40	22	26	13.1	17.1	50	54	244	264
	30	18	22	11.8	15.8	41	45	233	253
	20	16	20	10.9	14.9	33	37	222	242
	10	13	17	10.4	14.4	25	29	212	232
	0	11	15	10.0	14.0	17	21	202	222
	-10	9	13	9.9	13.9	9	13	192	212

*Outdoor section only

NOTE: On heating connect service gauges to the connecting tubing suction line fitting for the hi-side pressure and to the compressor suction line gauge port for low-side pressure readings.

NOTE: To correct for voltage at the unit terminals, multiply the tabulated current by a factor from the table below:
208 volts - 1.10
220 volts - 1.05
230 volts - 1.0
240 volts - 0.95

Figure 6–23 Unit operating characteristics—Heating (Courtesy Addison Products Co.)

REVIEW QUESTIONS

1. When a unit starts, the voltage at the compressor contactor should not drop more than _____ volts.
2. Which operating mode should be used to perform final checkout tests on the system?
3. With the unit in the cooling mode, when the indoor temperature drops to 75°F, the temperature of the air off the inside coil should be 55°F. True or False?
4. When measuring the supply and return air temperatures, temperature readings at the supply and return air grills are accurate enough. True or False?
5. Allowing standard pressure drop in the 100' long suction line, what is the operating superheat of the evaporator coil with R-22 refrigerant in the unit and the unit in the cooling mode? The suction pressure at the outdoor unit is 68 psig and the temperature of the vapor line at the coil is 50°F.
6. The compressor discharge pressure is 234 psig and the liquid off the condenser coil is 105°F. What is the amount of subcooling of the liquid R-22 refrigerant off the condenser coil?
7. The system in Problem 6 is a standard system. Is the refrigerant charge:
   a. Correct
   b. Undercharged
   c. Overcharged
8. Name the three methods used to obtain the correct refrigerant charge for a capillary tube system.
9. Which of the refrigerant charging methods in Problem 8 is the most accurate?
10. The heat pump is operating in the cooling mode. The air entering the outdoor coil is 98°F. The average temperature of the air leaving the coil is 110°F. The CFM rating of the outdoor section is 2,670 CFM. The unit is drawing 14 amperes at 232 volts. What is the cooling capacity of the system?
11. The unit in Problem 10 has a factory rating of 30,000 BTUH at 95°F. Is the unit operating in an acceptable capacity range?
12. The velocity of the air through an 8" × 20" duct is 1,050 ft./min. What is the CFM through the duct?
13. An electric air handler drawing 62.5 amperes with an applied voltage of 240 volts produces an air temperature rise through the unit of 28°F. What is the CFM of air through the unit?
14. In the heating mode, the 4 HP unit in Problem 13 produces a temperature rise through the inside coil of 21°F. What is the gross capacity of the unit?
15. With a 45°F outdoor ambient temperature and the gross capacity determined in Problem 14, is there any problem with the unit?
16. With a 17°F outdoor ambient temperature and the gross capacity determined in Problem 14, is there any problem with the unit?
17. A fossil fuel unit is operating on 1,050 BTU/cu. ft. natural gas of 0.07 specific gravity. The average time per revolution of the ½ cu. ft. test dial on the gas meter is 19 seconds. The unit is up to standard factory efficiency rating of 80%. The temperature rise of the air through the unit is 72°F. What is the CFM of air through the unit?
18. An oil-fired unit using a 0.80 GPH nozzle at 100 psig nozzle pressure is operating at 75% efficiency. Producing a 72°F temperature rise in the air through the unit, what is the CFM of the air through the unit?

7 Troubleshooting—Refrigeration

7–1 GENERAL

The majority of the problems that are encountered in air-to-air heat pump systems are the same as in refrigeration and air conditioning systems. Because the air-to-air heat pump is capable of operating in both cooling and heating modes and also encounters frost problems, both the electrical and refrigeration systems have been further complicated to handle these functions. In this text, the major concentration will be on those problems and solutions that apply to air-to-air heat pump systems.

In Section 1, Chapter 2, Testing Instruments and Equipment, the various test equipment that applies to heat pump systems was discussed. Many of the items listed have common usage when servicing heating and air conditioning systems as well as heat pumps. When used in conjunction with heat pumps, the basic heating and air conditioning test equipment will apply. References to such items will be made as various problems and solutions are discussed.

To properly diagnose operational problems in heat pump systems, four pressure readings and eight or more temperature readings are required.

Pressure Readings

The four pressure readings that are required are:

1. Liquid line pressure
2. Vapor line pressure
3. Compressor suction pressure
4. Compressor discharge pressure

Liquid Line Pressure. The pressure in the liquid line of an air-to-air heating and cooling heat pump system is the liquid pressure in the line located between the pressure-reducing/check-valve combination (Trombone) on each of the heat exchanger coils. When the check valves in the system are operating properly, the liquid line pressure will be within 2 to 5 psig of the compressor discharge pressure. This is the normal flow resistance through the condenser coil in either of the operating modes.

When the check valves malfunction, the results will be reflected in the line pressure. If the check valve that is supposed to open in the particular operating mode sticks closed, the liquid line pressure will be measured between two pressure-reducing devices. The check valve did not open and allow the refrigerant to bypass the pressure-reducing device. The liquid line pressure will be considerably below the condensing pressure.

If the check valve that is supposed to close in the particular operating mode does not close, the system will not have a pressure-reducing device in the flowpath. This will cause flooding of the evaporator.

Vapor Line Pressure. The pressure in the vapor line will depend upon the operating mode of the unit. In the cooling mode, the vapor line pressure will be within 5 psig of the suction pressure. In the heating mode, the vapor line pressure will be within 5 psig of the compressor discharge pressure. Because the vapor line is subject to high compressor pressures in the heating mode, a 0 to 500 psig standard gauge is used to measure this pressure.

Compressor Suction Pressure. The correct place to measure the compressor suction pressure is in the "suction line" between the reversing valve and the compressor suction vapor connection. This is the only true suction line in the unit. Refrigerant vapor travels in one direction only—to the compressor—regardless of the operating mode.

A compound gauge is used to read the suction pressure.

Compressor Discharge Pressure. The correct place to measure the compressor discharge pressure is in the line between the compressor discharge connection and the reversing valve. This "hot gas line" carries vapor in one direction only—from the compressor to the reversing valve.

The vapor line will carry refrigerant vapor and the liquid line will carry liquid refrigerant in either direction, depending upon the system operating mode.

Temperature Readings

A minimum of eight temperature readings is required to analyze problems in air-to-air heat pump systems.

1. Return air DB temperature. A sling psychrometer is used to determine the DB temperature and % RH in order to determine the total heat content of the return air and the heat load on the evaporator coil. This information is needed to determine the desired ΔT°F of the air through the coil.

2. Return air WB temperature.
3. Air temperature leaving the evaporator coil. This is necessary to determine the actual ΔT°F of the air through the coil.
4. Suction line temepratture at the evaporator outlet connection. This is used in determining the evaporator operating superheat. The location of the thermometer on the system will change with the change in operating mode from cooling to heating and reverse.
5. Liquid line temperature at the outlet of the condenser coil. This temperature reading is used to determine the system subcooling. This test is used primarily in the cooling mode.
6. Temperature of the air entering the condenser. This temperature reading is the average of the temperature readings of the air entering the condenser air openings on the outside section in the cooling mode. In the heating mode, this temperature reading is the temperature of the return air to the inside condenser coil.
7. Three or more thermometers located in the discharge air from the outdoor section. These readings are used to get the average air temperature of the discharge air to measure the temperature rise in the cooling mode or temperature drop in the heating mode.
8. Suction line temperature between the reversing valve and the compressor suction inlet connection. This temperature reading is used to check for possible hot gas bypass in the case of suspected reversing valve problems.

All these pressure and temperature readings are necessary to analyze problems in the heat pump operation. Most problems in the heat pump system can be compared to problems in refrigeration as well as air conditioning systems. Heat pumps are different because they can develop problems in the heating mode, the cooling mode, or both.

Problem Categories

Heat pump problems, however, as in refrigeration or air conditioning systems, can be divided into three categories:

1. Air system problems
2. Refrigeration system problems
3. Electric control problems

7–2 AIR SYSTEM PROBLEMS

If it is assumed that the CFM through the evaporator was set when the unit was installed, air problems will show up as an increase in the ΔT°F across the coil. It is not possible for the heat transfer capacity of the system to increase. The air quantity can only decrease. Therefore, the increase in the ΔT°F across the coil can only be due to a decrease in the CFM of air through the evaporator coil.

In the heating mode, the reduction in CFM will cause a rise in compressor discharge pressure and an increase in operating cost. If the reduction becomes great enough, the unit will cut out on the high head pressure control, bringing in the lockout system and lighting the emergency light on the thermostat. Immediate service is needed to prevent damage to the motor-compressor assembly.

Air Filters

The heat pump system operates year-round, which necessitates more frequent air filter changes. To keep operating costs low and promote better area conditions, the air filters should be changed at least every 60 days.

The best solution to air filter problems is the installation of an electrostatic air cleaner. Electrostatic air cleaners are easily cleaned and their operating air flow resistance is very low. Coil surfaces also remain cleaner.

Blower Motor and Drive

The blower motor and drive should be inspected at least on an annual basis. Motor and blower bearings should be lubricated once per year with not more than 10 drops of an electric motor oil. *DO NOT USE AUTOMOBILE MOTOR OIL.* Additives in automobile motor oil will clog bearing oil feed systems and cause bearing burnout.

Duct Restrictions

In the cooling mode, closing off supply grills in unused rooms will have the effect of increasing the temperature drop across the evaporator coil and could result in ice formation on the coil. In extreme cases, liquid refrigerant return to the compressor could cause damage.

In the heating mode, the reaction is more severe. Only a small reduction in CFM will have a large effect upon the compressor operation. High temperatures and pressures are reached very quickly. The effect is multiplied very rapidly when outdoor temperatures are above 60°F.

7–3 HEAT PUMP REFRIGERATION SYSTEM PROBLEMS

As in refrigeration and air conditioning systems, refrigeration system problems can be divided into two categories:

1. Refrigerant quantity
2. Refrigerant flow rate

Refrigerant Quantity

In a heat pump system, the refrigerant quantity will affect the system operation in both operating modes. In the cooling mode, the reduction in cooling capacity will reduce the temperature drop through the evaporator. In the heating mode, the temperature rise through the condenser will decrease.

In either case, the system has lost refrigerant via a leak. The leak must be found and repaired, and the system evacuated and recharged.

Starting from an evacuated system, the refrigerant is weighed in. The correct charge is calculated as outlined under Section 2, Chapter 6, Startup, Checkout, and Adjustment.

Refrigerant Flow Rate

In Chapter 1, Basic Principles, the refrigerant flow systems of the heat pump were discussed. In this discussion, the fact that different flow circuits are used for each operating mode was emphasized. This is important when attempting to analyze flow problems.

The first question is: does the problem exist in only one of the operating modes or both? If only one, which one? This can help to point out the location of the particular problem.

In the cooling mode, the heat pump system operates the same as an air conditioning system. Refrigerant flow is from the compressor, through the reversing valve, to the outdoor coil, through the check valve around the pressure-reducing device, through the liquid line, through the pressure-reducing device on the indoor evaporator coil, reversing valve, and accumulator, and return to the compressor.

In the heating mode, the operation is reversed. The outdoor coil is the evaporator and the indoor coil is the condenser. Refrigerant flow is from the compressor, through the reversing valve, to the indoor condenser coil, through the check valve around the pressure-reducing device, through the liquid line, through the pressure-reducing device on the outdoor evaporator coil, through the coil, reversing valve, and accumulator, and return to the compressor.

What is the difference in the operation of the two systems? The difference is the use of the pressure-reducing device and evaporator and bypassing the pressure-reducing device for the condenser. The action is reversed by the action of the reversing valve under control of the electrical control system.

7–4 CHECK VALVES

Check valves can cause problems by sticking in a given position, either open or closed. The position the check valve is stuck in will only show up in one operating mode and not the other. This is why it is so important to operate the system in both modes to isolate the problem.

For example, suppose the check valve on the outdoor coil is stuck closed. Which operating mode will encounter the problem? The question really is in which mode is the check valve required to be open and allow the refrigerant to bypass the pressure-reducing device?

The pressure-reducing device has to be bypassed when the companion coil is used as a condenser. In this case, the outdoor coil is the condenser in the cooling mode.

How would the stuck-closed check valve affect the operation of the refrigeration system? The stuck check valve forces the liquid refrigerant to flow through the pressure-reducing device on the condenser coil before entering the liquid line to travel to the pressure-reducing device on the indoor coil. This has the effect of practically doubling the pressure drop of the liquid refrigerant before it enters the evaporator. Thus:

1. Suction pressure will be extremely low.
2. Evaporator superheat will be high.
3. Liquid refrigerant subcooling will be high. Liquid refrigerant will build up in the condenser.
4. Compressor discharge pressure will be low. The amount of heat picked up in the evaporator will be low. The system will reach a very low thermal balance and the condensing temperature (discharge pressure) will be low.

The first reaction to these problem symptoms is that the system is low on refrigerant. *Before any refrigerant is added,* operate the system in the opposite mode. Good performance in the opposite mode means the system is correct except for a device that was supposed to open in one mode but not in the other and failed to function properly.

The indication of this problem is twofold.

1. The liquid refrigerant subcooling is high.
2. The pressure in the liquid line is considerably lower than 5 psig below the compressor discharge pressure measured at the hot gas line between the compressor and reversing valve.

Suppose the check valve on the indoor coil is stuck open. In which operating mode will the problem surface? The cooling mode. In this mode, the check valve is supposed to close, forcing the refrigerant through the pressure-reducing device. With the check valve open, both check valves are now open and no pressure drop occurs between the condenser and the evaporator. The refrigerant boiling point is not reduced and there is little cooling. Suction pressure is high, liquid subcooling is zero, and liquid refrigerant is flooding back to the reversing valve, accumulator, and compressor. However, the system will operate in the heating mode because the check valve has to be open in the heating mode.

The problems are the opposite when the check valve on the outdoor coil gives problems. If the check valve is stuck closed, the problem will take place in the cooling mode. Suction pressure and discharge pressure will be low. The liquid line pressure will be considerably lower than the compressor discharge pressure. *Do not add refrigerant until the system has been operated and checked in the opposite mode.* Even though outside temperatures are above normal heating operating requirements, the system will still reflect the operating condition of the check valve. However, the system should not be operated too long in the heating mode as the compressor discharge pressure and amperage draw will rise very rapidly.

If the check valve is stuck open, the problem will take place in the heating mode. There will be no pressure and boiling point drop in the outside evaporator. There will be very little heat off the inside condenser and the accumulator, reversing valve, and compressor will be flooded with liquid refrigerant.

Check Valve Repair

Before replacing the check valve, which means reclaiming the system refrigerant and charging the refrigerant after the repair, try using a magnet to free the valve. Placing a magnet against the inlet end of a stuck shut check valve and moving the magnet to the opposite end will generally force the valve flapper or ball to leave the seat and free the system. If the valve is stuck open, start the magnet at the outlet end of the valve and move it to the inlet end.

If this procedure is not successful, the valve must be replaced. To reduce the possibility of future problems, always replace with a ball-type valve. Before installing the valve, make sure the ball is loose in the guide cage. Shake the valve. If it rattles, use it. If it does not rattle, return it to your supplier for another. To save time and travel, shake it when you obtain it to make sure it is good.

When installing the valve, be careful not to overheat the valve body and warp the ball cage. Use thermomastic on the body of the valve when making the solder connections.

7–5 REVERSING VALVES

As explained in Chapter 2, Component Parts, Reversing Valves, the reversing valve is a pilot-valve-operated slide action main valve. The entire operation depends upon a pressure difference between the compressor suction and discharge pressures of 75 psig or higher.

To obtain proper operation of the reversing valve, the system must have a complete refrigerant charge and operation in either mode to have at least the minimum pressure difference of 75 psig developed between the suction and discharge pressures. All testing must be done with the system operating.

Problems in reversing valves are electrical and mechanical.

Electrical Problems

Electrical problems are confined to the electrical solenoid coil. When voltage is applied to the coil, a click should be heard from the pilot valve plunger changing position. If no click is heard, check across the coil lead connections to make sure voltage has been applied to the coil circuit. If power is supplied to the solenoid coil, the coil pull test should be done. Remove the nut and spacer washer holding the coil on the plunger housing. Pull the coil partially off the housing. A definite drag should be felt from the magnetic pull of the valve coil. Do not pull the coil completely off the housing, as it will quickly overheat.

If a drag is felt, the coil is active and the pilot valve is stuck. Replace the coil and tighten it into place with the spacer washer and nut. Cycle the power to the coil several times to try to free the pilot valve plunger. If this is not successful, reclaim the refrigerant charge, replace the valve, evacuate, and recharge the system.

When the coil is removed and a magnetic pull is not felt, the coil is dead. Check the coil for continuity using an ohmmeter. Check the coil connections and leads for opens. Coil replacement is required for an open coil. Leads can be replaced.

Mechanical Problems

Mechanical problems in the valve consist of failure of either the pilot valve or main slide valve to shift. When power is applied to the coil and a pilot valve click is heard, *if the pressure difference between suction and discharge pressures is more than 75 psig,* the main slide valve should change position. If the valve does not shift after several on/off cycles of the power to the coil, the valve should be replaced.

However, the slide valve may be in the midposition with hot gas bypassing to the suction side directly through the valve. This can happen if the valve is energized immediately upon the start of unit operation. This means that the valve is free to slide whenever the pressure difference is high enough. The slide action could be so slow that when it reaches the midposition, the pressure difference loss due to the bypass of the vapor can cause the slide to remain in the midposition.

To cause the slide to move off dead center, the system pressure has to be raised. The easiest way is to put the system in the cooling mode. Block off the condenser air inlets with plastic on the inlet side of the outdoor coil face. Allow the unit to operate until the head pressure reaches the equivalent of 130°F condensing temperature. Keep the unit operating and cycle the valve several times. This will usually cause the valve slide to move and put the valve into operation. If this procedure is not successful, the valve will have to be replaced.

Replacing Reversing Valves

When replacing reversing valves, there are several rules to follow:

1. Always replace the valve with one of comparable size. A smaller valve would put too much vapor flow resistance into the system. A larger valve could cause sluggish valve operation and future problems.
2. The valve must always be mounted in a horizontal position with the pilot valve above the main body of the valve. This is to keep oil from gathering in the pilot valve and the connecting lines. Oil logging in these parts can produce very sluggish valve operation and possible valve failure.
3. The valve body must never be subjected to temperatures above 250°F. The use of thermomastic material on the valve body when soldering the tube connections is an absolute must. The surest way to require a second valve change is to overheat the first replacement valve.

7–6 STANDARD REFRIGERATION SYSTEM PROBLEMS

The discussion to this point has primarily been on problems that are restricted to heat pumps that will occur in either the heating or cooling mode or both. The system could have any of the standard refrigeration system problems that may occur in either the heating or cooling modes or both.

The eleven probable causes of problems with the refrigeration system are shown in Figures 7–1 A and B. The measurement of the temperatures and pressures that are used in the symptoms versus probable cause tables are determined from the four pressures and ten or more temperature readings discussed in Chapter 6 under the topic Performance Check.

CAPILLARY AND RESTRICTOR SYSTEMS

PROBABLE CAUSE	LOW SIDE (SUCTION) PRESSURE psig	D.X. COIL SUPERHEAT °F	HIGH SIDE (HOT GAS) PRESSURE psig	CONDENSER LIQUID SUBCOOLING °F	COND. UNIT AMPERAGE DRAW AMPS.
1. INSUFFICIENT OR UNBALANCED LOAD	LOW	LOW	LOW	NORMAL	LOW
2. EXCESSIVE LOAD	HIGH	HIGH	HIGH	NORMAL	HIGH
3. LOW AMBIENT (COND. ENTERING AIR°F)	LOW	HIGH	LOW	NORMAL	LOW
4. HIGH AMBIENT (COND. ENTERING AIR °F)	HIGH	HIGH	HIGH	NORMAL	HIGH
5. REFRIGERANT UNDERCHARGE	LOW	HIGH	LOW	LOW	LOW
6. REFRIGERANT OVERCHARGE	HIGH	LOW	HIGH	HIGH	HIGH
7. LIQUID LINE RESTRICTION	LOW	HIGH	HIGH	HIGH	LOW
8. PLUGGED CAPILLARY TUBE	LOW	HIGH	HIGH	HIGH	LOW
9. SUCTION LINE RESTRICTION	LOW	LOW	LOW	NORMAL	LOW
10. HOT GAS LINE RESTRICTION	HIGH	LOW	HIGH	NORMAL	HIGH
11. INEFFICIENT COMPRESSOR	HIGH	LOW	LOW	LOW	LOW

Figure 7–1A Symptoms versus probable causes for capillary and restrictor systems

T. X. VALVE SYSTEMS

PROBABLE CAUSE	LOW SIDE (SUCTION) PRESSURE psig	D.X. COIL SUPERHEAT °F	HIGH SIDE (HOT GAS) PRESSURE psig	CONDENSER LIQUID SUBCOOLING °F	COND. UNIT AMPERAGE DRAW AMPS.
1. INSUFFICIENT OR UNBALANCED LOAD	LOW	LOW	LOW	NORMAL	LOW
2. EXCESSIVE LOAD	HIGH	NORMAL	HIGH	NORMAL	HIGH
3. LOW AMBIENT (COND. ENTERING AIR°F)	LOW	HIGH	LOW	NORMAL	LOW
4. HIGH AMBIENT (COND. ENTERING AIR °F)	HIGH	NORMAL	HIGH	NORMAL	HIGH
5. REFRIGERANT UNDERCHARGE	LOW	HIGH	LOW	LOW	LOW
6. REFRIGERANT OVERCHARGE	HIGH	NORMAL	HIGH	HIGH	HIGH
7. LIQUID LINE RESTRICTION	LOW	HIGH	HIGH	HIGH	LOW
8. T.X. VALVE POWER ELEMENT DEAD	LOW	HIGH	LOW	HIGH	LOW
9. SUCTION LINE RESTRICTION	LOW	HIGH	LOW	HIGH	LOW
10. HOT GAS LINE RESTRICTION	HIGH	HIGH	HIGH	NORMAL	HIGH
11. INEFFICIENT COMPRESSOR	HIGH	HIGH	LOW	LOW	LOW

Figure 7–1B Symptoms versus probable causes for T.X. valve systems

Keep in mind that regardless of what mode the heat pump system is operating in, the eleven symptoms in Figures 7-1 A and B are the indicators that are read for troubleshooting. To attempt to give continuity to the problem analysis to promote the saving of time in the service and repair process, we start with the most probable cause "Insufficient Air to the Evaporator" and end with the least encountered "Inefficient Compressor."

In the horizontal line of each probable cause is the resulting reaction to each of the symptoms. These reactions are listed as low, high, or normal. The use of the word *normal* is not to be used as a "guesstimation." The "normal" must actually be determined to be able to classify the symptom. Determining the "normal" for each of these symptoms was discussed earlier in this chapter. This should be reviewed as part of this discussion.

Insufficient Air Through the Evaporator

Cooling Mode. Insufficient air over the inside evaporator coil will produce a larger than normal temperature drop through the coil. The heat transfer ability of the system attempts to remain constant. Therefore, as the CFM of air through the evaporator is reduced, more total heat is removed from each CFM. This increases the ΔT°F of the air through the coil. Air filter restrictions, duct restrictions, and blowers and drives were discussed in Chapter 6.

The reduced load on the evaporator will reduce the suction pressure below normal. The refrigerant vaporization rate in the coil will be reduced. As a result, the point of vaporization travels toward the coil outlet. In extreme cases, it will leave the coil and liquid refrigerant will enter the suction line. Compressor damage could result.

The compressor discharge is lowered. The reduction in heat absorbed in the evaporator reduces the heat load on the condenser. Not as much split is required for heat transfer and the condensing temperature (compressor discharge pressure) drops below normal.

The amount of liquid refrigerant in the condenser will stay about the same so any change in the amount of subcooling will be small. The compressor amperage draw will be below normal due to the reduced vapor pumping requirements.

Heating Mode. The CFM of air through the evaporator (the outdoor coil) is set by factory design. Only two sources of problems apply here.

1. Dirty coil surfaces. Dirt and debris accumulation reduces the CFM through the coil.
2. Inside fan motor. Excessive amperage draw of the fan motor can cause the motor to cut out on the internal overload. This cuts the fan operation and the unit will cut off on compressor high-pressure lockout.

This is a difficult problem to diagnose because by the time a service technician arrives, the motor has cooled and the overload cutout has reset. Check the motor amperage draw to determine if the motor is overloaded.

Unbalanced Load

An unbalanced load on the evaporator coil will cause the heat absorbing ability of the coil to drop. The ΔT°F of the air through the coil will be lower than normal. The capacity is reduced because some of the circuits in the coil are overloaded while others are lightly loaded. The ΔT°F of the air through the overload circuits will be low due to the small reduction in air temperature as it passes over the coil.

The lightly loaded circuits allow liquid refrigerant to leave the circuit and enter the suction line. This reduces the amount of heat absorption into the refrigerant in the coil as well as reduces the system net capacity by forcing heat absorption in the suction side of the system.

T.X. Valve Systems. In a T.X. valve system, the liquid refrigerant from the lightly loaded circuit(s) passing the feeler bulb of the T.X. valve will cause the valve to close down. This reduces the suction pressure (refrigerant boiling point in the coil) and reduces the system net capacity. The suction pressure will be lower than normal. The compressor discharge pressure will be low due to the reduced heat load on the condenser.

The condenser liquid subcooling will be higher than normal. This is a result of the reduced liquid refrigerant demand by the T.X. valve and a buildup of liquid refrigerant reserve in the condenser.

The compressor vapor load is reduced, which results in a reduction of the compressor amperage draw (LOW).

Capillary Tube or Restrictor Systems. In systems using capillary tubes or restrictors, the unbalanced load will cause a reduction in the amount of heat absorbed in the evaporator coil. The lightly loaded circuits will not evaporate the liquid refrigerant provided and will flood liquid refrigerant into the suction line. This will further reduce the system net capacity.

Low side (suction) pressure will be low due to reduced load on the evaporator coil. Compressor discharge pressure will be low because of the reduced amount of vapor to be handled. The condenser does not require as high a split to extract the necessary amount of heat. The condenser liquid subcooling will remain approximately the same. Compressor amperage draw will be low because the vapor load on the compressor has been reduced.

Excessive Load

An excessive load either of high CFM of air or high total heat content of the air will have the opposite effect of insufficient load. The temperature drop of the air through the evaporator coil will be low. With high CFM through the coil, less heat will be removed from each cubic foot of air. With a higher total heat content in the air, more heat per pound of air means less temperature reduction per pound. Suction pressure, compressor discharge pressure, and compressor amperage draw will be high due to the additional load on the system. The condenser liquid subcooling will be about normal.

Cooling Mode. In the cooling mode, both the CFM of air to the inside evaporator coil as well as the total heat content will affect the operation of the system. It is not reasonable to expect the two factors to reach such a magnitude to cause unit cutoff. The conditions of the air off the coil will not be as desired until the area temperature and so on are reduced to a reasonable coil load. Again, no adverse effect on the system can be expected.

Heating Mode. In the heating mode, the total heat content of the outdoor air into the evaporator can have a decided effect on the system. In the heating mode, the system is designed to operate in temperature conditions when heating is required. This means an outdoor ambient temperature of 65° F or below.

Operating at ambient temperatures of 70° F or higher will quickly overload the system and cause cutoff on the high-pressure lockout system.

Low Ambient Air to the Condenser

Operating a heat pump in the cooling mode at outdoor ambient temperatures below normal operating range produces compressor discharge pressures that are too low for satisfactory operation. This reduction in compressor discharge pressure will reduce the amount of liquid refrigerant supplied to the evaporator coil. This in turn reduces the amount of vapor produced in the coil and reduces the coil heat absorbing ability.

The ΔT°F of the air through the coil will be low due to the coil capacity reduction. The suction pressure will be low. The balance between the ability of the coil to produce vapor and the compressor to pump vapor will produce a lower coil operating (suction) pressure. The compressor discharge pressure will be low because the compressor is not pumping as much vapor.

The refrigerant quantity has not changed, only the flow rate has decreased. The liquid subcooling will be on the high side of the normal range. Compressor amperage draw will be normal. The compressor is not working as hard.

Capillary Tubes and Restrictors. Changes in operating characteristics due to ambient temperature changes are more radical in systems using fixed orifice pressure-reducing devices such as capillary tubes or restrictors than in systems using T.X. valves.

In fixed orifice type systems, the peak net capacity is at design conditions of 95°F. As the outdoor ambient temperatures drop, a gradual reduction in capacity occurs until the outdoor ambient reaches a low of 65°F. From this temperature on down, the reduction in capacity becomes very radical. To operate this type of system below 65°F requires some means of maintaining the proper compressor discharge pressure. Low ambient control can be achieved via cycling condenser fan motor, dampers in the air stream, and so on.

T.X. Valve Systems. Systems using T.X. valves will tolerate a greater reduction in compressor discharge pressure than fixed orifice systems. As the compressor discharge pressure is reduced, reducing the refrigerant flow, the T.X. valve will modulate open to compensate and keep up the refrigerant flow rate. However, when the pressure difference between inlet and outlet across the valve drops under 100 psig, the valve reaches the fully open position. Any further reduction in compressor discharge pressure reduces the system capacity very radically.

Where fixed orifice systems require compressor discharge pressure control below 65°F, a T.X. valve system will usually operate with good net capacity down to 35°F before pressure-maintaining controls are needed.

High Ambient Air to the Condenser

The high limit of condenser entering air temperatures are 115°F in the cooling mode and 70° F in the heating mode. In either mode, the higher condensing temperature required to put heat energy into this high-temperature air results in excessive compressor discharge pressures. The higher the condenser ambient air temperature, the higher the compressor discharge pressure.

The ΔT°F of the air through the evaporator will drop due to loss in compressor pumping efficiency, and increase in flash gas quantity in the coil due to the higher temepratura liquid off the condenser. Suction pressure will rise due to the greater refrigerant flow as well as the rise in compressor discharge pressure.

The amount of rise in suction pressure compared to discharge pressure will be different between systems using T.X. valves and those using capillary tubes or restrictors.

T.X. Valve Systems. As the liquid pressure rises, the T.X. valve will modulate down to maintain coil operating temperatures. The reduction in compressor pumping capacity will tend to counteract this and a rise in suction pressure will result.

Capillary Tube or Restrictor Systems. The flow rate through capillary tubes and restrictors is totally dependent upon the pressure across the device. As a result, a rise in compressor discharge pressure greatly increases the flow rate of the refrigerant into the coil. This produces a higher coil operating (suction) pressure.

The evaporator coil superheat will decrease as the compressor discharge pressure rises. In T.X. valve systems, the valve will maintain the coil superheat within the operating pressure range of the valve. The superheat reduction as compared to compressor discharge rise will be small. In the capillary tube or restrictor system, the change is very radical.

Figure 7–2 shows a table of superheat temperatures at various evaporator air inlet WB temperatures versus condenser air inlet DB temperatures. A direct comparison can be made between 65°F DB and 115°F DB condenser air inlet temperature at the same conditions of evaporator air inlet temperature of 70°F WB.

At 65°F condenser air inlet DB, the superheat is 30°F. At 115°F, the superheat drops to 8°F. This is totally a result of the rise in compressor discharge pressure forcing a higher flow rate of liquid refrigerant into the evaporator coil.

The condensing temperature of the compressor discharge vapor and the temperature of the liquid off the condenser will rise. The difference between the two, the subcooling, will remain within the normal range.

Compressor amperage draw will be higher due to the compressor pumping against higher pressures requiring more power.

Refrigerant Undercharge

Less refrigerant in the system means less liquid refrigerant fed to the coil, less vapor developed (lower suction pressure), less vapor pumped (low compressor discharge pressure), and less work done by the compressor (low amperage draw). The ΔT°F of the air through the evaporator coil will be lower than normal. The evaporator coil superheat will be high.

Due to the lack of liquid refrigerant in the condenser, the amount of subcooling will decrease until no liquid refrigerant will be built up in the condenser. All pressures as well as amperage draw will be low.

Refrigerant Overcharge

An overcharge of refrigerant in a heat pump system will affect the operational results depending upon the type of pressure-reducing device used and the amount of overcharge.

T.X. Valve Systems. In heat pump systems using single-acting T.X. valves (valves that flow in one direction only), the valve will react to the higher head pressure developed from the excessive refrigerant in the condenser. In this situation, the following pressure and temperature reactions will occur:

1. The ΔT°F of the air through the coil will drop due to the drop in the system net capacity.

SUPERHEAT TABLE

CONDENSER AIR INLET TEMP.—°F DB	EVAPORATOR AIR INLET TEMP.—°F WB											
	54	56	58	60	62	64	66	68	70	72	74	76
60	13	17	18	20	24	26	28	30				
65	11	13	17	17	18	22	25	28	30			
70	8	11	12	14	16	18	22	25	28	30		
75		7	10	12	14	16	18	23	26	28	30	
80			6	8	12	14	16	18	23	27	28	30
85				6	8	12	14	17	20	25	27	28
90					6	9	12	15	18	22	25	28
95						7	11	13	16	20	23	27
100							8	11	14	18	20	25
105							6	8	12	15	19	24
110								7	11	14	18	23
115									8	13	16	21

Figure 7–2 Superheat table (Courtesy Addison Products Co.)

2. The low side (suction) pressure will be higher. The additional pressure on the liquid refrigerant entering the T.X. valve will tend to increase the valve capacity.
3. The T.X. valve will maintain the same coil superheat at the higher coil operating temperature until the extra compressor discharge pressure becomes excessive. This would be at about 50% overcharge.
4. Compressor discharge pressure will be higher due to the reduction in heat transfer surface from the additional refrigerant occupying additional tube volume.
5. Condenser liquid subcooling will be higher. More liquid refrigerant in the condenser means more exposure to heat transfer surface to lower the temperature of the liquid.
6. Compressor amperage draw will be high due to the additional work required by the compressor against the higher discharge pressure.

T.X. valve systems using a reverse-flow valve that has a fixed reverse-flow orifice must be treated as a fixed orifice or restrictor system.

Capillary Tube or Restrictor Systems. The charge tolerance in a capillary or restrictor system is ±1 oz. of refrigerant. Any excessive amount of refrigerant, other than what the evaporator will hold in the off portion of the operating cycle, will cause liquid refrigerant runout into the suction line at startup. Because there is no means of stopping the refrigerant flow when the cycle is off, the refrigerant continues to flow into the evaporator until the pressures in the system reach a balance. The correct refrigerant charge will fill the evaporator with no excess.

The correct method of charging the system was covered in Chapter 6.

As would be expected, suction pressure, compressor discharge pressure, condenser liquid subcooling, and compressor amperage draw will be high. The ΔT°F of air though the coil will be low because the system net capacity will be low. Evaporator superheat will be 0°F due to excess liquid refrigerant flow out of the coil.

Liquid Line Restriction

A restriction in the liquid line will reduce the refrigerant flow to the evaporator. As a result of the reduced amount of refrigerant flow to the coil:

1. The ΔT°F of the air through the coil will be low.
2. The suction pressure will be low.
3. The evaporator coil superheat will be high.
4. The compressor discharge pressure will drop even though the condenser contains more liquid. The reduction in pumped vapor by the compressor will lower rather than raise the condensing temperature needed to transfer the reduced amount of heat.
5. The first reaction to the low suction and discharge pressure will be "shortage of refrigerant." Before adding any refrigerant, *always check the liquid subcooling off the condenser*. If it is low, there is a refrigerant shortage. If it is high, look for a flow resistance in the liquid line. The liquid subcooling will be high because the liquid refrigerant will accumulate in the condenser.

The question of where the restriction could be was discussed under "Check Valves" earlier in this chapter. The system should be operated in both modes to determine if a single function restriction exists.

If the restriction occurs in both operating modes, that portion of the liquid line between the inner and outer trombones is the prime suspect. Plugged filter-driers, kinked lines, or an oversupply of solder in solder joints should be investigated.

If quick couple connectors are used in the line installation, an improperly opened fitting diaphragm can cause a piece of the diaphragm to block the refrigerant flow through the fitting.

Plugged Capillary Tubes or Restrictor

The reaction to this problem will be the same as the liquid line restriction except it will be in only one of the operating modes. If it is in the cooling mode, it will be on the inside coil. If it is in the heating mode, it will be on the outside coil.

1. ΔT°F of the air through the coil will be low.
2. Suction pressure will be low because of the reduced refrigerant supply to the coil.
3. Evaporator superheat will be high because of reduced refrigerant supply to the coil.
4. Compressor discharge pressure will be low because there is less vapor to pump.
5. Condenser liquid subcooling will be high. Liquid refrigerant will back up in the condenser.
6. Compressor amperage draw will be low.

To determine which capillary tubes are plugged, shut off the air through the evaporator. Operate the compressor until the evaporator has accumulated a fair amount of frost in the fin area. Part of the coil will be clear of frost because that circuit(s) is not receiving any refrigerant. The capillary tubes or tubes or restrictors are plugged.

In high humidity areas, operate the unit with the coil exposed to see which sections form condensation on the coil fins. The section(s) that remains dry has a plugged capillary(s).

Suction Line Restriction

A suction line restriction reduces the flow of vapor from the evaporator to the compressor. This raises the coil operating pressure and coil operating temperature:

1. The ΔT°F of the air through the coil will be low.
2. The low side (suction) pressure will be extremely low.
3. Evaporator superheat is 0°F. Refrigerant is backed up in the evaporator due to the reduction in the vaporizing rate due to the rise in the coil operating temperature and pressure.
4. Compressor discharge pressure will be low. The heat supply to the compressor is low and the compressor work is reduced.
5. Condenser subcooling is low. The liquid refrigerant has accumulated in the evaporator coil.
6. Compressor amperage draw is low.

To make positive identification of the problem, it is necessary to install a gauge pressure tap on the outlet manifold of the evaporator. This will reveal the operation pressure on both ends of the suction line. The difference in pressure between these two points should not be over 5 psig.

Hot Gas Line Restriction

The compressor must develop a higher discharge pressure to force vapor into the condenser. The entire system will have a lower net capacity at a higher amperage draw:

1. The ΔT°F of the air through the coil will be low.
2. Low side (suction) pressure will be high. The low compressor pumping capacity will result in a higher pressure balance between the evaporator and the compressor.
3. Evaporator coil superheat will be low. The coil cannot evaporator the liquid refrigerant at the desired rate due to the reduced pumping rate of the compressor.
4. If compressor discharge pressure is measured at the liquid line outlet of the condenser, it will be low. If it is measured at the compressor hot gas outlet, it will be high. If may be necessary to install an extra gauge pressure tap in the hot gas line at the compressor discharge outlet to measure the pressure at both places. The difference in the pressure readings should not be more than 5 psig.
5. Regardless of where the compressor discharge pressure is measured, the compressor amperage draw will be high. If the liquid line pressure tap shows low head pressure but the compressor amperage draw is high, suspect a hot gas line restriction.

Inefficient Compressor

In those cases where the compressor reaches the worn stage in which pumping capacity is reduced, the total system capacity reduces:

1. The ΔT°F of the air through the coil is low.
2. The low side (suction) pressure is high.
3. The evaporator coil superheat is low.
4. The compressor discharge pressure is low.
5. The condenser liquid subcooling is low. The extra refrigerant is in the evaporator.
6. The compressor amperage draw is low because the compressor is doing less work. *Before attempting to change the compressor, check its performance by shutting off the air over the evaporator coil.* The coil should frost over. If it does not, change the compressor.

If the system has service valves, a pump down can be done on the compressor. The compressor should be able to develop a 15" vacuum or lower.

REVIEW QUESTIONS

1. In the heating mode, the pressure in the vapor line should be within 5 psig of the:
   a. Compressor suction pressure
   b. Compressor discharge pressure
   c. Liquid refrigerant pressure
2. In the cooling mode, if the suction pressure is 57 psig, the pressure in the vapor line should not be over _____ psig.
3. Name the three categories of heat pump problems.
4. A heat pump operating in the heating mode cuts off on the safety lockout when the outdoor temperature exceeds 48°F. What is the most likely cause of the problem?
5. The system operates satisfactorily in the cooling mode but will not produce sufficient capacity in the heating mode. Pressure readings in the heating mode show both the suction and discharge pressures are low and subcooling is high. What is the most likely cause of the problem?
6. In order for a reversing valve to operate properly, the minimum pressure differential across the valve is:
   a. 75 psig
   b. 100 psig
   c. 125 psig
   d. 150 psig
7. When installing a reversing valve, in the soldering process, the maximum temperature the valve body will tolerate is:
   a. 200°F
   b. 250°F
   c. 300°F
   d. 350°F
8. The minimum outdoor ambient temperature at which a capillary tube heat pump system, in the cooling mode, will produce satisfactory operating is:
   a. 0°F
   b. 17°F
   c. 35°F
   d. 45°F
   e. 65°F
9. The minimum outdoor ambient temperature at which a T.X. valve heat pump system will produce satisfactory operation is:
   a. 0°F
   b. 17°F
   c. 35°F
   d. 45°F
   e. 5°F
10. With a capillary tube system operating in the cooling mode, as the outdoor ambient temperature rises, the superheat of the inside coil drops. True or False?
11. With a T.X. valve system operating in the cooling mode, as the outdoor ambient temperature rises, the superheat of the inside coil drops. True or False?
12. Starting with the correct charge of refrigerant in the system, what effect would the following changes have on the amount of subcooling?
    a. Loss of refrigerant
    b. Addition of refrigerant
    c. Check valve sticking closed
    d. Check valve sticking open
    e. Plugged screen
    f. Suction line restriction
    g. Broken valve in the compressor
    h. Suction line restriction
    i. Broken valve in the compressor

8 Troubleshooting—Electrical

In addition to the control systems that are used in refrigeration and air conditioning systems to perform the standard on/off functions, additional controls are required to perform those functions that are necessary to obtain heat pump operation. These can be categorized as follows:

1. Conditioned area temperature—heating mode
2. Conditioned area temperature—cooling mode
3. Changeover between heating and cooling modes
4. Control of defrost function
5. System protection

8–1 AREA TEMPERATURE CONTROL

The first three categories use temperature controls to perform the control function. Category 1, heating mode, and category 2, cooling mode, functions are usually incorporated into a single device. Category 3, heating mode/cooling mode changeover, is incorporated into the same device in those heat pump systems using electric auxiliary heat. However, in add-on units to fossil fuel heating systems, the changeover may be under the control of a separate thermostat. In this case, the changeover function is eliminated from the conditioned area thermostat. This control was discussed under Indoor Section, Vertical Up Air Flow—Control Wiring as well as Add-On Systems Control Wiring.

To diagnose problems in these categories, further discussion of the functions of these controls is necessary. Because of the possible variety of control methods that apply, it is necessary to separate the two different types of application—standard System or Add-On System.

Standard System

When the heat pump system is installed using electric auxiliary heat (standard system), practically all manufacturers incorporate area temperature control and operating mode selection into the conditioned area thermostat.

Control Functions

Figure 8–1 shows a typical heat pump thermostat with the eight features provided to perform the eight necessary control functions:

1. System totally off (feature D, system switch).
2. System on "heating only" (feature D, system switch in conjunction with feature A, heating temperature selector).

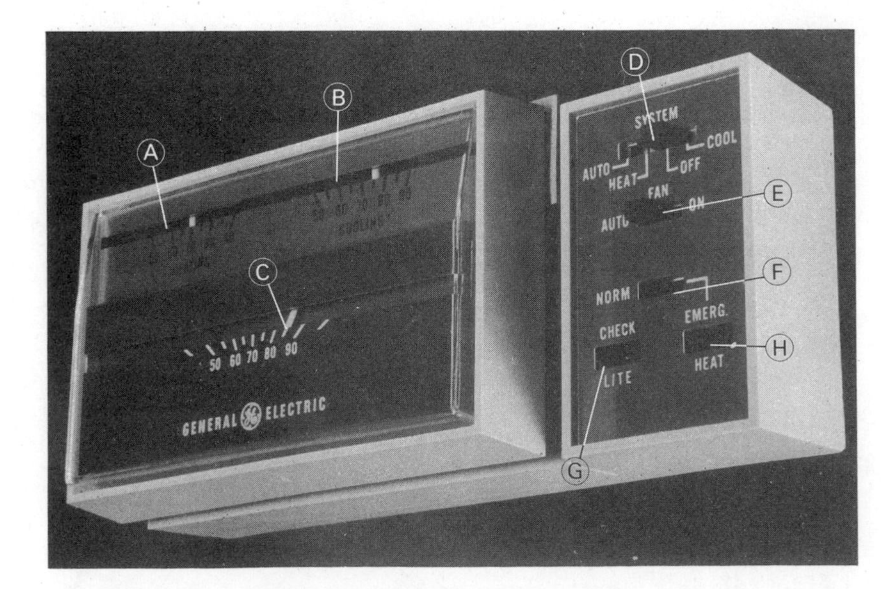

(A) Heating temperature selector
(B) Cooling temperature selector
(C) Temperature indicator
(D) Systems switch
(E) Fan switch
(F) Normal/Emergency heat switch
(G) Check-Lite
(H) Emergency heat light

Figure 8–1 Heat pump thermostat (Courtesy Addison Products Co.)

3. System with two stages of heat (feature A, heating temperature selector, controls both stages of heat with a fixed difference of set points of 2°F).
4. System on "cooling only" (feature D, System switch, in conjunction with feature B, cooling temperature selector).
5. System on "automatic changeover" (feature D, system switch in conjunction with both feature A, heating temperature selection, and feature B, cooling temperature selection. Feature B is a two-stage control. The first stage controls the reversing valve for system changeover and the second stage controls the area temperature in the cooling mode. A fixed differential of 3°F is used between the two stages.
6. System on "Emergency Heat." Feature F, normal/emergency heat switch, is used to accomplish the changeover from heat pump operation to auxiliary heat operation controlled by the first stage of the heating control. The "Emergency Heat Lite" (feature H) is used in conjunction with the changeover switch to provide a constant reminder to the area occupant that emergency heat is being used.
7. Room air circulation only. This is a standard feature of all air conditioning and heat pump systems. Continuous air circulation (CAC) is accomplished by the setting of the fan switch (feature E). With this switch in the "Auto" position, the inside blower cycles on and off with the refrigeration system. In the "On" position, the inside blower operates continuously, even if the system switch (feature D) is in the "Off" position.
8. Trouble indication. Not all heat pump thermostats incorporate feature G (check lite). When the system uses a high pressure lockout relay to provide the manual reset high-pressure protection required by UL, a check light in the thermostat is provided to indicate if this problem in the protection system has functioned.

Heat Pump Wiring Diagram

Figure 8–2 shows an area thermostat connected into a typical heat pump system using a two-stage electric auxiliary heat package. The top of the diagram shows the high-voltage control system on the outdoor section. The lower section shows the high-voltage control section of the indoor section. The middle section shows the area thermostat connected to the low-voltage control section (above the double arrowheads) and the low-voltage control section of the indoor section (below the arrowheads).

Trouble Diagnosis

To provide a basis for trouble diagnoses, a discussion of each function on the control system is needed.

System Totally Off. In this switch position, power is supplied to the thermostat via the connecting wire from the Red (R) lead and terminal from the 24 V control power supply. The circuits in the thermostat assembly have been emphasized (see Figure 8–3).

Tracing the energized circuits through the thermostat assembly shows:

1. 24-volt power directly to the selector switch section controlling the heating thermostat contacts (both HTG-1 and HTG-2).
2. Power through the emergency switch in the "Norm" position to the "Auto" and "Cool" sections of the season switch.
3. Power directly to the "On" side of the "Fan" operation selector switch. This is to provide continuous operation of the inside fan even if the season switch is in the "Off" position.
4. Power through the "Check Lite" to the open contact of the lockout relay. This provides operation of the indicator light if the lockout system has been actuated.

In all four circuits, no action takes place because the circuits have not been completed through their respective loads and to the other side of the power source.

System on Heating Only. In this season switch position, the contacts in the "Heat" position of the season switch have been closed (see Figure 8–4). Power is now carried to the first stage heating (HTG-1) thermostat contact. When this contact closes upon a demand for heat in the conditioned area, power is supplied to two loads:

1. Through the "Auto" circuit of the "Fan" switch, through the fan relay coil (FR) and to the other side of the power supply. This completes this circuit and energizes the fan relay coil. Normal function of the relay is to close the normally open contact in the fan relay. This supplies power to the blower motor in the indoor section and air movement should result.
2. Power is supplied from the HTG-1 thermostat contact through the "Norm" contact assembly of the "Emergency Heat" switch. This supplies power through terminal "W1" on the thermostat and terminal "Y" in the outdoor section. From terminal "Y," power flows to the compressor contactor coil (CC), through the normally closed contacts of both the high-pressure cutout control and the lockout relay. From here, the power flow is through the common terminal (C1) of the outdoor section, to the common terminal (C1) of the indoor section and to the other side of the 24-volt power source.

The system now operates under the control of the first stage heating contact (HTG-1).

System with Two Stages of Heat. Figure 8–5 shows the control circuit through the second stage heat thermostat. These circuits are in addition to those shown in Figure 8–4. Power flows through the HTG-2 contact into two separate heating control circuits. From terminal "W2" on the thermostat, the circuit is to terminal "W1" in the indoor section. From here the circuit divides.

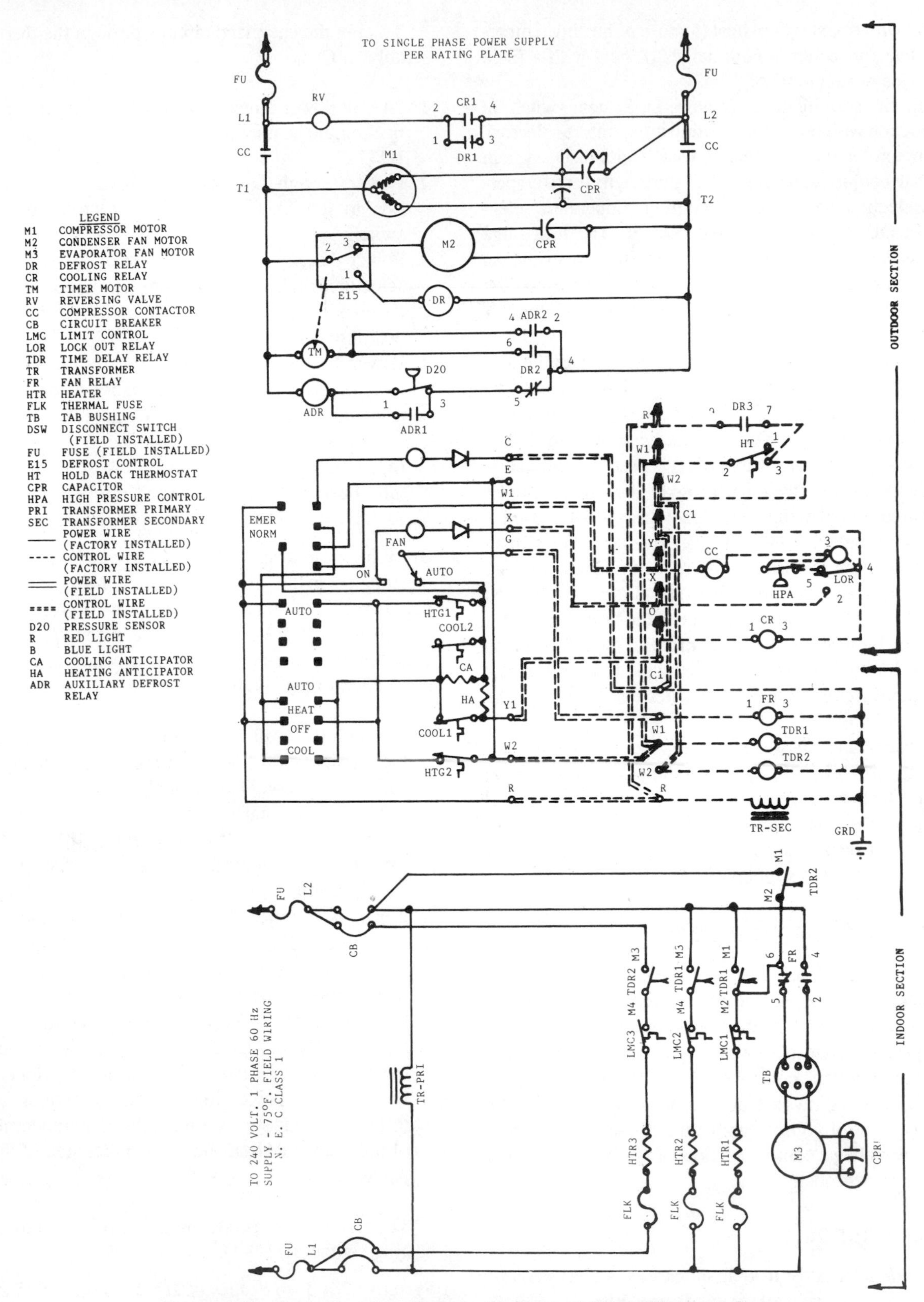

Figure 8–2 System wiring diagram (Courtesy Addison Products Co.)

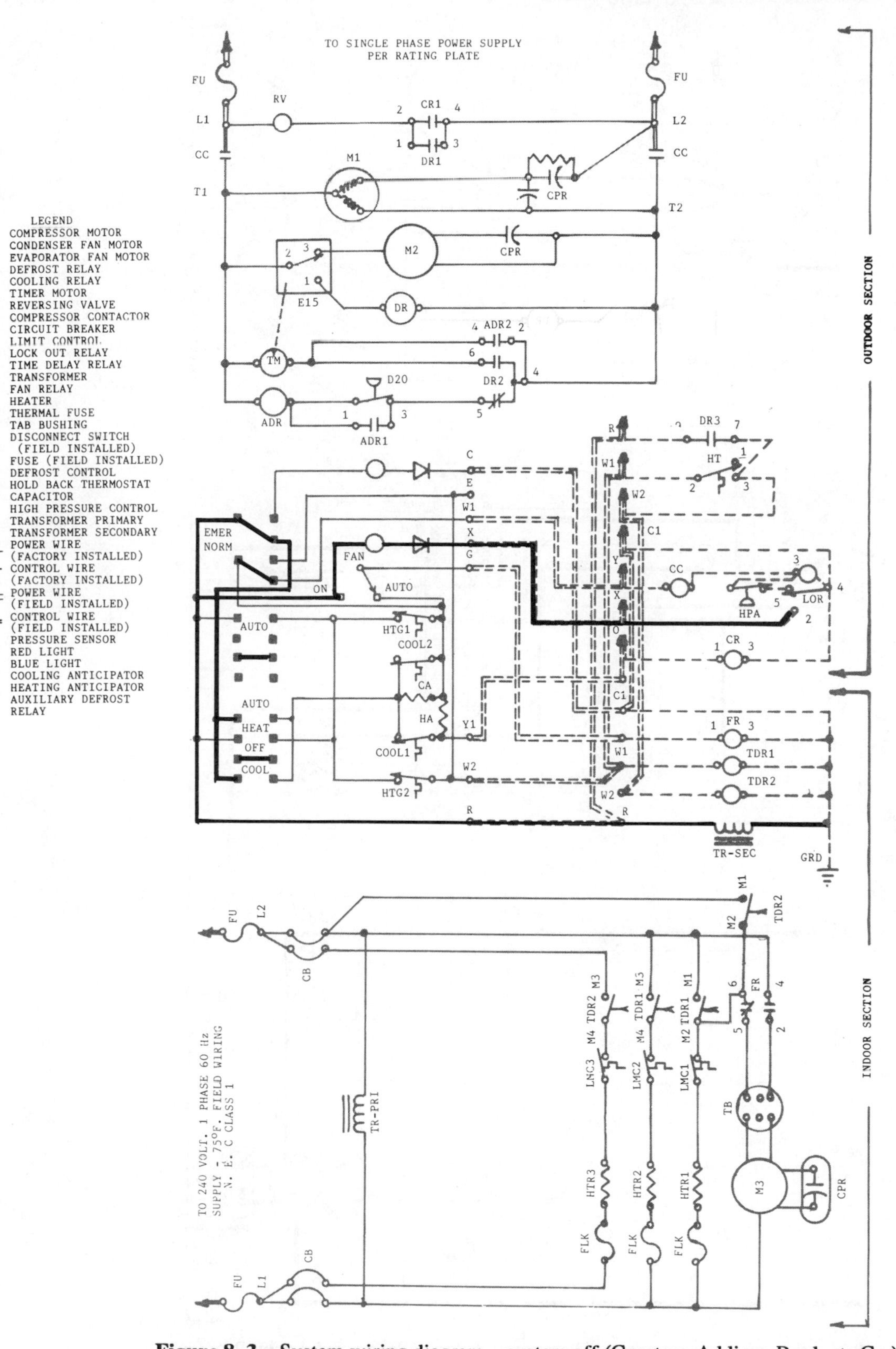

Figure 8–3 System wiring diagram—system off (Courtesy Addison Products Co.)

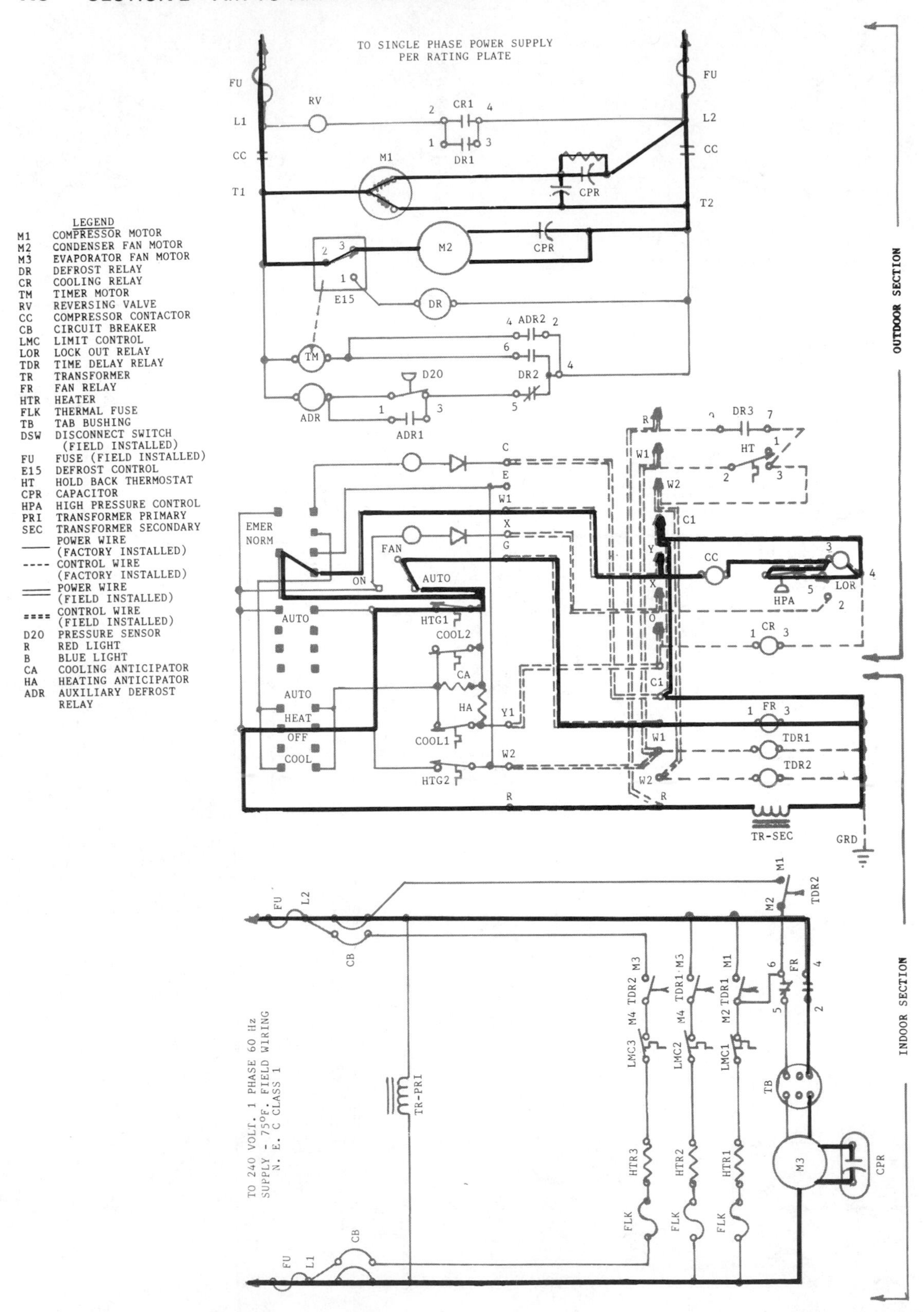

Figure 8–4 System wiring diagram—heating first stage (Courtesy Addison Products Co.)

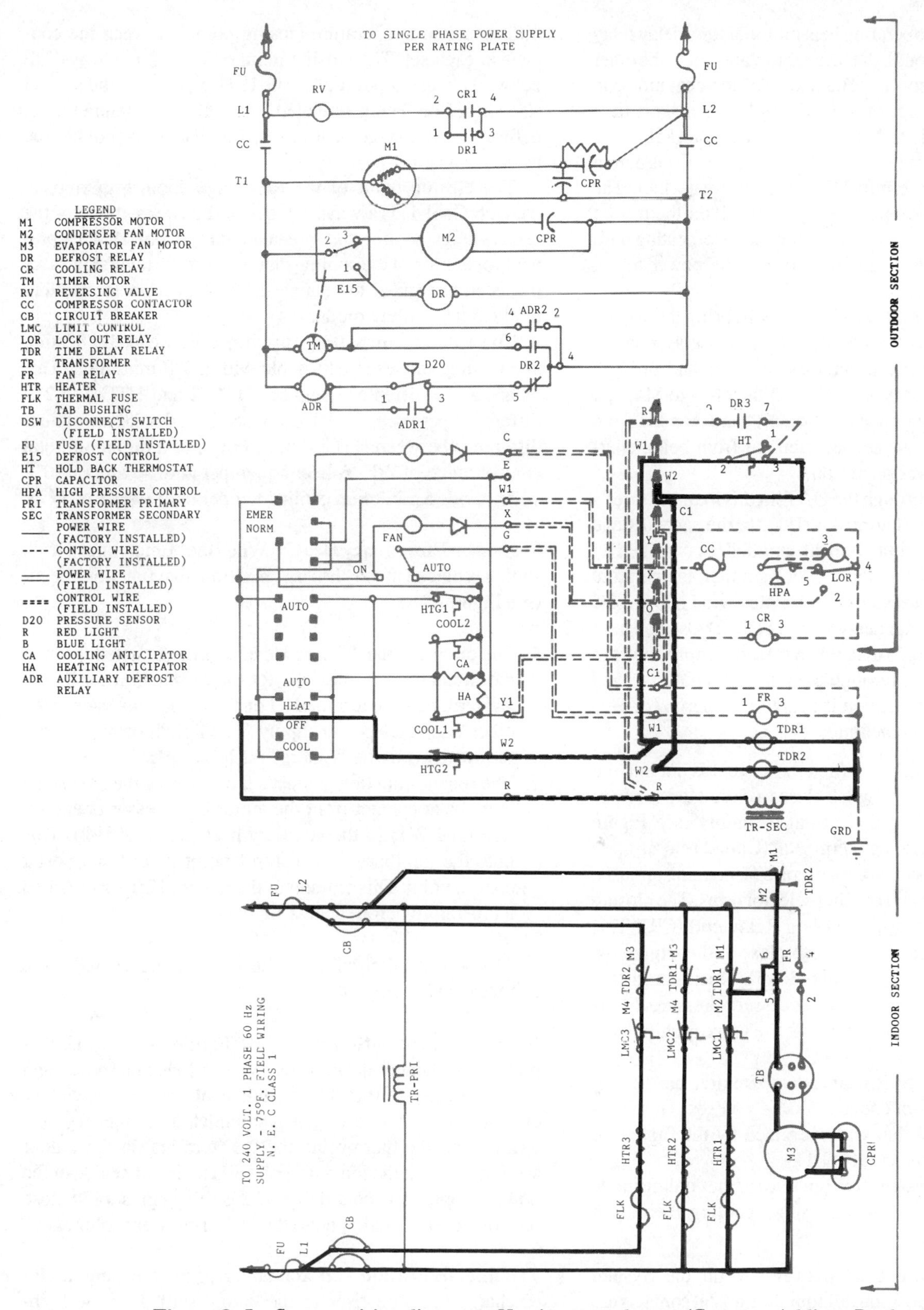

Figure 8–5 System wiring diagram—Heating second stage (Courtesy Addison Products Co.)

1. Power flows to the operating heat motor of time delay relay No. 1 (TDR-1), through the heat motor, and out to the other side of the power source. The action of the heat motor in TDR-1 closes contact M1 to M2, and after a short time delay, contact M3 to M4. The closing of contact M1 to M2 energizes the circuit to heater No. 1 (HTR-1). This action also puts power to terminal 6 of the fan relay (FR). The purpose of this is to keep the fan operating, if the thermostat were to be satisfied or shut off, to prevent overheating until the contact in TDR-1 has opened and the power to the heater has shut off.

   The closing of contact M3 to M4 will bring the second heater element into the circuit. The time delay relay is constructed to close the contacts in order and open in reverse order—close M1 to M2 and then M3 to M4, and then in reverse order, open M3 from M4 and then M1 from M2. This prevents the second element from being energized without blower operation.
2. Power is supplied through the OAT, in this manufacturer's unit, the Holdback Thermostat (HT), to the second stage electric heat. The amount of heat supplied is controlled by a thermostat measuring the outdoor ambient temperature and set at the second balance point plus wind allowance. The second stage time delay relay (TDR-2) also has a set of contacts (M1 to M2) that, when closed, supply power to the sustaining circuit through the fan relay contact (5 and 6) to keep blower operation if heater 3 is energized. This also is to prevent overheating.

System on Cooling Only. When the system is put into the cooling mode, the System Switch closes a circuit to put power to the two-stage cooling thermostat assembly (see Figure 8–6). The first stage cooling thermostat (Cool-1) closes upon a 3°F or more rise in the area temperature, depending upon the settings of the heating and cooling selector arms. The closing of this contact energizes the Cooling Relay coil (CR). This causes the cooling contact (CR1) to close and energize the reversing valve pilot valve solenoid coil. Operation of the refrigeration system supplying the necessary pressure difference across the reversing valve causes the valve to shift to the cooling mode.

With another 2°F rise in area temperature, the second stage cooling thermostat contact (Cool-2) closes. This energizes the same circuit that was energized by the first stage heating thermostat contact (HTG-1).

The refrigeration system now operates in the cooling mode under the control of the second stage cooling thermostat contact (Cool-2).

System on Automatic Changeover. With the System Switch in the "Auto" position, all four thermostat contacts are energized. The heating contacts (HTG-1 and HTG-2) are energized through the lower section of the season switch, the cooling contacts (Cool-1 and Cool-2) through the upper single contact. This allows the unit to operate under control of any of the contacts (see Figure 8–7).

There is no possibility of overlapping of the control circuits in a properly functioning thermostat as there are mechanical differences in temperature closing points between the contacts in each set. The closing point of HTG-2 is always 2°F below the closing point of HTG-1. This prevents the second stage heat from being activated, on a fall in area temperature, before the first stage heat has had a chance to provide the heating capacity.

The closing point of the first stage cooling thermostat contact (Cool-1) is always 2°F below the closing point of the second stage cooling thermostat contact (Cool-2). This prevents operation of the cooling mode before the reversing valve has been energized to change the system from the heating mode to the cooling mode.

The temperature setting difference between the heating and cooling modes is adjustable with a 3°F minimum. This means a 2°F difference between HTG-2 and HTG-1, a 2°F difference between Cool-1 and Cool-2, and a 3°F minimum difference between HTG-1 and Cool-1, or a minimum total control range of 7°F. A heating temperature setting of 70°F will provide a minimum cooling temperature setting of 77°F.

System on Emergency Heat. When the "Emergency Heat" switch is placed in the "Emerg" position, two functions occur (see Figure 8–8).

1. The circuit to the "Emerg Heat" light is actuated through the circuit through the light (R), the current limiting diode, and terminal C to terminal C1 on the indoor unit back to the other side of the power supply. This light is on all the time the switch is in the "Emerg" position.
2. The second function switches the action of the first stage thermostat contact from the motor compressor operation (terminal W1) to the auxiliary heat (terminal W2). This puts the auxiliary heat control point at the thermostat control point. This eliminates the required 2°F drop in area temperature before heat is supplied.

The additional circuitry is the same as the second stage operation in Figure 8–4.

Room Air Circulation Only. In Figure 8–9, we see that power is obtained ahead of the Season Selector switch and routed directly to the "On" terminal of the fan switch. By closing this switch, a circuit is completed through the "G" terminal on the thermostat, the "G" terminal in the indoor section, through the fan relay coil (FR) to the other side of the power supply. The energizing of this relay coil should close the contact (FR) and operate the indoor blower motor.

Trouble Indication. If for any reason, the compressor discharge pressure rises to the cutout setting of the high-pressure control (HPA), the contact in this control opens. This forces the compressor contactor operating current to flow through the lockout relay coil.

Figure 8–10 shows this situation. The HPA contact is open and current flow is through the compressor contactor coil (CC) to terminal 3 on the lockout relay, through the relay coil to terminal 4, and return to the other side of the power supply.

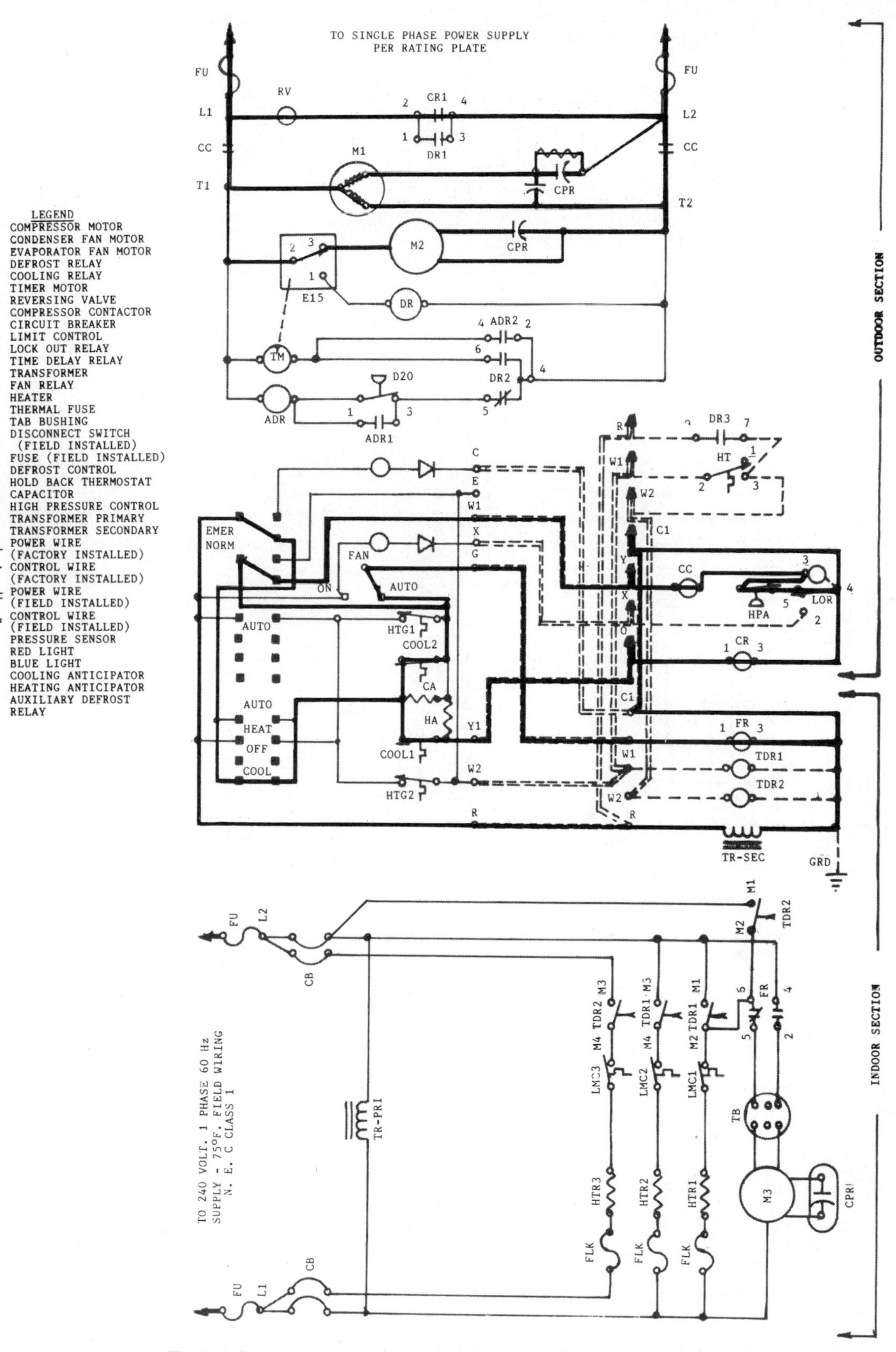

Figure 8–6 System wiring diagram—cooling (Courtesy Addison Products Co.)

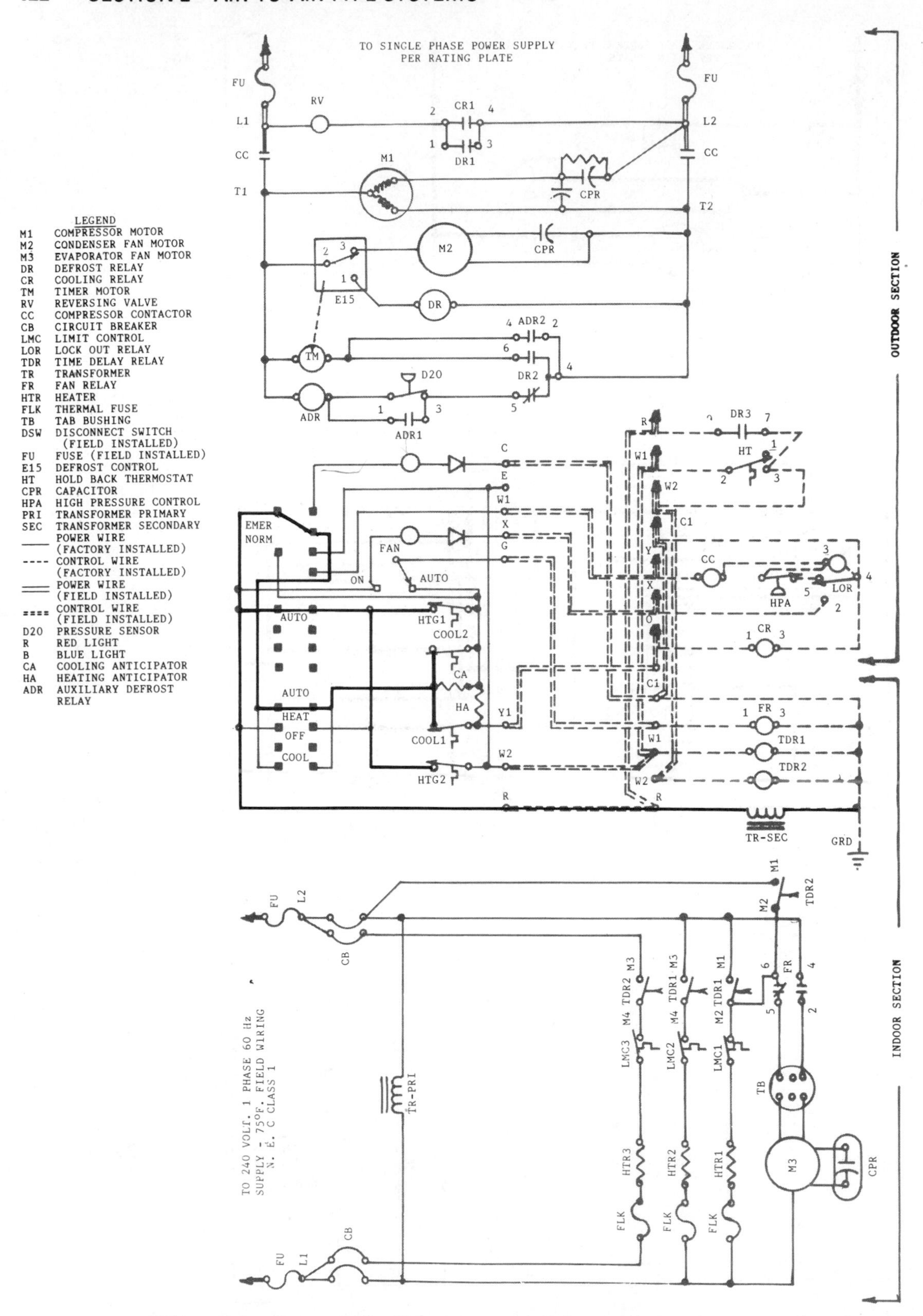

Figure 8–7 System wiring diagram—automatic changeover (Courtesy Addison Products Co.)

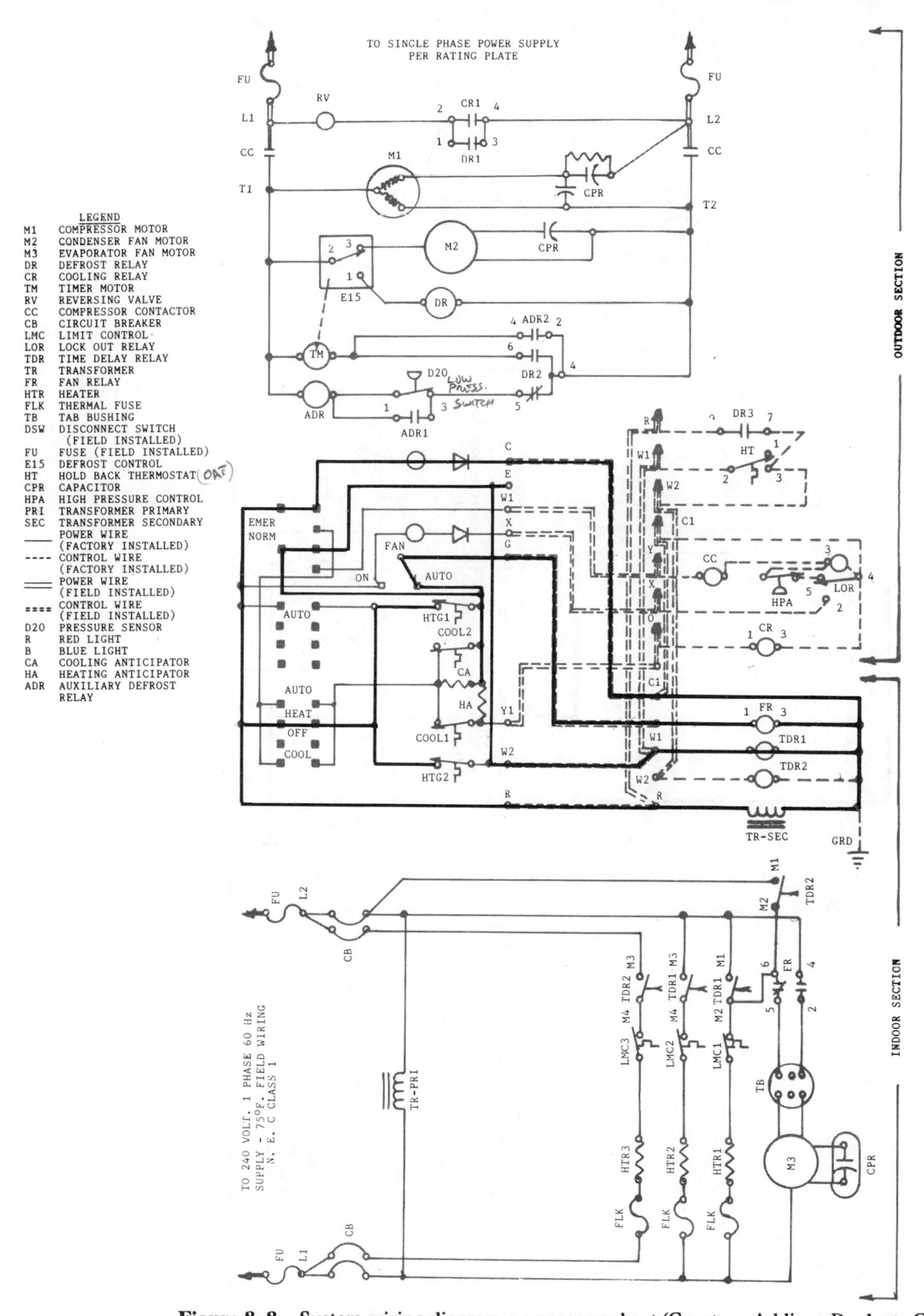

Figure 8–8 System wiring diagram—emergency heat (Courtesy Addison Products Co.)

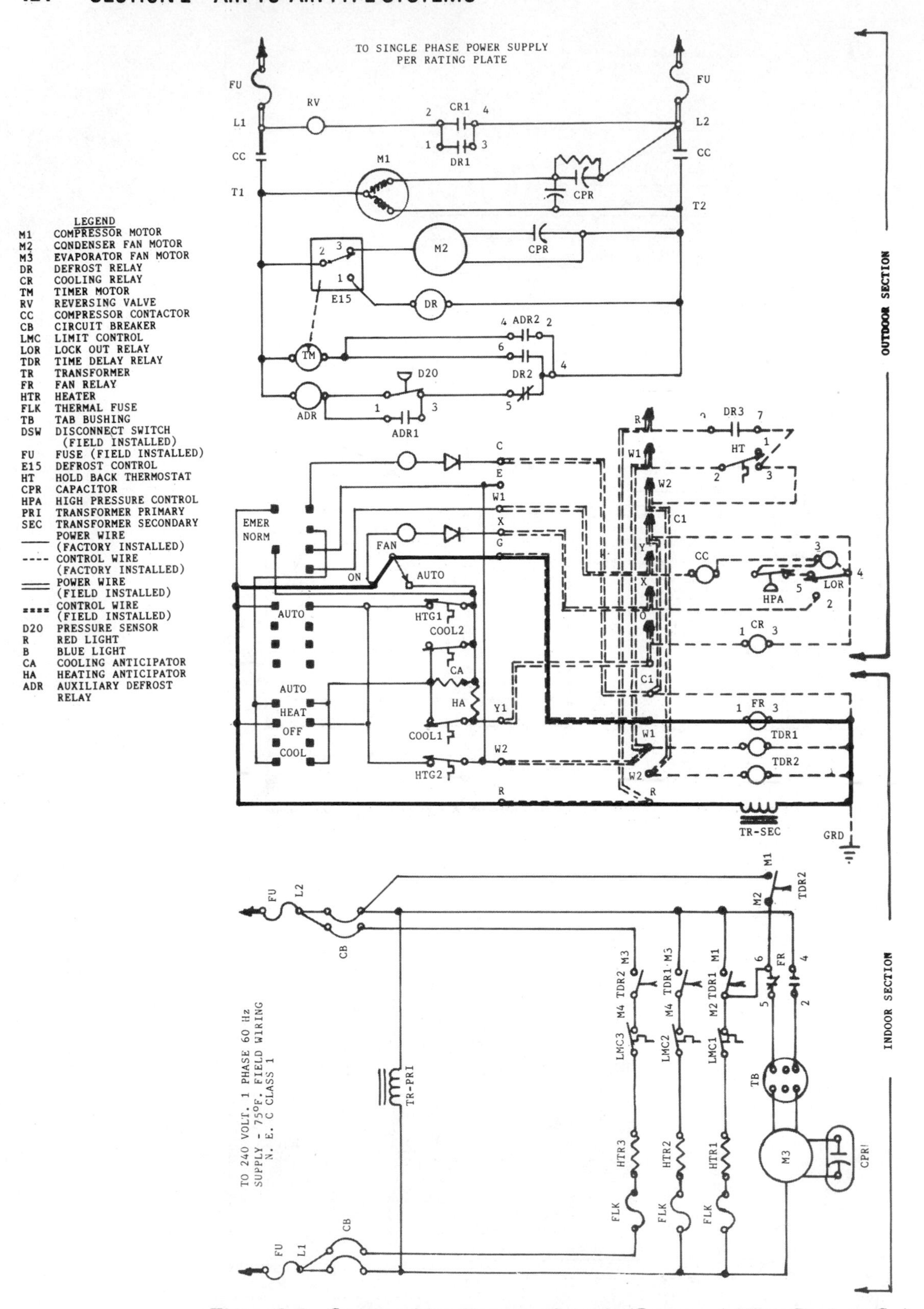

Figure 8–9 System wiring diagram—fan only (Courtesy Addison Products Co.)

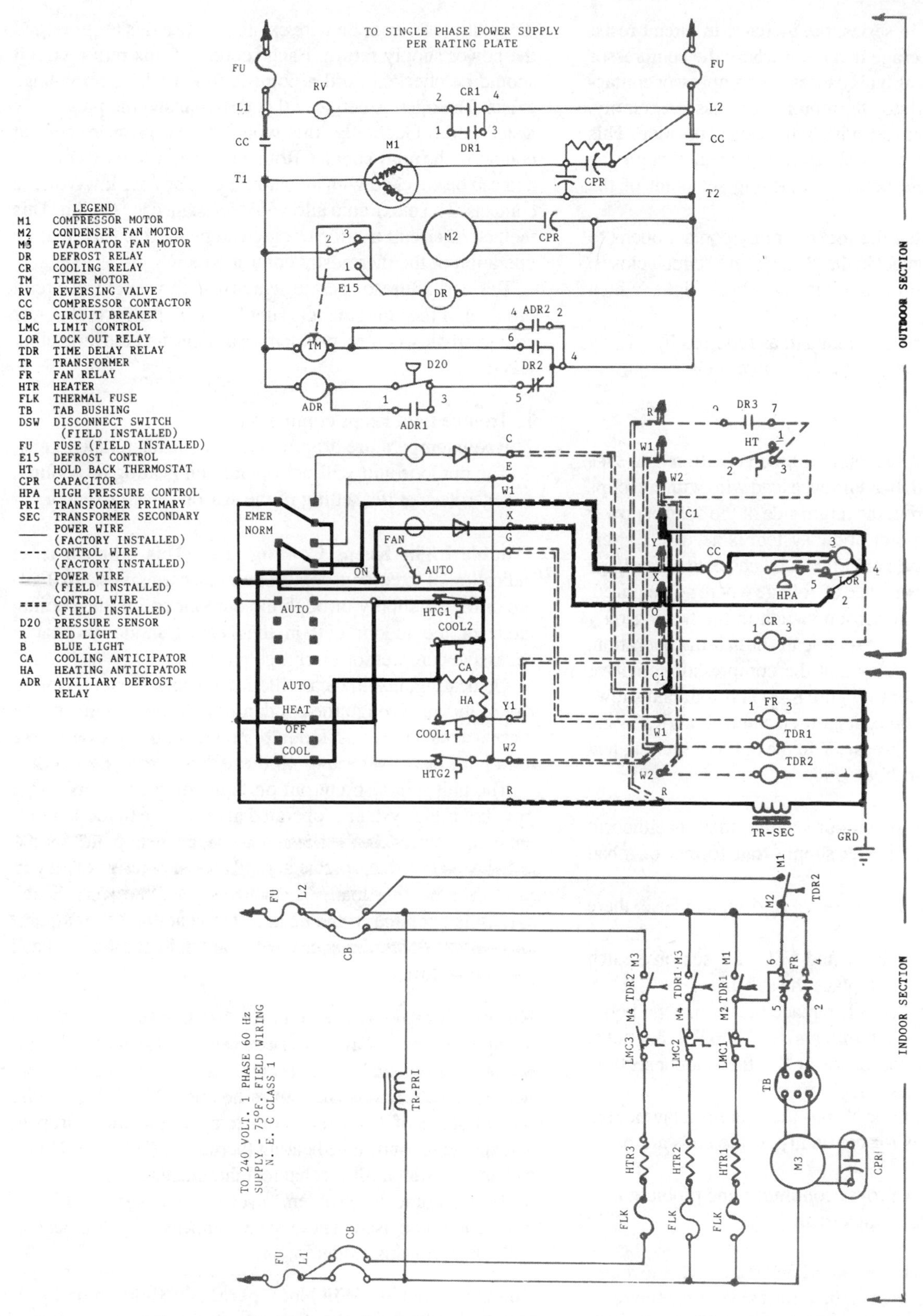

Figure 8–10 System wiring diagram—lockout safety circuit (Courtesy Addison Products Co.)

With the two coils in series, the increase in circuit resistance reduces the amperage to a point where the compressor contactor will not hold in. This causes the compressor contactor contacts to open and stop the motor-compressor assembly. The lockout relay, however, stays in the closed position. This prevents the compressor from operating even though the discharge pressure falls below the closing set point of the high-pressure control.

At the same time that the lockout relay contact opens to prevent recycling, a contact in the "Check Lite" circuit closes. This causes current flow through the check light, which indicates that a problem exists.

Resetting of this circuit is manual, as required by UL, by breaking the power supply to the lockout circuit.

Problem Analysis

When analyzing problems, a voltmeter with a 0 to 24+ voltage range is needed. In addition, a lead wire with end clips long enough to reach from the return side of the control power supply to the furthest point of the system is needed. By this method, each section of a circuit can be checked for voltage where required or the existence of voltage where not required.

For example, with the season switch in the heating only position, and the thermostat set to call for heating operation, both the inside blower relay and the compressor contactor should be actuated. Refer to Figure 8–3. This diagram shows that 24-volt power should be measured at terminals R, G, and W1 on the thermostat sub-base, Y in the outdoor section, and G in the indoor section. If we find:

1. No voltage at R on the indoor section, there is either no power to the control power supply transformer or a bad transformer.
2. No power to terminal R on the thermostat sub-base, there is an open cable wire.
3. No power to terminals G and W1, the season switch contacts are defective. Replace the sub-base.
4. Power at G but none at W1, replace the thermostat body.
5. Power at the thermostat terminals R, G, and W1, but not at terminal G in the indoor section or Y in the outdoor section, a cable problem exists.
6. Power at terminals G or Y, but there is no relay action, check the relay coils for continuity. If the coils are open, replace the relay.
7. The relay coils show proper continuity, the problem may be in the return side of the circuit.

The fan relay circuit has no additional controls in the return side. Therefore, the voltage at the return side of the fan relay coil should be 0 volts because this terminal is connected directly to the return side of the power supply. If a voltage exists at this terminal, an open wire exists between the relay and the power supply. Replace the wire.

The return side of the compressor contactor coil circuit contains the lockout relay protection system. Therefore, if a voltage is measured at the compressor contactor terminal on the return side, an open wire exists between this terminal and the power supply return. Each section of this return circuit should be checked until a point is found where no voltage exists. The open is between this terminal and the previously tested point. Generally, this would be the pressure control contact or the relay contact. Both are normally closed contacts and can be checked with an ohmmeter. They are low-voltage contacts. The maximum allowable resistance is 1/2 ohm. This method of circuit trouble tracing can be used in each feature operation of the thermostat control circuits.

Besides failure to operate, which usually is a control circuit problem, if the equipment is supplied with the correct power characteristics, control problems can show up in several other ways:

1. Trouble light keeps coming on.
2. Room temperature drops at certain outside temperatures.
3. The outdoor unit will not operate on heating or cooling, regardless of the setting of the season selector switch.

Trouble Light Keeps Coming On. This problem is an indication of excessive compressor discharge pressure. Failure of the air supply through the outdoor coil in the cooling mode or the indoor coil in the heating mode will cause excessive compressor discharge pressure.

On heating, install clean filters and check the temperature rise of the air through the condenser coil. If the temperature rise exceeds 35°F at 70°F outdoor temperature, check for the reason for the reduction of air through the condenser coil.

The unit will also cut out on high compressor discharge pressure if the system is operated at excessive outdoor ambient temperatures. In the heating mode, the heat pump should only be expected to operate at outdoor temperatures that will produce a heating load; i.e., below 65°F. Therefore, if the system is operated in the heating mode at outdoor temperatures above 65°F, the excessive load could cause high head pressure cutoff.

Room Temperature Drops at Certain Outdoor Temperatures. This room temperature drop of 2°F will occur at the second balance point of the system when the heat pump will not handle the entire heat loss. To bring on the second stage of heat, the room temperature must drop to actuate the second stage heating thermostat (HTG-2). This is a characteristic of all mechanical thermostats.

To overcome this problem, an electronic thermostat (solid state) should be used. These have compensating circuits that will adjust for this situation.

The Outdoor Unit Will Not Operate on Heating or Cooling Regardless of the Setting of the Season Selector Switch. Throughout this text, constant reference has been made to the matching of terminal designations on the various components in the system. There are no standard wiring designations. Each manufacturer develops its own to fit its particular designs or needs. Therefore, it is imperative that the manufacturer's wiring instructions are followed.

The most common occurrence is matching the W[1] terminal of the thermostat to the W1 terminal of the auxiliary heat. This operates the auxiliary heat in both the heating and cooling modes. To operate the outdoor (high side) section, the W1 terminal of the thermostat must be connected to the Y terminal of the outdoor section.

8–2 DEFROST SYSTEMS

The operation of the various types of defrost systems was covered in Chapter 3, Component Parts. Problems with the various systems can be put in the following categories:

1. Unit will not initiate the defrost mode.
2. Unit will not terminate the defrost mode.
3. Unit has unnecessary or nuisance defrost cycles.
4. Unit does not complete the defrost function. Frost and/or ice is left on the outside coil.

The solution to the four categories of problems will vary with each type of defrost system. Some of the problems may have several possible solutions while others may have only one. To reduce the necessity for cross reference, each type of defrost system has been considered separately.

8–3 TEMPERATURE INITIATION/ TEMPERATURE TERMINATION

Mechanical Type

Figure 8–11 shows the first mechanical defrost control using the temperature initiation/temperature termination principle. This control uses the temperature of the outside evaporator coil liquid refrigerant inlet to determine if the coil is at a temperature low enough to form and retain frost. This temperature set point is usually 26°F. The large feeler bulb on the tube on the right side of the control in Figure 8–11 is the low-temperature sensing bulb.

Figure 8–12 shows this control mounted in a package unit showing the feeler bulb locations. The air sensing tube (straight tube type) is wound into a helix shape for maximum sensitivity. Mounted behind the air inlet grill, it measures the temperature

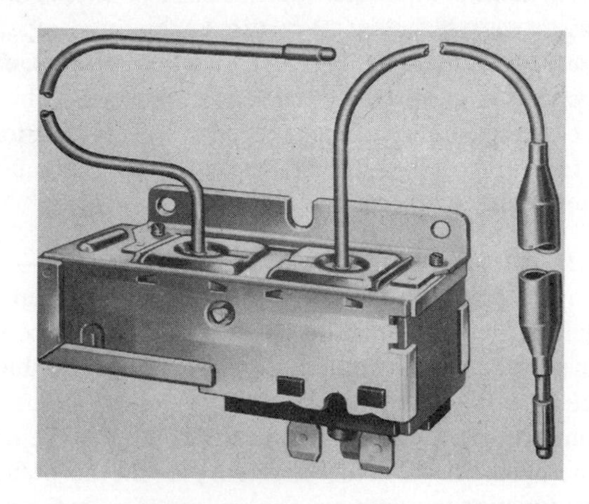

Figure 8–11 D52 defrost control (Courtesy Ranco Inc.)

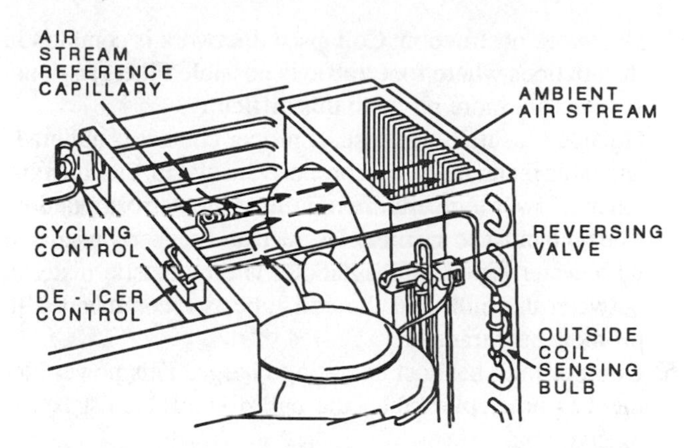

Figure 8–12 D52 defrost control in a package unit (Courtesy Ranco Inc.)

of the outdoor air as it enters the unit. This control operates by measuring the difference in temperature between the entering air and the coil operating temperature. The basic unit design under normal clear coil operating conditions uses a temperature difference between air and coil of 15°F at 40°F outdoor temperature down to 10°F at 10°F outdoor temperature.

Because the possibility of frost formation decreases considerably above outdoor temperatures of 45°F, the control is designed to prevent nuisance defrost functions above 48°F air temperature.

As the outdoor evaporator coil gathers an insulating coat of frost, the temperature difference between air and coil automatically increases to try to maintain the heat removal capacity. When the insulating effect reaches a temperature difference increase of 10°F, the control initiates the defrost function. When the outdoor coil reaches 55°F, during the defrost function, the control reverses and the system returns to the heating mode.

Unit Will Not Initiate the Defrost Function. Because this control depends upon the operating temperature of the outdoor evaporator coil, anything that affects this operating temperature affects the initiation of the defrost function.

Shortage of Air Through the Inside Condenser Coil. Any reduction in the air quantity through the condenser will raise the condensing temperature required to return the refrigerant vapor to the liquid state. The rise in condensing temperature along with its corresponding rise in pressure, will force an increase in liquid refrigerant flow into the outdoor evaporator coil. In addition, the higher compressor discharge pressure reduces the compressor pumping capacity.

Both factors force the formation of extra frost insulation effect to bring the coil temperature down to the initiating temperature to start the defrost function.

Reduction of CFM through the indoor coil can be caused by:

1. Clogged filters. These should be changed at least every 60 days.
2. Closed supply grills. Isolating part of the conditioned rooms by closing supply grills will cause defrost control failure.

3. Ductwork obstruction. Collapsed ductwork is common in installations where foot traffic is possible Fiberglass duct systems are more prone to this difficulty.
4. Outside sensing bulb loose. If proper contact is not made and if the feeler bulb is not properly insulated, the coil must reach a lower temperature to initiate the defrost function. The bulb must be securely fastened and properly insulated with water repellent insulation. Thermomastic material between the bulb and the coil tube is recommended to promote heat transfer.
5. Outdoor bulb has lost operating charge. This power element is not replaceable; the entire control must be replaced.
6. Overcharge of refrigerant. The effect of an overcharge of refrigerant is the same as the reduction of air through the inside condenser coil. The higher compressor discharge pressure causes the higher evaporator pressure and operating temperature. Make sure the system is operating with the correct refrigerant charge.

Unit Will Not Terminate the Defrost Function. This problem occurs because the contact locking device in the control sticks and does not release the contact arm. Control replacement is necessary.

Unit Has Unnecessary or Nuisance Defrost Cycles. This problem does not present itself with this defrost control.

Unit Does Not Complete the Defrost Function—Frost and/or Ice Remain on the Coil. Premature shut down of the system by the satisfaction of the area thermostat because of excessive reheat capacity will result in this problem. The solution is to reduce the amount of reheat BTUH so that it is equal to or less than the sensible heat capacity of the refrigeration system.

Electronic (Solid State) Type

Figure 8–13 shows a typical solid state temperature initiation/temperature termination (demand defrost) defrost control. This control system consists of a printed circuit board with solid state integrated components for the temperature-sensing functions as well as switching components to control the defrost relay circuit.

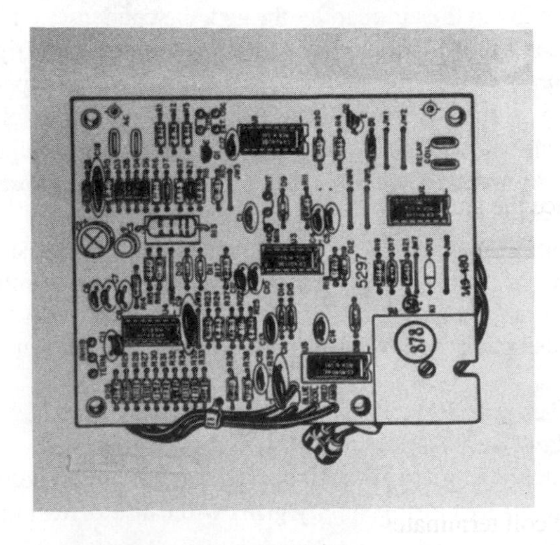

Figure 8–13 Solid state defrost control (Courtesy Rheem Air Conditioning Division)

The defrost control board has two permanently attached thermistors, which are the temperature sensors. One thermistor is an air type mounted in the air stream entering the condenser air circuit. The other is a specially formed device to be fastened to the evaporator coil downstream from the pressure-reducing device. Each location has been precisely selected by the manufacturer. Any replacement should find the new thermistors in the same locations.

Figure 8–13 shows the coil sensor secured by a clamp to the copper tube of the coil. The clamp must hold the thermistor assembly securely in position with only sufficient pressure to retain the thermistor in position. Thermomastic material in the clamp voids and waterproof insulation are required.

This control is assembled with predetermined resistances in the thermistors and leads. Therefore, the leads cannot be altered in any way. The circuit board and thermistors must be matched as an assembly.

Unit Will Not Initiate the Defrost Function. Make sure the board is supplied the required voltage. There will be two test terminals for measuring the input voltage. These will vary in location and designation with the various type boards. Make sure the thermistor fastened to the evaporator tube is properly secured and insulated.

If both items check out, substitute another board and thermistor assembly. The systems are not repairable.

The electronic control is subject to the same problems as the mechanical control. The CFM of air through the inside condenser coil must be correct to keep the compressor discharge pressure correct. The refrigerant operating charge must also be correct.

Unit Will Not Terminate the Defrost Function. Check the system by substituting another board and thermistor assemblies.

Unit Has Unnecessary or Nuisance Defrost Cycles. Most solid state demand defrost controls have a built-in timer circuit to monitor the action of the coil sensor thermistor. The thermistor must be below the required operating temperature for a time period before the defrost function is initiated. This time period is usually between 4 and 6 minutes. This is to reduce the possibility of short repetitive defrost periods. If such repetitive defrost periods are encountered, substitute another board and thermistor assemblies.

Unit Does Not Complete the Defrost Function—Frost and/or Ice Remain on the Coil. This system will terminate the defrost function upon loss of unit operation. Therefore, reheat must be kept to a minimum to reduce the possibility of the area thermostat terminating the system operation.

Some control systems use an extra contact in the defrost relay to bypass the area thermostat to prevent this premature defrost function termination. The amount of reheat should be limited to prevent excessive room temperature rise and unnecessary operating cost.

8–4 PRESSURE INITIATION/ TEMPERATURE TERMINATION

Mechanical Type

The pressure initiated/temperature terminated defrost control measures the air flow resistance through the outdoor evaporator coil. If the air flow resistance is sufficient because of frost buildup, the pressure drop across the diaphragm switch shown in Figure 8–14 closes. This control contains both the defrost initiation function, which is controlled by pressure difference, and the defrost termination, which is based on coil temperature.

The SPDT switch controls the outdoor fan motor through the normally closed contact and the defrost relay through the normally open contact. The common of the two contact sets is connected to the source of power. This system requires a 240-volt defrost relay, the same operating voltage as the outdoor fan motor.

The coil sensing bulb is connected to a bellows pressure point that operates against the terminating lever in the control. When the temperature sensing bulb is above 32°F, the pressure in the bellows forces the pressure point against the terminating lever, which prevents the diaphragm switch from acting. When the temperature drops below 32°F, the pressure in the bellows is released and the pressure switch is free to move.

The diaphragm in the pressure switch measures the pressure across the outdoor evaporator coil. This is a measurement of the air flow resistance through the coil.

On a blow-through type unit, the low-pressure connection measures atmospheric pressure while the high-pressure connection measures the pressure in the fan discharge area ahead of the evaporator coil. On draw-through units, the connections are reversed. The high-pressure connection measures atmospheric pressure while the low-pressure connection measures

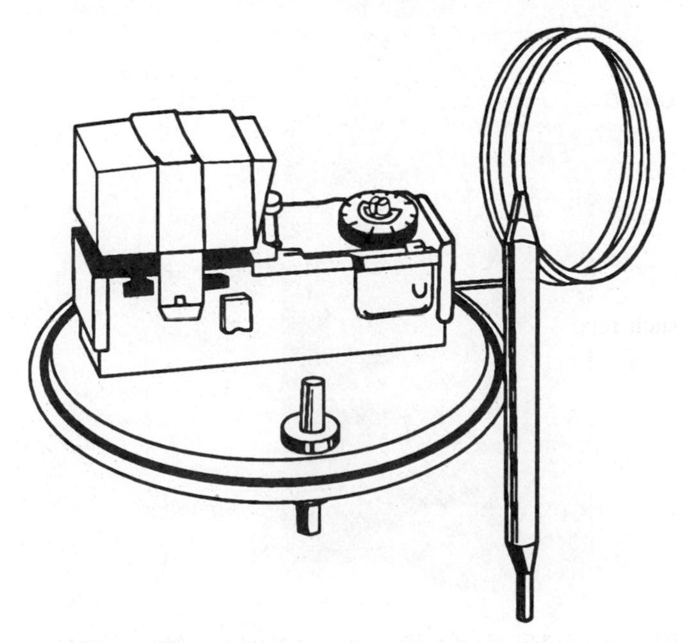

Figure 8–14 D52 defrost control—pressure actuated (Courtesy Ranco Inc.)

the pressure between the evaporator air outlet and the condenser fan inlet.

Most units use an initiation pressure of 0.40 inches WC to 0.65 inches WC. The initiation pressure selection point usually requires that 70% to 80% of the coil face area be covered with frost.

With the coil temperature below 32°F, when the coil air flow resistance reaches the control set point, the initiation lever in the control responds to the action of the pressure diaphragm and reverses the switch contacts. This stops the outdoor fan and energizes the defrost relay. The system is put into the defrost mode.

When the outdoor coil reaches the termination set point (55°F to 85°F, depending upon the unit design), the pressure increase in the bellows moves the termination lever to reverse the switch action. The defrost relay is de-energized and the outdoor fan motor is energized. The unit is back in the heating mode.

Unit Will Not Initiate the Defrost Mode. Two things must happen before the defrost mode can be initiated.

1. The coil temperature-sensing bulb must be at a temperature below 32°F. Make sure the bulb is securely fastened to the coil tube and is insulated. Using thermomastic material between the bulb and the coil tube is recommended.
2. The diaphragm must sense the pressures across the coil. Make sure all pressure connection tubes are clean with no kinks. Insects can plug these tubes and prevent the defrost function.

Unit Will Not Terminate the Defrost Mode. Check the control operation by substituting another control. If the feeler bulb of the control is tight and the coil temperature is above the set point, seized contact points require control replacement.

Unit Has Unnecessary or Nuisance Defrost Cycles. This control system is subject to erratic wind pressures, which may cause frequent short cycling of the defrost control system. If this is a major problem for the particular application, changing to the solid state demand control will help to reduce this problem.

Unit Does Not Complete the Defrost Function—Frost and/or Ice Remain on the Coil. This is usually caused by an excessive amount of reheat during the defrost function. Reduce the amount of reheat to not exceed the sensible heat capacity of the refrigeration system.

Electronic (Solid State) Demand Defrost Control

Figure 8–15 shows an electronic (solid state) demand defrost control. This control also operates upon the change in air flow resistance through the evaporator coil to initiate the defrost function. A thermistor (temperature sensor) attached to the coil terminates the defrost mode. This thermistor also prevents the system from considering a defrost function if the coil temperature is over the initiating set point.

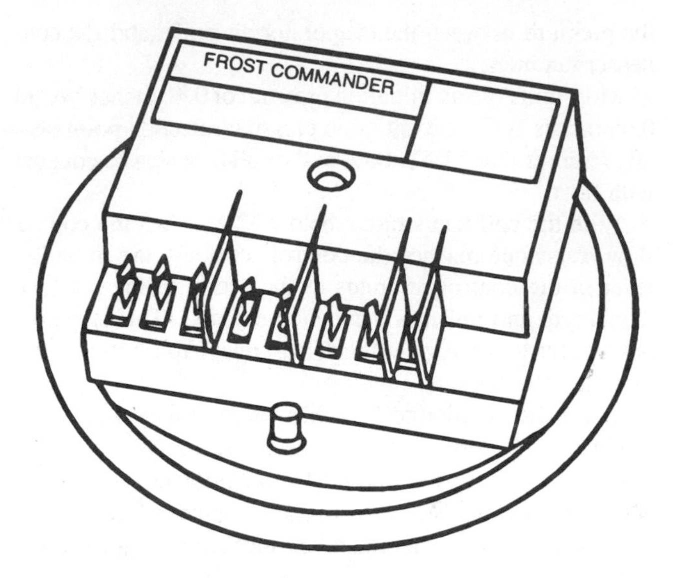

Figure 8–15 Solid state pressure-actuated defrost control (Courtesy Ranco Inc.)

The set points of the control in Figure 8–15 are 27°F initiation and 55°F termination. Each system manufacturer selects the set points for its design. When controls are replaced, these specifications should be followed.

When the amount of frost accumulates on the coil to a point of 70% to 80% coverage (0.20" WC), the control closes a pressure-actuated circuit. Unlike a mechanical contact set, this circuit must be activated for a continuous period of approximately 20 seconds before the defrost function is initiated. This prevents nuisance "wind gust" defrost cycles.

The same solid state circuit board has a time override feature. This period of 10 minutes is the maximum time the control will allow the system to be in the defrost mode, regardless of the evaporator coil temperature.

When power is applied to the board through the action of the area thermostat calling for heat and starting the heating function, the board must be energized for a period of 20 to 25 minutes before a defrost function can be initiated. This prevents short defrost cycles in the event of unit shutdown before the defrost function is completed.

By means of electrical circuitry, the faults of the mechanical control are overcome.

Unit Will Not Initiate the Defrost Function. The coil temperature must be below the set point of the control and the coil temperature sensor must be securely fastened in place and properly insulated.

The pressure taps from the high- and low-pressure connections must be clear and not kinked. Do not attempt to clear these tubes by using your breath. Moisture can condense in the tubes and freeze, thus blocking the tubes and compounding the problem. After removing the tubes from the control taps, use a short burst of pressure from your nitrogen cylinder.

Substitute another control to check the control operation. These controls are not field repairable.

Unit Will Not Terminate the Defrost Mode. The control has circuitry that limits the defrost mode to a 10-minute period. If the unit will not terminate after the 10-minute period is over, substitute another control.

Unit Has Unnecessary or Nuisance Defrost Cycles. This problem has not arisen with this control.

Unit Terminates Before the Defrost Period is Completed—Frost and/or Ice Remain on the Coil. This can be a temporary problem during periods of extremely high humidity, but it will clear after such weather conditions clear. If icing conditions continue, check to make sure the terminating thermostat is at the correct location according to the manufacturer's instructions. If the location is correct, replace the control.

8–5 TIME INITIATION/TEMPERATURE TERMINATION

Mechanical Type

The mechanical time initiation/temperature termination defrost control uses a motor-driven cam that operates in conjunction with a temperature-limiting control. The combination is supplied power from the load side of the compressor contactor.

A defrost timer is shown in Figure 8–16 mounted in the upper right-hand corner in a typical control box in an outdoor section of a heat pump system. The timer operates in conjunction with the terminating thermostat located on the evaporator coil, after the pressure-reducing device. This location is behind the panel holding the rating plate.

The terminating thermostat (Figure 8–17) is mounted against one of the tubes off the pressure-reducing device to measure the boiling point temperature of the refrigerant entering the coil. At the design closing temperature, selected

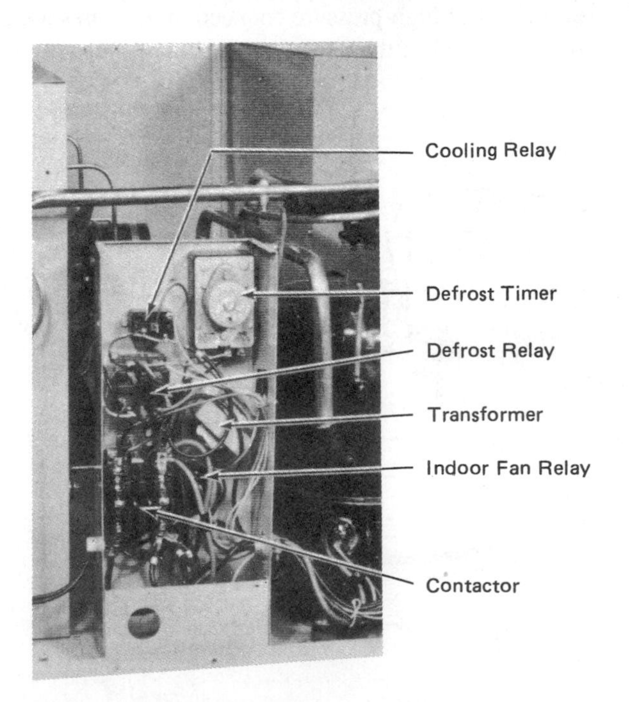

Figure 8–16 Typical heat pump electrical box (Courtesy Addison Products Co.)

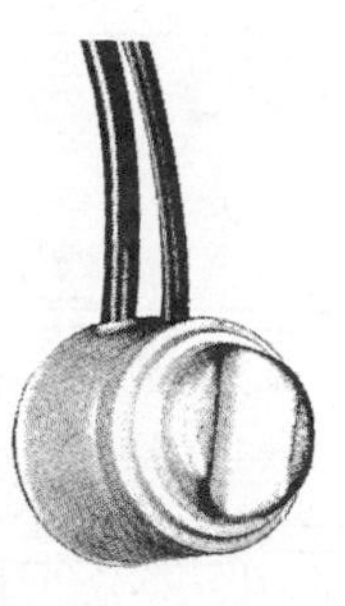

Figure 8–17 Terminating thermostat (Courtesy Addison Products Co.)

by the manufacturer, this control closes and allows the defrost timer to operate when supplied power. This thermostat also is the terminating thermostat, opening the defrost control circuit when the coil reaches the termination set point. This control usually has a set point to close between 26°F and 32°F and open at 45°F to 65°F. Each manufacturer sets these control set points according to the unit design. For best operation, substitution of this thermostat must be with one with the same set points.

Depending on the cam arrangement used, the timer will initiate the defrost function every 30 to 90 minutes of compressor operation when the coil temperature is below the initiating temperature of the terminating thermostat.

Figure 8–18 shows the cam arrangement for both 30-minute and 90-minute operation. In this illustration, the three-leaf contact arrangement is shown. The leaf and cam arrangement is in the normal operating position—both the No. 2 and No. 3 leaves are riding on the top of the cam. The insulator push rod is holding the contacts open between leaves No. 1 and No. 2. Leaf No. 3 is slightly shorter than leaf No. 2. As the cam turns clockwise, leaf No. 3 falls into the notch. This relieves the upward pressure in leaf No. 1. It drops and the contacts close. This energizes the defrost relay.

The defrost relay has a normally open contact that closes a circuit around the timer contact. This holds the relay closed when the timer cam rotates further, allowing leaf No. 2 to fall. This action opens the contact between leaf No. 1 and leaf No. 2.

The defrost function continues until the power to the defrost relay is interrupted. At this point, the defrost relay drops open and the system returns to the heating mode. This power interruption normally is caused by the terminating thermostat when the outdoor evaporator reaches the thermostat set point. *Any power interruption will terminate the defrost function.*

Figure 8–19 shows a wiring diagram of an outdoor section using a defrost timer in conjunction with a defrost relay. Tracing the circuit, we find that terminals 3 and 4 in the timer motor are connected in parallel with terminals 4 and 6 in the defrost relay. The closing of this contact between 4 and 6 in the defrost relay shorts out the contact in the timer motor.

The power is taken off terminal T1 on the compressor motor contactor, through the fuse protection and the defrost terminating thermostat to terminal 1 of the timer motor and the defrost relay coil. From terminal 1 in the defrost timer through the timer motor to terminal 4, the circuit goes to terminal 4 of the defrost relay and back to terminal T2 of the contactor. This puts line voltage power across the timer motor to operate it when both the contactor and terminating thermostat are closed.

This circuit remains active as long as the compressor contacts and the defrost terminating thermostat contacts are closed.

90 MIN. CAM AS SUPPLIED

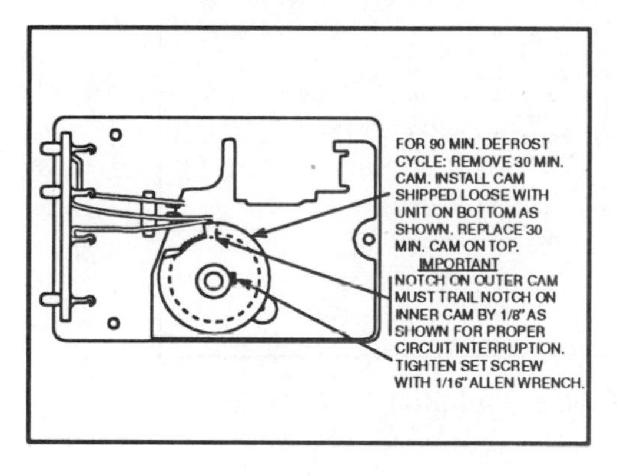

30 MIN. CAM

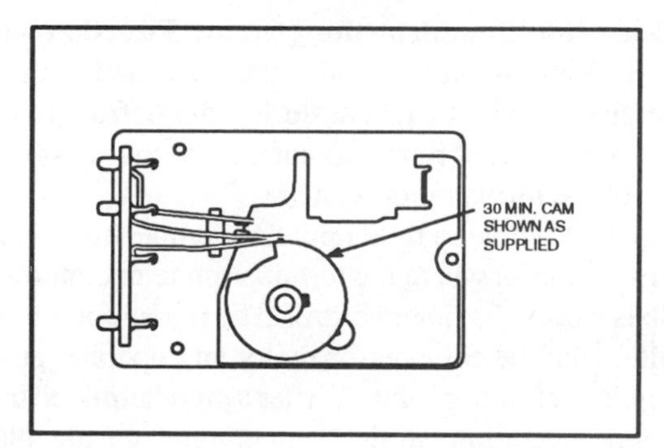

Figure 8–18 Timer motor and cam assembly (Courtesy Addison Products Co.)

Unit Will Not Initiate the Defrost Function. The most common cause of failure to initiate the defrost function is failure of the defrost terminating thermostat to reach the closing set point. Therefore:

1. Make sure the thermostat is mounted securely on the evaporator coil and properly insulated.
2. Make sure that the correct amount of air is passing through the inside condenser coil. Check the compressor discharge pressure.
3. Check the refrigerant charge. An overcharge of refrigerant will raise the compressor discharge pressure and evaporator coil operating temperature.
4. Check the timer motor for proper running operation.

Unit Will Not Terminate the Defrost Mode. The unit will remain in the defrost mode when:

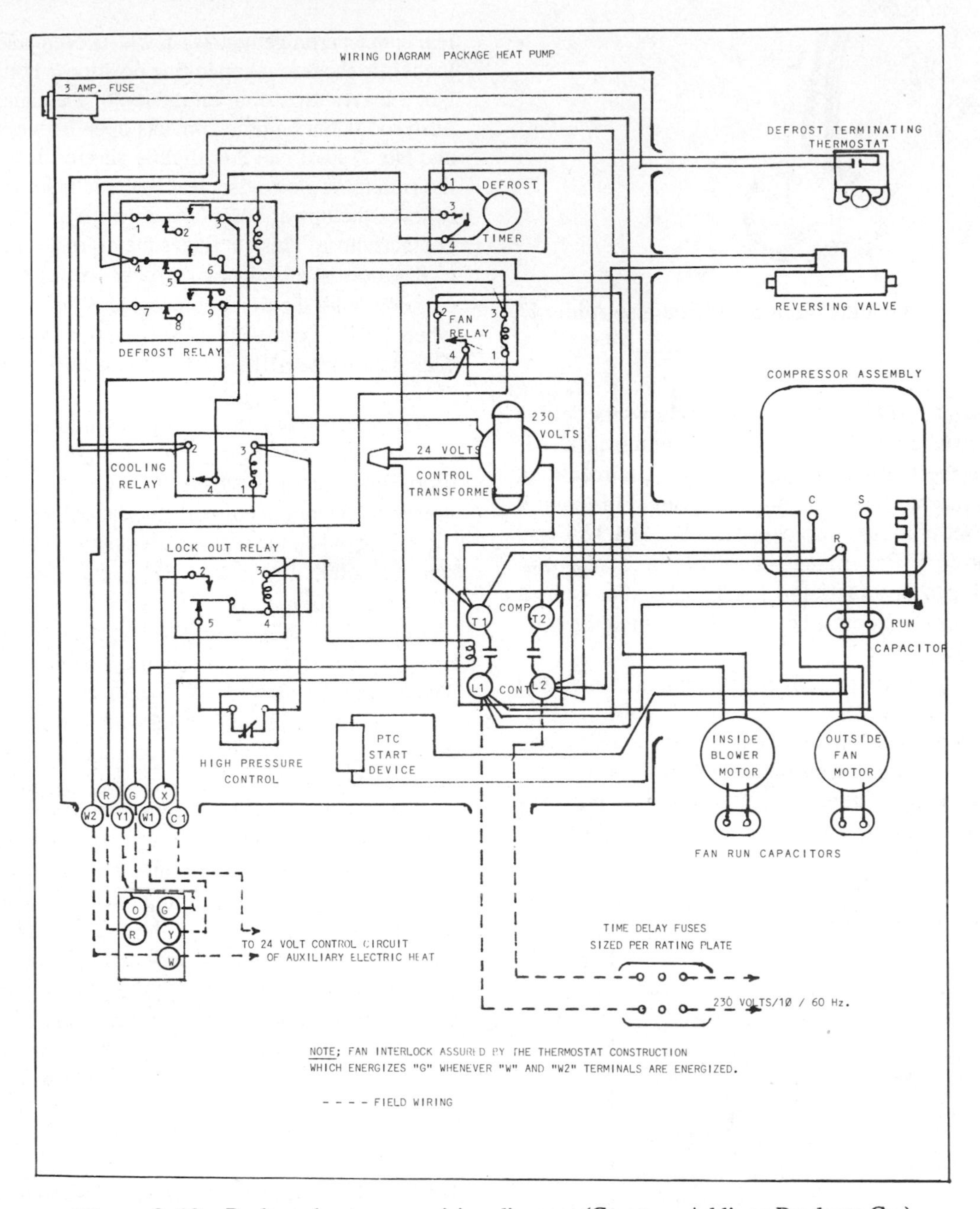

Figure 8–19 Package heat pump wiring diagram (Courtesy Addison Products Co.)

1. The outdoor coil will not reach termination temperature. Extreme outdoor ambient temperatures and high wind speed through the coil can prevent the coil termination temperature. Substitution of an electronic type that has a defrost limiting circuit should take care of the problem.
2. The defrost timer stalls. Check the timer for proper cam operation. Substitute a new timer for performance check.

Unit Has Unnecessary or Nuisance Defrost Cycles. Add the second cam to convert from 30 minute to 90 minute operating time period.

Unit Does Not Complete the Defrost Function—Frost and/or Ice Remain on the Coil. This is usually caused by an excessive amount of reheat during the defrost function. Reduce the amount of reheat to not exceed the sensible heat capacity of the refrigeration system.

Figure 8–20 shows a time initiation/temperature termination defrost control where the termination temperature-sensing bulb is part of the timer control. The timer motor is wired in parallel with the compressor motor and operates anytime the compressor motor operates. After a predetermined time of compressor operation, if the temperature-sensing bulb is

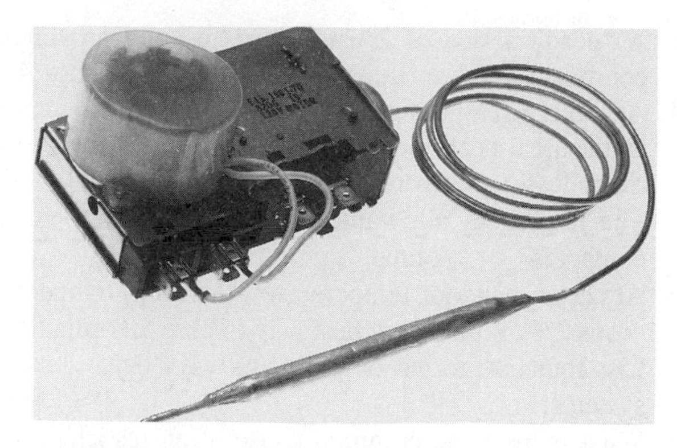

Figure 8–20 E15 time/temperature defrost control (Courtesy Ranco Inc.)

below the initiation temperature set point, the control will initiate a defrost function.

The basic problem is that the defrost cycle timing is fixed by the operation of the timer motor and cams. Therefore, this control does not allow for variance in the rate of frost formation.

During periods of 35°F to 45°F outdoor ambient temperatures with very high relative humidity, the defrost rate may not be sufficient to maintain a clean coil. During periods of very low humidity, the defrost function may occur even though it is not needed. If these characteristics are objectionable, a solid state (electronic) control should be substituted.

Electronic (Solid State) Type

Figure 8–21 shows an electronic (solid state) time initiation/temperature termination defrost control. Using the same type of termination thermostat as is used with the mechanical timer and thermostat combination (see Figure 8–17), this board will initiate the defrost function on a predetermined time period of compressor operation and the termination thermostat below the closing set point. The timer circuit cycles with the compressor operation. However, an integrated hold circuit accumulates the compressor operating time. The time between defrost functions is selected by switching a flexible lead and spade terminal between contact points of 50, 70, or 90 minutes.

This control, like the mechanical type, has an override circuit built into the board. This means that the control automatically terminates the defrost function at the end of a ten-minute operating period if the termination thermostat does not open. Therefore, strong wind through the outdoor evaporator coil has limited effect on the control.

Problems with this control will be the same as with the mechanical type.

1. The termination thermostat must be securely fastened and insulated.
2. Condenser cfm and refrigerant charge must be correct.
3. The amount of reheat must be limited to the sensible heat capacity of the refrigeration system.

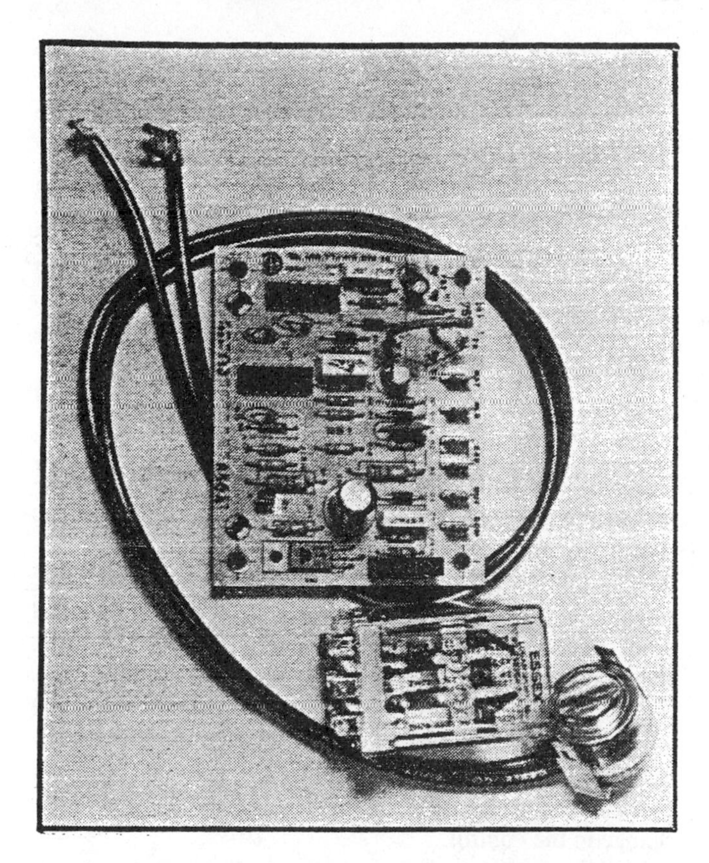

Figure 8–21 Solid state defrost control board (Courtesy Ranco Inc.)

8–6 PRESSURE-TIME INITIATION/ TEMPERATURE TERMINATION

The major fault of the time initiation/temperature termination mechanical control is the operation of the timer motor to measure compressor running time. This does not take into consideration the amount of frost on the coil. Thus, the differences in humidity level in the outside air have little effect on the defrost function timing.

To overcome this, the pressure-time initiation/temperature termination control uses a static pressure (Ps) switch to measure the air flow resistance through the coil to control the timer motor. This control, therefore, limits the timer motor operation only to the condition of a coil covered with frost and operating below the termination thermostat set point.

The Ranco D20 defrost pressure switch is used in conjunction with the Ranco E15 defrost timer to initiate the defrost function. A complete operating description of this control setup was discussed in Section 1, Chapter 3, Components.

When the outdoor evaporator coil is below the closing temperature set point of the termination thermostat, any closing of the pressure switch causes the timer motor to operate. This occurs only when power is available because the compressor motor is operating. After a sufficient period of operating time to cause the cam to turn to drop the follower into the initiation slot, the defrost function is initiated. Termination is by the rise in the feeler bulb temperature to force the control switch out of the defrost mode into the heating mode.

The major advantage of this control is the fact that the defrost function can only be initiated when the coil is sufficiently restricted with frost to raise the static pressure (Ps) drop across the coil to the closing point of the switch contacts. This control system also has the advantage of automatic termination of the defrost period regardless of the outdoor coil temperature. Therefore, short unnecessary defrost periods are eliminated. The effect of cold heavy wind on the unit is minimized.

Problems with the control, however, can be the same as with the pressure initiated/temperature terminated control.

Unit Will Not Initiate the Defrost Function. Any restriction in the pressure-sensing tubes will cause failure of the control to initiate the defrost function. The temperature-sensing bulb of the control must be securely fastened to the evaporator coil tube and thoroughly insulated with waterproof insulation. Thermomastic material between the bulb and tube is a must.

The bulb must also be mounted in a reverse vertical position with the connecting capillary tube out the bottom of the bulb. This is to prevent the operating charge in the power element from gathering in the bulb and reducing the pressure change in the control.

Unit Will Not Terminate the Defrost Mode. This problem occurs when the switch in the control seizes, the operating cams stick, or there is a defective timer motor. Replace the control.

Unit Does Not Completely Defrost the Coil—Frost and/or Ice Remain on the Coil. The system uses a directly controlled defrost relay to keep the unit in the defrost mode until the coil temperature reaches the termination temperature set point. Any power interruption only stops the timer motor until power is resumed. The system will then automatically resume the defrost function until the proper coil temperature is reached. Though it is not critical, the amount of reheat should be kept to a minimum to prevent delays in the heating of the conditioned area.

Discussion of trouble diagnosis in each control type assumes that the proper power is available to the unit as well as the control system. This text covers the various basic types of defrost controls. To cover all makes and models is not possible in a single text. The service technician should gather and retain manufacturers' literature and specification sheets as controls are developed and marketed.

REVIEW QUESTIONS

1. What is the easiest test for an inefficient compressor?
2. When the thermostat heat-cool season switch is placed in the "cool" position, the compressor runs constantly regardless of the setting of the "cool" temperature indicator. What is the most likely cause of the problem?
3. A "dead" period of 2°F to 4°F is usually provided in combination thermostats. What is this time factor for?
4. A cooling temperature setting of 78°F on a standard mechanical heat-cool thermostat will result in a maximum heating temperature setting of _____°F.
5. The maximum allowable resistance of a low-voltage contact is _____ ohms.
6. At certain outdoor temperatures, the room temperature drops 2°F. What is the best way to cure this situation?
7. List four categories of problems with defrost control systems.
8. What is the most common cause of an ice ring on the bottom of the outside coil?
9. What is the most common cause of improper defrost operation of a temperature initiation/temperature termination defrost control?
10. What is the most common cause of the problem in Problem 9?
11. What is the most common cause of failure to initiate the defrost mode in a pressure initiation/temperature termination defrost control system?
12. In a pressure initiation/temperature termination defrost control system, what two things must happen before the defrost function will start?
13. In order to initiate the defrost mode with the pressure initiation/temperature termination defrost control, how much of the outdoor coil needs to be blocked off on most units?
    a. 50% to 60%
    b. 60% to 70%
    c. 70% to 80%
    d. 80% to 90%
    e. 90% to 100%
14. What is the most common cause of failure to initiate the defrost mode in a time initiation/temperature termination defrost control system?
15. When the defrost timer contacts close and initiate the defrost mode, what holds the control in the defrost mode when the timer contacts open?
16. The amount of reheat used in the defrost mode must not exceed the net cooling capacity of the heat pump. True or False?
17. Name the four functions that the defrost system must perform to properly operate the defrost mode.

Section THREE
Liquid-to-Air Type Systems

Liquid-to-Air 9

9–1 GENERAL

With the rise in popularity of the air-to-air heat pump, a demand developed in areas of extreme temperature swings for a better heat source. The most desirable heat source would be one that would retain a constant temperature year-round. The readily available heat source that comes the closest to meeting this requirement is the underground water supply or "Earth Aquifer." The water in the earth at a level of 50 feet or more below grade level varies very little in temperature. Figure 9–1 is a line graph showing the average water temperatures at the 50-foot to 150-foot level throughout the continental United States.

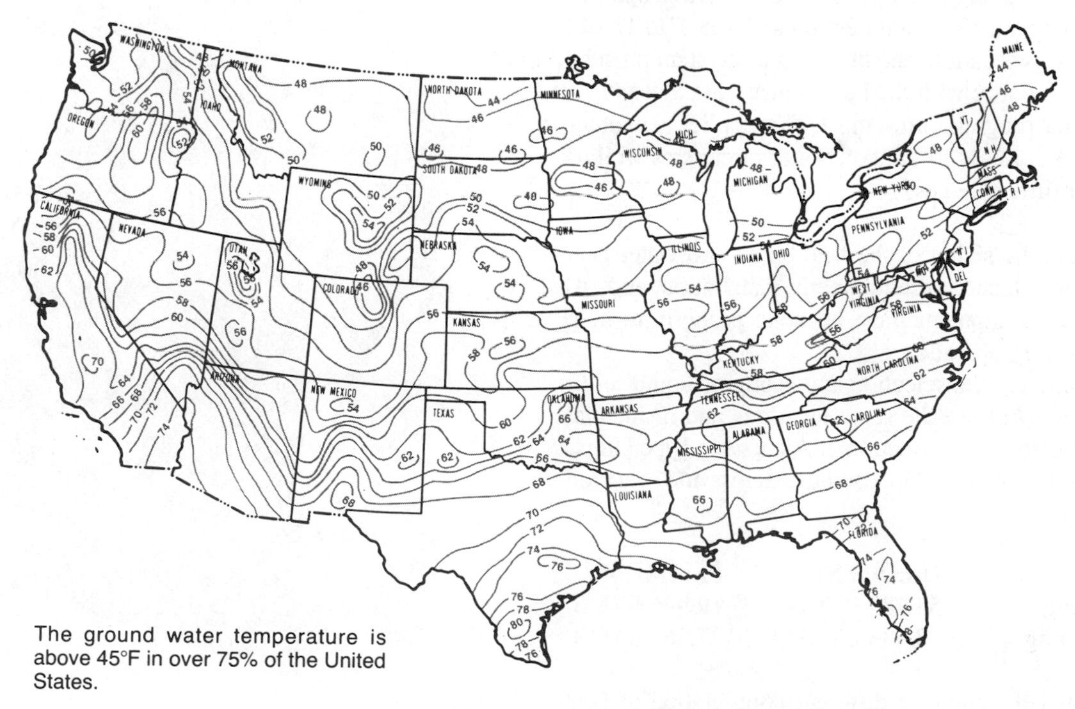

Figure 9–1 Groundwater and well water temperature—50' to 150' in depth (Courtesy Bard Manufacturing Co.)

In all of the states where an adequate supply of water is available, the average temperature of the water is such that it can be used as a heat source for heating operation or as a heat sink for cooling operation. To use this availability, liquid-to-air heating and cooling heat pumps have been developed. A variety of ways of using the ground water have been developed. These include:

1. Well water source—open circuit
2. Well water source—closed circuit
3. Buried closed loop
4. Water reservoir

Each of these will be discussed in Chapter 11, Types of Liquid Heat Sources and Sinks.

The basic fundamentals of a liquid-to-air heat pump system were discussed in Chapter 1, Basic Fundamentals. This described the fundamental system of refrigerant flow for both the heating and cooling modes. In Section 1, Chapter 3, Components, liquid-to-refrigerant heat exchangers were discussed. These sections should be reviewed. Each type of equipment will be reviewed in each text category.

9–2 SIZING

The rules for sizing liquid-to-air heat pumps are the same as for air-to-air heat pumps:

1. The unit must be sized to handle the cooling load.
2. Enough auxiliary heat must be included to handle the entire heat loss in the event of heat pump system malfunction.
3. The air distribution system must be draft free in spite of handling supply air with temperatures of 105°F to 110°F.
4. Installation, evacuation, and charging the system must be done in accordance with the best industry practices.
5. Maintenance programs are a must. This applies to maintenance of cleanliness in the coil water circuits as well as regular air filter changes.

Each step in the sizing of the heat pump and water heat source system, selection of equipment, installation, and adjustment of the equipment must be done properly to help ensure satisfactory operation.

When discussing the selection and sizing of liquid-to-air heat pumps, we will use the same locations and residential heat loss and gain used in Section 2, Air-to-Air Type Units.

Repeating the calculated loads, the heat loss and heat gain were:

	Omaha, NB	**Phoenix, AZ**
Winter Heating	53,782 BTUH	19,996 BTUH
Summer Cooling	18,833 BTUH	27,265 BTUH

Each location will require a different combination of heat pump and auxiliary heat because of the different heating and cooling load ratios.

As explained in Section 2, the heat loss and heat gain calculations were made according to the recommended practices outlined in the Environmental Systems Library published by the Air Conditioning Contractors Association (ACCA). It is highly recommended that these manuals be used, as they are constantly being updated as new information is being developed.

9–3 EQUIPMENT (SPLIT SYSTEMS)

Liquid-to-air heat pumps are made in the same style and configurations as air-to-air heat pumps. Both split systems using high-side and low-side sections and connected by refrigerant lines as well as package systems are marketed.

High Side Section

Figure 9–2 shows the enclosed cabinet of the high side section. Access panels close the front of the cabinet and the major portion of the top. The permanent portion of the top cover contains connections to the liquid and vapor lines, supply and discharge water, and water connections for the domestic water heater.

A popular feature with liquid-to-air heat pumps is the ability to heat domestic water with waste heat from the heat pump. This is especially desirable during the summer cooling season where all the heat picked up from the conditioned areas is wasted. Some of this wasted heat is recovered to heat the domestic hot water supply.

Figure 9–3 shows the refrigerant flow cycle of a liquid-to-air heat pump with the domestic hot water exchanger in the

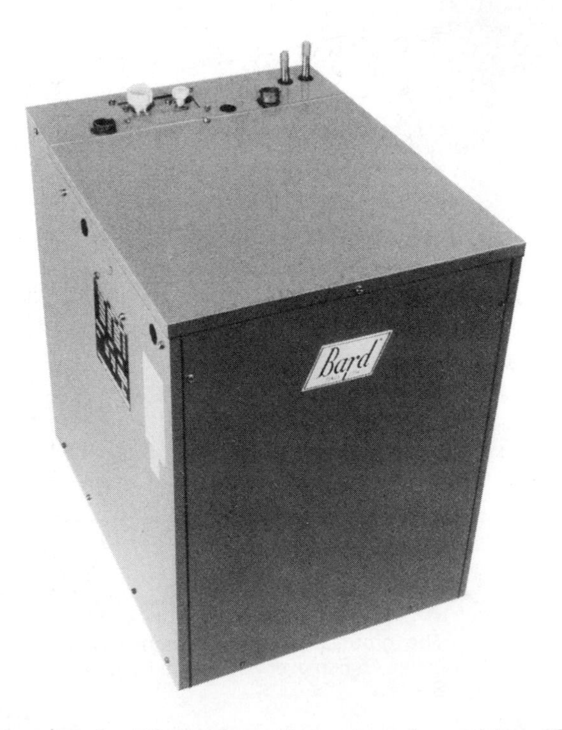

Figure 9–2 High side section—complete cabinet (Courtesy Bard Manufacturing Co.)

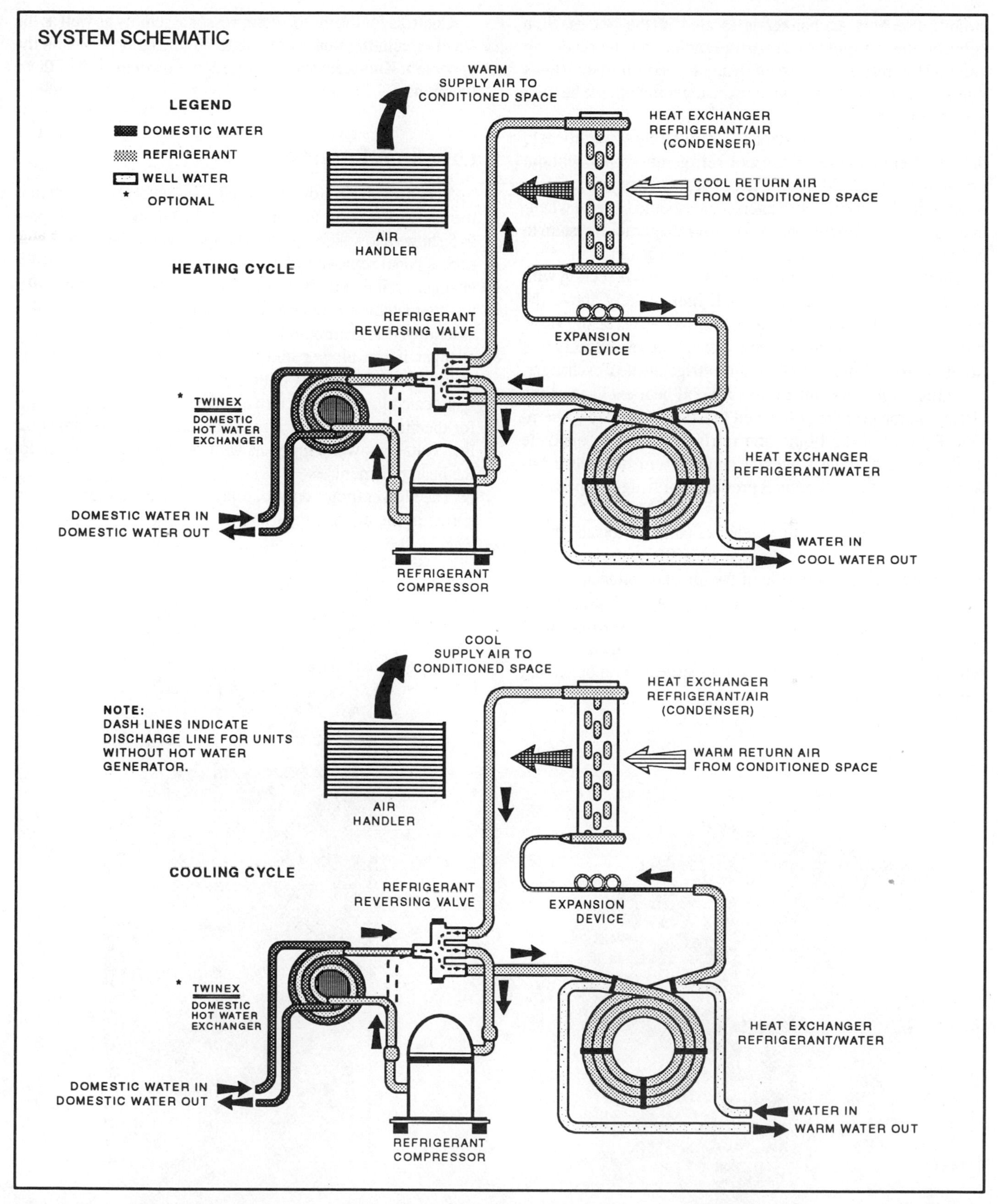

Figure 9–3 Refrigeration and water circuits (Courtesy Friedrich Air Conditioning and Refrigeration Co., Climate Master Division)

circuit. The heat exchanger is located in the hot gas line between the compressor discharge outlet and the reversing valve. Hot refrigerant vapor from the compressor flows through the domestic hot water exchanger in both the heating and cooling modes.

This heat exchanger is sized to remove only enough heat to drop the temperature of the hot refrigerant vapor without producing condensed liquid. These heat exchangers are usually referred to as desuperheaters. Care should be used whenever adding a desuperheater coil to any refrigeration system to ensure against overcapacity of the desuperheater coil.

Figure 9–4 shows the interior of the high side section with the front and top panels removed. Immediately above the compressor is the reversing valve. Behind the reversing valve is the pressure-reducing device and check valve assembly (the trombone) on the high side liquid-to-refrigerant heat exchanger.

Using a tube-in-a-tube (coaxial) coil heat exchanger, the circuits in the coil are connected to have counterflow operation. By having the liquid and refrigerant flow in opposite directions, the highest average temperature difference between liquid and refrigerant is produced with higher operating efficiency.

A separate tube-in-a-tube heat exchanger is located to the right of the motor-compressor assembly. This tube has the hot gas connection from the side of the top end connected to the reversing valve. The water circulating pump discharge is into the top connection of the coil. This coil also operates on a counterflow principle. The water enters the same end of the tube-in-the-tube assembly as the desuperheated vapor leaving the tube assembly.

Figure 9–4 Interior view of the high side (Courtesy Bard Manufacturing Co.)

Controls for the motor-compressor assembly as well as the water circulating pump are located in the control box on the upper left. Knockouts are provided for power to each of these functions.

Low Side Section

The low side section of a liquid-to-air heat pump system is the same as used in the air-to-air heat pump system. Figure 9–5 shows a low side section of a split system. With the front access panel removed, the indoor air-to-refrigerant heat exchanger is shown in the bottom section. This heat exchanger has its matching pressure-reducing device and check valve assembly (the trombone). Immediately above the heat exchanger is the blower and motor assembly for circulating conditioned air to the conditioned area.

Located in the top section are the electric heater elements for the auxiliary heat supply. This section also contains the low-voltage power supply, as well as sequencer controls for the heater elements.

The blower motor control relay is located in the upper left corner of the main compartment.

Figure 9–5 Low side section—air handler (Courtesy Bard Manufacturing Co.)

Refrigerant Lines

The liquid and vapor lines used to connect the two sections to complete the system are standard refrigerant lines. The vapor line is insulated to maintain system efficiency. In the case of the illustrated equipment, quick-couple fittings using precharged lines are used. Other systems, however, may use field-soldered lines as well as flare-type connections. These installations require evacuation after installation and leak testing. Industry use of precharged lines is increasing but they have not received 100% acceptance.

The split system has the versatility of installation because the low side section may be installed with vertical up, vertical down, or horizontal air flow. The high side section is not classified as the "Outdoor" section as installation of this equipment outdoors is not recommended. The outdoor atmosphere is not needed for operation and the danger of freeze up would exist. Therefore, the high side section does not have to be in the conditioned area, but it must be located where the surrounding temperature will not drop below 32°F.

9–4 ADD-ON SYSTEMS

Liquid-to-air heat pumps are marketed for add-on type applications in the same manner as air-to-air systems. Figure 9–6 shows the heat exchanger assembly that would be added to a gas-fired or oil-fired heating unit. This heat exchanger has the necessary pressure-reducing device and check valve assembly (the trombone) as well as condensate pan and drain connection and refrigerant line connections.

Location of this device requires the same restrictions as adding air conditioning to any heating system. The heat exchanger is located downstream from the heating unit heat exchanger. This is to prevent damage to the heat exchanger in the heating unit.

This application also requires the complete changeover from heat pump to auxiliary heat at the application balance point. This will be covered in the discussion on Balance Point in this chapter.

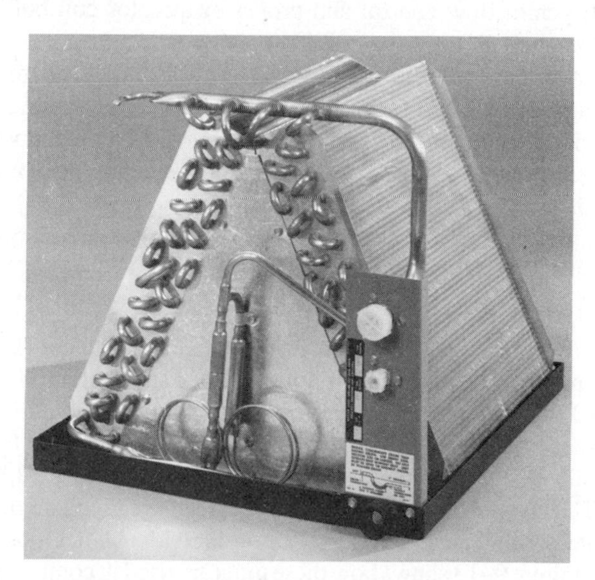

Figure 9–6 Heat pump "A" coil (Courtesy Bard Manufacturing Co.)

9–5 EQUIPMENT (PACKAGE SYSTEMS)

The requirement of locating the high side section of the heat pump in a nonfreeze area promoted the design and marketing of the complete package unit before the split unit. All manufacturers of liquid-to-air heat pumps produce package systems. Only a few have added the split system to their line of products.

Package units are being marketed in a wide variety of products, from standard packages for domestic installations to large packages for industrial use. Some specialty applications are also being marketed.

The major design fault with the package unit is that the motor-compressor assembly will operate in only one position. It cannot be inverted or laid horizontal. Therefore, each package design has a single application conformity.

Vertical Type

Figure 9–7 shows a vertical-type package unit with the complete cabinet enclosed. The air filter on the right side is the return air connection with the supply air out the top.

Figure 9–8 shows the same package unit with the panels removed. The upper and lower compartments are supported by an airtight panel as the upper section is part of the conditioned air circulating system. On the right is the air-to-refrigerant heat exchanger with the condensate pan beneath it.

Figure 9–7 Vertical package unit with the complete cabinet enclosed (Courtesy Bard Manufacturing Co.)

Figure 9–8 Interior view of vertical package unit (Courtesy Bard Manufacturing Co.)

In the top is the blower and motor assembly for air circulation. Beneath the blower assembly is the control panel with the low-voltage power supply. Alongside the control panel is the circulating pump for the domestic hot water heating circuit.

In the lower compartment are the motor-compressor assembly and the main liquid-to-refrigerant heat exchanger to the right of the compartment. The hot water desuperheating coil is located on the floor of the compartment under the motor-compressor assembly. The supply and discharge water connections are shown on the coaxial heat exchanger. The connections for the domestic hot water circuit are on the far side of the illustration.

Horizontal Type

To fill the need for application to attic or crawl space installations, the horizontal-type package unit was developed.

Figure 9–9 shows a horizontal design package unit with return and supply air connections on the same side of the unit. For versatility, these duct connections can be relocated. The return air connection with the air filter can be in either the front or rear panel. The blower assembly can be relocated to give any of the options for the supply duct connections.

Figure 9–10 illustrates the versatility that this manufacturer supplies in the unit design. This is a common practice in the marketing of these units.

Figure 9–11 shows the unit with the panels removed. The lower compartment contains the motor-compressor assembly, the liquid-to-refrigerant coaxial heat exchanger, domes-

Figure 9–9 Horizontal package unit (Courtesy Bard Manufacturing Co.)

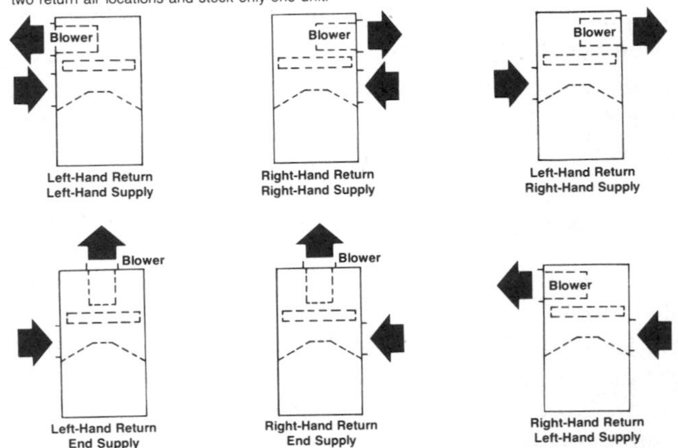

Figure 9–10 Duct connection flexibility in the horizontal package unit (Courtesy Bard Manufacturing Co.)

tic hot water coil, and circulating pump and accumulator. The right-hand compartment contains the air-to-refrigerant heat exchanger, pressure-reducing device, air circulating blower assembly, and the control panel. The pressure-reducing device is a bi-flow T. X. valve. This device is designed to provide refrigerant flow control and proper evaporator coil boiling points in both operating modes.

A liquid-to-air heat pump, with a controlled condensing temperature in the heating mode of 105°F, operates within the same pressure ranges in both the heating and cooling modes. Therefore, the use of a single pressure-reducing device for both functions is common practice.

Design Variety

Liquid-to-air heat pumps have developed into a vast variety of design and application, from the small residence to large commercial applications. Figure 9–12 is an illustration of a package terminal unit used in multiples in high rise buildings, motels, office buildings, and so on. These units, sized from 6,600 BTUH cooling and 6,400 BTUH heating to 14,800 BTUH cooling and 14,000 BTUH heating, are used for conditioning small areas such as offices or individual hotel rooms.

Figure 9–13 shows how these units are used in conjunction with large units to handle the conditioning requirements of a large building. This system shows two heat energy sources: a

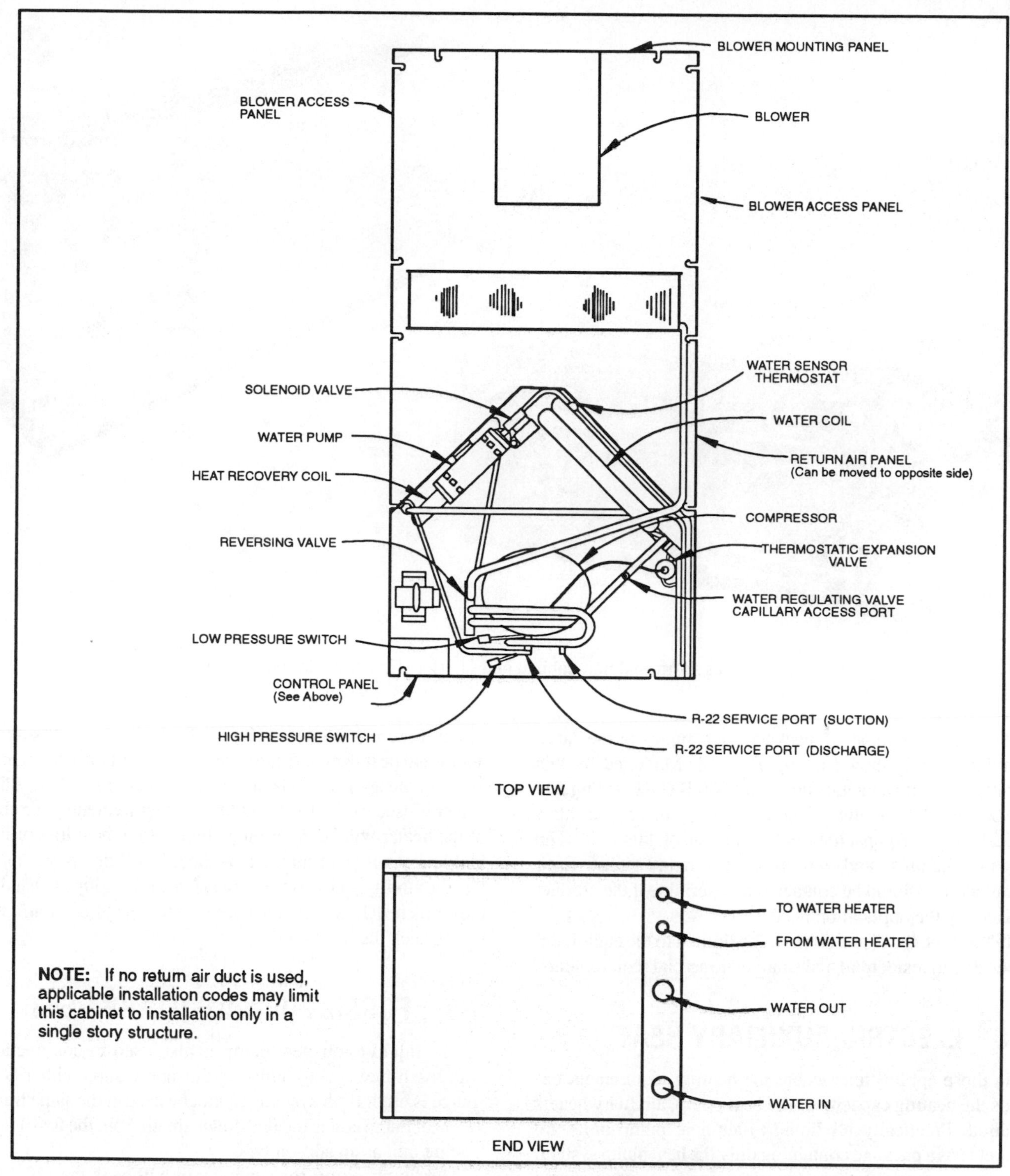

Figure 9–11 Interior view of the horizontal package unit (Courtesy Bard Manufacturing Co.)

hot water heater operating in conjunction with a solar energy system. The cooling energy heat sink is the evaporative cooling tower. This is an excellent way to handle the various heating and cooling loads of a large building. Each area or zone is conditioned by its own individual system.

With a central hydronic loop circulating water, each unit removes heat from or adds heat to the central loop, depending upon the individual room requirements. Such systems require the skill of professional engineers for successful application of all the types of equipment involved.

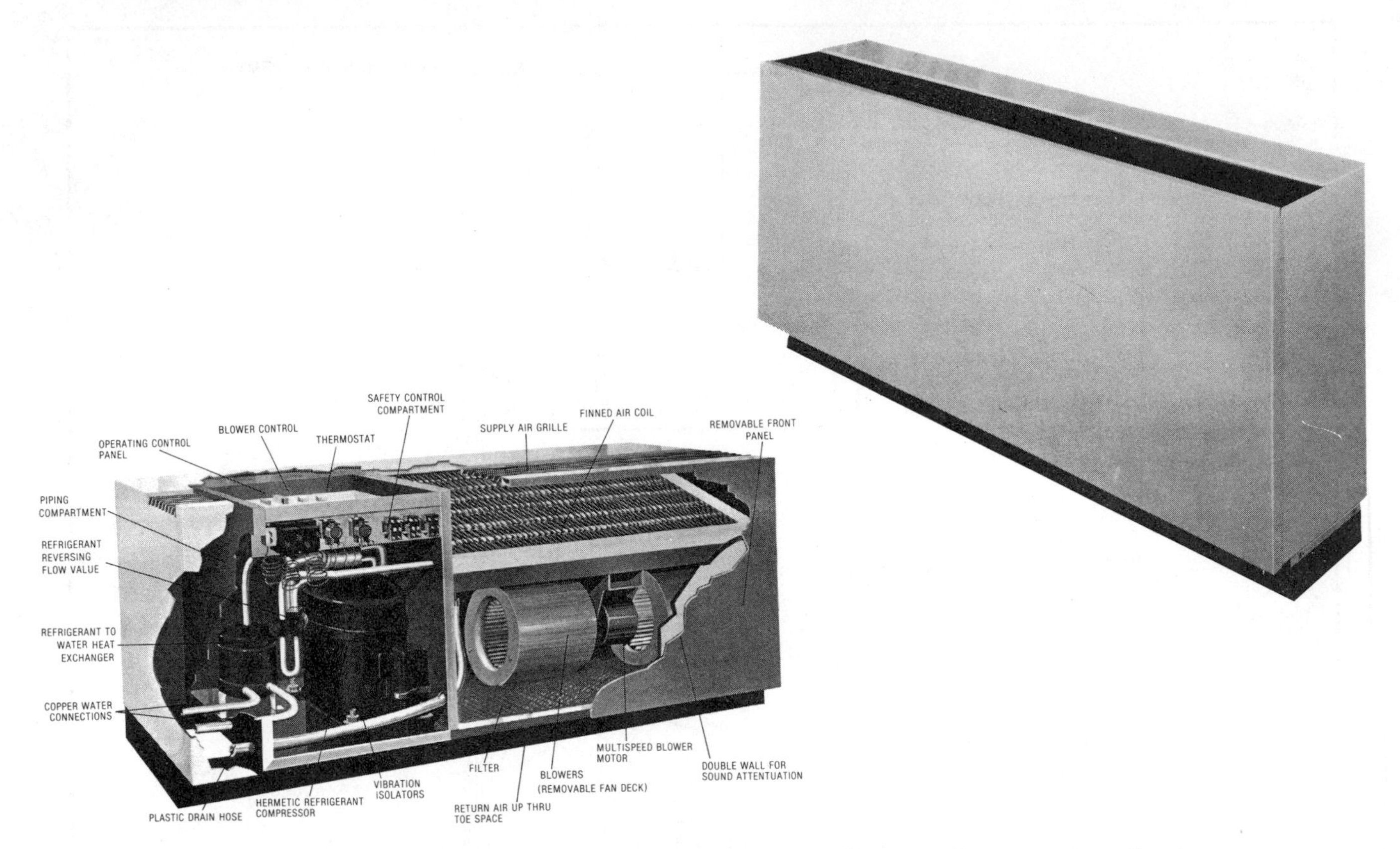

Figure 9–12 Package terminal unit cabinet and interior (Courtesy Command-Aire Corp.)

The other extreme of package heat pump is the large commercial unit shown in Figure 9–14. Marketed by this manufacturer in capacities up to 337,000 BTUH heating and 449,000 BTUH cooling (40 nominal tons), this represents a typical range of capacities of equipment of this type. The range of features and options are too many to list. Each manufacturer should be contacted to determine if the product will satisfy the application need.

In this text, the discussion will be limited to the equipment applicable to residential and small commercial requirements.

9–6 ELECTRIC AUXILIARY HEAT

In those applications where the heating requirement exceeds the heating capacity of the heat pump, auxiliary heat is required. Practically all liquid-to-air heat pump units are marketed as a package containing only the heat pump system. This is the opposite from the air-to-air heat pump where multiple options of electric heat sizes are available for insertion as an integral part of the unit.

With the liquid-to-air heat pump, the auxiliary heat is supplied by the use of electric duct heaters. Located in the supply duct, 4 feet (48") downstream from the unit, a duct heater, Figure 9–15, would be used. This heater will supply additional heat over and above the capacity of the heat pump.

After the application balance point is determined, the duct heater is used to handle the additional requirements. The size of the heater will be determined by the total heat loss of the building. As in air-to-air systems, electric utilities may require that the auxiliary electric heat be sized large enough to handle the entire heat loss in the event of heat pump system malfunction. Check the local codes.

9–7 FOSSIL FUEL AUXILIARY HEAT

The liquid-to-air heat pump is also used in conjunction with fossil fuel heating units. In this application, either heat source is used. Both systems cannot be used at the same time. The heat pump coil is located downstream from the fossil fuel heating unit as an add-on type.

The supply air off the heating unit will be in the 130°F to 150°F range. This high an air temperature entering the condenser coil of the heat pump, with the heat pump operating, will result in system damage. Therefore, the system must operate with either system operating, not both at the same time.

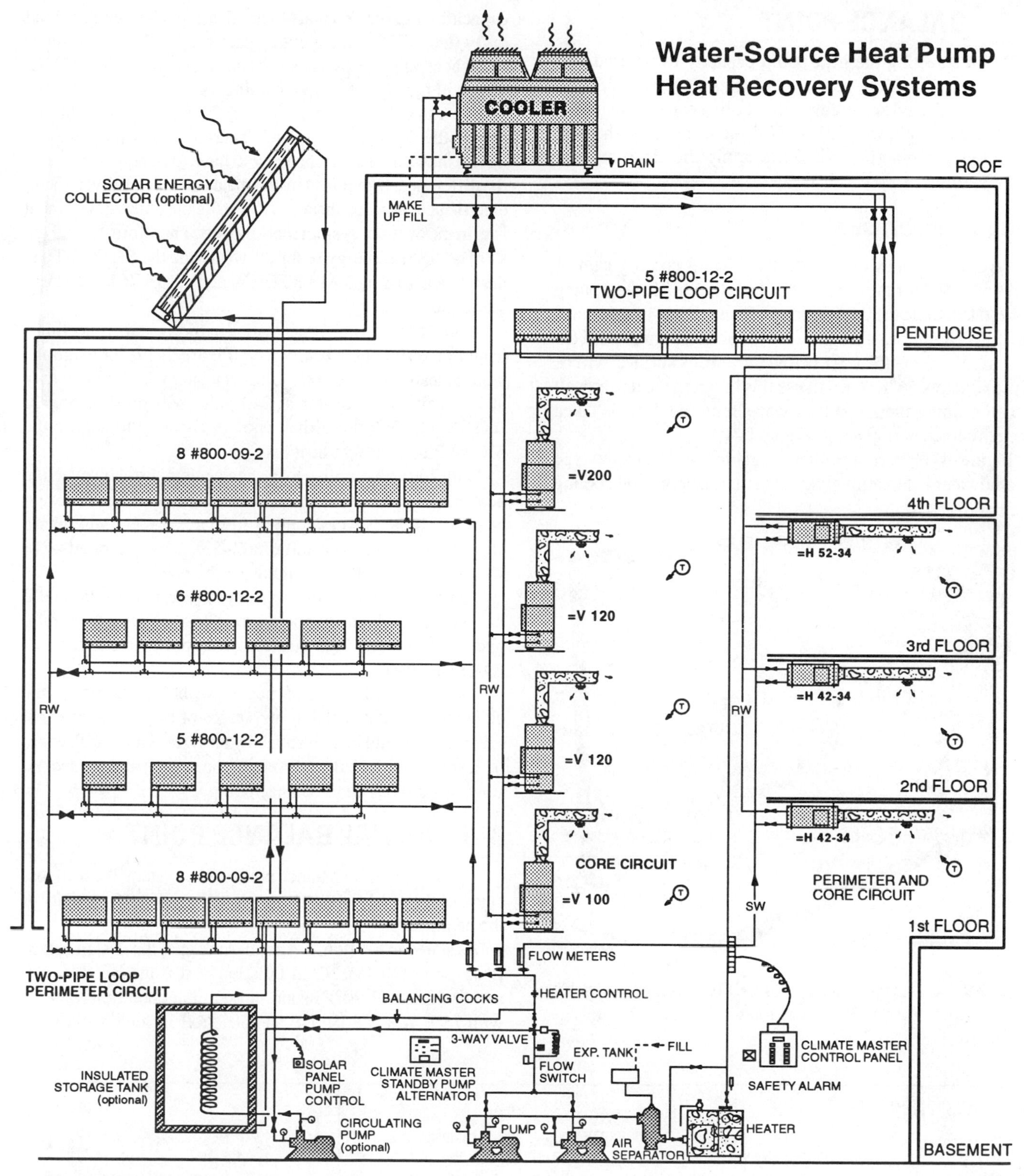

Figure 9–13 Commercial heat recovery system (Courtesy Friedrich Air Conditioning and Refrigeration Co., Climate Master Division)

9–8 BALANCE POINT

When selecting the heat pump capacity to be used in a particular application, rule No. 1 states that the unit must always be selected according to the cooling capacity required. For sizing the unit, we will use the same location and heating and cooling requirements used to apply the air-to-air heat pump systems.

Omaha, Nebraska

With a heat loss of 53,782 BTUH and cooling gain of 19,996 BTUH for the Omaha, NB installation, the heat pump is sized to handle the cooling load with auxiliary heat to make up the difference. The ground water temperature in the Omaha, NB area, 15' to 50' below the earth's surface, will be approximately 56°F. We will use this in sizing the unit as well as determining the amount of water required. This is taken from the line chart given in Figure 9–1.

Figure 9–16 is a specification table for a package-type groundwater heat pump giving the rated heating and cooling capacities. For the Omaha, NB application with a cooling load of 19,996 BTUH, we would select the WPH22–1 groundwater heat pump, which has a cooling rating of 21,400 BTUH. This will handle the design cooling load.

The rated heating capacity of 27,600 BTUH will not handle the design heating load. Therefore, auxiliary electric heat in the form of a duct heater will have to be added to supply the difference. If required by the local codes or utility, enough electric heat must be installed to handle the entire design heat loss in the event of system malfunction. Therefore, to find the kWh of electric energy required, we divide the 53,782 BTUH design heat loss by 3,413 BTU/kW and get 15.76 kW to meet the requirement.

A duct heater would be installed in the system. Selecting a heater of sufficient capacity from Figure 9–17, for example, would mean using the 7603–52 duct heater with 19.2 capacity at 240 volts. This heater at 240 volts will provide 65,510 BTUH, which is more than enough capacity to supply the 53,782 BTUH design heat loss.

With 208 volts applied to the heater elements, however, the heat output is 49,133 BTUH. This will not handle the design heat loss. If the power supply is from a 208 volt, three-phase supply, the next larger unit (7602–53), which is rated at 18 kWh or 61,416 BTUH, will have to be used.

To prevent the electric auxiliary heat from operating until the outside ambient temperature is low enough to require operation to handle the heat loss, outdoor ambient controls (OAT) are used.

Using the 7603–52 duct heater, with four elements, the elements would be used in two stages of two elements each. Each stage would be controlled by an OAT. One OAT would be set at the initial balance point of the system and the second stage set at the second balance point.

Figure 9–14 Large commercial unit (Courtesy Friedrich Air Conditioning and Refrigeration Co., Climate Master Division)

9–9 INITIAL BALANCE POINT

To find the initial balance point of the system, the heat loss curve of the conditioned area is plotted against the capacity of the heat pump.

As stated in Section 2, Air-To-Air Type, the design conditions for the Omaha, NB installation were established at –5°F in the winter with 70°F inside and summer conditions of 95°F with 75°F WB at 80°F DB and 50% RH inside conditions.

DUCT HEATERS

				MINIMUM	WIRE SIZE*		MAX.	DIMENSIONS			
PART NO.	PH	VOLTS	KW	AMPACITY	Cu	A1	FUSE	H&A	W	B	C
8604-067	1	240	4.8	25	#10	#8	25	8	12	17	6
8604-068	1	240	9.6	50	#6	#4	50	8	16	17	6
8604-069**	1	240	15.0	79	#3	#1	80	12	18	27	12
8604-070**	1	240	19.2	100	#1	#0	100	12	18	27	12

* Use wire suitable for at least 90°C. **Fused units (over 48 amperes).
NOTE: All duct heaters are supplied with backup protection and internal fusing as required by NEC.

Figure 9–15 Duct heater specifications (Courtesy Addison Products Co.)

The calculated heat loss of 53,782 BTUH is plotted on the graph in Figure 9–18 at the –5°F ambient temperature. A straight line is drawn from this point to the 65°F point on the outside ambient temperature scale. The inside heating temperature used for calculation is 70°F. This temperature is not used for calculating the balance points. Internal heat sources such as lights, refrigerators, and so on will usually supply enough heat energy to handle the heat loss down to 65°F outdoor ambient temperature. Therefore, to promote accuracy in unit application, the 65°F outdoor temperature is considered the starting point for unit operation requirement.

With the heat loss at the range of outdoor temperatures encountered plotted on the graph, the heating capacity of the heat pump is also plotted on the same graph. The heating capacity of a liquid-to-air heat pump remains constant and is not affected by outdoor air temperatures. The use of regulating valves on the water supply keeps a constant condensing temperature of 105°F. The quantity of water used will vary with the water temperature, but the condensing temperature (compressor discharge pressure) remains constant. Therefore, the unit heating capacity remains constant.

WPH, YHP — GROUND WATER HEAT PUMPS

Model	Rated Cooling	Rated Heating	Electrical
WPH22-1	21,400	27,600	1-60-208/230
WPH28-1	28,600	34,600	1-60-208/230
WPH34-1	33,300	44,700	1-60-208/230
WPH44-1	41,300	51,800	1-60-208/230
WPH44-3	41,300	51,800	3-60-208/230
WPH54-1	50,600	65,800	1-60-208/230
WPH54-3	50,600	65,800	3-60-208/230
WPH64-1	63,000	73,200	1-60-208/230
WPH64-3	63,000	73,200	3-60-208/230
WPH64-4	63,000	73,200	3-60-460
YHP44-1	—	51,000	1-60-208/230

Figure 9–16 Rating table for ground water heat pumps (Courtesy Addison Products Co.)

The unit used in our example has a heating capacity of 27,600 BTUH. This unit capacity has been plotted on the graph in Figure 9–18. Where the two lines cross is the initial balance point. For this application, it is 28°F. This means that the liquid-to-air heat pump will handle the heat loss down to 28°F. At this temperature, the unit will run continuously.

9–10 SECOND BALANCE POINT

Any further drop in outdoor temperature will allow the conditioned area temperature to drop unless additional heating capacity is introduced. The original heat loss was calculated using factors based on 15 mph wind against the building. To allow for higher wind speeds with increased wind effect on the heat loss, 3°F is added to the initial balance point to determine the first OAT setting. In this application, the first OAT would be set at 31°F.

The closing of the contacts in the first OAT will energize the first stage of the auxiliary heat and supply 7.2 kWh or 32,755 BTUH. Added to the capacity of the liquid-to-air heat pump, a total of 60,355 BTUH is supplied to the conditioned area. Figure 9–19 shows the additional heating capacity of the first stage auxiliary heat added to the heating capacity of the heat pump.

As the outdoor temperature drops, the heat loss of the conditioned area increases. The system capacity, however, remains constant and the system capacity line parallels the heat pump capacity line. Where the system capacity line and the heat loss line cross is the "second balance point." In this application, we see that the second balance point is at –13°F. The median of annual extreme temperatures for the Omaha, NB area is –12°F.

The second stage of auxiliary heat is available, with the second OAT set at –10°F (–13°F + 3°F wind effect), when

			TOTAL HEATER CAPACITY 208V		TOTAL HEATER CAPACITY 240V (480V)		NO.			
MODEL NO.	FOR USE WITH	POWER SUPPLY	KW	BTUH	KW	BTUH	OF ELEMENTS	SEQ. STEPS	OUTDOOR THERMOSTAT	OPTIONAL REMOTE PANELS MODEL NO. / QTY.
7603-49	All Single Phase	1-60-208/240	3.6	12,283	4.8	16,378	1	1	None	7603-66/1
7603-50	All Single Phase	1-60-208/240	7.2	24,566	9.6	32,755	2	1	None	7603-67/1
7603-51	All Single Phase	1-60-208/240	10.8	36,850	14.4	49,133	3	2	7603-70	7603-66/1 Plus 7603-67/1
7603-52	2-5 Ton Single Phase	1-60-208/240	14.4	49,133	19.2	65,510	4	2	7603-70	7603-67/2
7603-53	2½-5 Ton Single Phase	1-60-208/240	18.0	61,416	24.0	81,888	5	3	7603-71	7603-66/1 Plus 7603-67/2
7603-54	3-5 Ton Single Phase	1-60-208-240	21.6	73,699	28.8	98,266	6	3	7603-71	7603-67/3
7603-55	4-5 Ton Single Phase	1-60-208-240	25.2	85,982	33.6	114,643	7	4	7603-71	7603-67/3 Plus 7603-66/1
7603-56	3-5 Ton Three Phase	3-60-208/240	7.2	24,566	9.6	32,755	2	1	None	7603-68/1
7603-57*	3-5 Ton Three Phase	3-60-208/240	10.8	36,850	14.4	49,133	3	2	7603-70	7603-68/1
7603-58	3-5 Ton Three Phase	3-60-208/240	14.4	49,133	19.2	65,510	4	2	7603-70	7603-68/2
7603-59	3-5 Ton Three Phase	3-60-208/240	18.0	61,416	24.0	81,888	5	3	7603-71	7603-68/2
7603-60*	3-5 Ton Three Phase	3-60-208/240	21.6	73,699	28.8	98,266	6	3	7603-71	7603-68/2
7603-61	4 & 5 Ton Three Phase	3-60-208/240	25,2	85,982	33.6	114,643	7	4	7603-71	7603-68/3
7603-62*	4 & 5 Ton 460V/3 Phase	3-60-480	—	—	11.8	40,262	3	1	None	Not Available
7603-63*	4 & 5 Ton 460V/3 Phase	3-60-480	—	—	19.9	67,899	3	1	None	Not Available
7603-64*	4 & 5 Ton 460V/3 Phase	3-60-480	—	—	23,6	80,523	6	1	7603-70	Not Available
7603-65*	4 & 5 Ton 460V/3 Phase	3-60-480	—	—	31.7	108,160	6	1	7603-70	Not Available

*These heater kits have equal phase loading and will not cause phase unbalance

NOTE: Heater kits are not fused and require field installation of over current protection.

Figure 9–17 Auxiliary electric heater kit specifications (Courtesy Addison Products Co.)

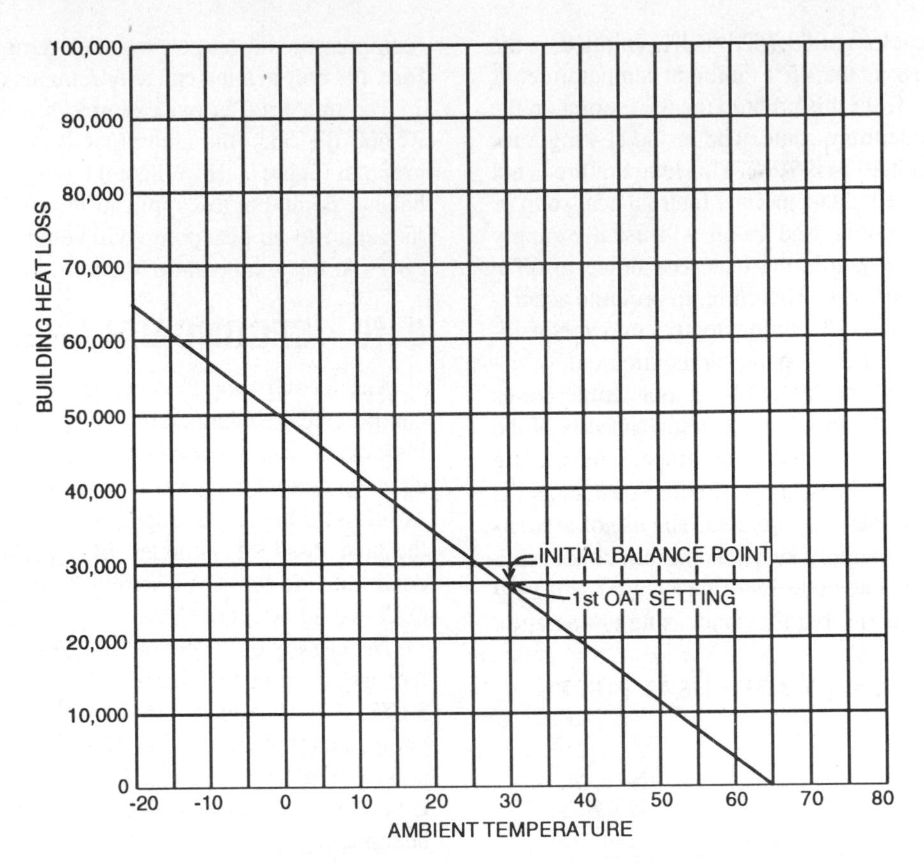

Figure 9–18 Balance point—Step 1

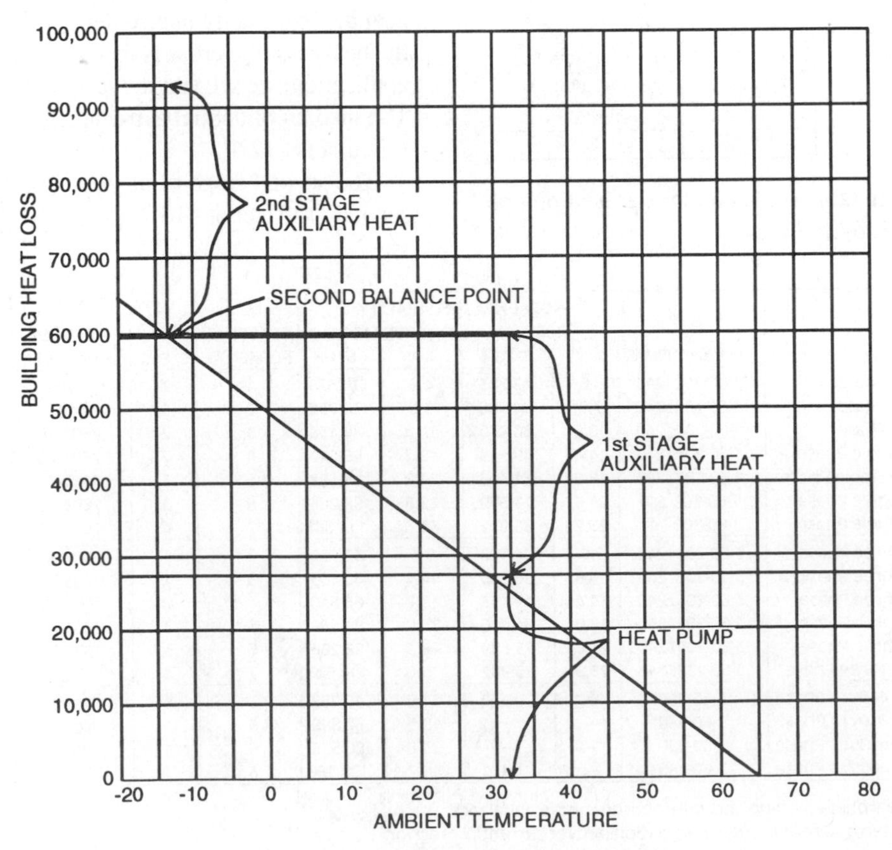

Figure 9–19 Balance point—Step 2

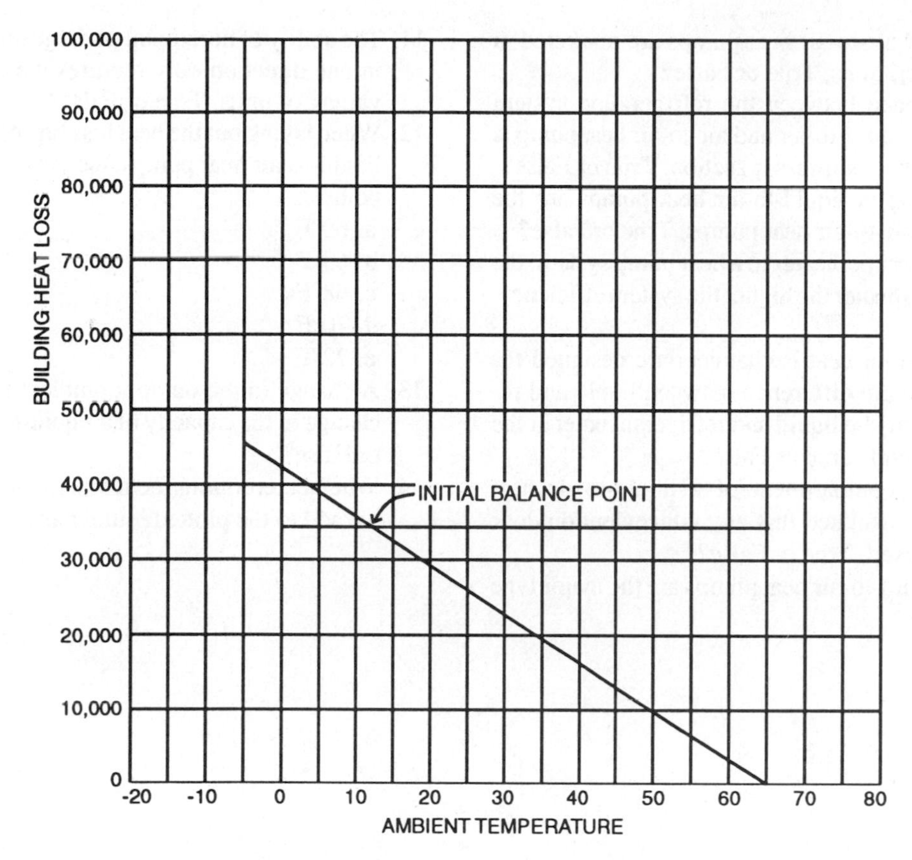

Figure 9–20 Balance point—Step 3

needed. The usage will be very small because of the low possibility of need. This second stage must be installed, however, to meet the extreme conditions and utility requirements.

Phoenix, Arizona

With a heat loss of 19,996 BTUH and a cooling gain of 27,265 BTUH, the heat pump sized to handle the cooling load will have the capacity to handle the winter heating load. From the specification table in Figure 9–16, we would select the WPH28–1 unit with a rated capacity of 28,600 BTUH. We do not need to make allowance for outdoor DB temperatures above the 105°F design temperature.

The use of groundwater as a heat source and sink will provide water at an average temperature of 70°F from 50' to 150' below ground level. The use of water from surface supplies such as lakes and streams is not recommended unless the water depth is 20' or more.

If an auxiliary electric duct heater must be installed to handle the entire heating requirement in the event of unit malfunction, a 7603–50 (Figure 9–17) duct heater with 7.2 kWh or 32,755 BTUH capacity is required. This is a two-stage element unit and could be installed with outdoor ambient thermostats (OAT) for two-stage operation. To see if this is necessary, the balance point of the heat pump versus the heat loss should be plotted.

The design conditions for the Phoenix, AZ installation were established at +34°F with 70°F inside in the winter and summer weather conditions of 105°F DB with 71°F WB at 50% RH. Using the point of 19,996 BTUH (20,000 BTUH) on the graph in Figure 9–20 at the +34°F design temperature, a line is drawn through this point from +65°F on the ambient temperature scale. The line is extended to an arbitrary point of 0°F which is 20°F below the lowest temperature recorded in the Phoenix, AZ area.

The heating capacity of the heat pump (34,600 BTUH) is plotted as a horizontal line at the 34,600 BTUH point on the building heat loss scale. Where the heat loss line and the unit capacity line cross, at +12°F, is the initial balance point. The possibility of the use of the auxiliary electric heat in normal operation will be very remote. Therefore, no outdoor ambient thermostat will be required. The auxiliary electric heat will only be used in conjunction with the emergency heat circuit of the thermostat.

REVIEW QUESTIONS

1. Aquifer temperatures 50' or more in depth will change very little with the seasons. True or False?
2. List four of the most popular methods of using groundwater as a heat source.
3. List the five basic rules for sizing liquid-to-air heat pumps.

4. Liquid-to-air and air-to-air heat pumps are marketed in the same configurations. True or False?
5. The only difference between the refrigeration system components in a liquid-to-air and air-to-air heat pump is the coil used in the compressor section. True or False?
6. The controls used in liquid-to-air heat pumps are the same as used in air-to-air heat pumps. True or False?
7. When adding a desuperheater to a heat pump system, the larger the desuperheater the higher the system efficiency. True or False?
8. Liquid-to-refrigerant heat exchangers are designed for constant temperature difference between liquid and refrigerant by having the liquid and refrigerant enter at the same end of the coil. True or False?
9. The compressor compartment of a liquid-to-air heat pump is so well insulated that any convenient outdoor location can be used. True or False?
10. Package unit liquid-to-air heat pumps are the major type marketed. Why?
11. The ability of horizontal package units to handle air flow in one direction only requires the marketing of a large variety of units. True or False?
12. When laying out the heat loss line of a heating load using liquid-to-air heat pumps, the inside temperature starting point is:
    a. 60°F
    b. 65°F
    c. 68°F
    d. 70°F
    e. 72°F
13. A change in the outdoor ambient temperature causes a change in the capacity of a liquid-to-air heat pump. True or False?
14. When determining the temperature setting of the OAT, we add to the plotted initial balance point. Why?

10 LIQUID HEAT SOURCE

10–1 GENERAL

Though it is possible to find heat pump applications that use liquids other than water, such as Glycol-Water solutions, waste heat from oil coolers, and so on, the most predominate source of heat is groundwater. This text will, therefore, be limited to the use of groundwater and buried water coils as the heat source or heat sink.

Groundwater is used as a heat source and heat sink by six different methods:

1. Stored water quantities
2. Drilled well—surface return
3. Drilled well—drilled well return
4. Horizontal pipe ground loop
5. Vertical pipe ground loop
6. Solar storage

Before selecting the type of supply system to be used, several factors must be determined:

1. The quantity of water required by the heat pump in gallons per minute (GPM). This will depend upon the temperature of the supply water and the unit size.
2. The temperature of the available water supply during the peak requirements in both the heating season (the coldest portion of the winter) and the cooling season (the hottest portion of the summer).
3. The quality of the water available.

These three factors must be considered when selecting the water source as well as the method of disposal.

10–2 WATER QUANTITY

The most important of the three factors is the quantity of water available to the system. All heat pump manufacturers list the water requirements in their specification literature for use in sizing the supply system. For example, Figure 10–1 is a table of capacity and efficiency ratings published by one manufacturer on two sizes of their range of units. Capacities for both cooling and heating are given for three different water flow rates at supply water temperatures from 45°F to 80°F in 5°F increments. A single cooling rating with 85°F water is given. No capacity rating on heating is given at 85°F water as this water temperature is too high for the heating cycle. This could cause cutoff on high head pressure lockout.

The rated capacity of the two units is given at ARI Standard 325–82 for groundwater heat pumps—70°F supply water at 5 GPM flow rate and with 50°F supply water at 5 GPM flow rate. Comparison of various units should be made at these conditions.

If the actual rated quantity of water required for the unit is not known, an average figure of 3 GPM per 12,000 BTUH cooling capacity or 3.5 GPM per 12,000 BTUH heating capacity can be used.

If the water supply is taken from the same source as domestic water, the entire water quantity demand of the building along with the heat pump must be taken into consideration. This will be discussed in more detail under the various types of water sources.

10–3 WATER TEMPERATURE

Figure 10–2 shows the average groundwater and well water temperatures at depths ranging from 50 to 150 feet throughout the United States. Using a 10°F drop in temperature of the water as it flows through the coaxial heat exchanger on the heating cycle, only in the extreme northern areas of the country would care have to be used in selecting the type of equipment for the particular installation. Some manufacturers market units that will operate with less than 10°F drop of the water through the coil and still produce enough heat transfer in the coaxial coil to produce boiling points above 32°F.

Working the unit too close to the freezing point of water in the coaxial coil could result in coil damage. Usually a boiling point of 28°F or lower in the refrigerant portion of the coil is needed to produce coil freeze up. This situation is serious enough to require a boiling point of not less than 34°F for best equipment protection.

This problem is not very widespread when using a well. However, extreme care must be exercised if the water source is a lake or stream or buried ground loop.

10–4 WATER QUALITY

All water contains a certain amount of minerals and other impurities, which may or may not affect the performance of the heat pump system. A good rule to follow is if you can drink the water, it is a satisfactory water supply for the heat pump system. If it is not suitable for drinking, special water treatment may be required to keep maintenance requirements of the system to a minimum.

CAPACITY AND EFFICIENCY RATINGS ①

MODELS WQS30 - WQSD30

	COOLING					HEATING				
WATER TEMP. °F	† GPM	CAPACITY	WATTS	EER*	EER ADJ**	GPM	CAPACITY	WATTS	COP*	COP ADJ**
	4	32,500	2,100	14.0	12.6	5	26,300	2,400	3.2	2.8
45	5	31,700	2,050	14.0	12.2	7	26,500	2,400	3.2	2.8
	7	30,800	2,000	14.2	11.7	8	27,100	2,450	3.2	2.7
	4	32,100	2,200	14.0	12.6	4	27,500	2,500	3.2	2.9
50	②5	31,600	2,150	14.1	12.4	5	28,200	2,500	3.3	3.0
	7	30,900	2,100	14.2	11.9	7	28,700	2,500	3.4	2.9
	4	31,700	2,300	13.7	12.4	4	29,000	2,500	3.4	3.1
55	5	31,400	2,250	13.9	12.2	5	29,600	2,500	3.5	3.1
	7	31,000	2,200	13.9	11.6	7	30,000	2,550	3.5	3.0
	4	31,200	2,400	13.2	12.0	4	30,300	2,550	3.5	3.2
60	5	31,200	2,350	13.4	11.9	5	31,000	2,550	3.6	3.2
	7	31,100	2,300	13.5	11.4	7	31,300	2,600	3.5	3.0
	4	30,800	2,400	13.1	11.9	4	31,600	2,600	3.6	3.3
65	5	31,100	2,450	12.7	11.4	5	32,200	2,600	3.6	3.3
	7	31,200	2,350	13.1	11.2	7	32,500	2,650	3.6	3.1
	4	30,400	2,600	11.9	10.9	4	32,900	2,600	3.7	3.4
70	②5	31,000	2,550	12.0	10.7	5	33,400	2,650	3.7	3.3
	7	31,300	2,450	12.4	10.6	7	33,700	2,700	3.7	3.2
	4	29,400	2,650	11.5	10.5	4	34,100	2,650	3.8	3.5
75	5	29,900	2,600	11.5	10.4	5	34,600	2,650	3.8	3.4
	7	30,300	2,550	11.7	10.0	7	34,900	2,700	3.8	3.3
	4	28,500	2,750	10.8	10.0	4	35,000	2,650	3.9	3.6
80	5	28,800	2,700	10.9	9.8	5	35,500	2,700	3.9	3.5
	7	29,300	2,650	10.9	9.5	7	36,000	2,700	3.9	3.4
	4	27,600	2,850	10.3	9.5					
85	5	27,800	2,800	10.3	9.3					
	7	29,000	2,700	10.5	9.1					

MODELS WQS36 - WQSD36

	COOLING					HEATING				
WATER TEMP. °F	† GPM	CAPACITY	WATTS	EER*	EER ADJ**	† GPM	CAPACITY	WATTS	COP*	COP ADJ**
	4	36,600	2,800	13.1	12.0	4	31,300	3,100	3.0	2.8
45	5	35,500	2,700	13.2	11.8	5	32,300	3,150	3.0	2.7
	7	31,200	2,500	13.0	11.2	7	33,600	3,200	3.1	2.7
	4	36,500	2,900	12.6	11.6	4	33,300	3,150	3.1	2.9
50	②5	35,600	2,800	12.7	11.5	5	34,300	3,200	3.1	2.9
	7	32,000	2,600	12.5	10.8	7	35,200	3,250	3.2	2.8
	4	36,500	3,000	12.2	11.3	4	35,600	3,200	3.3	3.0
55	5	35,700	2,900	12.3	11.1	5	36,400	3,250	3.3	3.0
	7	32,800	2,700	12.2	10.5	7	37,000	3,300	3.3	2.9
	4	36,500	3,100	11.8	10.9	4	37,500	3,250	3.4	3.1
60	5	35,800	3,000	11.8	10.7	5	38,300	3,300	3.4	3.1
	7	33,500	2,850	11.7	10.2	7	38,700	3,350	3.4	3.0
	4	36,400	3,200	11.4	10.6	4	39,000	3,300	3.5	3.2
65	5	35,900	3,100	11.5	10.4	5	39,900	3,350	3.5	3.2
	7	34,300	3,000	11.3	9.9	7	40,200	3,350	3.5	3.1
	4	36,400	3,300	11.0	10.3	4	40,500	3,350	3.5	3.3
70	②5	36,000	3,200	11.3	10.3	5	41,100	3,400	3.5	3.3
	7	35,000	3,100	11.1	9.7	7	41,600	3,450	3.5	3.2
	4	34,900	3,400	10.7	10.0	4	41,700	3,400	3.6	3.4
75	5	34,800	3,300	11.2	10.2	5	42,100	3,450	3.6	3.3
	7	34,200	3,200	10.9	9.7	7	42,800	3,500	3.6	3.2
	4	33,500	3,500	10.4	9.7	4	42,500	3,500	3.6	3.3
80	5	33,500	3,400	11.0	10.1	5	43,000	3,550	3.6	3.3
	7	33,400	3,350	10.7	9.5	7	43,500	3,550	3.6	3.2
	4	32,100	3,600	10.1	9.5					
85	5	32,400	3,500	10.9	10.0					
	7	32,600	3,450	10.6	9.4					

*Unit only rating — without well water pump watts included.
**Unit rating plus estimated water well pump power requirements of 60 watts per GPM.

†Rated water flow: 5 GPM.
①Applicable when matched with H3AQ1 and B36EHQ1 air handlers at rated air flow.
②Tested in accordance with ARI Standard 325-82 for ground water.

Figure 10–1 Capacity and efficiency rating from manufacturer's literature (Courtesy Bard Manufacturing Co.)

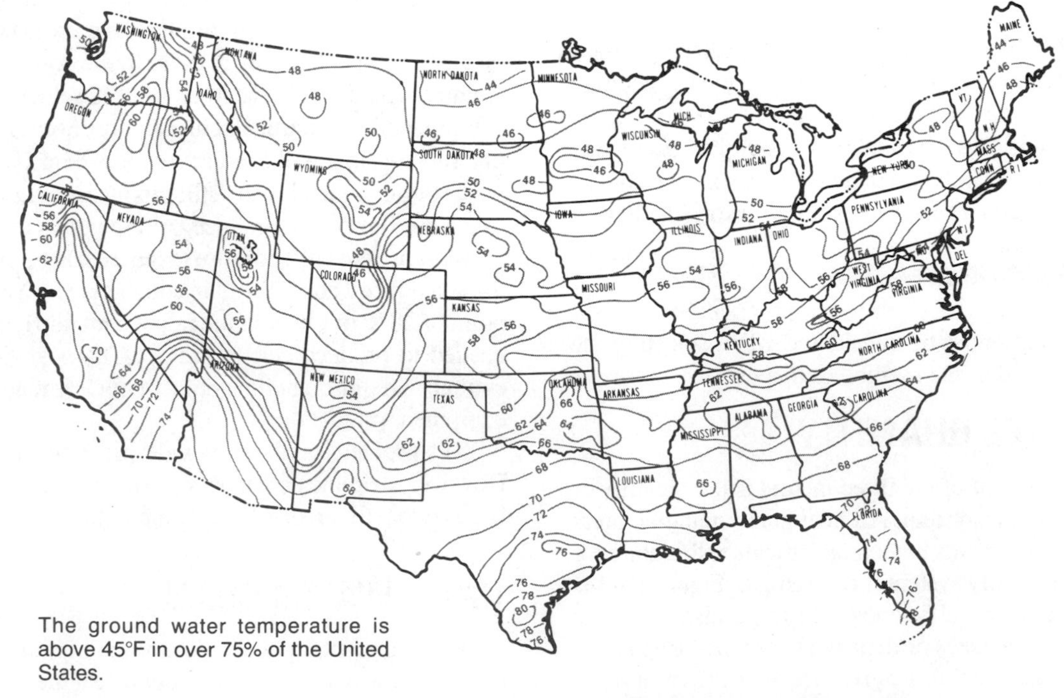

Figure 10–2 Groundwater and well water temperatures at 50' to 150' in depth (Courtesy Bard Manufacturing Co.)

Regardless of the water quantity, if it contains sand or other particles, a filter must be used to prevent erosion of the heat exchanger coil and fouling of the water flow control valves.

After flowing through the liquid-to-refrigerant heat exchanger, supply water has to be disposed of. To conserve water, the disposal means must be designed to replenish the source from which the water was obtained.

10–5 RETURN WATER

The most popular method of water disposal, with the widest application possibility, is returning the water directly back to the source of supply. This method of disposal is called *reinjecting*. This method of disposal created the term *return water*.

The three factors of supply water—quantity, temperature, and quality—must also be applied to the return water.

Quantity

Care must be taken to assure there will be no change in the quantity. To assure a continuing supply of water from a source, the return system above ground or below ground must be adequately sized to assure complete return of the water to the source. This will be discussed in more detail when each type of water source is covered.

Temperature

The temperature of the water will drop from 5°F to 15°F as it flows through the liquid-to-refrigerant heat exchanger. To prevent the temperature of the return water from affecting the temperature of the supply water, there must be adequate distance of water storage either above or below ground to accomplish this.

Quality

The only change in the water as it passes through the equipment is change in temperature. Therefore, the water that is returned to the water source is as safe for human consumption as the water that came from the water source. The unit will not add to or detract from the water quality.

Any break in the water supply system will only create a water spill with no creation of a health hazard. If a leak develops in the liquid-to-refrigerant heat exchanger, it will be possible for refrigerant and refrigeration oil to enter the return water system.

Refrigerant oil is a highly refined mineral oil with a lower level of impurities than mineral oil used for human consumption. The refrigerants used in heat pumps are man-made refrigerants that meet U.L. standards of being stable, nontoxic, noncorrosive, nonirritating, and nonflammable. Therefore, any release of these materials will not affect the safe and sanitary condition of the water.

10–6 SCALING

A serious effect of water quality is the effect the mineral level in the water will have on the performance of the liquid-to-refrigerant heat exchanger.

The first step in minimizing the effect of the formation of mineral scale in the water circuit of the coil is the use of a cupronickel water coil. The cupronickel surface has less ability to catch and retain minerals than copper. This is not the single answer to the problem.

Scaling is the process by which minerals contained in water precipitate out of solution and build up on the inside surfaces of pipes and valves. As scale builds up, heat transfer ability goes down and the resistance to water flow through the pipes increases. The result is decreased water flow and heat transfer capacity. If nothing is done, scale will eventually fill the pipes and stop the water flow.

The minerals that combine to form scale are normally found in any water taken directly from the earth. They are held in suspension in the water by the carbon dioxide (CO_2) in the water. If the water temperature rises excessively and/or the pressure on the water drops suddenly, the suspended minerals are released from solution and form carbonate scale. Two important things must be done on the water to minimize scale formation:

1. Maintain the water pressure.
2. Limit the temperature rise.

Maintain the Water Pressure

In order to maintain the pressure on the water in the coil, all controls, shut-off valves, and so on are located on the outlet side of the coil. If the regulation system is such that the flow regulation system will allow pressure loss, a slow-acting motorized valve is also recommended in the return side of the system. The slow-acting motorized valve will assure sustaining the pressure in the coil without any severe pressure changes.

Limit the Temperature Rise

This rule applies to the system in the cooling mode. The small temperature change made in the water in the heating mode will have little effect on scale formation. This is also a temperature drop that tends to improve the mineral-retaining ability of the water.

In the cooling mode, heat is added to the water, which raises the temperature. To minimize scaling, the temperature rise in the water should not exceed 20°F.

10–7 CORROSION

Corrosion is the effect by which pipes, valves, fittings, and so on in the water system are eroded or dissolved by compounds suspended in the water. These may be either acid or base materials, both of which have the same effect on pipe materials.

Most groundwater is highly corrosive. To protect against corrosion, two things can be done:

1. Use cupronickel liquid-to-refrigerant heat exchangers.
2. Use PVC or polybutylene pipe and fittings that will not corrode.

Galvanic Action

Another form of corrosion is galvanic action. This is the destruction of one metal by the generation of electrical energy transfer when dissimilar metals are joined. To help prevent this action, iron or galvanized pipe and fittings should not be used. The preferred materials are PVC, polybutylene, polyethylene, or rubber. If metal pipe must be used, copper is the preferred metal.

Encrustation

Encrustation is the buildup of a slimy orange-brown deposit. It is caused by iron bacteria and is as effective as scale in reducing heat transfer capacity. Maintaining line pressure and a closed system (no exposure of the water to atmosphere) will help to minimize the growth of iron bacteria.

If iron bacteria is in the water, periodic cleaning of the water circuit in the coil, controls, and piping circuit is required. Flushing with a phosphoric acid (food grade acid) solution will remove the iron bacteria deposits as well as scale.

Each treatment, however, will also remove a quantity of the pipe itself. To maximize pipe life, cupronickel coils must be used when iron bacteria is encountered. The safest thing to do is use cupronickel coils in all installations.

The peculiarities of each type of water supply and disposal means will be discussed in the chapters on each type of system.

10–8 STORAGE WATER QUANTITY

Lakes and ponds can provide a low cost source of water for heating and cooling. Direct use of the water from the bottom of the lake or pond is not recommended. Unless a filtration system is used, algae and turbid water can quickly foul up the liquid-to-refrigerant heat exchanger.

To provide filtered water, a dry well should be dug alongside the lake or pond to be used as the water source. Figure 10–3 shows a submersible pump used to remove water from a dug dry well located alongside the water source. Normal

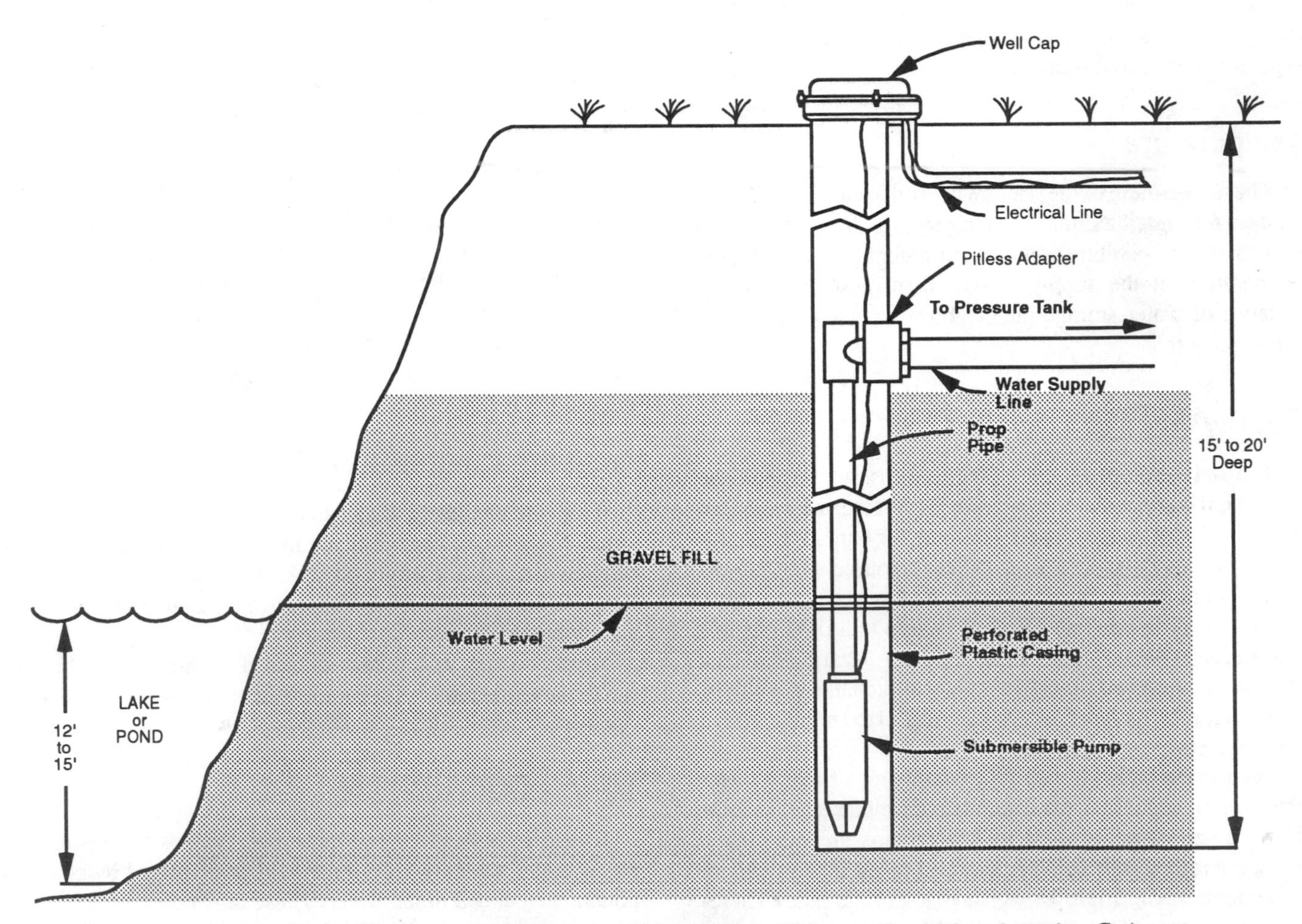

Figure 10–3 Supply well for lake or pond installation (Courtesy Bard Manufacturing Co.)

procedure is to dig a 15- to 20-foot hole as close to the body of water as possible without causing collapse from the water pressure in the lake or pond. The hole should be deep enough to prevent further digging due to the water fill rate into the hole. A minimum standard is to have the bottom of the hole 12 to 15 feet below the low water level of the lake or pond.

Once excavated, a perforated *plastic* casing is installed with gravel back fill placed around the casing. The gravel bed will be the filter to improve the water quality.

10–9 PROCEDURE RULES

There are several rules that must be followed when installing a system of this type:

1. The lake or pond must be at least 1 acre (40,000 ft.^{2}) of surface for each 50,000 BTUH of heat pump cooling capacity. Another way to figure the water source size is it must have a surface area twice the cubic foot size of the area to be conditioned.
2. The average depth of the supply source must be 5 feet or more with some of the area of water depth of 15 feet or more.
3. A submersible pump should be used. Jet pumps and suction pumps use a higher amount of electrical energy per quantity of liquid pumped.
4. The pump must be sized to deliver the required amount of water for the unit. Manufacturers' specifications will have the GPM of water required depending upon the water temperature and other factors.
5. A pressure tank to supply constant pressure must be used. Use of the pump alone will promote scaling as well as cause erratic system operation. Use a pressure control on the supply tank to operate the pump as required. A pump cycling with the unit will promote problems of operation as well as scaling.
6. The sizing of the supply and return piping system is very important to compensate for friction loss in pressure in the system. This is especially important if the water supply system must be over 200 ft. in length.
7. All water lines must be below the minimum water level of the lake or pond. This will prevent drain out in the off cycle and prevent loss of water supply at startup until the water pump can fill the supply system. *The water lines must be installed below the frost line.*
8. Using 4-in. field tile (rigid plastic or corrugated), the return line discharge location must be located at least 100 ft. from the dry well location. Metal pipe should never be used for this application.
9. To ensure good drainage, especially in cold locations where freeze up is possible, the return line should be installed with a minimum slope of 2" in 100'. This will minimize ice buildup where the return line ends above the frost line.
10. The discharge end of the return line must be located high enough above the high water level of the lake or pond to prevent water backing up into the pipe and freezing.
11. Where local weather conditions prevent use of a gravity drain system, standard plastic pipe installed below the low water level of the lake or pond *and below the frost line* can be used to carry return water into the water source. ***Warning:*** *The ice located above such a water outlet will tend to be thinner than the rest of the ice span due to the motion of the water in this location.*

It must also be noted that water temperatures at the bottom of an ice-covered pond can drop down close to freezing, around 36°F to 38°F, which is too cold for standard liquid-to-air heat pumps. Standard units are designed to operate with a minimum supply water temperature of 45°F or higher. Therefore, for this application, a unit that will operate with a low water temperature down to 35°F is recommended. This type of supply system prevents the use of antifreeze in the water.

REVIEW QUESTIONS

1. List the six methods of using groundwater as a heat source and heat sink.
2. Name the three factors that must be considered when selecting the type of heat source supply system.
3. A supply water temperature of 85°F is used to rate a liquid-to-air heat pump in both the heating and cooling modes. True or False?
4. What is an average figure to use for GPM water flow rate for:
   a. Heating________GPM/12,000 BTUH
   b. Cooling________GPM/12,000 BTUH
5. Most manufacturers use what temperature drop through the groundwater to refrigerant coil when rating the unit?
6. What is the minimum refrigerant boiling point in the groundwater to refrigerant coil?
7. What is meant by the term *reinjecting*?
8. List two ways to limit the formation of scale in the groundwater system.
9. In order to maintain the pressure on the water in the coil, all controls are located in the inlet or outlet side of the coil circuit?
10. To minimize scaling, the maximum temperature rise of the water through the coil is:
    a. 30°F
    b. 25°F
    c. 20°F
    d. 15°F
    e. 10°F
11. List two ways to protect against corrosion of pipes by the groundwater.
12. The destruction of one metal by the generation of electrical energy transfer when two metals are joined is called ____________________.

13. The buildup of deposits in the pipe by iron bacteria is called ______________.
14. Water taken from a lake or pond must be filtered. What is the most practical way to accomplish this?
15. When using a lake as a water source, the area of the surface of the lake must be __________ for each 50,000 BTUH heat pump cooling capacity.
16. When using a lake as a water source, the average depth of the lake must exceed __________ feet.
17. When using a lake as a water source, the lake must have a depth of __________ feet or more.
18. Three types of pumps, centrifugal, suction, and jet types, are approved for heat pump applications. True or False?
19. A pressure tank to maintain a constant water pressure is a requirement. True or False?
20. Air relief valves are provided in the water supply system to permit drain back during the off cycle of the unit. True or False?

11 TYPES OF LIQUID HEAT SOURCES/SINKS

11–1 GENERAL

Water in the ground is found in spaces within the sand, ground, and rock under the earth's surface. To some extent there is groundwater everywhere. The quantity, depth, temperature, and quality vary widely. These factors must be taken into consideration in the selection, sizing, and installation of a liquid-to-air heat pump using groundwater as a heat source or sink. The well driller in the proposed location of the heat pump should be consulted on these factors to avoid problems with the installation.

11–2 TERMINOLOGY

The following terms must be understood when discussing well-type supply systems.

- *Aquifer*. The water-bearing formation below the surface of the earth is called the aquifer. This body of water is contained in the sand, gravel, dirt, rock, and so on. It varies in depth and area depending upon the type of material holding the water as well as the geographical location.
- *Consolidated formation*. All materials below the surface of the earth contain some water. Consolidated formation materials, which are solid rock (granite, limestone, sandstone, shale, and so on), only have water in cracks and fissures. This type of material does not contain an adequate supply of water for heat pump operation. This material, however, has a high rate of emissivity—the ability to take on or give off heat energy. The most prominent application in this type of material is the geothermal well.
- *Unconsolidated formation*. A mixture of loose or granulated materials (sand, gravel, soft clay, soil, and so on) that contains large quantities of easily obtainable water.
- *Water table*. The surface of the aquifer. In an underground aquifer, the water table is the level of the water in the water-bearing unconsolidated formation that is closest to the surface of the earth.
- *Static water level*. The level to which water will rise in a well casing when no water is being drawn. The static water level is at the surface of the water column contained in the well.
- *Pumping water level*. The level to which the water level drops when the pump is delivering full load water quantity.
- *Drawdown*. The drawdown is the difference in water level in the well casing between the static water level and the pumping water level.
- *Specific capacity*. Specific capacity of a well is the quantity of water flow in gallons per minute for each foot of drawdown.
- *Cone of depression*. The drop in water level that surrounds a well casing caused by the well drawing water from the aquifer. The point of the cone is the pumping water level.
- *Cone of impression*. The mounding of the water surrounding the well casing when water is forced into the aquifer. The more difficult the process of absorption of the water into the aquifer, the higher the cone of impression. If the static water level in the aquifer is near the surface and the cone of impression rises high enough, the return well will overflow. Because of this possibility, all return wells should be designed to handle overflow in a controlled way.

11–3 DRILLED SUPPLY WELLS

The most common type of well is the drilled well. Drilled wells are small diameter holes bored into the ground by a screw-type shaft attached to a drilling rig. Using a stream of water to flush out the material loosened by the drill, the hole can be bored to any depth to find groundwater. Usually the economics of drilling cost as well as the cost of lifting the water from the well limits the well depth to 200 feet.

Once the hole is drilled, a pipe of noncorrosive material (PVC or steel) is inserted. This pipe, called the well casing, protects the well from collapsing, allows the insertion of the submersible pump to a position below the operating well level, and allows a free flow of water from the ground into the well casing. The various components of a typical supply well are pictured in Figure 11–1.

Proper sizing and testing of the supply well is absolutely necessary for successful operation of the heat pump system. A well supplying a liquid-to-air heat pump must be able to produce an adequate supply of water for an extended period of time at a demand rate equal to the total anticipated water quantity requirement for the conditioned area. The supply time can be for a period of 20 hours out of the 24 hours in a day up to 10 days at a time. The water use is big for a residence and may add up to double that of the heat pump alone.

Figure 11–2 shows the domestic water use of various household items with a column showing the peak demand allowance for the pump in GPM and the individual fixture flow rate in GPM. This table is used to determine the household water needs in gallons per minute (GPM).

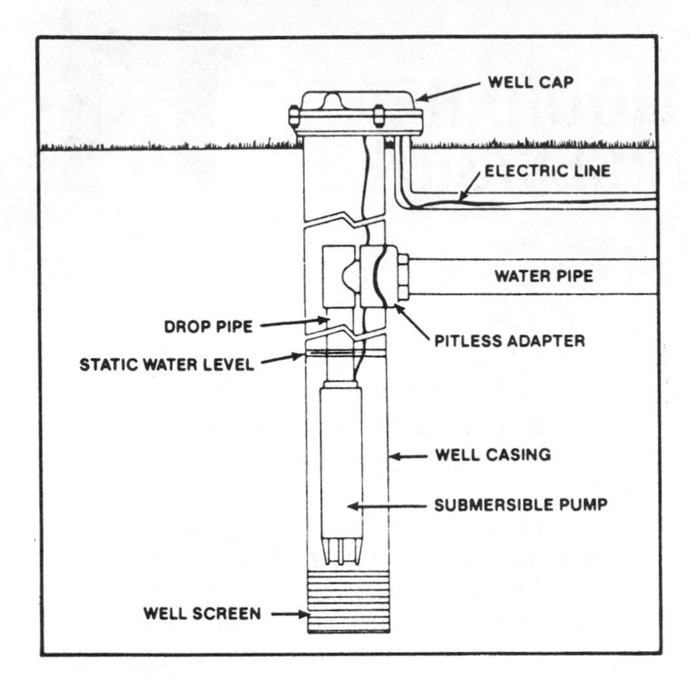

Figure 11–1 Basic drilled well (Courtesy Mammoth, A Nortek Co.)

Domestic Water Uses	Peak Demand Allowance for Pump GPM	Individual Fixture Flow Rate GPM
Household Uses		
Bathtub or tub-and-shower combination	2.00	8.0
Shower only	1.00	4.0
Lavatory	.50	2.0
Toilet—flush tank	.75	3.0
Sink, kitchen, including garbage disposal	1.00	4.0
Dishwasher	.50	2.0
Laundry sink	1.50	6.0
Clothes washer	2.00	8.0
Irrigation, Cleaning, and Miscellaneous		
Lawn irrigation	2.50	5.0
Garden irrigation (per sprinkler)	2.50	5.0
Automobile washing	2.50	5.0
Tractor and equipment washing	2.50	5.0
Flushing driveways and walkways	5.00	10.0
Cleaning milking equipment and milk storage tank	4.00	8.0
Hose cleaning barn floors, ramps, etc.	5.00	10.0
Swimming pool (initial filling)	2.50	5.0

Figure 11–2 Domestic water use table (Courtesy Bard Manufacturing Co.)

To use the table, the number and type of fixtures in the conditioned area are listed, along with their peak demand allowance. The actual usage of each fixture is larger than the peak demand. However, not all fixtures are used at the same time. The peak demand allowance takes the probable usage time and combination into account to figure the total volume of water needed at one time. This total plus the heat pump requirement will be the load on the well pump.

The sizing of the well pump and supply system will be different for each type of application and will be discussed under each of the types.

11–4 RETURN WATER

Groundwater that has passed through the heat pump has been changed in temperature only. Regardless, the return water must be disposed of according to local regulations.

Surface Water Disposal

Figure 11–3 shows the method of disposing of the return water into a pond, lake, or stream. The discharge pipe must be sized for minimum friction loss (4" diameter or larger) and have a drop to the disposal area of not less than 2" in 10'.

The discharge end of the pipe must be high enough above the maximum water level in the pond, lake, or stream to prevent water backing up into the pipe and freezing. A screen should be installed in the end of the pipe to prevent the entrance of frogs, algae, and so on.

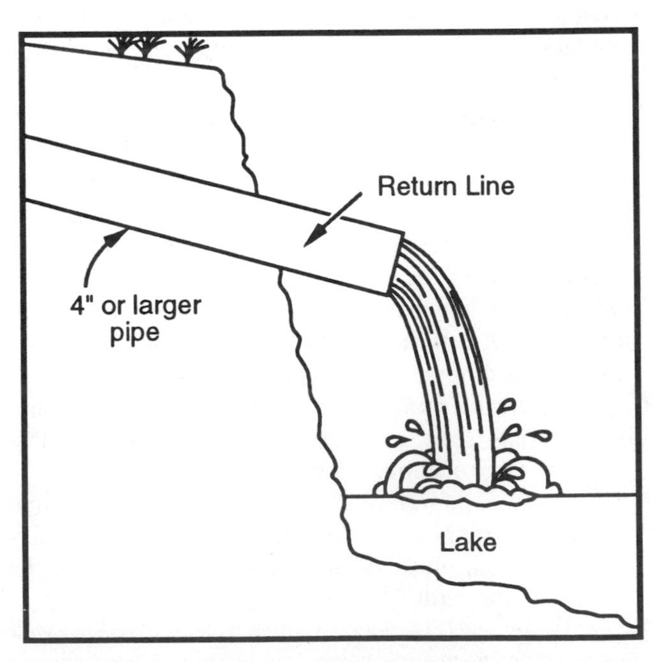

Figure 11–3 Return line discharge into a pond, lake, or stream (Courtesy Bard Manufacturing Co.)

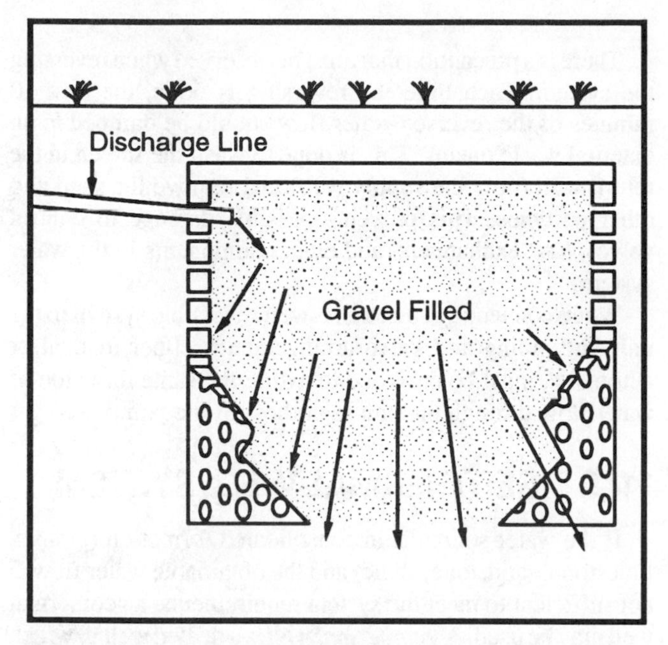

Figure 11–4 Dry well discharge (Courtesy Bard Manufacturing Co.)

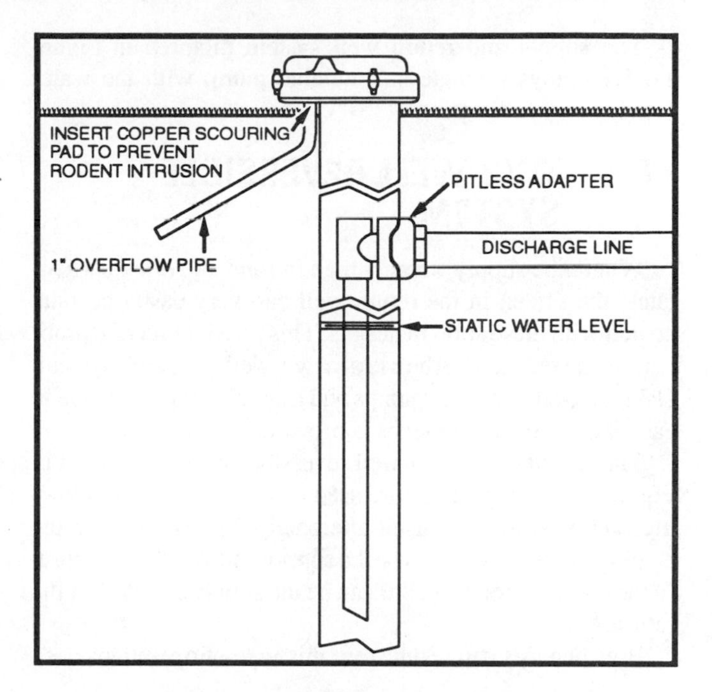

Figure 11–5 Discharge well designed for controlled overflow (Courtesy Mammoth, A Nortek Co.)

Dry Well Disposal

Figure 11–4 shows a cross section of a typical construction of a dry well for return water disposal. The dry well is basically a large hole in the ground that is filled with gravel. Return water flows into the hole and through the gravel to return to the aquifer. This method of return water disposal is restricted to unconsolidated formations of sand or gravel that are very porous.

Installations of this type in more dense unconsolidated formations such as soil, soft clay, and so on will not carry the water away from the well at the desired rate. Before a decision is made to use a dry well, a percolation test of the ground is necessary. A reliable well driller should be contacted on this problem.

Field Tile Disposal

Perforated plastic field tile is commonly used in installations where sufficient land is available for water disposal. This is the favorite method of water disposal in heavy clay soils. The tile is buried from 3 to 7 feet deep below the frost line. With this disposal method, return water is discharged into the tile, through which it eventually seeps to an area where it is carried back into the aquifer.

The number of feet of perforated plastic tile required for the installation will depend upon the ability of the earth to absorb water. To obtain this information necessary to determine this, contact the septic tank system contractor in the neighborhood. That person will want to know the total quantity of water to be handled in gallons per hour. This amount is obtained from the manufacturer's specifications for the unit under consideration.

Return Well Disposal

A return well is a supply well in reverse. The parts of a return well are shown in Figure 11–5. When designing an installation using a two-well system (supply and return well) several things must be considered:

1. The wells must be placed far enough apart to reduce the possibility of the temperature of the return water affecting the temperature of the supply water. This amount of separation will depend upon the type of material in the aquifer as well as the water storage capacity of the aquifer. An acceptable rule is 100' or more separation between the wells.
2. The size of the return well must be greater than the size of the supply well. Generally, the sizes are figured on a two-to-one basis, the return well being two times the size of the supply well.
3. The screen size in each well is also figured on a two-to-one basis. The return well has two times the screen free area of the supply well.
4. To prevent air being trapped in the water, which promotes encrustation action, the end of the return pipe must be below the static water level in the well casing.

The supply and return well system pictured in Figure 11–5 employs a single submersible pump with the water flowing in one direction.

11–5 TWO-WELL REVERSIBLE SYSTEM

Where the supply water is high in sand and other particulates, the screen in the return well can very easily become coated with these solid materials. This possible service problem can be reduced by the use of a two-well reversible system. Using two submersible pumps and a supply and return line in each well, a flushing means is provided.

The layout of the two-well reversible system is shown in Figure 11–6. Two pumps are used to reverse the flow, allowing each well to clean itself alternately. For example, in the heating mode, Well A can be the supply and Well B the return. In the cooling mode, Well B can be the supply and Well A the return.

Five benefits can result from this type of operation:

1. There is a temperature buildup in the cooling mode return that will increase the operating efficiency when it is the supply well in the heating mode.
2. The reverse action also applies in the cooling mode due to the supply well being cooled by the heat pump in the heating mode.
3. Water return to the aquifer is more effective when the return area of the aquifer is redeveloped each year.
4. The wells may be spaced closer together.
5. Each well is sized as a return well.

There is a precaution that must be observed when reversing the system. Each time the reversing is done, the first 10 minutes of the reversed water flow should be dumped to an external waste means. This is done to wash the screen in the return well (now the supply well) and remove the sand and other particles. This prevents possible damage to control valves, heat exchangers, and other components in the water system.

When connecting such a two-well reversible system to the unit, the piping and valve arrangements, either manual or automatic, must be designed to assure the same direction of water flow through the unit regardless of the pump used.

11–6 GEOTHERMAL WELL SYSTEM

If the water source is in consolidated formation (granite, limestone, sandstone, shale) and the obtainable water flow is not sufficient to meet the system requirements, a geothermal well may be used. A *geothermal well* is a drilled well in which the return water is returned to the same well from which the supply water is taken. In a geothermal well, the water is only the carrying agent to remove heat from the underground mass for heating or supply heat to the underground mass for cooling. The earth strata material is the heat source or sink depending upon the operating mode.

There are three types of geothermal wells:

1. Dedicated geothermal well
2. Domestic geothermal well
3. Sand geothermal well

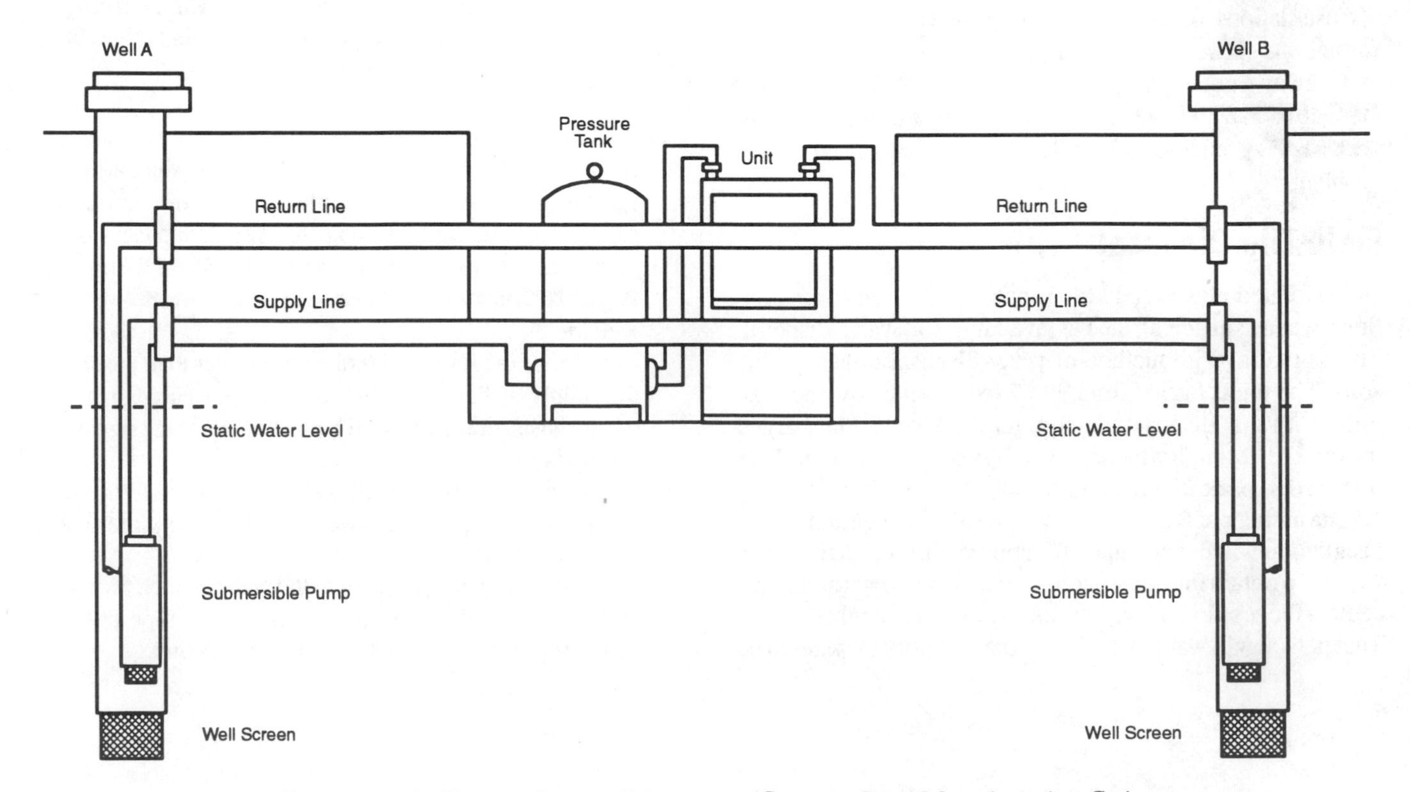

Figure 11–6 Two-well reversible system (Courtesy Bard Manufacturing Co.)

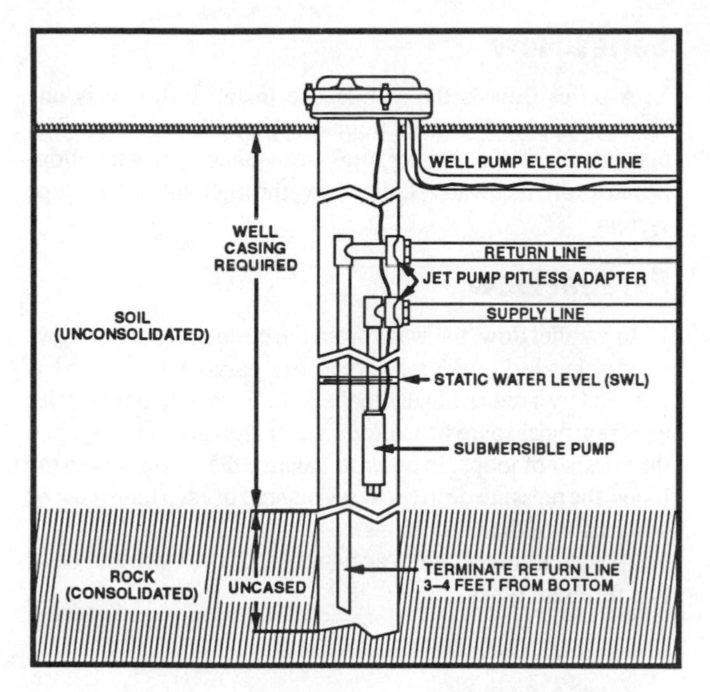

Figure 11–7 Dedicated geothermal well (Courtesy Mammoth, A Nortek Co.)

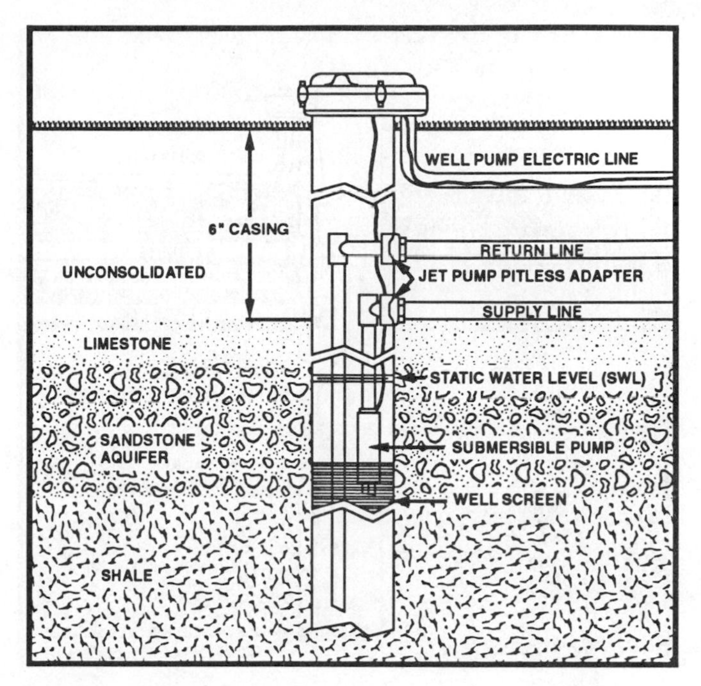

Figure 11–8 Domestic geothermal well (Courtesy Mammoth, A Nortek Co.)

In the first two types, the water is drawn from the top of the well below the pumping water level and returned to the bottom of the well. The water flows up from the return outlet to the supply inlet to lose or gain heat, depending upon the operating mode.

In the third type, sand geothermal well, the action is reversed. The water is drawn from the bottom of the well and returned to the top.

Dedicated Geothermal Well

The various parts of a dedicated geothermal well are shown in Figure 11–7. A 6" diameter well casing is required to accommodate the submersible pump as well as the return water pipe. This casing goes down as far as and into the consolidated formation far enough to create a seal. The object is to keep loose material from the unconsolidated surface material from falling into and clogging the well.

No well casing is used in the consolidated formation for maximum heat transfer capacity. The submersible pump is set 3 to 5 feet below the top of the water column in the well and the return water pipe drops to within 3 to 4 feet of the bottom of the well.

Standard practice in the industry when figuring the depth of the well is 150' deep for each 12,000 BTUH of system cooling requirement. The 24,000 BTUH unit in the Omaha, NB application would require a well 300' deep. In the Phoenix, AZ area, the well would be 450' deep.

Because of the weight of the pipe and to promote longer pipe life, 80 PVC pipe is recommended for installation in the well.

Domestic Geothermal Well

In a domestic geothermal well (Figure 11–8), the water supply is for both the heat pump and domestic use. Because water will be drawn from the well for domestic use, the consolidated formation has to contain limestone or sandstone aquifer from which the domestic water can be obtained.

The well is a drilled well with a 6" well casing that penetrates through the less dense material and enters to seal the more dense material. The well is then drilled the full depth of 150' per 12,000 BTUH of cooling capacity of the unit.

The well casing, however, has two well screens to take water from the aquifer material in the less dense water bearing level. The major supply screen is located at the base of the water bearing strata and a purge screen is located near the top of the water-bearing strata. The pump is located between the two screens below the pumping water level.

Sand Geothermal Well

A sand geothermal well is rarely used as its use is limited to unconsolidated aquifers where the sand and/or gravel layers holding the water are 75' or more in depth. Where possible, the use of this type geothermal well is recommended to save drilling and pipe cost. See Figure 11–9.

The well casing size can be 4" or 5", depending upon the size of the submersible pump used. Also, the well does not depend upon heat transfer through the well casing. The water from the system is removed from the aquifer and returned to the aquifer.

The submersible pump is located in the bottom of the well below the "packer" plate that separates the supply and return water.

The well outlet screen at the top of the aquifer has two to three times the free area of the intake screen at the bottom of the well casing. This is because an aquifer will receive water at only 75% to 80% of the rate at which it gives up water. This means that the return side of the water transfer means to the

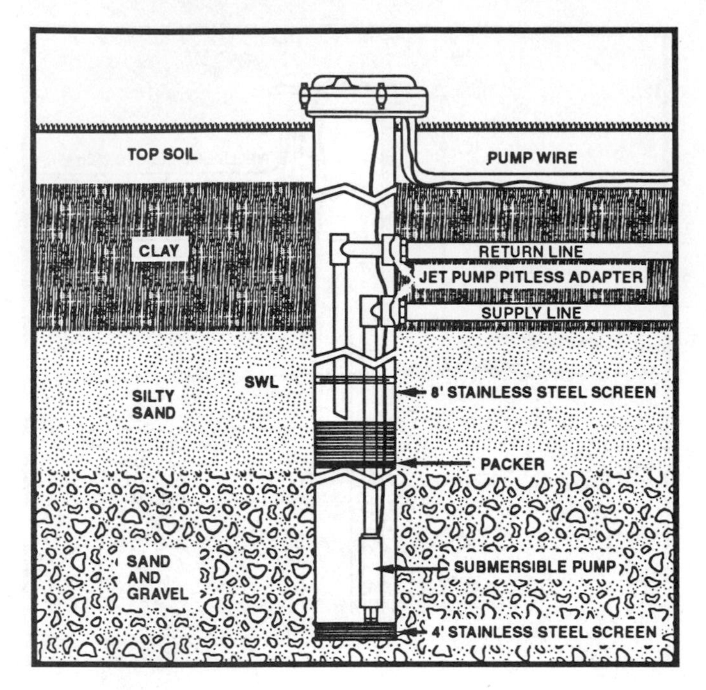

Figure 11–9 Sand geothermal well (Courtesy Mammoth, A Nortek Co.)

aquifer must be larger than the supply. The upper screen must also be located below the static water level to prevent contact with air in the soil and prevent encrustation on the screen.

Both the pump and return pipe openings are located immediately above their respective screens.

11–7 GROUND LOOPS

Where the supply of groundwater or the ability to dispose of return water is not sufficient for liquid-to-air heat pump application or where regulations prohibit the use of the aquifer water source, water circulated through a closed pipe system buried in the ground can be used as a heat source and heat sink. Heat is transferred to and from the soil surrounding the pipes due to the temperature difference between liquid in the pipe and the surrounding strata. The pipe system may be buried in either a horizontal or vertical configuration deep enough to be in moist earth for best heat transfer.

These ground loops, sometimes referred to as earth couplings, are generally categorized into six basic systems:

1. Single-layer horizontal
2. Two-layer horizontal
3. Four-layer horizontal
4. Single U-bend vertical
5. Double U-bend vertical
6. Solar storage

The types of systems are also divided into the piping arrangement used—series flow or parallel flow. Each type of flow has its advantages and disadvantages.

Series Flow

A series flow is the simplest to install in that it is one continuous pipe from the circulating pump to the return. The disadvantage is a greater flow resistance (pressure drop) because all the water must flow through the entire pipe system.

Parallel Flow

In parallel flow, the water from the circulating pump flows from a header, divides between the parallel loops, and is returned by a return header to the unit. Each loop only carries a proportional share of the total water flow, depending upon the number of loops. In order to balance the flow through the loops, the pressure drop (flow resistance) of each loop must be within 5% of the other loops.

Header arrangement to accomplish this balance can be of two different types:

1. Close header
2. Reverse return header

Close Header. Figure 11–10 shows a line drawing piping arrangement of a close header vertical loop system. The vertical loops are shown in a circle around the header setup with a minimum of 15' spacing between each vertical loop. For the sake of illustration, the header connections are pictured as separated. All the loop connections are connected to supply and return header manifolds, which are located adjacent to each other with only 15" to 20" of earth between to keep heat transfer between the supply and return pipes to a minimum. The total pipe lengths as well as total equivalent length of each loop must be equal to each of the other loops.

Reverse Return Header. The reverse return header vertical loop system (Figure 11–11) uses two interconnecting headers connecting the supply and return of each vertical

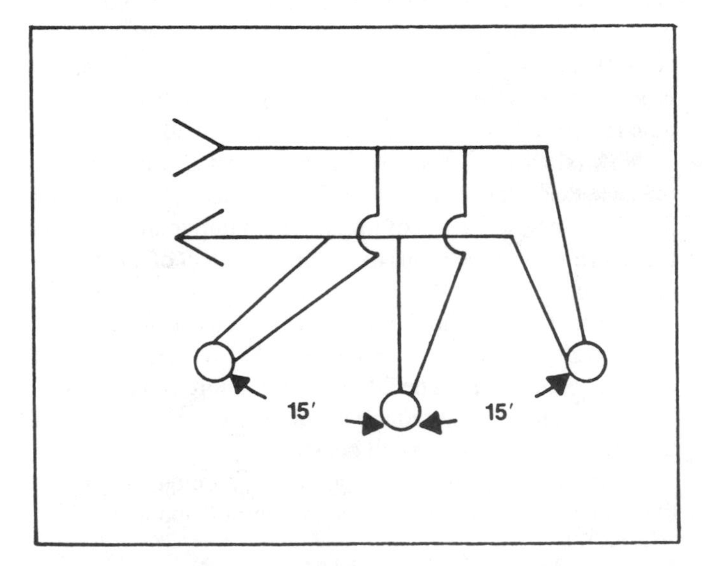

Figure 11–10 Close header vertical well (Courtesy Mammoth, A Nortek Co.)

loop. Each vertical loop, spaced a minimum of 15' apart, is connected to the headers with equal pipe length and equal equivalent length of the fittings. To balance the total resistance of the header system, water is supplied directly into the supply header closest to the unit. Return water is taken from the far end of the return header and returned to the unit. Thus, there is a balanced flow through the vertical loops.

Single-layer Horizontal

The single-layer horizontal ground loop shown in Figure 11–12 is the simplest type of system to install. The loop is a single length of plastic pipe laid in a single trench and then backfilled. The water flows through a single layer of pipe in a single loop. All the water used for the heat source or sink flows through the one pipe.

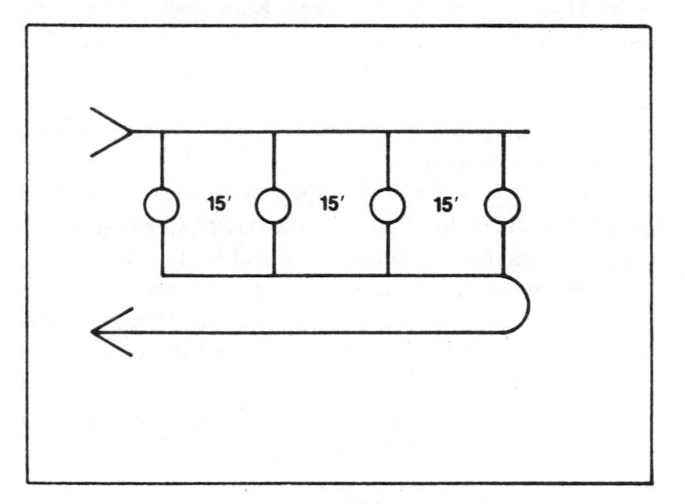

Figure 11–11 Reverse return header vertical well (Courtesy Mammoth, A Nortek Co.)

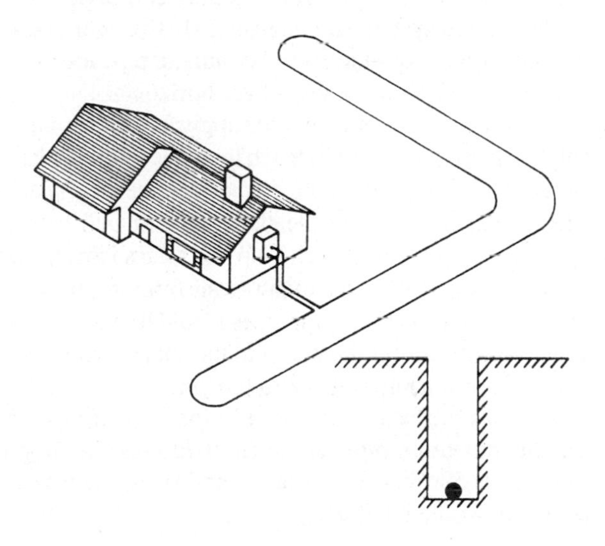

Earth Coil Type:	Horizontal-Single Layer
Water Flow:	Series
Typical Pipe Size:	1 1/2 to 2 inches
Nominal Length:	350 to 500 feet/ton
Burial Depth:	4 to 6 feet

Figure 11–12 Single-layer horizontal ground loop (Courtesy Mammoth, A Nortek Co.)

Using 1 ½" to 2" pipe buried from 4 to 6 feet below grade, between 350 and 500 feet of pipe is required for each 12,000 BTUH cooling capacity. Because of the pipe length involved, this system is used on the smaller size units and/or where trench area is not limited. With a single-layer system, the length of the trench involved is the same as the length of the pipe.

This system does have the advantage that it can be installed by the use of a vibratory plow as shown in Figure 11–13. The vibratory plow wedges a path through the earth and inserts the pipe in one operation. A trench is not needed and no backfilling is done. This type of installation of the pipe is most effective in rock-free soil where straight runs of longer length can be made.

This type of installation is limited to the single-pipe system with a maximum depth of 30" to 36". Where sharp turns or 180° bends are required, a backhoe or hand digging is required.

Two-layer Horizontal

In the two-layer horizontal system, the single-loop system is laid with two pipes in the same trench (Figure 11–14). This system requires a double backfill. The pipe is laid out with half the pipe length laid on the floor of the trench. After the first layer of backfill is in place, the second or return half of the loop is laid on the top of the first backfill. The backfill is then completed.

For the purpose of illustration, the return loop is pictured in a vertical position. Actually, the loop is laid in a hand- or backhoe-dug enlarged area at the end of the loop. The pipe is then laid for a circle return to avoid kinks. The backfill is done by hand to create a rise from the lower pipe to the upper pipe. This will allow air to leave the bend and prevent flow blockage.

The wider trench required for this installation can be done by a chain trencher (Figure 11–15) if the soil is dense enough to retain walls until the backfill is completed. In loose sandy

Figure 11–13 A vibratory plow (Courtesy Mammoth, A Nortek Co.)

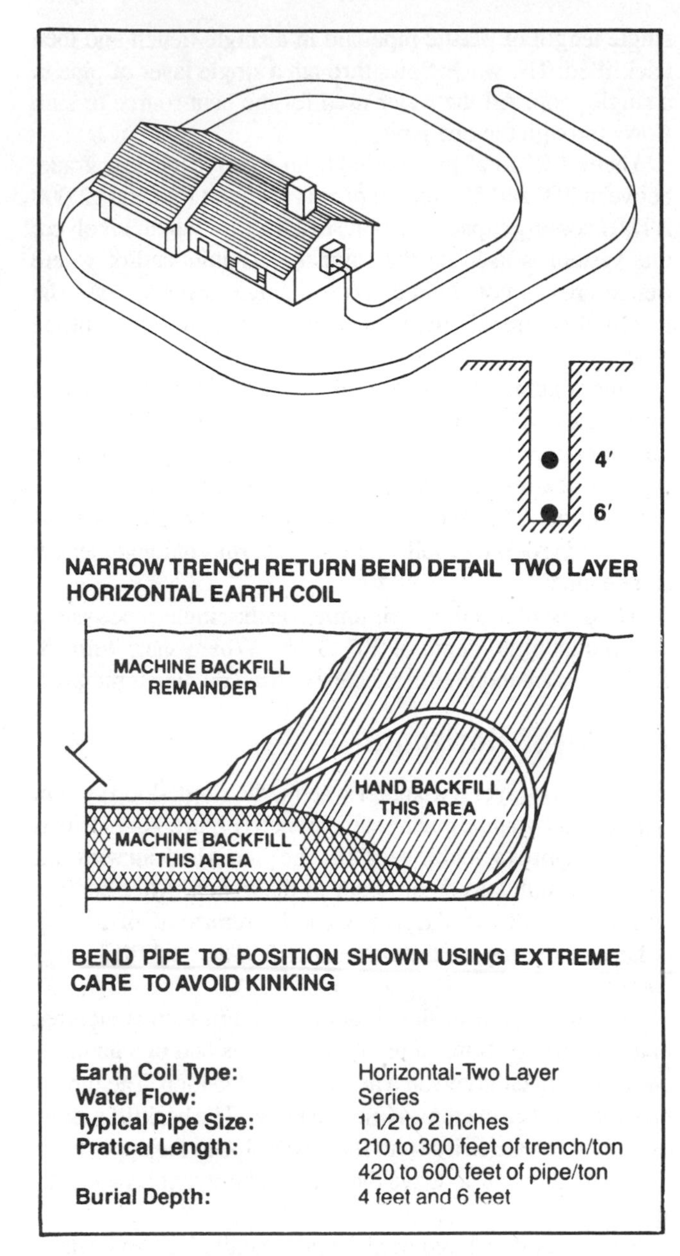

Figure 11–14 Two-layer horizontal ground loop (Courtesy Mammoth, A Nortek Co.)

Figure 11–15 A chain trencher (Courtesy Mammoth, A Nortek Co.)

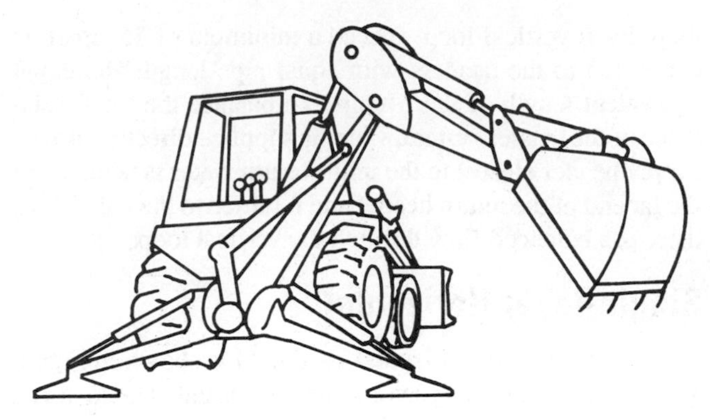

Figure 11–16 A backhoe (Courtesy Mammoth, A Nortek Co.)

soil, a wider trench would be required and a backhoe (Figure 11–16) would be used.

This system is still a single-pipe system where all the heat source/sink water flows through the one pipe. Because two pipes are affecting the temperature and heat content of the earth next to the pipe loop, more pipe is required to get the heat transfer—420 to 600 feet of pipe per 12,000 BTUH cooling capacity. This will require 210 to 300 feet of trench per 12,000 BTUH cooling capacity.

With the two-pipe system, the lower pipe must be at least 6 feet in the ground with 2 feet minimum between the pipes.

Four-layer Horizontal

In areas where the pipes will be laid in heavy wet soil with a high rate of heat absorption or rejection (emissivity), a four-layer horizontal system can be installed. This will allow the minimum space requirement to lay out the pipe loops.

Figure 11–17 shows a four-layer horizontal loop system laid out in two loops with vertical supply and return headers. Using a 6-feet deep trench for each loop, the loops are laid out at layers of 6', 5', 4', and 3' below grade (12" vertical spacing).

The connections to the headers are arranged to supply water to the lower half of each loop and return from the other half of the loop. This helps to purge air from the system and prevent air locks. Typical pipe sizes would be 1 ½" to 2" pipe for the pipe lines from the pump to the supply manifold and from the return manifold back to the unit.

The loops are usually ¾" to 1" pipe size. If ¾" pipe is used, the maximum pipe length (to hold pressure drop to a minimum) is 500' per loop. For 1" pipe sizes, the maximum pipe length would be 750".

Single U-bend Vertical

Where trench area is at a minimum, vertical loops are used to gain additional pipe surface to earth exposure. Figure 11–18 shows a single U-bend vertical system using series flow between three vertical loops.

Using a drilled or bored well of a diameter five times the diameter of the pipe used for the loop system, each well is drilled to the required depth for the system capacity. The total

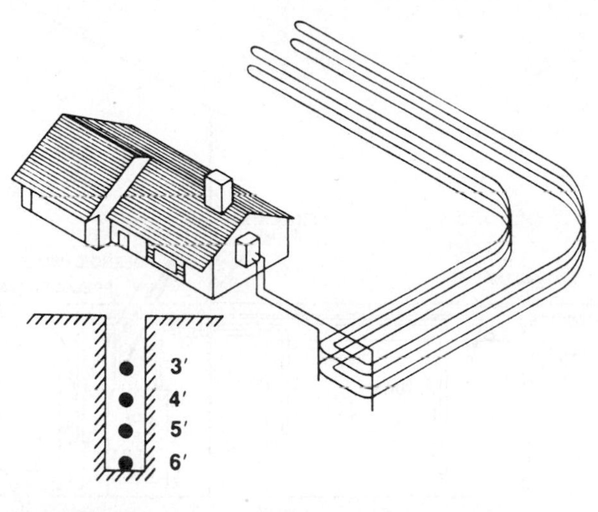

Earth Coil Type:	Horizontal–Four Layer
Water Flow:	Parallel
Typical Pipe Size:	Parallel loops 3⁄4 to 1 inches headers 1 1⁄2 to 2 inches
Parallel Pipe Length:	500 Ft. Max. Pipe Length (3⁄4 inch) 750 Ft. Max. Pipe Length (1 inch)
Burial Depth:	6 feet, 12 inch spacing

Figure 11–17 Four-layer, horizontal multiple ground loop (Courtesy Mammoth, A Nortek Co.)

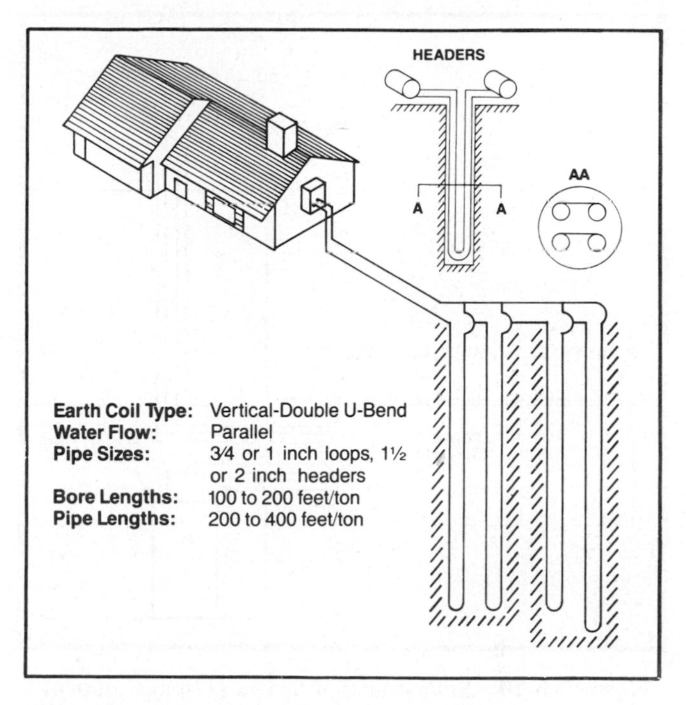

Figure 11–19 Vertical U-bend—parallel flow (Courtesy Mammoth, A Nortek Co.)

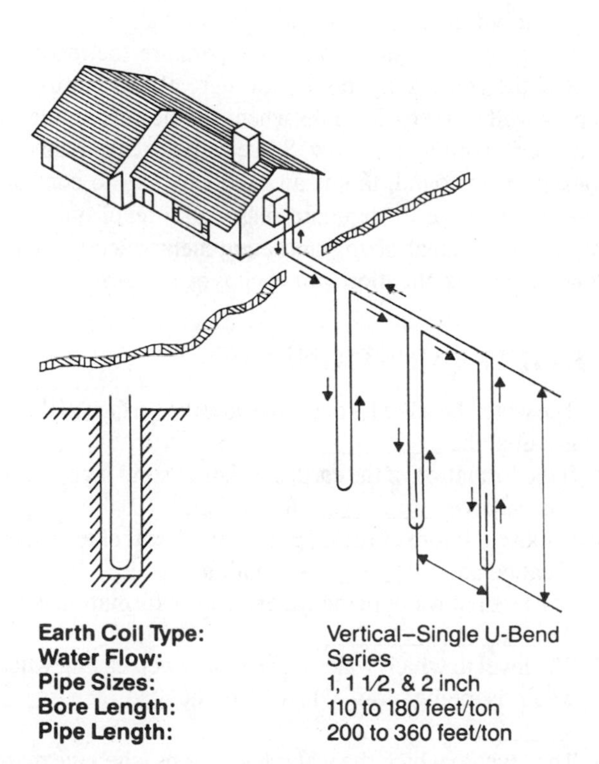

Figure 11–18 Vertical U-bend—series flow (Courtesy Mammoth, A Nortek Co.)

well depth is 110 to 180 feet per 12,000 BTUH cooling capacity with 200 to 360 feet of pipe per 12,000 BTUH cooling capacity. The depth and number of wells will be determined by the type of earth material encountered and the cost of drilling.

Though shown as a series flow (Figure 11–18), the piping arrangement can be parallel flow using the piping arrangement shown in Figure 11–19. The parallel piping arrangement will result in lower overall flow resistance though more pipe is required.

Double U-bend Vertical

The double U-bend vertical system is shown in Figure 11–19. This system uses two loops in each well. Though the number of wells is reduced, each well is of a larger diameter (10 times the pipe diameter), with two loops in each well.

This system is best used in soils where the well depth is under 50' in heavy wet soil with a high emissivity rate (ability to absorb or reject heat energy). It is the most difficult system to install and balance.

11–8 RESERVOIR SYSTEM

A reservoir system is a closed geothermal loop system using a reservoir storage system for the heat energy heat source or heat sink. Using a quantity of water to store heat energy during the cooling mode and supply energy in the heating mode, the reservoir system will operate without an outside heat source or heat sink when the heating and cooling requirements balance in a 24-hour period. For those periods

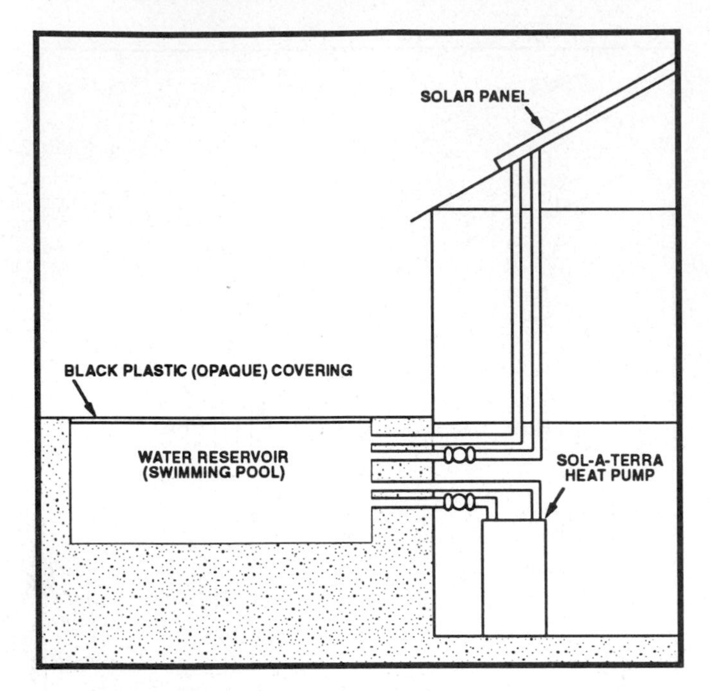

Figure 11–20 Solar-heated reservoir system (Courtesy Mammoth, A Nortek Co.)

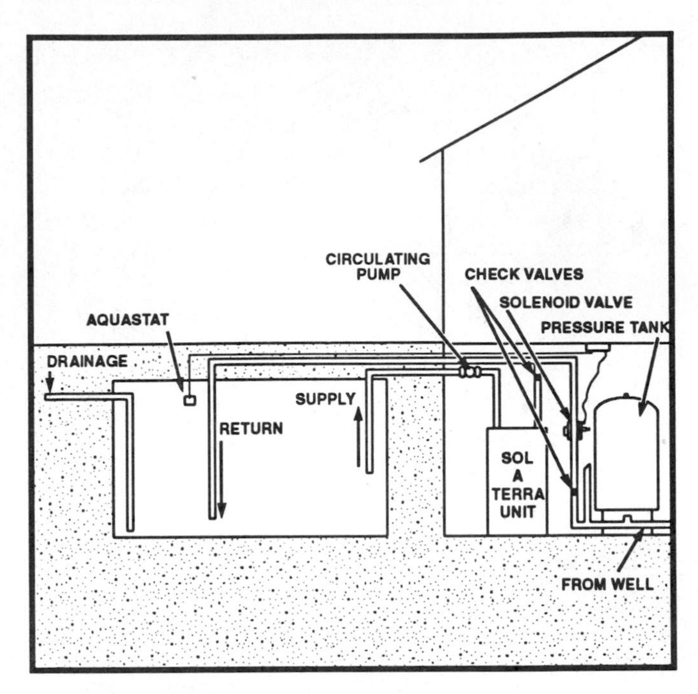

Figure 11–21 Buried tank reservoir system—Well water makeup (Courtesy Mammoth, A Nortek Co.)

when the loads are unbalanced, two methods are used to provide heat source or sink capacity:

1. Solar cell and evaporative tower
2. Well water supply

Solar Cell and Evaporative Tower

Figure 11–20 shows the use of a swimming pool as a reservoir and a solar cell for additional heating capacity. In this installation, the use of a black opaque cover for the pool increases the solar heat gain. The solar system is controlled to keep the water temperature within the recommended water temperature range for the heat pump (45°F to 70°F). For cooling operation, an evaporative cooling tower is used in conjunction with the swimming pool to maintain water temperatures below 80°F for supply to the heat pump.

Application of this type of system is very limited for single-unit systems. The area of application would be where the heating load is the minor load and possible sunshine hours are such that considerable solar boost is available. Summer weather conditions must be such that evaporative cooling towers operate at full capacity. This requires a hot, dry climate.

For multiple-unit systems that handle different exposures in the same building, this open pool system can save water costs during balanced load periods.

Well Water Supply

Figure 11–21 shows a buried-tank, water reservoir closed loop system. This system is applicable to single-unit systems to reduce the total yearly water consumption. With a 3,000-gallon tank buried in the ground (two 1,500-gallon tanks are usually paired to make the 3,000-gallon volume), the water reservoir will receive or release 25,020 BTU per each degree of water temperature change. 3,000 gallons times 8.4 pounds per gallon times 1 BTU per pound equals 25,020 BTU. Working on a 10°F temperature change in the water, 25,020 BTU/°F ×<10°F = 250,200 BTU of storage heat in the water reservoir before an outside source is needed.

Using two aquastats (water temperature thermostats) to control the water temperature, a water valve is controlled to supply well water to the tanks whenever the tank temperature becomes too high to too low. Since well water remains fairly constant year-round, this is an excellent way to control the tank temperature to maintain reasonably level unit performance. The actual sizing and equipment selection will be done under "Application" for each type system.

REVIEW QUESTIONS

1. The water-bearing formation below the surface of the earth is called the ________.
2. If the formation of the earth consists of solid layers of rock, it is called a ________ formation.
3. A loose mixture of rocks, gravel, sand, soft clay, and so on is called an ________ formation.
4. The level of water in the water-bearing formation is called the ________ ________.
5. The level to which water will rise in a well casing when no water is being drawn is called the ________ ________ ________.
6. The level to which the water level drops when the pump is delivering full-load water quantity is called the ________ ________ ________.

7. The difference between the water level standing in a well and the water level with the pump running at full capacity is called the _______ _______.
8. The drop in water level that surrounds a well casing caused by the pump removing water from the well is called the _____ __________.
9. The mounding of the water surrounding the well casing when water is forced into the ground surrounding the well casing is called the _______ ___ __________.
10. The most common type of well is the driven well. True or False?
11. The actual water usage is the amount of water a fixture will use in a 24-hour period. True or False?
12. The peak demand allowance is the maximum amount of water the fixture will require. True or False?
13. Before a decision to use a dry well for water disposal can be made, a __________ test must be performed.
14. An acceptable minimum distance for separation of supply and return wells is:
    a. 50'
    b. 75'
    c. 100'
    d. 125'
    e. 150'
15. When calculating the size of supply and return wells, which well is larger and by what ratio?
16. How far down in the return well must the return pipe extend?
17. When is the use of a two-well reversible system advisable?
18. When reversing the two-well system, the first 10 minutes of water pumping must have the water go to a waste. Why?
19. A drilled well in which the return water is returned to the same well casing the supply is obtained from is called a __________ well.
20. List the three types of drilled wells described in Problem 19.
21. In which of the three wells listed in Problem 20 is the water drawn from the top of the well and returned to the bottom?
22. When drilling a well through solid layers of rock, a perforated well casing is placed in the drilled hole to ensure water flow through the rock level. True or False?
23. The well casing has two well screens with the pump located between the screens. This set is used in what type of well?
24. Name the two types of configurations in the buried ground loop systems.
25. Ground loop systems need only be buried deep enough to be protected from surface traffic. True or False?
26. Name the two different piping arrangements used in ground loop systems.
27. If two or more loops are used in the system, the water flow between the pipes is accomplished by either of two types of header arrangements. Name them.
28. When installing return bends in a ground loop system, the major concern is to keep the pipe bend water flow resistances as low as possible. True or False?
29. The diameter of a drilled or bored well casing is a minimum of _____ times the diameter of the system water pipe.
30. A closed geothermal loop system using a water storage device is called a __________ __________.

12 APPLICATION

12–1 GENERAL

The application of liquid-to-air heat pumps will differ from air-to-air heat pumps only in the sizing of the liquid supply and disposal systems. Unit selection is found in Chapter 9 and descriptions of the various types of liquid heat source/sink systems are given in Chapter 11. The major discussion in this chapter will be on the equipment selection and sizing of the various types of liquid heat sources.

The first liquid-to-air systems installed used a source of water and a disposal means. This required a source from which to remove the water; a bored or drilled well for supply water and an open or well disposal. In all such systems, a common means of handling the water exists. The water must be pumped, carried to the unit, through the unit, and be disposed of. To do this, certain components are common to all these systems.

12–2 WATER PUMP

Three types of water pumps are marketed: suction, jet, and submersible. Suction pumps are not applicable for lifting water over an atmosphere of pressure of 27' of water height. In fact, the efficiency of such a pump is so low at 15' lift that the pump output is very low and the operating cost high. Jet pumps lift water higher than suction pumps, but they require the movement of water several times the actual amount available to the supply system.

The operating cost of both suction and jet pumps is considerably higher than the operating cost of the submersible pump. Therefore, they are not cost effective. The use of suction and jet pumps is not recommended in liquid-to-air systems.

The submersible pump is located at some point in the aquifer, usually at the bottom of the bored or drilled well. This places a positive load on the pump with pumping resistance primarily on the discharge side. Intake flow resistances into the pump are low enough that pump cavitation or pumping loss is practically nonexistent.

The first step in selecting the size of the pump is to determine the volume of water required from the well. This has to be more accurate than a "standard" or "rule of thumb." If the pump is undersized, inadequate water supply to the unit will cause system operating problems. If the pump is oversized, the pump will short cycle, resulting in very short pump and control life.

To determine the volume of water the pump must handle, the total load requirement on the pump must be determined. The load may be the liquid-to-air unit only (dedicated system) or it may include the building requirements as well (combined system).

Dedicated System

If the well water system is dedicated to the heat pump only, with no additional connected loads, the pump may be sized to supply the precise flow required by the heat pump only. The quantity of water required to handle the heat pump in either the heating mode or cooling mode, whichever is the larger, will depend upon the temperature of the groundwater. For example, for the Omaha, NB application, using the WPH22–1 unit selected for the cooling load of 19,996 BTUH, the quantity of water required would be selected from the application rating and water flow requirement tables covering this unit. These are published in the manufacturer's literature.

Figure 12–1 shows the water requirements for the WPH22–1 unit for both cooling and heating with entering water temperatures (EWT) from 45°F to 90°F. Using the groundwater temperature map in Figure 10–2, we determine that the water temperature in the Omaha, NB area is approximately 53°F. Using the 50°F temperature for heating and 55°F temperature for cooling, we find that we need a maximum flow rate of 4 GPM for cooling and 8 GPM for heating. We size the pump for the 8 GPM heating requirement.

For the Phoenix, AZ application, the groundwater temperature is 72°F. This means that for a WPH22–1H unit in the Phoenix, AZ area, using 75°F water for cooling will require 7 GPM. Using 75°F water for heating will require 5 GPM. The pump is sized for the cooling requirement of 7 GPM.

Combined System

In a combined system, the well pump supplies water for both the heat pump and for household use. Therefore, a decision must be made to provide enough water volume for either the total of the unit and the domestic use or the greater of the two.

COOLING

WPH22-1H

E.W.T.	GPM	BTUH TOTAL	BTUH SENS.	WATTS	EER W/OUT PUMP	EER WITH PUMP	HEAT REJ.
45	2.0	24800	16700	1730	14.3	13.4	29800
	3.0	25700	17100	1660	15.5	14.0	30500
	4.0	26100	17300	1630	16.0	14.0	30800
50	2.0	24200	16500	1770	13.7	12.8	29300
	3.0	25100	16900	1710	14.7	13.3	30000
	4.0	25500	17000	1670	15.3	13.4	30300
55	2.0	23700	16300	1820	13.0	12.2	29000
	3.0	24500	16600	1750	14.0	12.7	29600
	4.0	24900	16800	1720	14.5	12.7	29900
60	3.0	23900	16400	1800	13.3	12.1	29100
	4.0	24300	16500	1770	13.7	12.1	29400
	5.0	24500	16600	1750	14.0	12.0	29600
65	3.0	23300	16100	1850	12.6	11.5	28700
	4.0	23700	16300	1820	13.0	11.5	29000
	5.0	23900	16400	1790	13.4	11.4	29100
70	3.0	22700	15900	1900	11.9	10.9	28300
	5.0	23300	16100	1840	12.7	10.9	28700
	7.0	23600	16300	1820	13.0	10.5	28900
75	3.0	22100	15600	1950	11.3	10.4	27900
	5.0	22700	15900	1900	11.9	10.3	28300
	7.0	22900	16000	1880	12.2	10.0	28400
80	4.0	21900	15500	1960	11.1	10.0	27700
	6.0	22300	15600	1920	11.6	9.8	28000
	8.0	22400	15700	1910	11.7	9.4	28100
85	4.0	21200	15300	2020	10.5	9.4	27200
	6.0	21600	15400	1990	10.9	9.2	27500
	8.0	21700	15500	1970	11.0	8.9	27500
90	5.0	20800	15100	2060	10.1	8.8	26900
	7.0	21000	15200	2040	10.3	8.5	27100
	9.0	21100	15300	2030	10.4	8.2	27100

HEATING

WPH22-1H

E.W.T.	GPM	BTUH TOTAL	WATTS	COP W/OUT PUMP	COP WITH PUMP	HEAT OF ABSORP.
45	6.0	19700	1780	3.2	2.7	14500
	8.0	20200	1800	3.3	2.6	15000
	10.0	20400	1820	3.3	2.5	15100
50	4.0	20200	1820	3.3	2.9	14900
	6.0	21300	1860	3.4	2.8	15800
	8.0	21800	1890	3.4	2.7	16200
55	4.0	22100	1910	3.4	3.0	16500
	6.0	23100	1950	3.5	2.9	17300
	8.0	23600	1980	3.5	2.8	17700
60	4.0	23700	1980	3.5	3.1	17800
	6.0	24800	2040	3.6	3.0	18700
	8.0	25400	2070	3.6	2.9	19200
65	4.0	24900	2040	3.6	3.2	18800
	6.0	26300	2120	3.6	3.1	19900
	8.0	26700	2130	3.7	3.0	20300
70	3.0	25400	2060	3.6	3.3	19300
	5.0	27400	2180	3.7	3.2	20900
	7.0	28300	2220	3.7	3.1	21600
75	3.0	27000	2160	3.7	3.4	20500
	4.0	28200	2210	3.7	3.4	21600
	5.0	29000	2260	3.8	3.3	22200
80	2.0	27000	2160	3.7	3.5	20500
	3.0	29100	2270	3.8	3.5	22300
	4.0	30500	2340	3.8	3.5	23400

Figure 12–1 Application ratings and water flow requirements (Courtesy Addison Products Co.)

12–3 PUMP SIZE

To calculate the domestic use, the "Peak Demand Allowance for Pump" column in Figure 12–2 is used. Two different categories of appliances or procedures that require water are listed along with the normal peak demand on the pump. Each fixture and procedure has a higher individual flow rate. Because not all would be used or done at the same time, an average usage is figured. For example, we will assume that the residence has the following:

1 tub and shower combination	2.00 GPM
1 shower only	1.00 GPM
2 lavatories	1.00 GPM
2 toilet flush tanks	1.50 GPM
1 kitchen sink	1.00 GPM
1 dishwasher	0.50 GPM
1 clothes washer	2.00 GPM
Total	9.00 GPM

This residence in a city area would demand a 9.00 GPM water volume to satisfy the normal household load.

If this is the Omaha, NB application, the heat pump will need 8 GPM plus the domestic use of 9 GPM for a total flow rate requirement of 17 GPM. The supply well pump should be sized to deliver 17 GPM.

Domestic Water Uses	Peak Demand Allowance for Pump GPM	Individual Fixture Flow Rate GPM
Household Uses		
Bathtub or tub-and-shower combination	2.00	8.0
Shower only	1.00	4.0
Lavatory	.50	2.0
Toilet—flush tank	.75	3.0
Sink, kitchen, including garbage disposal	1.00	4.0
Dishwasher	.50	2.0
Laundry sink	1.50	6.0
Clothes washer	2.00	8.0
Irrigation, Cleaning, and Miscellaneous		
Lawn irrigation	2.50	5.0
Garden irrigation (per sprinkler)	2.50	5.0
Automobile washing	2.50	5.0
Tractor and equipment washing	2.50	5.0
Flushing driveways and walkways	5.00	10.0
Cleaning milking equipment and milk storage tank	4.00	8.0
Hose cleaning barn floors, ramps, etc.	5.00	10.0
Swimming pool (initial filling)	2.50	5.0

Figure 12–2 Domestic water use table (Courtesy Bard Manufacturing Co.)

12–4 HEAD PRESSURE

The pump delivers the desired quantity of water against the resistance of the system to accept the water. The system resistance is called the pump head pressure, or the pump head. This head pressure is the total pressure that the pump must work against to deliver the required amount of water through the system.

The physical properties of the system that affect pump head are lift, friction (including all piping, unit, controls, fittings, and so on), and pressure resistance of the pressure tank. The three sources of head pressure are added together to determine the total head pressure the pump must work against. This head pressure is expressed as feet of water column (ft. H_2O or ft. WC). When the total volume of water required and the "feet of head" are calculated, the properly sized well pump is selected using the pump manufacturer's literature. Pumps are sized using "feet of head," not psig. Therefore, pressure (psig) is converted by multiplying the pressure required by 2.31 feet of head/1 psig.

Lift

Lift is the work the pump must do against the pull of gravity to push the water uphill from the pump in the well to the level of the pressure tank. The lift is made up of the total weight of the water contained in the supply pipe on the uphill side of the system.

Friction

Friction is the flow resistance of the pipe, valves, fittings, and so on that the pump must overcome to deliver the required GPM.

Back Pressure

Back pressure is the pressure in the pressure tank that opposes the flow of water that raises the pressure in the tank. As the tank fills, the tank pressure increases because the air space above the bladder in the tank is forced upward and the air volume is reduced. This action also puts more pressure on the water quantity in the tank, which increases the back pressure. This requires a greater effort by the pump to overcome the back pressure. Pressure tank controls are commonly set for 30 psig on and 50 psig off set points.

12–5 PRESSURE TANK SELECTION

A pressure storage tank is a requirement to provide the water to the system upon demand without short cycling the well pump. A pressure tank is always required for satisfactory system operation. Even in a dedicated system, a pressure storage tank must be used.

The pressure storage tank holds the water in the tank under pressure of the air chamber in the tank. This allows the supply of water to the system when the pump is off. The tank supplies the water to the system on demand, while the pump acts as a makeup supply to the tank.

Because the pressure tank replaces the well pump during the pump off mode, it must be capable of supplying the required volume of water from the minimum to the maximum system requirements. Therefore, to reduce the short cycling of the well pump, the pressure tank must be large enough to require a minimum of one minute for the well pump to raise the tank pressure to the off set point.

Nominal Capacity (Gal.)	Drawdown (Gal.) 20–40 PSI	30–50 PSI	40–60 PSI
20	7	6	5
30	10	9	8
40	14	12	10
80	28	25	21
120	41	35	31
200	69	60	52
270	93	80	70

Figure 12–3 Pressure tank chart (Courtesy Mammoth, a Nortek Co.)

Pressure Tank Size

The tank must also be large enough to be able to supply the water volume load for at least two minutes. Within these requirements, the well pump will operate with longer on and off periods of time. This extends the pump life as well as helps to keep the electrical energy requirements to a minimum. As a general rule, when sizing the pressure tank, the larger the tank the better. However, the purchase cost of the tank acts as a deterrent to oversizing.

Figure 12–3 is a table of the "drawdown" capacity of different size tanks from 20 gallons to 270 gallons at three different pressure ranges—20 to 40, 30 to 50, and 40 to 60 psig. The 30 to 50 psig range is the most commonly used range. The *drawdown* is the amount of water the tank will supply per minute into the system within each pressure range.

In the Omaha, NB application, where the total flow requirement is 18 GPM, an 80 gallon tank with a 25 GPM drawdown would be selected.

Pressure Tank Type

The introduction of air into the water promotes scaling and encrustation in the system. Therefore, the old type well system that used a single tank with the air and water in direct contact is not recommended. In addition, the action of the air in the water promotes the electrolysis action in the system, which causes pipe and fitting failure.

The tank that must be used is a rubber bladder type. A rubber diaphragm in the center of the tank separates the air and water and prevents absorption of the air into the water. Use of this type of tank allows the use of spring-loaded intake valves at the inlet of the submersible pump to maintain pressure in the supply system.

Figure 12–4 shows the operation of a bladder-type tank. When first installed (A), the bladder is forced to the bottom of

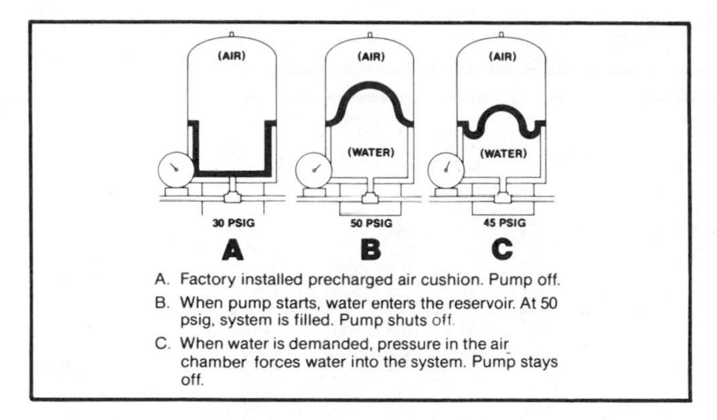

Figure 12–4 Operation of a pressure tank (Courtesy Mammoth, a Nortek Co.)

the tank by the air charge in the tank. This air charge is the minimum set point of the control.

When the tank is filled to the pressure control Off set point (B), the bladder is in the upward extended position. When the system demands a water flow, the bladder lowers from the effect of the water pressure drop. This maintains an equal pressure above and below the bladder. This action continues until the control On set point is reached, starting the pump and the fill portion of the operating cycle.

Short Cycling

A major factor in selection of the pump and pressure tank sizes is the length of the pump operating cycle. Although the average life of a submersible well pump is seven years, this can be shortened considerably by operating the pump on a short operating cycle. The standard for judging if the pump is subject to short cycling is if the pump restarts less than one minute after shutdown. If this occurs, the pump is seriously short cycling. A pump will last considerably longer if it runs continuously than if it repeatedly starts and stops.

A continuously operating pump is not desirable as it is difficult to match pump capacity to a system water demand. For best results, the pump and tank combination should be sized to operate continuously with maximum flow requirement on the system. This not only reduces the initial purchase cost of the pump since a smaller pump can be used, it also helps to reduce the system operating cost. The smaller pump draws less wattage. The net result is a properly sized system at the lowest cost to the installer and the greatest savings to the system owner.

12–6 PIPE MATERIAL

The use of iron pipe or galvanized steel pipe in a groundwater heat pump system is *not recommended.* The use of such materials promotes electrolysis action as well as scale formation. The recommended material for pipe and fittings is PVC, polybutylene, and rubber. Copper is permissible but not desirable.

The piping between the pump and the pressure tank that is buried underground must be either PVC or polybutylene and must be buried below the frost line or at a minimum of four feet below grade, whichever is deeper. The same material is used to connect to the pressure tank. From the pressure tank to the heat pump, any of the recommended materials can be used.

Pipe Sizing

Part of the system flow resistance or friction is the loss in feet of head of the water flow through the pipe. This friction loss depends upon the size of the pipe and the amount of water in GPM the pipe is expected to handle. The larger the pipe, the lower the friction loss. The more the expected water flow rate, the higher the friction loss.

Friction loss is measured in feet of head per 100' of pipe (ft. WC/100'). Figure 12–5 is a table of the friction loss through plastic pipe in ft. WC/100'. Six sizes of pipe from 3/8" inside diameter (ID) to 2" inside diameter (ID) are listed. (In plumbing work, the pipe size is given in inside diameter (ID) dimensions. In refrigeration work, tubing is sized by outside diameter (OD) dimensions. The water quantities are shown from 1 GPM to 50 GPM. Where blank spaces are shown, the pipe size/GPM combinations are not recommended. This pipe friction loss table will be used when designing each type of system.

Fitting Friction Loss

The total feet of pipe through which the pump must force the water is made up of the actual physical length of pipe plus the equivalent length of pipe of the fittings and other controls. In addition, the friction loss through the liquid-to-refrigerant heat exchanger must be included.

Figure 12–6 is a table of the friction loss through fittings and ball valves expressed in the equivalent feet of pipe length. For example, a 90° standard plastic elbow in a 1" OD size has a friction loss equal to 6' of straight pipe. A threaded adapter of the same material and size has a friction loss equal to 3' of straight pipe.

Only ball-type shut-off valves should be used. These straight flow-through valves have a much lower friction loss than globe or gate valves. A 1" ball valve has a friction loss equal to 4' of straight pipe.

If one elbow adds the equivalent of 6' of pipe, 10 elbows add up to 60' of straight pipe. The objective of the pipe layout is to use as few fittings as possible to keep the total friction loss to a minimum. This will be discussed in detail when laying out each type of system.

12–7 ACCESSORIES

Figure 12–7 shows some of the accessories that are used with liquid-to-air heat pumps. Not all of these would be used on the same installation. For example, the combination of the slow-closing motorized valve with the flow meter and dis-

PLASTIC PIPE:

FRICTION LOSS PER 100 FT.

GPM	GPH	3/8"		1/2"		3/4"		1"		1¼"		1½"	
		Ft.	Lbs.	Ft.	Lbs.	Ft.	Lbs.	Ft.	Lbs.	Ft.	Lbs.	Ft.	Lbs.
1	60	4.25	1.85	1.38	.60	.356	.155	.11	.048				
2	120	15.13	6.58	4.83	2.10	1.21	.526	.38	.164	.10	.044		
3	180	31.97	13.9	9.96	4.33	2.51	1.09	.77	.336	.21	.090	.10	.043
4	240	54.97	23.9	17.07	7.42	4.21	1.83	1.30	.565	.35	.150	.16	.071
5	300	84.41	36.7	25.76	11.2	6.33	2.75	1.92	.835	.51	.223	.24	.104
6	360			36.34	15.8	8.83	3.84	2.69	1.17	.71	.309	.33	.145
8	480			63.71	27.7	15.18	6.60	4.58	1.99	1.19	.518	.55	.241
10	600			97.52	42.4	25.98	11.27	6.88	2.99	1.78	.774	.83	.361
15	900					49.68	21.6	14.63	6.36	3.75	1.63	1.74	.755
20	1,200					86.94	37.8	25.07	10.9	6.39	2.78	2.94	1.28
25	1,500							38.41	16.7	9.71	4.22	4.44	1.93
30	1,800									13.62	5.92	6.26	2.72
35	2,100									18.17	7.90	8.37	3.64
40	2,400									23.55	10.24	10.70	4.65
45	2,700									29.44	12.80	13.46	5.85
50	3,000											16.45	7.15
60	3,600											23.48	10.21

Figure 12–5 Friction loss through pipe (Courtesy Gould Pumps Inc.)

Fitting Type	Material	Nominal Pipe/Fitting Size (" ID) 1/2	3/4	1	11/4	11/2	2	21/2
Insert Coupling	Plastic	3	3	3	3	3	3	3
Threaded Adapter	Copper	1	1	1	1	1	1	1
	Plastic	3	3	3	3	3	3	3
90° Standard	Copper	2	3	3	4	4	5	6
Elbow	Plastic	4	5	6	7	8	9	10
Standard Tee	Copper	1	2	2	3	3	4	5
(Flow-Thru Run)	Plastic	4	4	4	5	6	7	8
Standard Tee	Copper	4	5	6	8	9	11	14
(Flow-Thru Side)	Plastic	7	8	9	12	13	17	20
Gate or Ball Value	Brass	2	3	4	5	6	7	8

Figure 12–6 Friction loss through fittings and valves (ft. hd/100') (Courtesy Mammoth, a Nortek Co.)

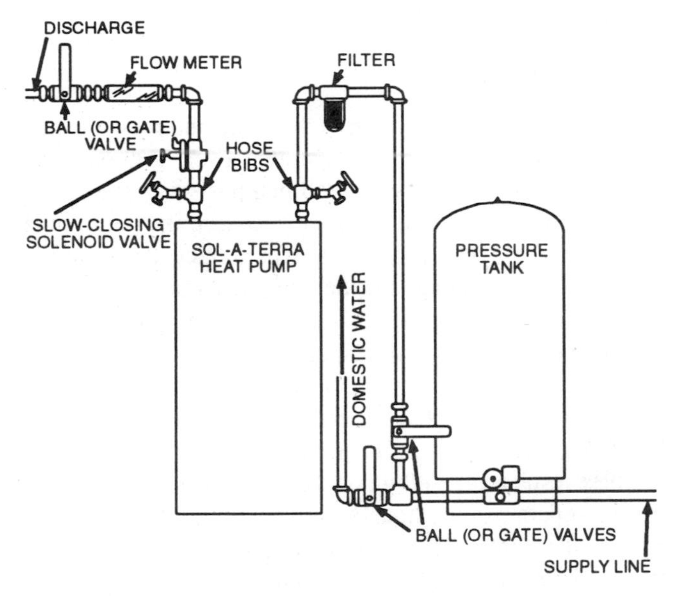

Figure 12–7 Typical piping accessories (Courtesy Mammoth, a Nortek Co.)

charge ball valve is a means of setting the water flow to the correct amount.

Instead of using the discharge ball valve to set the flow rate, a constant flow valve could be used. The discharge ball valve would then be used only for isolating the unit from the piping system for service maintenance.

Water regulating valves controlling head pressure in the cooling mode and suction pressure in the heating mode are also used. This is the preferred method as it controls the water flow more accurately when water temperatures and pressures change.

Although it is not required, a slow-acting motorized valve is sometimes used for positive shut off in conjunction with water regulating valves.

Hose bibs are required for service connections when descaling the water circuit in the unit.

Filters are required when large particles are encountered in the water supply. Sand and gravel tend to destroy the regulating parts of control valves as well as plug the return well screens. Installation of filters large enough to handle the water flow rate requirements per season is recommended. When large enough filters are used, the filters need only be changed twice per year.

Slow-acting or Motorized Valves

If the water flow control system is regulated by set devices such as a constant flow valve or set ball valve, water will flow continuously through the unit even in the off portion of the operating cycle. To prevent this unnecessary water flow, a slow-acting or motorized valve should be used. The valve must be of a slow opening/closing type to prevent sudden release or buildup of pressure in the unit water circuit. Figure 12–8 is an illustration of a slow-acting valve used for this purpose.

The valve is wired across the control circuit of the motor-compressor assembly. This cycles the valve with the compressor. The valve must be located in the discharge side of the unit water system. This prevents loss of water pressure in the unit, which promotes scale deposits on the pipe and tube surfaces. The slow-acting valve also prevents water hammer with its accompanying noise and potential damage.

Motorized valves are full-turn valves. The water flow turns 90° through the control mechanism and then turns another 90° to exit from the valve body. This is two 90° turns through the valve body. As a result, the friction loss through this device is considerably higher than with ordinary fittings.

The amount of water that is required also affects the friction loss. For example, the friction loss through a 1" motorized valve carrying 10 GPM is equal to 1 psig or 2.5 equivalent feet of straight pipe (see Figure 12–9). If 20 GPM is required, the friction loss increases to 7.6 equivalent feet of straight pipe.

When the entire piping system is completed and calculated, the use of multiple controls to regulate the water flow may be necessary. This will be determined when laying out the piping for each type of system.

Figure 12–8 Slow-acting motorized valve (Courtesy Erie Manufacturing Co.)

Shut-off Valves

Shut-off valves must be installed across any portion of the system that must be isolated for service maintenance. For example, they must be installed on the inlet and outlet of the unit to facilitate coil cleaning, and on the inlet and outlet of any

	VALVE SIZE INCHES	WATER CAPACITIES IN GALLONS PER MINUTE											
		CV	PRESSURE DROP ACROSS VALVE (PSI)										
		1	2	3	4	5	10	15	20	25	30	40	50
BRONZE NPTF BODY	1/2	3	4	5	6	6.5	9	11.5	13	15	16	18	21
	3/4	6	8	10	12	13	18	23	26	30	32	37	42
	1	11	15	19	22	24	34	42	49	55	60	69	77
	1-1/4	17	24	29	34	38	53	65	76	85	93	107	120
	1-1/2	28	39	48	56	62	88	108	125	140	153	177	197
	2	46	65	79	92	102	145	178	205	230	251	290	325
	2-1/2	65	91	112	130	145	205	251	290	325	356	411	459
	3	90	127	155	180	201	284	348	402	450	492	569	636
IRON FLANGE BODY	3	69	97	119	138	154	218	267	308	345	377	436	487
	4	125	176	216	250	279	395	484	559	625	684	790	883
	6	330	466	571	660	737	1043	1278	1475	1650	1807	2087	2333

Figure 12–9 Pressure drop across water valves (Courtesy Erie Manufacturing Co.)

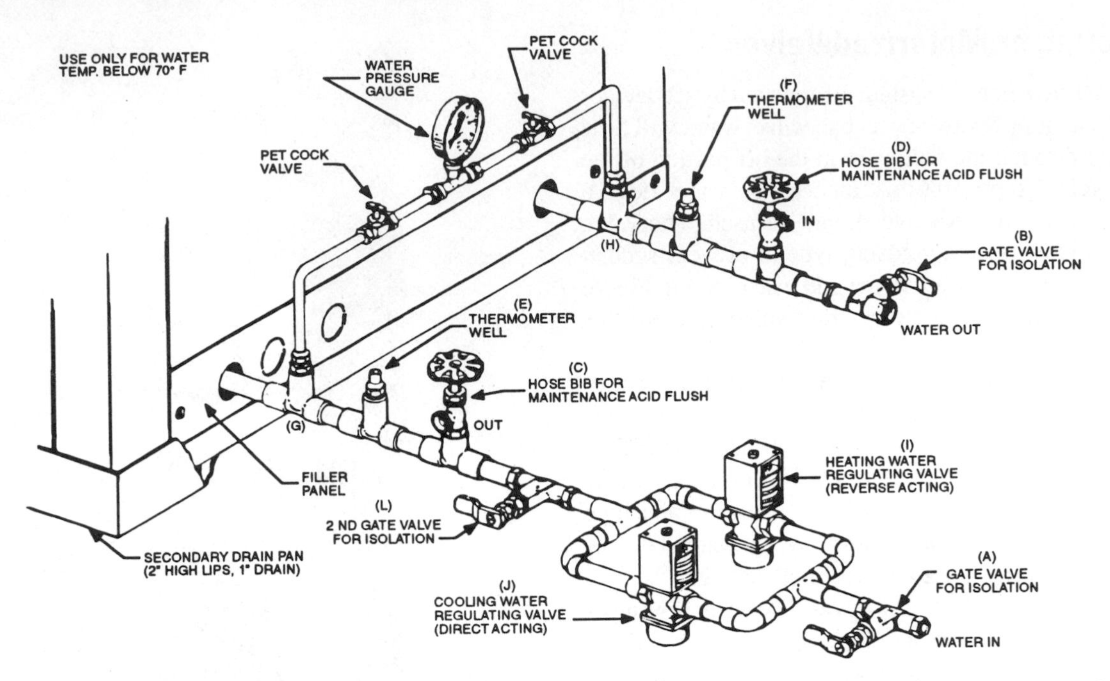

Figure 12–10 Typical piping arrangement using flow-regulating valves (Courtesy Addison Products Co.)

filter or controls. This allows filter or control change without excessive water loss. Figure 12–10 shows the piping arrangement recommended by one manufacturer for the use of water-regulating valves as well as valves for isolation of the unit for acid flushing of the heat exchanger coil when required.

Ball valves or gate valves must be used because of their straight-through flow pattern. The friction loss through a gate or ball valve is given in the table in Figure 12–5. For example, a 1" gate or ball valve has a flow friction loss equal to 4 equivalent feet of straight pipe. The loss through the three valves in Figure 12–10 would be 12 equivalent feet of straight pipe.

Pressure-regulating Valves

When water-regulating valves are used to control the water flow according to the unit requirements, water pressure and water temperature, two valves are used (see Figure 12–10). A direct-acting valve is used to control the unit condensing pressure and temperature in the cooling mode. As the condensing pressure and temperature rise, the valve opens to allow an increased water flow. On pressure and temperature fall, the valve closes. The predominant setting for the valve is to maintain 105°F condensing temperature.

A reverse-acting valve is used to control the evaporator operating pressure and boiling point in the heating mode. As the evaporator pressure and boiling point fall, the valve opens to increase the water flow and heat energy available. When the evaporator pressure and boiling point increase, the valve modulates closed. The predominant setting of this valve is to maintain a 40°F boiling point in the evaporator.

Each manufacturer has its own method of checking and adjusting these valves. The manufacturer's instructions should be followed for best results.

Figure 12–11 shows the pressure drop at various flow rates through water-regulating valves. To convert to equivalent feet of water, 1 psig = 2.31 feet of water. For example, the 1" valve with 14 GPM flow rate would have a 2.7 psig pressure drop × 2.31 feet of water/1 psig or a friction loss equal to 6.24 equivalent feet of straight pipe.

Constant Flow Valves

A constant flow valve is an automatic valve installed on the discharge side of the heat pump unit. It maintains the flow rate through the unit within ±10% of the set amount even though the water pressure will vary between the control on and off set points.

Figure 12–12 shows the flow friction loss in both psig and ft. H_2O for all GPM flow quantities. For purposes of estimation of the total feet of head for pump sizing, the constant flow valve is rated at 34.7 ft. H_2O.

Strainers

Strainers should always be installed in the supply line to the unit to catch any sand or particles in the water. The best location in a well supply system is in the line between the pump and the tank. This prevents material buildup in the pressure tank as well as the rest of the system.

A Y-type filter is recommended for easy access to replace the filter material. A good filter material is a rolled piece of plastic window screen.

Drain Cocks or Hose Bibs

Valve taps into the water system are required for treatment of the coils in the heat pump. In time, all units will require descaling treatment. To make the system access a simple

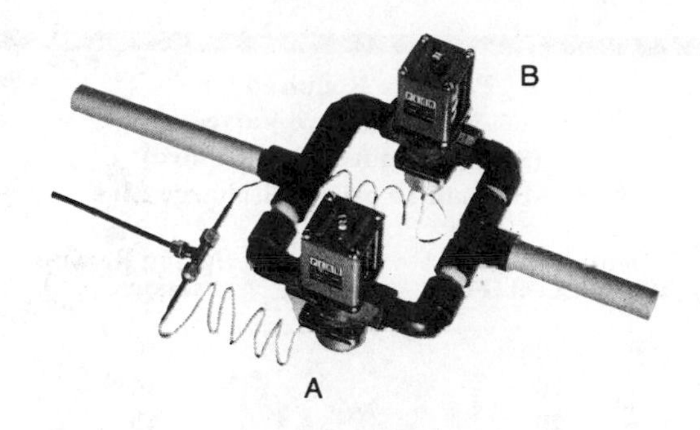

Model	Valve Size	A Cooling Valve	B Heating Valve	PRESSURE DROP—WATER REGULATING VALVES										
				GPM	2	4	6	8	10	14	18	22	26	30
WRH/HRH 22, 28, 34	½"	V46AB-1	V46NB-2	PSI	.3	1.4	3.1	5.5	8.7	17.0				
WRH/HRH 44, 54, 64	¾"	V46AC-1	V46NC-2	PSI			1.2	2.1	3.3	6.5	10.7	16.0		
WRH/HRH 54, 64,	1"	V46AD-1	V46ND-2	PSI					1.4	2.7	4.5	6.7	9.4	12.5

Figure 12–11 Pressure drop through water-regulating valves (Courtesy Addison Products Co.)

Friction Loss Through Constant Flow Values		
All GPM Ratings	PSI 15	Ft. H_2O 34.7

Figure 12–12 Friction loss through constant flow valves (Courtesy Bard Manufacturing Co.)

process, drain cocks or hose bibs are installed in the piping circuit. Figures 12–7 and 12–10 show where these connecting valves are installed. Although they will not be needed for immediate use, it is a serious mistake not to include them in the original installation.

In ground loop installations, these valves are required on both sides of the shut-off valves—on the unit side for treatment of the inside coils and on the loop side for purging the ground loop.

P/T Ports

Figure 12–13 shows a typical arrangement for inclusion of P/T ports into the system for service testing and adjusting. A P/T port is a fitting that contains a neoprene rubber insert

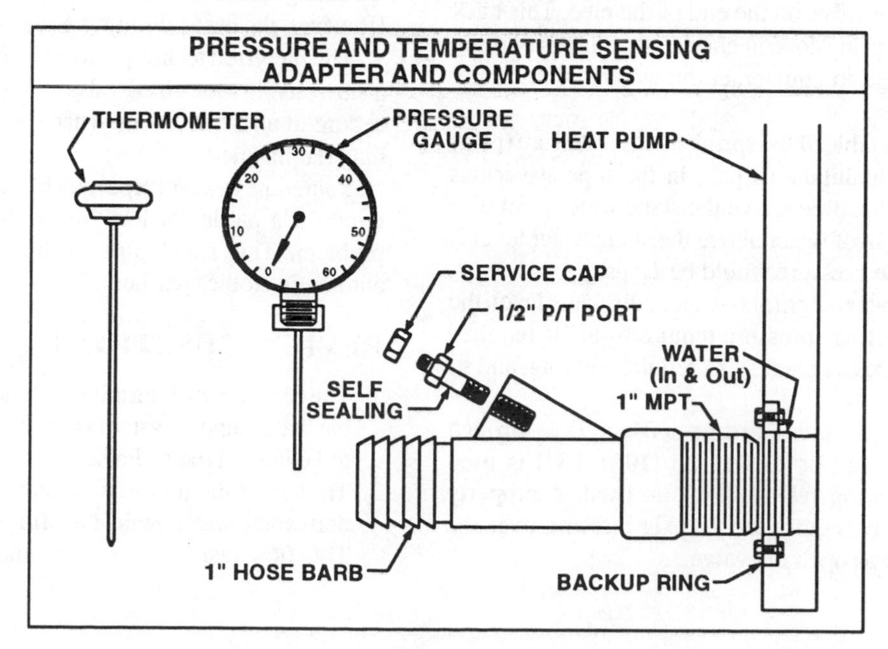

Figure 12–13 P/T port and accessories (Courtesy Bard Manufacturing Co.)

collar designed to admit a test probe. The test probe may be part of a dial or digital thermometer or a probe adapter for a pressure gauge. Insertion of the probe through the rubber collar allows access to the water circuit without leakage of water past the probe. The pressure limit on the seal is 100 psig. When the probe is removed, the collar closes and the water pressure enforces the seal. An exterior cap protects the port when not in use.

Use of these ports is highly recommended as it greatly reduces the time for attaching instruments and does not require reducing the pressure in the system.

Flow Meters

A flow meter located in the discharge line from the unit is a valuable addition to the system (see Figure 12–7). Use of the flow meter will enable the setting of the required GPM of water through the unit. Otherwise, a calibrated bucket and stop watch are required.

Use of the bucket and stop watch means the addition of a means of directing the water flow from the disposal means to the bucket for a period of time. The cost of a flow meter is not much more investment than the fittings, valves, and so on to provide the same test. Including the time involved to set up and conduct the flow test, the flow meter is more economical.

Spring-loaded Check Valves

In systems using a return well for water disposal, a means of preventing air entrapment in the system must be provided by maintaining pressure in that portion of the system after the cutoff controls. If this pressure is allowed to drop to atmospheric pressure, air entrapment, which stimulates encrustation, can occur.

To prevent the air entrapment, the drop pipe in the return well must terminate below the static water line in the well and have a back pressure valve on the end of the pipe. This back pressure valve is a spring-loaded check valve that has spring pressure high enough to counteract the weight of the water plus 10 psig.

Figure 12–14 is a table of the spring release pressure (psig) that is required to maintain 10 psig in the pipe at various heights of water in the pipe above the static water level. For example, if the height of water above the static water level is 0', the spring release pressure would be 10 psig.

The higher the water height above the static water level, the greater the spring release pressure required—at 10' requires 14 psig, at 50' requires 31 psig, at 90' requires 49 psig, and so on.

When calculating the pressure drop or friction loss through the valve, only 10 psig or 23.1 ft. hd (10 × 2.31) is used regardless of the spring release pressure used. A properly sized valve will only require the 10 psig increase over the weight of the water to open the valve.

Pressure Required On Back Pressure Valve (Spring-Loaded Check Valve) To Maintain 10 psi on Discharge Line

Depth To Static Water Level (Feet)	Spring Release Pressure (psi)
0	10
10	14
20	18
30	23
40	27
50	31
60	36
70	40
80	45
90	49
100	53

Figure 12–14 Spring-loaded check valve—spring pressure requirement (Courtesy Bard Manufacturing Co.)

12–8 PROBLEMS TO AVOID

Water pipes inside the house, regardless of the material used in the pipe, tend to condense moisture on the outside (sweat). Therefore, the pipes must be insulated where the condensate off the pipe will be a nuisance or cause damage. Some recommended insulating materials are spiral wound fiberglass, insulating tapes, or foam tubes.

Another serious problem that must be avoided is water hammer. *Water hammer* is the excessive vibration caused by pressure buildup when the flow of water is stopped very rapidly. Using rubber bushings between the pipe and the mounting means will reduce the objectionable vibration noise. However, the best solution is to eliminate the water hammer.

The preferred technique to prevent water hammer is to use a slow-acting motorized valve. This should be standard procedure in all installations where water under pressure is the material handled.

Sometimes a stand pipe may be incorporated into the water circuit if a particular location in the system is creating the problem. This stand pipe is the same type as used with automatic clothes washers.

REVIEW QUESTIONS

1. The first step in determining the size of a pump needed for the water supply system is to determine the size of the pipe to be used. True or False?
2. To deliver the required amount of water, the pump must deliver the water against the flow resistance of the system. This flow resistance is called the _______ _______.

3. Name the three physical properties that affect the flow resistance of the system.
4. To convert pressure in psig to feet of head (ft. hd) the pressure figure is multiplied by:
   a. 1.897
   b. 2.159
   c. 2.31
   d. 2.861
   e. 3.307
5. The total weight of water contained in the supply pipe on the uphill side of the system is called the _______.
6. The pressure in the pressure tank that opposes the flow of water is called the _______ _______.
7. The most common pressure control settings for water system controls are:
   a. 40 psig on; 60 psig off
   b. 10 psig on; 40 psig off
   c. 20 psig on; 50 psig off
   d. 30 psig on; 40 psig off
   e. 30 psig on; 50 psig off
8. If a pressure tank is sized correctly, the tank will supply the required volume of water to the system for a period of _____ minutes before the pump starts.
9. The amount of water the pressure tank will supply to the system between the Off and On set points of the control is called the _______ _______ _______.
10. The use of iron pipe or galvanized steel pipe is not recommended. Why?
11. List three of the recommended materials for pipes and fittings.
12. The friction loss of pipe depends upon what two factors?
13. Shut-off isolation valves are required in the water circuit. True or False?
14. When water-regulating valves are used in the cooling mode, is the valve used a direct-acting or reverse-acting type?
15. The predominant setting of the valve in Problem 14 is to maintain a _____°F condensing temperature.
16. When water-regulating valves are used, in the heating mode, is the valve a direct-acting or reverse-acting type?
17. The predominant setting of the valve in Problem 16 is to maintain a _____°F boiling point in the evaporator.
18. To protect the unit from sand and other particles in the water, a filter must be installed between the tank and the unit. True or False?
19. A spring-loaded check valve with a 35 psig spring release pressure is installed at the bottom of the return pipe. When the unit is operating, what is the friction loss through the valve in ft. hd?
20. What is the best way to prevent water hammer?

13 SIZING WELL SUPPLY SYSTEMS

13–1 GENERAL

In this chapter we will attempt to illustrate the steps needed to design the various types of systems using wells for supply and various types of disposal systems. Not all types of systems are equally adaptable to every installation. For example, a system using a drilled well supply and surface disposal would not be recommended in Minnesota where the outdoor temperature reaches –40°F and disposal ponds or streams freeze over. When the frost line reaches 6' below grade, surface water disposal becomes difficult.

On the other hand, in Florida, where water is available 5' to 10' down and surface disposal in a swimming pool or pond is available, the drilled well and surface disposal system is ideal. In Phoenix, AZ, where water is in short supply, ground loops are preferred over any supply/disposal system. Therefore, to illustrate each type of system, a region where the system has a reasonable chance for performing satisfactorily will be used in the examples. The same size unit will be used for each system.

As explained previously, the amount of water needed for the heat pump will depend upon the temperature of the supply water. Therefore, the geographical location of the installation is important to find out the average water temperature available. Figure 13–1 is a map of groundwater temperatures ranging from 50' to 150' below grade throughout the United States. This map gives the water temperatures available to the heat pump when delivering the heat pump GPM requirements.

Using the heat pump information plus the information on the system components, each system is calculated for pipe sizes, pump sizes, and pressure tank sizes. For the purpose of consistency, most of the calculating is done in feet of head (FTHD) or feet of water (ft. H_2O), which is a unit of pressure. Both terms are used interchangeably.

This pressure is called pump head. *Pump head* (PH) is the sum of the pressure exerted by gravity (feet of lift), pressure exerted by the pressure tank, and pressure loss due to friction through the piping system, the pipe, fittings, controls, and so on.

These factors are calculated for each type of system to determine the pump size, pressure tank size, pipe size, and control set points.

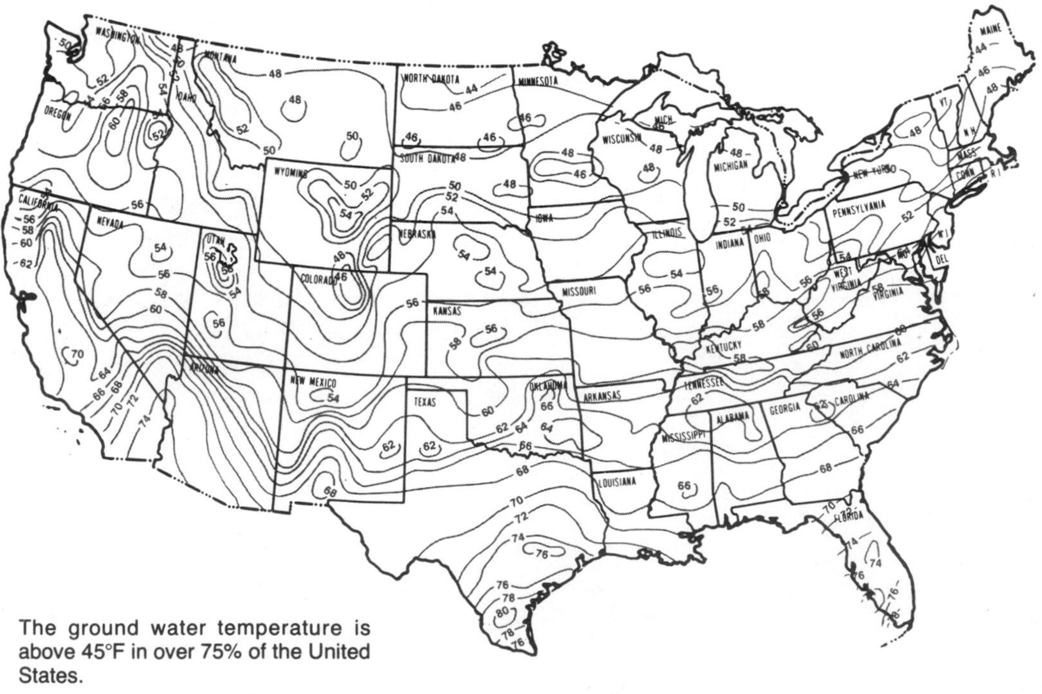

Figure 13–1 Ground and well water temperatures—50' to 150' in depth (Courtesy Bard Manufacturing Co.)

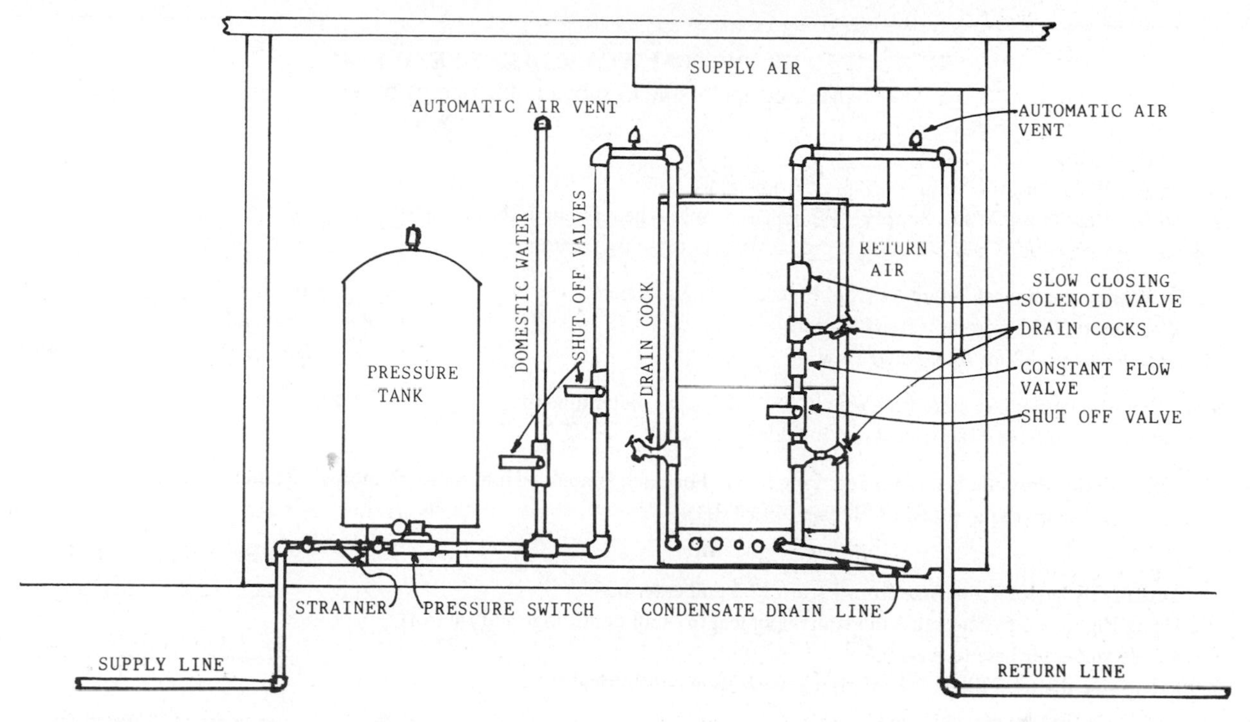

Figure 13–2 Typical piping to a water source heat pump (Courtesy Bard Manufacturing Co.)

Figure 13–2 shows a typical piping arrangement to a liquid-to-air heat pump unit. This setup will be used as the basic setup for each of the following examples covering the various types of systems using a well for supply and surface, dry well, or field tile for disposal. In each design section, when reference is made to the control and piping detail, this illustration is used. A complete picture of each type of system will be discussed under each design.

13–2 WATER SYSTEM WORKSHEET

When calculating the pump, pressure tank, and pipe sizes for the system, the worksheet in Figure 13–3 is used. For calculation consistency, most of the calculations are done in feet of head (FTHD), which is a unit of pressure. This pressure is the pump head.

Using the worksheet in Figure 13–3, an explanation of each step involved in the sizing process follows.

Well Pump Sizing

Divide the water piping system into branches, labeling each branch. The two basic branches are A and B. One or more branch C may be in the system. In the case of branch C, label as C1, C2, C3, and so on.

Branch A is the prime source of water, the pipe system from the discharge of the submersible pump to the pressure tank. Branch B is the water piping circuit from the pressure tank, through the heat pump unit and controls (valves, controls, and so on), and the piping from the unit to the water disposal means. Branch(es) C is the pipe system supplying water from the pressure tank to the domestic supply throughout the building. Most domestic systems require a minimum pressure of 30 psig (69 ft. hd). If the branch supplies an excessive load, it will require actually calculating the total branch pressure drop.

If the domestic supply system has more than one branch, these should be individually labeled as C1, C2, C3, and so on. The total GPM requirement of all the branches is used to find the total domestic GPM.

Step 1. Domestic Need. In a dedicated system, the heat pump is the only load on the water system; therefore, step 2 is not needed. The domestic system has the domestic load as part of the total GPM requirement. Therefore, determine the domestic water needs in GPM from Figure 11–2. To use the water usage table in Figure 11–2, list the number and type of fixture in the domestic load, such as sinks, toilets, and so on. Add up the peak demand allowance for each fixture. The actual usage (Individual Fixture Flow Rate) will be larger than the peak demand GPM. However, not all fixtures in the domestic load will require water at the same time. The pump will not be required to supply the total possible volume of water at one time, so an average figure is used. For example:

Quantity	*Item*	*Peak GPM*
1	Tub and shower combination	2.0
2	Lavatories 1@	2.0
2	Toilets .75@	1.5
1	Kitchen sink	1.0
1	Laundry sink	1.5
1	Clothes washer	2.0
	Total GPM	9.0

This total is entered on line 1 of the worksheet. If more than one branch C is involved, all fixtures on all branches C are included.

WATER SYSTEM WORKSHEET SE3WKS-1

(Method applicable to submersible pumps only)

A. Well Pump Sizing

Branch A—Well Pump—Piping from pump to pressure tank.
Branch B—Heat Pump Water Supply—Piping from tank to heat pump to drain.
Branch C—Domestic Water Supply—Piping from tank to house fixtures.

1. Determine the household needs from Figure 12-2. Enter here.	Branch C.	________	GPM
2. Heat pump GPM from unit specifications.	Branch B.	________	GPM
3. Add lines 1 and 2 for total water flow.	Branch A.	________	GPM

Note: If piping layout has more branches (C1, C2, C3, etc.), determine the flow rate for these from Figure 12-2 and include in the total.

B. Determine Water Pressure Requirements Pipe Sizing For Each Branch—Household plumbing, Branch B, may be assumed to have a total pressure requirement of 30 psig (69 FTHD).

	Br. A	Br. B	
4. Tentatively select a pipe size from Figure 12-5 and enter here.	________	________	ID
5. Using Figure 12-6, determine the equivalent length of all the fittings and shut-off valves and enter here (equiv. ft.)	________	________	EF
6. Determine the total lineal feet of pipe in each branch. (physical ft.)	________	________	ft.
7. Add lines 5 and 6 and enter here.	________	________	
8. From Figure 12-5, for pipe size and GPM needed for each branch, determine each branch friction loss. (FTHD/100')	________	________	
9. Multiply line 8 by line 7; divide by 100 to determine the total branch friction loss. (FTHD/100')	________	________	
10. From the heat pump specifications, enter the unit pressure drop. (FTHD)	________	________	
11. Pressure drop through the controls from Figures 12-9, 12-11, and 12-12. (FTHD)	________	________	
12. Calculate the total pressure drop. Branch A pressure drop is the same as line 9. Enter Branch A here. (FTHD)	________		
Branch B, by adding lines 9, 10, and 11 (FTHD)	________	________	
Branch C.	________	________	
13. Multiply line 12 by 0.433 to convert to psig. Enter here.	________	________	

14. From the piping layout, determine parallel flow among the branches. Beginning at the well pump, add the friction loss in psig for the well pump branch (Branch A) to the branch having the higher pressure, Branch B or C.
Note: If more than three branches are required by the piping layout, select that branch that has the highest pressure drop and add this pressure drop to Branch A. Enter on line 14 the number obtained as total pressure loss due to pipe friction. ________ (psig)
15. Add 20 psig to line 14 to obtain the pressure control cutout set point. ________ (psig)
16. Multiply line 15 by 2.31 to convert to FTHD. Enter here. ________ (FTHD)
17. Pump requirement.

________ GPM at ________ (FTHD)

Vertical distance to water in well ________ ft. lift
TOTAL PRESSURE ________ (FTHD)

C. Pressure Tank Sizing (Bladder or diaphragm types only)

18. Enter desired minimum off time of the well pump in minutes. (Never less than two minutes.)	________	minutes
19. Enter the pressure control cutin set point.	________	psig
20. Enter pressure cutout set point from line 15.	________	psig
21. Multipy line 3 by line 18 to determine the minimum drawdown.	________	GPM

22. Refer to Figure 12-3 or the pressure tank specifications for a specific model of tank(s) to select the nominal capacity of tank needed, using information from lines 19, 20, and 21.

Figure 13–3 Water system worksheet—SE3WKS-1

Step 2. Determine the GPM water flow required by the heat pump to be installed. This information from the manufacturer's literature will be determined by the temperature of the groundwater used. Figure 13–1 is a map of groundwater temperatures throughout the U.S. From this map, the approximate groundwater temperature can be determined.

Using the manufacturer's literature (see Figure 10–1), the GPM is determined by the water temperature available and the heating and/or cooling capacity needed. For example, if 50°F groundwater is available and 28,700 BTUH heating capacity is required, the unit will need 8 GPM. This information is entered on line 2 of the worksheet.

Step 3. Add up the total water requirement for the heat pump plus any building requirement. This is the total water flow the pump *will be required to deliver to the system.*

In this example, the 9 GPM requirement for the building (domestic load) plus the 8 GPM requirement for the heat pump = 17 GPM total requirement. This is the GPM of water supplied by the pump through branch A to the pressure tank.

Determine Water Pressure Requirements for Pipe Sizing of Each Branch

Step 4. Tentatively select a pipe size for each branch, based on the GPM requirement for that branch. Select a pipe size from Figure 12–5 that will result in a friction loss of less than 10 FTHD.

In this example, the 17 GPM in branch A will require 1-1/4" pipe for a friction loss of 4.81 FTHD. The pipe size between the pump and tank is 1-1/4". Branch B will carry 8 GPM and can be 1" for a friction loss of 4.58 FTHD. Branch C, supplying the domestic requirement of 9 GPM, will be 1" with a friction loss of 5.73 FTHD.

When selecting pipe sizes, if the friction loss is 7 or higher, it is advisable to use the next larger pipe. The actual pipe loss is a small part of the total, but it should be kept as low as possible.

The pipe sizes for branches A and B are recorded on the worksheet on line 4.

Step 5. Tabulate all the fittings in branches A and B and record on line 5. The fittings should be listed by type and size. For example, in Figure 13–2, the following fitting tabulation would result:

Type	Quantity	Size	Equivalent Length	TEL
Branch A				
90° Elbow	2	1¼"	7'	14'
	Total Equivalent Length			14'
Branch B				
90° Elbow	6	1"	6'	36'
Tee	4	1"	4'	16'
Shut-off Valve	2	1"	4'	8'
	Total Equivalent Length			60'

The totals of 14 EL for branch A and 60 EL for branch B are entered on line 5 of the worksheet.

Step 6. Determine the total lineal feet of pipe in each branch. For branch A, determine the total lineal (actual) feet of pipe length from the submersible pump in the well to the pressure tank. The point of measurement can be the tee in the line where the tank is connected. It may be necessary to drop a weighted line down the well casing to the top of the pump to measure this pipe length. For estimating purposes, find out the static water level from the local well driller and add 5' to this distance.

For branch B, determine the total pipe length from the tank connection tee to the inlet pipe connection on the heat pump plus the lineal (actual) length of the return pipe from the outlet connection to the end of the discharge pipe in the pond, dry well, or field tile.

The total of each branch is recorded on line 6.

Step 7. Adding the total equivalent pipe lengths for the fittings and shut-off valves on line 5 and the total lineal feet of pipe in each branch on line 6 gives the total equivalent pipe lengths for each branch. These are recorded on line 7.

Step 8. From Figure 12–5, determine the friction loss per 100 feet of pipe for the GPM required in each branch. For example, in branch A, using 1-1/4" pipe to carry 17 GPM, the pipe friction loss would be 4.81 FTHD/100'. In branch B, using 1" pipe to carry 8 GPM, the pipe friction loss would be 4.58 FTHD/100'. The friction losses for branches A and B are recorded on line 8.

Step 9. Multiply line 8 (branch friction loss per 100') by the total length of the branch (lineal feet plus equivalent feet) and divide by 100'. This will determine the total branch friction loss per 100' of length.

$$\text{TBFL}/100' = \frac{\text{BFL}/100' \times \text{TEF}}{100'}$$

where:

TBFL = Total branch friction loss—the total friction loss of pipe, fittings, and controls in the branch (FTHD).

BFL/100' = The friction loss (FTHD) per 100' of pipe based on the size of the pipe and the GPM of water carried.

TEF = Total equivalent feet length of the branch, including both the physical (lineal) feet length of the pipe plus the loss through the fittings, controls, and so on.

100' = All losses are based on 100' of pipe. Therefore, the total loss through the branch is reduced to a figure per 100'.

These totals are recorded on line 9.

WRH										
22	G.P.M.	2	3	4	5	6	7	8	9	10
	PSI	.9	1.5	2.3	3.0	4.0	5.0	6.1	7.2	8.3
28	G.P.M.	2	3	4	5	6	7	8	9	10
	PSI	.7	1.2	1.8	2.5	3.2	4.0	5.0	6.0	7.0
34	G.P.M.	3	5	7	9	11	13			
	PSI	.8	1.5	2.8	4.0	5.5	7.1			
44	G.P.M.	4	6	8	10	12	14			
	PSI	1.5	2.6	4.0	5.8	6.8	8.4			
54	G.P.M.	5	6	8	10	12	14	16		
	PSI	1.3	1.7	2.5	3.6	5.0	6.5	8.6		
64	G.P.M.	5	6	8	10	12	14	16		
	PSI	1.5	1.9	3.0	4.8	7.2	9.5	12.5		

Figure 13–4 Unit heat exchanger pressure drop (Courtesy Addison Products Co.)

Step 10. From the manufacturer's specification literature, the pressure drop based on the GPM required is listed. In our example, we assumed that the unit selected would be the WPH22–1H in Figure 13–4. At the required 8 GPM, the water coil in the unit would have a pressure drop of 6.1 psig or 14.09 FTHD. The FTHD for the unit is recorded in the branch B column of line 10.

Step 11. Using Figures 12–9 and 12–12 or Figure 12–11, tabulate the total equivalent length of pipe loss through the controls used in the heat pump circuit. For example, the following would be listed from Figure 12–5:

Item	Quantity	Size	FTHD	Total FTHD
Constant flow valve	1	1"	34.7	34.7
S.C. motorized valve	1	1"	3.6	3.6
Strainer	1	1"	2.0	2.0
	Total Equivalent Length			40.3

If water-regulating valves are used instead of the constant flow valves, the FTHD for the valves will be approximately 3.5 FTHD (1" valve with 8 GPM flow). (See Figure 12–11.) The chart in Figure 12–11 does not list a valve for less than 10.6 GPM flow. However, because the pipe requirement is for 1" pipe, reducing the pipe size to ¾" valve will produce considerably more flow resistance than if left the same pipe size. The recommendation is to use valves the same size as the pipe size with 1" maximum. For larger pipe sizes, multiple valves in parallel should be used.

If the discharge water disposal is to a return well, Figure 13–5, a spring-loaded check valve is required on the end of the return pipe to maintain a 10 psig minimum pressure in the return pipe. The equivalent pipe resistance of this valve is 10 psig × 2.31 or 23.1 FTHD. This must be added into the total for branch B.

The total equivalent pipe length totals for branches A and B are entered on line 11.

Step 12. Calculate the total pressure drop, in FTHD, of branch B by adding lines 9, 10, and 11. This will give the pressure drop in FTHD through the heat pump coil and the controls. Record this total in column B on line 12. The FTHD loss for column A is the same figure as line 9.

Step 13. The FTHD friction loss of branches A and B of line 12 are multiplied by 0.433 (psig/FTHD) to convert these figures to psig. This information is recorded on line 13.

Step 14. From the piping layout, determine the lowest pressure that could be used in the system and still have adequate water flow through the fixtures with the heat pump operating at the same time.

In a dedicated system, the pressure loss through the system applies to only the heat pump circuit. In a domestic system, branch A psig loss plus the largest psig in the other branches is used as the total psig loss.

From the piping layout, determine the psig loss using each of the branches in conjunction with branch A. Branch C, the domestic supply line, has an assumed pressure drop of 30 psig. If, however, more than one branch C is involved, the psig loss of each branch must be calculated and the branch total must be used.

Beginning at the well pump, add the friction loss in psig for the well pump branch A to the higher pressure drop and record on line 14.

Step 15. To determine the hp needed for the pump, the pump is sized on the highest pressure the pump must operate against. This is done by adding 20 psig (the pressure control differential) to the cutin set point from line 14. This cutout pressure set point is recorded on line 15.

Step 16. Convert the cutout pressure to FTHD by multiplying FTHD by 2.31 FTHD/1 psig and record on line 16.

Step 17. To determine the pump capacity needed for the system (pumps are rated by GPM output per FTHD), three items of information are needed:

1. The highest flow rate required has been recorded on line 3 (branch A—the pipe system from the pump to the pressure tank).
2. Total system flow resistance in FTHD. This amount has been recorded on line 16.

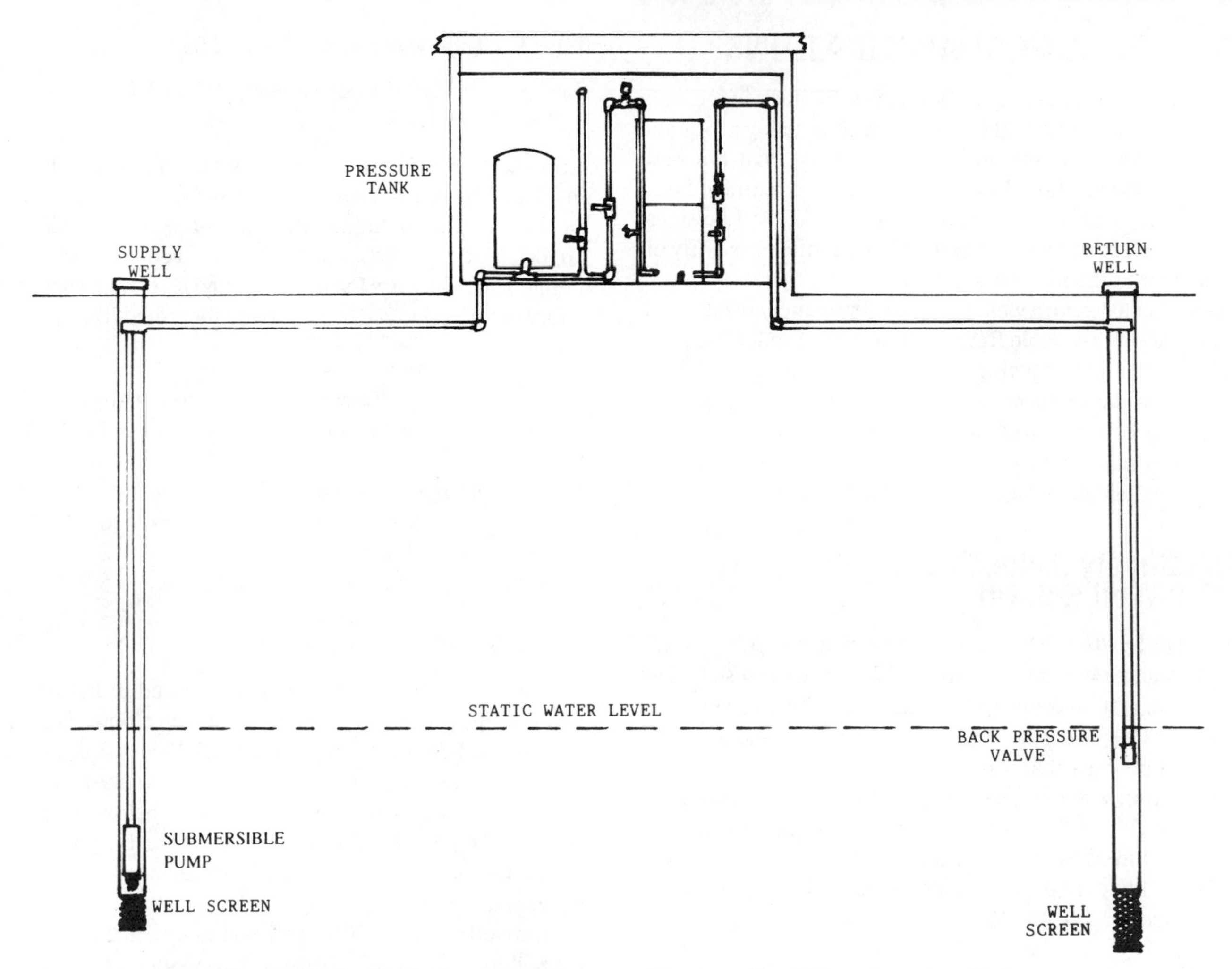

Figure 13–5 Typical return well system (Courtesy Bard Manufacturing Co.)

3. Vertical distance to the water in the well. This distance is from the height of the water in the pressure tank to the pumping water level in the well.

The pumping water level in a well will depend upon the amount of water pumped versus the ability of the water to flow from the surrounding aquifer. This information must be obtained from your local well driller who is familiar with the aquifer condition in the area.

When measuring the height of the water, all elevations must be taken into account. If the pressure tank is above grade, in the first floor of the building, the distance from the grade level down to the pumping water level plus the height of the water in the tank (2/3 of the tank height) is the total lift.

If the pressure tank is in the basement of the building or in a well chamber below frost line, the distance between the pumping water level and grade level minus the distance from 2/3 of the tank height to grade level is used. This information is recorded on the second line (ft. lift) of line 17. The FTHD and ft. lift added together is the total pressure (FTHD) the pump must operate against.

Pressure Tank Sizing (Bladder or Diaphragm Type Must Be Used)

Step 18. Enter the desired minimum off time of the well in minutes. To get maximum pump life, the off time must be two minutes or more. Enter this time on line 18.

Step 19. Enter the pressure control cutin point on line 19.

Step 20. Enter the pressure control cutout set point from line 15. A recommendation of 20 psig is used as the differential for the pressure control.

Step 21. Determine the minimum drawdown the pressure tank must have by determining the total quantity of water the tank must deliver before the pump starts. This is found by multiplying the GPM required (line 3) by the minimum off time selected for pump operation.

Enter this total minimum drawdown on line 21.

Step 22. Refer to Figure 12–3 to determine the nominal capacity (gallons) pressure tank required.

13–3 PRACTICAL APPLICATIONS

To illustrate the use of the worksheet for each of the well supply systems, the calculations will be done using a location for water temperatures and disposal means that are best adapted to the location. For example, an open return to lake, pond, or drywell would be used from Florida to Louisiana where good groundwater is available with little possibility of freeze over of the receiving means.

Disposal to a return well is used predominately in the north where water is available from shallow wells (under 100') but a winter frost level develops.

Where water levels are in the 25'+ level but the quantity is minimal, geothermal wells are recommended. These types of systems are calculated using the same liquid-to-air heat pump in all the applications. The locations will vary.

Well Supply (Lake, Pond, or Dry Well Return)

Domestic System. We will assume that the WPH22–1H liquid-to-air heat pump (see Figure 12–1) is to be installed in a domestic system in the gulf coastal area using a dry well for disposal. This could be a pond or lake. The aquifer is of sufficient porosity that water returns very easily.

The return water is sometimes used for irrigation via lawn sprinklers. When these are used, a pressure loss of 20 psig must be allowed for sprinkler operation.

The building to be used is a residence having the following fixtures (see Figure 11–2).

Quantity	Item	PDA
1	Tub and shower combination	2.0
1	Shower only	1.0
4	Lavatory 0.5 @	2.0
2	Toilet flush tank 0.75 @	1.5
1	Kitchen sink (including garbage disposal)	1.0
1	Dishwasher	0.5
1	Laundry sink	1.5
1	Clothes washer	2.0
	Total gallons per minute	11.5

Local conditions. We will assume that the location is north central Florida with groundwater temperature of 70°F. The aquifer is 6' below grade and a normal well can be expected to have a 2' drawdown. The building is slab construction with the liquid-to-air heat pump and pressure tank located in the garage area. The drilled well is 50' east of the building and the dry well is 50' west of the building.

Using the Water System Worksheet in Figure 13–6, the step calculations would be:

Line 1. Household needs (Figure 12–2); Branch C—11.5 GPM.

Line 2. Heat pump need (70°F water)—5.0 GPM.

Line 3. Total water need—16.5 GPM.

Line 4. 16.5 GPM with less than 10 FTHD/100'—1-1/4" ID See Figure 12–5.

Line 5. This system is using a well supply and dry well disposal. Figure 13–2 is a typical piping arrangement from the submersible pump through the liquid-to-air heat pump to the atmospheric type disposal means (dry well or pond). Using Figure 13–2, which is located at ground level with supply and return pipes buried 24" in the ground for protection, the fitting count is:

Quantity	Item	Size	IEL	TEL
2	90° Elbow	1¼"	7	14
1	Tee	1¼"	5	5
Branch A total equivalent length				19
6	90° Elbow	1"	6	36
8	Tee	1"	2	16
1	Strainer	1"	5	5
3	Ball valve	1"	4	12
Branch B total equivalent length				69

Line 6. Total lineal feet of branches A and B. In branch A, vertical pipe runs from the pump to the top elbow. The pump is located 6' below the static water level. With a 2' drawdown, the pump is located 4' below the pumping water level.

As previously stated, the aquifer is 6' below grade. The horizontal pipe is 2' below grade. Therefore, the pump to top elbow distance is 6' below static water level to 4' above static water level of 10'.

The well is located 50' from the building and the pipe enters the building through the floor at the pressure tank. This adds 50' horizontal plus 2' vertical or 52' to the 10' well pipe. The total of branch A is 62 lineal feet.

In branch B, the heat pump is located adjacent to the pressure tank with an elbow in the line from the tee at the tank to the first shut-off valve. This makes this pipe approximately 5' long through the fittings, controls, and so on.

The return line drops through the floor (2') and 50' to the dry well. With approximately 2' for fittings, controls, and so on, plus 2' vertical and 50' horizontal, branch B is 59'.

Line 7. Total lineal feet and equivalent feet in each branch. Branch A has 62 lineal feet and 19 equivalent feet for a total of 81'. Branch B has 59 lineal feet and 101 equivalent feet for a total of 118'. These are recorded on line 7.

Line 8. Branch friction loss/100'.

Branch A: 1-1/4" pipe with 16.5 GPM = 4.0 FTHD/100'
Branch B: 1" pipe with 5.0 GPM = 1.92 FTHD/100'

Line 9. Determine the total branch friction loss:

$$\text{Branch A} = \frac{4.0\ \text{FTHD/100'} \times 81}{100'} = 3.24\ \text{FTHD}$$

$$\text{Branch B} = \frac{1.92\ \text{FTHD/100'} \times 116'}{100'} = 2.15\ \text{FHTD}$$

WATER SYSTEM WORKSHEET SE3WKS-1
(Method applicable to submersible pumps only)

A. Well Pump Sizing

Branch A—Well Pump—Piping from pump to pressure tank.
Branch B—Heat Pump Water Supply—Piping from tank to heat pump to drain.
Branch C—Domestic Water Supply—Piping from tank to house fixtures.

1. Determine the household needs from Figure 12-2. Enter here.	Branch C.	11.5	GPM
2. Heat pump GPM from unit specifications.	Branch B.	5.0	GPM
3. Add lines 1 and 2 for total water flow.	Branch A.	16.5	GPM

Note: If piping layout has more branches (C1, C2, C3, etc.), determine the flow rate for these from Figure 12-2 and include in the total.

B. Determine Water Pressure Requirements Pipe Sizing For Each Branch—Household plumbing, Branch B, may be assumed to have a total pressure requirement of 30 psig (69 FTHD).

	Br. A	Br. B	
4. Tentatively select a pipe size from Figure 12-5 and enter here.	1¼"	1¼"	ID
5. Using Figure 12-6, determine the equivalent length of all the fittings and shut-off valves and enter here (equiv. ft.)	19	69	EF
6. Determine the total lineal feet of pipe in each branch. (physical ft.)	62	59	ft.
7. Add lines 5 and 6 and enter here.	81	118	
8. From Figure 12-5, for pipe size and GPM needed for each branch, determine each branch friction loss. (FTHD/100')	4.0	1.92	
9. Multiply line 8 by line 7; divide by 100 to determine the total branch friction loss. (FTHD/100')	3.24	2.15	
10. From the heat pump specifications, enter the unit pressure drop. (FTHD)		6.93	
11. Pressure drop through the controls from Figures 12-9, 12-11, and 12-12. (FTHD)		1.40	
12. Calculate the total pressure drop. Branch A pressure drop is the same as line 9. Enter Branch A here. (FTHD)	3.24		
Branch B, by adding lines 9, 10, and 11 (FTHD)		14.78	
Branch C.		69.30	
13. Multiply line 12 by 0.433 to convert to psig. Enter here.		1.40	
Branch B		6.40	
Branch C		30.00	
14. From the piping layout, determine parallel flow among the branches. Beginning at the well pump, add the friction loss in psig for the well pump branch (Branch A) to the branch having the higher pressure, Branch B or C. *Note*: If more than three branches are required by the piping layout, select that branch that has the highest pressure drop and add this pressure drop to Branch A. Enter on line 14 the number obtained as total pressure loss due to pipe friction.	31.40	(psig)	
15. Add 20 psig to line 14 to obtain the pressure control cutout set point.	51.40	(psig)	
16. Multiply line 15 by 2.31 to convert to FTHD. Enter here.	118.73	(FTHD)	
17. Pump requirement. 16.5 GPM at	118.73	(FTHD)	
Vertical distance to water in well	9.0	ft. lift	
TOTAL PRESSURE	127.73	(FTHD)	

C. Pressure Tank Sizing (Bladder or diaphragm types only)

18. Enter desired minimum off time of the well pump in minutes. (Never less than two minutes.)	2	minutes
19. Enter the pressure control cutin set point.	30.00	psig
20. Enter pressure cutout set point from line 15.	50.00	psig
21. Multipy line 3 by line 18 to determine the minimum drawdown.	33	GPM
22. Refer to Figure 12-3 or the pressure tank specifications for a specific model of tank(s) to select the nominal capacity of tank needed, using information from lines 19, 20, and 21.	Tank size	120

Figure 13–6 Drilled well and dry well

Line 10. Unit pressure drop. Figure 13–4 shows the unit selected has a water tube pressure drop with a 5 GPM flow rate of 3 psig. Therefore, the 3 psig × 2.31 FTHD/1 psig = 6.93 FTHD.

Line 11. Pressure drop through the controls. This installation is using water-regulating valves for controlling discharge pressure in the cooling mode and suction pressure in the heating mode. The GPM is more for cooling than for heating. Therefore, we will use the cooling mode requirement for calculation.

The only control is the direct-acting cooling water regulating valve. (see Figure 12–11), with 1.40 FTHD at 10 GPM flow rate.

Line 12. Total pressure drop in the branches.

Branch A: 3.24 FTHD + 0 + 0 = 3.24 FTHD
Branch B: 3,828 FTHD + 11.55 FTHD + 3.23 FTHD = 14.78 FTHD
Branch C: 30 psig × 2.3 FTHD/1 psig = 69.3 FTHD

Line 13. Convert branch FTHD to psig.

Branch A: 3.24 FTHD × 0.433 = 1.40 psig
Branch B: 14.78 FTHD × 0.433 = 6.40 psig
Branch C: 69.3 FTHD = 0.433 = 30.00 psig

Line 14. Determine the higher parallel flow among the branches.

Branch A (1.40 psig) + Branch B (6.40 psig) = 7.80 psig drop
Branch A (1.40 psig) + Branch C (30.00 psig) = 31.40 psig drop

The combination of branches A and C is the larger of the two combinations. This figure is used in the calculation of line 15.

Line 15. 31.40 psig (line 14) plus 20 psig = 51.40 psig.

This is the cutout set point of the water pressure control.

Line 16. 51.40 psig × 2.31 FTHD/1 psig = 118.73 FTHD.

Line 17. The pump must lift the water from the pumping water level to the height of 2/3 of the tank height. This is a lift of 6' in the well plus 3' in the tank or 9'.

The total FTHD the pump must work against is 118.73 + 9' or a total of 127.73 FTHD, producing a flow rate of 16.5 GPM.

Pressure Tank Sizing

Line 18. The minimum off time is 2 minutes.

Line 19. The pressure control cutin set point is 31.40 psig (line 15). The practical setting would be 30 psig.

Line 20. The pressure control cutout set point is 51.40 psig (line 15). The practical setting would be 50 psig.

Line 21. Minimum drawdown = 16.5 GPM (line 3) × 2 minutes (line 18) or 33 gallons in the off mode.

Line 22. The pressure tank size from Figure 12–3 is 120 gallon capacity.

Well Supply (Lake, Pond, or Dry Well Return)

Dedicated System. If the groundwater in the location of the liquid-to-air heat pump is not potable (drinkable), the domestic water supply would be from a separate source. The well pump would be used to supply the heat pump only, and the system would be classified as a *dedicated* system.

Using the Water System Worksheet in Figure 13–7, the step calculations would be:

Line 1. Line 1 covering branch C is not used in a dedicated system as there is no domestic load.

Line 2. Heat pump need (70°F water)—5.0 GPM.

Line 3. Total water need—5.0 GPM.

Line 4. 5 GPM with less than 10 FTHD/100'—1" (see Figure 12–5).

Line 5. This system uses a well supply and dry well disposal. Figure 13–2 is a typical piping arrangement from the submersible pump through the liquid-to-air heat pump to the atmospheric type disposal means (dry well or pond). Using Figure 13–2, located at ground level and supply and return pipes buried 2' in the ground for protection, the fitting count would be:

	Quantity	Item	Size	IEL	TEL
Branch A					
	2	90° Elbow	1"	3	6
	1	Tee	1"	4	4
		Total equivalent length			10
Branch B					
	6	90° Elbow	1"	6	36
	8	Tee	1"	2	16
	1	Strainer	1"	5	5
	3	Ball valve	1"	4	12
		Total equivalent length			69

Line 6. Total lineal feet of branches A and B. In branch A, the vertical pipe runs from the pump to the top elbow. The pump is located 6' below the static water level. With a 2' drawdown, the pump is located 4' below the pumping water level.

As previously stated, the aquifer is 6' below grade. The horizontal pipe is 2' below grade. Therefore, the pump to top elbow distance is 6' below the static water level to 4' above the static water level or 10'.

The well is located 50' from the building and the pipe enters the building through the floor at the pressure tank. This adds

WATER SYSTEM WORKSHEET SE3WKS-1
(Method applicable to submersible pumps only)

A. Well Pump Sizing

Branch A—Well Pump—Piping from pump to pressure tank.
Branch B—Heat Pump Water Supply—Piping from tank to heat pump to drain.
Branch C—Domestic Water Supply—Piping from tank to house fixtures.

1. Determine the household needs from Figure 12-2. Enter here.	Branch C.	0.0	GPM
2. Heat pump GPM from unit specifications.	Branch B.	5.0	GPM
3. Add lines 1 and 2 for total water flow.	Branch A.	5.0	GPM

Note: If piping layout has more branches (C1, C2, C3, etc.), determine the flow rate for these from Figure 12-2 and include in the total.

B. Determine Water Pressure Requirements Pipe Sizing For Each Branch—Household plumbing, Branch B, may be assumed to have a total pressure requirement of 30 psig (69 FTHD).

	Br. A	Br. B	
4. Tentatively select a pipe size from Figure 12-5 and enter here.	1"	1"	ID
5. Using Figure 12-6, determine the equivalent length of all the fittings and shut-off valves and enter here (equiv. ft.)	10	69	EF
6. Determine the total lineal feet of pipe in each branch. (physical ft.)	62	59	ft.
7. Add lines 5 and 6 and enter here.	81	118	
8. From Figure 12-5, for pipe size and GPM needed for each branch, determine each branch friction loss. (FTHD/100')	1.8	1.8	
9. Multiply line 8 by line 7; divide by 100 to determine the total branch friction loss. (FTHD/100')	1.458	2.124	
10. From the heat pump specifications, enter the unit pressure drop. (FTHD)		6.93	
11. Pressure drop through the controls from Figures 12-9, 12-11, and 12-12. (FTHD)		1.40	
12. Calculate the total pressure drop. Branch A pressure drop is the same as line 9. Enter Branch A here. (FTHD)	1.48		
Branch B, by adding lines 9, 10, and 11 (FTHD)		10.454	
13. Multiply line 12 by 0.433 to convert to psig. Enter here.	0.641	4.526	

14. From the piping layout, determine parallel flow among the branches. Beginning at the well pump, add the friction loss in psig for the well pump branch (Branch A) to the branch having the higher pressure, Branch B or C. *Note*: If more than three branches are required by the piping layout, select that branch that has the highest pressure drop and add this pressure drop to Branch A. Enter on line 14 the number obtained as total pressure loss due to pipe friction.	20.00	(psig)
15. Add 20 psig to line 14 to obtain the pressure control cutout set point.	40.00	(psig)
16. Multiply line 15 by 2.31 to convert to FTHD. Enter here.	92.44	(FTHD)
17. Pump requirement. 5.0 GPM at	92.44	(FTHD)
Vertical distance to water in well	9.00	ft. lift
TOTAL PRESSURE	101.44	(FTHD)

C. Pressure Tank Sizing (Bladder or diaphragm types only)

18. Enter desired minimum off time of the well pump in minutes. (Never less than two minutes.)	2	minutes
19. Enter the pressure control cutin set point.	20.00	psig
20. Enter pressure cutout set point from line 15.	40.00	psig
21. Multipy line 3 by line 18 to determine the minimum drawdown.	10.00	GPM
22. Refer to Figure 12-3 or the pressure tank specifications for a specific model of tank(s) to select the nominal capacity of tank needed, using information from lines 19, 20, and 21. Tank size	30	gal.

Figure 13–7 Dedicated system

50' horizontal plus 2' vertical or 52' to the 10' well pipe. The total of branch A is 62 lineal (physical) feet.

In branch B, the heat pump is located adjacent to the pressure tank with an elbow in the line from the tee at the tank to the first shut-off valve. This makes this pipe approximately 5' long through the fittings, controls, and so on.

The return line drops through the floor 2' and 50' to the dry well. With approximately 2' for fittings, controls, and so on, plus 2' vertical and 50' horizontal, branch B is 59'.

Line 7. Total lineal (physical) feet in each branch. Branch A has 62 lineal feet and 19 equivalent feet for a total of 81'. Branch B has 59 lineal feet and 69 equivalent feet for a total of 118'. These are recorded on line 7.

Line 8. Determine the friction loss (FTHD/100') of each branch used. See Figure 12–5.

Branch A: 1" pipe with 5 GPM = 1.8 FTHD/100'
Branch B: 1" pipe with 5 GPM = 1.8 FTHD/100'

Line 9. Determine the total branch friction loss per 100' of pipe:

$$\text{Branch A} = \frac{1.8 \text{ FTHD/100'} \times 81}{100'} = 1.458 \text{ FTHD}$$

$$\text{Branch B} = \frac{1.8 \text{ FTHD/100'} \times 118}{100'} = 2.124 \text{ FTHD}$$

Line 10. Unit pressure drop. Figure 13–4 shows the unit selected has a water tube pressure drop with a 5 GPM flow rate of 3 psig. Therefore, the 3 psig × 2.31 FTHD/1 psig = 6.93 FTHD.

Line 11. Pressure drop through the controls. This installation is using water-regulating valves to control the discharge pressure in the cooling mode and suction pressure in the heating mode. The GPM is more for cooling than for heating. Therefore, we will use the cooling mode requirement for calculation.

The only control is the direct-acting cooling water regulating valve (see Figure 12–11), with a flow resistance of less than 1.4 FTHD at the 5 GPM flow rate.

Line 12. Branch C is not included. Therefore, only branches A and B are considered.

Branch A: 1.48 FTHD + 0 + 0 = 1.48 FTHD
Branch B: 2.124 FTHD + 6.93 FTHD + 1.40 FTHD = 10.454 FTHD

Line 13. Convert branch FTHD to psig.

Branch A: 1.48 × 0.433 = 0.641 psig.
Branch B: 10.454 × 0.433 = 4.526 psig
Branch C is not used.

Line 14. Determine the higher parallel flow among the branches. Only branches A and B are used.

Branch A (0.641 psig) + Branch B (4.526 psig) = 5.167 psig

Line 15. 5.167 psig (line 14) plus 20 psig = 25.167 psig.

The calculated cutin and cutout set points for the control are 5.167 psig cutin and 25.167 psig cutout. The actual control set point range would be 20 psig cutin and 40 psig cutout.

Line 16. 40 psig × 2.31 FTHD/1 psig = 92.44 FTHD.

Line 17. The pump must lift the water from the pumping water level to the height of 2/3 of the pressure tank height. This is a lift of 6' in the well plus 3' to the tank or 9'.

The total FTHD the pump must work against is 92.4 FTHD (pipe loss) + 9' (lift height) or 101.40 FTHD, producing a flow rate of 5 GPM.

Pressure Tank Sizing

Line 18. The minimum pump off time is 2 minutes.

Line 19. The calculated pressure control cutout set point is 5.167 psig. The practical set point is 20 psig.

Line 20. The calculated pressure control cutout set point is 61.4 psig. The practical set point is 40 psig.

Line 21. Minimum drawdown = 5.0 GPM × 2 minutes = 10 gallons pressure tank drawdown. The pressure tank from Figure 12–3 would be a 30-gallon capacity.

Well Supply (Return Well)

Dedicated System. A two-well or return-well system is used where winter temperatures are low enough to produce a ground frost line 2 feet or more below grade. Figure 13–5 shows the supply well on the left and return well on the right. Each well is located 50' from the building and 100' apart. We will assume that this installation is in Omaha, NB with a winter time design temperature of 15°F.

The supply and return lines must be buried a minimum of 4' below grade level. In this area, the water is expected to be 50°F at the 50' to 150' level (see Figure 13–1). Assume that the static water level in the aquifer is 50' below grade with ground porosity that produces 4' of drawdown at full pumping rate. This establishes the water pumping level at 54' below grade.

The unit that will be used on this installation is the WPH22–1H (Figure 12–1) with auxiliary electric heat.

Well Pump Sizing

The piping detail of the unit is shown in Figure 13–8. Branch A is the well pump piping from the pump to the pressure tank. Branch B is the piping from the pressure tank through the heat pump to the return well. Branch C is not used; this is a dedicated system. Using the Water System Worksheet (Figure 13–9), the steps involved would be:

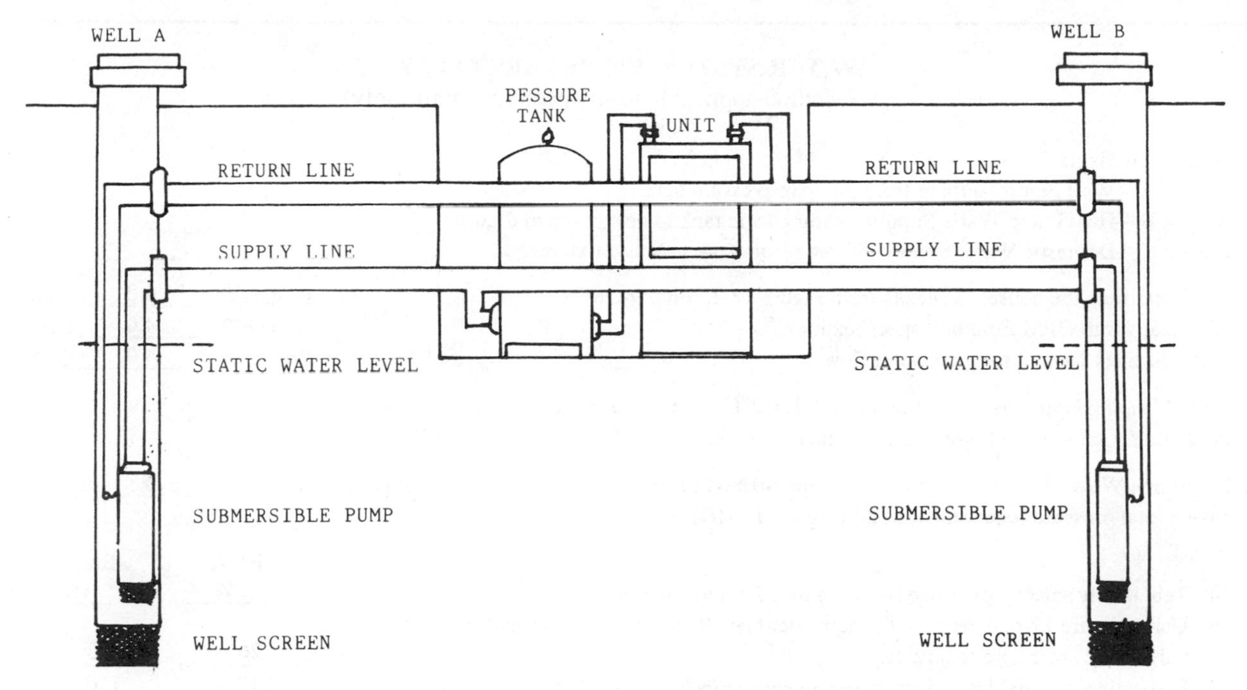

Figure 13–8 Dedicated two-well system (Courtesy Mammoth, a Nortek Co.)

Line 1. Not needed—dedicated system.

Line 2. Heat pump GPM from unit specifications using 50°F water (see Figure 12–1). The unit will require 2 GPM for the cooling mode and 4 GPM for the heating mode. The heating mode amount (branch B: 4 GPM) will be used.

Line 3. Total water flow in branch A—4 GPM. Since this is a dedicated system, there is no domestic load.

Line 4. Tentative pipe size from Figure 12–5.

Branch A: 4 GPM—¾" ID
Branch B: 4 GPM—¾" ID

Line 5. Using Figure 12–6, determine the equivalent length of all fittings and shut-off valves.

Branch A: Pump to pressure tank (Figures 13–2 and 13–5).

Quantity	Item	Size	I FTHD	T FTHD
2	Ball valve	¾"	3	6
3	90° Elbow	¾"	5	15
1	Strainer	¾"	5	5
	Total equivalent length (FTHD)			26

Branch B: Pressure tank through heat pump to the return well aquifer (Figures 13–2 and 13–5).

Quantity	Item	Size	I FTHD	T FTHD
7	90° Elbow	¾" ID	5	35
6	Tee	¾" ID	4	24
	(Includes 2 tees for P/T plugs.)			
2	Ball valve	¾"	3	6
	Total equivalent length (FTHD)			65

Line 6. Determine the total lineal (physical) feet of pipe in each branch. In branch A, the pump is 58' below grade (4' below the pumping water level) supplying a horizontal pipe buried 4' below grade level. The vertical riser pipe is 54' long. The horizontal pipe to the building is 50' long.

The pipe enters the building 4' above the floor with the pressure tank connection 1' above the floor. From the wall to the pressure tank connection is 3'. Therefore: 54' + 50' + 3' + 3' = 110 lineal feet in branch A.

In branch B, the pipe from the pressure tank to the unit uses an inverted loop as high as the supply and return pipes, 3' vertical distance. The outlet rises from the unit to the discharge horizontal line (3') plus the discharge line (50') to the return well plus the drop in the well add up to the lineal length of branch B. Notice that the drop pipe in the return well has a back pressure valve installed and the drop pipe end is below the static water level.

The total lineal feet of branch B is 2' horizontal + 3' vertical up + 1' horizontal to the unit plus 3' vertical + 50' horizontal + 51' vertical down in the return well for a total lineal feet length of 110'.

Line 7. The total equivalent feet length (TEL) of each branch in the system is the equivalent length of the fittings plus the lineal feet of pipe. Branch A: Fittings (26 FTHD) plus pipe length (110') totals up to 136 FTHD. Branch B: Fittings (65 FTHD) plus the pipe lineal length (110') totals up to 175 FTHD.

Line 8. The friction loss per 100' equivalent feet of pipe is based on 3/4" plastic pipe at the 4 GPM flow rate. From Figure 12–5, the figure is 3.7 FTHD/100' in both branches A and B.

WATER SYSTEM WORKSHEET SE3WKS-1
(Method applicable to submersible pumps only)

A. Well Pump Sizing

Branch A—Well Pump—Piping from pump to pressure tank.
Branch B—Heat Pump Water Supply—Piping from tank to heat pump to drain.
Branch C—Domestic Water Supply—Piping from tank to house fixtures.

1. Determine the household needs from Figure 12-2. Enter here.	Branch C.	0.0	GPM
2. Heat pump GPM from unit specifications.	Branch B.	4.0	GPM
3. Add lines 1 and 2 for total water flow.	Branch A.	4.0	GPM

Note: If piping layout has more branches (C1, C2, C3, etc.), determine the flow rate for these from Figure 12-2 and include in the total.

B. Determine Water Pressure Requirements Pipe Sizing For Each Branch—Household plumbing, Branch B, may be assumed to have a total pressure requirement of 30 psig (69 FTHD).

Line	Br. A	Br. B	
4. Tentatively select a pipe size from Figure 12-5 and enter here.	¾"	¾"	ID
5. Using Figure 12-6, determine the equivalent length of all the fittings and shut-off valves and enter here (equiv. ft.)	26	65	EF
6. Determine the total lineal feet of pipe in each branch. (physical ft.)	110	110	ft.
7. Add lines 5 and 6 and enter here.	136	175	
8. From Figure 12-5, for pipe size and GPM needed for each branch, determine each branch friction loss. (FTHD/100')	3.7	3.7	
9. Multiply line 8 by line 7; divide by 100 to determine the total branch friction loss. (FTHD/100')	5.032	6.475	
10. From the heat pump specifications, enter the unit pressure drop. (FTHD)		5.3	
11. Pressure drop through the controls from Figures 12-9, 12-11, and 12-12. (FTHD)		60.11	
12. Calculate the total pressure drop. Branch A pressure drop is the same as line 9. Enter Branch A here. (FTHD)		5.032	
Branch B, by adding lines 9, 10, and 11 (FTHD)		71.885	
13. Multiply line 12 by 0.433 to convert to psig. Enter here.	2.18	31.13	

14. From the piping layout, determine parallel flow among the branches. Beginning at the well pump, add the friction loss in psig for the well pump branch (Branch A) to the branch having the higher pressure, Branch B or C.
Note: If more than three branches are required by the piping layout, select that branch that has the highest pressure drop and add this pressure drop to Branch A. Enter on line 14 the number obtained as total pressure loss due to pipe friction. 33.31 (psig)
15. Add 20 psig to line 14 to obtain the pressure control cutout set point. 53.31 (psig)
16. Multiply line 15 by 2.31 to convert to FTHD. Enter here. 123.15 (FTHD)
17. Pump requirement.

4.0 GPM at 123.15 (FTHD)

Vertical distance to water in well 49.0 ft. lift

TOTAL PRESSURE 172.15 (FTHD)

C. Pressure Tank Sizing (Bladder or diaphragm types only)

18. Enter desired minimum off time of the well pump in minutes. (Never less than two minutes.)	2	minutes
19. Enter the pressure control cutin set point.	35.00	psig
20. Enter pressure cutout set point from line 15.	55.00	psig
21. Multipy line 3 by line 18 to determine the minimum drawdown.	8	GPM
22. Refer to Figure 12-3 or the pressure tank specifications for a specific model of tank(s) to select the nominal capacity of tank needed, using information from lines 19, 20, and 21. Tank size	30	gal.

Figure 13–9 Dedicated two-well system

Line 9. Determine the total branch friction loss:

$$TBEL = \frac{TEL \times FTHD/100'}{100'}$$

Branch A:

$$TBEL = \frac{136 \times 3.7}{100'} = 5.032 \text{ FTHD}$$

Branch B:

$$TBEL = \frac{175 \times 3.7}{100'} = 6.475 \text{ FTHD}$$

Line 10. From the heat pump specifications, enter the unit pressure drop (FTHD). Figure 13–4 states that the WPH22–1H unit has a water side pressure of 2.3 psig with a water flow of 4 GPM. Converted to FTHD, 2.3 psig × 2.31 FTHD/1 psig = 5.3 FTHD.

Line 11. The pressure drop in FTHD through the controls, from Figures 13–1 and 13–4 add up to:

Quantity	Item	FTHD
1	Constant flow valve (Figure 12–11)	34.7
1	S, C. Solenoid valve (Figure 12–8) 1 psig × 2.31 =	2.31
1	Back pressure valve (Figure 12–13) 10 psig × 2.31 =	23.1
	Total FTHD through controls	60.11

Line 12. Determine the total pressure drop through the system—branch A and branch B.

Branch A: From line 9 = 5.032 FTHD
Branch B: From lines 9, 10, and 11 = 71.885 FTHD

Line 13. Convert the branch friction loss in FTHD to pressure loss.

Branch A: 5.032 FTHD × 0.433 psig/1 FTHD = 2.18 psig
Branch B: 71.885 FTHD × 0.433 psig/1 FTHD = 31.13 psig

Line 14. Determine the total friction loss of the system in psig.

Branch A (2.18 psig) + Branch B (31.13 psig) = 33.31 psig total system pressure loss.

Line 15. Add 20 psig to obtain the pressure control cutout point. 33.31 psig + 20 psig = 53.31 psig. The practical set point is a pressure control cutout set point of 55 psig.

Line 16. Determine the maximum FTHD the pump must operate against. 55.31 psig × 2.31 FTHD/1 psig = 123.15 FTHD.

Line 17. The pump must deliver a maximum of 4 GPM against 123.15 FTHD with a 49' vertical lift. Therefore, the pump must be rated to deliver 4 GPM against 172.15 FTHD (123.15 FTHD pipe plus 49 FTHD vertical lift).

Pressure Tank Sizing

Line 18. The minimum off time of the well pump is 2 minutes.

Line 19. The pressure control cutin set point is 35 psig.

Line 20. Pressure control cutout set point is 55 psig.

Line 21. Pressure tank drawdown = 2 minutes × 4 gallons = 8 gallons drawdown. The nominal capacity of the pressure tank (Figure 12–3) with the 10-gallon drawdown in the 40–60 psig range is 30 gallons.

In this application, we used 3/4" plastic pipe with a 1/2 hp pump with 3/4" fittings plus a slow-acting motorized valve for positive water flow control and a constant flow valve for water quantity regulation. This control system was used as an example to point out the friction losses involved. The preferred system uses two pressure-actuated regulating valves, which produce lower system pressure loss.

Well Supply (Two-Well Reversible—Dedicated System)

When calculating the pump and pipe size for a two-well reversible system, the procedure is the same as for a well supply—well return system except that:

1. Each system must be calculated separately. Do not assume equal conditions. Changes in piping setup as well as ground conditions can vary between the two well locations (see Figure 11–8).
2. Each system will have two extra ball valves (four total) in each circulation system.

The extra ball valves are for diverting water flow through the heat pump coil in the same direction, regardless of the pumping system used.

Do not attempt to make the system into an automatic changeover system without a means of timer delay for a flushing period. When changeover is made, the unit must be operated with an open hose bib or drain cock from the return side to an external disposal means for a period of not less than 10 minutes. The purpose is to flush the well screen on the operating pump side.

Sand and particles accumulate in the return screen under normal operating conditions. The flushing action is needed to keep system contamination to a minimum.

13–4 GEOTHERMAL WELL

When there is not enough groundwater in the aquifer or if the water is contained in consolidated formations, and is an unsatisfactory supply or disposal means for a liquid-to-air heat pump using an open loop system, a closed loop or geothermal well system is used. In a geothermal well system, the aquifer formation is used as a heat source and heat sink in the operating modes. As water is circulated from the well, through the unit, and returned to the same well, heat is transferred between the water and the material surrounding the well shaft.

Basic Geophysical Principles

To understand the geothermal heat exchange system, two basic principles must be understood:

1. Geothermal temperature gradient
2. Thermal conductivity

Geothermal Temperature Gradient. This is the increase in the earth's temperature as the distance below grade level (earth's surface) increases. The deeper the hole, the higher the temperature at the bottom of the hole.

The geothermal gradient begins at approximately 100' below grade (ground) level. The approximate rise in ground temperature is 3°F for each 100' increase in depth. For example, Figure 13–10 shows the average geothermal temperature gradient in the New London, Connecticut area. The water temperatures encountered at the various depth levels are an average of the temperatures found in 10 test wells drilled into the granite rock of New London County. The temperature of well water in the New London County area ranges between 52°F and 54°F.

Thermal Conductivity. This is the rate at which heat tends to flow through a material. This flow rate is expressed in BTU per square foot of the material surface times the temperature difference per thickness in feet of the material. The thermal conductivity of a material or combination of materials is referred to as the *K* factor. The formula for determining the *K* factor is:

$$K = \text{BTU/ft.}^2 \times \Delta°\text{F/ft. thickness}$$

The higher the *K* factor, the faster heat energy will flow through the material.

Depth (Ft)	Temperature (°F)
0	25.0
50	51.0
100	52.6
150	54.2
200	55.7
250	57.3
300	58.8
350	60.4
400	61.9
450	63.5
500	65.0

Figure 13–10 Average geothermal temperature gradient, New London, CT (Courtesy Mammoth, a Nortek Co.)

The thermal conductivity (*K* factor) for various materials is given in Figure 13–11. From this table, you can see that wool, the least dense material, has a *K* factor of only 0.021 BTUH/ft.^{2}/1°F/1 ft. thickness. The other extreme is copper, the most dense material, which has a *K* factor of 225 BTUH/ft.^{2}/1°F/1 ft. thickness.

The material encountered in geothermal wells varies from a *K* factor of 0.75 for limestone to a *K* factor of 2.0 for granite. From these factors, we see that for a given heat transfer requirement, we would not need as much heat transfer surface if the well were drilled in granite ($K = 2.00$) as we would in limestone ($K = 0.75$). Therefore, if the entire well is limestone, we would need 2.67 times the pipe heat transfer surface than if it were drilled in granite.

Because the earth layers vary over a wide range, calculating the exact surface needed would be very complicated. Therefore, the industry has developed standards based upon experience in all types of earth formations:

1. A standard geothermal well uses a 6" casing.
2. Well depth is 150' of wetted hole per 12,000 BTUH unit capacity. The *wetted* hole portion of the well is that portion of the casing below the static water level in the aquifer.

Using these two standards, it is apparent that this application is for heat pumps of 60,000 BTUH or less. Even at 60,000 BTUH load requirement, the well drilled to 750' below the static water level can be uneconomical. In these cases, the ground loop system would be a better application.

Dedicated Geothermal Well

Figure 13–12 shows the component parts of a dedicated geothermal well. Both supply with pump and the return pipe are in the same casing. Using jet pump pitless adapters in the casing shell, the supply and return vertical lines are connected to their matching horizontal lines to and from the building. The submersible pump is located 6' below the static water level to allow for any possible water loss from the well during

Material	Conductivity *K*
Wool	0.021
Dry topsoil	0.075
PVC	0.084
Polybutylene	0.125
Sand	0.188
Gravel	0.220
Wet topsoil (42% water by volume)	0.620
Limestone	0.750
Concrete	1.000
Sandstone	1.060
Granite	2.000
Steel	36.000
Copper	225.000

Figure 13–11 Thermal conductivity factor of various materials (Courtesy Mammoth, a Nortek Co.)

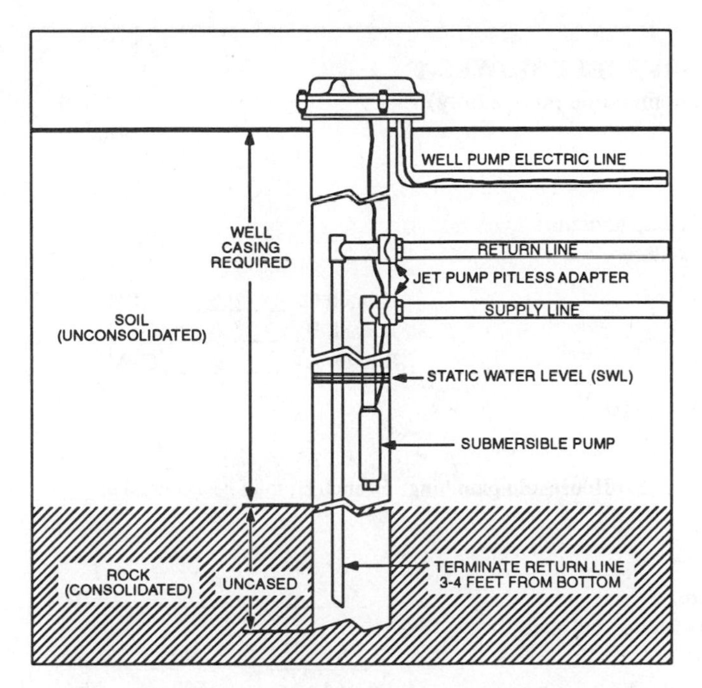

Figure 13–12 Dedicated geothermal well (Courtesy Mammoth, a Nortek Co.)

the operating mode. The return line terminates 3' to 4' from the bottom of the well. The water circulation in the well is from the bottom up, losing or gaining heat as the unit requires.

Notice that the actual steel pipe casing only extends into the consolidated formation far enough to create a seal. This prevents water and debris from being drawn into the well casing during the operating mode.

The total depth of the well drilling will depend upon the depth of the static water level in the aquifer and the unit capacity—either heating or cooling, whichever is larger.

Dedicated Geothermal Well System Size

To determine the pump and pipe sizes, we need to know the unit capacity requirements, both heating and cooling, and the anticipated ground temperature. For our example, we will assume that this unit is to be located in the Phoenix, AZ area where the aquifer water is very limited.

The unit size for the Phoenix, AZ application is a WPH28-1H (Figure 13–13), which has a cooling capacity of 28,000 BTUH and a heating capacity of 34,600 BTUH. The heating capacity of 34,600 BTUH is used to size the well components.

In the Phoenix, AZ area, the temperature of the water at the 50' to 150' level can be expected to be in the 70°F range. (See Figure 13–1.) When the unit is operating, an average temperature difference between the ground temperature and water temperature through the unit can be expected to be 15°F. This means an average temperature of 55°F in the heating mode and 85°F in the cooling mode.

Using the water temperatures for determining the GPM requirement in Figure 13–13, we find that 3.0 GPM is needed for heating (38,500 BTUH with 55°F water) and 5.0 GPM for cooling (28,900 BTUH with 85°F water). We will size the water supply system on the cooling requirement of 5 GPM.

WPH28-1H

E.W.T.	GPM	BTUH TOTAL	BTUH SENS.	WATTS	EER W/OUT PMP	EER WITH PUMP	HEAT REJ.
	2.0	31600	22100	2300	13.7	13.1	38500
45	3.0	33100	22400	2180	15.2	14.0	39500
	4.0	33800	22700	2130	15.9	14.3	40100
	2.0	30900	21900	2350	13.1	12.5	37900
50	3.0	32400	22200	2240	14.5	13.4	39000
	4.0	33100	22400	2180	15.2	13.7	39500
	3.0	31600	21900	2300	13.7	12.7	38500
55	4.0	32400	22200	2240	14.5	13.1	39100
	5.0	32900	22400	2220	14.9	13.2	39500
	3.0	31000	21500	2350	13.2	12.3	38000
60	4.0	31600	21800	2300	13.7	12.4	38500
	5.0	32100	22000	2260	14.2	12.5	38800
	4.0	30800	21600	2360	13.1	11.8	37900
65	6.0	31500	21900	2300	13.7	12.1	38400
	8.0	31900	22100	2280	14.0	11.6	38700
	4.0	30100	21300	2420	12.4	11.3	37400
70	6.0	30800	21600	2370	13.0	11.3	37900
	8.0	31100	21800	2340	13.3	11.0	38100
	5.0	29700	21000	2450	12.1	10.8	37100
75	7.0	30100	21200	2420	12.4	10.6	37400
	9.0	30400	21400	2400	12.7	10.3	37600
	5.0	28900	20700	2520	11.5	10.3	36500
80	7.0	29300	20900	2490	11.8	10.1	36800
	9.0	29500	21100	2470	11.9	9.8	36900
	5.0	28100	20400	2590	10.8	9.7	35900
85	7.0	28600	20600	2550	11.2	9.6	36300
	9.0	28800	20800	2540	11.3	9.4	36500
	6.0	27500	20100	2640	10.4	9.2	35500
90	8.0	27900	20300	2600	10.7	9.1	35800
	10.0	28100	20400	2580	10.9	8.8	35900

WPH28-1H

E.W.T.	GPM	BTUH TOTAL	WATTS	COP W/OUT PUMP	COP WITH PUMP	HEAT OF ABSORP.
	4.0	22900	2140	3.1	2.8	16500
45	7.0	24400	2200	3.2	2.7	16800
	10.0	25100	2240	3.3	2.6	17400
	4.0	24800	2220	3.3	3.0	17200
50	7.0	26500	2300	3.4	2.9	18600
	10.0	27300	2340	3.4	2.7	19300
	4.0	26800	2310	3.4	3.1	18900
55	7.0	28600	2400	3.5	3.0	20400
	10.0	29400	2440	3.5	2.8	21000
	3.0	27400	2340	3.4	3.2	19400
60	5.0	29700	2450	3.5	3.2	21300
	8.0	31100	2520	3.6	3.0	22400
	3.0	28900	2410	3.5	3.3	20600
65	5.0	31400	2520	3.6	3.3	22700
	8.0	33000	2600	3.7	3.1	24100
	3.0	30700	2500	3.6	3.4	22100
70	5.0	33400	2620	3.7	3.4	24400
	8.0	35100	2700	3.8	3.2	25800
	2.0	29700	2440	3.6	3.4	21300
75	4.0	34300	2660	3.8	3.5	25200
	6.0	36200	2750	3.9	3.6	26800
	2.0	31500	2520	3.7	3.5	22800
80	4.0	36400	2760	3.9	3.6	26900
	6.0	38500	2860	4.0	3.6	28000

Figure 13–13 WPH28-H application ratings and water flow specifications (Courtesy Addison Products Co.)

The physical requirements of pump and piping are:

1. The well is located 30' from the building.
2. The static water level is 25' below the grade level.
3. The well depth will be based on the standard of 150' of wetted pipe per 12,000 BTUH in the heating mode.

$$\text{Length of wetted pipe} = \frac{150' \times 38,500 \text{ BTUH}}{12,000 \text{ BTUH}} = 481.25'$$

Well Pump Sizing

Using the water system worksheet in Figure 13–14, the well pump and pressure tank are sized using the steps as given on the worksheet.

WATER SYSTEM WORKSHEET SE3WKS-1
(Method applicable to submersible pumps only)

A. Well Pump Sizing

Branch A—Well Pump—Piping from pump to pressure tank.
Branch B—Heat Pump Water Supply—Piping from tank to heat pump to drain.
Branch C—Domestic Water Supply—Piping from tank to house fixtures.

1. Determine the household needs from Figure 12-2. Enter here.	Branch C.	0.0	GPM
2. Heat pump GPM from unit specifications.	Branch B.	5.0	GPM
3. Add lines 1 and 2 for total water flow.	Branch A.	5.0	GPM

Note: If piping layout has more branches (C1, C2, C3, etc.), determine the flow rate for these from Figure 12-2 and include in the total.

B. Determine Water Pressure Requirements Pipe Sizing For Each Branch—Household plumbing, Branch B, may be assumed to have a total pressure requirement of 30 psig (69 FTHD).

	Br. A	Br. B	
4. Tentatively select a pipe size from Figure 12-5 and enter here.	1"	1"	ID
5. Using Figure 12-6, determine the equivalent length of all the fittings and shut-off valves and enter here (equiv. ft.)	31	90	EF
6. Determine the total lineal feet of pipe in each branch. (physical ft.)	64	529	ft.
7. Add lines 5 and 6 and enter here.	92	619	
8. From Figure 12-5, for pipe size and GPM needed for each branch, determine each branch friction loss. (FTHD/100')	1.8	1.8	
9. Multiply line 8 by line 7; divide by 100 to determine the total branch friction loss. (FTHD/100')	1.656	11.142	
10. From the heat pump specifications, enter the unit pressure drop. (FTHD)		5.775	
11. Pressure drop through the controls from Figures 12-9, 12-11, and 12-12. (FTHD)		24.5	
12. Calculate the total pressure drop. Branch A pressure drop is the same as line 9. Enter Branch A here. (FTHD)	1.656		
Branch B, by adding lines 9, 10, and 11 (FTHD)		41.42	
13. Multiply line 12 by 0.433 to convert to psig. Enter here.	0.72	17.93	

14. From the piping layout, determine parallel flow among the branches. Beginning at the well pump, add the friction loss in psig for the well pump branch (Branch A) to the branch having the higher pressure, Branch B or C. *Note*: If more than three branches are required by the piping layout, select that branch that has the highest pressure drop and add this pressure drop to Branch A. Enter on line 14 the number obtained as total pressure loss due to pipe friction.	18.65	(psig)
15. Add 20 psig to line 14 to obtain the pressure control cutout set point.	40.00	(psig)
16. Multiply line 15 by 2.31 to convert to FTHD. Enter here.	92.4	(FTHD)
17. Pump requirement. 5.0 GPM at	92.4	(FTHD)
Vertical distance to water in well	21.0	ft. lift
TOTAL PRESSURE	113.4	(FTHD)

C. Pressure Tank Sizing (Bladder or diaphragm types only)

18. Enter desired minimum off time of the well pump in minutes. (Never less than two minutes.)	2	minutes
19. Enter the pressure control cutin set point.	20.00	psig
20. Enter pressure cutout set point from line 15.	40.00	psig
21. Multipy line 3 by line 18 to determine the minimum drawdown.	10	GPM

22. Refer to Figure 12-3 or the pressure tank specifications for a specific model of tank(s) to select the nominal capacity of tank needed, using information from lines 19, 20, and 21.

Figure 13–14 Dedicated geothermal well

Step 1. Determine the household needs. This is a dedicated geothermal well, so there is no household load.

Step 2. Enter heat pump GPM from the unit specifications: 3GPM heating or 5 GPM cooling. We will use the larger of the two, 5 GPM.

Step 3. Add lines 1 and 2 for the total water flow requirement, 5 GPM

Step 4. Tentatively select a pipe size from Figure 12–5 and enter here. Based on the 5 GPM flow rate:

Branch A: 1" (1.8 FTHD/100')
Branch B: 1" (1.8 FTHD/100')

Step 5. Determine the equivalent length of all the fittings and shut-off valves and enter here. To determine the fittings, valves, and so on, use the piping layout in Figure 12–10.

Branch A: Well pump to pressure tank.

Quantity	Item	FTHD @	Total FTHD
3	90° Elbow	6	18
2	Shut-off valve	4	8
1	Strainer	5	5
		Total FTHD	31

Branch B: Pressure tank to unit and unit to bottom of well.

Quantity	Item	FTHD @	Total FTHD
7	90° Elbows	6	42
9	Tee	4	36
3	Shut-off valve	4	12
		Total FTHD	90

Step 6. Determine the total lineal feet of pipe in each branch (physical length in feet). In branch A, the pump is located 6' below the static water level, which is 25' below grade level. This puts the pump 31' below grade level. The horizontal supply pipes to the building are 4' below grade. The area does not have a "frost line" situation. The 4' below grade level is only for protection of the pipes from potential damage as well as to reduce the summer sun effect on the system temperature. Therefore, the vertical riser pipe is 31' below grade level (pump location) less the 4' of pipe protection of 27' of vertical length. The horizontal pipes are 30' in length (pump to building) plus the 5' from the horizontal depth to the tank connection in the building or 35 feet. Allowing a horizontal 2 feet for connection to the pressure tank means the total physical length of Branch A is 64 lineal feet.

The pipe length in branch B is from the pressure tank to the heat pump unit (through the inlet connections) and from the heat pump (through the outlet connections and controls) to the bottom of the return well. From the pressure tank to the unit is approximately 3'. From the unit through the fittings and controls to the vertical pipe through the floor of the room is approximately 5'. The vertical pipe through the floor (5') plus the 35' of distance to the well head means 40' of pipe from the horizontal control portion to the well head. The calculated depth of the well is 481'. Therefore, the total physical length of the return pipe is 3' + 5' + 40' + 481' = 529'.

Step 7. Find the TEF (Total Equivalent Feet) of branches A and B by adding the TEF line 5 and the total physical length (lineal feet) in line 6.

Branch A: 32 TEF + 64 lineal feet = 92 TEF
Branch B: 90 TEF + 529 lineal feet = 619 TEF

Step 8. Determine the pipe loss in FTHD/100' for the GPM and pipe size in each branch. We have tentatively selected 1" pipe with a FTHD/100' of 1.8 FTHD.

Step 9. Determine the total branch friction loss in FTHD/100'.

Branch A:

$$\text{T FTHD} = \frac{92' \times 1.8}{100'} = 1.656 \text{ FTHD/100'}$$

Branch B:

$$\text{T FTHD} = \frac{619 \times 1.8}{100'} = 11.142 \text{ FTHD/100'}$$

Step 10. From the heat pump specifications, enter the unit pressure drop (FTHD). The WPH28–1H selected for this application using the 5 GPM required during the cooling mode (Figure 13–13) will have a pressure drop of 2.5 psig (Figure 13–4). Therefore, the pressure drop of 2.5 psig × 2.31 FTHD/1 psig = FTHD.

Step 11. Determine the pressure drop through the controls. The water flow control in the cooling mode is a 1' pressure-regulating valve with a pressure drop at the 5 GPM flow rate of less than 1.4 (Figure 12–11). In this application, a spring-loaded check valve is required at the base of the return pipe to keep the pressure up in the system and prevent encrustation of the minerals in the water. The spring-loaded check valve will have to have a spring tension strong enough to hold against the weight of water in the 27' of vertical height of water in the return pipe above the static water level plus 10 psig. This will require a spring pressure of 21.69 psig (22 psig). However, the flow resistance to the system is still the 10 psig opening pressure. The FTHD flow resistance of the spring-loaded check valve is 10 psig × 2.31 FTHD/1 psig or 23.1 FTHD. Therefore, the regulating valve pressure loss of 1.4 FTHD plus the spring-loaded check valve pressure loss of 23.1 FTHD results in a 24.5 FTHD pressure loss for the controls.

Step 12. Calculate the total pressure drop of branch B by adding lines 9, 10, and 11.

$$11.142 + 5.775 + 24.5 = 41.42 \text{ FTHD}$$

The total pressure drop in branch A (1.656 FTHD) was determined in line 9.

Step 13. Convert the FTHD pressure loss in each branch to psig by multiplying the FTHD in each branch by 0.433 psig/1 FTHD.

Branch A: 1.656 FTHD × 0.433 psig/1 FTHD = 0.72 psig
Branch B: 41.39 FTHD × 0.433 psig/1 FTHD = 17.93 psig

Step 14. Determine the total pipe friction of the system. In this dedicated system, only branches A and B are used. No domestic load is included. Therefore, branch(es) C is not needed. The 0.72 psig loss in branch A plus the 17.92 psig loss in branch B adds up to a system pressure loss of 18.65 psig.

Step 15. Determine the operating control point by adding the control differential of 20 psig to the system pressure loss of 18.65 psig. The pressure control cutout set point should be 38.65 psig. The practical setting would be 40 psig.

Step 16. Determine the maximum friction loss in FTHD the pump will encounter by multiplying the pressure control cutout set point by 2.31 FTHD/1 psig.

$$40 \text{ psig} \times 2.31 \text{ FTHD/1 psig} = 92.4 \text{ FTHD}$$

Step 17. Determine the total FTHD that the pump must operate against to deliver the required flow rate of 5 GPM at 92.4 FTHD. The vertical lift distance in the well is 21' (static water level 25' below grade minus the 4' distance of the horizontal pipes below grade). Therefore, the total pressure in FTHD against the pump is 92.4 plus 21 or 113.4 FTHD.

Using the 5 GPM flow rate required and the 113.4 FTHD flow resistance, the pump size and hp required are selected from the pump manufacturer's specification literature.

Pressure Tank Sizing

Step 18. Enter the desired minimum off time of the well pump in minutes—2.

Step 19. Enter the pressure control cutin set point. 40 psig cutout set point – 20 psig differential = 20 psig cutin set point.

Step 20. Enter the pressure control cutout set point from line 15—40 psig.

Step 21. Determine the minimum drawdown GPM required for the pressure tank. The system flow rate of 5 GPM (line 3) multiplied by the minimum off time of 2 minutes = 10 GPM. The tank size selected from Figure 12–3 or the tank manufacturer's specification literature would be a 30-gallon nominal capacity.

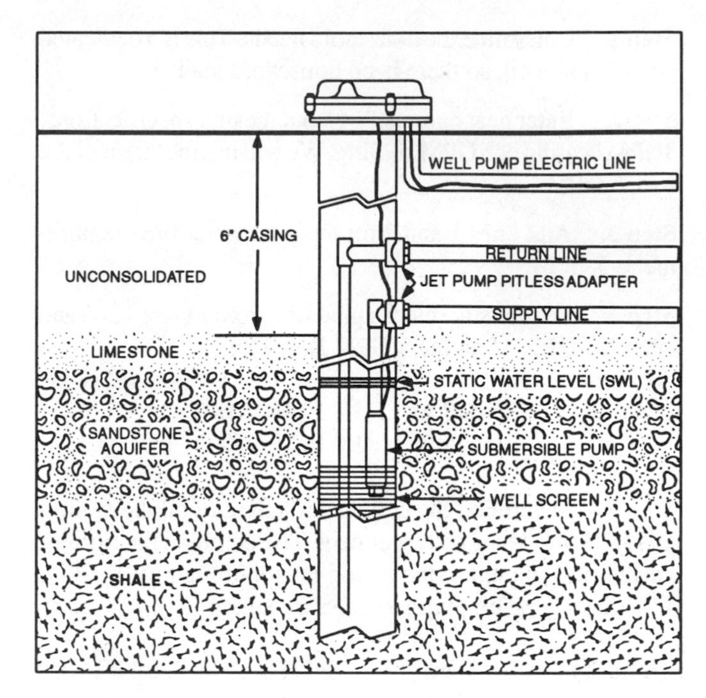

Figure 13–15 Domestic geothermal well (Courtesy Mammoth, a Nortek Co.)

Therefore, in this application, we will use a WPH28–1H liquid-to-air heat pump, supplied 5 GPM by a submersible pump, in a dedicated geothermal well drilled to a depth of 481' of wetted pipe plus the 25' that the static water level is below grade for a drilled depth of 506'.

Domestic Geothermal Well

Figure 13–15 shows the well casing arrangement of a domestic geothermal well. All components are the same as in a dedicated geothermal well except:

1. A well screen is part of the well casing for intake of the water required for domestic or building use.
2. This type well is restricted to only potable water supply.
3. The well screen must be located in semiconsolidated material such as limestone, sandstone, or other semisolid materials.

Location of this screen in unconsolidated material such as sand, gravel, or mud will cause material deposits at the bottom of the drilled well. This will very rapidly destroy the effectiveness of the well. For these reasons, the use of a domestic geothermal well is not recommended.

Well and Pump Size

The well and pump are sized using the same procedure as for a dedicated geothermal well.

The exception is the branches (C) that must be taken into the total GPM flow requirement.

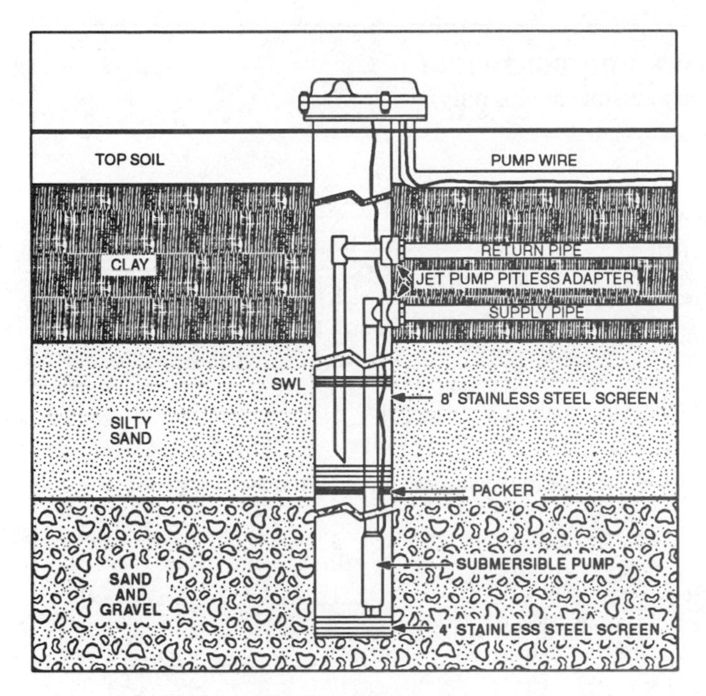

Figure 13–16 Sand geothermal well (Courtesy Mammoth, a Nortek Co.)

Step 1. A building GPM is included.

Step 2. A decision is made to obtain the maximum flow requirement. This will be branches A plus B or A plus C, whichever is greater.

Sand Geothermal Well

In those rare earth formations where the aquifer water is contained in a sand and gravel formation of 75' or more thickness, a sand geothermal well can be used to reduce drilling costs. Figure 13–16 shows the component parts of a sand geothermal well. This type well is the reverse of a standard geothermal well. The water is pumped from the bottom of the well and returned to the top of the well. A seal or "packer" plate is placed below the return section to prevent flow directly between the return and the supply pump. The water is forced to leave the well casing and mix with the water in the aquifer.

Flowing from the 8' stainless steel return screen down through the aquifer, the water gives up or takes on heat energy, depending upon the operating mode of the system. The entire aquifer is the heat source or heat sink. The top of the upper return screen must be located at least 4' below the static water level to prevent air intake and encrustation. Although it is not shown, a spring-loaded check valve should be used on the end of the return pipe to prevent loss in the system below 10 psig.

The submersible pump is located above the 4' intake screen. Note that the return screen requires twice the area of the intake screen. The screens must be located as far apart as possible (at the top and bottom of the aquifer) so as to use as much of the aquifer as possible for the heat source/sink activity. The lower end of the well casing should be located close to the lower edge of the aquifer without entering the consolidated formation under the aquifer.

Due to the fact that a considerable amount of silt and sand is moved through this system, use as a domestic geothermal well is not recommended unless large filters are installed in the domestic water supply.

Sand Geothermal Well and Pump Size

To determine the well and pump size, we need to know the unit capacity requirements, both heating and cooling, and the anticipated ground temperature. For our example, we will assume that the unit is located in the Sacramento, CA area, with a groundwater temperature at the 50' to 150' level of 64°F (see Figure 13–1). We will assume that the application has a cooling capacity requirement of 27,265 BTUH and a heating capacity requirement of 19,996 BTUH.

The application is located in the Sacramento river valley where the underground river aquifer is 85' of sand and gravel on top of a layer of granite rock. The top of the aquifer is 15' below grade level. The well is located 20' from the building with the horizontal supply and return pipes 2' below grade level for protection. There are no frost line problems in this area.

The unit for this application is a WPH28–1H with a rated cooling capacity of 28,600 BTUH and a heating capacity of 34,600 BTUH. Using the table in Figure 13–13, we find that the WPH28–1H unit needs 4 GPM of 65°F water to produce 31,000 BTUH cooling capacity. This will leave the rest of the system requirement with a very small reserve capacity. Using 6 GPM to produce 31,600 BTUH cooling capacity will produce 10% reserve capacity. We will use the 6 GPM requirement. The system will supply the heating requirement with reserve capacity.

Using the worksheet in Figure 13–17, the well pump and pressure tank is sized using the following steps:

Step 1. Determine the household needs. In this installation, the system will be used for the heat pump unit only; it is a dedicated type system.

Step 2. Determine the heat pump GPM requirement for the application location groundwater temperature of 65°F, from the unit manufacturer's specification literature (see Figure 13–13).

Step 3. Add lines 1 and 2 for total water flow—6 GPM.

Step 4. Tentatively select a pipe size to have a friction loss of less than 10 FTHD for the required GPM flow rate (see Figure 12–5). In branch A, 3/4" pipe with an 8.0 FTHD/100' loss could be used. However, 1" pipe with 1.8 FTHD/100' loss will produce considerably less friction loss through the piping system. 1" pipe is recommended.

Step 5. Determine the equivalent length of all the fittings and shut-off valves and enter the total in equivalent feet. To determine the fittings, valves, and so on, use the piping layout in Figure 12–10.

WATER SYSTEM WORKSHEET SE3WKS-1
(Method applicable to submersible pumps only)

A. Well Pump Sizing

Branch A—Well Pump—Piping from pump to pressure tank.
Branch B—Heat Pump Water Supply—Piping from tank to heat pump to drain.
Branch C—Domestic Water Supply—Piping from tank to house fixtures.

1. Determine the household needs from Figure 12-2. Enter here.	Branch C.	0.0	GPM
2. Heat pump GPM from unit specifications.	Branch B.	6.0	GPM
3. Add lines 1 and 2 for total water flow.	Branch A.	6.0	GPM

Note: If piping layout has more branches (C1, C2, C3, etc.), determine the flow rate for these from Figure 12-2 and include in the total.

B. Determine Water Pressure Requirements Pipe Sizing For Each Branch—Household plumbing, Branch B, may be assumed to have a total pressure requirement of 30 psig (69 FTHD).

	Br. A	Br. B	
4. Tentatively select a pipe size from Figure 12-5 and enter here.	1"	1"	ID
5. Using Figure 12-6, determine the equivalent length of all the fittings and shut-off valves and enter here (equiv. ft.)	31	93.5	EF
6. Determine the total lineal feet of pipe in each branch. (physical ft.)	117	49	ft.
7. Add lines 5 and 6 and enter here.	148	142.5	
8. From Figure 12-5, for pipe size and GPM needed for each branch, determine each branch friction loss. (FTHD/100')	2.5	2.5	
9. Multiply line 8 by line 7; divide by 100 to determine the total branch friction loss. (FTHD/100')	3.7	3.56	
10. From the heat pump specifications, enter the unit pressure drop. (FTHD)		7.39	
11. Pressure drop through the controls from Figures 12-9, 12-11, and 12-12. (FTHD)		24.5	
12. Calculate the total pressure drop. Branch A pressure drop is the same as line 9. Enter Branch A here. (FTHD)	3.7		
Branch B, by adding lines 9, 10, and 11 (FTHD)		35.45	
13. Multiply line 12 by 0.433 to convert to psig. Enter here.	1.6	15.35	

14. From the piping layout, determine parallel flow among the branches. Beginning at the well pump, add the friction loss in psig for the well pump branch (Branch A) to the branch having the higher pressure, Branch B or C. *Note*: If more than three branches are required by the piping layout, select that branch that has the highest pressure drop and add this pressure drop to Branch A. Enter on line 14 the number obtained as total pressure loss due to pipe friction.	16.95	(psig)
15. Add 20 psig to line 14 to obtain the pressure control cutout set point.	40.00	(psig)
16. Multiply line 15 by 2.31 to convert to FTHD. Enter here.	92.4	(FTHD)
17. Pump requirement. 6.0 GPM at	92.4	(FTHD)
Vertical distance to water in well	13.0	ft. lift
TOTAL PRESSURE	105.4	(FTHD)

C. Pressure Tank Sizing (Bladder or diaphragm types only)

18. Enter desired minimum off time of the well pump in minutes. (Never less than two minutes.)	2	minutes
19. Enter the pressure control cutin set point.	20.00	psig
20. Enter pressure cutout set point from line 15.	40.00	psig
21. Multipy line 3 by line 18 to determine the minimum drawdown.	12	GPM

22. Refer to Figure 12-3 or the pressure tank specifications for a specific model of tank(s) to select the nominal capacity of tank needed, using information from lines 19, 20, and 21.

Figure 13–17 Geothermal well pump and pressure tank sizing

Quantity	Item	FTHD @	T FTHD
Branch A: Well pump to pressure tank			
3	90° Elbow	6	18
2	Shut-off valve	4	8
1	Strainer	5	5
	Total FTHD	31	
Branch B: Pressure tank to unit and from unit to top of well			
7	90° Elbows	6	42
9	Tee	4	36
3	Shut-off valves	4	12
1	S.L. check valve	3.5	3.5
	Total FTHD	93.5	

Step 6. Determine the total lineal feet of pipe in each branch (physical length). In branch A, the pump is located 5' above the bottom of the sand and gravel aquifer, which is 100' below grade level. The horizontal supply and return pipes are buried 2' below grade level for protection. Therefore, the vertical pipe from the pump to the adapter in the well casing is 93'. The horizontal pipe to the building is 20' plus 1' to the tank tee connection. Branch A equivalent feet = 93' + 20' + 3' + 1' = 117'.

The pipe length in branch B is from the pressure tank to the heat pump unit through the inlet connections, from the heat pump unit to the top of the sand geothermal well. From the pressure tank to the unit is approximately 5'. From the unit, through fittings, controls, and so on, to the floor of the room is approximately 5'. The vertical pipe through the floor (2') plus the 20' of horizontal pipe to the well head equals 22'.

The spring-loaded check valve located on the end of the return line is 4' below the upper level of the aquifer. The top of the aquifer is 15' below grade level. Therefore, the check valve is located 19' below grade level minus the 2' the horizontal pipes are below grade level. The vertical pipe is 17' long. Branch B equivalent feet = 5' + 5' + 22' + 17' = 49'.

Step 7. Find the TEF (total equivalent feet) of branches A and B by adding the TEF in line 5 to the lineal feet of each branch in line 6.

Branch A: 117 TEF + 31 lineal feet = 138 TEF
Branch B: 93.5 TEF + 49 lineal feet = 142.5 TEF

Step 8. Determine the pipe loss in FTHD/100' for the GPM and pipe size in each branch. We have selected 1" pipe with a FTHD/100' of 2.5.

Step 9. Determine the total branch friction loss in FTHD/100'.

Branch A:

$$\text{FTHD} = \frac{148 \times 2.5}{100'} = 3.7 \text{ FTHD/100'}$$

Branch B:

$$\text{FTHD} = \frac{142.5 \times 2.5}{100'} = 3.56 \text{ FTHD/100'}$$

Step 10. From the heat pump specification literature, enter the unit pressure drop (FTHD). The WPH28–1H unit selected for this application using the 6 GPM required during the cooling mode (Figure 13–13) will have a pressure drop of 3.2 psig (Figure 13–4). Therefore, the pressure drop of 3.2 psig × 2.31 FTHD/1 psig = 7.39 FTHD.

Step 11. Determine the pressure drop through the controls. The water flow control in the cooling mode is a 1' pressure-regulating valve with a pressure drop at the 6 GPM flow rate of 1.4 psig (Figure 12–11). In this application, a spring-loaded check valve is required at the base of the return pipe. This valve will have to have a spring strong enough to hold against the weight of the water in the return pipe up to the height of the water in the pressure tank. This water height is 15' below grade to 5' to the top of the water in the pressure tank or 20'. The weight of the water in psig = 20 FTHD × 0.433 psig/FTHD or 8.66 psig.

The water weight of 8.66 psig plus the 10 psig opening pressure means that the spring-loaded check valve requires a spring pressure of 18.66 psig. However, the flow resistance through the valve is only 10 psig (opening pressure) × 2.31 FTHD/1 psig or 23.1 FTHD. The regulating valve pressure loss of 1.4 FTHD plus the spring-loaded check valve of 23.1 FTHD is a total of 24.5 FTHD loss through the controls.

Step 12. The total pressure drop of branch A (3.7 FTHD) was determined in line 9. Calculate the total pressure drop of branch B by adding lines 9, 10, and 11 (FTHD). 3.56 + 7.39 + 24.5 = 35.45 FTHD.

Step 13. Convert the FTHD pressure loss in each branch to psig pressure loss by multiplying the FTHD of each branch by 0.433 psig/1 FTHD.

Branch A: 3.7 FTHD × 0.433 psig/1 FTHD = 1.6 psig
Branch B: 35.45 FTHD × 0.433 psig/1 FTHD = 15.35 psig

Step 14. Determine the total pipe friction loss of the system. In this dedicated system, only branches A and B are used. No domestic load is included. Therefore, no branch(es) C is included.

Branch A (1.60 psig) + Branch B (15.35 psig) = 16.95 psig.

Step 15. Determine the operating cutout control point by adding the control differential pressure of 20 psig to the system pressure loss.

20 psig + 16.95 psig =
36.95 calculated control cutout set point.

The practical setting would be 40 psig.

Step 16. Determine the maximum friction loss in FTHD the pump will encounter at the pressure control cutout set point.

2.31 FTHD/1 psig × 40 psig = 92.4 FTHD

Step 17. Determine the total FTHD that the pump must operate against to deliver the required flow rate. The required flow rate is 6 GPM against a friction loss of 92.4 FTHD. The vertical lift distance in the well is 13' (15' from the grade level to top of the aquifer minus 2' for pipe burial). Therefore, the total pressure in FTHD against the pump is 92.4 plus 13 = 105.4 FTHD. Using the 6 GPM flow rate required and the 105.4 FTHD flow resistance, the pump size and hp required are selected from the pump manufacturer's specification literature.

Pressure Tank Sizing

Step 18. Enter the desired minimum off time of the well in minutes—2.

Step 19. Enter the pressure control cutin set point. 40 psig cutout set point – 20 psig differential = 20 psig cutin set point.

Step 20. Enter the pressure control cutout set point from line 15—40 psig.

Step 21. Determine the minimum drawdown GPM required from the pressure tank. The system flow rate of 6 GPM × 2 minutes drawdown time = 12 GPM minimum drawdown.

The tank size selection from Figure 12–3 or the manufacturer's specification literature would be an 80-gallon nominal capacity. The 40-gallon size would not satisfy the 2-minute minimum time requirement.

This application will require a WPH28–1H liquid-to-air heat pump supplied 6 GPM of 64°F water by a 1/2 hp pump in a dedicated sand geothermal well drilled to a depth of 100' to the lower edge of the aquifer.

REVIEW QUESTIONS

1. When selecting pipe sizes, what is the maximum friction loss per pipe size?
2. When calculating the total lineal feet of pipe length in the system, the vertical lift in the well is equal to the drilled depth of the well. True or False?
3. List three items of information needed to determine the pump capacity needed for the system (GPM output/FTHD).
4. The increase in temperature of the earth as the distance below grade increases is called the _____ _____ _____.
5. The approximate rise in temperature of the earth is _____°F per each 100' below grade level.
6. The rate at which heat tends to flow through a material is called its _______ _______.
7. A well drilled through granite will not have as high a heat transfer rate as a well drilled through limestone. True or False?
8. What section of a well is classified as the "wetted portion"?
9. The water height in the return pipe is 38'. What opening pressure is required in the spring-loaded check valve at the base of the pipe?
10. The system pressure loss is 53 FTHD. What cutin and cutout pressures would be used to set the pump pressure control?
11. The system water flow rate required is 9 GPM. What is the minimum tank drawdown that can be used when sizing the pressure tank?

14 SIZING GROUND LOOP SYSTEMS

14-1 GENERAL

Ground loop systems operate on a different basis than a drilled well system. In drilled well systems, the water in the aquifer is the heat source/heat sink material. In ground loop systems, the earth mass itself is the heat source/heat sink material. As a result of the temperature difference between the liquid circulating in the pipe and the earth material that must be created and maintained to get the desired heat transfer rate, the heat pump must operate with circulating liquid temperatures as low as 25°F and as high as 100°F.

When selecting the heat pump for these ground loop applications, it is extremely important that the unit will operate within the range of these minimum and maximum temperatures. Some unit models on the market have much smaller operating ranges such as 45°F to 85°F or 60°F to 85°F entering liquid temperatures. These units will not operate satisfactorily on ground loop applications.

Heat is transferred between the unit and the earth by a closed loop fluid coil buried in the earth. A pump is used to circulate this fluid. Plastic pipe placed in horizontal trenches or vertical drilled holes is used to carry the heat transfer fluid. Horizontal pipe is usually buried from 3' to 6' below grade in the northern hemisphere. Burying the pipe deep reduces the ability of the sun to recharge the heat energy used in the winter.

Because of the shallow depths used, antifreeze is used in the liquid circulating system to prevent freeze up of the water in the system and to allow the system to gain capacity and efficiency by using the latent heat removed from the earth material when the water in the soil is frozen. The antifreeze materials used are usually propylene glycol or calcium chloride.

In the southern hemisphere, the pipes are buried from 4' to 12' below grade. This is to reduce the effect of the high soil surface temperatures from the summer sun on the system performance.

Vertical drilled holes may be up to 300 feet deep. This will depend upon the type of material encountered during the drilling operation. Multiple holes may be more desirable than one deep hole.

The pipe length used in either type of system may vary from 150' to 1,000' per 12,000 BTUH of unit capacity, depending upon such things as:

1. UC (unit capacity—cooling)
2. UH (unit capacity—heating)
3. COPc (coeffecience of performance—cooling)
4. COPh (coefficience of performance—heating)
5. Tm (mean annual earth temperature)
6. EWTc (entering water temperature—cooling)
7. EWTh (entering water temperature—heating)
8. Rs—soil resistance (varies with type of soil)
9. Rp—pipe resistance (resistance to heat flow through the wall of the pipe)
10. Type of earth coil
11. Depth of the earth coil
12. Fc (unit cooling run factor)
13. Fh (unit heating run factor)

14-2 HEAT PUMP SELECTION

Two important factors must be considered before the size or capacity of a heat pump can be determined and a selection made.

1. An accurate heat gain and heat loss of the building must be determined. This has been discussed in previous chapters. It is well, however, to stress the need for accuracy in these calculations. The ACCA Environmental Systems Library Manuals H, J, and N are highly recommended for this procedure. The bin method of energy calculation in Manual J is used in the calculation of pipe lengths in the ground loop.
2. The minimum and maximum earth temperatures at the depth the closed loop coil is to be located must be determined. These are necessary to select the water temperature range the unit must operate against.

As with air-to-air heat pumps, the unit is selected to satisfy the cooling capacity required. If the cooling load is less than the heating load, supplemental heat is used to make up the difference. If the cooling capacity required exceeds the heating capacity required, no supplemental heat is needed. The utility requirements, however, apply to liquid-to-air units as well as air-to-air units. This means that auxiliary heat capable of handling the entire design heat loss of the building must be included in the installation.

The sensible output capacity of the unit in the cooling mode should not be less than the calculated total sensible load nor should it exceed the calculated sensible load by more than 25%. The corresponding latent capacity of the unit in the cooling mode should not be less than the calculated total latent load. The equipment sensible and latent capacities should be determined from the manufacturer's specification literature. These capacities will depend upon the local groundwater temperatures as well as design conditions expected.

COPc (Coefficience of Performance—Cooling)

The coefficience of performance (COP) of a unit is described as the BTUH cooling of the unit divided by the electrical energy needed to produce the capacity. The cooling COP of the unit is the BTUH cooling capacity of the unit divided by the heat energy equivalent of the electrical energy needed.

$$\text{Cooling COP} = \frac{\text{BTUH Unit Capacity}}{\text{Watts Input} \times 3.413 \text{ BTU/Watt}}$$

The manufacturer may rate its units by the EER (energy efficiency ratio) rating method. The EER rating is the total BTUH cooling capacity divided by the watts of electrical energy. To convert the EER rating to COP rating, divide the EER rating by 3.413.

COPh (Coefficience of Performance—Heating)

The heating coefficience of performance (COPh) is the heating capacity of the unit divided by BTU equivalent of the watts input.

$$\text{Heating COP} = \frac{\text{BTUH Unit Heating Capacity}}{\text{Watts Input} \times 3.413 \text{ BTU/Watt}}$$

Tm (Mean Annual Earth Temperature)

The mean annual earth temperature can be assumed to equal the well water temperature from a well 50' to 150' deep. It can be approximated by adding 2°F to the mean annual air temperature. The table in Figure 14–1 gives the mean annual

		T_m (°F) – A_s (°F) – T_o (Days)				
		T_m	A_s	T_o	HDD	CDD
AL	Birmingham	65	19	31	2710	1928
	Montgomery	67	18	31	2250	2238
AZ	Phoenix	73	23	33	1680	3508
	Tucson	70	18	34	1700	2896
AR	Little Rock	64	21	32	3170	1925
CA	Los Angeles	64	7	54	1960	1185
	Merced	68	25	33	2470	1200
	San Diego	64	7	54	1500	722
CO	Colo Springs	51	21	36	6410	461
	Denver	52	22	37	6150	625
	Grand Junc.	55	25	32	5660	1140
DC	Washington	57	22	36	4240	1517
FL	Appalachicola	70	15	36	1308	2663
	Jacksonville	71	14	32	1230	3059
GA	Atlanta	62	19	32	2990	1589
	Augusta	65	18	30	2410	1995
	Macon	67	17	33	2160	2294
ID	Boise	53	21	34	5830	714
	Idaho Falls	46	23	34	7890	286
IL	Chicago	51	25	37	6160	925
	E. St. Louis	57	24	34	4900	1475
	Urbana	53	26	35	6000	664
IN	Fort Wayne	53	24	35	6220	748
	Indianapolis	55	24	34	5630	974
	South Bend	52	25	37	6460	695
IA	Des Moines	52	28	35	6610	928
	Sioux	51	29	34	6960	932
KS	Dodge City	57	25	35	4986	1411
	Topeka	56	26	35	5210	1361
KY	Louisville	60	22	33	4610	1268
LA	Lake Charles	70	16	32	1490	2739
	New Orleans	70	15	32	1400	2706
	Shreveport	66	19	32	2160	2538
ME	Portland	48	22	39	7570	252
MA	Plymouth	51	21	43	5630	661
MI	Battle Creek	50	24	35	6580	628
	Detroit	50	25	39	6290	743
	Sau St Mar.	42	26	40	9050	139
MN	Duluth	41	28	37	9890	176
	Int. Falls	39	31	34	10600	176
	Minneapolis	47	29	35	8250	527

Figure 14–1 Earth temperature data for U.S. Cities (Courtesy Bard Manufacturing Co.)

		T_m (°F)	A_s (°F)	T_o (Days)		
		T_m	A_s	T_o	HDD	CDD
MS	Biloxi	70	17	32	1500	2793
	Columbus	65	19	32	2890	2039
	Jackson	67	18	31	2260	2321
MO	Columbia	57	24	35	5070	1269
	Kansas City	58	26	35	4750	1420
	Springfield	58	23	34	4900	1382
MT	Billings	49	23	37	7150	498
	Great Falls	48	23	36	7670	339
	Missoula	46	21	32	8000	188
NB	Grand Isla.	52	27	35	6440	1036
	Lincoln	53	28	34	6050	1148
	N. Platte	51	26	35	6680	802
NV	Ely	47	22	35	7710	207
	Las Vegas	69	23	32	2610	2946
	Winnemucca	52	22	33	6760	407
NJ	Trenton	55	22	38	4980	968
NM	Albuquerque	59	22	31	4250	1394
	Roswell	63	22	30	3680	1560
NY	Albany	50	25	38	6900	574
	Binghamton	48	24	38	7340	369
	Niagara Falls	50	24	24	6688	549
	Syracuse	50	24	38	6720	591
NC	New Bern	65	17	35	2400	1964
	Greensboro	60	20	31	3810	1341
ND	Bismark	44	31	33	8960	487
	Grand Forks	42	33	35	9930	400
	Williston	45	29	34	9180	450
OH	Akron	52	23	37	6140	634
	Columbus	55	22	34	5670	809
	Dayton	56	24	35	5620	936
	Toledo	51	25	36	6430	685
OK	Altus	65	24	33	3190	2347
	Ok. City	62	23	34	3700	2068
	Tulsa	62	23	34	3730	1949
OR	Astoria	53	9	45	5190	13
	Medford	55	17	34	4880	562
	Portland	54	13	37	4635	300
PA	Middleton	55	23	35	5280	1025
	Philadelphia	55	22	34	5180	277
	Pittsburgh	52	23	36	5950	647
	Wilkes-Barre	52	23	36	6160	608
SC	Charleston	66	16	32	2146	2078
	Greenville	62	19	33	3070	1573
	Sumpter	65	18	32	2453	2078
SD	Huron	47	30	35	8220	716
	Rapid City	50	25	38	7370	661
TN	Bristol	59	20	32	4143	1107
	Knoxville	61	21	31	3510	1569
	Memphis	63	21	32	3210	2029
	Nashville	60	21	32	3610	1694
TX	El Paso	66	20	28	2680	2098
	Ft. Worth	68	21	34	2390	2587
	Houston	71	16	33	1410	2889
	San Antonio	72	16	32	1560	2994
UT	Salt Lake	53	24	35	5990	927
VT	Burlington	46	26	37	8030	396
VA	Norfolk	61	20	37	3440	1707
	Richmond	60	19	33	3910	1353
	Roanoke	59	20	33	4150	1030
WA	Moses Lake	53	23	29	5145	707
	Seattle	53	12	36	4424	162
	Spokane	49	21	32	6770	388
WV	Charleston	58	20	33	4510	1055
	Elkins	52	20	32	5680	389
WI	Green Bay	46	26	37	8100	389
	Madison	49	27	36	7720	460
WY	Casper	49	24	37	7510	458
	Cheyene	48	21	39	7370	327
	Lander	46	24	38	7870	383
	Sheridan	48	24	35	7740	446

HDD—Heating Degree Days CDD—Cooling Degree Days

Figure 14–1 (cont.) Earth temperature data for U.S. cities (Courtesy Bard Manufacturing Co.)

earth temperatures (Tm) for 110 cities throughout the United States. These can be used to approximate locations within the vicinity of the listed cities.

The mean annual earth temperature (Tm) is the mean average temperature of the earth as it changes throughout the year. The surface temperature has an effect on the earth temperature but the greater effect is from the rays of the sun.

The earth has a dampening effect on the temperature changes because of its ability to take on or give off heat, depending upon the density and moisture content of the earth material, called the diffusivity of the earth material, as well as the distance from the surface. As the depth increases, the annual swing decreases.

The soil swing curves in Figure 14–2 show the effect of air and solar changes on the soil from the surface down to the 12' level. The curves show surface, 2', 5', and 12' levels. These curves also show the time of the year that the minimum (Tmin) and maximum (Tmax) temperatures will be reached. For example, the minimum surface temperature (Tmin) can be expected to be reached on approximately the 35th day of the year. Some weather condition variations can vary this date by a ±3 day span.

The storage effect of the earth will cause a delay of 14 days to reach minimum temperatures at the 2' level, 35 days at the 5' level, and 83 days at the 12' level. The storage capacity of the earth is demonstrated by the minimum temperature variations of 23°F below the mean temperature on the surface to 5°F below the mean temperature at the 12' level.

In the cooling season, reverse temperatures occur with the higher surface temperature (Tmax) occurring about August 6th, ±3 days. Again, as the depth below the surface increases, the maximum temperature (Tmax) is decreased as well as delayed. At the 2' level, the Tm decreases by 17°F and is delayed approximately 14 days (August 20th). At the 12' depth, the maximum temperature (Tmax) occurs around Oct. 29th, and reaches a high of 5°F above the mean annual temperature (Tm).

Figure 14–2 shows the curves for an average soil density and moisture content (difusivity). To allow for soils of various degrees of thermal difusivity, the curves are pictured in Figure 14–3. From Figure 14–3, the minimum and maximum temperatures that can be expected for various depths can be determined for three types of soil.

Using this information, the depth that the earth loop must be installed can be determined to stay within the entering water temperature (EWTc and EWTh) requirements of the heat pump. This becomes part of the steps used to determine the pipe length and depth for each application.

EWTc (Entering Water Temperature—Cooling)

The entering water temperature to the unit, the temperature of the water leaving the earth coil, will be higher than the normal temperature of the earth. This is due to the heat

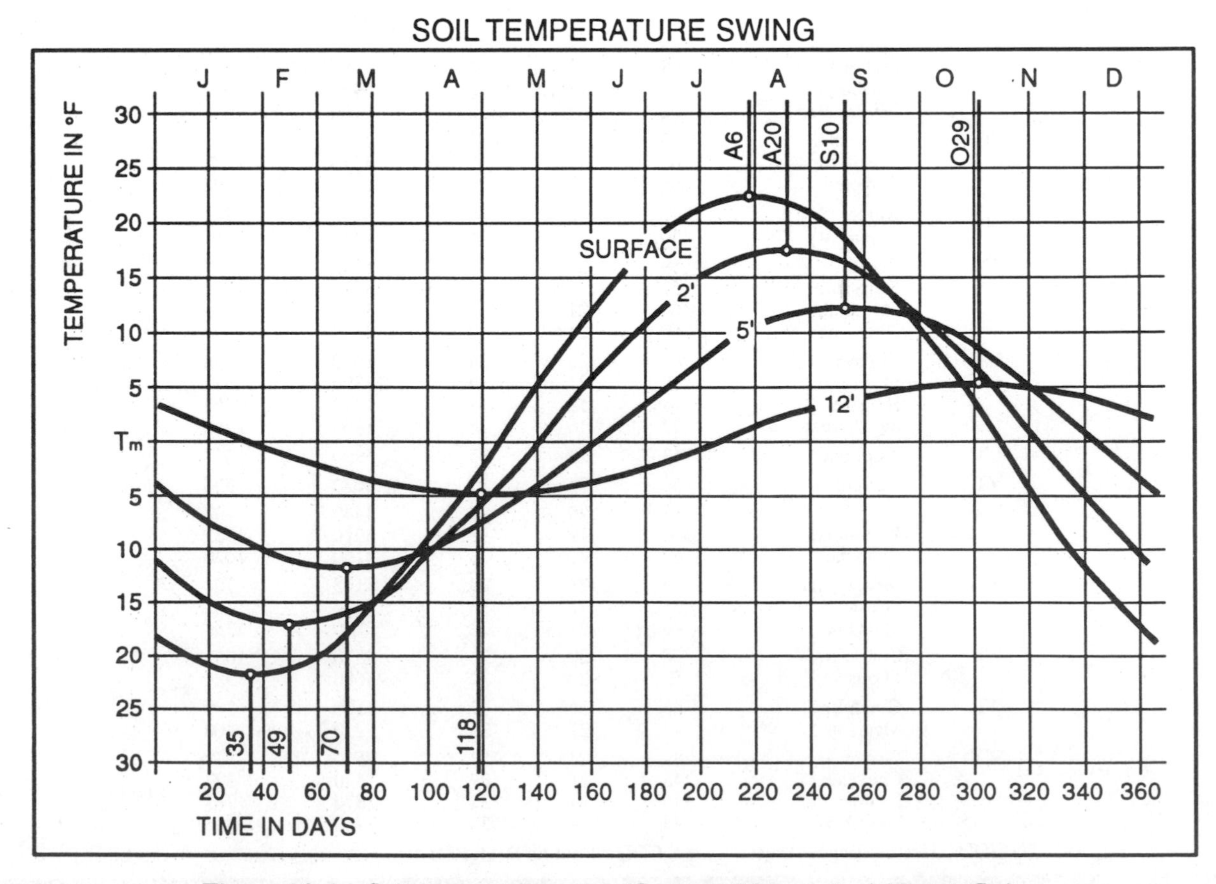

Figure 14–2 Soil temperature swing (Courtesy Mammoth, A Nortek Co.)

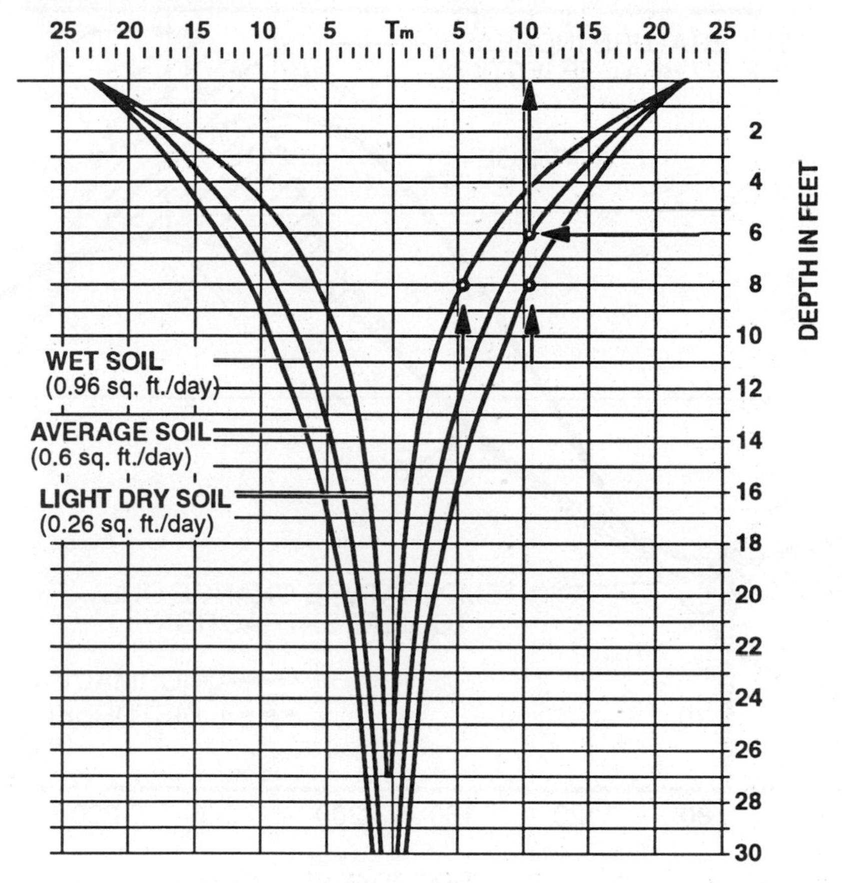

Figure 14–3 Comparison of Tm at varying depths for different soils (Courtesy Mammoth, A Nortek Co.)

rejection from the circulating water to the earth. A temperature difference is needed to get heat transfer.

The amount of this temperature difference will depend upon the BTU of rejection as well as the amount of pipe surface to transfer the heat. A balance of these two factors must be arrived at to get the desired transfer rate without exceeding the water temperature limits of the unit as well as to keep the amount of pipe to a minimum.

EWTh (Entering Water Temperature—Heating)

The same effect occurs during the heating mode. The earth is cooled by the extraction heat removal. The change, however, is considerably less because the heat of extraction is the net capacity of the unit while the heat of rejection is the total or gross capacity of the unit.

Based strictly on the difference between the heat quantities, it would appear that the cooling mode will require more pipe surface. However, when all factors are considered, this does not always apply.

Disturbed Earth Effect

Figure 14–4 shows the effect that the heat of rejection and extraction has on the earth coil temperature in the cooling and heating modes. The "Disturbed Earth" title is used to denote heat content change, which affects the earth temperature. Again, these temperature curves are based on average soil difusivity.

This application is in Omaha, NB, where the groundwater temperature is 54°F (see Figure 13–1), the same as the Lincoln, NB arca. The table in Figure 14–1 gives the Mean annual earth temperature (Tm) as 53°F. In the heating mode, with a disturbed earth drop of 6°F and a 10°F ΔT between the earth and ground loop water, the unit entering water temperature (EWTh) would be 37°F. In the cooling mode, with a disturbed earth temperature rise of 15°F and a 10°F ΔT between the earth and ground loop liquid, the unit entering water temperature would be 79°F.

Rs (Soil resistance)

Soil resistance is the resistance to heat flow through the soil. A light dry soil will not carry heat energy as rapidly as a dense moist soil. In addition, the depth of the pipe below the surface, the distance between pipes, and the size and number of pipes involved in the ground loop all have an effect on the soil resistance.

Figure 14–5 shows a table of soil resistances for heavy damp soil, heavy dry soil, and light damp soil for various pipe sizes from 3/4" to 2" used in single loop, double loop, and four-pipe multiple loop as well as vertical loop in rock or heavy damp soil. For example, a single-loop pipe system of a

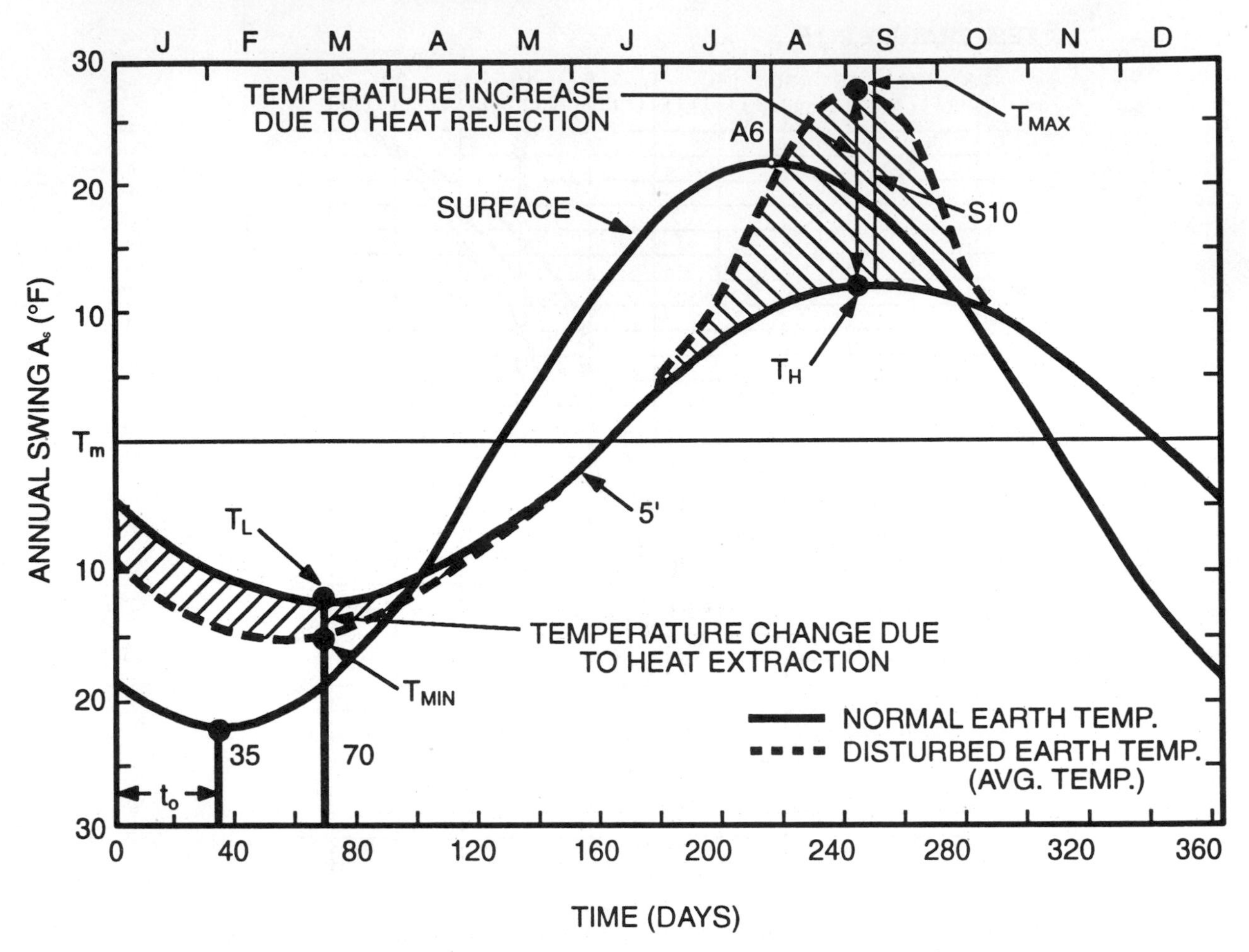

Figure 14–4 Disturbed earth temperature (Courtesy Mammoth, A Nortek Co.)

PIPE SIZE	Rs (HEAVY SOIL - DAMP) / Rs (HEAVY SOIL - DRY OR LIGHT SOIL - DAMP)										Rs (ROCK) / Rs (HS-DAMP)
	3	4	5	6	3, 5	4, 6	3, 4, 5, 6	4, 5, 6, 7	3, 4, 1	3, 4, 2	
3/4	1.02 / 1.38	1.06 / 1.44	1.09 / 1.47	1.11 / 1.49	1.31 / 1.77	1.37 / 1.84	2.05 / 2.75	2.15 / 2.86	2.11 / 2.85	1.88 / 2.53	0.60 / 1.06
1	0.97 / 1.32	1.02 / 1.37	1.04 / 1.40	1.06 / 1.42	1.26 / 1.70	1.32 / 1.77	2.00 / 2.88	2.10 / 2.79	2.07 / 2.78	1.84 / 2.47	0.57 / 1.01
1-1/4	0.92 / 1.25	0.97 / 1.31	0.99 / 1.34	1.01 / 1.36	1.22 / 1.63	1.27 / 1.70	1.96 / 2.61	2.05 / 2.72	2.02 / 2.71	1.79 / 2.40	0.54 / 0.96
1-1/2	0.89 / 1.21	0.94 / 1.27	0.97 / 1.30	0.98 / 1.32	1.19 / 1.59	1.25 / 1.66	1.92 / 2.57	2.02 / 2.68	1.99 / 2.67	1.76 / 2.36	0.53 / 0.94
2	0.85 / 1.15	0.89 / 1.20	0.92 / 1.24	0.94 / 1.26	1.14 / 1.53	1.20 / 1.60	1.88 / 2.51	1.98 / 2.62	1.94 / 2.61	1.71 / 2.29	0.50 / 0.89

Figure 14–5 (Rs) resistance value of soil (Courtesy Bard Manufacturing Co.)

1" size at 3' depth would have a soil resistance (Rs) of 0.97 in heavy damp soil or a higher resistance of 1.32 in heavy dry soil or light damp soil. The higher figure designates a higher resistance to heat flow—less ability to transfer heat. If this same pipe were buried 4', the Rs would increase to 1.02 for heavy damp soil and 1.37 for heavy dry or light damp soil.

This table uses averages for soil conditions rather than the calculation steps for specific conditions. The error rate using this table will usually be less than 3%. The final calculation may result in pipe length with less than 3' in 100' error.

Rp (Pipe Resistance)

Industry experience has produced information on the types of material recommended for the buried pipe loops used in earth loop systems. Plastic materials suffer the least from earth material corrosive effect and result in the longest pipe life. Figure 14–6 shows four different materials that are used in the manufacture of pipe for ground loop systems.

1. Polyethylene—Schedule 40
2. Polyethylene—Schedule 11
3. Polybutylene—Schedule 17
4. Polybutylene—Schedule 13.5

The resistance to heat transfer through these materials, the Rp (pipe resistance factor), is given in Figure 14–6. Note that two different resistances are given: Rp with the pipe in a horizontal position and Rpe with the pipe in a vertical position. These factors are used in the formulas for calculating the length of pipe needed for the earth/liquid heat exchanger.

Type of Earth Coil

The type of earth coil to be used will affect the pipe length involved. When only a single pipe is used in a trench, the highest heat transfer rate per foot of pipe occurs. The greater the amount of trenching involved, the greater the area required for the layout.

Figure 14–7 shows the description of a typical horizontal series system using one pipe in a continuous trench. The maximum recommended heat pump size on a single loop system is 60,000 BTUH, which means a maximum pipe length of 2,500'. With only one pipe in the trench, burial depth would be 3.5' in northern climates and 6' in southern climates.

Figure 14–8 shows a typical horizontal series system using a double back loop with two pipes in the trench. Because having two pipes in the same trench increases the soil resistance, a longer length of pipe is required to get the desired heat transfer rate. In this case, the practical length of pipe increases from 350' to 500' per 12,000 BTUH to 420' to 600' per 12,000 BTUH. The trench length is reduced from the single pipe trench at 350' to 500' per 12,000 BTUH to 210' to 300' per 12,000 BTUH. The two recommended pipe depths and 3' and 5' in the northern regions and 4' to 6' in the southern regions.

The third type of horizontal system is the multiple-level or parallel system (Figure 14–9). The example shown is a four-pipe stack in a narrow trench. Soil resistances in Figure 14–5 also show the four pipes in double stacks on each side of a wide trench. In a stacked narrow trench, pipe depths are 6', 5', 4', and 3' for the northern regions and 7', 6', 5', and 4' for southern regions. In the double-stack wider trench, the pipe depths would be the same as for a single-loop double back system — 3' and 5' in the northern regions and 4' and 6' in the southern regions.

PIPE SIZE	Rp / Rpe: PE SCH 40	PE SDR-11	PB SDR-17	PB SDR-13.5
3/4	0.17 / .116	↑	↑	↑
1	.159 / .109			
1-1/4	.130 / .089	.144 / .096	0.16 / 0.11	0.20 / 0.14
1-1/2	.117 / .080			
2	.098 / .068	↓	↓	↓

Figure 14–6 (Rp) resistance value of pipe (Courtesy Bard Manufacturing Co.)

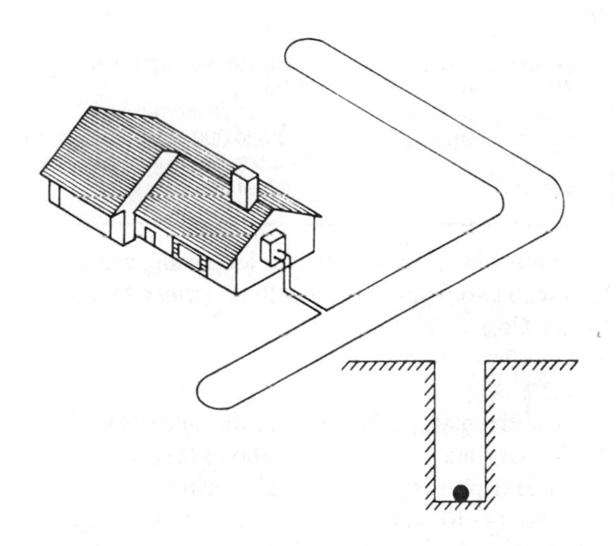

Earth Coil Type: Horizontal-Single Layer Series
Water Flow:
Typical Pipe Size: 1 1/2 to 2 inches
Nominal Length: 350 to 500 feet/ton
Burial Depth: 4 to 6 feet

Figure 14–7 Horizontal series loop with one branch in one trench (Courtesy Mammoth, A Nortek Co.)

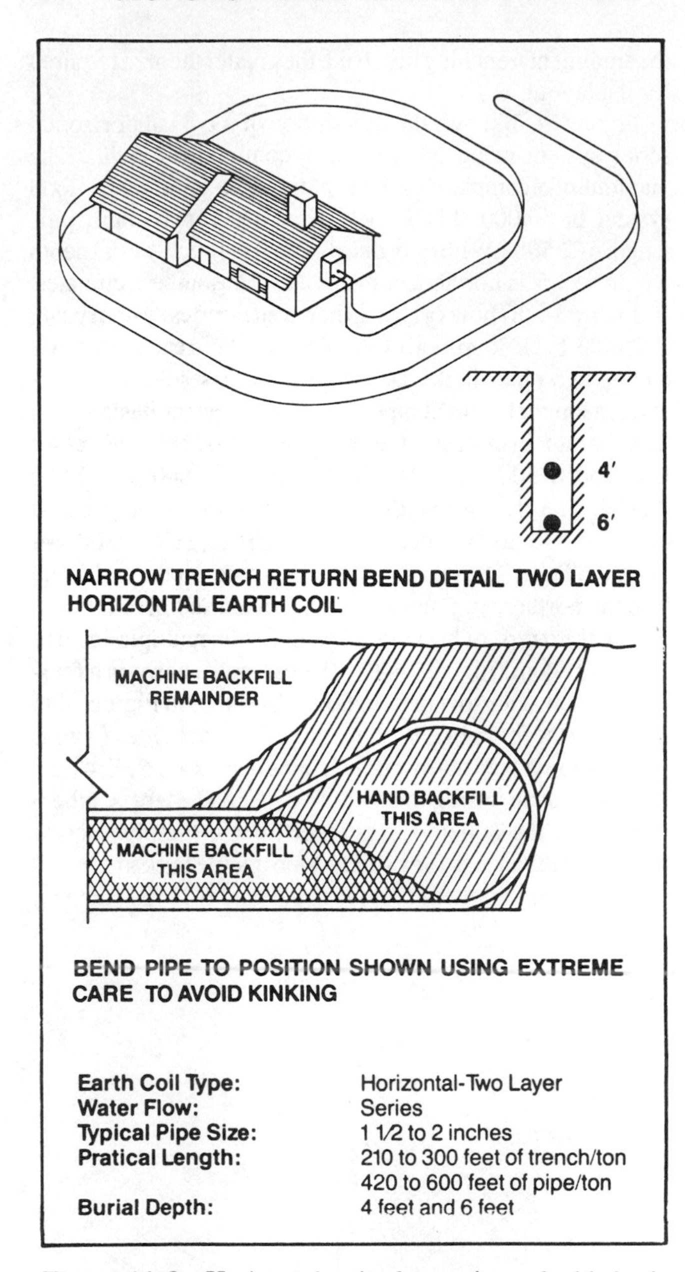

Figure 14–8 Horizontal series loop using a double back loop with two pipes in the trench (Courtesy Mammoth, A Nortek Co.)

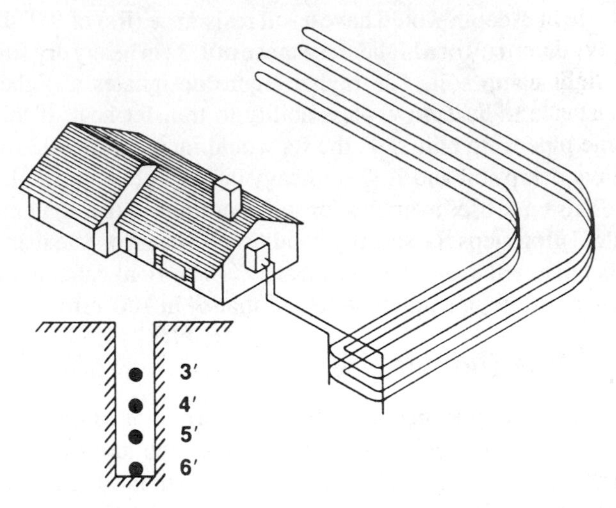

Earth Coil Type:	Horizontal–Four Layer
Water Flow:	Parallel
Typical Pipe Size:	Parallel loops 3/4 to 1 inches headers 1 1/2 to 2 inches
Parallel Pipe Length:	500 Ft. Max. Pipe Length (3/4 inch) 750 Ft. Max. Pipe Length (1 inch)
Burial Depth:	6 feet, 12 inch spacing

Figure 14–9 Horizontal multilevel parallel system (Courtesy Mammoth, A Nortek Co.)

The horizontal spacing between the pipe stacks should not be less than 1' and 2'. Figure 14–5 shows the difference in pipe resistance (Rp) using the 1' and 2' spacing.

Figure 14–10 shows the vertical series system using a continuous series circuit. The bored or drilled hole total hole length is 100' to 175' per 12,000 BTUH, which would require 200' to 350' of pipe. Using approximately 140' of pipe per 12,000 BTUH, a single bore depth should not be more than 300'.

The minimum space between the bored holes is 10' for heavy wet soil and 15' for heavy dry soil or light wet soil. The

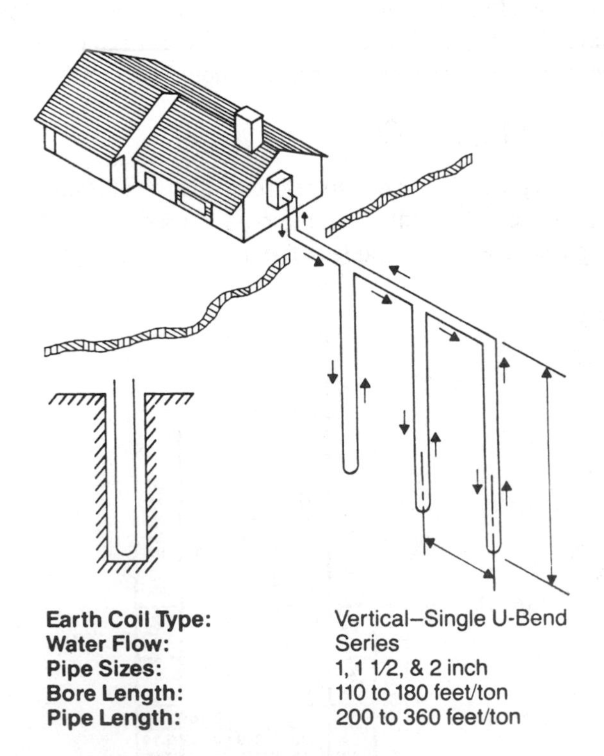

Earth Coil Type:	Vertical–Single U-Bend
Water Flow:	Series
Pipe Sizes:	1, 1 1/2, & 2 inch
Bore Length:	110 to 180 feet/ton
Pipe Length:	200 to 360 feet/ton

Figure 14–10 Vertical series system (Courtesy Mammoth, A Nortek Co.)

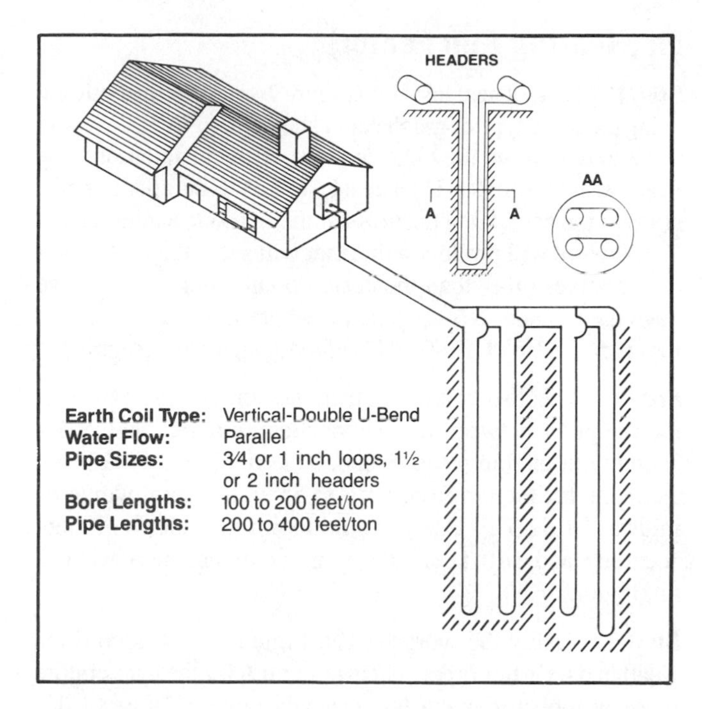

Figure 14–11 Vertical parallel system (Courtesy Mammoth, A Nortek Co.)

space between the bored holes should not exceed double the minimum as this is a waste of pipe length and unnecessary flow resistance through longer pipe.

Figure 14–11 shows a vertical system with parallel flow through the vertical loops. Bore and pipe length are approximately the same. However, the advantage of the parallel system is a lower flow rate through each loop. Therefore, smaller pipe with thinner walls which have less heat transfer resistance (Rp) can be used.

Proper design to assure equal flow rates through the loops is very important to obtain the highest operating efficiency.

Depth of Earth Coil

In the illustrations of each type of coil layout, average soil depths are given. In some cases, however, it is necessary to deviate from these to assume that the entering water temperatures to the unit will not exceed the manufacturer's recommendations. Water temperatures can be critical to the operation of the system, especially in the heating mode in the northern regions.

Practically all heat pump manufacturers rate their units in the Tmax of 45°F to 90°F in the cooling mode and a Tmin of 45°F to 80°F in the heating mode. Some manufacturers rate the same units or manufacture a low temperature unit that will operate to a Tmin of 25°F. To operate the higher temperature unit in the Tmin of 25°F range, two restrictions are imposed:

1. If a low-pressure fixed-setting control switch is used in the unit, it must be removed and an adjustable switch must be substituted to be able to reach the lower cutout settings.
2. A mixture of water and antifreeze must be used in the closed loop system to protect the refrigerant-to-liquid heat exchanger coil in the unit from freezing damage. If there is any possibility of loop liquid temperatures dropping below 35°F, the antifreeze solution must be included.

Remember: When using 20% propylene-glycol antifreeze solution in the ground loop system, the minimum water flow rate selected for the heat pump must be increased 40% to obtain the same heat transfer rate. For example, if a 6 GPM flow rate for water is needed, the flow rate used to size the pump when the 20% antifreeze solution is used would be 6 GPM × 1.4 or 7.2 GPM.

The pipe flow resistance or piping head loss also increases with the use of the antifreeze solution. The 20% propylene-glycol solution increases the flow resistance by 36%. Flow resistance for water × 1.36 = Flow resistance for the solution.

To determine the pipe depth required for the loop, the anticipated ground temperature swing must be determined. Based on the minimum and maximum entering water temperatures to the unit (EWTc and EWTh) and the mean annual earth temperature (Tm) of the area, the temperature change curves in Figures 14–3 and 14–4 are used to determine the required depth of the pipe below grade level.

For example, if the unit is located in Phoenix, AZ, the mean annual earth temperature (Tm) from Figure 14–1 would be 73°F. With an anticipated 15°F rise in disturbed earth temperature, to supply water within the maximum unit EWTc of 80°F, the maximum deviation in the water from the Tm temperature would be 7°F. Using the curves in Figure 14–3, the pipe would have to be 8' in the ground.

If the unit location were in Grand Forks, ND, where the Tm is 42°F, the cooling mode operation would result in a 42°F Tm plus the 15°F disturbed earth differential or 52°F earth temperature. The unit could use a 20°F differential between the earth temperature and the EWTc for the cooling mode operation to result in a unit EWTc of 72°F. The recommendation would be 15°FΔT resulting in an EWTc of 72°F. This would mean a pipe depth of 4' in wet soil to 2' in dry soil. Average soil depth would be 4'.

The same process is used when determining the pipe depth for the heating mode. The disturbed earth factor for heating is less than cooling—approximately 6°F. For the Phoenix, AZ application, the heating pipe depth is determined from the 73°F Tm less the disturbed earth factor of 6°F (67°F) less the 10°F temperature difference between the earth and water, resulting in and EWTh to the unit of 57°F. This is within the unit design range and would not require an antifreeze solution in the loop.

For the heating mode, wet soil would require 8' depth, dry soil a 4' depth, and average soil, a 6' depth. Because the cooling load is the major load, the cooling depth requirement would be used.

In the Grand Forks, ND application, the 42°F Tm less the 6°F disturbed earth factor means an earth temperature of 36°F. Using 10°FΔT between the earth and water means a liquid temperature of 26°F. To keep the operating temperature above 32°F, a 5°FΔT would have to be used. This means a pipe depth of 15' in wet soil, 8' in light dry soil, and 12' in average soil. Antifreeze solution would be used to allow operation of the system below the 32°F liquid temperature. A 10°FΔT would allow a pipe depth of 8' in wet soil, 4' in light dry soil, and 6' in average soil. Use of the antifreeze solution will require adjustment in the pipe flow resistance factor.

Run Factor

The run factor is the percent of the time the unit can be expected to operate to handle the heating load during the coldest month (January) and the warmest month (August).

Both run factors must be calculated to determine the pipe length required in each of the operating modes. The longer pipe requirement is used in the system design. To determine the run factor, the average heat gain per hour is calculated and compared to the unit capacity. The formula used is:

$$\text{Run Factor} = \frac{\text{Average Heat Gain or Loss/Hour}}{\text{Unit Capacity in BTUH}}$$

The recommended method of calculating the average heat gain or loss per hour is the Bin Method of calculating annual energy requirements outlined in Manual J of the ACCA Environmental Systems Library. These manuals are highly recommended for use in this process due to their completeness as well as constant update as new information becomes available.

The tables in Figure 14–12 are the bin weather data for 158 cities in the United States. The tables show the average number of hours throughout the year that the outdoor temperature is in each temperature bin. For example, in Akron, Ohio, the outdoor temperature will be between 50°F and 55°F for an average of 684 hours each year. The extremes of temperature for the Akron, Ohio area are –10°F to –5°F for four hours each year to +95°F to +100°F for four hours each year. The asterisks in the adjoining columns indicate that the outdoor temperatures may have reached these extremes but for a period of less than one hour.

This time/temperature information is used to establish the hourly heat loss or gain in each temperature category to arrive at an annual heating or cooling energy requirement. After the annual energy requirement is determined, an average hourly requirement is determined and compared to the unit heating or cooling capacity. This determines the possible percentage of time the unit must operate to handle the heating and cooling requirements. These percentages are called the heating and cooling run factors.

To use the bin method of calculating the run factors, the following steps are required.

Fh (Heating Run Factor)

Step 1. Draw a heat loss line (Figure 14–13) for the building using the same type of graph paper shown in Figures 9–18 and 9–19 to determine the balance point of the installation. The heat loss line from 0 BTUH heat loss at 65°F outdoor temperature to the calculated heat loss at the outdoor winter design temperature will produce a line that will show the heat loss at temperatures other than the design points. For example, we have the Omaha, NB application where the calculated heat loss is 57,782 BTUH at –3°F winter design temperature.

Step 2. Locate points on the heat loss line (Figure 14–13) at the midpoint between the outdoor temperatures that correspond with the bin weather data table in Figure 14–12. For example, for the bin from 50°F to 55°F, the point would be midway at 52.5°F. Using the midpoint provides sufficient accuracy and still keeps the number of calculations to a minimum.

Step 3. Using the worksheet in Figure 14–14, record the weather data in hours per year in column A for the temperature bins that apply to Omaha, NB. From the table in Figure 14–12, we see that the lowest temperature bin involved in the Omaha, NB area is –15°F to –10°F. The outdoor temperature will be in the range of –15°F to –10°F for an average of 6 hours per year. This is recorded in column A on the –15°F to –10°F line.

From this point, the average hours per year are recorded in each of the temperature bins—21 hours in the –10°F to –5°F bin, 54 hours in the –5°F to 0°F bin, and so on until the +60°F to +65°F bin is reached. The +65° F to +70°F bin is not used in either the heating or cooling calculations. Though the heat loss is calculated at 70°F inside design temperature, no additional heat is needed down to +65°F outside temperature. Inside heat sources will supply the heating requirements.

Step 4. From the heat loss line on the graph in Figure 14–13, the building heat loss in MBTUH (1,000 BTU per hour) is determined from the left scale of the graph and recorded on the line of the appropriate temperature bin. For example, the heat loss at –15°F to –10°F (read at –12.5°F) is 66,000 BTUH. This is recorded as 60 MBTUH to reduce the size of the totals that have to be handled.

At –10°F to –5°F (–7.5°F) the heat loss is 61.8 MBTUH, at –5°F to 0°F (–2.5°F) the heat loss is 57.5 MBTUH, and so on until all the heat loss in MBTUH is recorded for each of the temperature bins involved.

Step 5. Find the total heat energy lost in MBTUH per year for each temperature bin by multiplying the hours in column A by the MBTUH in column B and recording the results in column C. For example, the 6 hours in column A times the 66 MBTUH in column B equals a total of 396 MBTUH per year in this temperature bin. This calculation is made for each of the temperature bins.

Step 6. Add the total heat loss in all the temperature bins (column C) and record the total in box D.

Bin Weather Data

OUTDOOR TEMPERATURE (F)	-35 to -30	-30 to -25	-25 to -20	-20 to -15	-15 to -10	-10 to -5	-5 to 0	0 to 5	5 to 10	10 to 15	15 to 20	20 to 25	25 to 30	30 to 35	35 to 40	40 to 45	45 to 50	50 to 55	55 to 60	60 to 65	65 to 70	70 to 75	75 to 80	80 to 85	85 to 90	90 to 95	95 to 100	100 to 105	105 to 110	110 to 115	OUTDOOR TEMPERATURE (F)
Akron, Ohio					•	4	10	17	44	127	214	422	654	819	706	680	605	684	736	762	801	639	454	261	112	30	6	•			Akron, Ohio
Albany, N. Y.				1	2	2	20	62	90	174	254	376	552	814	751	677	694	694	720	769	741	559	387	230	115	31	10				Albany, N. Y.
Albuquerque, N.M.							1	2	4	19	76	142	319	523	666	714	710	712	678	783	838	780	629	506	364	228	66	3			Albuquerque, N. M.
Allentown, Pa.						1	3	7	23	61	129	248	501	851	848	712	661	702	792	773	812	657	471	305	153	43	12	•			Allentown, Pa.
Amarillo, Tex.					1	2	1	5	34	52	115	191	315	499	595	612	684	714	671	752	824	804	607	475	381	288	129	24	1		Amarillo, Tex.
Anchorage, Alaska			5	28	46	94	124	211	285	376	560	662	795	831	713	637	741	995	930	499	180	50	8								Anchorage, Alaska
Atlanta, Ga.								2	•	6	14	20	82	255	455	627	651	700	817	908	978	1188	880	620	361	172	23	2			Atlanta, Ga.
Augusta, Ga.										2	9	35	94	210	368	497	570	666	785	875	957	1167	959	652	531	274	88	17	1		Augusta, Ga.
Austin, Tex.									•	2	15	13	51	78	220	355	472	567	717	768	1032	1105	1251	819	600	438	215	46	1		Austin, Tex.
Bakersfield, California													7	77	247	541	746	908	977	966	898	831	742	613	474	371	254	100	18	•	Bakersfield, California
Baltimore, Md.							•	2	7	43	89	184	328	642	755	770	673	683	696	729	794	883	655	438	263	109	23	2			Baltimore, Md.
Baton Rouge, La.										1	5	6	41	103	199	336	482	598	664	828	992	1346	1307	826	622	376	50	2			Baton Rouge, La.
Billings, Mont.		1	3	30	51	71	81	89	121	163	236	372	522	730	785	781	721	741	749	686	564	447	340	242	144	76	17	3			Billings, Mont.
Binghamton, New York				1	1	5	23	67	156	263	350	515	568	873	666	598	595	673	694	808	781	548	318	145	31	2					Binghamton, New York
Birmingham, Ala.									3	4	13	48	111	273	423	489	581	621	767	875	1044	1166	915	631	471	256	57	8	1		Birmingham, Ala.
Bismarck, N.D.	7	21	41	71	102	161	214	278	304	349	358	436	558	702	575	482	501	549	612	632	560	442	346	233	139	68	15	2	•		Bismarck, N.D.
Boise, Idaho							4	2	18	47	142	272	498	854	881	865	809	790	748	663	609	484	376	307	214	135	39	6			Boise, Idaho
Boston, Mass.					•	1	9	4	35	74	151	256	429	674	848	828	757	766	781	804	819	676	433	245	127	93	10	•			Boston, Mass.
Brownsville, Tex.												1	13	9	25	73	187	270	390	564	963	1419	1901	1579	862	489	20				Brownsville, Tex.
Buffalo, N.Y.						•	2	19	66	125	230	427	605	821	796	684	643	727	752	771	755	621	396	232	76	15	1				Buffalo, N.Y.
Burbank, Calif.													•	10	83	292	661	1186	1562	1562	1163	808	565	431	257	129	48	9	3		Burbank, Calif.
Burlington, Ia.					3	13	29	55	108	161	233	356	542	797	708	572	485	528	613	714	753	743	584	377	199	95	31	1			Burlington, Ia.
Burlington, Vermont		1	2	5	17	39	81	135	216	272	332	491	561	752	716	637	603	655	694	703	670	573	362	189	53	9	1				Burlington, Vermont
Calgary, Alta.			12	35	75	109	171	194	224	303	335	436	619	820	838	781	797	839	698	558	594	273	173	89	28	5					Calgary, Alta.
Casper, Wyoming			2	3	15	30	45	73	116	200	324	495	683	806	831	782	670	606	642	592	532	423	347	283	201	66	3				Casper, Wyoming
Charleston, S.C.											3	14	58	150	308	433	546	633	792	1006	1134	1311	1156	736	436	148	34	1			Charleston, S.C.
Charleston, W. V.							1	7	22	73	135	252	356	630	633	607	667	661	689	767	949	912	606	471	270	57	3				Charleston, W. V.
Charlotte, N. C.									2	5	23	64	166	360	515	634	684	730	752	839	908	1115	747	567	397	203	52	5			Charlotte, N. C.
Chattanooga, Tenn.								1	4	10	32	87	215	423	538	631	681	696	746	776	930	986	713	535	388	260	87	16	1		Chattanooga, Tenn.
Cheyenne, Wyo.			1	5	4	13	34	55	92	177	279	499	608	797	810	883	765	769	771	637	514	409	301	214	93	32	3	•			Cheyenne, Wyo.
Chicago, Ill.					4	11	18	63	79	117	207	337	517	842	826	616	561	617	581	676	767	697	531	355	209	96	32	4			Chicago, Ill.
Cincinnati, Ohio					1	2	6	10	36	60	101	218	475	709	692	642	619	661	692	738	855	822	613	431	250	105	24	2			Cincinnati, Ohio
Cleveland, Ohio						1	23	9	49	109	198	346	581	809	760	614	615	633	645	733	831	721	562	311	155	56	8	1			Cleveland, Ohio
Cold Bay, Alaska							5	39	39	132	303	476	673	1245	1755	1327	1484	1042	211	31	6	2	•								Cold Bay, Alaska
Colorado Springs, Col.				1	1	7	18	43	75	136	262	448	626	755	673	678	725	760	805	784	600	477	397	276	176	46	1				Colorado Springs, Col.
Columbia, S.C.										1	11	36	115	286	394	496	564	661	778	838	941	1126	964	655	502	279	93	17	2		Columbia, S.C.
Columbus, Ohio					4	5	6	24	25	71	136	177	537	739	740	661	645	655	701	737	814	741	558	392	212	83	18	1			Columbus, Ohio
Corpus Christi, Texas											•	3	9	27	83	180	302	444	551	748	1041	1175	1538	1408	785	436	36	1			Corpus Christi, Texas
Dallas, Tex.									1	4	17	34	91	231	371	504	576	629	656	693	795	831	942	880	659	493	273	79	9	•	Dallas, Tex.
Dayton, Ohio				•	1	5	16	23	55	99	182	309	558	786	698	601	576	580	627	717	832	817	607	402	202	65	8	1			Dayton, Ohio
Denver, Colorado			1	•	1	6	22	36	78	119	216	359	553	721	717	692	704	678	731	783	684	549	437	332	236	103	10	1			Denver, Colorado
Des Moines, Ia.					9	26	56	81	131	195	280	409	583	763	630	492	515	521	597	694	783	709	562	378	223	94	30	5	•		Des Moines, Ia.
Detroit, Michigan						1	4	17	61	131	248	377	618	884	808	595	566	592	633	695	783	721	516	314	148	47	9	•			Detroit, Michigan
Duluth, Minn.	2	9	20	50	142	131	190	229	284	373	499	617	638	766	585	528	605	699	733	633	458	323	194	86	23	2	•				Duluth, Minn.
Edmondton, Alta.	4	10	19	64	105	154	196	281	333	338	404	426	571	707	644	515	680	814	750	597	423	283	182	78	23	4					Edmondton, Alta.
El Paso, Tex.									•	3	8	35	90	205	342	514	590	712	759	839	893	970	865	740	586	406	204	42	2		El Paso, Tex.
Evansville, Ind.			•	1	•	1	4	9	20	47	81	146	380	654	683	658	643	678	658	670	735	811	721	530	346	199	83	12			Evansville, Ind.
Fairbanks, Alaska	121	126	159	206	270	332	401	401	379	401	447	457	455	495	409	429	513	637	658	515	361	202	118	54	14	1					Fairbanks, Alaska
Fargo, N.D.	2	19	33	70	124	182	237	255	274	360	385	439	578	657	569	445	449	513	641	680	626	505	362	216	108	30	6				Fargo, N.D.
Flint, Michigan			•	1	2	11	34	75	142	208	347	487	707	863	675	565	597	574	638	746	745	588	411	253	88	11	•				Flint, Michigan
Fort Wayne, Ind.				•	2	5	15	32	61	101	183	366	615	890	754	639	586	613	664	682	762	667	502	353	180	80	15	1			Fort Wayne, Ind.
Fort Worth, Texas									1	3	12	44	132	294	422	538	591	648	622	689	774	889	982	788	596	440	246	54	3		Fort Worth, Texas
Fresno, California												•	34	168	426	673	952	1036	1006	921	803	709	607	490	392	297	192	56	5		Fresno, California
Galveston, Tex.											1	5	10	27	47	115	279	486	706	1195	1236	885	1129	1738	848	56	1				Galveston, Tex.
Grand Rapids, Mich.				1	1	2	6	28	67	157	256	454	701	983	794	589	564	589	671	729	718	594	415	280	122	38	6	•			Grand Rapids, Mich.
OUTDOOR TEMPERATURE (F)	-35 to -30	-30 to -25	-25 to -20	-20 to -15	-15 to -10	-10 to -5	-5 to 0	0 to 5	5 to 10	10 to 15	15 to 20	20 to 25	25 to 30	30 to 35	35 to 40	40 to 45	45 to 50	50 to 55	55 to 60	60 to 65	65 to 70	70 to 75	75 to 80	80 to 85	85 to 90	90 to 95	95 to 100	100 to 105	105 to 110	110 to 115	OUTDOOR TEMPERATURE (F)

*Less than one hour

AVERAGE NUMBER OF HOURS EACH TEMPERATURE SHOWN OCCURS IN A YEAR

Source of data: Lennox Ind.

Figure 14–12 Bin weather data (Courtesy Air Conditioning Contractors of America)

(Continued)

OUTDOOR TEMPERATURE (F)	-35 to -30	-30 to -25	-25 to -20	-20 to -15	-15 to -10	-10 to -5	-5 to 0	0 to 5	5 to 10	10 to 15	15 to 20	20 to 25	25 to 30	30 to 35	35 to 40	40 to 45	45 to 50	50 to 55	55 to 60	60 to 65	65 to 70	70 to 75	75 to 80	80 to 85	85 to 90	90 to 95	95 to 100	100 to 105	105 to 110	110 to 115	OUTDOOR TEMPERATURE (F)
Great Falls, Mont.	1	5	15	55	49	58	97	92	112	154	231	383	553	712	809	822	847	884	805	635	501	374	270	160	86	36	5	1			Great Falls, Mont.
Green Bay, Wisconsin				3	19	42	95	160	231	321	373	515	689	820	649	542	522	598	720	758	658	474	331	176	6	9					Green Bay, Wisconsin
Greensboro, N.C.									*	15	36	92	261	466	614	621	686	721	817	866	963	955	659	504	318	135	33	2			Greensboro, N.C.
Halifax, N.S.						1	12	44	81	174	245	353	528	959	998	799	809	884	941	956	547	251	105	29	2						Halifax, N.S.
Harrisburg, Pa.								3	9	33	87	170	393	766	901	758	682	685	757	753	823	790	545	345	189	61	14	1			Harrisburg, Pa.
Hartford, Connecticut				*	2	3	11	33	77	153	233	370	552	825	807	683	575	649	752	751	755	617	419	274	118	29	3				Hartford, Connecticut
Helena, Mont.	1	10	28	61	69	73	99	86	153	211	286	389	645	809	900	826	776	761	690	560	465	344	234	164	92	29	5				Helena, Mont.
Hilo, Hawaii																			11	361	2272	3001	2126	975	24						Hilo, Hawaii
Honolulu, Hawaii																			6	122	722	2789	3458	1552	117	1					Honolulu, Hawaii
Houston, Tex.											4	9	25	50	100	220	396	553	684	796	1069	1263	1621	922	676	323	53	1			Houston, Tex.
Huron, South Dakota		1	7	21	56	83	145	208	262	305	419	476	571	652	574	502	488	513	569	614	624	554	443	318	205	103	47	9	*		Huron, South Dakota
Indianapolis, Ind.				1	2	3	7	28	53	74	135	279	543	752	751	653	614	620	601	708	763	760	604	420	238	102	23	2			Indianapolis, Ind.
Jackson, Miss.								1	3	3	4	17	64	176	325	485	557	605	669	810	989	1196	1051	719	546	383	143	16			Jackson, Miss.
Jacksonville, Fla.												1	10	52	133	262	350	468	688	895	979	1329	1658	949	607	316	70	2			Jacksonville, Fla.
Kansas City, Mo.						*	7	23	51	93	147	251	389	600	643	660	581	598	617	593	740	767	687	569	367	221	107	36	8	1	Kansas City, Mo.
King Salmon, Alaska	4	22	43	87	144	221	222	250	237	259	323	435	628	994	1047	872	923	984	595	289	131	43	13	3							King Salmon, Alaska
Knoxville, Tenn.									5	15	26	60	209	442	565	639	698	676	709	847	935	1042	793	578	335	156	32	3			Knoxville, Tenn.
LaCrosse, Wis.	1	1	8	19	29	52	97	140	188	272	305	489	600	784	654	518	481	520	655	738	754	638	458	225	103	32	2				LaCrosse, Wis.
LaGuardia, N.Y.								1	6	22	69	139	262	507	804	825	814	788	765	771	842	882	659	368	170	51	14	2			LaGuardia, N.Y.
Lake Charles, La.										1	5	6	14	58	126	303	450	622	717	838	1038	1302	1444	834	640	327	40	2			Lake Charles, La.
Lansing, Mich.					*	2	12	33	76	158	299	451	734	920	741	575	592	628	696	706	725	596	409	256	121	30	3				Lansing, Mich.
Laredo, Texas											*	1	8	27	82	192	343	471	563	687	838	958	1346	1161	801	626	469	189	6		Laredo, Texas
Las Vegas, Nevada											1	7	44	194	396	591	716	769	786	699	644	651	669	678	602	474	431	301	101	16	Las Vegas, Nevada
Lexington, Kentucky						*	7	16	35	80	144	238	441	654	627	629	611	644	656	710	898	957	630	464	263	62	2				Lexington, Kentucky
Little Rock, Ark.							1	1	2	6	23	41	136	306	469	603	702	652	704	782	847	996	964	691	497	288	100	11			Little Rock, Ark.
Los Angeles, Calif.														1	15	151	532	1183	2130	2331	1458	670	202	68	18	4	1	*			Los Angeles, Calif.
Louisville, Ky.				*	*	2	2	8	13	37	66	137	324	631	703	646	663	650	700	702	757	831	717	538	350	209	64	14			Louisville, Ky.
Lubbock, Texas						*	1	5	7	33	86	180	346	490	546	620	618	642	700	688	829	833	708	544	447	322	109	14	2		Lubbock, Texas
Macon, Georgia										3	11	39	109	210	362	487	545	668	755	800	905	1239	995	665	538	354	85	3			Macon, Georgia
Madison, Wis.	1	1	1	6	13	31	60	93	171	215	307	459	659	896	690	539	528	565	636	724	680	586	455	264	131	42	5				Madison, Wis.
Medford, Ore.									*	1	7	48	224	660	984	1143	1086	1000	901	715	566	443	329	274	207	124	55	6			Medford, Ore.
Memphis, Tenn.						1	*	1	4	7	12	54	157	345	518	595	645	659	716	788	855	926	902	672	471	297	120	17	1		Memphis, Tenn.
Miami, Fla.															4	26	71	147	277	452	810	1708	2463	1795	888	125	2				Miami, Fla.
Midland, Texas									1	5	23	80	163	337	451	602	669	631	678	720	793	914	865	673	532	426	177	28	1		Midland, Texas
Milwaukee, Wis.			1	2	4	18	47	83	116	176	285	421	459	913	774	611	591	585	634	749	753	597	390	226	96	32	6	*			Milwaukee, Wis.
Minneapolis, Minn.		3	8	50	46	72	124	177	219	288	357	526	598	666	623	504	468	522	609	735	713	620	432	268	128	33	2				Minneapolis, Minn.
Missoula, Mont.				2	9	19	38	43	89	170	284	465	736	1016	952	848	812	797	666	553	418	320	223	159	97	41	7	*			Missoula, Mont.
Mobile, Ala.										1	4	5	23	74	180	370	488	567	657	898	1135	1486	1338	785	523	220	38	2			Mobile, Ala.
Moline, Ill.				*	8	13	39	66	111	175	247	376	551	827	714	551	537	551	591	684	744	694	574	387	209	88	19				Moline, Ill.
Montgomery, Ala.										3	5	17	63	159	329	455	567	663	708	799	965	1235	1121	751	500	332	89	7			Montgomery, Ala.
Montreal, Que.			3	4	25	79	107	180	280	318	355	437	523	719	706	555	548	589	666	693	740	593	388	103	43	3					Montreal, Que.
Nashville, Tenn.						2	2	2	5	22	43	89	241	444	562	643	601	644	729	836	876	919	804	583	404	220	76	13	2		Nashville, Tenn.
Newark, New Jersey								2	11	38	109	202	360	637	846	784	701	692	697	755	814	819	615	383	200	79	21	2	*		Newark, New Jersey
New Orleans, La.											1	4	7	18	94	143	448	556	719	844	1047	1259	1848	966	606	259	16	1			New Orleans, La.
New York, New York								1	10	26	2	188	330	603	858	838	796	722	745	754	877	926	604	263	96	28	5	*			New York, New York
Norfolk, Va.											6	38	108	312	543	737	782	681	795	858	995	1113	909	481	273	113	20	1			Norfolk, Va.
Oklahoma City, Okla.							*	2	19	41	66	135	232	401	498	609	684	683	725	802	769	932	847	606	421	240	89	20	1		Oklahoma City, Okla.
Oakland, California													*	20	155	509	971	1858	2431	1498	756	339	135	61	26	8	2	*	*		Oakland, California
Omaha, Neb.				*	6	21	54	88	127	180	261	374	521	683	672	584	567	527	544	637	743	703	579	435	263	134	45	14	1		Omaha, Neb.
Orlando, Florida													4	30	84	156	245	377	564	830	1123	1680	1732	1004	666						Orlando, Florida
Philadelphia, Pa.									9	32	100	189	335	654	818	758	701	663	710	735	808	864	655	420	225	74	17	1			Philadelphia, Pa.
Phoenix, Arizona													8	57	182	391	540	659	769	767	776	762	774	798	698	611	507	324	129	15	Phoenix, Arizona
Pittsburgh, Penn.						1	7	30	60	159	233	360	569	774	688	631	587	637	678	799	910	722	503	311	98	12					Pittsburgh, Penn.
Pocatello, Idaho					4	10	34	58	94	151	250	462	678	891	835	743	703	688	638	585	536	467	359	283	189	88	17	*			Pocatello, Idaho
Portland, Me.					1	9	15	64	117	190	271	384	573	750	841	805	810	773	857	808	642	387	263	129	61	14	*				Portland, Me.
OUTDOOR TEMPERATURE (F)	-35 to -30	-30 to -25	-25 to -20	-20 to -15	-15 to -10	-10 to -5	-5 to 0	0 to 5	5 to 10	10 to 15	15 to 20	20 to 25	25 to 30	30 to 35	35 to 40	40 to 45	45 to 50	50 to 55	55 to 60	60 to 65	65 to 70	70 to 75	75 to 80	80 to 85	85 to 90	90 to 95	95 to 100	100 to 105	105 to 110	110 to 115	OUTDOOR TEMPERATURE (F)

*Less than one hour

AVERAGE NUMBER OF HOURS EACH TEMPERATURE SHOWN OCCURS IN A YEAR

Source of data: Lennox Ind.

Figure 14–12 (cont.) Bin weather data (Courtesy Air Conditioning Contractors of America)

(Continued)

OUTDOOR TEMPERATURE (F)	-35 to -30	-30 to -25	-25 to -20	-20 to -15	-15 to -10	-10 to -5	-5 to 0	0 to 5	5 to 10	10 to 15	15 to 20	20 to 25	25 to 30	30 to 35	35 to 40	40 to 45	45 to 50	50 to 55	55 to 60	60 to 65	65 to 70	70 to 75	75 to 80	80 to 85	85 to 90	90 to 95	95 to 100	100 to 105	105 to 110	110 to 115	OUTDOOR TEMPERATURE (F)
Portland, Ore.							1	2	4	15	31	52	97	301	702	1291	1313	1312	1359	959	546	367	208	128	53	21	3				Portland, Ore.
Providence, R.I.						1	12	12	34	66	134	257	418	709	801	690	822	824	827	844	803	664	419	228	73	22	2				Providence, R.I.
Pueblo, Colo.	*	1	1	1	1	8	26	39	52	117	203	365	496	570	556	664	674	677	708	769	714	587	484	383	296	210	68	10			Pueblo, Colo.
Raleigh, N.C.										7	23	72	197	383	520	589	668	728	813	915	968	1049	726	518	356	174	49	5			Raleigh, N.C.
Rapid City, S. D.			1	5	10	46	79	128	194	246	273	420	594	742	694	675	621	632	666	655	590	488	370	275	198	108	52	10	1		Rapid City, S. D.
Regina Sask.	12	27	51	99	168	221	238	281	330	349	441	533	563	648	559	499	530	616	693	533	481	386	282	168	92	47	10				Regina Sask.
Reno, Nevada						1	4	15	37	101	227	387	530	733	829	890	909	845	690	572	477	418	371	333	243	120	35	2			Reno, Nevada
Richmond, Va.										13	58	108	269	497	617	688	695	710	759	812	854	914	726	483	330	163	59	12			Richmond, Va.
Roanoke, Virginia								1	11	28	86	186	307	533	702	710	712	695	704	758	924	928	620	459	305	91	8				Roanoke, Virginia
Rochester, N.Y.					*	1	6	30	69	135	233	403	606	823	816	695	661	691	746	750	712	590	378	241	125	43	9				Rochester, N.Y.
Sacramento, Calif.													8	93	355	701	1049	1298	1329	1071	773	630	486	375	276	192	93	34	5	*	Sacramento, Calif.
Salem, Oregon									*	3	20	46	172	435	707	1100	1398	1352	1163	833	540	405	261	169	100	47	14	5	*		Salem, Oregon
Salt Lake City, Utah							1	16	46	90	150	308	572	842	841	725	652	690	661	632	641	588	461	363	272	150	43	3			Salt Lake City, Utah
San Antonio, Tex.									1	1	8	15	36	81	161	340	407	529	675	853	1019	1180	1310	863	623	427	215	22			San Antonio, Tex.
San Diego, Calif.															11	88	393	972	1863	2389	1821	832	270	99	22	4	*				San Diego, Calif.
San Francisco, Calif.														15	121	517	1319	2469	2278	1174	571	192	65	31	10	2					San Francisco, Calif.
San Juan, P.R.																				*	158	1450	3780	2521	837	19					San Juan, P.R.
Savannah, Ga.											2	9	35	106	235	367	496	586	736	935	1054	1313	1279	820	488	206	66	8			Savannah, Ga.
Scranton, Pennsylvania						2	9	31	78	178	264	392	575	848	805	628	592	629	719	784	804	666	413	254	83	14					Scranton, Pennsylvania
Seattle, Wash.										5	21	42	133	536	1121	1605	1337	1295	1132	699	406	233	105	36	17	4	1				Seattle, Wash.
Shreveport, La.								1	1	3	8	25	43	161	306	464	584	658	695	811	934	1065	1112	767	583	394	133	18	*		Shreveport, La.
Sioux City, Ia.			*	2	14	38	73	97	152	247	361	468	595	732	642	550	523	527	575	642	701	618	507	365	207	97	26	4	1		Sioux City, Ia.
Sioux Falls, S.D.		*	2	16	43	59	102	152	208	293	448	520	585	712	625	501	498	522	605	669	684	566	443	277	155	67	7	1			Sioux Falls, S.D.
South Bend, Indiana				*	1	7	26	47	81	166	250	449	661	870	694	564	526	567	608	728	806	698	507	318	150	41	3				South Bend, Indiana
Spokane, Wash.				*	3	8	16	29	52	91	153	302	625	1060	974	853	805	786	715	633	525	414	294	212	136	66	16	1			Spokane, Wash.
Springfield, Ill.					3	3	20	34	68	88	150	273	531	832	688	623	588	569	573	651	757	725	645	486	257	152	39	7	1	*	Springfield, Ill.
Springfield, Missouri						1	4	14	29	68	139	235	437	588	621	616	621	602	620	759	846	876	647	475	331	169	60	6	2	*	Springfield, Missouri
St. John, N.B.				2	14	42	97	128	201	264	335	427	669	986	784	692	796	941	1022	887	390	207	79	11	1						St. John, N.B.
Syracuse, N. Y.				1	1	7	10	59	79	141	241	370	564	800	752	708	669	709	753	742	752	604	406	255	109	28	4				Syracuse, N. Y.
Tallahassee, Florida												9	57	126	219	331	428	568	760	895	1061	1618	1275	739	530	149	4				Tallahassee, Florida
Tampa, Fla.													1	10	48	137	216	345	570	877	1187	1387	1910	1126	752	195	6				Tampa, Fla.
Toledo, Ohio						7	12	28	50	113	214	337	597	920	811	640	567	611	661	716	782	655	491	306	169	61	18	*			Toledo, Ohio
Topeka, Kan.						2	10	28	58	105	173	280	500	681	654	601	581	590	707	625	729	723	649	490	334	214	110	34	7		Topeka, Kan.
Toronto, Ont.					2	14	45	98	195	291	339	413	600	872	745	616	577	581	700	691	701	577	387	224	84	14					Toronto, Ont.
Tucson, Arizona												2	25	98	248	417	598	716	800	763	781	870	959	777	656	520	357	152	31	*	Tucson, Arizona
Tulsa, Oklahoma							1	2	12	29	75	159	265	438	535	611	637	622	636	671	752	838	816	649	481	333	149	43	10	1	Tulsa, Oklahoma
Vancouver, B.C.										5	13	21	110	386	753	1364	1618	1321	1217	936	559	310	108	18	3	1					Vancouver, B.C.
Waco, Texas										1	3	24	84	216	354	501	558	651	622	701	830	909	1101	829	612	482	249	42	1		Waco, Texas
Wake Island, Pacific																					5	621	3336	3944	863	1					Wake Island, Pacific
Washington D.C.										11	48	104	213	532	768	798	715	690	723	745	823	947	735	496	284	101	31	1			Washington D.C.
West Palm Beach, Fla.														2	18	57	115	202	291	455	835	1672	2413	1664	860	183	3				West Palm Beach, Fla.
Wichita Falls, Texas								*	4	10	27	114	228	388	497	581	636	627	606	677	714	784	825	741	560	406	260	83	2		Wichita Falls, Texas
Wichita, Kan.						1	3	22	47	78	140	236	349	549	614	618	658	669	670	669	757	768	673	503	341	213	105	47	11	2	Wichita, Kan.
Wilmington, Del.							*	1	9	39	107	200	369	682	816	752	708	668	731	721	804	849	630	390	209	66	16	2			Wilmington, Del.
Winnipeg, Man.	8	17	60	134	187	248	278	303	333	361	406	454	505	667	532	451	476	552	637	632	702	424	290	152	65	17	2				Winnipeg, Man.
Winston Salem, N.C.								1	4	18	43	116	245	456	622	708	704	710	694	801	907	1086	703	518	323	105	7				Winston Salem, N.C.
Yakima, Wash.				1	1	1	6	7	22	50	118	304	629	904	807	841	876	823	752	694	582	450	342	271	165	86	31	2			Yakima, Wash.
Youngstown, Ohio						3	10	31	63	158	242	420	645	840	681	645	620	664	698	773	821	641	421	266	102	21	2				Youngstown, Ohio
OUTDOOR TEMPERATURE (F)	-35 to -30	-30 to -25	-25 to -20	-20 to -15	-15 to -10	-10 to -5	-5 to 0	0 to 5	5 to 10	10 to 15	15 to 20	20 to 25	25 to 30	30 to 35	35 to 40	40 to 45	45 to 50	50 to 55	55 to 60	60 to 65	65 to 70	70 to 75	75 to 80	80 to 85	85 to 90	90 to 95	95 to 100	100 to 105	105 to 110	110 to 115	OUTDOOR TEMPERATURE (F)

*Less than one hour

AVERAGE NUMBER OF HOURS EACH TEMPERATURE SHOWN OCCURS IN A YEAR

Source of data: Lennox Ind.

Figure 14–12 (cont.) Bin weather data (Courtesy Air Conditioning Contractors of America)

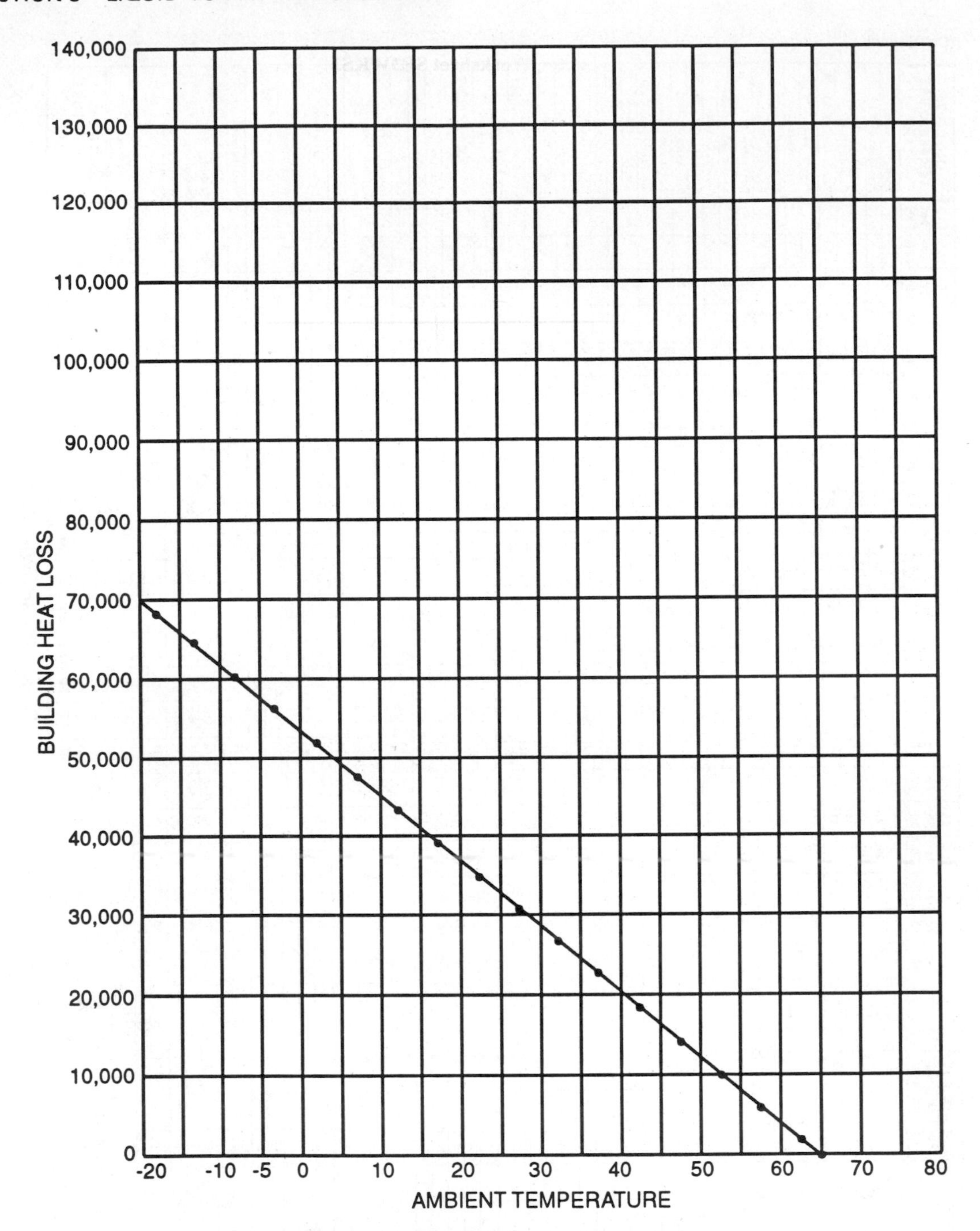

Figure 14–13 Building heat loss graph

Step 7. Add the number of weather data hours in the temperature bins involved (column A) and record the total in box E.

Step 8. Divide the total MBTUH in box D by the total number of hours in box E and record the results in box F. This will be the average MBTUH per hour. Only use the decimal figures to three places.

Step 9. Find the average BTUH heat loss per hour by multiplying the MBTUH figure in box F by 1,000. Record this average BTUH figure in box G.

Step 10. From the manufacturer's specifications literature covering the unit selected, determine the heating capacity of the unit for the EWTh that will be supplied by the ground loop system. See EWTh (Entering Water Temperature—Heating)

With an estimated 37°F EWTh, the unit must be capable of handling entering water temperatures below 45°F and an antifreeze solution must be used in the ground loop.

With an estimated cooling load of 19,996 BTUH, a WPH22-1H unit from the specifications in the table in Figure 14–15, would deliver 16,400 BTUH at 6 GPM, 16,800 BTUH at 8 GPM, or 17,100 BTUH at 10 GPM. For estimating purposes,

Run Factor Worksheet SE3WKS3

Name John H. Doe
Location Omaha, NB

Temperature Bin	(A) Weather Data (HR/YR)	(B) Loss/Gain (MBTUH PER HR)	(C) Total Heat Loss/Gain A × B (MBTUH)	(D) Calculations
−35 to −30				(D) Heating
−30 to −25				Total Heat Loss
−25 to −20				(in 1,000)
−20 to −15				129,171.1
−15 to −10	6	66.0	366.0	(E) Total Hours
−10 to −5	21	61.8	1297.8	Per Season
−5 to 0	54	57.5	3105.0	5,866
0 to +5	98	53.0	5194.0	(F) Avg. BTU/HR
+5 to +10	127	47.7	6057.9	(in 1,000)
+10 to +15	190	44.3	8417.0	22.020
+15 to +20	261	40.0	10440.0	(G) Avg. BTU/HR
+20 to +25	374	35.7	13351.8	22,020
+25 to +30	521	31.8	16567.8	(H) Unit Heating
+30 to +35	683	27.5	18782.5	Capacity (BTUH)
+35 to +40	672	23.5	15792.0	16,400
+40 to +45	584	19.0	11096.0	(I) Fh = 1.0
+45 to +50	567	14.8	8391.6	(J) Cooling
+50 to +55	527	10.2	9375.4	Total heat Gain
+55 to +60	544	6.2	3372.8	(in 1,000)
+60 to +65	637	1.0	637.0	20,956
+65 to +70	—	—	—	(K) Total Hours
+70 to +75	703	2.7	1898.1	Per Season
+75 to +80	579	7.5	4342.5	2,174
+80 to +85	453	12.5	5437.5	(L) Avg. BTU/HR
+85 to +90	263	17.4	4576.2	(in 1,000)
+90 to +95	134	22.3	2988.2	9.639
+95 to +100	45	27.3	1228.5	(M) Avg. BTU/HR
+100 to +105	14	32.0	448.0	9,639
+105 to +110	1	37.0	37.0	(N) Unit Cooling
+110 to +115				Capacity (BTUH)
+115 to +120				29,500
+120 to +125				(O) Fc 0.33

Figure 14–14 Worksheet

we will use the smaller 6 GPM with unit heating capacity at 16,400 BTUH. This unit capacity is recorded in box H.

Step 11. Determine the heating run factor by dividing the average heat loss per hour in BTUH (box G) by the unit heating capacity in BTUH. The result is the percentage of the time the unit will have to operate to handle the heating requirement. This percentage is recorded in box I.

Because the unit cannot operate more than 100% of the time, auxiliary heat has to be used. The heating run factor (Fh) for this application would be 1.0 (100%). If the average BTUH heating load is less than the unit heating capacity, the percentage figure will be less than 100% and a number smaller than 1.0 will be used. For example, calculating the Fh factor for the Phoenix, AZ application that has been previously discussed would produce an Fh factor of 0.29. The average BTUH heating load calculates out at 8,477 BTUH with a unit heating capacity of 29,700 BTUH. The 8,447 BTUH heating load divided by the 29,700 BTUH unit capacity results in a run time of 29% or an Fh of 0.29.

Fc (Cooling Run Factor)

The steps involved in calculating the Fc (cooling run factor) are the same as for calculating the Fh (heating run factor).

WPH 22-1H

EWT	GPM	BTUH TOTAL	WATTS	COP W/OUT PUMP	HEAT OF ABSORP.
	6	13500	1480	2.7	9300
25	8	13800	1490	2.7	9600
	10	14000	1500	2.7	9800
	6	15000	1550	2.8	10500
30	8	15300	1570	2.9	10800
	10	15500	1580	2.9	11000
	6	16400	1620	3.0	11800
35	8	16800	1640	3.0	12100
	10	17100	1660	3.0	12300
	6	18000	1700	3.1	13100
40	8	18400	1720	3.1	13500
	10	18700	1740	3.2	13700

WPH 44-1H

EWT	GPM	BTUH TOTAL	WATTS	COP W/OUT PUMP	HEAT OF ABSORP.
	8	24900	2540	2.8	17700
25	12	25800	2580	2.8	18500
	16	26300	2600	2.8	18900
	8	27700	2660	3.0	20100
30	12	28600	2700	3.0	21000
	16	29200	2720	3.1	21400
	8	30600	2790	3.2	22600
35	12	31700	2840	3.2	23600
	16	32300	2870	3.3	24100
	8	33300	2900	3.3	25000
40	12	34600	2960	3.4	26000
	16	35200	2990	3.4	26600

WPH 28-1H

EWT	GPM	BTUH TOTAL	WATTS	COP W/OUT PUMP	HEAT OF ABSORP.
	6	17000	1840	2.7	11700
25	8	17500	1860	2.7	12100
	10	17900	1880	2.8	12300
	6	18900	1920	2.9	13300
30	8	19400	1940	2.9	13700
	10	19800	1960	2.9	14000
	6	20800	2000	3.0	14900
35	8	21400	2030	3.1	15400
	10	21800	2050	3.1	15700
	6	22700	2090	3.2	16600
40	8	23300	2120	3.2	17100
	10	23700	2140	3.2	17400

WPH 54-1H

EWT	GPM	BTUH TOTAL	WATTS	COP W/OUT PUMP	HEAT OF ABSORP.
	10	30900	3250	2.8	20600
25	15	32000	3300	2.8	21600
	20	32700	3330	2.9	22200
	10	34200	3400	3.0	23500
30	15	35500	3460	3.0	24700
	20	36200	3500	3.0	25200
	10	37600	3550	3.1	26300
35	15	39100	3610	3.2	27600
	20	39800	3650	3.2	28300
	10	41100	3700	3.3	29300
40	15	42700	3780	3.3	30700
	20	43500	3820	3.4	31400

WPH 34-1H

EWT	GPM	BTUH TOTAL	WATTS	COP W/OUT PUMP	HEAT OF ABSORP.
	8	22900	2290	2.9	15900
25	10	23400	2310	3.0	16300
	12	23700	2330	3.0	16600
	8	25300	2400	3.1	17900
30	10	25800	2420	3.1	18300
	12	26200	2440	3.1	18600
	8	27800	2510	3.2	20000
35	10	28300	2540	3.3	20400
	12	28700	2550	3.3	20800
	8	30300	2630	3.4	22100
40	10	30800	2660	3.4	22600
	12	31200	2670	3.4	22900

WPH 64-1H

EWT	GPM	BTUH TOTAL	WATTS	COP W/OUT PUMP	HEAT OF ABSORP.
	10	38000	4510	2.5	24100
25	15	39400	4580	2.5	25200
	20	40200	4620	2.5	25900
	10	41900	4700	2.6	27300
30	15	43300	4770	2.7	28600
	20	44200	4820	2.7	29300
	10	45800	4880	2.8	30600
35	15	47400	4950	2.8	32000
	20	48400	5000	2.8	32800
	10	49700	5070	2.9	34064
40	15	51500	5150	2.9	35400
	20	52500	5200	3.0	36400

(1) Antifreeze solution must be used to protect down to 0°F, or at least to 20° below lowest entering water temperature.

(2) The fixed set point low-pressure control must be replaced with an adjustable low-pressure control.

(3) The above data takes in to account the loss in heat transfer due to the use of antifreeze solution in the loop.

Figure 14–15 Unit application to earth-coupled closed loop systems (Courtesy Addison Products Co.)

Step 1. Draw a heat gain line for the building using a graph paper similar to the one in Figure 14–16 for the application. Using the Omaha, NB installation, the heat gain line is based on 0 BTUH at +70°F outside temperature and 19,996 BTUH at the +91°F cooling outside design temperature. According to the bin weather data in Figure 14–12, the outdoor ambient temperature will reach the +100°F to +105°F bin temperature. Therefore, the heat gain line must reach the +105°F bin temperature line.

Step 2. Locate the points on the heat gain line at the midpoint temperature positions in the bin temperature ranges that correspond with the bin weather data in Figure 14–12.

Step 3. Using the lower portion of the worksheet in Figure 14–14, record the weather data in hours per year in column A for the temperature bins that apply to the Omaha, NB area. From the table in Figure 14–12, we see that the highest temperature involved in the Omaha, NB area is in the +105°F to +110°F bin—1hour. This is recorded in column A in the +105°F to +110°F bin. From this point, the average hours per year are recorded in each of the temperature bins until the +65°F to +70°F bin is reached. This bin is not used; neither heating nor cooling is normally required in this range.

Step 4. From the heat gain line in the graph in Figure 14–16, the building heat gain in MBTUH (1,000 BTUH) is determined from the scale on the left of the graph and recorded in column B of the worksheet (Figure 14–14) in the appropriate temperature bin. For example, the heat gain 105°F to 110°F (read at 107.5°F) is 36,700 BTUH. This is recorded as 36.7 MBTUH to reduce the size of the total numbers that have to be handled. At +100°F to +105°F (–102.5°F), the heat gain is

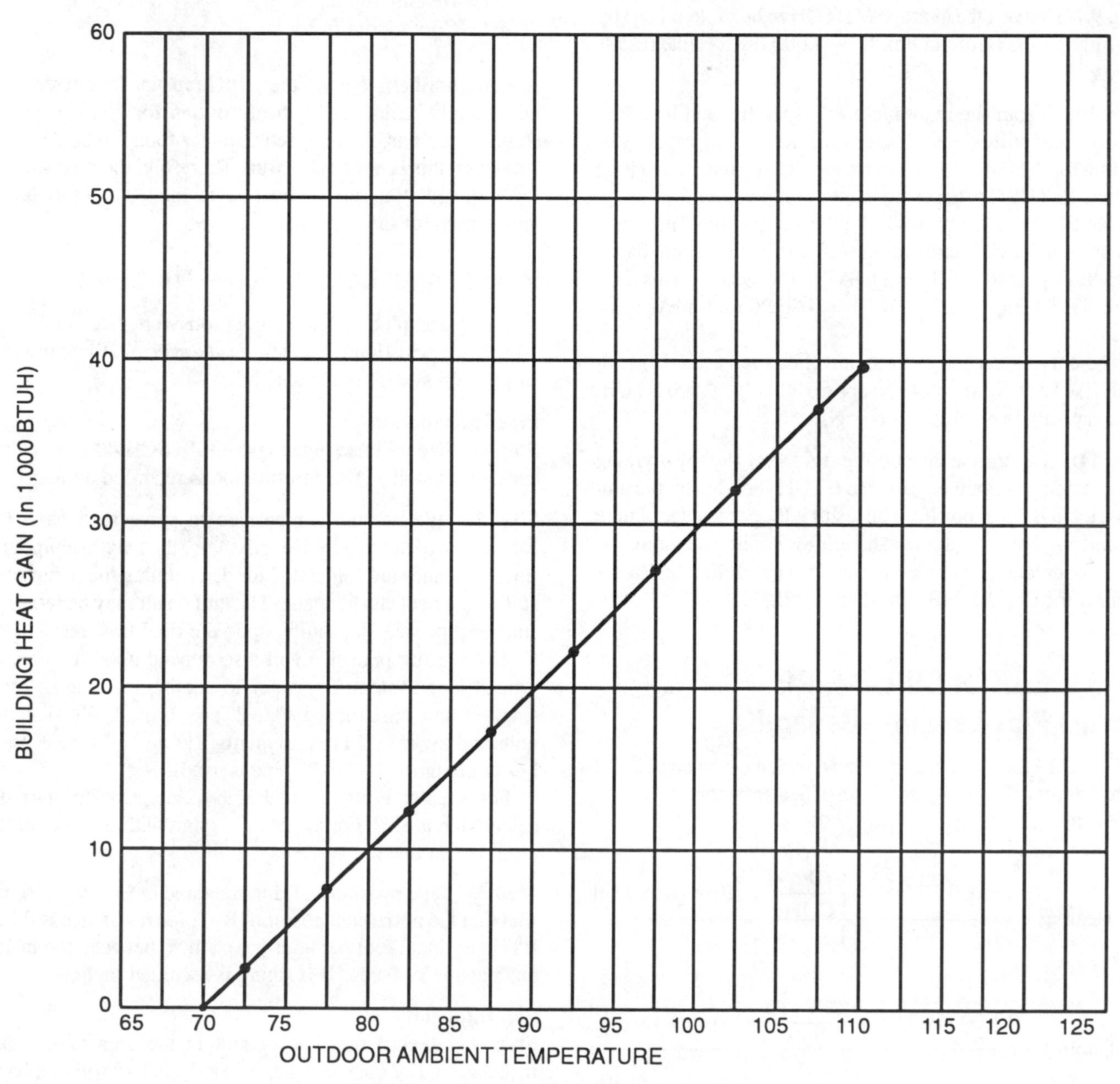

Figure 14–16 Building heat gain line graph

31.7, at +95°F to +100° F, 27.0, and so on until all the heat gain figures have been recorded in column B for each of the temperature bins involved.

Step 5. Find the total heat gain energy in MBTUH per year for each temperature bin by multiplying the hours in column A by the MBTUH in column B and recording the results in column C. This calculation is made for each of the temperature bins.

Step 6. Add the number of heat gain MBTUH in the temperature bins involved and record the total in box J.

Step 7. Add the number of weather data hours in the temperature bins involved and record the total in box K.

Step 8. Divide the total MBTUH in box J by the total number of hours in box K and record the results in box L. This is the average MBTUH.

Step 9. Convert the average MBTUH in box L to BTUH by multiplying the figure in box K by 1,000. Record the results in box M.

Step 10. From the manufacturer's specification literature covering the unit selected, determine the cooling capacity of the unit for the EWTc that will be supplied by the ground loop system. See EWTc, Entering Water Temperature—Cooling.

The EWTc to the unit will be the sum of the (Tm) mean annual earth temperature plus the (DEF) disturbed earth factor of 15°FΔT plus the 10°F design ΔT of the system. This adds up to 53°F (Tm) + 15°F (DEF) + 10°FΔT for an EWTc of 78°F.

The cooling capacity of the unit (See Figure 13–13) with 78°F EWTc at GPM would be approximately 29,500 BTUH. This capacity is recorded in box N.

Step 11. Determine the cooling run factor (Fc) by dividing the average heat gain per hour in BTUH (box M) by the unit capacity in BTUH (box N). The result is the percentage of time the unit will have to operate to handle the cooling requirement. This percentage is recorded in box O. The cooling run factor (Fc) for the Omaha, NB installation is 0.33.

14–3 EARTH COIL—PIPE

Design Pipe Length—General

Using the factors that have been covered in Chapter 13, the formulas for calculating the length of heat transfer pipe are the following:

$$\text{Lh Heating} = \frac{12{,}000\ \text{BTUH} \times \frac{\text{COPh - 1}}{\text{COPh}} \times [\text{Rp} + (\text{Rs} \times \text{Fh})]}{[\text{T}_\text{L} - \text{Tmin}]}$$

$$\text{Lc Cooling} = \frac{12{,}000\ \text{BTUH} \times \frac{\text{COPc} + 1}{\text{COPc}} \times [\text{Rp} + (\text{Rs} \times \text{Fc})]}{[\text{Tmax} - \text{Th}]}$$

where:

Lc = Length of pipe needed for each 12,000 BTUH cooling capacity
Lh = Length of pipe needed for each 12,000 BTUH heating capacity
COPh = Heating COP of unit selected
COPc = Cooling COP of unit selected
Rs = Soil resistance
Rp = Pipe resistance
T_l = Low soil temperature at low point day of the year—To
Th = High soil temperature at peak day of the year—To + 180 days
Tmin = The design minimum entering water temperature to the unit (EWTh)
Tmax = The design maximum entering water temperature to the unit (EWTc)
Fh = Heating run factor
Fc = Cooling run factor

In the northern regions, the earth loop pipe length required will usually be longer for heating than for cooling. In the southern regions, the reverse is usually found to be the case. However, this is not always true. Therefore, the pipe lengths for both operating modes must be calculated and the longer length used for the application.

Pipe Length Calculation—Northern

Using the pipe length calculation worksheet in Figure 14–17, we will figure the heat exchanger pipe length for a single loop system in Omaha, NB.

Pipe Information

Step 1. We will use pipe material PE SCH 40 (Polyethylene-Schedule 40). This information is recorded on line 1.

Step 2. We assume a pipe size to serve as a basis for calculation of the required length. After the length of pipe and flow resistance are calculated for determining the pump size, and adjustment can be made. The end result may be larger or smaller pipe size, depending upon the final flow resistance.

The starting pipe size will also depend upon the type of ground loop system. For a single continuous loop or for a single double back loop, 1¼" to 2" pipe is used. With smaller units and lower GPM requirements, 1¼" to 1½" pipe is used. For larger units, 1¾" to 2" pipe is used.

For smaller units, a good size assumption to start the calculation is 1½". For larger units, start with 2". Record the pipe size on line 2.

Step 3. Pipe resistance. From the table in Figure 14–6, we find that the horizontal position (Rp) pipe resistance is 0.117 BTU per lineal foot per hour per each °F between the inside and outside surfaces. This figure is recorded on line 3.

Soil Information

Step 4. Record the type of soil in the area of the unit installation. For our example, we find a heavy soil that has a

Pipe Length Calculation Worksheet SE3WKS-2
Horizontal Earth Coils
Data Collection

Job Name John H. Doe — Pipe Length—Cooling 410 Ft.
Location Omaha, NB — Pipe Length—Heating 527 Ft.

A. Pipe Information
1. Pipe Material Polyethylene-Schedule 40
2. Pipe Size 1½"
3. Pipe Resistance (Rp) 0.117 (Figure 14-6)

B. Soil Information
4. Soil Type Heavy Soil-Damp
5. No. of pipes in trench 1
6. Horizontal spacing between pipes 0
7. Soil Resistance (Rs) 0.98 (Figure 14-5)

C. Location Information
8. Mean earth temperature (Figure 14-1) Tm 53 °F
9. Soil Surface temperature (Figure 14-1) As 28 °F
10. Soil temperature variation (maximum 15°F) STV 10 °F
(If multiple pipes, average the temperatures at each pipe depth.)
11. Horizontal earth coil depth (Figure 14-3) D 8 ft.
(Average depth for multiple coils.)
12. High soil temperature (Tm + STV) Th 63 °F
13. Low soil temperature (Tm – STV) T_L 43 °F

D. Heat Pump Information
14. Highest EWTc temperature (Tmax) (Figure 12-1) 80 °F
15. Lowest EWTh temperature (Tmin) (no antifreeze) (Figure 12-1) 45 °F
16. Lowest EWTh temperature (Tmin) (20% antifreeze)(Figure 14-15) 25 °F
17. Unit cooling capacity (UC) at 6 GPM at Tmax 22,300 BTUH (Figure 12-1)
18. Unit heating capacity (UH) (no antifreeze) at 6 GPM at Tmin 19,700 BTUH (Figure 14-15)
19. Unit heating capacity (UH) (20% antifreeze) at 6 GPM at Tmin = 13,500 BTUH
20. Unit cooling COPc at Tmax 80 °F at 6 GPM 3.4 (Figure 12-1)
21. Unit heating COPh (no antifreeze) at Tmin 45 °F at 6 GPM 3.2 (Figure 12-1)
22. Unit heating COPh (20% antifreeze) at Tmin 25 °F at 6 GPM 2.7 (Figure 14-15)

E. Unit Run Factor
23. Cooling outside design temperature 91 °F
24. Cooling design heat gain 19,996 BTUH
25. Heating design outside temperature –3 °F
26. Heating design heat loss 53,702 BTUH
27. Cooling run factor (Fc) (Figure 14-14) 0.33
28. Heating run factor (Fh) (Figure 14-14) 1.00

F. Ground Loop Heat Exchanger Length (This does not include supply and return pipes or pipe headers.)
Cooling
Pipe Length/12,000 BTUH Cooling Capacity

29. $$\text{Pipe Length} = \frac{12{,}000\ \text{BTUH} \times \dfrac{\text{COP}+1}{\text{COP}} \times [\text{Rp} + (\text{Rs} \times \text{Fc})]}{(\text{Tmax} - \text{Th})}$$

Figure 14–17 Pipe length calculation worksheet

(Insert values from information section)

30. $$PLc = \frac{12{,}000\ BTUH \times \frac{3.4+1}{3.4} \times [0.117 + (0.98 \times 0.33)]}{(80-63)}$$

31. $$\text{Solve (A)} = \frac{COP+1}{COP} = \frac{3.4+1}{3.4} = \frac{4.4}{3.4} = 1.29$$

32. $$PLc = \frac{12{,}000\ BTUH \times (A)1.29 \times [0.117 + (0.98 \times 0.33)]}{(80-63)}$$

33. Solve (B) = (Rs × Fc) = (0.98 × 0.33) = 0.323

34. $$PLc = \frac{12{,}000\ BTUH \times (A)1.29 \times [0.117 + (B)0.323]}{(80-63)}$$

35. Solve (C) = [Rp + (B)] = (0.117 + (B)0.323] = 0.44

36. $$PLc = \frac{12{,}000\ BTUH \times (A)1.29 \times (C)0.44}{(80-63)}$$

37. Solve (D) = (Tmax – Th) = (80 – 63) = 17°F

38. $$PLc = \frac{12{,}000\ BTUH \times (A)1.29 \times (C)0.44}{(D)17} = 400.65\ Ft.$$

39. $$\text{Total } PLc = \frac{UC\ 22{,}300\ BTUH \times 400.65/12{,}000\ BTUH}{12{,}000} = 744.5\ Ft.$$

40. Cooling heat exchanger pipe length = 745 Ft.
Heating (No antifreeze)

41. $$\text{Pipe Length} = \frac{12{,}000\ BTUH \times \frac{COP-1}{COP} \times [Rp + (Rs \times Fh)]}{(T_L - Tmin)}$$

42. $$PLh = \frac{12{,}000\ BTUH \times \frac{3.2-1}{3.2} \times [0.117 + (0.98 \times 1.00)]}{(43-45)}$$

43. $$\text{Solve (A)} = \frac{COP-1}{COP} = \frac{3.2-1}{3.2} = \frac{2.2}{3.2} = 0.6875$$

44. $$PLh = \frac{12{,}000\ BTUH \times (A)\ 0.6875 \times [0.117(0.98 \times 1.00)]}{(43-45)}$$

45. Solve (B) = (Rs × Fh) = (0.98 × 1.00) = 0.98

46. $$PLh = \frac{12{,}000\ BTUH \times (A)0.6875 \times [0.117 + (B)0.98]}{(43-45)}$$

47. Solve (C) = [Rp +(B)] = [0.117 + 0.98] = 1.097

48. $$PLh = \frac{12{,}000\ BTUH \times (A)0.6875 \times (C)1.097}{(43-45)}$$

49. Solve (D) = (T_L – Tmin) = (43 – 45) = –2 (not usable)

50. $$PLh = \frac{12{,}000\ BTUH \times (A)---- \times (C)----}{(D)} = ----\ Ft.$$

51. $$\text{Total } PLh = \frac{UH----BTUH \times ---Ft./12{,}000\ BTUH}{12{,}000} = ----\ Ft.$$

Figure 14–17 (cont.) Pipe length calculation worksheet

Heading (20% antifreeze)

52. $$\text{Pipe Length} = \frac{12{,}000 \text{ BTUH} \times \frac{COP-1}{COP} \times [Rp + (Rs \times Fh)]}{(T_L - Tmin)}$$

53. $$PLh = \frac{12{,}000 \text{ BTUH} \times \frac{2.7-1}{2.7} \times [0.117(0.98 \times 1.00)]}{(43-25)}$$

54. $$\text{Solve (A)} = \frac{COP-1}{COP} = \frac{2.7-1}{2.7} = \frac{1.7}{2.7} = 0.63$$

55. $$PLh = \frac{12{,}000 \text{ BTUH} \times (A)0.63 \times [0.177 + (0.98 \times 1.00)]}{(43-25)}$$

56. Solve (B) = (Rs × Fh) = (0.98 × 1.00) = 0.98

57. $$PLh = \frac{12{,}000 \text{ BTUH} \times (A)0.63 \times [0.117 + (B)0.98]}{(43-25)}$$

58. Solve (C) = [Rp + (B)] = [0.117 + 0.98] = 1.097

59. $$PLh = \frac{12{,}000 \text{ BTUH} \times (A)0.63 \times (C)1.097}{(43-25)}$$

60. Solve (D) = $(T_L - Tmin)$ = (43 – 25) =18

61. $$PLh = \frac{12{,}000 \text{ BTUH} \times (A)0.64 \times (C)1.097}{(D)18} = 468 \text{ Ft.}/12{,}000 \text{ BTUH}$$

62. $$TPLh = \frac{UH13{,}500 \text{ BTUH} \times 468 \text{ Ft.}/12{,}000 \text{ BTUH}}{12{,}000} = 526.5 \text{ Ft.}$$

Heating heat exchanger pipe length is ___527___ Ft.

Figure 14–17 (cont.) Pipe length calculation worksheet

good water content (damp). This was discussed previously in this chapter.

Step 5. No. of pipes in the trench. In Figure 14–5, the effect on the soil resistance because of the number of pipes in the trench as well as the pipe arrangement, is shown in the columns for the various pipe sizes. For example, for 1½" pipe, the "heavy soil-damp" soil resistance is 0.89 to 0.98, depending upon the depth of the pipe below grade level. If two pipes are in the trench, the soil resistance is from 1.19 to 1.25.

Step 6. Four pipes in a vertical stack configuration increases the resistance to 1.92 and 2.02. Four pipes two layers of two pipes each one foot apart, has a soil resistance of 1.99. If the pipes in the horizontal layers are 2 feet apart, the soil resistance drops to 1.78.

These changes in soil resistance demonstrate the effect that the heat energy from each pipe has on the other pipes. Close proximity of the pipes requires a higher pipe temperature to transfer the given heat energy. This is reflected in the soil resistance factor.

Step 7. From the information determined in steps 4, 5, and 6, the soil resistance factor is determined from the table in Figure 14–5. For the Omaha, NB installation using 1½" diameter pipe in a single loop with one pipe in the trench at a depth of 6' in heavy, damp soil, the factor is 0.98. This is recorded on line 7 of the worksheet.

Location Information

To determine the earth temperatures that will be encountered during the heating and cooling operating modes, information about the earth conditions in the area of the installation are recorded in lines 8 through 13.

Step 8. Mean annual earth temperature (Tm). The mean annual earth temperature (Tm) can be assumed to equal the well water temperature from a well 50' to 150' deep. The Tm figure for various cities in the United States is found in the Tm column in Figure 14–1. For cities not listed, the city closest in location or matching weather conditions should be used. For example, for Omaha, NB, we would use the Tm figure of 53 for Lincoln, NB. This is recorded on line 8.

Step 9. Soil surface temperature (As). The annual soil surface temperature swing (As) is found in the second column of Figure 14–1. This is the temperature change that occurs at the surface of the earth throughout the year. The surface temperature swing can be plotted. For example, the curves in

Figure 14–2 illustrate a surface temperature swing of 44°F–22°F above the mean temperature during the summer to 22°F below the mean temperature in the winter. As the depth of the temperature test location increases, the capacitor effect of the earth material delays the thermal change and reduces the temperature swing.

The As factor is recorded on line 9. For the Omaha, NB area, this is 28°F.

Step 10. Soil temperature variation (STV). This is an assumed figure which is a balance between the allowable earth temperature and the depth of the pipe location which affects installation costs. A small pipe depth means a wider earth temperature swing (STV), which affects the unit performance. The result would be lower water temperatures in the winter and higher water temperatures in the summer. A larger pipe depth means closer temperature variation but more installation cost.

Experience limits the soil temperature variation (STV) to 15°F maximum, with a preferred allowance of 10°F. This allows the temperature surrounding the pipe to rise 10°F above the mean annual soil temperature (Tm) in the cooling mode and drop 10°F below the mean annual soil temperature (Tm) in the heating mode.

From the curves in Figure 14–3 for wet soil, average soil, and dry soil, we can determine the desired pipe depth to produce the desired soil temperature variation (STV).

For example, for the heavy, damp soil (wet soil) in the Omaha, NB area, to limit the STV to 10°F, the pipe depth must be 8'.

Light soil will require more pipe, but the pipe can be buried at a reduced depth. The objective is to balance the pipe quantity, pipe depth (installation cost), and unit performance for the soil and weather conditions encountered.

Step 11. Horizontal earth coil depth (D). From Figure 14–3, we have determined that the pipe will be buried to a depth (D) of 8'. If the installation requires two pipes in the same trench, the average depth will be 8'—one at 7' and one at 9', with a minimum of 2' between.

A four-pipe layer system would not be recommended as the top pipe would be 5' from the surface with the others spaced at 2' intervals—7', 9', and 11'. The deeper trench could raise installation costs. However, in small pipe field areas, this may be necessary.

The average pipe depth is recorded on line 11.

Step 12. High soil temperature (Tm + STV). The high soil temperature is equal to the mean annual soil temperature plus the design soil temperature variation (STV). In our example, this is a Tm of 53°F plus an STV of 10°F for a high soil temperature (Th) of 63°F. This is recorded on line 12.

Step 13. Low soil temperature (T_L). The low soil temperature (T_L) is equal to the mean annual soil temperature (Tm) minus the design soil temperature variation (STV). In our example, this is a Tm of 53°F minus an STV of 10°F for a low soil temperature of 43°F.

Heat Pump Information

Steps 14 through 22 are used to record the information from the manufacturer's specification literature for the heat pump unit selected for the application. The maximum and minimum water temperatures the unit will operate against satisfactorily as well as the water flow rate, whether or not antifreeze is required, and the unit performance characteristics are recorded.

Step 14. Highest cooling entering water temperature (EWTc). The unit highest cooling entering water temperature, classified as Tmax, is entered on line 14. For example, this is taken from Figure 12–1 for the WPH-22-1H unit and we are limiting it to 80°F. This is done because we, from experience, know that the water temperatures in the Omaha, NB area have a low summertime temperature. When we figure the unit for the higher water temperatures in the Phoenix, AZ area, we will have to use the 90°F rating.

The Tmax of 80°F is recorded on line 14.

Step 15. Lowest entering water temperature (EWTh) without antifreeze. With no antifreeze in the water system, the lowest water temperature the unit will tolerate is 45°F. (See Figure 12–1.) Therefore, the Tmin in this situation is 45°F. This is recorded on line 15.

Step 16. Lowest entering water temperature (EWTh) with 20% antifreeze mixture. Using a mixture of 20% antifreeze in the earth loop system, the unit will tolerate an EWTh to a Tmin of 25°F. This is recorded on line 16.

Step 17. Unit cooling capacity (UC). Using the Tmax temperature selected in step 14 and the unit capacity table in Figure 12–1, we find that we can get 21,900 BTUH capacity at 4 GPM, 22,300 BTUH at 6 GPM, and 22,400 BTUH at 8 GPM. The estimated cooling load is 19,996 BTUH.

Using a 10% tolerance, we need a capacity of 21,996 BTUH. To accomplish a capacity close to this without oversizing the pump, we will use the 6 GPM figure. The UC 22,300 BTUH is recorded on line 17.

Step 18. Unit heating capacity (no antifreeze) (UH). Using the 6 GPM selected for cooling at the 45°F Tmin, we determine that, from Figure 12–1, the unit will deliver a UH of 19,700 BTUH. This is recorded on line 18.

Step 19. Unit heat capacity (UH) (20% antifreeze). Using the 20% antifreeze solution, the unit will tolerate an EWTh of 25°F. With the 6 GMP flow rate (Figure 14–15), the unit will deliver 13,500 BTUH. This is recorded on line 19.

Step 20. Unit cooling COPc at Tmax. The COPc (coefficience of performance) of the unit selected from Figure 12–1, at the Tmax of 80°F with the 6 GPM water flowrate is 3.4.

The efficiency rating in the table is given as the EER rating (energy efficiency rating) of 11.6 without the water circulating pump. To convert the EER rating to COP rating, the EER rating is divided to 3.413. Therefore, the EER rating of 11.6 divided by 3.413 results in a COP rating of 3.4. This is recorded on line 20.

Step 21. Unit heating COPh rating (no antifreeze). The unit, not using antifreeze, requires a Tmin of 45°F. With a 6 GPM water flow rate, the unit has a COPh rating, without the circulating water pump, of 3.2 (see Figure 12–1). This is recorded on line 21.

Step 22. Unit heat COPh (20% antifreeze). The Tmin rating (minimum entering water temperature), using 20% antifreeze, is 25°F. With the 6 GPM flow rate, the COPh without the circulating pump is 2.7 (see Figure 14–15). This is recorded on line 22.

Unit Run Factor

The unit run factor, the percentage of time the unit can be expected to operate to handle the cooling or heating load, is calculated using the method outlined previously. The method uses the worksheet in Figure 14–14.

To obtain the information to use the run factor calculation method, the design temperatures and heat gain and loss figures are recorded on lines 23, 24, 25, and 26.

Step 23. Cooling outdoor design temperature. This was established when the ACCA manual J and J–1 worksheets were used to calculate the cooling gain. For the Omaha, NB area, the cooling outdoor design temperature was established at +91°F. This is recorded on line 23.

Step 24. Cooling design heat gain. The cooling design heat gain was determined to be 19,996 BTUH. This is recorded on line 24.

Step 25. Heating design outdoor temperature. For the Omaha, NB area, Manual J lists the heating outdoor design temperature as –3°F. This is recorded on line 25.

Step 26. Heating design heat loss. The heating design heat loss was determined to be 53,702 BTUH. It is recorded on line 26.

Step 27. Cooling run factor. Using the worksheet in Figure 14–14, the cooling run factor for the Omaha, NB application was determined to be 0.33. This is recorded on line 27.

Step 28. Heating run factor. Using the worksheet in Figure 14–14, the heating run factor for the Omaha, NB area was determined to be 1.35. The unit cannot run more than 100% of the time so a factor of 1.0 is used and recorded on line 28.

Ground Loop Heat Exchanger Length

The remaining steps are mathematical calculations to determine the amount of buried pipe needed to get the required heat transfer rate in both the heating and cooling operating modes. To determine if antifreeze solution is needed, the heating pipe length is calculated for each condition. Experience in a given location will tend to reduce these calculation steps. The first time through, however, each step must be taken.

The calculation method determines the amount of pipe needed for both heating and cooling based on 12,000 BTUH heat transfer. After this pipe length is determined, the total pipe length is adjusted according to the actual cooling and heating capacities of the unit.

Cooling—Pipe Length

Step 29. Calculation formula. Step 29 shows the overall formula used to calculate the pipe length for cooling (PLc) for each 12,000 BTUH of unit cooling capacity. Reviewing the formula shows the figures used with the various abbreviations that are to be used in this formula. There are several substeps in the calculations and each of these will be handled in turn.

Step 30. The formula with the actual values substituted for the abbreviations. To help keep the process in sequence, all values are carried through the various steps. These values from steps 1 through 28 are based on the Omaha, NB application.

Step 31. Solve (A). To reduce the complication of the numerator in the formula, solve the $\frac{COP+1}{COP}$ function. The result of this calculation becomes the figure for (A), 1.29.

Step 32. Substitute the figure for (A) in the formula.

Step 33. Solve (B). The figures (Rs × Fc) are multiplied to remove the parentheses. Substituting the figures for Rs (0.98) and Fc (0.33), these figures multiplied together result in the (B) figure of 0.323.

Step 34. Substitute the (B) figure of 0.323 in the formula instead of the (×) parentheses block.

Step 35. Solve (C). The figures for Rp and (B) are substituted for the [Rp + (B)]. This results in a calculation of (0.117 + 0.323), which adds up to a (C) of 0.44.

Step 36. The (C) result is placed in the formula in place of the [0.117 + (B) 0.323].

Step 37. Solve (D). The formula denominator has a set of parentheses enclosing the (Tmax – Th). The actual figures have been carried throughout the calculation steps. The calculation for (D) to remove the parentheses is Tmax – Th or 80°F – 63°F for a total of 17°F.

Step 38. Final formula. In step 38, we find the formula has been resolved to three figures in the numerator (above the line) and one in the denominator (below the line). Therefore, 12,000 BTUH times 1.29 times 0.44 results in 6811.2. This figure (the numerator) divided by 17 (the denominator) results in a pipe length per 12,000 BTUH cooling capacity of 400.65 feet.

Step 39. Total cooling pipe length. To determine the total pipe length required in the heat exchanger section of the buried pipe loop, we adjust the pipe length per 12,000 BTUH by the ratio of the unit cooling capacity compared to 12,000 BTUH. This ratio is the unit cooling capacity divided by 12,000 BTUH or a ratio of 23,300 divided by 12,000 or 1.8583. This ratio times the pipe length per 12,000 BTUH gives the total heat exchanger pipe length required.

Therefore, the unit cooling capacity (22,300 BTUH) times the pipe length per 12,000 BTUH (400.65) and divided by 12,000 BTUH gives a total heat exchanger pipe length for cooling of 745.5 ft.

Step 40. To reduce complications in the calculation process, use the closest whole number (746 ft.).

Heating (no antifreeze)
To calculate the heating heat exchanger pipe length, the same basic formula and steps to solve it are used.

Step 41. Pipe length formula—heating—no antifreeze. The items that change are the COP, T_L, and Tmin. These are recorded in lines 21, 13, and 15.

Step 42. Pipe length formula—heating—no antifreeze. Substitute the actual factors for the abbreviations.

Step 43. Solve (A). The heating COP at 45°F EWT with 6 GPM (3.2) minus 1 divided by the heating COP (3.2) equals 0.6875. This is recorded on line 43.

Step 44. Insert the factor found in step 43 into the formula in the (A) space.

Step 45. Solve (B). The figures for Rs and Fh are multiplied to remove the parentheses. In our example, the Rs factor of 0.98 and the Fh factor of 1.00 are multiplied to get the (B) factor of 0.98. This is recorded on line 45.

Step 46. Substitute the (B) factor into the basic formula to remove the parentheses.

Step 47. Solve (C). The Rp factor of 0.117 and the (B) factor of 0.98 are added together to get the (C) factor of 1.097. This is recorded on line 47.

Step 48. Substitute the (C) factor for the Rp + (B) and remove the brackets. This is recorded on line 48.

Step 49. Solve the (D). The minimum entering water temperature to the unit, EWTc (Tmin), is subtracted from the low water temperature of the system (T_L). In this Omaha, NB application with a Tmin of 45°F and a T_L if 43°F, the result is –2°F. This is not operationally possible. Therefore, the unit that has a 45°F minimum EWTc cannot be used in this application. In this application, steps 50 and 51 cannot be used.

Heating (20% Antifreeze)
The unit that can be used at water temperatures (T_L) below 45°F, will have different Tmin, UH, and COP factors. These have been recorded on lines 16, 18, and 22. Using the same step-by-step procedure as in steps 41 to 51 for heating—no antifreeze, the steps from 52 to 62 are done using the appropriate factors.

Step 52. The basic formula.

Step 53. Substitute the numerical values of each of the abbreviations.

Step 54. Solve for (A).

Step 55. Substitute the value for (A) into the formula to remove the fraction in the numerator of the formula.

Step 56. Solve for (B).

Step 57. Substitute the value for (B) into the formula to remove the parentheses.

Step 58. Solve for (C).

Step 59. Insert the value found for (C) into the formula to remove the brackets.

Step 60. Solve for (D). In our example, we find that with a unit low entering water temperature of 25°F (Tmin) and the unit entering low water temperature (T_L) of 43°F, we will have an 18°F difference in temperature between the soil and the 20% antifreeze mixture for heat transfer. This value is recorded on line 60.

Step 61. With all factors substituted, the formula can be solved. 12,000 BTUH times (A) 0.64 times (C) 1.097 divided by (D) 18 equals 468 Ft. of pipe per 12,000 BTUH of unit capacity.

Step 62. To find the total pipe length needed (TPLH) for heating, the length per 12,000 BTUH is adjusted by the ratio of 12,000 BTUH to the heating capacity of the unit at the EWTh and GPM. In this application, it is the unit heating capacity (UH) at 25°F EWT and 6 GPM times 468 Ft. per 12,000 BTUH divided by 12,000 BTUH. This determines that the total pipe length of the heat exchanger for the heating operation (TPLH) is 526.5 ft. The heat exchanger pipe length is 527 ft.

The longer of the pipe lengths for cooling or heating is used in the application. In the Omaha, NB application, the pipe length would be a TPLh of 527 ft.

Pipe Length Calculation—Southern

Using the same worksheet (Figure 14–17) to calculate the ground loop heat exchange pipe length, we find that the depth the pipe is buried has a great effect on the amount of pipe needed for the necessary heat rejection/absorption capacity. Figure 14–18 shows the calculation steps needed to determine the TPLc and TPLh in the Phoenix, AZ area. For a comparison between the Omaha, NB and Phoenix, AZ applications, the same pipe material and size are used in both places.

Other factors, such as ground temperature, pipe depth, and so on will change. With the 73°F mean annual temperature (Tm) in the Phoenix, AZ area (Figure 14–1), we know that the 20% antifreeze solution will not be needed. 73°F less the maximum soil temperature variation of 15°F means a minimum soil temperature of 58°F. This puts the lowest EWTh above the 36°F limit. Therefore, the antifreeze calculations are not needed.

Pipe Information

Step 1. Polyethylene schedule 40 pipe will be installed.

Step 2. Pipe size— 1½".

Step 3. Pipe resistance (Rp)—0.117 (Figure 14–6).

Pipe Length Calculation Worksheet SE3WKS-2
Horizontal Earth Coils
Data Collection

Job Name John H. Doe Pipe Length—Cooling 2,842 Ft.
Location Phoenix, AZ Pipe Length—Heating 393 Ft.

A. Pipe Information
1. Pipe Material Polyethylene-Schedule 40
2. Pipe Size 1½"
3. Pipe Resistance (Rp) 0.117 (Figure 14-6)

B. Soil Information
4. Soil Type Heavy Soil-Dry (Average)
5. No. of pipes in trench 1
6. Horizontal spacing between pipes 0
7. Soil Resistance (Rs) 1.21 (Figure 14-5)

C. Location Information
8. Mean earth temperature (Figure 14-1) Tm 73 °F
9. Soil Surface temperature (Figure 14-1) As 23 °F
10. Soil temperature variation (maximum 15°F) STV 8 °F
(If multiple pipes, average the temperatures at each pipe depth.)
11. Horizontal earth coil depth (Figure 14-3) D 8 Ft.
(Average depth for multiple coils.)
12. High soil temperature (Tm + STV) Th 81 °F
13. Low soil temperature (Tm – STV) T_L 65 °F

D. Heat Pump Information
14. Highest EWTc temperature (Tmax) (Figure 12-1) 90 °F
15. Lowest EWTh temperature (Tmin) (no antifreeze) (Figure 12-1) 45 °F
16. Lowest EWTh temperature (Tmin) (20% antifreeze)(Figure 14-15) 25 °F
17. Unit cooling capacity (UC) at 6 GPM at Tmax 27,000 BTUH (Figure 12-1)
18. Unit heating capacity (UH) (no antifreeze) at 6 GPM at Tmin 22,900 BTUH (Figure 14-15)
19. Unit heating capacity (UH) (20% antifreeze) at 6 GPM at Tmin = ---- BTUH
20. Unit cooling COPc at Tmax 90 °F at 6 GPM 3.05 (Figure 12-1)
21. Unit heating COPh (no antifreeze) at Tmin 45 °F at 6 GPM 3.15 (Figure 12-1)
22. Unit heating COPh (20% antifreeze) at Tmin 25 °F at 6 GPM ---- (Figure 14-15)

E. Unit Run Factor
23. Cooling outside design temperature 104 °F
24. Cooling design heat gain 27,265 BTUH
25. Heating design outside temperature 34 °F
26. Heating design heat loss 19,996 BTUH
27. Cooling run factor (Fc) (Figures 14-19 and 14–20) 0.46
28. Heating run factor (Fh) (Figures 14-19 and 14–20) 0.32

F. Ground Loop Heat Exchanger Length (This does not include supply and return pipes or pipe headers.)
Cooling
Pipe Length/12,000 BTUH Cooling Capacity

29. Pipe Length $= \dfrac{12,000 \text{ BTUH} \times \dfrac{\text{COP}+1}{\text{COP}} \times [\text{Rp} + (\text{Rs} \times \text{Fc})]}{(\text{Tmax} - \text{Th})}$

Figure 14–18 Pipe length calculation worksheet

(Insert values from information section)

30. $PLc = \dfrac{12{,}000\ BTUH \times \dfrac{3.05+1}{3.05} \times [0.117 + (1.21 \times 0.46)]}{(90-81)}$

31. Solve $(A) = \dfrac{COP+1}{COP} = \dfrac{3.05+1}{3.05} = \dfrac{4.05}{3.05} = 1.328$

32. $PLc = \dfrac{12{,}000\ BTUH \times (A)1.328 \times [0.117 + (1.21 \times 0.46)]}{(90-81)}$

33. Solve (B) = (Rs × Fc) = (1.21 × 0.46) = 0.557

34. $PLc = \dfrac{12{,}000\ BTUH \times (A)1.21 \times [0.117 + (B)0.557]}{(90-81)}$

35. Solve (C) = [Rp + (B)] = [0.117 + 0.557] = 0.674

36. $PLc = \dfrac{12{,}000\ BTUH \times (A)1.328 \times (C)0.674}{(90-81)}$

37. Solve (D) = (Tmax – Th) = (90 – 81) = 9°F

38. $PLc = \dfrac{12{,}000\ BTUH \times (A)1.328 \times (C)0.674}{(D)9} = 1{,}195\ Ft.$

39. $Total\ PLc = \dfrac{UC\ 22{,}300\ BTUH \times 1{,}195/12{,}000\ BTUH}{12{,}000} = 2{,}738.5\ Ft.$

40. Cooling heat exchanger pipe length = 2,739 Ft.
Heating (No antifreeze)

41. $Pipe\ Length = \dfrac{12{,}000\ BTUH \times \dfrac{COP-1}{COP} \times [Rp + (Rs \times Fh)]}{(T_L - Tmin)}$

42. $PLh = \dfrac{12{,}000\ BTUH \times \dfrac{3.15-1}{3.15} \times [0.117(1.21 \times 0.32)]}{(65-45)}$

43. Solve $(A) = \dfrac{COP-1}{COP} = \dfrac{3.15-1}{3.15} = \dfrac{2.15}{3.15} = 0.682$

44. $PLh = \dfrac{12{,}000\ BTUH \times (A)\ 0.682 \times [0.117(1.21 \times 0.32)]}{(65-45)}$

45. Solve (B) = (Rs × Fh) = (1.21 × 0.32) = 0.3872

46. $PLh = \dfrac{12{,}000\ BTUH \times (A)0.682 \times [0.117 + (B)0.3872]}{(65-45)}$

47. Solve (C) = [Rp +(B)] = [0.117 + 0.3872] = 0.5042

48. $PLh = \dfrac{12{,}000\ BTUH \times (A)0.682 \times (C)0.5042}{(65-45)}$

49. Solve (D) = (T_L – Tmin) = (65 – 45) = 20

50. $PLh = \dfrac{12{,}000\ BTUH \times (A)0.682 \times (C)0.5042}{(D)20} = 206\ Ft.$

51. $Total\ PLh = \dfrac{UH\ 22{,}900\ BTUH \times 206\ Ft/12{,}000\ BTUH}{12{,}000} = 393\ Ft.$

Figure 14–18 (cont.) Pipe length calculation worksheet

Soil Information

Step 4. Heavy soil-dry. This could be classified as average soil to determine the Tm at various depths. (Figure 14–3).

Step 5. The same type loop system figured for the Omaha, NB application will be figured—a single loop system with one pipe in the trench.

Step 6. Single pipe—no horizontal spacing.

Step 7. Soil Resistance (Rs)—1.21 (Figure 14–5).

Location Information

Step 8. The mean annual earth temperature (Tm) (Figure 14–1) is 73°F.

Step 9. The soil surface temperature swing (As) in the Phoenix, AZ area is 23°F. (Figure 14–1).

Step 10. Soil temperature variation (STV). This factor has the greatest effect upon the cost of the installation. To reduce the amount of pipe involved, a smaller STV is used. However, the smaller STV means a deeper trench for the pipe location. The cost of more pipe in a shallower trench versus less pipe in a deeper trench must be considered.

Experience shows that an 8' depth of pipe is the most practical if a horizontal ground loop system is to be used. In this calculation, we will use the 8' depth. This will result in an STV of 8°F for the average soil (Figure 14–3).

Step 11. Horizontal earth coil depth. We have selected 8' as an average depth.

Step 12. High soil temperature (Th). Tm + STV = 73°F + 8°F = 81°F.

Step 13. Low soil temperature (T_L). Tm – STV = 73°F – 8°F = 65°F.

Heat Pump Information

Step 14. The highest EWTc temperature (Tmax) (Figure 13–13) from the manufacturer's specification literature is 90°F.

Step 15. The lowest EWTh temperature (Tmin) (no antifreeze) (Figure 13–13) from the manufacturer's specification literature is 45°F.

Step 16. The lowest EWTh temperature (Tmin) (20% antifreeze solution) (Figure 14–15) from the manufacturer's specification literature is 25°F.

Step 17. Unit cooling capacity (UC) with 6 GMP flow rate at Tmax of 85°F—27,500 BTUH.

Step 18. Unit heating capacity (UH) (no antifreeze) at 6 GPM flow rate at Tmin of 45°F—19,700 BTUH(see Figure 13–13).

Step 19. Unit heating capacity (UH) at Tmin of 25°F—not used.

Step 20. Unit cooling coefficience of performance (COPc) at Tmax of 90°F at 6 GPM flow rate—3.05 (see Figure 13–13). 10.4 EER divided by 3.413 BTU/Watt = 3.05.

Step 21. Unit heating coefficience of performance (COPh) (no antifreeze) at Tmin of 45°F at 6 GPM flow rate—3.15.

Step 22. Unit heating coefficience of performance (COPh) (20% antifreeze)—not used.

Unit Run Factor

Using the worksheet shown in Figure 14–14, the cooling and heating run factors have been worked out in Figures 14–19 and 14–20.

Step 23. Cooling outdoor design temperature—from the ACCA Environmental Systems Library, Manual J, 104°F.

Step 24. Cooling design heat gain—from the calculation method in Manual J, 27,265 BTUH.

Step 25. Heating design outdoor temperature—from ACCA Manual J, 34°F.

Step 26. Heating design heat loss—from the calculation method in ACCA Manual J—19,996 BTUH.

Step 27. Cooling run factor (Fc) (Figures 14–19 and 14–20) —0.46.

Step 28. Heating run factor (Fh) (Figures 14–19 and 14–20) —0.32

Ground loop heat exchanger length.

Cooling

Step 29. Pipe length (PLc)—cooling/12,000 BTUH basic formula.

Step 30. Basic formula with all numerical values in place.

Step 31. Solve for (A)—1.328.

Step 32. Basic formula with (A) inserted.

Step 33. Solve for (B)—0.557

Step 34. Basic formula with (A) and (B) inserted.

Step 35. Solve for (C)—0.674

Step 36. Basic formula with (A) and (C) inserted.

Step 37. Solve for (D)—9.

Step 38. Basic formula with final (A), (C), and (D) values inserted. Performing the mathematic calculations reveals the PLc per 12,000 BTUH is 1,195 ft.

Step 39. Adjusting the PLc per 12,000 BTUH to TPLc for 27,500 BTUH = 2,738.5 ft.

Step 40. TPLc for cooling is a minimum of 2,739 ft.

Heating (No Antifreeze)

Step 41. PLh/12,000 BTUH basic formula.

Step 42. Basic formula with values inserted.

Step 43. Solve for (A)—0.682.

Step 44. Basic formula with value for (A) inserted.

Step 45. Solve for (B)—0.3872.

Step 46. Basic formula with values for (A) and (B) inserted.

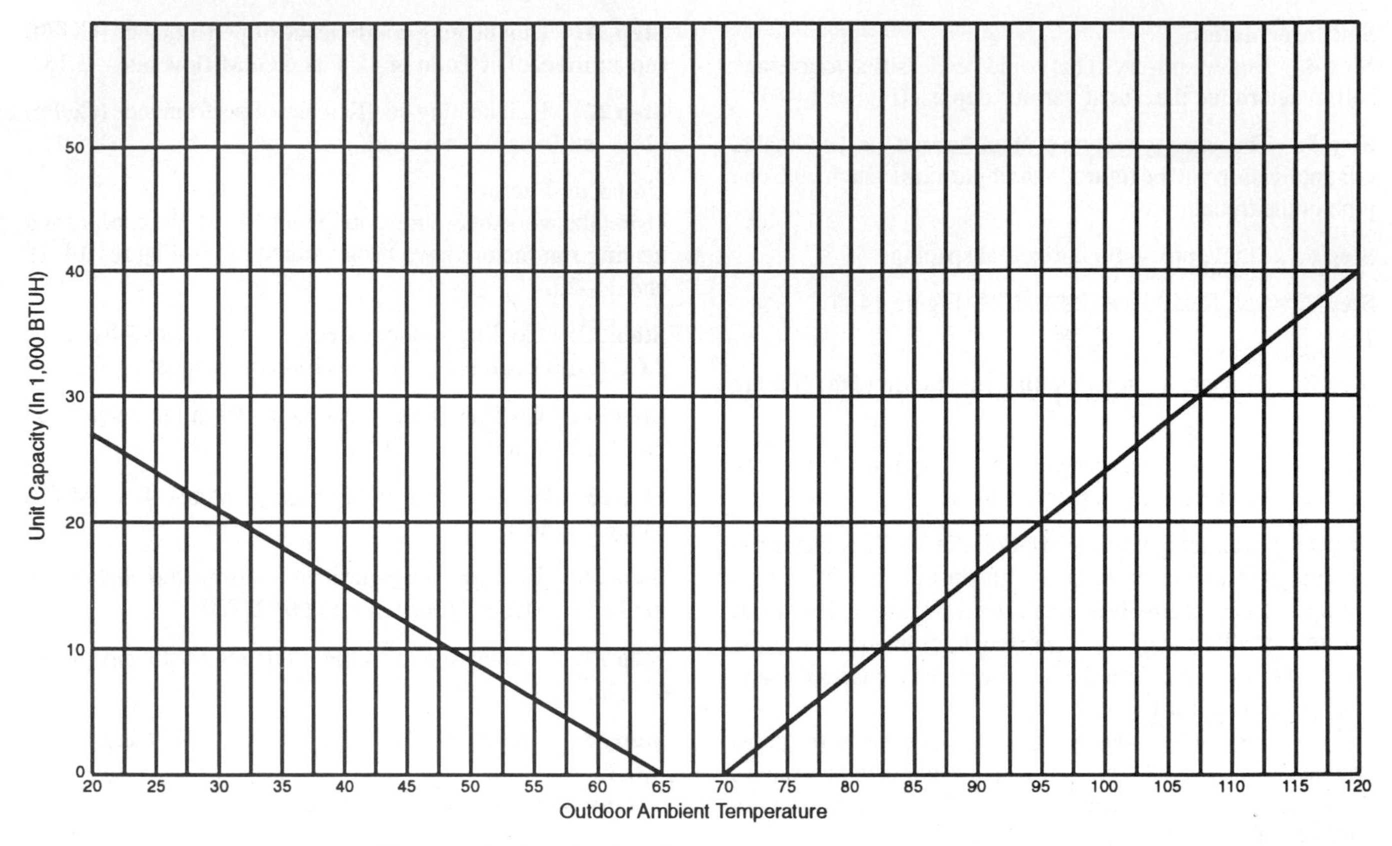

Figure 14–19 Heat loss/heat gain curve graph—southern

Step 47. Solve for (C)—0.504.

Step 48. Basic formula with values for (A) and (C) inserted.

Step 49. Solve for (D)—20.

Step 50. Basic formula with values for (A), (C), and (D) inserted. Solving the formula results in 206 ft. of pipe per 12,000 BTUH heating load.

Step 51. Adjusting the 206 ft. of pipe per 12,000 BTUH to the unit capacity of 22,900 BTUH results in a heat exchanger pipe length requirement of 393 ft.

Because the weather and ground temperatures are high enough to eliminate the need for antifreeze solution, steps 52 through 62 are not used.

The longer of the heating and cooling pipe lengths is used for the application. In this example, the cooling TPLc of 2,739' is used. The final decision must be based on the per foot cost of deep trenching with less trench and pipe distance versus shallow trenching with more trench and pipe distance. This usually works out to a balance of an 8' deep trench.

14–4 MULTIPLE PIPE SYSTEM—PARALLEL SYSTEM

To reduce the amount of trenching required, the pipes can be arranged in two horizontal layers using four pipes or two U-bend pipes. Each U-bend pipe will have the flow enter through the top pipe and leave through the bottom pipe. This is to get counterflow heat exchange between the fluid in the pipe and the earth material. The warmer liquid will meet the warmer earth material for as close to even ΔT°F as possible.

The liquid flow will be in two parallel circuits 2' apart. To calculate the pipe and trench lengths, the worksheet in Figure 14–21 is used. The cooling and heating run factors have already been calculated using the heat loss/gain curves in Figure 14–19 and the worksheet in Figure 14–20.

We will use the same earth temperatures and materials that were used to calculate the single loop system. The average depth of the pipes will be 8' for direct comparison to the single loop system.

It is not necessary to use as large a pipe, however, as only 50% of the liquid will flow through each pipe. The recommended pipe size for this system is ¾" to 1". We will use 1" pipe at an average depth of 8'.

Pipe Information

Step 1. Pipe material is Polyethylene Schedule 40.

Step 2. Pipe size is 1".

Step 3. Pipe resistance (Rp) is 0.159 (Figure 14–6).

Soil Information

Step 4. Soil type is Heavy Soil—Dry (Average).

Step 5. This installation will have four pipes in the same trench in two vertical loop circuits.

Run Factor Worksheet SE3WKS3

Name John H. Doe
Location Omaha, NB

Temperature Bin	(A) Weather Data (HR/YR)	(B) Loss/Gain (MBTUH Per HR)	(C) Total Heat A × B (MBTUH)	Calculations
−35 to −30				(D) Heating
−30 to −25				Total Heat Loss
−25 to −20				(in 1,000)
−20 to −15				24,876.6
−15 to −10				(E) Total Hours
−10 to −5				Per Season
−5 to 0				3,373
0 to +5				(F) Avg. BTU/HR
+5 to +10				(in 1,000)
+10 to +15				7.373
+15 to +20				(G) Avg. BTU/HR
+20 to +25				7,373
+25 to +30	8	22.5	180	(H) Unit Heating
+30 to +35	57	19.9	1,134.5	Capacity (BTUH)
+35 to +40	182	16.6	3,021.2	22,900
+40 to +45	391	13.6	5,317.6	(I) Fh = 0.32
+45 to +50	540	10.5	5,670.0	(J) Cooling
+50 to +55	659	7.5	4,942.5	Total heat Gain
+55 to +60	769	4.5	3,460.5	(in 1,000)
+60 to +65	767	1.5	1,150.5	58,914
+65 to +70	—	—	—	(K) Total Hours
+70 to +75	762	1.5	1,150.5	Per Season
+75 to +80	774	6.0	1,524.0	4,619
+80 to +85	798	10.0	7,980.0	(L) Avg. BTU/HR
+85 to +90	698	14.0	4,619.0	(in 1,000)
+90 to +95	611	18.0	10,998.0	12.755
+95 to +100	507	22.0	11,154.0	(M) Avg. BTU/HR
+100 to +105	324	26.0	8,424.0	12,755
+105 to +110	129	30.0	3,870.0	(N) Unit Cooling
+110 to +115	15	34.0	510.0	Capacity (BTUH)
+115 to +120	1	38.0	38.0	27,500
+120 to +125	—	—	—	(O) Fc 0.46

Figure 14–20 Fh and Fc run factor worksheet—southern

Step 6. The horizontal spacing between the loops is 2'.

Step 7. Soil resistance for this pipe arrangement is 2.47 (Figure 14–5).

Location Information

Step 8. Mean earth temperature (Tm) is 73°F (Figure 14–1).

Step 9. Soil temperature swing (As) is 23°F (Figure 14–1).

Step 10. Soil temperature variation (STV) is 8°F.

Step 11. Horizontal earth coil depth (D) (Average) is 8'.

Step 12. High soil temperature (Th) (Tm + STV) is 81°F.

Step 13. Low soil temperature (T_L) (TM – STV) is 65°F.

Heat Pump Information

Step 14. Highest EWTc temperature (Tmax) is 90°F.

Step 15. Lowest EWTh temperature (Tmin) (no antifreeze) is 45°F.

Step 16. Lowest EWTh temperature (Tmin) (20% antifreeze) is 25°F.

Step 17. Unit cooling capacity (UC) at 6 GPM at Tmax is 27,500 BTUH.

Step 18. Unit heating capacity (UH) (no antifreeze) at 6 GPM at Tmin is 22,900 BTUH.

Step 19. No antifreeze is required.

Pipe Length Calculation Worksheet SE3WKS-2
Horizontal Earth Coils
Data Collection

Job Name ___John H. Doe___ Pipe Length—Cooling ___11,064___ Ft.
Location ___Phoenix, AZ___ Pipe Length—Heating ___584___ Ft.

A. Pipe Information
1. Pipe Material ___Polyethylene-Schedule 40___
2. Pipe Size ___1"___
3. Pipe Resistance (Rp) ___0.159 (Figure 14-6)___

B. Soil Information
4. Soil Type ___Heavy Soil-Dry (Average)___
5. No. of pipes in trench ___4___
6. Horizontal spacing between pipes ___2___ ft
7. Soil Resistance (Rs) ___2.47___ (Figure 14-5)

C. Location Information
8. Mean earth temperature (Figure 14-1) Tm ___73___ °F
9. Soil Surface temperature (Figure 14-1) As ___23___ °F
10. Soil temperature variation (maximum 15°F) STV ___8___ °F
(If multiple pipes, average the temperatures at each pipe depth.)
11. Horizontal earth coil depth (Figure 14-3) D ___8___ ft.
(Average depth for multiple coils.)
12. High soil temperature (Tm + STV) Th ___81___ °F
13. Low soil temperature (Tm – STV) T_L ___65___ °F

D. Heat Pump Information
14. Highest EWTc temperature (Tmax) (Figure 12-1) ___90___ °F
15. Lowest EWTh temperature (Tmin) (no antifreeze) (Figure 12-1) ___45___ °F
16. Lowest EWTh temperature (Tmin) (20% antifreeze)(Figure 14-15) ___25___ °F
17. Unit cooling capacity (UC) at ___6___ GPM at Tmax ___27,500___ BTUH (Figure 12-1)
18. Unit heating capacity (UH) (no antifreeze) at ___6___ GPM at Tmin ___22,900___ BTUH (Figure 14-15)
19. Unit heating capacity (UH) (20% antifreeze) at ___---___ GPM at Tmin = ___----___ BTUH
20. Unit cooling COPc at Tmax ___90___ °F at ___6___ GPM ___3.05___ (Figure 12-1)
21. Unit heating COPh (no antifreeze) at Tmin ___+45___ °F at ___6___ GPM ___3.15___ (Figure 12-1)
22. Unit heating COPh (20% antifreeze) at Tmin ___+25___ °F at ___6___ GPM ___----___ (Figure 14-15)

E. Unit Run Factor
23. Cooling outside design temperature ___104___ °F
24. Cooling design heat gain ___27,265___ BTUH
25. Heating design outside temperature ___34___ °F
26. Heating design heat loss ___19,996___ BTUH
27. Cooling run factor (Fc) (Figure 14–20) ___0.46___
28. Heating run factor (Fh) (Figure 14–20) ___0.32___

F. Ground Loop Heat Exchanger Length (This does not include supply and return pipes or pipe headers.)
Cooling
Pipe Length/12,000 BTUH Cooling Capacity

29. $$\text{Pipe Length} = \frac{12{,}000\ \text{BTUH} \times \frac{\text{COP}+1}{\text{COP}} \times [\text{Rp} + (\text{Rs} \times \text{Fc})]}{(\text{Tmax} - \text{Th})}$$

Figure 14–21 Pipe length calculation for parallel coils—southern

(Insert values from information section)

30. $$\text{PLc} = \frac{12,000\ \text{BTUH} \times \frac{3.05+1}{3.05} \times [0.159 + (2.47 \times 0.46)]}{(90-81)}$$

31. $$\text{Solve (A)} = \frac{\text{COP}+1}{\text{COP}} = \frac{3.05+1}{3.05} = \frac{4.05}{3.05} = 1.328$$

32. $$\text{PLc} = \frac{12,000\ \text{BTUH} \times (\text{A})1.328 \times [0.159 + (2.47 \times 0.46)]}{(90-81)}$$

33. Solve (B) = (Rs × Fc) = (2.47 × 0.46) = 1.136

34. $$\text{PLc} = \frac{12,000\ \text{BTUH} \times (\text{A})1.328 \times [0.159 + (\text{B})1.136]}{(90-81)}$$

35. Solve (C) = [Rp + (B)] = [0.159 + 1.136] = 2.726

36. $$\text{PLc} = \frac{12,000\ \text{BTUH} \times (\text{A})1.328 \times (\text{C})2.726}{(90-81)}$$

37. Solve (D) = (Tmax – Th) = (90 – 81) = 9°F

38. $$\text{PLc} = \frac{12,000\ \text{BTUH} \times (\text{A})1.328 \times (\text{C})2.726}{(\text{D})9} = 4,826\ \text{Ft.}$$

39. $$\text{Total PLc} = \frac{\text{UC } 27,500\ \text{BTUH} \times 4,826/12,000\ \text{BTUH}}{12,000} = 11,061.5\ \text{Ft.}$$

40. Cooling heat exchanger pipe length = 11,062 Ft.
Maximum loop length – 750' = 8 loops of 692 Ft. each.
Heating (No antifreeze)

41. $$\text{Pipe Length} = \frac{12,000\ \text{BTUH} \times \frac{\text{COP}-1}{\text{COP}} \times [\text{Rp} + (\text{Rs} \times \text{Fh})]}{(T_L - \text{Tmin})}$$

42. $$\text{PLh} = \frac{12,000\ \text{BTUH} \times \frac{3.15-1}{3.15} \times [0.159 + (1.84 \times 0.32)]}{(65-45)}$$

43. $$\text{Solve (A)} = \frac{\text{COP}-1}{\text{COP}} = \frac{3.15-1}{3.15} = \frac{2.15}{3.15} = 0.682$$

44. $$\text{PLh} = \frac{12,000\ \text{BTUH} \times (\text{A})\ 0.682 \times [0.159 + (1.84 \times 0.32)]}{(65-45)}$$

45. Solve (B) = (Rs × Fh) = (1.84 × 0.32) = 0.589

46. $$\text{PLh} = \frac{12,000\ \text{BTUH} \times (\text{A})0.682 \times [0.159 + (\text{B})0.589]}{(65-45)}$$

47. Solve (C) = [Rp +(B)] = [0.159 + 0.589] = 0.748

48. $$\text{PLh} = \frac{12,000\ \text{BTUH} \times (\text{A})0.682 \times (\text{C})0.748}{(65-45)}$$

49. Solve (D) = (T_L – Tmin) = (65 – 45) = 20

50. $$\text{PLh} = \frac{12,000\ \text{BTUH} \times (\text{A})0.682 \times (\text{C})0.748}{(\text{D})20} = 306\ \text{Ft.}$$

51. $$\text{Total PLh} = \frac{\text{UH } 22,900\ \text{BTUH} \times 306\ \text{Ft}/12,000\ \text{BTUH}}{12,000} = 584\ \text{Ft.}$$

Figure 14–21 (cont.) Pipe length calculation for parallel coils—southern

Step 20. Unit cooling coefficience of performance (COPc) at Tmax 90°F at 6 GPM is 3.05.

Step 21. Unit heating coefficience of performance (COPh) at Tmin 45°F at 6 GPM is 3.15.

Step 22. No antifreeze is required.

Step 23. Cooling outside design temperature is 104°F.

Step 24. Cooling design heat gain is 27,265 BTUH.

Step 25. Heating design outside temperature is 34°F.

Step 26. Heating design heat loss is 19,996 BTUH.

Step 27. Cooling run factor (Fc) (Figures 14–19 and 14–20) is 0.46.

Step 28. Heating run factor (Fh) (Figures 14–19 and 14–20) is 0.32.

Ground Loop Heat Exchanger Length

Step 29. Basic formula for pipe length calculation.

Step 30. Formula with all values in place.

Step 31. Solve for (A)—1.328.

Step 32. Basic formula with (A) in inserted.

Step 33. Solve for (B)—1.136.

Step 34. Basic formula with (B) inserted.

Step 35. Solve for (C)—2.726.

Step 36. Basic formula with (C) inserted.

Step 37. Solve for (D)—9°F.

Step 38. Basic formula with (A), (C), and (D) inserted. This produces a pipe length per 12,000 BTUH of 4826.8'.

Step 39. Solve for pipe length per the unit cooling (UC) capacity—11,061.5'.

Step 40. The total cooling heat exchanger length is 11,062'.

With a 750' maximum per loop length, this would mean a minimum of 11,062' total pipe length divided by the 750' maximum per loop or 14.77 loops. With two loops per trench, 16 loops would be the minimum number of loops in 8 trenches, each 692' long, 8' deep, and 2' wide.

The cost of such an installation would not be practical, which rules out the single loop, single loop, double back, and multiple loop system for the Phoenix, AZ application. All three systems are very practical in areas of lower ground temperatures. The pipe length calculations for each type of system should be done for the area in question to make a decision on the type of system to use.

14–5 VERTICAL (SERIES) SYSTEM

Figure 14–10 shows a series system using U-tube heat exchanger pipes in vertical bore holes. This system is a single series circuit with all the liquid flowing through the single circuit. Using the worksheet in Figure 14–22, we need to calculate the total length of heat exchanger pipe needed to handle the heating and cooling capacity of the unit selected. We will assume that this is the Phoenix, AZ application using the same calculated heat gain (27,265 BTUH) and heat loss (19,996 BTUH).

For the installation, we will use 1½" Polyethylene Schedule 50 plastic pipe with butt fusion welds. These will be discussed under Chapter 15, Installation. The calculations show that 2,672' of the 1½" plastic pipe will be installed in 5 bore holes 4¾" in diameter and 267' deep.

The table in Figure 14–23 shows the minimum diameter bore hole required for single U-bend and double U-bend (two U bends per hole) for plastic pipe from ¾" to 2" diameter. The use of double U-bend (two U bends per bore hole) is limited to bore holes in rock where the high cost of boring the hole justifies the use of more pipe in each bore hole. The use of the double U-bend installation is also limited to areas where the mean annual earth temperature (Tm) is 60°F or below. With the 73°F Tm in the Phoenix, AZ area, the single U-bend system is used with 1½" pipe in 4¾" bore holes.

Vertical Pipe Length Calculation

Using the completed worksheet in Figure 14–22, the calculations are done step by step on the worksheet.

Pipe Information

Step 1. Pipe material is Polyethylene Schedule 40.

Step 2. Pipe size is 1½".

Step 3. Pipe resistance. The symbol for pipe resistance changes for vertical pipe; it is Rpe. Using the pipe resistance table in Figure 14–6, the pipe resistance for the various size pipes and pipe material are shown below the line in each box. For example, the pipe resistance for 1½" P E SCH–40 pipe is 0.080 when installed in a vertical position.

Soil Information

Step 4. Heavy soil—damp. The aquifer will be encountered between 60' and 300' below the surface, depending upon the location. All soil encountered is classified as Damp.

Step 5. Soil resistance is found in the right hand column of Figure 14–5. The figure for 1½" pipe found under the line Rs (HS-DAMP) is 0.94.

Location Information

Step 6. The mean annual earth temperature (Tm) is the same as the aquifer temperature at the 50' to 150' level. These are found in Figure 14–1 in the first (Tm) column. For Phoenix, AZ this is listed at 73°F.

Step 7. Disturbed earth variation. The temperature rise or fall of the earth temperature next to the heat exchanger pipe will vary with the depth of the pipe. This can vary from 3°F to 18°F, depending on the distance below grade. To reduce the calculation complications, an average of 10°F is used. This means that the earth temperature next to the pipe will rise an average of 10°F in the cooling mode and drop an average of

Pipe Length Calculation Worksheet SE3WKS-4
Vertical Earth Coils
Data Collection

Job Name John H. Doe Pipe Length—Cooling 2,672 Ft.
Location Phoenix, AZ Pipe Length—Heating 330 Ft.
Number of boreholes 5 Sizes 4¾" Depth 267 Ft.

A. Pipe Information
1. Pipe Material Polyethylene-Schedule 40
2. Pipe Size 1½"
3. Pipe Resistance (Rpe) 0.080 (Figure 14-6)

B. Soil Information
4. Soil Type Heavy Soil-Damp
5. Soil Resistance (Rs) 0.94 (Figure 14-5)

C. Location Information
6. Mean earth temperature (50' to 150' depth) (Figure 14-1) Tm 73 °F
7. Disturbed earth variation DEV 10 °F
8. High earth temperature (Tm + DEVc) Th 83 °F
9. Low earth temperature (Tm – DEVh) T_L 63 °F

D. Heat Pump Information
10. Highest EWTc temperature (Tmax) (Figure 12-1) 90 °F
11. Lowest EWTh temperature (Tmin) (no antifreeze) (Figure 12-1) 45 °F
12. Lowest EWTh temperature (Tmin) (20% antifreeze)(Figure 14-15) 25 °F
13. Unit cooling capacity (UC) at 6 GPM at Tmax 27,500 BTUH
14. Unit heating capacity (UH) (no antifreeze) at 6 GPM at Tmin 22,900 BTUH (Figure 14-15)
15. Unit heating capacity (UH) (20% antifreeze) at --- GPM at Tmin ----- BTUH
16. Unit cooling COPc at Tmax 90 °F at 6 GPM 3.05
17. Unit heating COPh (no antifreeze) at Tmin 45 °F at 6 GPM 3.15
18. Unit heating COPh (20% antifreeze) at Tmin --- °F at --- GPM ----

E. Unit Run Factor
19. Cooling outside design temperature 107 °F
20. Cooling design heat gain 27,265 BTUH
21. Heating design outside temperature 34 °F
22. Heating design heat loss 19,996 BTUH
23. Cooling run factor (Fc) (Figures 14–19 and 14–20) 0.46
24. Heating run factor (Fh) (Figures 14–19 and 14–20) 0.32

F. Vertical Ground Loop Heat Exchanger Length (This does not include supply and return pipes or pipe headers.)
Cooling
Pipe Length/12,000 BTUH Cooling Capacity

25. $$\text{Pipe Length} = \frac{12{,}000\ \text{BTUH} \times \frac{\text{COP}+1}{\text{COP}} \times [\text{Rp} + (\text{Rs} \times \text{Fc})]}{(\text{Tmax} - \text{Th})}$$

(Insert values from information section.)

26. $$\text{PLc} = \frac{12{,}000\ \text{BTUH} \times \frac{3.05+1}{3.05} \times [0.080 + (0.94 \times 0.46)]}{(90-83)}$$

Figure 14–22 Pipe length calculation for vertical coils—southern

27. Solve $(A) = \frac{COP+1}{COP} = \frac{3.05+1}{3.05} = \frac{4.05}{3.05} = 1.328$

28. $PLc = \frac{12{,}000\ BTUH \times (A)1.328 \times [0.080 + (0.94 \times 0.46)]}{(90-83)}$

29. Solve (B) = (Rs × Fc) = (0.94 × 0.46) = 0.4324

30. $PLc = \frac{12{,}000\ BTUH \times (A)1.328 \times [0.080 + (B)0.4324]}{(90-83)}$

31. Solve (C) = [Rp + (B)] = [0.080 + 0.4324] = 0.5124

32. $PLc = \frac{12{,}000\ BTUH \times (A)1.328 \times (C)0.5124}{(90-83)}$

33. Solve (D) = (Tmax – Th) = (90 – 83) = 7°F

34. $PLc = \frac{12{,}000\ BTUH \times (A)1.328 \times (C)0.5124}{(D)7} = 1{,}166\ Ft.$

35. $Total\ PLc = \frac{UC\ 27{,}500\ BTUH \times 1{,}166/12{,}000\ BTUH}{12{,}000} = 2{,}672\ Ft.$

36. Total bored hole length = 1,336 Ft.

Heating (No Antifreeze)

37. $Pipe\ Length = \frac{12{,}000\ BTUH \times \frac{COP-1}{COP} \times [Rp + (Rs \times Fh)]}{(T_L - Tmin)}$

38. $PLh = \frac{12{,}000\ BTUH \times \frac{3.15-1}{3.15} \times [0.080 + (0.94 \times 0.32)]}{(63-45)}$

39. Solve $(A) = \frac{COP-1}{COP} = \frac{3.15-1}{3.15} = \frac{2.15}{3.15} = 0.6825$

40. $PLh = \frac{12{,}000\ BTUH \times (A)\ 0.6825 \times [0.080 + (0.94 \times 0.32)]}{(63-45)}$

41. Solve (B) = (Rs × Fh) = (0.94 × 0.32) = 0.3008

42. $PLh = \frac{12{,}000\ BTUH \times (A)0.6825 \times [0.080 + (B)0.3008]}{(63-45)}$

43. Solve (C) = [Rp +(B)] = [0.080 + 0.3008] = 0.3808

44. $PLh = \frac{12{,}000\ BTUH \times (A)0.6825 \times (C)0.3808}{(63-45)}$

45. Solve (D) = (T_L – Tmin) = (63 – 45) = 18

46. $PLh = \frac{12{,}000\ BTUH \times (A)0.6825 \times (C)0.3808}{(D)18} = 173\ Ft.$

47. $Total\ PLh = \frac{UH\ 22{,}900\ BTUH \times 173\ Ft./12{,}000\ BTUH}{12{,}000} = 330\ Ft.$

48. Total bored hole length 165 Ft.

Figure 14–22 (cont.) Pipe length calculation for vertical coils—southern

MINIMUM DIAMETERS FOR BORE HOLES		
Nom Pipe Size	Single U-Bend	Double U-Bend
¾"	3¼"	4½"
1	3½"	5½"
1¼	4	5¾"
1½	4¾"	6
2	6	7

Figure 14–23 Bore hole sizes (Courtesy Mammoth, A Nortek Co.)

10°F in the heating mode. The high and low earth temperatures are adjusted accordingly.

Step 8. High earth temperature (Th) (Tm + DEVc). The mean earth temperature (Tm) plus the 10°F disturbed earth variation (DEVc) means a high earth temperature (Tm) of 73°F + 10°F or 83°F.

Step 9. Low earth temperature (T_L) (Tm – DEVh). The mean earth temperature (Tm) minus the DEVh means a low earth temperature of 73°F – 10°F or 63°F.

Heat Pump Information

Step 10. Highest entering water temperature (EWTc) for cooling from the manufacturer's specification literature (Tmax). From Figure 12–1, the manufacturer's specifications show 90°F.

Step 11. Lowest entering water temperature (EWTh) for heating (no antifreeze) (Tmin) from the manufacturer's specification literature (Figure 13–13) is 45°F.

Step 12. Lowest entering water temperature (EWTh) for heating (20% antifreeze solution) from the manufacturer's specification literature (Figure 14–15) is 25°F.

Step 13. Unit cooling capacity (UC) at 6 GPM at Tmax (90°F EWTc) is 27,500 BTUH. This is found in the manufacturer's specification literature (Figure 13–13).

Step 14. Unit heating capacity (UH) at 6 GPM at Tmin (no antifreeze) is 22,900 BTUH. This is found in the manufacturer's specification literature (Figure 14–15).

Step 15. Unit heating capacity (UH) at 6 GPM at Tmin (20% antifreeze solution). The ground water temperature is above 60°F and the use of antifreeze should not be needed.

Step 16. Unit coefficience of performance (COPc) at Tmax of 90°F with 6 GPM water flow rate is 3.05. The unit rating is 10.4 EER divided by 3.413 BTU/Watt equals a COP of 3.05.

Step 17. Unit coefficience of performance (COPh) (no antifreeze) at Tmin of 45°F with 6 GPM water flow rate is 3.15.

Step 18. (COPh) (20% antifreeze solution) at Tmin of 25°F. This is not needed in this calculation.

Unit Run Factor

The unit run factors are calculated using the bin temperature method found in the ACCA Manual J and the worksheet in Figure 14–20.

Step 19. Cooling outside design temperature from the ACCA manual J is 107°F.

Step 20. Cooling design heat gain is 27,265 BTUH.

Step 21. Heating outside design temperature is 34°F.

Step 22. Heating design heat loss is 19,996 BTUH.

Step 23. Cooling run factor (Fc) Figures 14–19 and 14–20 is 0.46.

Step 24. Heating run factor (Fh) Figures 14–19 and 14–20 is 0.32.

Ground Loop Heat Exchanger Length

The pipe length calculated will be for the U-bend sections in the bore holes. All supply and return pipe headers are included in the flow rate requirements for the pump size calculation but not as part of the heat exchanger surface.

Cooling

Step 25. Basic formula with abbreviations.

Step 26. Basic formula with all numerical values substituted.

Step 27. Solve for (A). The COPc of 3.05 + 1 (4.05) divided by the COPc of 3.05 = a COPc factor (A) of 1.328.

Step 28. Substitute the (A) factor into the formula.

Step 29. Solve for (B). The soil resistance (Rs) times the percentage of operation (Fc) equals the resistance factor (B) of 0.4324.

Step 30. Substitute the resistance factor (B) into the formula.

Step 31. Solve for (C). The pipe resistance plus the soil resistance factor add up to the heat travel resistance (C) between the water and the earth—0.5124.

Step 32. Substitute the total resistance factor (C) into the formula.

Step 33. Solve for (D). The temperature difference between the water and the earth in the cooling mode—90°F minus 83°F = 7°F.

Step 34. With all final factors for (A), (C), and (D) in the formula, 12,000 BTUH times the COPc factor (A) times the resistance factor (C) divided by the water-to-earth temperature difference factor equals the heat exchanger pipe length per 12,000 BTUH = 1,166 ft.

Step 35. The total heat exchanger pipe length for the WPH 28-1H unit with 27,500 BTUH capacity times the 590 ft. of pipe per 12,000 BTUH divided by 12,000 BTUH adjusts the heat exchanger pipe length to 2,672 ft.

Step 36. Single U-bend vertical loops use double pipe length per length of bore hole. The total bore hole depth is 1/2 of 2,672 ft. or 1,336 ft. The maximum bore hole depth is 300 ft. Therefore, 5 bore holes of 267 ft. are required to handle the cooling capacity.

Heating (No Antifreeze)

Step 37. Basic formula with abbreviations for heating mode calculation.

Step 38. Basic formula with numerical values substituted for symbols.

Step 39. Solve for (A). Heating COPh minus 1 divided by the heating COPh equals 0.6825.

Step 40. Basic formula with COPh factor substituted into the formula.

Step 41. Solve for (B). The soil resistance (Rs) times the run factor (Fh) equals the resistance factor (B)—0.3008.

Step 42. Basic formula with factors for (A) and (B) included.

Step 43. Solve for (C). The pipe flow resistance plus the soil percentage factor (B) add up to the total flow resistance between the water and the earth—0.3808.

Step 44. Basic formula with factors (A) and (C) included.

Step 45. Solve for (D). The temperature difference between the water and the earth in the heating mode (T_L) is 63°F minus 45°F equals 18°FΔT.

Step 46. Basic formula with factors (A), (C), and (D) included. Solving the equation of 12,000 BTUH times (A) 0.6825 times (C) 0.3808 divided by (D) 18°FΔT = 173 ft. In the heating mode, the system requires 173 ft. of heat exchanger pipe per 12,000 BTUH.

Step 47. The total length of heat exchanger pipe (TPLh) in the heating mode is the unit capacity of 22,900 BTUH times the length of pipe per 12,000 BTUH, 173 ft., divided by 12,000 BTUH. The length of heat exchanger pipe needed for the heating mode is 330 ft.

Step 48. The total bored hole length is ½ of 330 ft. or 165 ft. One bored hole 165 ft. deep would handle the heating load.

In all ground loop systems the pipe length and bored hole requirements depend upon the greater load. In the Omaha, NB application, the greater length is the heating load. In the Phoenix, AZ application, the pipe length selection is for the cooling load.

Vertical (Parallel) System

Figure 14–11 shows a vertical ground loop heat exchanger with the ground loops in parallel flow. As in horizontal multiple loop systems, the main advantage of this system is the use of multiple flow paths for the heat-carrying liquid. Instead of the liquid flowing through the one pipe, the quantity of liquid flow is divided among the parallel circuits. As a result, smaller pipe is used in the vertical loops. With the smaller pipe in the vertical loop, a smaller bore hole is used at lower bore cost. For example, the use of 1" plastic pipe in the vertical loop only requires a 3-1/2" bore hole. As the bore holes are filled after the U-bend loop is inserted, an oversized hole increases the quantity of fill. Fill material will be covered in Chapter 15, Installation.

Vertical (Parallel) Pipe Length Calculation

The steps for calculating the TPLc and TPLh of heat exchanger pipe loops have been worked out in Figure 14–24.

The step-by-step procedure is the same as for a vertical (series) system except the pipe size and resistance factor (Rpe) are different. The resistance factor for 1" vertical (Rpe) from Figure 14–6 is 0.109. Therefore, the smaller pipe with a higher resistance factor will result in a longer pipe length.

From the worksheet in Figure 14–24, we see that 2,799' of 1" heat exchanger pipe is required for the Phoenix, AZ application. This results in 5 bore holes, each 280' deep.

In the calculation examples given, a detailed step-by-step procedure was used. This may seem monotonous, but it should be followed until a good grasp of the basic steps is gained. Usually this will take fifteen or twenty application problems.

Experience will shorten the process but "short cuts" or "rule of thumb" calculation methods will only lead to problems.

REVIEW QUESTIONS

1. Liquid-to-air heat pumps connected to ground loops have to operate with circulating liquid temperatures as low as ____°F and as high as ____°F.
2. The usual pipe depth for horizontal loops is 3' to 6' below grade in the northern portions of the U.S. Why?
3. Name two reasons why antifreeze solutions are used in horizontal loop systems in the northern portions of the U.S.
4. The maximum recommended hole depth for vertically drilled holes for vertical loop systems is:
   a. 250'
   b. 300'
   c. 350'
   d. 400'
5. The difference between the mean annual earth temperature and the well water temperature will vary according to the season of the year. True or False?
6. Which has the greater effect on the earth temperature, the air temperature or sunshine?
7. The annual earth temperature swing decreases with the increase in the distance below grade level. True or False?
8. The heat storage effect of the earth is called the________ of the earth.
9. The amount of heat given off the buried coil during the operation in the cooling mode is called the ____ __ ____.
10. The amount of heat absorbed by the buried coil during operation in the heating mode is called the ____ __ ____.

Pipe Length Calculation Worksheet SE3WKS-4
Vertical Earth Coils
Data Collection

Job Name John H. Doe — Pipe Length—Cooling 2,799 Ft.
Location Phoenix, AZ — Pipe Length—Heating 356 Ft.
Number of boreholes 5 Sizes 3½" Depth 280 Ft.

A. Pipe Information
1. Pipe Material Polyethylene-Schedule 40
2. Pipe Size 1"
3. Pipe Resistance (Rpe) 0.109 (Figure 14-6)

B. Soil Information
4. Soil Type Heavy Soil-Damp
5. Soil Resistance (Rs) 0.94 (Figure 14-5)

C. Location Information
6. Mean earth temperature (50' to 150' depth) (Figure 14-1) Tm 73 °F
7. Distrubed earth variation DEV 10 °F
8. High earth temperature (Tm + DEVc) Th 83 °F
9. Low earth temperature (Tm – DEVh) T_L 63 °F

D. Heat Pump Information
10. Highest EWTc temperature (Tmax) (Figure 12-1) 90 °F
11. Lowest EWTh temperature (Tmin) (no antifreeze) (Figure 12-1) 45 °F
12. Lowest EWTh temperature (Tmin) (20% antifreeze)(Figure 14-15) 25 °F
13. Unit cooling capacity (UC) at 6 GPM at Tmax 27,500 BTUH
14. Unit heating capacity (UH) (no antifreeze) at 6 GPM at Tmin 22,900 BTUH (Figure 14-15)
15. Unit heating capacity (UH) (20% antifreeze) at --- GPM at Tmin ---- BTUH
16. Unit cooling COPc at Tmax 90 °F at 6 GPM 3.05
17. Unit heating COPh (no antifreeze) at Tmin 45 °F at 6 GPM 3.15
18. Unit heating COPh (20% antifreeze) at Tmin --- °F at --- GPM ----

E. Unit Run Factor
19. Cooling outside design temperature 107 °F
20. Cooling design heat gain 27,265 BTUH
21. Heating design outside temperature 34 °F
22. Heating design heat loss 19,996 BTUH
23. Cooling run factor (Fc) (Figures 14–19 and 14–20) 0.46
24. Heating run factor (Fh) (Figures 14–19 and 14–20) 0.32

F. Vertical Ground Loop Heat Exchanger Length (This does not include supply and return pipes or pipe headers.)
Cooling
Pipe Length/12,000 BTUH Cooling Capacity

25. $$\text{Pipe Length} = \frac{12{,}000\ \text{BTUH} \times \frac{\text{COP}+1}{\text{COP}} \times [\text{Rp} + (\text{Rs} \times \text{Fc})]}{(\text{Tmax} - \text{Th})}$$

(Insert values from information section)

26. $$\text{PLc} = \frac{12{,}000\ \text{BTUH} \times \frac{3.05+1}{3.05} \times [0.109 \times (0.94 \times 0.46)]}{(90-83)}$$

Figure 14–24 Pipe length calculation for vertical coils—southern

27. $\text{Solve (A)} = \frac{\text{COP}+1}{\text{COP}} = \frac{3.05+1}{3.05} = \frac{4.05}{3.05} = 1.328$

28. $\text{PLc} = \frac{12{,}000\ \text{BTUH} \times (\text{A})1.328 \times [0.109 + (0.94 \times 0.46)]}{(90-83)}$

29. Solve (B) = (Rs × Fc) = (0.94 × 0.46) = 0.4324

30. $\text{PLc} = \frac{12{,}000\ \text{BTUH} \times (\text{A})1.328 \times [0.109 + (\text{B})0.4324]}{(90-83)}$

31. Solve (C) = [Rp + (B)] = (0.109 + 0.4324] = 0.5414

32. $\text{PLc} = \frac{12{,}000\ \text{BTUH} \times (\text{A})0.109 \times (\text{C})0.5414}{(90-83)}$

33. Solve (D) = (Tmax – Th) = (90 – 83) = 7°F

34. $\text{PLc} = \frac{12{,}000\ \text{BTUH} \times (\text{A})1.328 \times (\text{C})0.5414}{(\text{D})7} = 1{,}232\ \text{Ft.}$

35. $\text{Total PLc} = \frac{\text{UC } 27{,}265\ \text{BTUH} \times 1{,}232/12{,}000\ \text{BTUH}}{12{,}000} = 2{,}799\ \text{Ft.}$

36. Cooling heat exchanger pipe length = 2,799 Ft.

Heating (No Antifreeze)

37. $\text{Pipe Length} = \frac{12{,}000\ \text{BTUH} \times \frac{\text{COP}-1}{\text{COP}} \times [\text{Rp} + (\text{Rs} \times \text{Fh})]}{(T_L - \text{Tmin})}$

38. $\text{PLh} = \frac{12{,}000\ \text{BTUH} \times \frac{3.15-1}{3.15} \times [0.109 + (0.94 \times 0.32)]}{(63-45)}$

39. $\text{Solve (A)} = \frac{\text{COP}-1}{\text{COP}} = \frac{3.15-1}{3.15} = \frac{2.15}{3.15} = 0.6825$

40. $\text{PLh} = \frac{12{,}000\ \text{BTUH} \times (\text{A})\ 0.6825 \times [0.109 + (0.94 \times 0.32)]}{(63-45)}$

41. Solve (B) = (Rs × Fh) = (0.94 × 0.32) = 0.3008

42. $\text{PLh} = \frac{12{,}000\ \text{BTUH} \times (\text{A})0.6825 \times [0.109 + (\text{B})0.3008]}{(63-45)}$

43. Solve (C) = [Rp +(B)] = [0.109 + 0.3008] = 0.4098

44. $\text{PLh} = \frac{12{,}000\ \text{BTUH} \times (\text{A})0.6825 \times (\text{C})0.4098}{(63-45)}$

45. Solve (D) = (T_L – Tmin) = (63 – 45) = 18

46. $\text{PLh} = \frac{12{,}000\ \text{BTUH} \times (\text{A})0.6825 \times (\text{C})0.4098}{(\text{D})18} = 186.5\ \text{Ft.}$

47. $\text{Total PLh} = \frac{\text{UH } 22{,}900\ \text{BTUH} \times 187\ \text{Ft.}/12{,}000\ \text{BTUH}}{12{,}000} = 356\ \text{Ft.}$

48. Total bored hole length 178 Ft.

Figure 14–24 (cont.) Pipe length calculation for vertical coils—southern

11. The change in the heat content of the earth during the operation in either the heating or cooling mode is called the ____ __ ______ .
12. Soil resistance is defined as __________ .
13. Which type of soil has the higher resistance to heat flow—heavy, wet soil or light, dry soil?
14. The heat transfer resistance of plastic pipe is constant throughout the length of the pipe, regardless of its position. True or False?
15. The minimum distance between bored or drilled holes in heavy, wet soil is:
    a. 10'
    b. 15'
    c. 20'
    d. 25'
16. The minimum distance between bored or drilled holes in heavy, dry soil is:
    a. 10'
    b. 10'
    c. 20'
    d. 25'
17. The minimum distance between bored or drilled holes in light, wet soil is:
    a. 10'
    b. 15'
    c. 20'
    d. 25'
18. To use a unit rated for liquid temperatures between 45°F and 90°F in the heating mode, in applications using liquid temperatures below 45°F, two restrictions must be followed. List them.
19. When selecting pumps, the flow rate required by the manufacturer can be used for systems containing antifreeze solutions. True or False?
20. The percentage of time the unit can be expected to operate during the coldest month (January) and warmest month (August) is called the _____ ______.
21. List five different pipe arrangements used in earth coil systems.
22. If the Fh for an application of a heat pump is 0.82, what amount of time is the unit expected to run during the month of January?
23. To convert from MBTUH to BTUH, multiply MBTUH by _____.
24. PVC pipe is not suitable for use in buried pipe loops. True or False?
25. For vertical pipe loop installations, the minimum depth per hole is:
    a. 60'
    b. 70'
    c. 80'
    d. 90'
    e. 100'
26. To ensure good contact between the pipe and the earth in vertical loops, the drill casing becomes part of the installation. True or False?

15 INSTALLATION

15–1 SITE PLAN

The layout of the building site must be done before any actual installation can begin. Many items of information are necessary before any bore holes or trenching can take place. The site plan must include the location of buried services to the building.

Figure 15–1 is a site plan taken from the design manual 2100–099F of Bard Manufacturing Co. of Bryan, Ohio. This is a typical site plan showing the location of the various building services. In a rural location, the services include buried electric, telephone, and propane gas lines as well as submersible well and septic tank and leach bed field.

Without a site plan, no one remembers the exact location of the utilities. Also, costs accelerate rapidly if any damage is done to the utilities. Regardless of the lack of memory, the damage is always the installer's fault. To assure that no damage is done to any of the services, the local locator services should be contacted. These services go by many different names such as Blue Stake, JULIE, Miss Dig, and so on. The local utilities as well as the city or county inspection department usually have the contact information. Damage to utilities usually carries a stiff fine as well as repair costs.

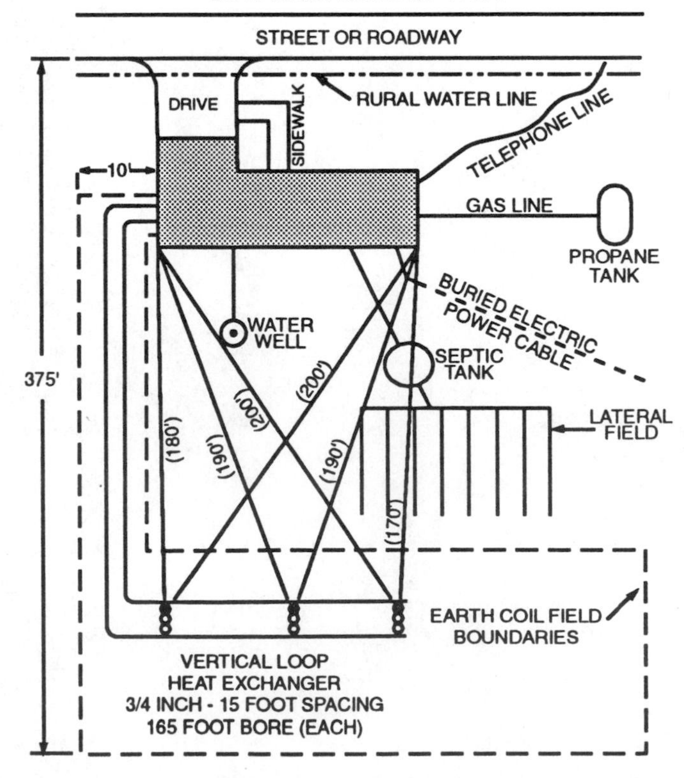

Figure 15–1 Site plan (Courtesy Bard Manufacturing Co.)

Local trenching contractors are usually well aware of the hazards and costs of indiscriminate digging in the area. The major worry is the "Do It Yourself" operator who does not have the necessary experience.

Remember: To reduce the possibility of encountering any buried utility service lines, a site plan is required.

15–2 PIPE MATERIAL

The types of pipe and pipe material used in ground loop systems are discussed in Chapters 12 and 14. A further discussion is in order to lead into the makeup of pipe fittings and joints.

The use of PVC pipe is not recommended for the buried portions of the ground loop systems. It may be used in the building for connecting the unit to the makeup water supply as well as connecting to the ground loop.

For the ground loop, use Polybutylene (PB) or Polyethylene (PE) in the horizontal coils, vertical U bends, and the supply and return line from the building.

Pipe Cleanliness

Pipe is much easier to keep clean than to clean after installation. When handling pipe coils, the pipe ends must be capped or taped. Trash, soil, or small animals or insects should not be allowed to enter the pipe. When loops are fabricated prior to installation, the loops are kept sealed.

Assembly

The technology of the assembly and installation of plastic pipe was developed by the natural gas and domestic well water distribution industries. Polybutylene and high-density Polyethylene are marketed with a guaranteed lifetime as high as 50 years.

Polybutylene. From the literature of Vanguard Plastics Inc. of McPherson, KS:

Question: What is Polybutylene?
Answer: When crude oil is refined to make everyday products such as gasoline and motor oil, unrefined "left over" material is collected. This material is used to make Polybutylene resin, which in turn is extruded into pipe.
Question: Since Polybutylene resin comes from crude oil, what about possible shortages in the future?
Answer: Remember, Polybutylene is a by-product of crude oil refining. As a result, no shortage of Polybutylene is anticipated in the foreseeable future.
Question: What's different about Polybutylene compared to other plastics and metals we use?
Answer: Polybutylene has excellent creep resistance as well as the highest resistance to stress cracking of all the Polyolefins. Polybutylene is the most flexible of the thermoplastics, yet it maintains the strength and durability to withstand shifting soil conditions or extreme temperature changes.

Since Polybutylene is inert, it will not rust, rot, or scale up like metal. It is not affected by electrolysis action.
Question: Why is some pipe blue and another gray? Are they different materials?
Answer: Both blue and gray are Polybutylene. The blue resin indicates cold water application while the gray resin will withstand higher temperature plumbing applications.
Question: Does Polybutylene create a health hazard?
Answer: All polybutylene piping has been approved by the National Sanitation Foundation for the use in potable water systems. There is no risk from Polybutylene pipe in open systems and closed loop systems do not involve discharge.
Question: What type of fastenings can be used?
Answer: Polybutylene pipe may be connected by using barbed insert IPS fittings with 300 Series stainless steel hose clamps or socket heat fusion.
Question: What is the strength limit?
Answer: Polybutylene thermoplastic has a high design stress that allows for continuous operation at 125 psig. It also has the strength to withstand excessive pressure surges.
Question: What are the operating temperature limits?
Answer: Polybutylene can withstand continuous use at temperatures as high as 110°F and intermittent use up to 120°F. Polybutylene offers limited protection against breakage due to freezing.

Polyethylene. From Orangeburg Industries Inc. of Ashville, NC, we learn that:

Polyethylene pipe meets or exceeds the American Society for Testing Materials (ASTM) standards covering Polyethylene plastic pipe. The standard for PE pipe is rated for use with water at 73.4°F at a hydrostatic design stress of 800 psig.

For installation, the tubing shall be installed in accordance with ASTM Standard D–2774–72 (Reapproved 1983) Standard Recommended Practice for Underground Installation of Thermoplastic Pressure piping.

Joining polyethylene pipe by heat fusion, installation practices standard ASTM D–2657 "Recommended Practices for Heat Joining of Thermoplastic Pipe and Fittings" should be followed.

Polyethylene pipe and tubing is designed for use as cold water supply lines or earth coupled and water source heat pump piping. Not recommended for hot water applications.

Polyethylene pipe and tubing shall not be installed in soil with solvents, fuel, organic compounds, or other detrimental materials which will cause permeation, corrosion, degradation, or structural failure of the material.

15–3 INSTALLATION OF PIPE

Horizontal Loops

Residential heat pump horizontal ground loop systems require installation functions that are different than standard installations.

1. Excavation or trenching or bore holes to receive the ground loops.
2. Joining the pipe and fittings to make up the loops.
3. Pressure testing the pipe sections before burial.
4. Proper connection to supply and return pipes into the building.
5. Backfilling of trenches or filling of bore holes.
6. Landscaping.
7. Cleaning and flushing the ground loop.
8. Charging the loop with water or antifreeze solution.
9. Proper piping to the heat pump unit.

Trenching. The various methods of trenching and the equipment involved are discussed in Chapter 11. The discussion involves the vibratory plow, the chain trencher, and the back hoe. The means used will depend upon the depth of the trench, the number of pipes in the hole, and the conditions of the soil.

Vibratory Plow. This device wedges a path through the earth and feeds the pipe into the slit as the machine moves forward. No actual trench is made. This method of application is effective in rock-free earth material in the straight-run portions of the ground loop. A 3' pipe depth is the maximum for most vibratory plow units.

The plow cannot be used for close turns or return bends. Therefore, a back hoe or handle shovel is required to dig the turnaround and connection locations. A further restriction is that the vibratory plow is limited to single pipe, series loop systems.

Chain Trencher. This device is capable of carving a groove in the earth with the width of the groove dependent upon the width of the chain buckets. According to the equipment manufacturers, a 4" wide chain is the most popular.

Some chain trenchers will reach depths of 9' to 12', but the most popular models have a 6' depth limit. Again, the soil must be rock free and of sufficient density to stand until the pipe has been installed before any wall collapse can occur.

Application can include reverse loop systems with the pipe stacked in the trench and the trench backfilled in the proper sequence of pipe and backfill to obtain the proper spacing of the pipes.

Back Hoe. This device is the most versatile of the three pieces of equipment. The back hoe is capable of trench preparation regardless of the soil conditions. Loose soil may require a wider trench to reach the desired depth.

Available in 1' to 3' wide buckets, the equipment will enable the trenching operation regardless of the soil conditions encountered.

For multiple loop systems, which require a wider trench, the back hoe with a front blade for backfilling is the most desirable equipment.

Vertical Loop Systems

Vertical loop systems are U-shaped pipe loops in holes drilled or bored into the earth. Holes to a maximum depth of 300' are used.

Boring holes for installing vertical U-bend earth loops is a different task than drilling a well for water. In a water well, there is no need to seal the bored hole walls to prevent water contamination because the water is to be pumped from the ground. The bored well for vertical loops is not intended to take water from the ground. Therefore, a seal is used to prevent water migration.

The vertical loop bore is easier as it is only necessary to have the hole open long enough to insert the U-bend loop. The bored hole will be filled after the loop is inserted. To obtain the highest heat transfer rate between the pipe and the earth requires an air- and void-free fill.

The depth of the well is not as important as getting the total wetted earth contact surface needed to accomplish the desired heat transfer rate. The main objective is to install a design length of heat exchanger pipe. The most economical boring procedure is to drill until the drill rate slows down and then set up and start another hole. However, the minimum depth per hole is 80'. The minimum distance between holes is 10'.

Using a wet rotary drill bit as shown in Figure 15–2, the drill rig rotates the drill while forcing a high-pressure liquid down to lubricate the drill as well as keep the bit teeth cool. Returning to the surface, the liquid carries the loosened earth material to the surface.

In most applications, water is used as the flushing agent. However, where the drilling operation is through layers of gravel or other loose materials, a drilling mud is used. Drilling mud, Bentonite clay, and/or other additives are used to seal the bored hole walls to reduce loss of drilling mud into the porous earth layers. If the earth porosity such as gravel and/or course sand is encountered, temporary steel casing can be used to keep the bored hole open until the U-bend heat exchanger pipe has been installed. After installation, the casing is pulled to increase the heat exchange efficiency between the pipe and the earth material.

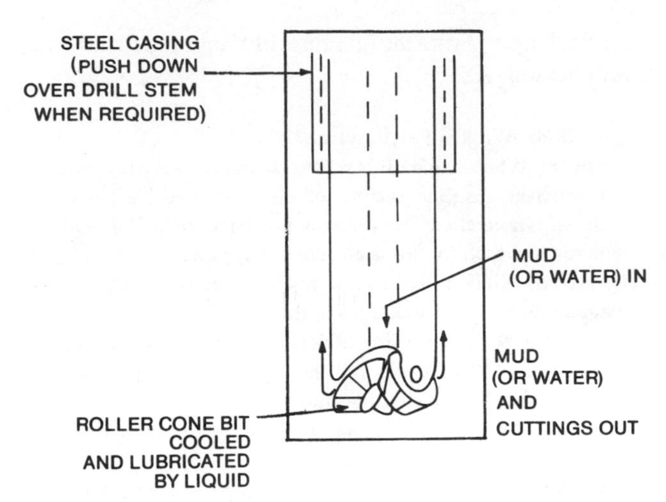

Figure 15–2 Wet rotary drill bit (Courtesy Mammoth, A Nortek Co.)

Three problems can be encountered in a wet rotary drilling operation:

1. Surface mess. The returning water has to be diverted to a settling pond or tanks to remove the solids before being returned to the well.
2. Pressure escape into the earth layers. This requires use of drill mud or steel casing.
3. Holding the hole open long enough to insert the heat exchanger loop. If experience dictates, temporary steel casing becomes standard practice.

Figure 15–3 shows a rotary, hollow-stem auger drill with cutting teeth and temporary plug at the bottom. The plug is in place during the drilling operation to prevent the drill from filling with the earth material. The drilling operation is dry with very little material at the surface. Some earth material will surface but the majority of the material will be pressed into the sides of the hole to compact and strengthen the sides. This process actually forms a well casing as the drill proceeds downward.

After the desired depth is reached, or if a rock or rock layer is encountered which terminates the well depth, the plug is removed using the long plug connector rod in the center of the drill stem.

With the plug removed, the U-bend pipe heat exchanger is dropped down the interior of the drill stem. The stem is then pulled out of the hole, leaving the U-bend pipe in place.

After either type of drilling operation, with the heat exchanger pipe loop in place, the bored or drilled hole is filled with Bentonite slurry or neat cement to ensure good wall contact between the pipe and the earth material. This addition to the drilled hole helps to seal the well to prevent water seeping from one aquifer to another as well as to improve the heat transfer rate.

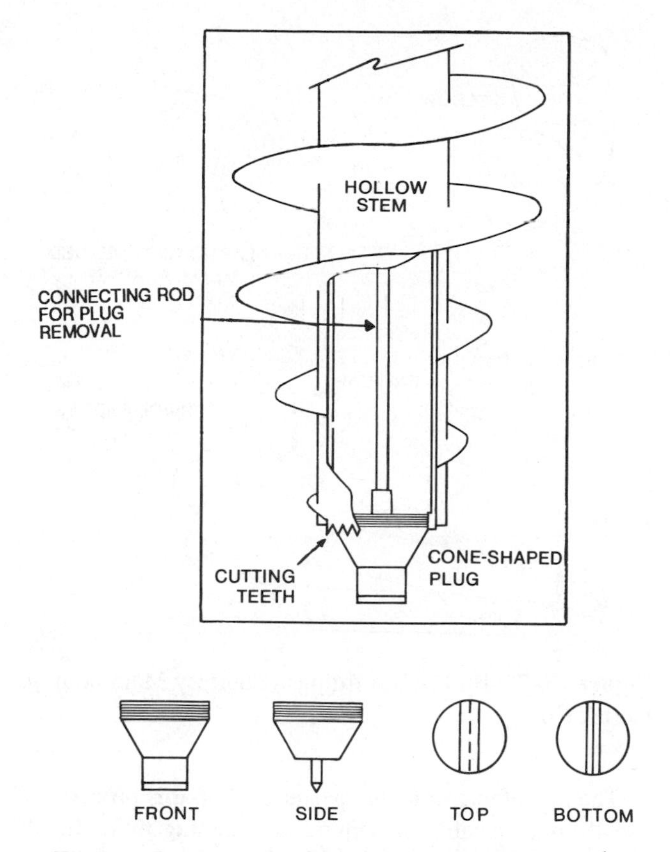

Figure 15–3 Dry rotary drill bit with hollow stem (Courtesy Mammoth, A Nortek Co.)

Joining Pipe and Fittings (Fusing)

To provide the highest reliability of the connections, all underground piping connections should be thermally fused (welded). Mechanical connections are not reliable enough for long underground loop life. The normal life of thermoplastic tube (Polyethylene and Polybutylene) is around 50 years. The connections should also have as long a life expectancy.

Two types of fusion techniques are used: socket fusion and butt fusion.

Socket Fusion

Polybutylene is best joined by socket fusion. This requires fittings in which the pipe is inserted. This process is similar to the installation of copper tube using soldered fittings.

Figure 15–4 shows some typical fittings used in the socket fusion process. The pipe and fittings are heated to the necessary fusion temperature before assembly. When the temperature is reached, the joint is quickly assembled and then allowed to cool for a specific time before pressure or tension is applied.

The fusion temperature and cooling time will depend upon the specific manufacturer's specifications. For example, the specification table (Figure 15–5) published by Orangeburg Industries Inc. covering their pipe products, shows the Melting Point (Vicat Softening Temperature) of 266°F. The pipe

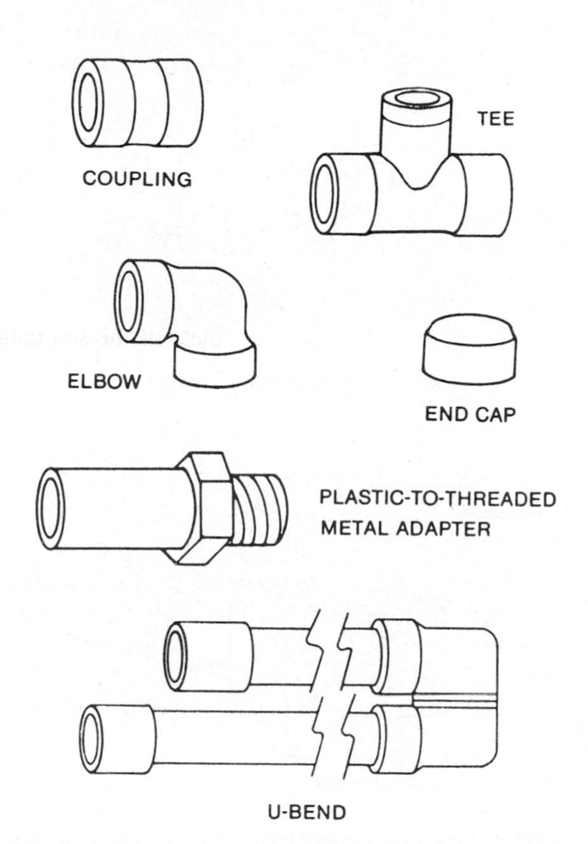

Figure 15–4 Socket fusion fittings (Courtesy Mammoth, A Nortek Co.)

Property	ASTM Reference	Nominal Value	Unit
Density (Pipe)	D-1505	0.955	g/cc
Melt Index, Condition E	D-1238	0.320	g/10 min.
Melting Point (Vicat Softening Temp)	D-1525	266	°F
Brittleness Temperature	D-746	< -103°F	°F
Thermal Expansion	D-696	0.00011	in/in/°F
Thermal Conductivity	C-177	2.77	BTU-In/Ft^3/hr/°F
"R" Value (Insulation) for 1" Thickness	—	0.3	R
Tensile Strength, yield (2.0 In/Min)	D-638	3425	PSI
Tensile Strength, ultimate (2.0 In/Min)	D-638	4000	PSI
Elongation (2.0 In/Min)	D-638	<800	Percent
Flexural Modulus	D-3350	136,000	PSI
Hydrostatic Design Basis (HDB)	D-2837	1600	PSI

Figure 15–5 Pipe specification table (Courtesy Orangeburg Industries Inc.)

heating tool is set to produce this pipe and fitting temperature to produce the bond that seals the socket joint.

The tool that is used for socket fusion (see Figure 15–6) provides for heating the faces of the pipe and fittings of various sizes as well as protection of the heater face plates when they are not in use. The steps needed for socket fusion are:

1. Select the correct pipe and fitting faces to be heated. Square cutting and trimming of the pipe end is not as critical in socket fusion as in butt fusion. The socket depth allows for some misalignment of the pipe end. Care should be used to cut as square a pipe end as possible to reduce the possibility of fusion problems. As in the assembly of any plastic product, a second chance for error is not available.

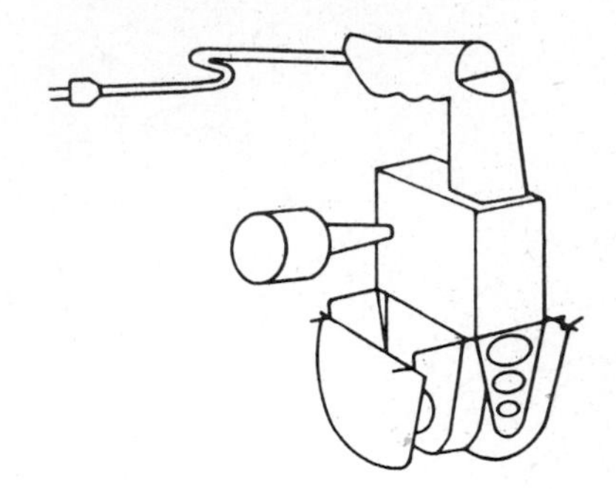

(A) SOCKET FACES BEING HEATED

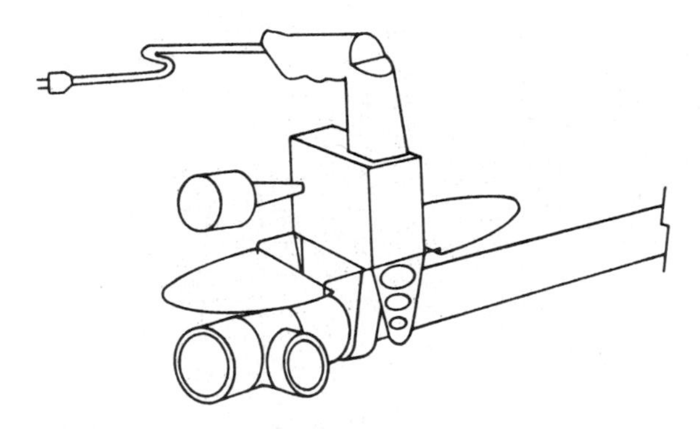

(B) TEE FITTING (FEMALE) AND PIPE END (MALE) BEING HEATED

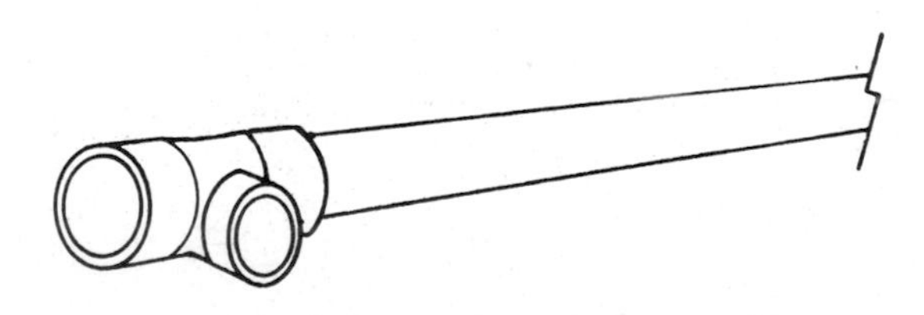

(C) TEE FITTING AND PIPE END FUSED

Figure 15–6 Steps of socket fusion process (Courtesy Mammoth, A Nortek Co.)

2. Heat the tool faces to the correct temperature. Using the required temperature from the pipe manufacturer's literature, set the tool thermostat control and allow full heating time of the tool faces. The pipe heating process should be as fast as possible to avoid excessive pipe collapse. The first step (A) in Figure 15–6 shows the position of the face plate guards while the tool is heating.
3. Hold the pieces to be fused against the tool faces until the proper temperature has been reached to plasticize the pipe and fitting faces. From the (B) illustration in Figure 15–6, both pipe and fitting are applied at the same time. Proper bond requires both the pipe and fitting faces be at the proper temperature.
4. Assemble the pipe/fitting joint. Remove the pipe and fitting joint from the tool and assemble immediately. The pipe material cools rapidly and too much drop in temperature will produce faulty joints. The joint is then allowed to cool and solidify.

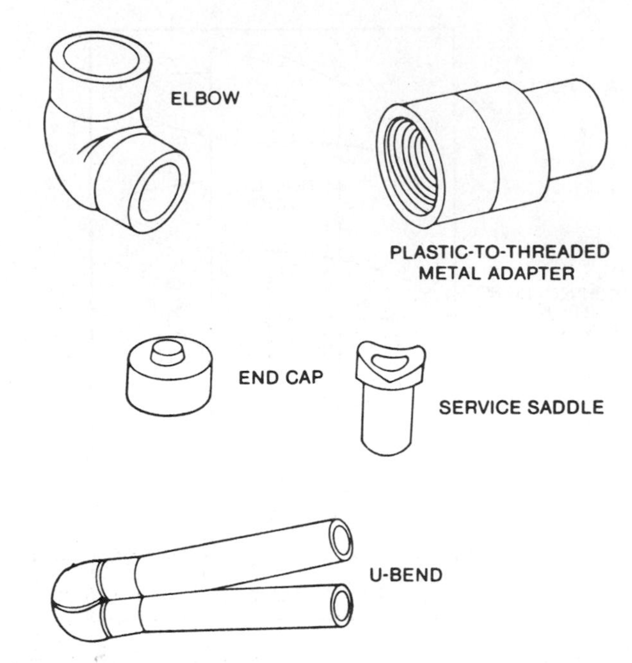

Figure 15–7 Butt fusion fittings (Courtesy Mammoth, A Nortek Co.)

The manufacturer of the heater used in this process will usually have a table of heating and cooling times for the various pipe and fitting materials. However, the pipe manufacturer's information should be used if it is available.

Item (C) in Figure 15–6 shows the assembled joint.

Butt Fusion

The butt fusion method of joining pipe is the most versatile. Figure 15–7 shows the fittings that are used in this process. Pipe ends are joined directly so no couplings are needed. A service saddle fused over a hole drilled in the pipe forms the necessary branch connection instead of using a tee fitting. Elbows are required if a sweeping turn of the pipe is not suitable.

Figure 15–8 shows a vertical U-bend heat exchanger butt fused into the supply and return pipes. Using service saddles to produce the branch connections and end caps to seal the supply and return headers, the only actual fittings used are the elbows, plastic-to-threaded metal pipe adapters, and the U-bend fitting at the bottom of the U-bend heat exchanger.

The procedure for butt fusing is the same as for socket fusing:

1. Select the correct pipe and fittings to be heated and joined.
2. Heat the faces of the tool to the correct temperature, depending upon the material in the pipe and fittings.
3. Hold the pieces to be fused against the tool faces until the plasticizing temperature is reached.
4. Assemble the pipe and fitting or two-pipe joint.

A different heater tool is used in the butt fusion process than in the socket fusion process. The butt fusion process requires square cutting and proper alignment of pipe ends to

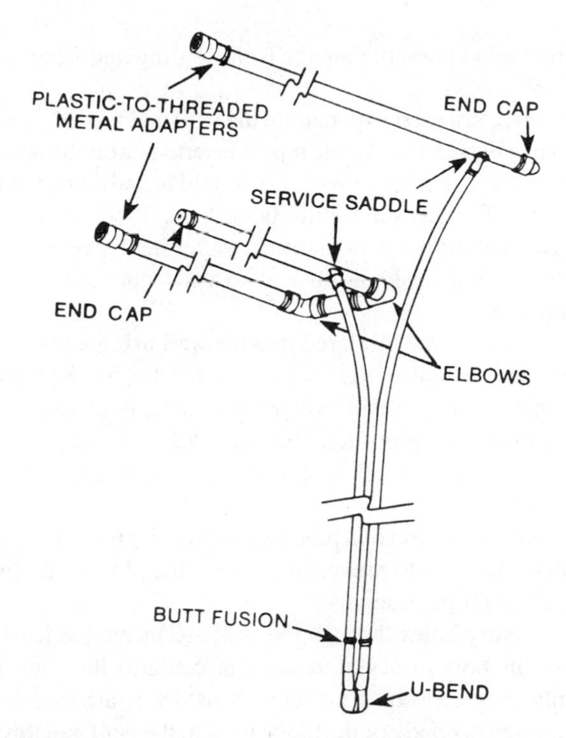

Figure 15–8 Butt fused system (Courtesy Mammoth, A Nortek Co.)

make a satisfactory butt fusion weld. Therefore, the tool used must have the ability to securely hold the parts in position as well as provide the ability to align the parts, trim and square the ends, and bring them together with enough pressure to preserve the joint until cooled.

Figure 15–9 shows the steps used in the butt fusion process. In step (A), the ends of the pipe are secured in place, trimmed to produce square faces, and aligned for proper joining. In step (B), the faces have been preheated and the pipe ends are forced into contact with the preheated faces of the tool to reach the proper plastizing temperature. In step (C), the pipe ends are forced together to make the joint and allowed to cool.

As with the socket fusion tool, accessories and fitting adapters are supplied with the tool for making the various types of butt joints.

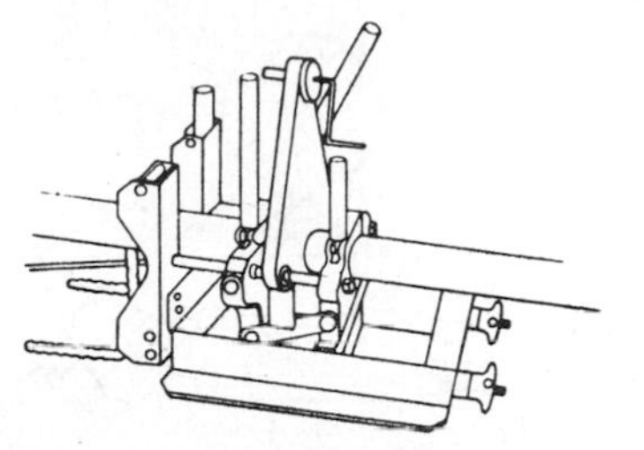

(A) SECURE, TRIM, AND ALIGN BOTH PIECES

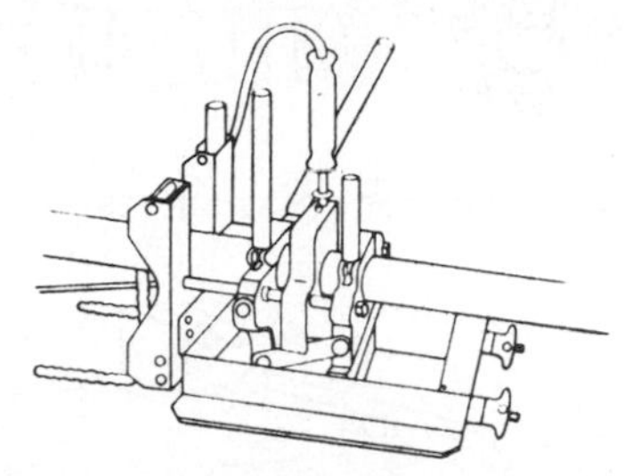

(B) HEAT THE TWO ENDS FOR A PRESCRIBED PERIOD OF TIME

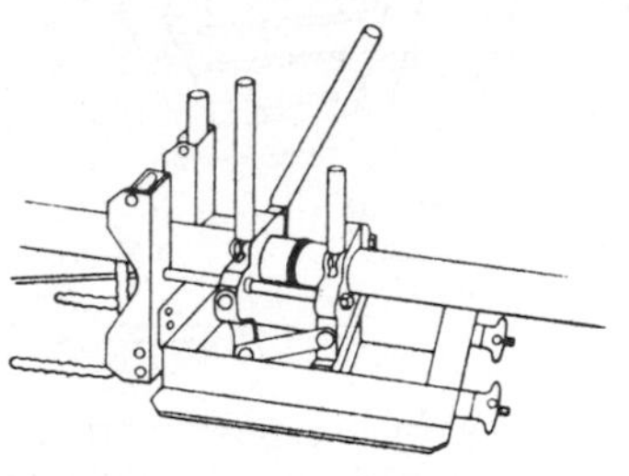

(C) BUTT THE PIECES TOGETHER TO FORM A DOUBLE-BEAD JOINT

Figure 15–9 Steps in butt fusion process (Courtesy Mammoth, A Nortek Co.)

Pressure Testing

ALWAYS PRESSURE TEST PIPE LOOPS BEFORE BURYING OR INSERTING IN BORED OR DRILLED WELLS.

For horizontal loops, the method of testing will depend upon the configuration of the pipe sections. Where all the pipes are at the same level, the pipe should be weighted down with a small amount of backfill to hold the pipe in place without covering the joints. Do not complete the backfilling process until the entire loop has been pressure tested, air has been purged out, and the system is connected to the heat pump and charged with liquid.

For multilevel configurations, or for situations where it is necessary to use the dug trench to receive the diggings of the next trench, each loop must be pressure tested before the next trench is dug. Before covering the manifold assemblies and the building entrance area, the entire heat exchanger loop must be pressure tested.

The proper design objective is to have the least number of fused joints in the system. The only requirement should be joints at the manifolds.

For vertical earth loops, the loops must be pressure tested before insertion in the wells. After all the vertical loops have been inserted in the wells and manifold connections have been made, the entire loop system is pressure tested at the unit before any backfilling or covering is done.

Figure 15–10 shows the tools and fittings used for pressure testing the heat exchanger loops. Two spline-type male insert fittings with stainless steel clamps are used to connect to the pipe. One fitting is capped or could have a capped Dill fitting with valve insert and cap. The second fitting has a pressure gauge and shut-off valve. The shut-off valve has a connection for connecting a portable air compressor.

With these fittings, the compressor is used to build up the air pressure in the pipe loop. The test pressure should be raised to a minimum of 50 psig. Leak testing can be done easily by using a soap solution and brush at each fused joint. No further work should be done until all joints have been tested and all leaks corrected.

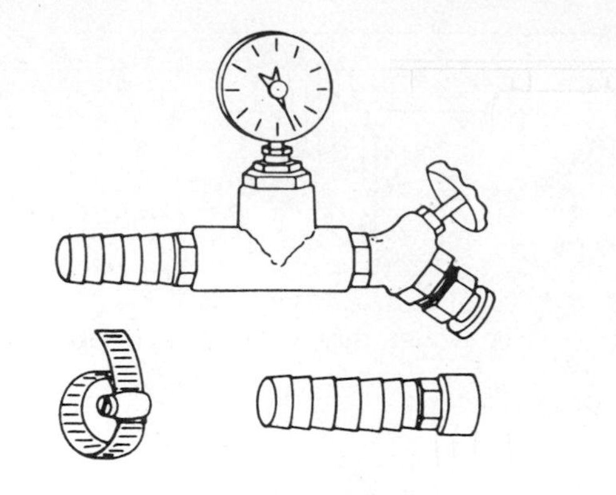

(A) FITTINGS FOR PRESSURE TESTING

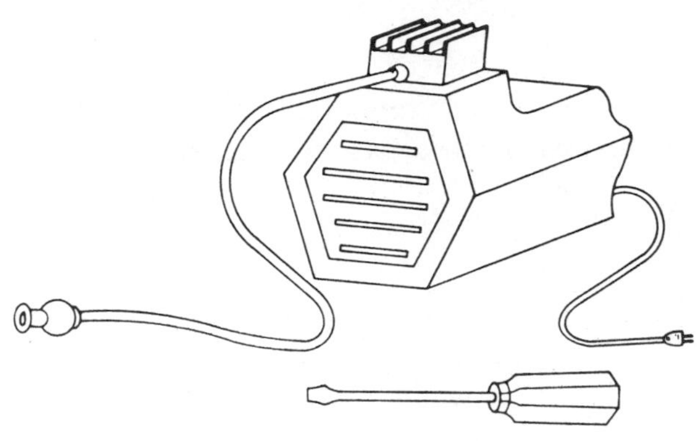

(B) TOOLS FOR PRESSURE TESTING

Figure 15–10 Tools and fittings used for pressure testing (Courtesy Mammoth, A Nortek Co.)

Entrance Pipes to the Building

It is very important that when the supply and return pipes are brought into the building, they are installed in a manner to prevent the entrance of moisture through the foundation of the building. When boring the necessary holes through the building foundation (see Figure 15–11), the holes must be properly positioned to prevent damage from settling and water penetration.

Sleeves are used to penetrate the wall and provide seals as shown in Figure 15–12. The pipe is inserted through the sleeve and sealed. The size of the sleeve should be only a slip fit over the pipe. The sleeves should be at least 6" longer than the thickness of the block or masonry wall. The sleeve is inserted flush with the inside wall and extends outward from the foundation.

The sleeves are mortared into the wall using a water-plug or water-resistant mortar. After the mortar has had proper curing time, the outside sleeve/wall area is coated with a foundation waterproofing. This provides moisture protection with a material that is flexible enough to stand against flexing or settling.

The construction of a pipe bed outside the foundation wall is very important to prevent damage to the pipe from settling as well as proper drainage.

The entry holes through the wall are located at least 15" above the bottom of the trench that contains the earth loop supply and return pipes. Located at least 18" apart, they should be through the body of the block used in the wall, not through any mortar joint (see Figure 15–13).

After the sleeves are inserted and properly sealed, a slope of backfill material up the wall to within 2" to 3" of the bottom of the pipes is put in place. This material will slope from the fill height to the trench bottom over a distance of 6' or more.

Pea gravel is then placed over the backfill to a depth to bring the top of the pea gravel level with the bottom of the sleeves. The pea gravel bed will extend about 4' beyond the base of backfill material. This is to provide positive moisture drain from the foundation.

When the pipes are inserted through the sleeves, the tight slip fit of the pipe in the sleeve, the extra length of sleeve, and the backfill and pea gravel help prevent shear damage to the pipe from any settling of the foundation wall. The space between the pipe and sleeve must also be filled with masonry waterproofing material (see Figure 15–14).

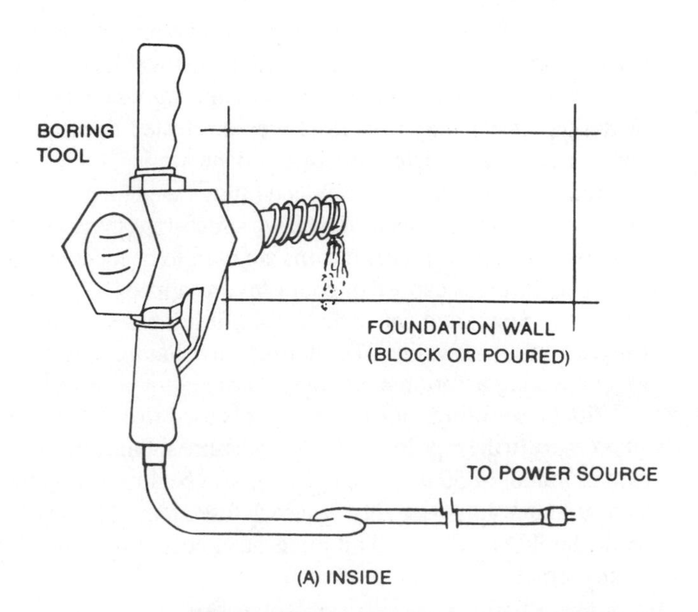

(A) INSIDE

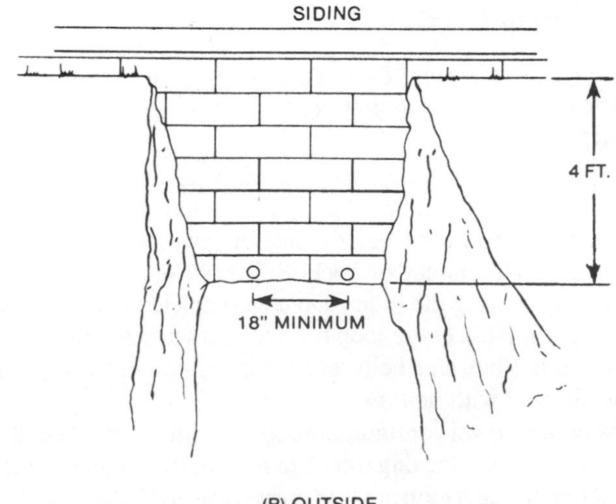

(B) OUTSIDE

Figure 15–11 Boring holes through foundation wall (Courtesy Mammoth, A Nortek Co.)

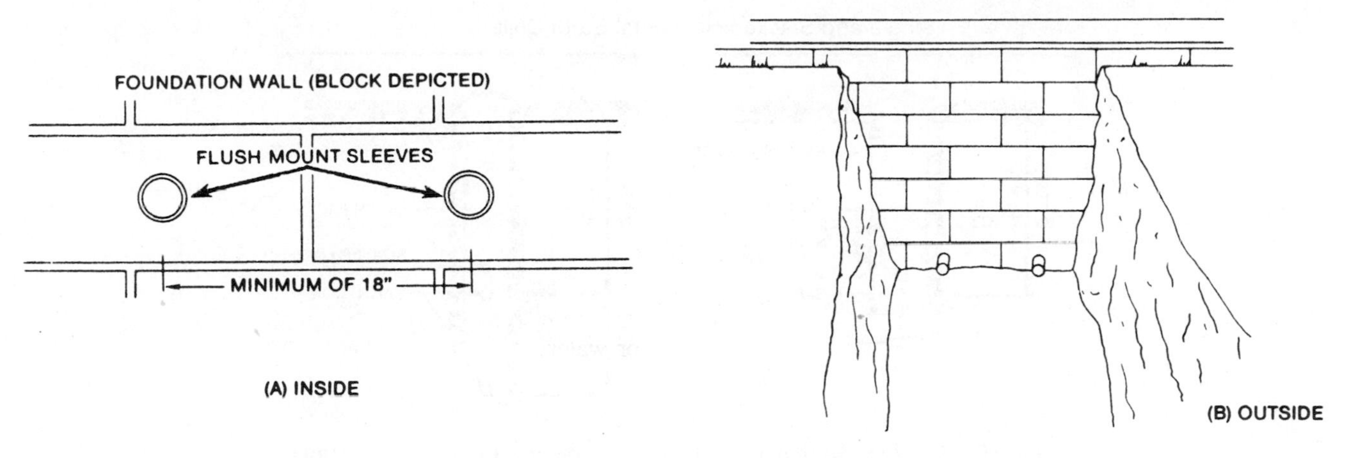

Figure 15–12 Installing sleeves (Courtesy Mammoth, A Nortek Co.)

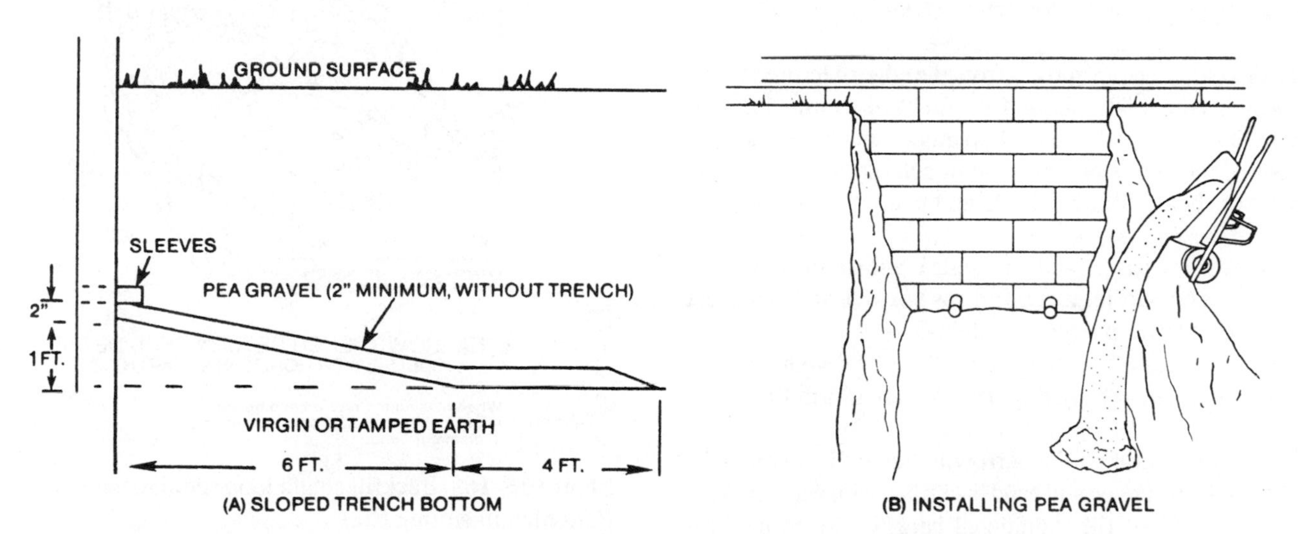

Figure 15–13 Preparation of pipe bed at foundation wall (Courtesy Mammoth, A Nortek Co.)

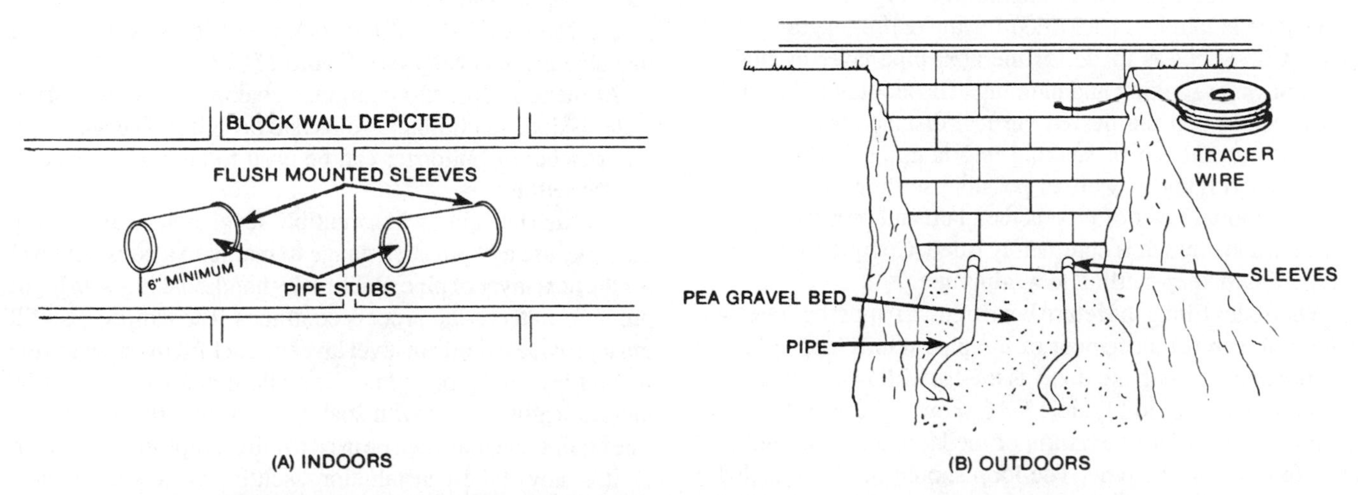

Figure 15–14 Inserting pipe through sleeves (Courtesy Mammoth, A Nortek Co.)

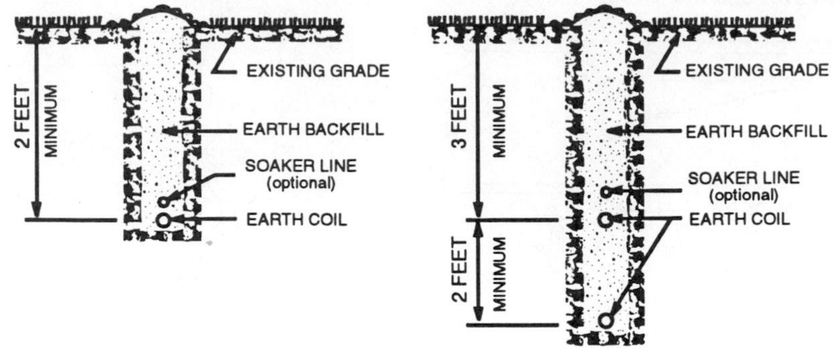

Single and Stacked Horizontal Earth Coils

EXISTING GRADE
2 FEET MINIMUM
EARTH BACKFILL
SOAKER LINE (optional)
EARTH COIL

EXISTING GRADE
3 FEET MINIMUM
EARTH BACKFILL
SOAKER LINE (optional)
EARTH COIL
2 FEET MINIMUM

Figure 15–15 Backfill in narrow trench (Courtesy Bard Manufacturing Co.)

15–4 BACKFILLING

Horizontal Loops

The backfilling is critical and its successful completion depends upon following correct procedures:

1. If rocks have been removed from the edge of the trench and the top of the spoils pile, and the dirt has not formed clumps due to rain, an experienced operator can angle blade the tops of the spoils pile into the trench on the first pass. A worker should follow closely and tamp the fines by hand and make sure no rocks fall in.
2. Several more passes with the angle blade should be used to make the backfilling as uniform as possible and to prevent air pockets over the pipe (bridging).
3. Several stages of tamping and rolling over with the trencher wheels may be necessary to obtain a firm fill.

In an installation with a narrow trench on level pipe, the pipe is laid in the trench and hand backfilled with water soaking for the first 12" of fill. Additional backfill is then installed mechanically, using high-pressure water soaking to eliminate bridging and voids. The left-hand illustration in Figure 15–15 shows the method of backfilling a single pipe trench. Extra mounding of the backfill is needed to compensate for settling.

In an installation with a narrow trench with multiple levels, as shown in the right-hand illustration of Figure 15–15, the lower pipe is laid and backfilled for the entire length of the loop. A level bed is made for the next pipe layer by high-pressure water soaking and tamping. The amount of backfill will depend upon the design vertical distance between the pipes. A careful job of soaking and tamping is needed to eliminate bridging as well as possible settling of the pipe. Each additional layer of pipe is then laid and properly backfilled until the trench is completely filled. Complete tamping is required to keep settling to a minimum.

When backfilling the return bend trench, it must be done by hand with careful placement of the pipe return loop. Water soaking and tamping are done as the backfill is installed.

The illustration in Figure 15–16 shows a vertical loop in the pipe. This is for illustration of the large loop required to prevent kinks in the pipe. The loop should be in an angled horizontal position from the lower layer to the upper layer.

Section View

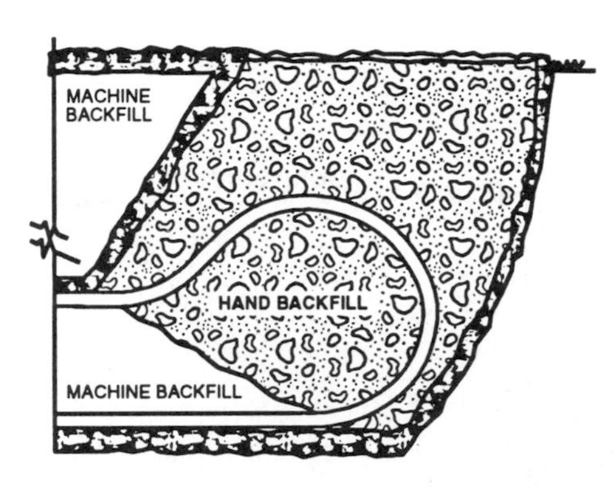

NARROW TRENCH RETURN BEND DETAIL
DOUBLE LAYER HORIZONTAL EARTH COIL

When making the return bend be careful not to kink the pipe. 2" pipe requires a 4' diameter bend.

Figure 15–16 Backfill single loop return bend (Courtesy Bard Manufacturing Co.)

This will prevent air entrapment in the loop and allow proper purging of the loop.

In a wide trench installation, where all the pipe is at one level, series loop or double back loop, the pipe is laid and backfill is hand shoveled every few yards to hold the pipe in place. The remainder of the rock-free backfill can then be installed mechanically (see Figure 15–17).

As the filling process continues, soaking via high-pressure water is done to eliminate voids and bridging. The weight of the trencher or bulldozer can be used to tamp the trench to reduce settling.

For each level in a wide, multiple-level trench, after laying the pipe, use a depth stick gauge to maintain a constant level for the next layer of pipe. Backfill by hand, soaking with high-pressure water as the process continues. The tamped backfill must provide a uniform level layer of backfill upon which the second layer of pipe is laid. The additional backfill can be mechanically moved with soaking until the trench is filled. The equipment may then be used to "tire tamp" the final layer.

If, in any of these installations, settling must be eliminated from the final landscaping, hydraulic or air-drive mechanical

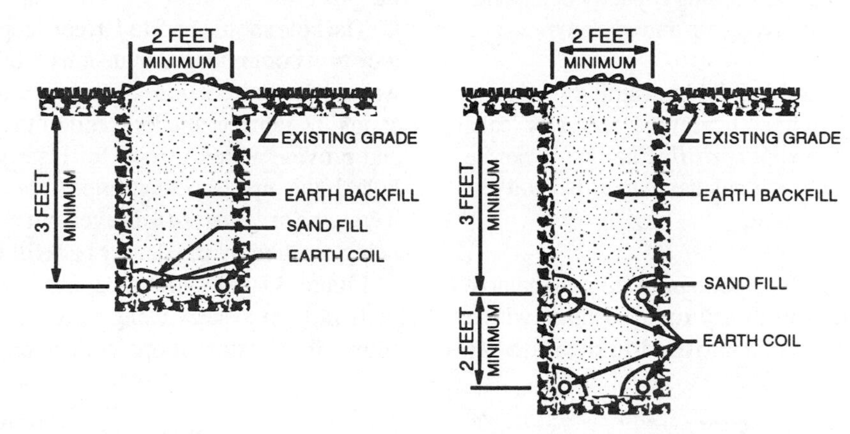

Figure 15–17 Backfill in wide trench (Courtesy Bard Manufacturing Co.)

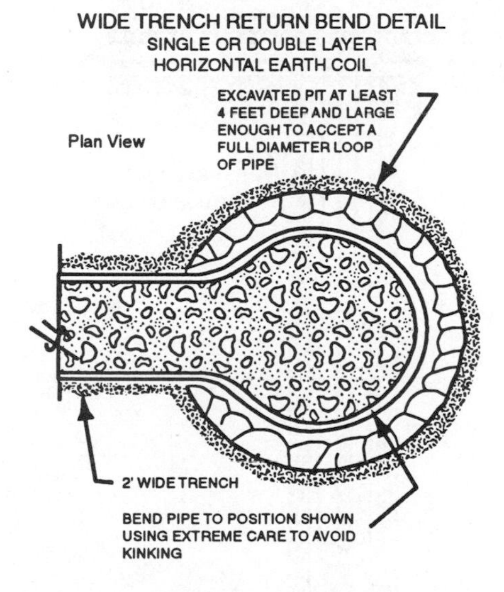

Figure 15–18 Backfill in wide trench return bend (Courtesy Bard Manufacturing Co.)

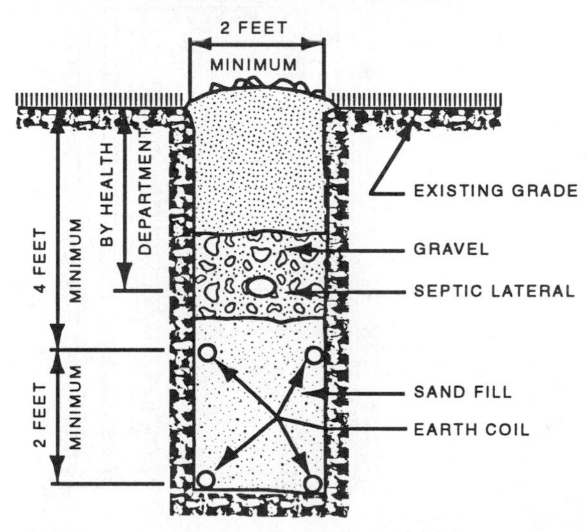

Figure 15–19 Wetting system to improve heat transfer efficiency (Courtesy Bard Manufacturing Co.)

tampers must be used. When using these machines, extreme care must be followed to prevent damage to the pipe loops.

Notice that in a wide trench installation, a sand fill around the pipe is recommended to reduce pipe damage.

Pipe returns use wider trenches to provide room for the return loops. Figure 15–18 shows the detail of the return bend pit. The pit is a large enough diameter to accommodate the pipe return bends without kinks. After the pipe is laid, the entire backfill must be done by hand to ensure proper positioning of the pipe and completely compacted fill. High-pressure water soak is required.

The thermal conductivity factor table of various materials in Figure 14–5 shows a conductivity for dry top soil of 0.075 while wet top soil is 0.620. If the soil is 42% water by volume, the ability to transfer heat is a little over 8 times as great. From this, we can see that a constant wetting system increases the heat transfer efficiency of the loop by better than 800%.

In applications where the building uses a septic tank and leach bed drainage field, the water from the septic tank is an excellent means of keeping the heat exchanger loop moist. Figure 15–19 shows the cross section of such an application. For the most effective moist fill, the heat exchanger loops are backfilled with sand to a height of 4" to 6" above the top of the heat exchanger loop. The leach bed laterals are then backfilled with gravel to a depth according to local health code. Above the gravel, the spoils from trenching are used to complete the backfill and seal the trench. This method of keeping the material around the pipe loops moist should be used whenever possible.

Vertical Loops

Additional considerations for backfilling of bored hole applications over and above those for horizontal loops are:

1. The conditions of the soil will determine if the wet rotary drill or dry rotary drill will be used for boring the hole. It will also determine the flushing material or grouting if a water seepage problem is encountered.

2. The depth and number of bore holes for vertical loops depends upon the heating and cooling capacity of the unit, the drilling rate, the site area, soil and rock types, and moisture level.
3. Each loop should be assembled, laid out straight, taped to reduce spring-back friction, and carefully tested for leaks and flow rate *before the hole is drilled* so that it can be lowered into the hole before it can cave in or the mud can settle to the bottom of the hole.

With wet rotary drilling, the pipe is inserted into the hole before the casing is pulled. With dry rotary drilling with a center plug bit, the pipe is inserted into the drill stem after the plug is removed. The drill stem is then pulled up from around the pipe.

The hole should be 5 to 10 feet deeper than the length of the loop to accommodate expansion of the loop. Fill the loop with water prior to insertion. A length of rebar or reinforcing rod should be taped to the loop end of the pipe to stiffen the pipe and provide added weight for insertion of the pipe into the hole. If the pipe is to be grouted in place, it *must be filled with water and pressurized* to a level that will prevent the pipe from being crushed by the denser backfill material.

Figure 15–20 shows the installation and backfill of a vertical loop system using bore holes. The top illustration shows the vertical loops connected in series. The bottom

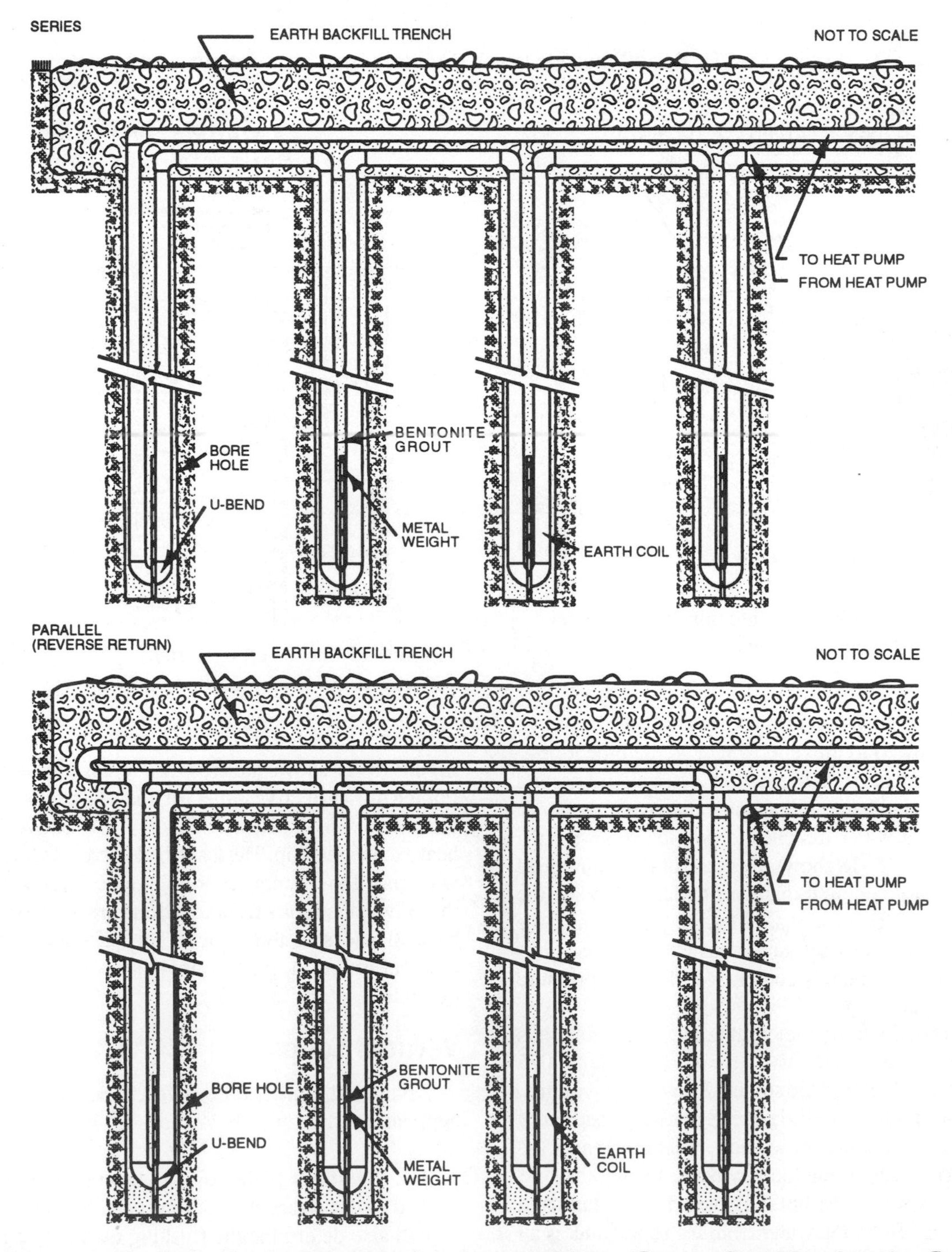

Figure 15–20 Series and parallel connections for vertical loops (Courtesy Bard Manufacturing Co.)

illustration shows the vertical loops connected in parallel to common headers. One header connects the vertical loops as it comes directly from the unit. The other header is connected to the far end of the connection manifold. This provides equal pipe distance and flow resistance for all the vertical loops.

Metal weights, which are taped to the loops, are used to provide pipe stiffening as well as to keep the pipe off the bottom of the hole until the grouting has been installed.

15–5 GROUTING

Bentonite grouting is the most popular for this use. Bentonite and water mixture will swell and adhere to all the surfaces in the well—pipe surface, well wall surface, and so on. It also remains wet and flexible. This means that the expansion and contraction of the pipe due to temperature changes does not affect surface contact. The heat transfer rate between pipe and earth material remains fairly constant.

Bentonite slurries also seal the walls of the bored hole to reduce any water leaks from the soil and the possibility of contamination between water layers in the earth.

For mixing the Bentonite slurry, the manufacturer's recommendations must be followed for the best heat transfer as well as quality of seal produced on the walls of the hole.

15–6 INSIDE PIPING

Figure 15–21 shows a typical ground loop to heat pump piping system. The essential items are pictured and referred to in the illustration.

Earth Loop Circuits. The pump must be connected so as to take the liquid from the earth loop and discharge it into the inlet end of the heat pump heat exchanger coil. This is to ensure that the liquid in the heat pump coil is at a positive pressure at all times.

A permanently installed flow meter is recommended to provide ease of adjusting for the desired flow rate as well as to check for any restrictions to flow developing in the system.

A safety flow switch must be installed ahead of the unit on the discharge side of the pump. This safety flow switch prevents unit operation in the event of circulating pump failure. The switch is a normally open contact in the control circuit to the compressor contactor. This prevents the contactor from closing until the proper flow pressure has been built up in the unit coil.

Thermometer wells are pictured; however, P/T plugs (see Chapter 12) are more desirable in these locations. P/T plugs allow for the checking of pressure drop through the unit as well as temperature change.

A drain valve and outlet must be provided for purging and service operations.

Fill Circuit. The liquid fill circuit is connected ahead of the pump. This circuit provides a source of water supply into the system with a means of controlling the system operating pressure.

Static pressure (pump off) in the loop should be above 10 psig. A minimum of 50 psig is recommended. To provide for pressure change compensation in the system due to temperature change of the liquid, a bladder-type expansion tank is connected into the system in the portion of the piping that is directly connected to the pipe loop.

For illustration convenience, the tank is shown connected to the top of the pipe. *The tank connection should come off the bottom of the horizontal supply pipe to prevent air in the water supply from entering the tank.* The air should be allowed to travel to the air vent riser stub so as to be vented out of the system automatically. The air purge operation of the pipe loop will eliminate the majority of the air in the pipe when filled. However, air in the water supply will be ejected from the water at a slow rate and must be vented.

If the system is to use an antifreeze solution, it is put into the system from pressurized drums through the brine charging valve. If no antifreeze solution is to be used, this valve assembly is not needed.

The piping inside the building may be copper tube and fittings or plastic. If plastic pipe and fittings are used, they must be the hot water (gray) Polybutylene, which will handle temperatures up to 190°F. All piping must be insulated to reduce heat loss as well as to eliminate condensation of moisture on the pipe.

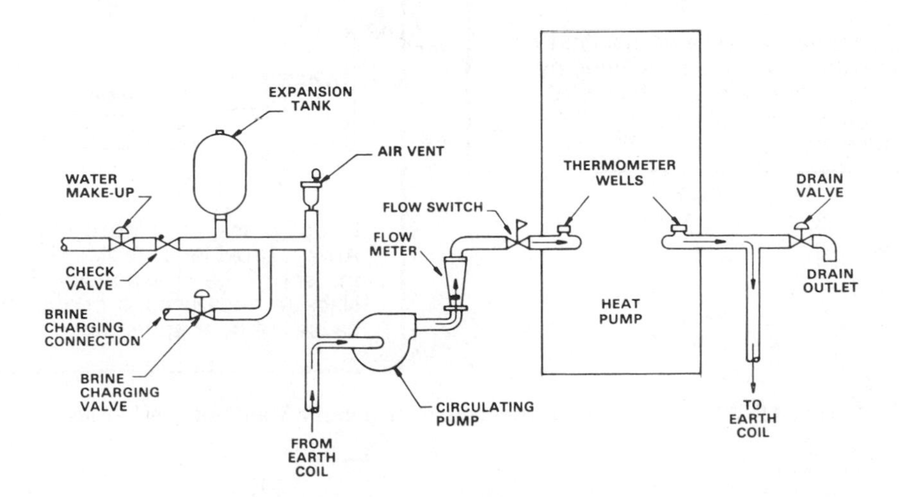

Figure 15–21 Typical ground loop to heat pump piping system (Courtesy Addison Products Co.)

Up to 25' of flexible hose may be used in connecting the pipe system to the unit. The advantage of using flexible hose is the reduction of unit vibration into the piping system. The hoses should be cut to the necessary length with as few bends as possible. Close bends increase pipe head loss so bends should be as wide as possible.

Figure 15–22 shows the possible problems with the use of flexible hose. Illustration (A) refers to twisting the hose to make connections. Illustration (B) shows the necessary precautions for installing the hose next to heat vents, furnace vents, and so on. These locations should be avoided. However, if the location is necessary, precautions to protect the hose must be taken.

Illustration (C) shows the most common problem. Saving on hose length only leads to trouble. The minimum radius for a bend in flexible hose is 10 times the outside diameter of the hose. This is a minimum distance; 15 to 20 times the diameter will produce less head loss if space permits.

Illustration (D) shows that there must always be some slack in the hose. The hose dimensions change with pressure and temperature. Some slack must be in the hose to allow for this.

15–7 TRACER SYSTEMS

A site plan such as is discussed in the beginning of this chapter is the ideal way to ensure the ground loop heat exchanger is locatable if needed in the future. Simply measuring the distances on the site plan provides the pipe joint, vertical well drillings, manifold connections, and other locations for excavation in case of service.

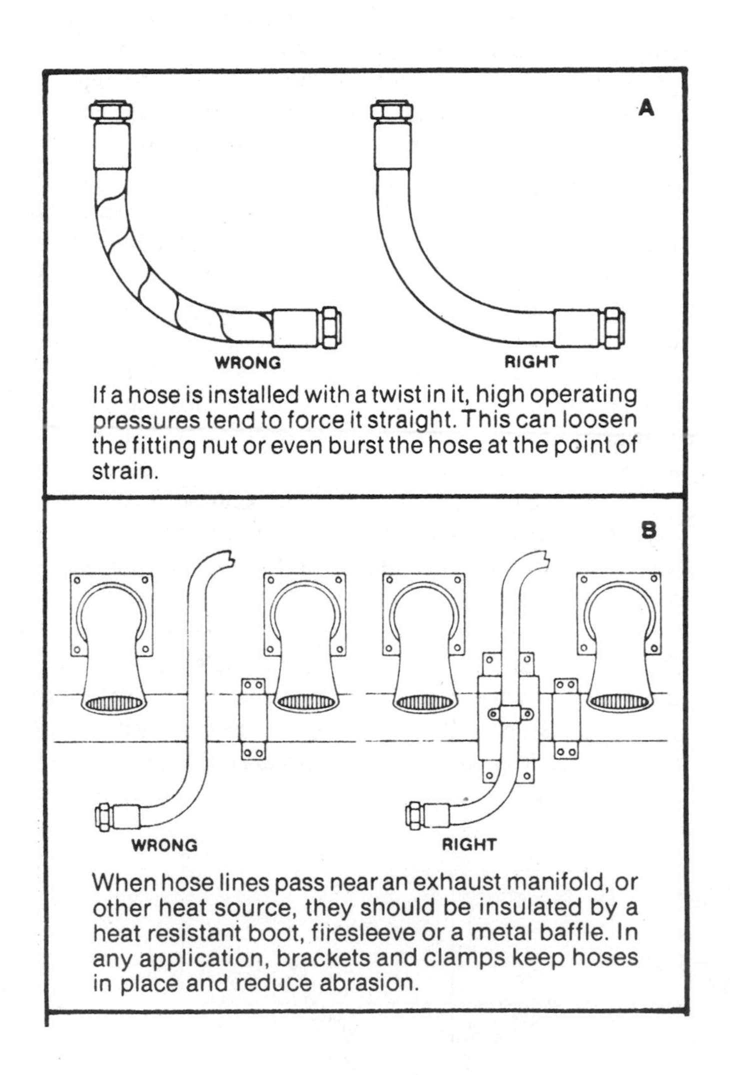

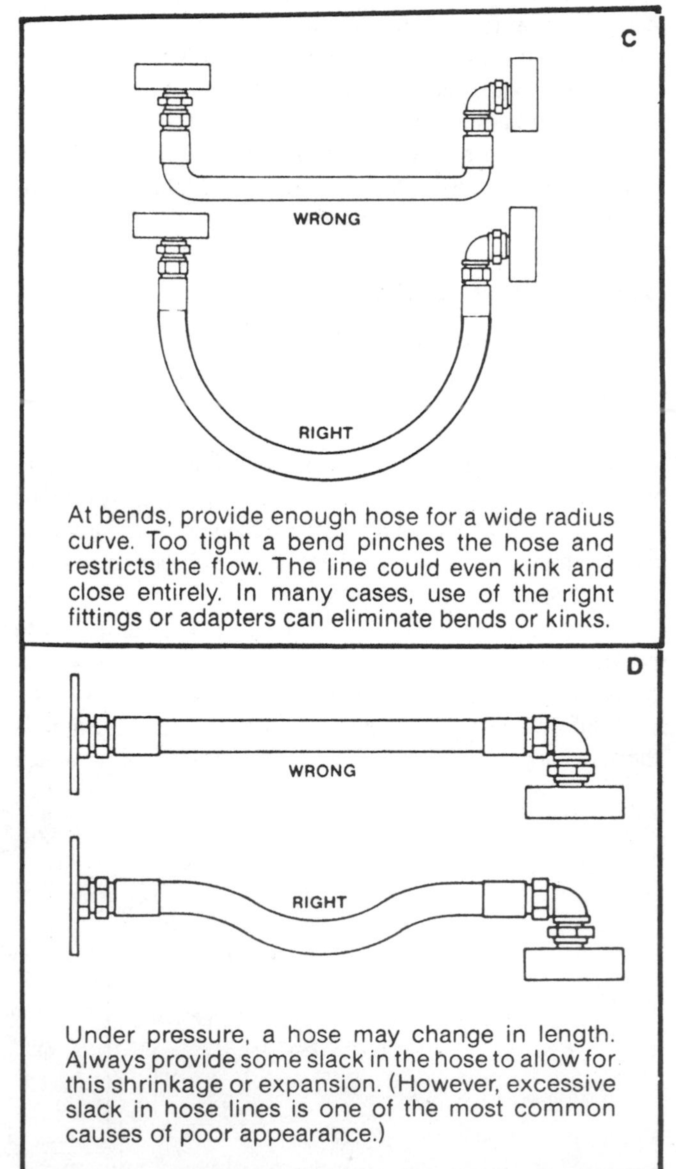

Figure 15–22 Flexible hose connections (Courtesy Mammoth, A Nortek Co.)

Site plans, however, have a way of disappearing with time. Therefore, a more permanent type of locating system is required. Using stainless steel rods and/or wire, a method of locating with portable metal detectors is used.

In Figure 15–14, the right-hand illustration shows a spool of tracer wire. This wire is laid approximately 18" below the finished grade level, directly above the ground loop as the trench is backfilled. Heavy stainless steel clothesline wire should be used for maximum wire life. **DO NOT USE ALUMINUM WIRE.** Aluminum wire is much more difficult to detect than stainless steel and has a shorter life expectancy. The earth salts along with ground moisture rapidly oxidize aluminum.

For vertical loop systems, lengths of steel rebar are located in each drilled hole as well as at the locations of pipe joints and manifold connections.

A tracer system is an absolute requirement in any ground loop heat exchanger installation.

REVIEW QUESTIONS

1. The most reliable connections used in earth coil piping are welded. True or False?
2. List two types of welded or fused joints.
3. When pressure testing the earth loop for leaks, the minimum pressure required is two atmospheres. True or False?
4. In earth loop circuits, where is the circulating pump always located?
5. The minimum pressure in the earth loop with the circulator off is:
   a. 10 psig
   b. 25 psig
   c. 50 psig
   d. 75 psig
6. The minimum radius for the bend in flexible hose is ___ times the diameter of the hose.
7. Locating a steel rod or wire above the buried earth coil is called a _____ _____.
8. The steel rod or wire is always buried approximately ____ inches below grade.

16 CIRCULATING PUMPS

16–1 GENERAL

The submersible pump used in well water supply systems, which consists of several stages of centrifugal impellers, is used to deliver water against high head resistance in GPM ratings up to 150 GPM. This system is classified as an open system because part of the water flow is subject to atmospheric pressures.

The ground loop system is a closed loop system where atmospheric pressure has no effect upon the system pressure or flow rate. The only two factors that affect the pump operation are the GPM flow rate and the flow resistance through the pipe measured in feet of head.

The pumps used in these systems are classified as *circulators*. Their only function is to circulate the liquid through the closed loop(s) at the desired flow rate. The only pressure differential across the pump is the flow resistance of the loop(s).

Figure 16–1 shows an open-motor circulator pump with flange connections into the piping system. Using a magnetic drive, the motor drives a magnetic sleeve that surrounds a magnetic iron block mounted on the same shaft as the pump impeller.

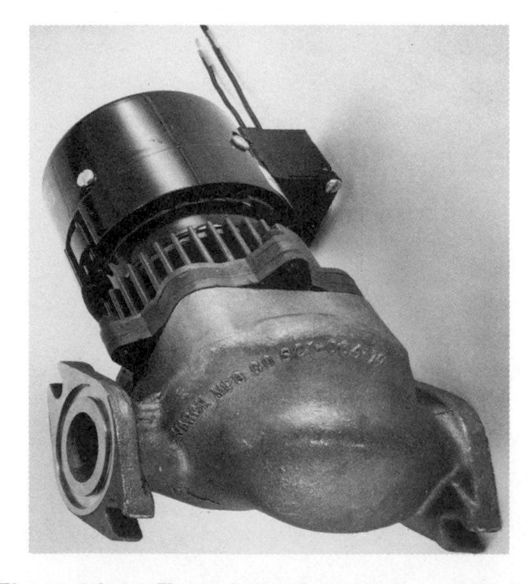

Figure 16–1 Typical circulating pump—March 5 (Courtesy March Manufacturing Co.)

Between the driving magnet and the driven block is an aluminum baffle. This baffle seals the pump and drive block into the liquid circuit and eliminates the need for a shaft seal. A ceramic shaft and bearing assembly is used in the pump portion for the longest bearing and shaft wear life.

Figure 16–2 shows a cut-away of a circulator pump of a different construction. In this pump, the entire rotor assembly of the motor is enclosed in the aluminum baffle for the liquid seal between the liquid system and the motor stator assembly. Ceramic shaft and bearings are also used for long wear life.

These types of construction, which do not use a rotary shaft seal on the motor shaft, will withstand system pressures up to 125 to 150 psig, depending upon the manufacturer's specifications.

Closed loop systems operate with a system static (pump off) pressure between 25 to 50 psig.

16–2 SPECIFICATIONS

Figure 16–3 is a typical specification table for a circulator. Five separate categories of pump sizes are listed from 0.01 (1/100) HP up to 0.20 (1/5) HP in motor size as well as two different AC voltages (115 and 24) and two different DC voltages (12 and 24). The pumping capacities vary from 4.5 GPM at 4.3 FTHD to 14.5 GPM at 31 FTHD.

These capacity ranges are used to select the pipe sizes in the loop. When sizing the loop, an arbitrary pipe size is selected as a start for the pump size calculation. If the end result of the FTHD calculation is higher than the pump range, it will be necessary to redesign the loop system to reduce the FTHD flow resistance.

The goal in the calculation process is to use as small a pump as possible and still get the flow rate through the pipe loop for the best heat transfer rate. After the flow rate and flow resistance in FTHD have been determined for the ground loop system, the pump curves for the pump selected are used to select the pump size (Model No.).

Figure 16–4 shows the performance curves for the circulator models listed in Figure 16–3. These curves show the pumping capacity of each model against varying flow resistances. For example, the Model 809 will deliver 4.46 GPM against 1.0 FTHD down to 0.26 GPM against 4.2 FTHD.

The largest pump (Model 830) will deliver 20.0 GPM against 2.5 FTHD down to 0.2 GPM against 32.0 FTHD.

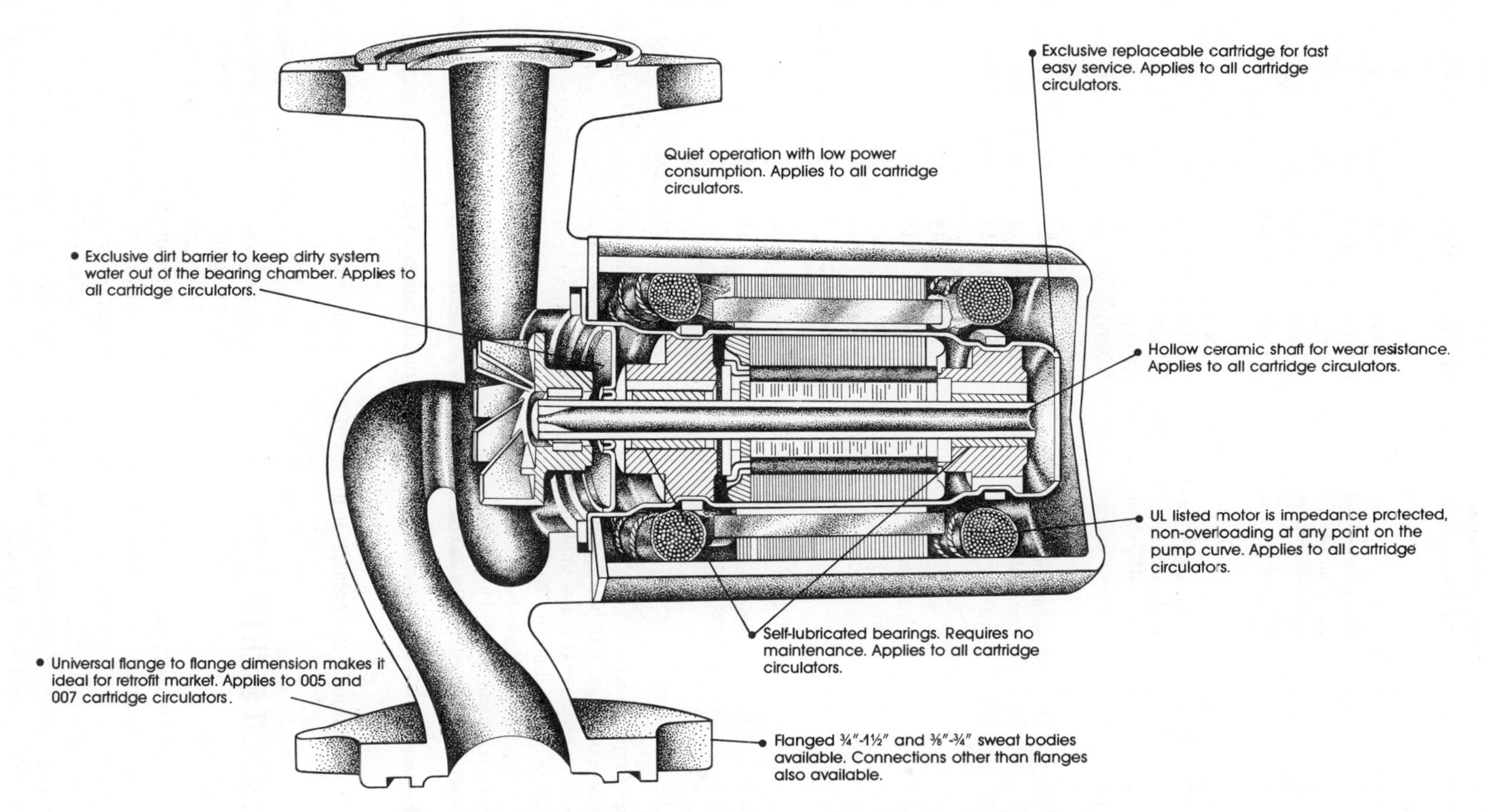

Figure 16–2 Typical circulating pump—Taco (Courtesy Taco, Inc.)

All pumps are rated for 150 psi internal pressure and 250°F water temperature.

Model	Max. Flow GPM	Max. Head Ft.	Intel &Outlet Configuration	HP	Motor♦ RPM	Volts	Hz	Amps	Watts	Materials in contact with solution
*809-BR★	4.5	4.3			1700	115	60	.4	30	
*809-PL	4.5	4.3			1700	115	60	.4	30	
*809-BR-C	4.5	4.3	Inline both ½" MPT center inlet ⅜" FPT & ½" MPT outlet		1700	115	60	.4	30	
*809-BR-24AC	4.5	4.3		1/100	1700	24	60	1.5	24	
809-BR-12DC	5.5	7.1			1950	12◊	D.C.	1.5	18□	Pump Casings: BR is Bronze. PL is Polysulfil Plastic. 316 Stainless Steel. Silicone Rubber "O" Ring, Ryton® and Teflon® Plastic Impeller assembly.
809-BR-24DC	5.5	7.1			1950	24◊	D.C.	.75	18□	
*809-HS-BR★	7.2	12.1			3400	115	60	1.2	90	
*809-HS-PL	7.2	12.1	Inline both ½" MPT center inlet ⅜" FPT & ½" MPT outlet		3400	115	60	1.2	90	
*809-HS-BR-C	7.2	12.1		1/25	3400	115	60	1.2	90	
809-HS-BR-12DC	7.5	15.5			3600	12	D.C.	3.8	48□	
809-HS-BR-24DC	7.5	15.5			3600	24	D.C.	1.9	48□	
815-BR★	8	18.6			3400	115	60	1.3	105	
815-PL	8	18.6	Inline both ½" MPT center inlet ⅜" FPT & ½" MPT outlet	1/25	3400	115	60	1.3	105	
815-BR-C	8	18.6			3400	115	60	1.3	105	
*821-CI	22	8.1			1700	115	60	1.8	110	
*821-CI-T	22	8.1	Inline Flange■ 90°, ¾" FPT		1700	115	60	1.8	110	CI is Cast Iron, flanged CI-T is Cast Iron, ¾" FPT. BR is Bronze, flanged. BR-T is Bronze, ¾" FPT. 316 Stainless Steel Rear Housing, Ryton Plastic Impeller, Carbon Bushing, Ceramic Thrust Washer and Nitrile Fibre Gasket.
*821-Br	22	8.1			1700	115	60	1.8	110	
*821-BR-T	22	8.1		1/20	1700	115	60	1.8	110	
821 Series any style	23	9.5			1800	12 or 24	D.C.	6.0/3.0	60□	
830-CI	14.5	31			3400	115	60	2.6	290	
830-CI-T	14.5	31	Inline Flange■ 90°, ¾" FPT		3400	115	60	2.6	290	
830-BR	14.5	31			3400	115	60	2.6	290	
830-BR-T	14.5	31		1/5	3400	115	60	2.6	290	
830 Series any style	13.5	32.0			3500	12 or 24	D.C.	24.0/12.0	220□	

★ BR housing available on special order in ① ¾" MPT; ② ⅝" flare tube connection; ③ smooth for solder connection.
♦ Motors come with sleeve bearings. Ball bearings available on special order. All motors have U.L. recognition.
∗ 809 A.C. and 821 A.C. pumps have U.L. recognition under File E43564.
◊ 809 D.C. motor brush life is a minimum of 20,000 hours.
■ ¾", 1", 1¼" & 1½" mating flanges available.
□ Watts draw based on open system. Closed loop system draws less wattage & amps.

Figure 16–3 Typical circulator specifications (Courtesy March Manufacturing Co.)

These capacity ranges are found in the greatest majority of loop systems.

16–3 CIRCULATOR SIZING

Single Loop System

To illustrate the steps used to calculate the circulator size, we will assume that we are installing a horizontal loop system such as shown in Figure 16–5. The pipe for the building, through the loop and return to the building, is installed in a continuous loop with butt-welded joints. Our previous calculation for this application shows 527' of heat exchanger ground loop for heating and 410' for cooling. We will use the longer length of 527' of 1-1/2" Polyethylene pipe in the heat exchanger portion.

To make the connections of the horizontal loop into the building, we will need an additional 10' of pipe (5' in the supply and 5' in the return). This is a total of 537' of the 1-1/2" size.

Inside the building, we will use 1" flexible hose to connect to the unit control manifold and circulator assembly. This will require approximately 10' of the 1" flexible hose.

The control and circulator assembly (Figure 15–21) consists of the circulator pump and flow safety switch as well as the flow meter, tees for the P/T plugs, and to the heat pump liquid-to-air coil inlet and outlet connections.

The unit that was selected for this application was the WPH22–1H (Figure 14–15) liquid-to-air heat pump with a flow rate requirement of 6 GPM. The pressure drop of the water coil in the unit at the 6 GPM flow rate is 4.0 psig (Figure 13–4).

Because the operating temperature of the water in the ground loop could drop below 32°F in the heating mode, an antifreeze solution must be used. This will increase the flow rate require-

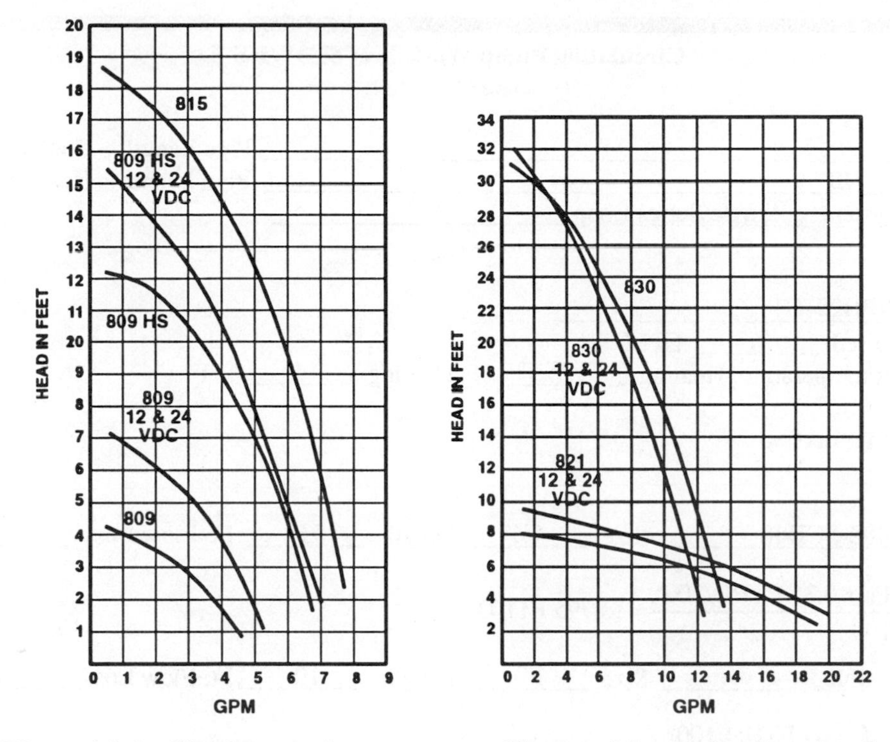

Figure 16–4 Circulator performance curves (Courtesy March Manufacturing Co.)

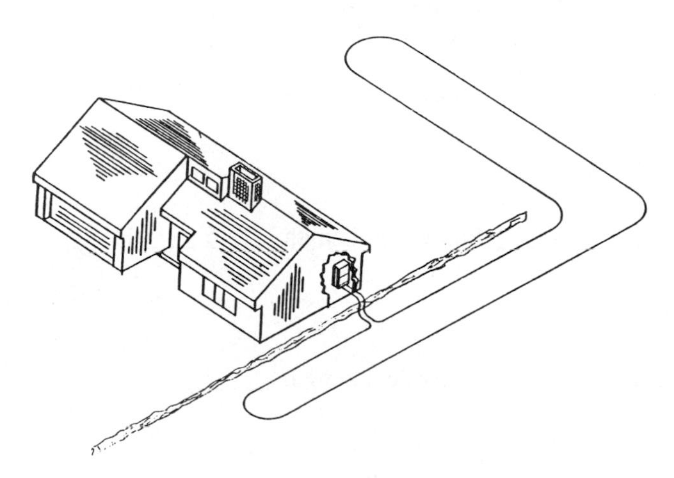

Figure 16–5 Horizontal loop with one pipe in trench (Courtesy Bard Manufacturing Co.)

ment by the factor for the antifreeze solution used. Figure 16–6 shows the flow adjustment factor multipliers that are used for the three types of antifreeze solutions used in closed loop earth coils where the ground temperature drops below 32°F. The particular antifreeze material used will depend upon the local codes that cover this application. The amount of antifreeze solution required is covered in Chapter 17.

Circulator Size Calculation

Using the worksheet shown in Figure 16–7, the following steps are done to determine the size and numbers of circulating pumps that are needed.

Step 1. Record the job name and location.

Step 2. Record the calculated pipe lengths of the buried heat exchanger pipe needed for both the heating and cooling modes. The greater of these two will be used to calculate the loop flow loss. The calculated lengths of pipe for the Omaha, NB application are 410 ft. for cooling and 527 ft. for heating.

Record the type of ground loop system to be installed.

Fluid	Multiplier
20% Propylene Glycol	1.36
20% Calcium Chloride	1.23
20% Methanol Alcohol	1.25

Figure 16–6 Correction factors for antifreeze solutions (Courtesy Bard Manufacturing Co.)

Step 3. Record the unit model number on line 1.

Step 4. Record the unit required flow rate on line 2. This is taken from the manufacturer's specification literature (see Figure 13–4).

Step 5. Record the heating and cooling minimum operating water temperatures the unit will encounter. This information is necessary to determine if an antifreeze solution must be used in the loop.

Pipe

Step 6. Record the various types of pipe used in the installation, the size, length, and flow loss per 100' of pipe. For example, the first pipe recorded is the ground loop pipe which is Polyethylene SCH–40 material, 1½" diameter, 537' in length, with a flow loss/100' of 0.35 FTHD (see Figure 16–8). As was previously discussed, 10' of this pipe has been added to connect from the building to the ground loop, making the

Circulating Pump Worksheet SE3WKS-5
Data Collection

Job Name John H. Doe — Pipe Length—Cooling 410 Ft.
Location Omaha, NB — Pipe Length—Heating 527 Ft.
Type of ground loop system Single Series Loop

A. Unit Information
1. Model No. WPH22-1H
2. Flow Rate Required 6 GPM
3. Operating water temperature: Heating 25 °F, Cooling 45 °F

Pipe and Fitting Information

B. Pipe (Figure 16-8)
Material PE3408-SCH40 Size 1½" Length 537 Ft. Flow Loss 0.35 FTHD/100'

$$\frac{\text{Pipe Length } 537' \text{ Ft.} \times 0.35 \text{ FTHD/100'}}{100} = 1.8795 \text{ FTHD}$$

Material Flexible Hose Size 1" Length 10 Ft. Flow Loss 2.69 FTHD/100'

$$\frac{\text{Pipe Length } 10' \text{ Ft.} \times 2.69 \text{ FTHD/100'}}{100} = 0.269 \text{ FTHD}$$

C. Fittings (Figure 16-9)
Type Tee Size 1" Quantity 4 Equivalent Length 9 Ft.

$$\frac{\text{Quan. } 4 \times \text{Equiv. Length } 9 \times 2.69 \text{ FTHD/100'}}{100} = 0.9684 \text{ FTHD}$$

Type E11–90° Size 1" Quantity 4 Equivalent Length 6 Ft.

$$\frac{\text{Quan. } 4 \times \text{Equiv. Length } 6 \times 2.69 \text{ FTHD/100'}}{100} = 0.6456 \text{ FTHD}$$

Type Coupling Size 1" Quantity 2 Equivalent Length 3 Ft.

$$\frac{\text{Quan. } 2 \times \text{Equiv. Length } 3 \times 2.69 \text{ FTHD/100'}}{100} = 0.1614 \text{ FTHD}$$

5. Total Fitting Flow Loss 1.7754 FTHD

D. Controls (Figures 12-7 and 12-24)
Type Flow Switch Size 1" Quantity 1 Equivalent Length 4 Ft.

$$\frac{\text{Quan. } 1 \times \text{Equiv. Length } 4 \times 2.69 \text{ FTHD/100'}}{100} = 0.0427 \text{ FTHD}$$

Type Flow Meter Size 1" Quantity 1 Equivalent Length 4 Ft.

$$\frac{\text{Quan. } 1 \times \text{Equiv. Length } 4 \times 2.69 \text{ FTHD/100'}}{100} = 0.0427 \text{ FTHD}$$

6. Total control flow loss 0.0854 FTHD

Figure 16–7 Circulator sizing

E. Pump Size Calculation

7. Unit water coil flow loss (From manufacturer's specification literature)
(Figures 13-4 and 14-15) ___4___ psig × 2.31 FTHD/1 psig = ___9.2400___ FTHD
8. Total pipe flow loss ___2.1485___ FTHD
9. Total fitting flow loss ___1.7754___ FTHD
10. Total control flow loss ___0.0854___ FTHD
11. Total flow loss ___13.2493___ FTHD
12. Antifreeze solution ___Propylene Glycol___
13. Antifreeze conversion factor ___1.36___ (Figure 16-6)
14. Unit Flow Rate ___6___ GPM × Antifreeze Conversion Factor ___1.36___ = Adjusted Flow Rate ___8.16___ GPM

F. Pump Selection

15. Circulating Pump Mfg. ___March___ and Model No. ___830–C1___
(Figures 16-3 and 16-4)
16. Number of pumps required ___1___

If multiple pumps in series are required, all pumps must be of the same make and model number to have the same performance characteristics.

Figure 16–7 (cont.) Circulator sizing

total pipe length 537'. This addition will vary with each pipe layout.

The second pipe recorded is the flexible rubber hose used to make the connections from the earth coil connections inside the building to the circulator, controls, and unit. This application requires 10' of 1" flexible hose with a flow loss of 2.69 per 100' (see Figure 16–7).

Step 7. Calculate the FTHD flow resistance for the first pipe length listed. 537' of pipe times 0.35 FTHD/100' divided by 100 equals 1.8795 FTHD for the 537'. This calculation is recorded in the formula below the pipe specifications.

Step 8. Repeat the calculation procedure for all additional pipe lengths listed. The second pipe length listed, the 10' of 1" flexible hose times the 2.69 FTHD/100' divided by 100 equals 0.269 FTHD for the 10' length. This is recorded in the formula listed below the specifications for the second pipe length.

In this application, additional pipe specification lines and formulas are not needed.

Step 9. Add the FTHD for each pipe length to obtain a total FTHD for all the pipe in the circuit. Record the result on line 4.

Fittings

Step 10. Record the various types of fittings, the quantity of each type, and the equivalent length of straight pipe of each type. In our example, using the illustration in Figure 16–9, we will have:

- Two 1" tees with end-to-side flow, each equal to 9' of straight pipe.
- Four 1", 90° elbows, each equal to 6' of straight pipe.
- Two 1" couplings, each equal to 3' of straight pipe. These are recorded on the specification lines under C. Fittings.

Step 11. Calculate the total FTHD the fittings will produce at the 6 GPM flow rate required by the unit. The calculation formula under each fitting type is used to find the total FTHD for the fitting type. For example: One 1" tee times 9 equivalent feet per tee times the 2.69 FTHD/100' for 1" pipe at 6 GPM divided by 100 equals 0.2421 FTHD.

Remember that the equivalent length of straight pipe for each fitting is based on pipe the same size and material as the fitting. The equivalent length of straight pipe for a plastic 1" tee is based on 1" plastic pipe. The FTHD/100' factor used in the calculation formula is the FTHD/100' of 1" plastic pipe at the 6 GPM flow rate or 0.35.

This process is repeated for each of the other types of fittings. A separate listing may have to be used if more than four types of fittings are used in the liquid flow circuit.

The Omaha, NB application has:

- Four 1" tees times 9 equivalent feet per tee times the 2.69 FTHD/100' for 1" pipe at 6 GPM divided by 100 equals 0.9684 FTHD for the four tees.
- Four 1", 90° Ell times 6 equivalent feet each times 2.69 FTHD/100' divided by 100 equals 0.6456 FTHD.
- Two 1" couplings times 3 equivalent feet each times 2.69 FTHD/100' divided by 100 equals 0.1614 FTHD.

Step 12. After the total FTHD for each type fitting has been determined, the totals for all fittings are added to determine the total fitting flow loss. The total for our example is 1.7754 FTHD.

PIPE SIZE AND MATERIAL		DI	Piping Feet of Head Loss at Different Flow Rates Per 100 Feet — CPM FLOW RATE 1	2	3	4	5	6	8	10	12	14
Connection Hose 1"		1.050	*	*	*	1.33	1.95	2.68	4.43	6.53	8.99	11.77
PVC ¾"—200 PSI			*	*	*	3.7	5.7	*	*	*	*	*
PVC 1"—200 PSI			*	*	*	1.0	1.9	2.7	4.2	6.3	8.9	11.8
Copper ¾"			*	*	*	4.3	6.3	*	*	*	*	*
Copper 1"			*	*	*	1.5	1.9	2.7	4.5	6.9	9.6	12.8
PE3408 (Polyethylene)		**DI**										
1. SDR-11	¾	0.860	0.31	1.03	2.07	3.41	5.03	*	*	*	*	*
2. SDR-11	1	1.077	0.11	0.36	0.71	1.18	1.73	2.38	3.92	*	*	*
3. SDR-11	1-¼	1.358	*	0.12	0.24	0.39	0.58	0.79	1.31	1.93	2.65	3.47
4. SDR-11	1-½	1.554	*	*	0.13	0.21	0.31	0.42	0.69	1.02	1.40	1.83
5. SDR-11	2	1.942	*	*	*	0.07	0.11	0.15	0.24	0.35	0.48	0.63
6. SCH 40	¾	0.824	0.38	1.26	2.54	4.18	6.16	8.46	*	*	*	*
7. SCH 40	1	1.049	0.12	0.40	0.81	1.33	1.96	2.69	4.45	*	*	*
8. SCH 40	1-¼	1.380	*	0.11	0.22	0.36	0.54	0.74	1.21	1.79	2.46	3.21
9. SCH 40	1-½	1.610	*	*	0.11	0.18	0.26	0.35	0.58	0.86	1.18	1.55
10. SCH 40	2	2.067	*	*	*	*	0.08	0.11	0.18	0.26	0.36	0.47
PB2110 (Polybutylene)		**DI**										
11. SDR-17, IPS	1-½	1.676	*	*	0.09	0.15	0.21	0.29	0.48	0.71	0.98	1.28
12. SDR-17, IPS	2	2.095	*	*	*	0.05	0.07	0.10	0.17	0.25	0.34	0.44
13. SDR-13.5, Cts	1	0.957	0.19	0.62	1.25	2.06	3.03	4.16	*	*	*	*
14. SDR-13.5, Cts	1-¼	1.171	*	0.24	0.48	0.79	1.17	1.60	2.64	*	*	*
15. SDR-13.5, Cts	1-½	1.385	*	0.11	0.22	0.36	0.53	0.72	1.19	1.76	2.41	3.2
16. SDR-13.5, Cts	2	1.811	*	*	0.06	0.10	0.15	0.20	0.33	0.49	0.68	0.88

Notes:
1. These head losses are for water at 40°F temperature.
2. Count each elbow, tee, reducer, air scoop, flow meter, ect., as 3 feet of equivalent pipe length and add to actual measured pipe length for total length.
3. To adjust the total earth loop piping head loss for other antifreezes and water solutions at 25°F, multiply pressure loss on line 2 for water by:

Propylene Glycol – 1.36, Calcium Chloride – 1.23, Methanol Alcohol – 1.25

Figure 16–8 Piping flow resistance in FTHD (Courtesy Bard Manufacturing Co.)

Friction Loss Through Fittings in Terms of Equivalent Feet of Pipe Length

Fitting Type	Material	Nominal Pipe/Fitting Size ("ID) ½	¾	1	1¼	1½	2	2½
Insert Coupling	Plastic	3	3	3	3	3	3	3
Threaded Adapter	Copper	1	1	1	1	1	1	1
	Plastic	3	3	3	3	3	3	3
90° Standard Elbow	Copper	2	3	3	4	4	5	6
	Plastic	4	5	6	7	8	9	10
Standard Tee (Flow Thru Run)	Copper	1	2	2	3	3	4	5
	Plastic	4	4	4	5	6	7	8
Standard Tee (Flow Thru Side)	Copper	4	5	6	8	9	11	14
	Plastic	7	8	9	12	13	17	20
Gate or Ball Valve	Brass	2	3	4	5	6	7	8

Figure 16–9 Friction loss through fittings (Courtesy Bard Manufacturing Co.)

Controls

The flow resistance of the controls that are in the liquid circuit must be included. In our example, we are using a flow safety switch and a flow meter. The flow safety switch, which has an equivalent length of 4 (same as a ball valve), produces a flow resistance of one device times 4 equivalent feet times 2.69 FTHD/100' divided by 100 equals 0.0427 FTHD.

The 1" flow meter with an equivalent length of 4 (same as a gate or ball valve) produces a flow resistance of one device times 4 equivalent feet times 2.69 FTHD/100' divided by 100 equals 0.0427 FTHD.

Step 13. The FTHD flow resistances for all the controls are added to determine the total control flow loss. The controls in our example add up to 0.0854 FTHD. This is recorded on line 6.

Pump size calculation

Step 14. The unit water coil loss for the required GPM flow rate is taken from the manufacturer's specification literature. For our example, the WPH22–1H unit selected (Figure 14–15) has a 4 psig flow loss through the water coil at the 6 GPM flow rate. This is recorded on line 7.

If the manufacturer's specification literature lists the coil pressure drop in psig, convert the pressure drop to FTHD by multiplying the psig by 2.31 FTHD/1 psig.

Step 15. Record on line 8 the total pipe flow FTHD from line 4.

Step 16. Record on line 9 the total fitting flow loss from line 5.

Step 17. Record on line 10 the total control flow loss from line 6.

Step 18. Add the FTHD losses of lines 7, 8, 9, and 10 and record the total on line 11.

Step 19. If the unit is expected to operate at liquid temperatures below 32°F, a 20% antifreeze solution must be used in the loop. Record on line 12 the type of antifreeze solution to be used. In our example, we will use a 20% Propylene Glycol solution.

Step 20. From the antifreeze correction factor table in Figure 16–6, record the correction factor for the antifreeze material selected. For Propylene Glycol, the factor is 1.36.

Step 21. Determine the antifreeze solution adjusted flow rate. The unit flow rate (6 GPM) times the antifreeze adjustment factor (1.36) equals the adjusted flow rate (8.16 GPM).

This means that the circulator must circulate 8.16 GPM to obtain the same heat transfer rate as a water-only flow rate of 6 GPM.

Step 22. Using the manufacturer's specification literature, select the circulator that will deliver the desired flow rate against the calculated FTHD.

Using Figure 16–4, we have selected a March circulator Model 821–C1. This will be a 1/20 HP motor operating on 115 volts with a 1.8 amperage draw. Pipe connections are flanges for easy removal for service.

Step 23. The number of circulators required is recorded on line 16. One circulator will handle the flow requirements for our example.

If the flow resistance is too great for the pump capacity, multiple circulators are used in series or stages. For example, if the total system flow resistance were 22 FTHD at the 8.16 GPM flow rate, the largest circulator would not satisfy the requirement. Two circulators in series would be used, with each circulator handling 50% of the flow resistance, or 11 FTHD at the 8.16 GPM flow rate. This would require two of the Model 830.

Looking at the capacity curves in Figure 16–4, we find that the Model 830 will deliver 8.16 GPM against 19.5 FTHD and the Model 821 will deliver 8.16 GPM against 6.5 FTHD. This adds up to 26 FTHD. It appears that these two circulators will handle the load.

MISMATCHING CIRCULATOR SIZES WILL CAUSE THE SMALLER CIRCULATOR TO OVERLOAD AND BURN OUT. When using multiple circulators or pumps in series stages, they must be of the same manufacturer and model number to get a balanced load handling.

The regulation of the flow rate when balancing the system will have the same effect on both circulators.

Multiple Loop Systems

When calculating the circulator size for a multiple loop system, only one of the loops is used in the calculation. A properly designed multiple loop system will have the same length of pipe, flow resistance, and flow rate in each loop. This is necessary for the highest heat transfer efficiency of the loop. Each loop must contain a proportional share of the pipe length and handle a proportionate share of the flow rate.

For example, in the Phoenix, AZ example, which is a vertical earth coil application with 5 vertical loops, the 2,799 ft. of pipe needed for the cooling mode means 560 ft. of pipe in each of 5 boreholes, which are 280 ft. plus 5 ft. expansion depth or 285 ft. deep.

The GPM flow rate through each coil must be above the minimum flow for the proper turbulence for the best heat transfer rate. Figure 16–10 is a table of the minimum flow for turbulence for pipe sizes from 3/4" to 2" with water and the antifreeze materials used in ground loops. For example, in the Phoenix, AZ application of 5 vertical loops with parallel flow using water above 32°F operation, the minimum GPM per loop is 1.4 GPM. With 5 loops, the minimum total flow rate would be 1.4 times 5 or 7 GPM.

If the system were to operate at 25°F and use 20% Propylene Glycol solution, the minimum flow rate would be 3.3

Nominal Pipe Size (Pipe ID)	Water at 40°F	Calcium Chloride 20% at 25°F	Propylene Glycol 20% at 25°F	Methanol 20% at 25°F
¾" (0.86)	1.1	2.2	3.3	2.4
1" (1.077)	1.4	2.8	4.1	3.1
1-¼" (1.380)	1.7	3.5	5.3	3.9
1-½" (1.676)	2.1	4.3	6.4	4.8
2" (2.095)	2.6	5.3	8.0	5.9

*For each separate loop.

Figure 16–10 Minimum flow for turbulence (GPM) (Courtesy Bard Manufacturing Co.)

GPM per loop or 3.3 × 5 or 16.5 GPM total. The circulator would have to be sized for the 16.5 GPM flow rate instead of the 7 GPM flow rate needed for the unit.

With a water flow rate of 7 GPM for the unit, each of the 5 vertical loops will carry ⅕ of the total flow rate or 1.4 GPM.

Using the piping arrangement in the bottom illustration of Figure 15–20 (parallel reverse return), only one of the loops is used to calculate the FTHD flow resistance.

Using the worksheet in Figure 16–11, the circulator size is calculated in the following steps:

Step 1. Record the job name and location.

Step 2. Record the heat exchanger pipe length required for both the cooling and heating modes. In our example, the application in Phoenix, AZ will require 2,799' of 1" PE3408–SCH40 plastic pipe for cooling and 356' for heating.

Step 3. Record the type of ground loop system to be installed. In our example, we will install 5 vertical coils connected for parallel flow. Each vertical coil will contain ⅕ of the total length of pipe or 2,800 divided by 5 equals 560'.

Unit Information

Step 4. Record the unit model number and system flow rate required. We have selected a WPH28–1H unit (see Figure 13–13), requiring a 7 GPM flow rate of 90°F water.

Step 5. Record the operating water temperature in both the heating and cooling modes. The WPH28–1H unit will operate at 45°F water for heating and 90°F water for cooling.

Pipe Information

Step 6. Record the information on the pipe used in one of the vertical coils. The flow resistance of the vertical coil is calculated by using the pipe length in the individual bore holes and a proportionate share of the water flow rate. The water flow rate for each vertical coil will be ⅕ of the total GPM (6 GPM) divided by 5 equals 1.2 GPM per coil.

BEFORE PROCEEDING, CHECK TO MAKE SURE THE FLOW RATE IS ABOVE THE MINIMUM TURBULENCE FLOW RATE. The table in Figure 16–10 shows the minimum turbulence flow rate for 1" plastic pipe containing water is 1.4 GPM. The operating flow rate must be increased to 7 GPM to meet the turbulence requirement.

Using the "Piping Head Loss" table in Figure 16–8, we find that the FTHD/100' for 1" PE3408–SCH40 pipe at 1.4 GPM flow rate is 0.232.

Step 7. Using the pipe length (560') in the single vertical coil and the 0.232 FTHD based on the 1" pipe with 1.4 GPM flow rate, calculate the FTHD for the 560' of pipe. The result is 1.2992 FTHD.

Step 8. Determine the length of the supply and return pipe and record on the second material line. For these pipes, we will use 1½" PE3408–SCH40 pipe.

The vertical bore holes are to be placed 15' apart, requiring 15' of horizontal pipe between the vertical coil connection tees. With 10' of pipe from the building to the first coil plus 15' between the vertical coils, 70' of pipe will be used in the supply pipe from the unit to the first vertical coil.

From the last vertical coil and return bend, another 70' of pipe will be in the return line to the unit. The horizontal lines in the return side of the vertical coils are not included as this would duplicate the pipe length.

To eliminate calculating the flow loss in each section of the header with proportionate amounts of water flow, the full flow rate is used in the pipe in one side of the header system. This will make the calculation slightly on the safe side.

In this application, the supply and return system will carry 7 GPM through the 140' of pipe. This will result in a FTHD/100' of 0.4650.

Step 9. Calculate the FTHD for the 140' of pipe. 140' × 0.4650 FTHD/100' divided by 100 equals 0.6510 FTHD.

Step 10. Record the information for the flexible hose used to connect the ground loop system to the heat pump unit. 10' of 1" flexible hose with a flow loss of 3.5550 FTHD/100' with 7 GPM flow rate.

Step 11. Calculate the FTHD loss for the 10' of flexible hose. 10' × 3.5550 FTHD/100' divided by 100 equals 0.3555 FTHD.

The remaining material and calculation lines are not needed in this application. If other types of pipe or hose were used, these lines would be used to determine the FTHD flow loss.

Step 12. Add all the FTHD flow resistances for all the pipes and record on line 4. Our example adds up to 5.5052 FTHD.

Fittings

We will use the right vertical coil in the lower illustration of Figure 15–20. Water flow will be from the heat pump unit, into the tee and out the branch (side flow), through the vertical loop (including two 90° Ells), through the 90° Ell at the top of the vertical loop, through three tees (flow thru), through two 90° Ells, to complete the header and return to the heat pump unit.

The fittings are:

- One 1½" tee with side flow with an equivalent length of 13' (see Figure 16–9).
- Three 1½" tees with flow thru with an equivalent length per fitting of 6'.
- Three 1½", 90° Ells with an equivalent length per fitting of 8, carrying 7 GPM.
- Two 1", 90° Ells with equivalent length per fitting of 6 carrying less than 3 GPM (1.4 GPM). The smallest pipe loss given in Figure 16–8 is 0.11 FTHD/100'

Step 13. This information is recorded on the first four lines.

Step 14. Calculate the FTHD for each of the first four fittings:

- Side-flow tee—One 1½" fitting times the equivalent length of 13 times 0.4650 FTHD/100' of equal pipe size at 7 GPM flow rate equals 0.0605 FTHD.

Circulating Pump Worksheet SE3WKS-5
Data Collection

Job Name John H. Doe — Pipe Length—Cooling 2,799 Ft.
Location Phoenix, AZ — Pipe Length—Heating 356 Ft.
Type of ground loop system Vertical Coil—Parallel Flow—5 bore holes

A. Unit Information
1. Model No. WPH28-1H
2. Flow Rate Required 7 GPM
3. Operating water temperature: Heating 45 °F, Cooling 90 °F

Pipe and Fitting Information

B. Pipe (Figure 16-8)
Material PE3408-SCH40 Size 1" Length 560 Ft. Flow Loss 0.232 FTHD/100'

$$\frac{\text{Pipe Length 560 Ft.} \times \text{0.232 FTHD/100'}}{100} = 1.2992 \text{ FTHD}$$

Material PE3408-SCH40 Size 1½" Length 140 Ft. Flow Loss 0.4650 FTHD/100'

$$\frac{\text{Pipe Length 140 Ft.} \times \text{0.4650 FTHD/100'}}{100} = 0.6510 \text{ FTHD}$$

Material Flexible Pipe Size 1" Length 10 Ft. Flow Loss 3.5550 FTHD/100'

$$\frac{\text{Pipe Length 10 Ft.} \times \text{3.5550 FTHD/100'}}{100} = 0.3555 \text{ FTHD}$$

4. Total Fitting Flow Loss 5.5052 FTHD

C. Fittings (Figure 16-9)
Type Side-Flow Tee Size 1½" Quantity 1 Equivalent Length 13 Ft.

$$\frac{\text{Quan. 1} \times \text{Equiv. Length 13} \times \text{0.4650 FTHD/100'}}{100} = 0.0605 \text{ FTHD}$$

Type Flow-Thru Tee Size 1½" Quantity 3 Equivalent Length 6 Ft.

$$\frac{\text{Quan. 3} \times \text{Equiv. Length 6} \times \text{0.4650 FTHD/100'}}{100} = 0.0837 \text{ FTHD}$$

Type 90°–Ell Size 1½" Quantity 3 Equivalent Length 8 Ft.

$$\frac{\text{Quan. 3} \times \text{Equiv. Length 8} \times \text{0.4650 FTHD/100'}}{100} = 0.1116 \text{ FTHD}$$

Type 90°–Ell Size 1" Quantity 2 Equivalent Length 6 Ft.

$$\frac{\text{Quan. 2} \times \text{Equiv. Length 6} \times \text{0.1100 FTHD/100'}}{100} = 0.0132 \text{ FTHD}$$

Type Flow-Thru Tee Size 1" Quantity 2 Equivalent Length 9 Ft.

$$\frac{\text{Quan. 2} \times \text{Equiv. Length 9} \times \text{3.5550 FTHD/100'}}{100} = 0.6399 \text{ FTHD}$$

Figure 16–11 Sizing a circulator for a vertical loop with parallel flow

Type 90°–Ell Size 1" Quantity 4 Equivalent Length 6 Ft.

$$\frac{\text{Quan. } 4 \times \text{Equiv. Length } 6 \times 3.5550 \text{ FTHD/100'}}{100} = 0.8532 \text{ FTHD}$$

Type Coupling Size 1" Quantity 2 Equivalent Length 3 Ft.

$$\frac{\text{Quan. } 2 \times \text{Equiv. Length } 3 \times 3.5550 \text{ FTHD/100'}}{100} = 0.2133 \text{ FTHD}$$

5. Total Fitting Flow Loss 1.9754 FTHD

D. Controls (Figures 12-7 and 12-24)

Type Flow Switch Size 1" Quantity 1 Equivalent Length 4 Ft.

$$\frac{\text{Quan. } 1 \times \text{Equiv. Length } 4 \times 3.5550 \text{ FTHD/100'}}{100} = 0.6399 \text{ FTHD}$$

Type Flow Meter Size 1" Quantity 1 Equivalent Length 4 Ft.

$$\frac{\text{Quan. } 1 \times \text{Equiv. Length } 4 \times 3.5550 \text{ FTHD/100'}}{100} = 0.6399 \text{ FTHD}$$

6. Total control flow loss 1.2798 FTHD

E. Pump Size Calculation.

7. Unit water coil flow loss (From Manufacturer's specification literature) (Figures 13-4 and 14-15) 4 psig × 2.31 FTHD/1 psig = 9.2400 FTHD
8. Total pipe flow loss 5.5052 FTHD
9. Total fitting flow loss 1.9754 FTHD
10. Total control flow loss 1.2798 FTHD
11. Total flow loss 18.0004 FTHD
12. Antifreeze Solution None
13. Antifreeze conversion factor ---- (Figure 16-6)
14. Unit Flow Rate 6 GPM × Antifreeze Conversion Factor ---- = Adjusted Flow Rate ---- GPM

F. Pump Selection

15. Circulating pump Mfg. March and Model No. 830 (Figures 16-3 and 16-4)
16. Number of pumps required 1

If multiple pumps in series are required, all pumps must be of the same make and model number to have the same performance characteristics.

Figure 16–11 (cont.) Sizing a circulator for a vertical loop with parallel flow

- Flow-thru tee—3 fittings times equivalent length of 6 times 0.4650 FTHD/100' of equal pipe size at 7 GPM flow rate equals 0.0837 FTHD.
- 90° Ell—3 fittings times equivalent length of 8 times 0.4650 FTHD/100' of equal size pipe at 1.4 GPM flow rate equals 0.1116 FTHD.

Step 15. The same control setup used in the single circuit horizontal loop will be used in this application. The fittings used are two 1" tees, four 1" Ells, and two 1" couplings. These are recorded on the next three fitting lines. If additional type and calculation lines are required, use a separate sheet.

- Flow-thru tee—Two 1" fittings with an equivalent length of 9' per fitting.
- 90° Ell—Four 1" fittings with an equivalent length of 6' per fitting.
- Coupling—Two 1" fittings with an equivalent length of 3' per fitting.

Step 16. Calculate the record the FTHD flow loss for each type fitting using the calculation formula below the type line.

- 1" flow-thru tee—Quantity 2 times equivalent length of 9 times 3.5550 (the FTHD/100' of 1" plastic pipe at 7 GPM flow rate) divided by 100 equals 0.6399 FTHD.
- 1" 90° Ell—Quantity 4 times equivalent length of 6 times 3.5550 divided by 100 equals 0.8532 FTHD.
- 1" Coupling—Quantity 2 times equivalent length of 3 times 3.5550 divided by 100 equals 0.2133 FTHD.

Step 17. Add all the FTHD figures for the fittings in the system and record on line 5. In our example, the total flow loss through the fittings is 1.9754 FTHD.

Controls

We are using the same controls as in the series loop. A flow safety switch and flow meter are the only controls necessary.

Step 18. The 1" flow safety switch with a flow rate of 7 GPM and an equivalent length of 4' produces a flow loss of 1 device times 4 equivalent feet times 3.5550 FTHD/100' divided by 100 equals 0.6399 FTHD.

The 1" flow meter with a flow rate of 7 GPM and an equivalent feet of 4 produces a flow loss of 1 device times 4 equivalent feet times 3.5550 FTHD/100' divided by 100 equals 0.6399 FTHD.

This information and the calculations are inserted in the controls section of the worksheet.

Step 19. The control flow losses in FTHD are added and the total is recorded on line 6.

Step 20. The unit coil flow loss is determined from the manufacturer's specification literature and recorded on line 7. For our example, the table in Figure 13–4 shows that the (WPH28–1H) unit has a pressure loss of 4 psig with a 7 GPM flow rate. This 4 psig pressure drop is recorded on line 7, changed to FTHD by multiplying by 2.31 FTHD/1 psig, and the 9.2400 FTHD is recorded on line 7.

Step 21. The total pipe flow loss of 5.5052 is transferred from line 4 to line 8.

Step 22. The total fitting flow loss of 1.9754 FTHD is transferred from line 5 to line 9.

Step 23. The total control flow loss of 1.2798 FTHD is transferred from line 6 to line 10.

Step 24. The total flow losses on lines 7, 8, 9, and 10 are added and the total is recorded on line 11 (18.0004 FTHD).

Step 25. The unit will operate with 45°F water in the heating mode. Therefore, no antifreeze solution will be used and no antifreeze adjustment will have to be made. The flow rate lines 12, 13, and 14 are not needed.

Pump Selection

Step 26. The pump selected will be sized on the 7 GPM flow rate against a total of 18 FTHD. From the pump curves in Figure 16–4, we select a March circulator Model 830.

Step 27. Only one circulator is needed for this application.

Throughout the discussion, the terms circulator, circulating pump, and pump have been used interchangeably. The reader must accept this usage as the terms are not standardized in the industry. The pumps used in closed loop systems are properly called circulators as their basic design is different from standard high-volume/high-pressure pumps. It is hoped that the use of the term *circulator* will become the predominate term when referring to the circulating device used in closed loop systems.

Only two types of systems have been used to demonstrate the circulator size calculations. The series circuit with all the liquid flowing through the one pipe uses all the basic calculation steps.

When the multiple pipe system is used, the exception to the basic steps is the proportioning of the liquid flow among the pipes. When proportioning the liquid flow rate among multiple pipes, the flow rate through each individual pipe must be kept above the minimum turbulation flow rate for the size pipe to assure proper heat transfer.

REVIEW QUESTIONS

1. If a circulator will not deliver the GPM needed against the system flow resistance, two or more circulators may be used in parallel to provide the necessary volume. True or False?
2. If the flow resistance is too high for a single circulator application, a combination of pumps may be used such as a 12 FTHD and an 8 FTHD circulator to supply a 20 FTHD requirement at the required flow rate. True or False?
3. When calculating the circulator size for a multiple loop system, all the loop resistances are calculated and an average resistance is used. True or False?
4. The multiple loop system has 4 loops. The unit flow rate required is 4 GPM. What will be the minimum flow rate that can be used to size the circulator?
5. If a 20% glycol solution were used in the loop in Problem 4, what would be the minimum GPM flow rate for sizing the circulator?

17 FLUSHING AND CHARGING THE LOOP

17–1 GENERAL

After the earth loop has been completed and pressure tested, it must be:

1. Completely flushed to clean all debris, dirt, and insects out of the system.
2. Filled to remove all the air from the system.
3. Pumped with antifreeze solution, if required, to obtain the proper concentration.
4. Operated to obtain the proper GPM flow rate.

ALL THIS IS DONE WITHOUT OPERATING THE HEAT PUMP UNIT.

Figure 17–1 shows an easily assembled unit for flushing and charging the system. This assembly is made up of a 1 hp pump mounted on a 55 gallon barrel water reservoir with flexible hoses to connect to the loop. A makeup garden hose from the building water supply is used to supply water for the flushing action. For final filling of the loop, a separate source of distilled or purified water is used.

The lift pipe from the reservoir to the pump has a large filter to protect the pump and keep the water clear before it enters the loop. The reservoir will catch all debris, dirt, and so on, and separate it from the water.

Pressure gauges have been installed at the outlet of the pump and the return connection of the loop. These are valuable in determining if there are restrictions in the loop system.

17–2 FLUSHING THE LOOP

Equipment for use in flushing the earth loop system is starting to be marketed. Figure 17–2 shows a typical flush cart that is marketed. The assembly contains the reservoir, pump, filter, gauges, control valves, and connection hoses. Mounted on a two-wheel dolly, it is easy to move and store.

Regardless of the type of equipment used, however, the pump must have enough capacity to provide a sufficient GPM flow rate that will purge the air and flush any material out of the pipe system. This will take a minimum water velocity of 2 feet per second. If the earth loop is a multiple system with parallel loops, the water must move through the parallel loops at a minimum of 2 feet per second.

Figure 17–3 shows a table of the minimum flow rates for proper purging for pipe sizes from ¾" to 2". With loop flow resistances as high as 35 FTHD and flow rates above 22 GPM, the pump motor size should be a minimum of ¾ hp, preferably 1 hp.

When the flushing connections from the reservoir assembly to the earth loop have been made, check to be sure all accessible fittings are secure and tight and any valves in the line are open. If the earth loop is connected to the heat pump unit, isolation valves in the lines to the unit should be closed to prevent debris from being flushed through the unit.

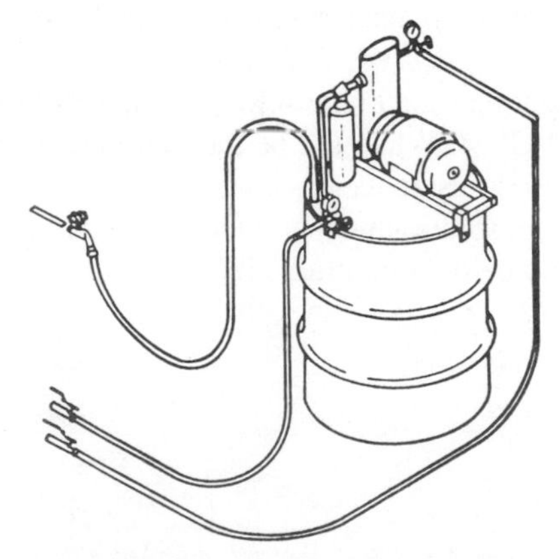

Figure 17–1 Equipment required for flushing and charging (Courtesy Mammoth, A Nortek Co.)

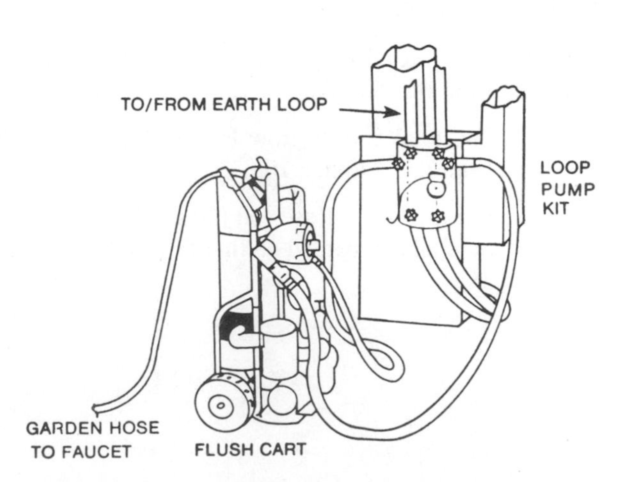

Figure 17–2 Manufactured flush cart assembly (Courtesy Mammoth, A Nortek Co.)

(PIPE ID) NOMINAL PIPE SIZE		FLOW RATE (GPM)
¾	(0.860)	3.6
1	(1.077)	5.7
1¼	(1.38)	9.4
1½	(1.676)	13.8
2	(2.095)	22.0

Figure 17–3 Minimum flow rate for purging earth coils (Courtesy Mammoth, A Nortek Co.)

Fill the reservoir from the water supply and start the pump. This will lower the water level in the reservoir very rapidly. The supply water should be continuous until the level in the reservoir stabilizes.

The liquid in the reservoir must be kept high enough to prevent air from being sucked into the pump inlet. To prevent this from happening, it may be necessary to cycle the pump to allow the water supply to catch up with the drop in water level.

After the level in the reservoir has stabilized and a steady stream of water is returning from the loop, without bursts of air, the pump should be operated for at least 20 minutes. This will help to ensure a thorough flushing of the loop.

At this time, a final leak test should be done on the loop. Establish a 50 psig water pressure on the loop by closing the water return valve and the pump discharge valve when the return pressure gauge reads 50 psig. The pump is cut off immediately to avoid overloading the pump.

Allow the system to stand with the 50 psig pressure for 15 minutes. In this period of time, the pressure in the system should not drop more than 5 psig. If there are no leaks in the system, the pressure will not drop.

If the pressure falls more than 5 psig, it may be necessary to dig holes at coupling or header locations to check for loose or failed connections. This will be easier if tracer wires or rods were installed. With all earth loop systems, check carefully for any visible signs of water leakage before digging or boring down to any coupling or header locations.

NOTE: For proper system operation, there must not be any leaks in the earth loop.

17–3 CHARGING THE LOOP

After the loop has been flushed and given a final pressure test for leaks, the system has to be filled with the solution that will remain in the system to carry the heat energy. To ensure long life of the loop as well as the unit coil, controls, and so on, the final water used should be distilled or purified water. Minerals and gases such as chlorine in the water cause corrosion and deposits that are not needed.

A double quantity of the distilled or purified water is needed in the charging process. The first quantity will be used to flush out the original flushing water from the system. After the original flushing is completed, the reservoir and pump system are drained and flushed clean and a new element is installed in the pump inlet filter.

If the earth loop has not been connected to the heat pump unit, it should be at this time. The access valves to the unit should be opened so that the entire system can be flushed with the final water.

The reservoir is filled with the distilled or purified water. With the pump operating, the first charge of final water is forced out through the system to force out the flushing water. To prevent contamination of the final water, the water from the system is routed to a waste system such as a lawn or garden or into the sewer system. The ejected water has not been contaminated so disposal is not environmentally harmful. The objective is merely to eject minerals and gases from the system.

To calculate the amount of the first charge of final water needed, the table in Figure 17–4 is used. This table gives the volume in gallons that 100' of each size pipe will hold. For example, the Omaha, NB application has 537' of 1½" Polyethylene SCH 40 pipe that holds 10.58 gallons per 100'. Therefore, 537' times 10.58 gal./100' divided by 100 = 56.86 gallons.

An arbitrary 2 gallons is added for the connecting lines, controls, and unit coil. This results in an initial charge of 56.8 + 2 or 58.8 gallons. To allow for some remainder in the flushing equipment, the charge would be rounded off to 60 gallons.

In the Phoenix, AZ application, 2,799 feet of 1" Polyethylene SCH 40 pipe and 70' of 1½" Polyethylene SCH 40 pipe are

Pipe Material	Nominal Pipe Size	Gallon Per 100' of Pipe
Polyethylene		
SDR-11	¾	3.02
SDR-11	1	4.73
SDR-11	1-¼	7.52
SDR-11	1-½	9.85
SDR-11	2	15.40
SCH 40	¾	2.77
SCH 40	1	4.49
SCH 40	1-¼	7.77
SCH 40	1-½	10.58
SCH 40	2	17.43
Polybutylene		
SDR-17 IPS	1-½	11.46
SDR-17 IPS	2	17.91
SDR-13.5 CTS	1	3.74
SDR-13.5 CTS	1-¼	5.59
SDR-13.5 CTS	1-½	7.83
SDR-13.5 CTS	2	13.38
Copper	1	4.3

Figure 17–4 Pipe capacity in gallons per 100' (Courtesy Bard Manufacturing Co.)

used. The 2,799' of 1" pipe contains 4.49 gallons per 100' and will require 2,799 times 4.49 divided by 100 or 125.68 gallons. The 70' of 1½" pipe contains 10.58 gallons per 100' and will require 70 times 10.58 divided by 100 or 7.4 gallons. 126 plus 7.4 plus 2 for the unit and controls adds up to a system charge of 135.4 or 136 gallons.

Fortunately no antifreeze is needed for this installation so the first final flush will usually be sufficient to remove the local water used for the original flushing operation. An additional 10% or 14 gallons of distilled or purified water should be used. After final flushing with the return water returning to the reservoir, the pump should be operated until clear water returns to the reservoir and all air has been purged.

17–4 ANTIFREEZE SOLUTIONS

In all areas except the deep South or the high temperature areas of the Southwest, an antifreeze solution will be required to prevent freeze up of the unit coil. The ground loop circulation liquid freezing point is the main determining factor for the strength of the antifreeze solution. The solution should not be any stronger than is necessary. The stronger the solution, the less the heat transfer ability. However, the solution must be strong enough to prevent freeze up.

The coldest expected temperature of the liquid leaving the unit coil in the heating mode determines the required freezing point of the earth loop liquid. The refrigerant temperature in the refrigerant side of the heat exchanger coil will be approximately 10°F below the leaving liquid temperature. Therefore, it is necessary to use an antifreeze solution with a concentration that will provide freeze-up protection for the minimum boiling point.

The selection of the ground loop antifreeze solution is based on many factors. These are listed in their order of importance:

1. Safety
2. Freezing point
3. Heat transfer characteristics (thermal conductivity)
4. Pumping requirements (frictional pressure drop determined by viscosity)
5. Cost (initial charge, make up, and inhibitor costs)
6. Corrosiveness
7. Physical and chemical compatibility with all parts of the system
8. Availability of material in the local area
9. Ease of transporting and storage
10. Long life

The choices of antifreeze materials are in three categories:

1. Salts—calcium chloride and sodium chloride
2. Glycols—ethylene and propylene
3. Alcohols—methyl, isopropyl, and ethyl

Some of the materials, such as the alcohols, are not desirable but are used. Therefore, they should be discussed.

Salt Antifreeze Solutions

The salts are safe, nontoxic, have good heat transfer characteristics, are low in cost, and have a long life. Their major drawback is that they are corrosive in the presence of air and most metals. Any liquid spills must be thoroughly flushed and cleaned up immediately because of the corrosive action of the solution.

They are considered nontoxic and environmentally safe. However, with the proper selection of metals in the system and complete purging of air out of the system, they can be used very successfully.

Calcium chloride brine is the second lowest cost brine, next to sodium chloride brine. Calcium chloride brines have been used down to –30°F. However, thermal and flow characteristics deteriorate rapidly below –5°F.

Calcium chloride is marketed in flake, solid, and solution forms. The flake form is used more than the other two. The flake form is available in type 1, which has a minimum weight of 77% calcium chloride, and type 2, which has a minimum weight of 94% calcium chloride. The type 2 material is the type recommended to keep foreign material in the solution to a minimum.

Both the calcium and sodium chloride brine solutions can cause considerable corrosion damage. Therefore, regular monitoring and maintenance are necessary.

Prevention of contact of the brine with oxygen and/or carbon dioxide helps prevent corrosion. Closed systems such as ground loop systems have a minimum of contact with gases. However, they still tend to have some corrosion. A strange thing about salt brine solutions is that strong solutions tend to be less corrosive than weak solutions.

Corrosion is best controlled by keeping contact with air to an absolute minimum and maintaining the pH factor of the solution between 7.7 and 8.0. A pH of 7.0 is neutral. Less than 7.0 is acid and more than 7.0 is alkaline. A slightly alkaline solution is best. Acidic, neutral, or strong alkaline solutions will be more corrosive.

Sodium dichromate inhibitor is considered to be the most economical and effective material for reducing the corrosive action of salt brines. Sodium dichromate is marketed as orange crystals that are easily dissolved in warm water. This material dissolves very slowly in cold water. Never add this material to the system without first dissolving it in warm water. The crystals will not dissolve easily and will plug valves and filters.

To use brine antifreeze solutions, the following procedures must be observed and followed:

1. All air must be purged from the system.
2. All air traps at the ends of headers and manifolds must be eliminated. Proper pipe layout and installation is essential.

3. Metal components with high zinc content must be eliminated. This includes and especially applies to circulators, fittings, and valves. Component selection with as much plastic content as possible is essential.
4. Any metals that are high in zinc content are to be avoided. If metal parts are necessary, 300 and 400 stainless steel, cupronickel, or other noble (cathodic) materials should be used.
5. Unit heat exchanger coils should be cupronickel rather than copper.

Following is a list of materials in the order of most desirable to least desirable to use:

1. 300 Series stainless steel
2. 400 Series stainless steel
3. Copper-nickel alloys
4. Bronze
5. Copper
6. Brass
7. Active stainless steel
8. Cast iron
9. Steel
10. Aluminum

The standard solution of calcium chloride and water will provide freeze protection down to +5°F. To calculate the pounds of type 2 calcium chloride flake needed to produce the solution, multiply the gallons of water needed for the earth loop by 2.0 pounds of salt per gallon.

In the Omaha, NB application with 60 gallons of water for the system charge, 60 gallons times 2.0 pounds/gallon = 120 pounds of type 2 calcium flake. Note that type 1 flake is not calculated. Use of this salt is not recommended.

The best application method is to prepare a second charge of the solution mixture made of the heated final charge water with the dissolved salt in a separate large drum or drums that can feed into the reservoir as needed. This solution is pumped through the system by the purge pump with the water in the ground loop going to waste until the brine solution fills the system. The pump should be operated for a sufficient length of time to remove any air that might have entered the system.

Be sure that the reserve tank and flush pump are flushed with clean water immediately after use for purging the brine solution. Any brine left in the flush pump will quickly destroy the pump.

Glycol Antifreeze Solutions

Glycols are safe though slightly toxic, are generally noncorrosive, have fair heat transfer characteristics, and are medium cost with limited life. However, for low-temperature operation, the glycols increase in viscosity requiring more pumping power with reduced heat transfer efficiency. This fact was demonstrated on the worksheet to determine the adjusted FTHD.

The use of propylene glycol antifreeze solution below 15°F liquid temperature results in excess viscosity, which results in excessively high FTHD flow loss and high pumping power requirements. The 15°F temperature is definitely the low limit for closed loop systems for the use of propylene glycol.

The quality of the water used to prepare propylene glycol solution is very important. The water must be distilled or purified (definitely "soft"), and have a low concentration of chloride and sulfate "ions."

Glycol inhibitors are usually added by the product manufacturer. Therefore, the product manufacturer used in the original charge must be used for any additional charge. Indiscriminate mixing of differently inhibited solutions must be avoided unless the different materials are known to be compatible.

Glycol solutions must be well on the alkaline side in the range of 8.8 to 9.2 pH. Solutions that have a pH factor below 7.5 must be replaced. Addition of an inhibitor cannot restore the proper pH to the solution.

The use of chromate treatment in a system using glycols is prohibited as it will result in a rapid buildup of sludge in the system.

Glycol solutions, even though inhibited, do not have an indefinite service life. The solution should be checked on a minimum of an annual basis. Weak solution should be removed and replaced with fresh solution.

Where the ground water at 100' depth or less is below 60°F, a 20% by volume solution of propylene glycol is required. This 20% by volume solution will provide freeze up protection down to 18°F liquid temperature.

To produce the 20% solution, a mixture in the proportions of 100 gallons of water to 20 gallons of propylene glycol is required. For the Omaha, NB application, the 60 gallons of distilled or purified water will require 12 gallons of propylene glycol. To determine the amount of glycol needed, multiply the gallons of water by 0.2.

To charge the glycol solution into the system, it may be added to the reservoir with the flush pump operating. The pump should operate until all the antifreeze material has had a chance to mix in the system. This must include the unit and controls along with the earth loop in the mixing process. Continue to operate the flush pump until all the air has been removed from the system.

Alcohol Antifreeze Solutions

If not properly handled, alcohols are not safe (they explode when mixed with air in vapor form and burn when a liquid exposed to air), have fair heat transfer characteristics, are medium cost, and have an indefinite life. Diluting the alcohol solution before delivery to the job site reduces the explosive hazard. The fact that they are noncorrosive helps their popularity. The many adverse qualities, however, reduce their use potential.

Methyl alcohol, sometimes referred to as methanol, wood alcohol, or carbinol, has been used as an antifreeze. The advantages of methanol solution are low cost, low corrosion activity, low viscosity, good thermal conductivity, and high heat transfer efficiency. The disadvantages of high volatility, high flammability, and high toxicity greatly outweigh the advantages. Pure methanol has a flash point of 54°F to 60°F, while a 30% methanol in water solution has a flash point of 75°F. A 19.41% by volume solution will have a freezing point of 15°F. The use of alcohols as an antifreeze in ground loop systems is not recommended.

REVIEW QUESTIONS

1. What is the minimum liquid velocity needed to purge the air from a ground loop system?
2. When determining the percentage of antifreeze in the liquid, the required freezing point of the earth loop liquid is the same as the average winter temperature for the area. True or False?
3. The minimum freezing temperature is not important because the more antifreeze used, the heavier the liquid, the higher the heat transfer rate. True or False?
4. Antifreeze materials that are used in earth loop systems are in three categories. Name them.
5. Which of the three types of antifreeze in Problem 4 is the most desirable to use?
6. When a solution is neither base nor acid, but neutral, it is said to have a pH number of _____.
7. As the solution becomes more acid, the pH number increases or decreases?
8. When using chloride materials for antifreeze solution, what one metal cannot be used in any portion of the system that will contact the chloride solution?
9. The lowest ground water temperature that can be used for a heat source without antifreeze in the earth coil is _____°F.
10. How many gallons of propylene glycol are needed to produce a 20% solution in an earth coil that holds 50 gallons of solution?
11. A 30% solution of methanol in water solution has a flash point of:
    a. 50°F
    b. 75°F
    c. 100°F
    d. 125°F

WASTE HEAT RECOVERY 18

18–1 GENERAL

Practically all liquid-to-air heat pump manufacturers market an option for the heat pump that is used to heat domestic water. Using the heat in the superheated refrigerant vapor off the compressor discharge, some of this heat is transferred to the domestic hot water system (see Figure 18–1). These units are known as desuperheaters, hot water heat exchangers, energy savers, and so on. Regardless of the name used, the device will reduce the cost of producing domestic hot water, especially if the regular hot water heater is an electrical type.

With the heat recovery coil in the refrigerant circuit between the compressor discharge outlet and the reversing valve, the coil receives the high-temperature superheated vapor from the compressor in both the heating and cooling modes (see Figure 18–2). In the heating mode, the hot water heat recovery coil reduces the heating capacity to the building, which may not save the desirable amount of energy that would be used to heat the water by normal means. In the cooling mode, the hot water coil acts as an additional condenser by desuperheating the hot vapor from the compressor before it reaches the refrigerant-to-liquid heat exchanger in the unit. This is a definite energy saver as this heat would normally be wasted to the earth.

This device is not advantageous in all applications. Where the cooling mode operation is minor, the small savings over conventional hot water heating means may not be justifiable. Where the cooling season predominates, however, the energy savings with this device makes this a desirable investment.

The refrigerant-to-water heat exchanger used in this application is not a single tube-in-a-tube construction. The construction is a three-tube type. This is classified as a tube-in-a-tube-in-a-tube assembly (see Figure 18–3). The inner tube is connected in the refrigerant system. The outer tube, the space between the second and third tubes (the outer shell) is connected into the water system.

The middle tube, the space between the refrigerant tube and the water tube, is open to the atmosphere. In the unlikely event that a leak should develop in the refrigerant tube, the water cannot enter the refrigerant circuit nor the refrigerant enter the water circuit. With a well supply and return system, no refrigerant will enter the water aquifer.

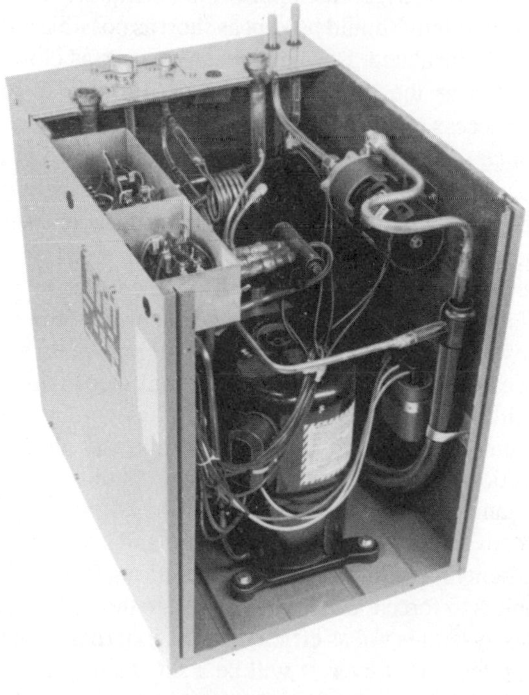

Figure 18–1 Remote high side with heat recovery system (Courtesy Bard Manufacturing Co.)

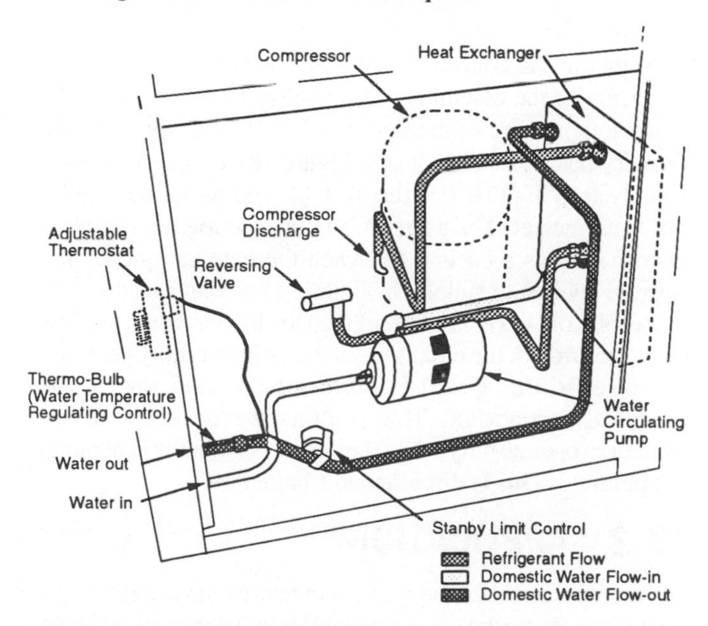

Figure 18–2 Basic refrigerant circuit with heat recovery/exchanger (Courtesy Friedrich Air Conditioning & Refrigeration Co., Climate Master Division)

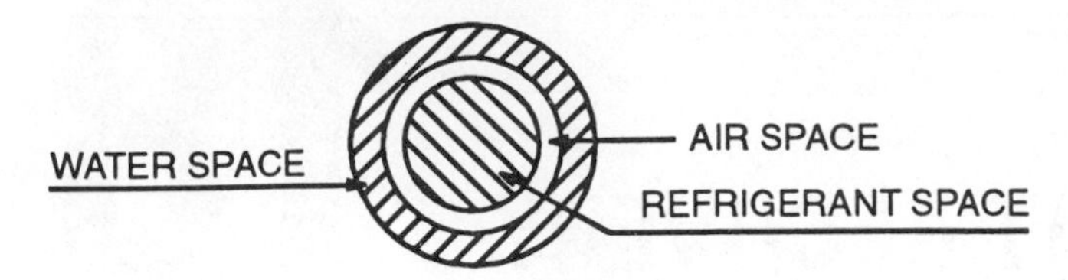

Figure 18–3 Cross section of energy saver heat exchanger (Tube in a tube in a tube)

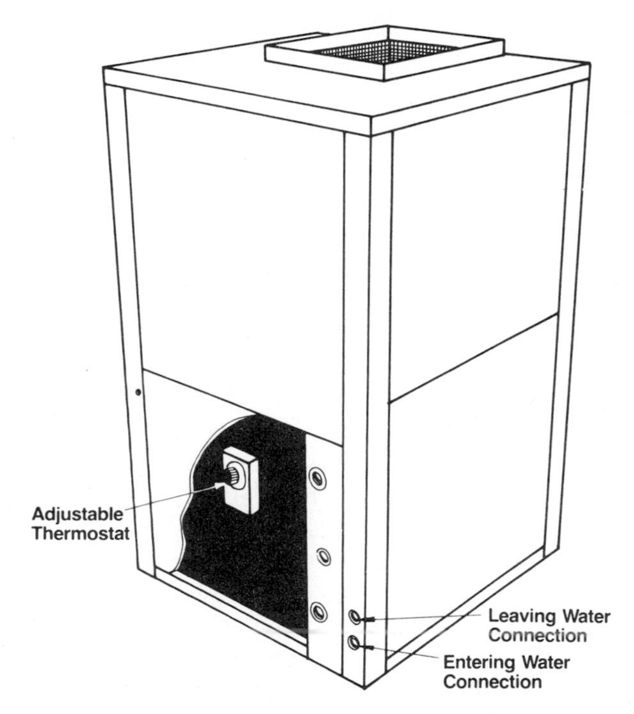

Figure 18–4 Adjustable thermostat for heat recovery operation (Courtesy Friedrich Air Conditioning and Refrigeration Co., Climate Master Division)

A thermostat control controls the operation of the circulator pump in the assembly. This control is used to limit the water temperature of the system when the heat pump is the primary source of energy (see Figure 18–4). This control is usually set 5°F to 10°F higher than the control for the regular hot water heater. The purpose is to use the regular source of energy only as a backup for when the heat pump does not operate enough to supply the building hot water load.

A solenoid valve is also included to close and prevent flow of water through the circuit when the system is not operating.

Power to this circuit is taken off the load side of the compressor contactor. This controls the operation of the system to operate only when the compressor in the heat pump is operating. This is discussed in Chapter 19.

18–2 APPLICATION

In order to realize the maximum energy savings from the heat recovery system, a second tank should be installed along with the main water heater for the building. This tank should be as large as space and economy will permit. In no case should it be sized to supply less than 50% of the daily hot water requirements for the occupants of the building. As a guide in estimating the daily hot water requirements, the U.S. Department of Energy recommends a figure of 16.07 gallons of hot water per day per individual. For example, a family of four would require 64.3 gallons per day. Therefore, the smallest tank to be used would be a 40 gallon capacity.

The most popular size, however, is the 80 gallon capacity. This tank is very popular in solar hot water systems and is built with special dip tube arrangements, without heaters, and has heavier insulation. However, a well-insulated electric water heater without the electric heating elements will also make a suitable storage tank.

Figure 18–5 shows a two-tank system. The domestic water supply enters the auxiliary tank to be heated by the heat recovery unit in the heat pump. The water enters the bottom of the tank via the "In" dip tube to prevent mixing the entering cold water with the top hot water.

Leaving the top of the auxiliary tank, the water flows into the bottom of the domestic water heater, again by a dip tube to prevent mixing. From this heater, it flows into the building distribution system.

The water flow circuit from the auxiliary tank through the heat recovery system is from a connection to what normally would be the drain cock connection. Connected to a tee between the tank and the drain cock, the drain cock is still the lowest port into the tank for the complete flushing and draining of the tank.

For art clarity, the pipes are shown in wide loop configuration. Actually, when determining the location of the storage tank and water heater, remember that the system functions with small temperature differences between the storage tank water and the refrigerant vapor off the compressor. For best results, all tubing should be kept as short as possible and must be heavily insulated. Locate the storage tank as close to the heat pump as the installation permits. Be sure to provide service access to all parts of the system.

Notice that isolating shut off valves are shown in all the lines as well as the inlet and outlet of the heat recovery system. These must be ball valves for minimum flow resistance. A strainer is also used in the line to the heat recovery unit to protect the pump from any particles in the water supply.

Where it is not possible to connect the water lines to make the flow connections to the hot water storage tank, a special fitting can be used in the bottom of the tank (see Figure 18–6). This fitting, called a coaxial fitting, has a tube-in-a-tube arrangement. The outer portion, which makes the physical connection to the tank, carries the cold water out of the bottom of the tank to the heat recovery unit. The hot water from the heat recovery unit enters the center tube and is forced upward by the bent tube inside the tank. The resulting flow of water in the tank is to force the hot water to rise to the top of the tank.

This system is not as efficient as the external connections in the piping. However, it will be a satisfactory connection means if the external connection system is not possible.

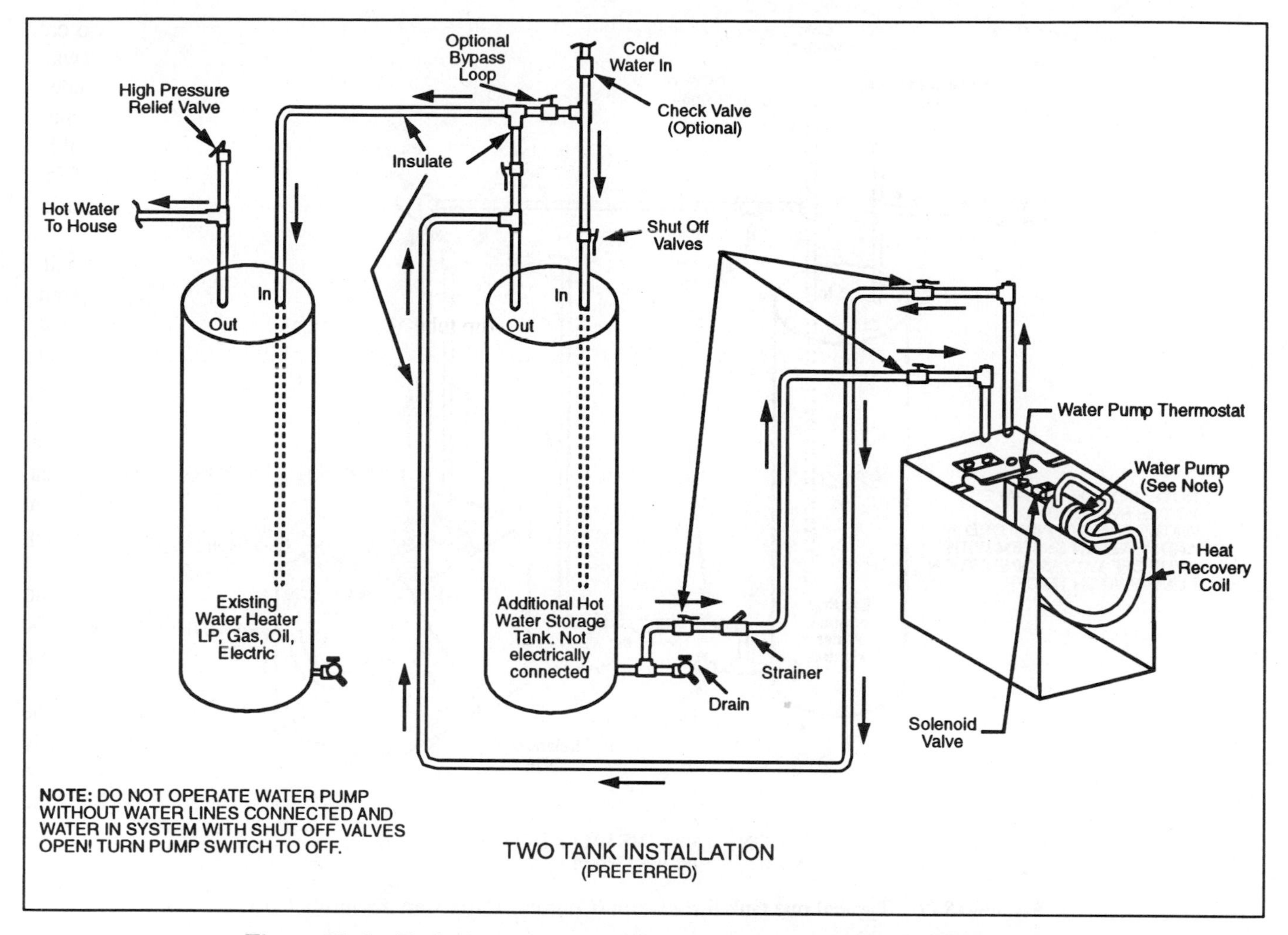

Figure 18–5 Typical two-tank installation (Courtesy Bard Manufacturing Co.)

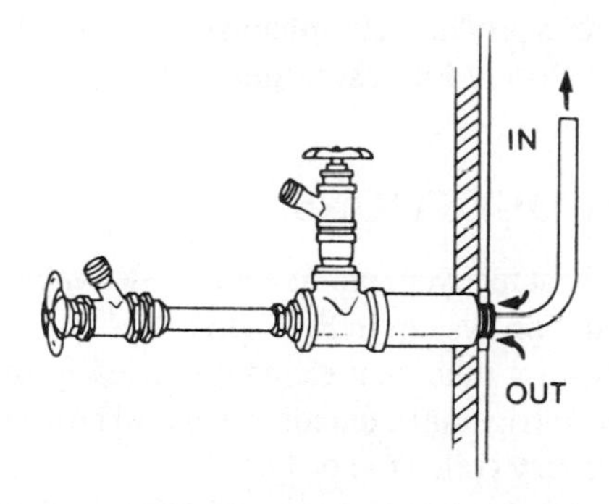

Figure 18–6 Coaxial tank fitting (Courtesy Mammoth, A Nortek Co.)

Where space or cost economy dictates only one hot water heater, a single tank installation is possible (Figure 18–7). The amount of savings, however, will not be as great.

The single hot water tank may be a new hot water heater, sized to handle 100% of the daily hot water requirements, or the existing water heater when the heat recovery system is added to the existing system. If the old water heater tank is used, it must be carefully inspected and cleaned. The tank should be drained and flushed until *all* loose sediment has been removed. This sediment, besides cutting down on the efficiency of the water heater and heat recovery unit, could damage the circulator or plug the strainer and stop the water flow. *A strainer is a must in all add-on systems.*

Single tank systems should be restricted to electric hot water heater tanks only. Fossil-fuel fired tanks should only be used with a two tank setup.

18–3 INSTALLATION

The installation of a heat recovery system is very simple in that standard piping is used. Plugged tees and access valves are installed for service accessibility. All shut-off valves are ball type.

The most important item is the local plumbing codes. All codes, local, state, or national, that apply to plumbing installations in the area must be followed. In practically all metropolitan areas, installation is restricted to a licensed plumber and inspection permits are required.

Where local and/or state codes do not apply, the B.O.A.C. Basic Codes published by the Building Officials and Code Administration, Inc. should be used for maximum safety and protection of the building water supply.

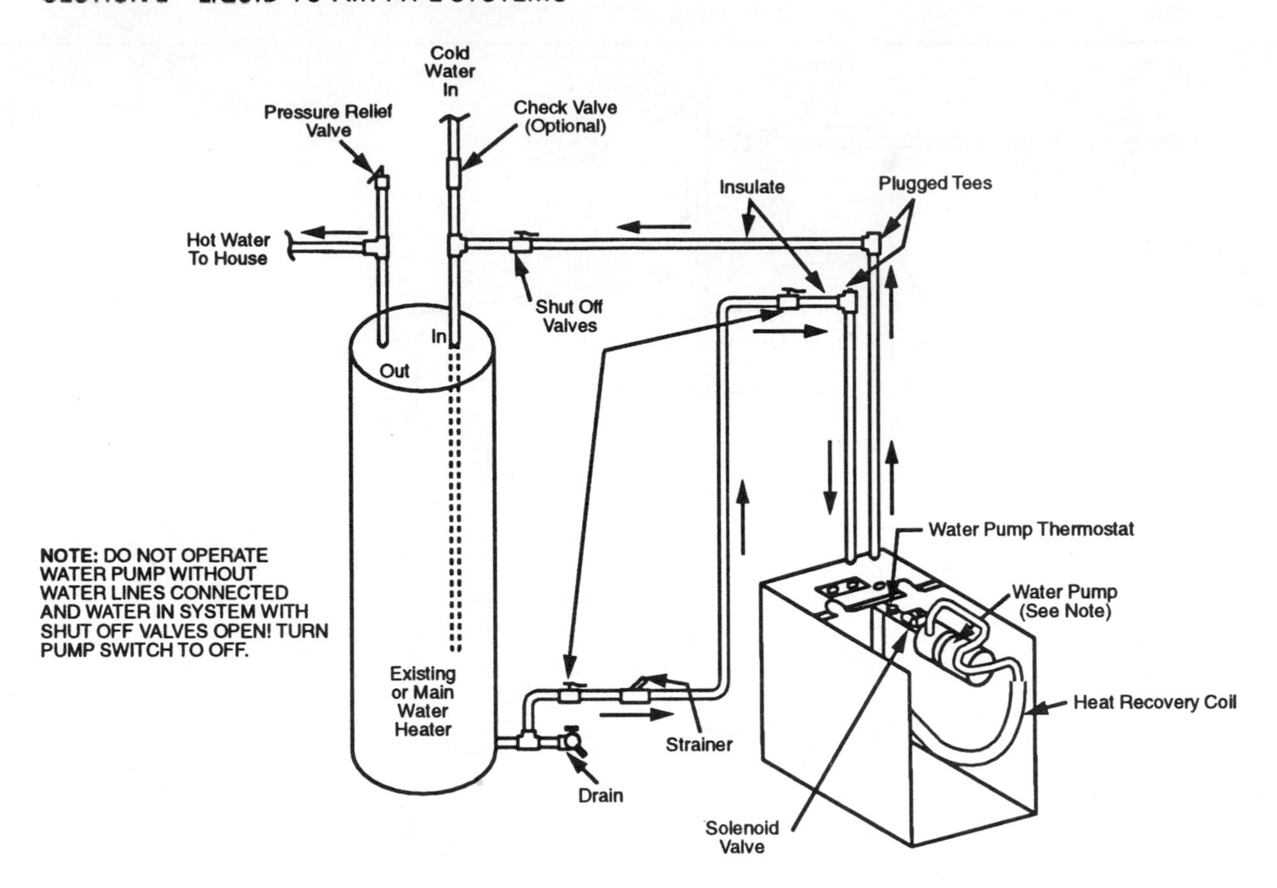

Figure 18–7 Typical one tank installation (Courtesy Bard Manufacturing Co.)

18–4 OPERATION OF THE HEAT RECOVERY UNIT

The circulating pump and solenoid valve are used in parallel with the heat pump compressor. No water can be circulated to the refrigerant-to-water coil unless the compressor is operating. The wiring circuit will be discussed in Chapter 19, Electrical, along with the electrical circuits of the heat pump.

A thermostat, which is attached to the pipe carrying the water from the storage tank, is wired in series and controls both the circulator and the solenoid valve. The thermostat is a normally closed type and opens when the temperature reaches the set point. This will stop the heating of the water by the recovery unit and is intended to prevent overheating the water.

The set point is determined by the amount of running time of the heat pump. It is usually set for a maximum water temperature of 150°F.

In climates that have high cooling requirements, the maximum setting is used for greatest economy. However, this applies to the two tank system. The maximum setting for the single tank system, for human safety, is 130°F.

Methods of startup and checkout will vary with each manufacturer's product. The manufacturer's literature should be followed closely for best results.

REVIEW QUESTIONS

1. A waste heat recovery unit is a highly desirable addition to a ground loop system in St. Paul, MN. True or False?
2. The tube-in-a-tube heat exchanger used in the unit as a liquid-to-refrigerant exchanger can also be used as a waste heat recovery coil. True or False?
3. If the regular water heater temperature control is set at 120°F, the waste heat recovery unit thermostat will be set at _____°F to _____°F.
4. When calculating the hot water requirement, the average gallon per day usage per individual is _____ gallons.
5. A special fitting that allows the supply and return connections to be made at a single port opening on the tank is called a _____ _____.
6. When the heat recovery unit is connected to a two tank system, the maximum thermostat setting is _____°F.
7. When the heat recovery unit is connected to a single tank system, the maximum thermostat setting is ____°F.

ELECTRICAL CONTROL CIRCUITS 19

19–1 GENERAL

The control of the liquid-to-air heat pump is essentially the same as the air-to-air units with the exception of the defrost function. The four essential functions are:

1. Conditioned area temperature—heating mode
2. Conditioned area temperature—cooling mode
3. Changeover between heating and cooling modes
4. System protection

19–2 ELECTRICAL CONTROL CIRCUITS

To illustrate these functions and the control and power circuits involved, Figures 19–1 through 19–8 are the wiring diagrams of a typical liquid-to-air heat pump unit. The thermostat in the illustrations is a Honeywell T87F3111 with a Q539J1089 sub-base. Howcvcr, many different types and/or makes of thermostat/sub-base combinations can be used as long as they supply the four basic functions.

The control manufacturer's literature is the best and final source of information as to the compatibility of the control to the particular unit. Before a strange brand or different control is used, the heat pump manufacturer should be consulted for approval.

Figure 19–1 shows the high- and low-voltage circuits that are active with the unit shut off. The high voltage is applied to the primary of the 240V/24V control transformer at all times. The 24V control voltage is applied to the "R" terminal of the thermostat sub-base and the inlet side of the selector switches.

Figure 19–2 shows the heat-off-cool season switch in the heat position. Power is now applied to the outer heat terminal of the single-pole, double-throw mercury contact. No action

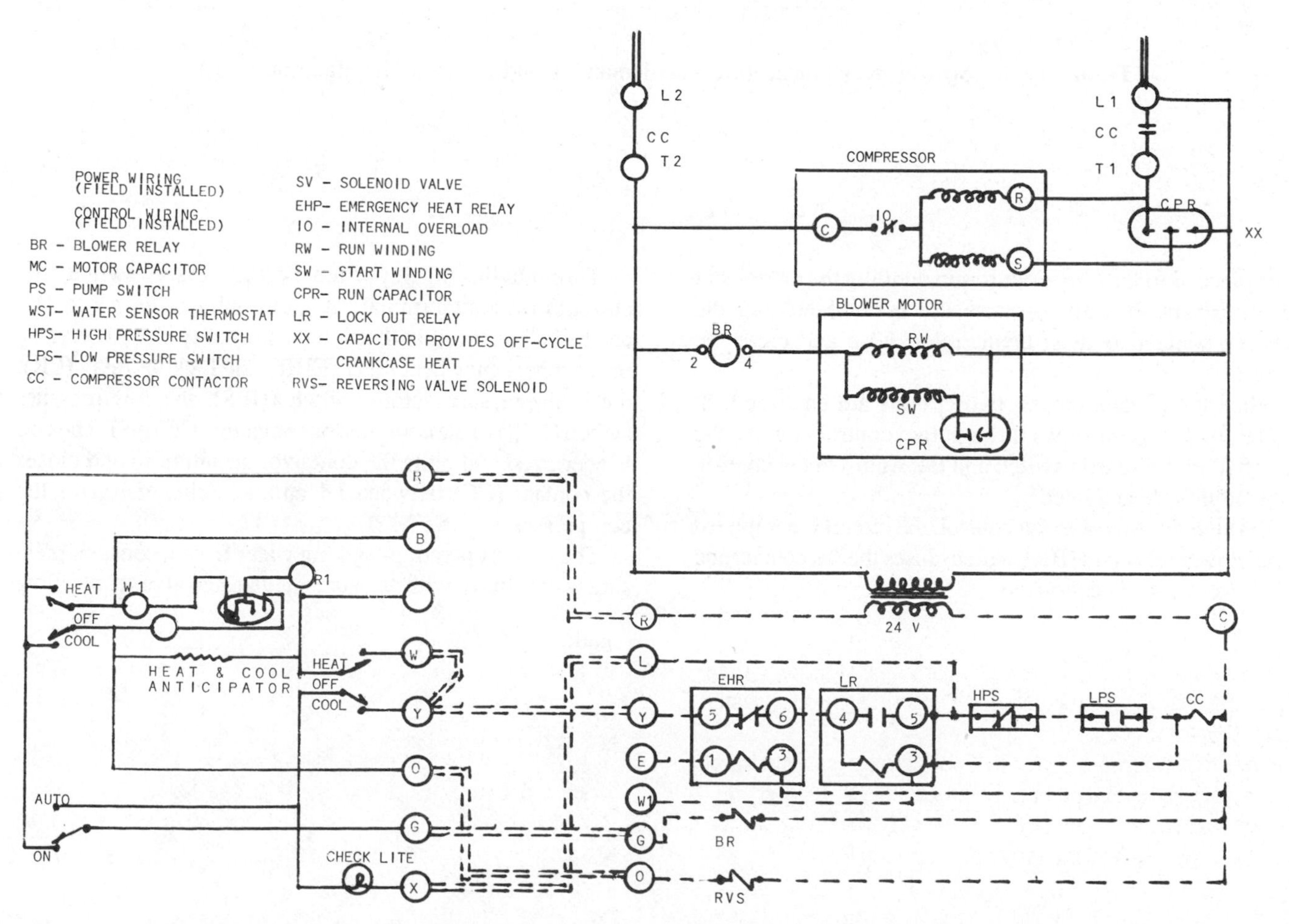

Figure 19–1 Basic control and power circuit—system off (Courtesy Bard Manufacturing Co.)

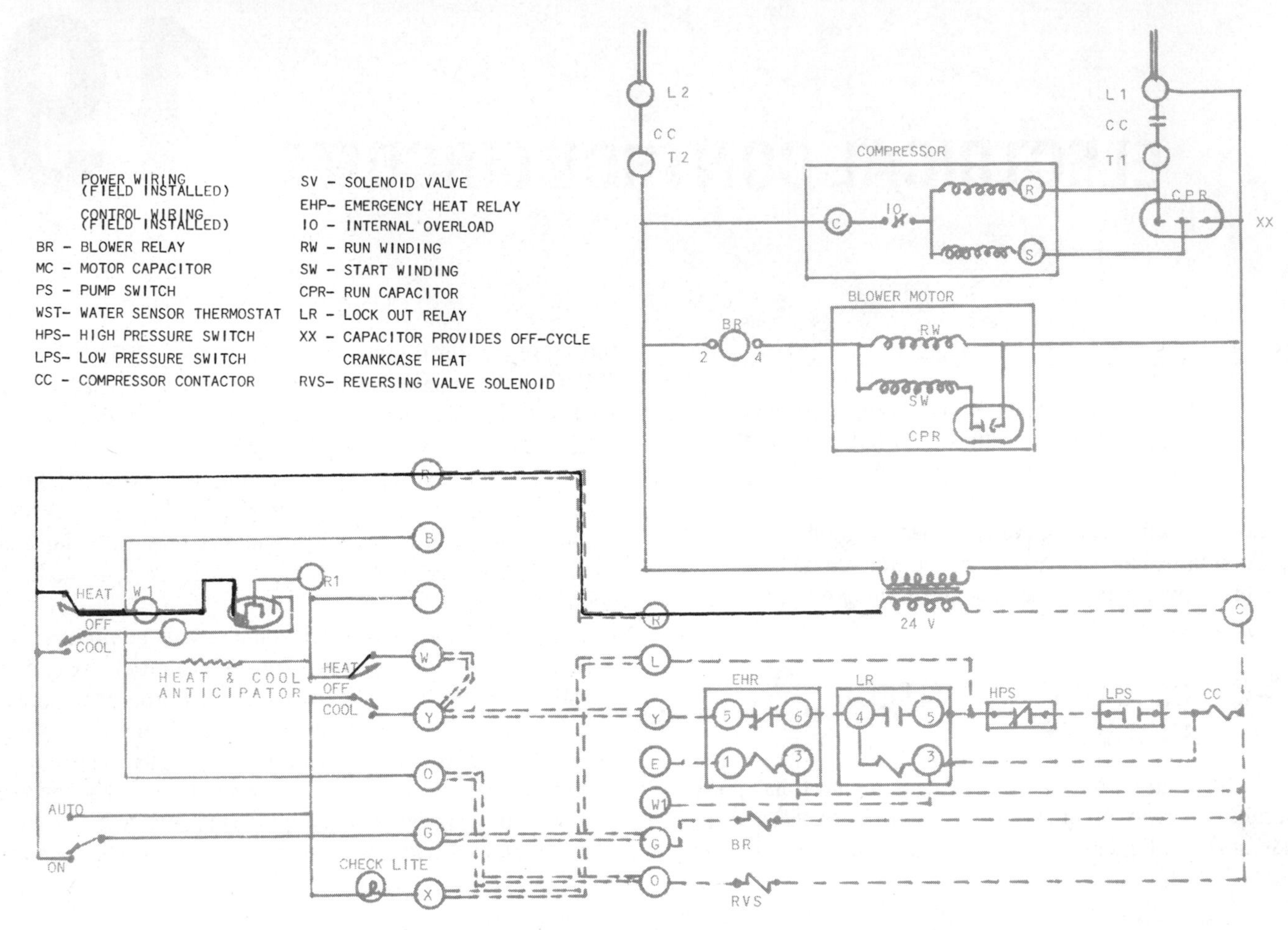

Figure 19–2 Season switch on heating—no demand (Courtesy Bard Manufacturing Co.)

takes place until the temperature surrounding the thermostat element drops enough to cause the bimetal holding the mercury contact to twist counterclockwise and close the contact.

When the contact closes, two circuits are energized. In Figure 19–3, current flows through two control circuits, the other half of the season switch and the Auto-On fan control. Both circuits are activated.

Through the Auto-On fan control, the circuit is completed to the blower relay coil (BR), which closes the BR contact and the blower motor is energized.

Through the other portion of the season switch, the current flows out terminal "W", through the jumper to "Y" and to "Y" on the unit, through the normally closed contacts of the emergency heat relay (EHR), the lockout relay (LR), the high-pressure cutout switch (HPS), the low-pressure switch (LPS) to the compressor contactor (CC) coil. The coil is energized and pulls the contactor armature in and closes the contact (CC) between L1 and T1. This energizes the compressor motor.

The unit is operating—taking water from a constant pressure water source with pressure-regulating valves controlling

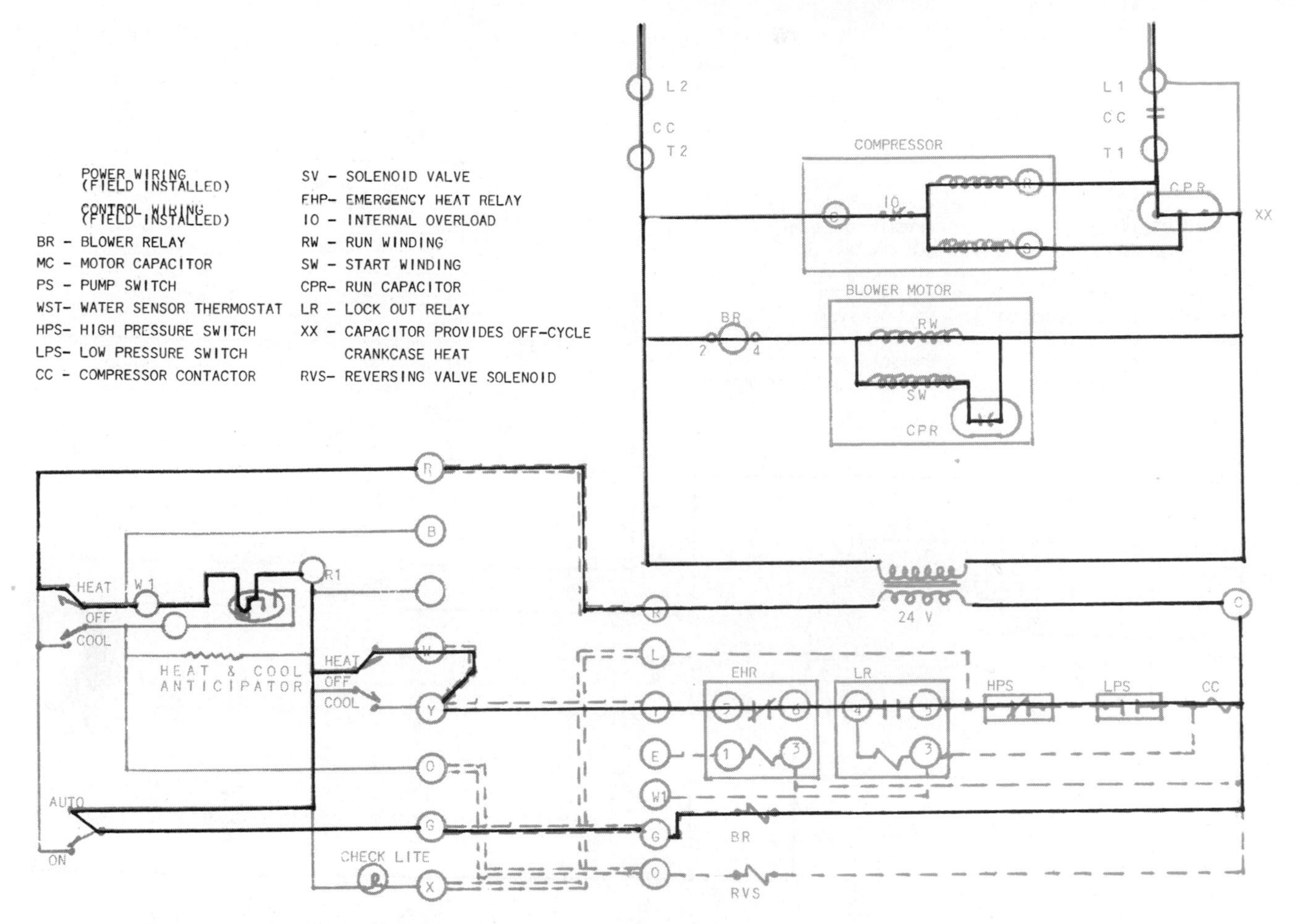

Figure 19–3 Season switch on heating—heating demand (Courtesy Bard Manufacturing Co.)

the water flow. Other types of water sources and ground loop circulator systems will be discussed later.

In Figure 19–4, the season switch is in the cool position without a demand for cooling operation. When the switch was put into the "cool" position, a circuit was made to the "O" terminal of the thermostat sub-base, through the thermostat cable to the "O" terminal on the unit, and to the solenoid coil of the reversing valve (RV). This action energizes the solenoid coil, which remains active as long as the season switch is in the "cool" position. The valve does not reverse, however, until a demand for cooling occurs.

In Figure 19–5, the rise in the temperature surrounding the bimetal element in the thermostat has caused the mercury contact bulb to tip in a clockwise rotation. This closes the circuit to terminal "Y" of the thermostat sub-base, terminal "Y" on the unit, through the relay contacts and pressure controls. The compressor contactor is energized and the contact (CC) closes. This operates the motor-compressor.

At the same time, the blower relay circuit is energized through the "G" circuit and the blower relay (BR) pulls in to energize the blower motor. As soon as the compressor has built sufficient pressure across the reversing valve, the valve

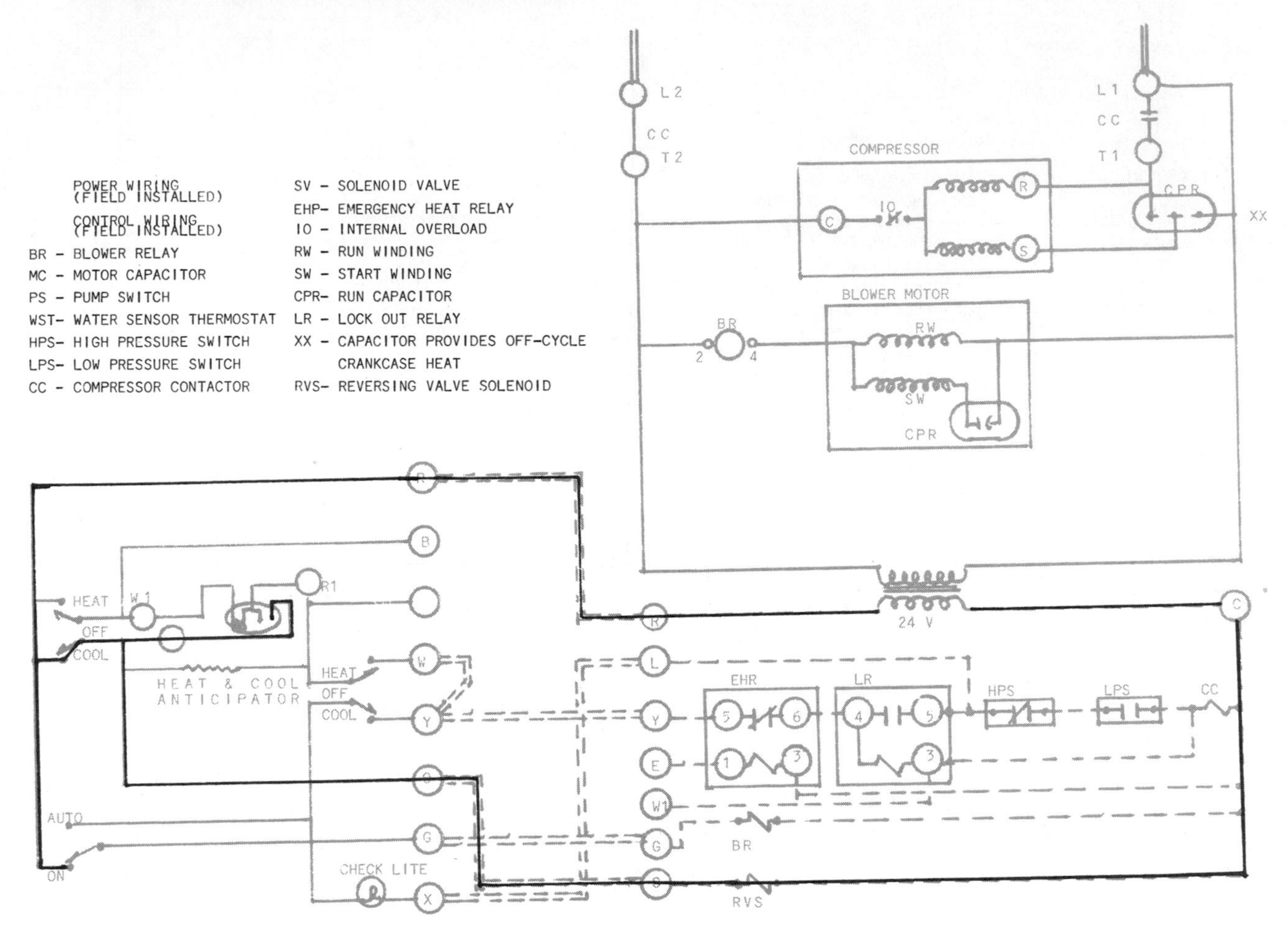

Figure 19–4 Season switch on cooling—no demand (Courtesy Bard Manufacturing Co.)

switches to the cooling mode. The valve will stay in this position, even though the motor-compressor assembly and blower motor cycle on the thermostat temperature control, until the season switch has been switched off the "cool" position.

The designer of this unit had the heating mode in mind as the greater requirement. Therefore, the solenoid valve is energized during the shorter cooling season.

In Figure 19–6, the thermostat sub-base is providing continuous blower operation without unit operation. This provides air circulation with heating or cooling. This action will also provide continuous air circulation (CAC) even though the unit cycles in either the heating or cooling mode.

Heat Recovery Control

When the unit contains a heat recovery unit, the heat recovery pump and flow control solenoid valve are connected across the power supply to the motor-compressor assembly on the load side of the compressor contactor contacts. This allows the heat recovery pump and flow control solenoid (SV) to be active only when the compressor is operating (see Figure 19–7).

The heat recovery pump and flow control solenoid are also controlled by a manual switch and thermostat. With the manual switch, the heat recovery pump and flow control

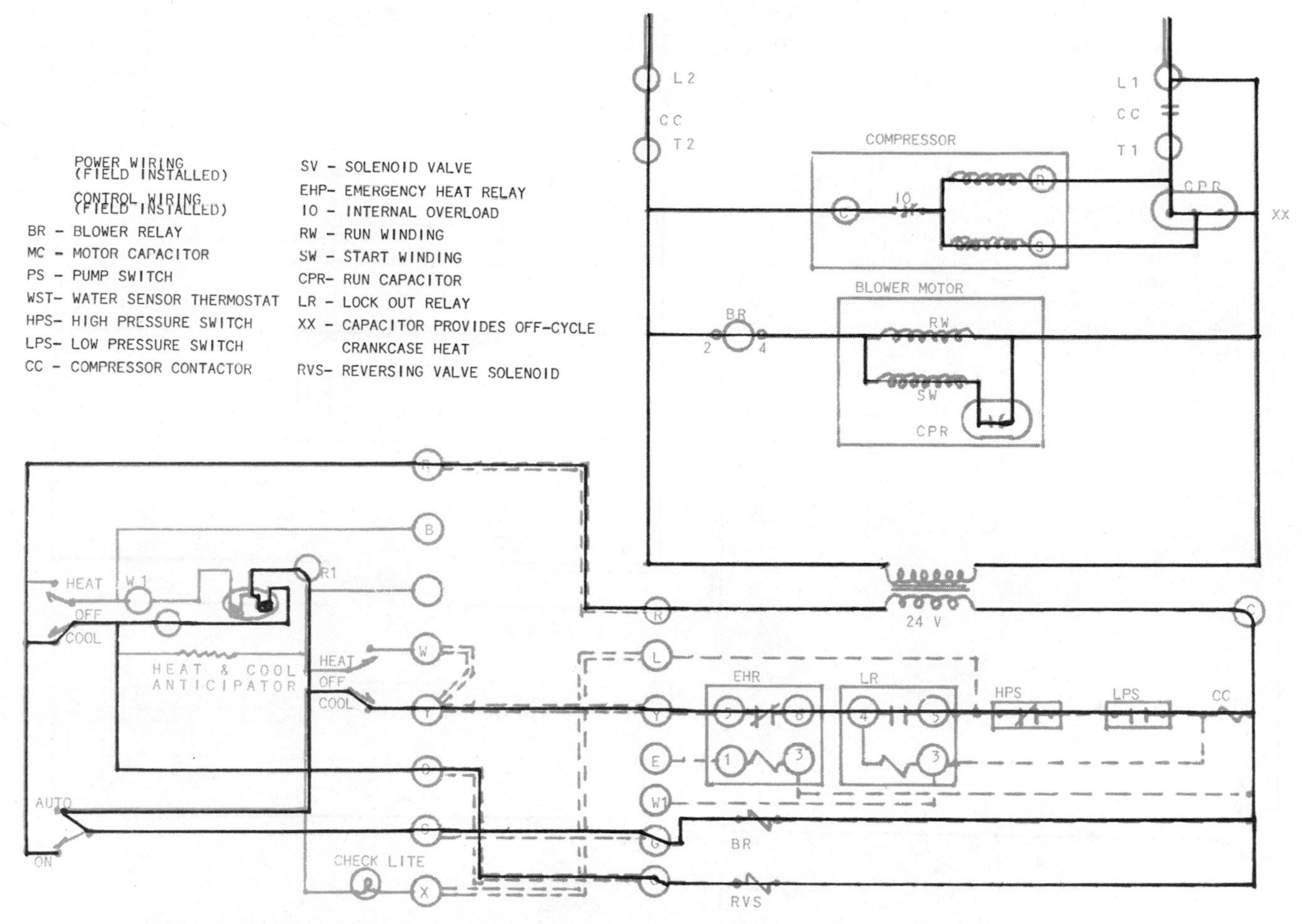

Figure 19–5 Season switch on cooling—cooling demand (Courtesy Bard Manufacturing Co.)

solenoid can be shut down without affecting the operation of the unit. This is a single-pole, manually operated switch.

To prevent too high a water temperature, a thermostat is mounted on the supply pipe from the tank to the pump and flow control solenoid. The thermostat is wired between the pump switch and the pump. As explained previously, this switch should have a top setting of 150°F on a two-tank system or 130°F on a single-tank system.

Lockout Safety Control

This unit is protected by the use of automatic reset high- and low-pressure controls and a lockout system (see Figure 19–8). The lockout relay system furnishes the manual reset protection required for UL approval. However, reset is possible from within the building by interrupting the power to the circuit.

If the high- or low-pressure controls operate to break the circuit to the compressor contactor coil, a circuit is created with the lockout relay coil and contactor coil in series.

The additional resistance of the lockout relay coil reduces the amperage through the circuit so low that the compressor contactor cannot remain in and it drops out and stops the motor-compressor assembly. This action is continuous as long as the thermostat calls for operation.

When the compressor stops and the system pressures equalize, if the system contains a sufficient quantity of refrigerant, the pressure switch will close. The compressor contac-

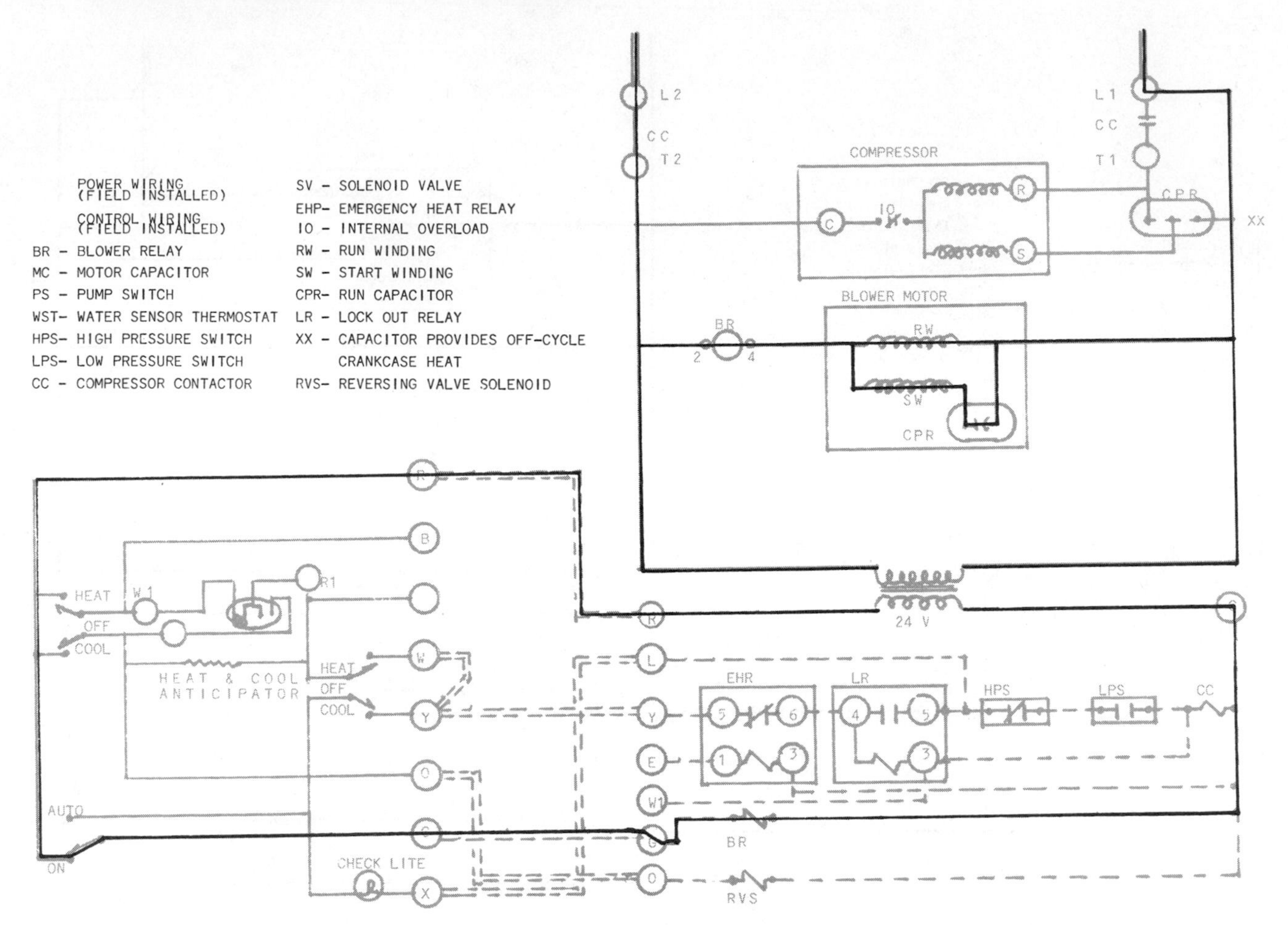

Figure 19–6 Season switch off—fan switch on (Courtesy Bard Manufacturing Co.)

tor will not be energized enough to close, but a circuit will be made through the pressure switches to the check-lite in the sub-base. This light will glow to tell the occupant of the building that the control system is in the lockout mode.

It is only necessary to interrupt the power to the lockout circuit to reset the system to put the unit back in operation. However, the cause of the lockout mode must be found and corrected. Continuously resetting the lockout system will only cause more serious damage to the system. The most common cause of lockout is low suction pressure caused by dirty air filters during the cooling mode or low water supply during the heating mode.

Auxiliary Solenoid Flow Control

Up to this point, the control system has been for a unit using a pressurized well water supply with the well pump operating off the tank pressure switch and the unit using water-regulating valves for control of suction pressure (cooling mode) and discharge pressure (heating mode). When the unit cycles off, the water flow stops due to the action of the water-regulating valves (see Figure 19–9).

If a constant flow valve or manual valves are used to regulate the water flow, an auxiliary flow control valve is required to stop the water flow when the unit cycles off. In our

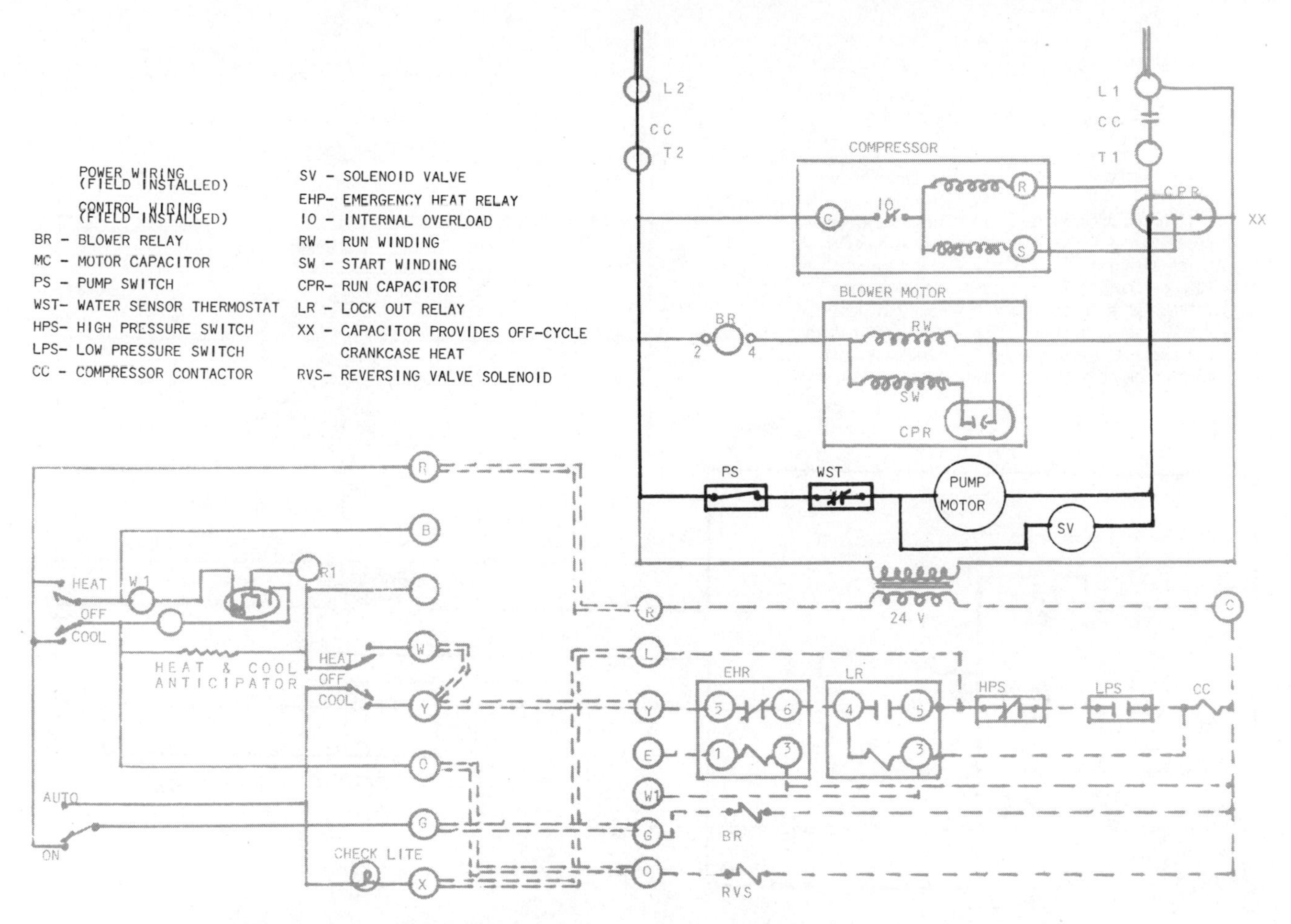

Figure 19–7 Heat recovery control circuit (Courtesy Bard Manufacturing Co.)

example circuit diagram, this valve is connected across terminals "W3" and "C". This energizes the valve whenever the compressor contactor coil (CC) is energized.

The reason for connecting directly across the compressor contactor coil (CC) is to put it into the lockout circuit. If the unit shuts down on lockout, the valve will close to prevent unnecessary water flow.

If unit problems require emergency heat through auxiliary heat units such as electric elements, gas, or oil fossil fuel unit operation, the shutdown of the unit via the emergency heat switch in the thermostat sub-base will also cut off unnecessary water flow.

Ground Loop Circulator Operation

When the unit is used with a buried earth loop, the circulator must be operating before the unit starts. Also, if the circulator or pump stops or the circulation GPM drops, the heat pump must cycle off. This safety feature is accomplished by having the thermostat control the circulator relay and the compressor motor contactor circuit (see Figure 19–10).

In the lead between the thermostat and unit, a flow switch is inserted in this circuit. Even though the thermostat is the master controller, the unit cannot operate unless the liquid flow rate through the ground loop is up to the desired amount.

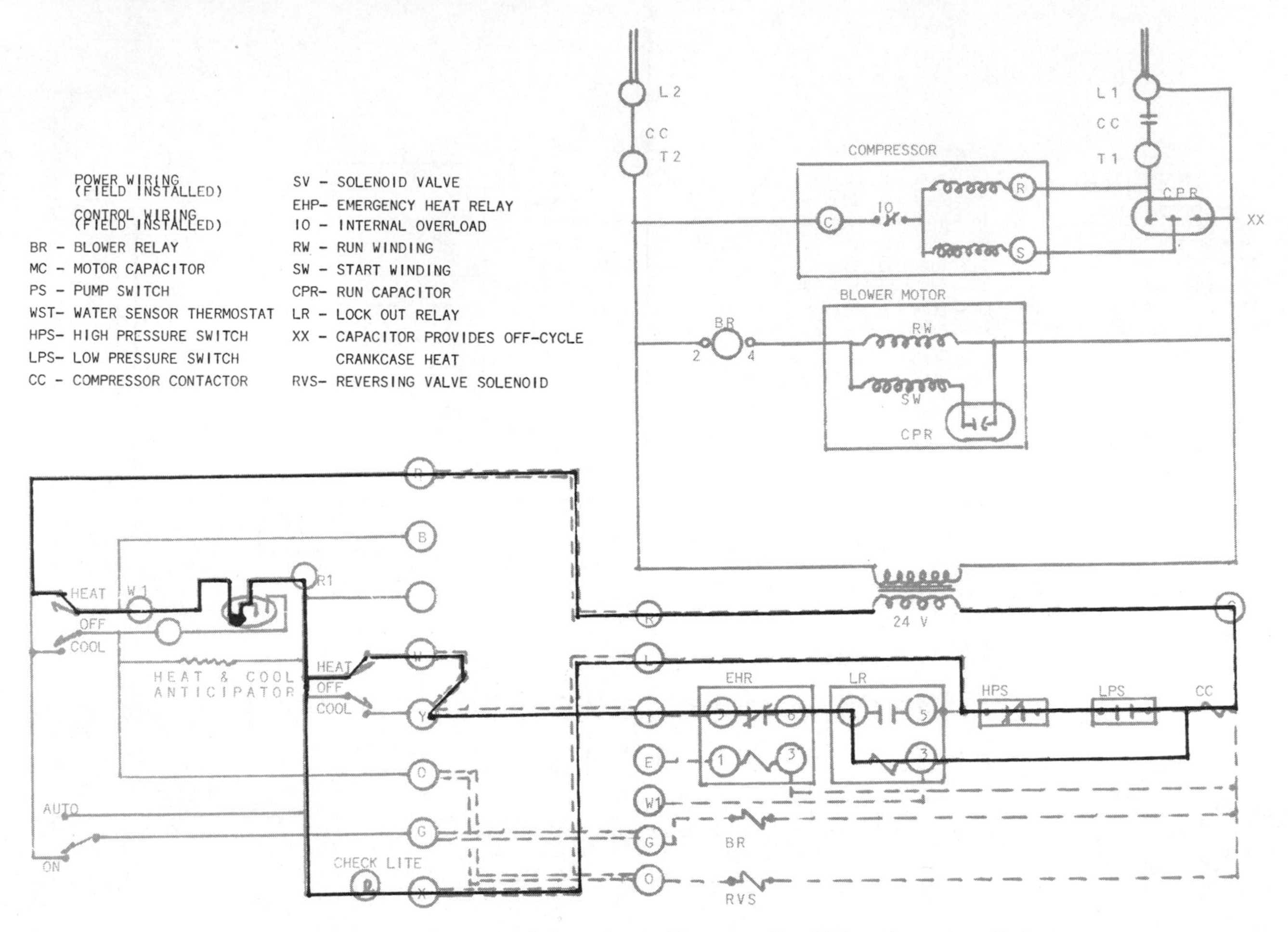

Figure 19–8 Lockout protection circuit (Courtesy Bard Manufacturing Co.)

The flow switch is set to close about 2 GPM below the design flow for the unit. An example of such a switch is the McDonnell and Miller flow switch Model FS4–3, which is mounted in the branch of a 1" tee. This switch is a SPDT switch using the normally open set of contacts to control the unit operation.

An attempt has been made here to illustrate the basic circuitry of the control system in liquid-to-air heat pumps. Sspecific information on particular controls and/or units should be obtained from the equipment manufacturer's literature.

19–3 POWER WIRING

The installation of electrical circuits is closely regulated by local, state, and national codes. Most areas require that such circuits be installed by licensed electricians.

Wiring of approved types and sizes is required to handle the electrical energy requirements of the system. The size of the wire used will depend upon the amperage draw of the unit supplied with power and the length of the power supply circuit.

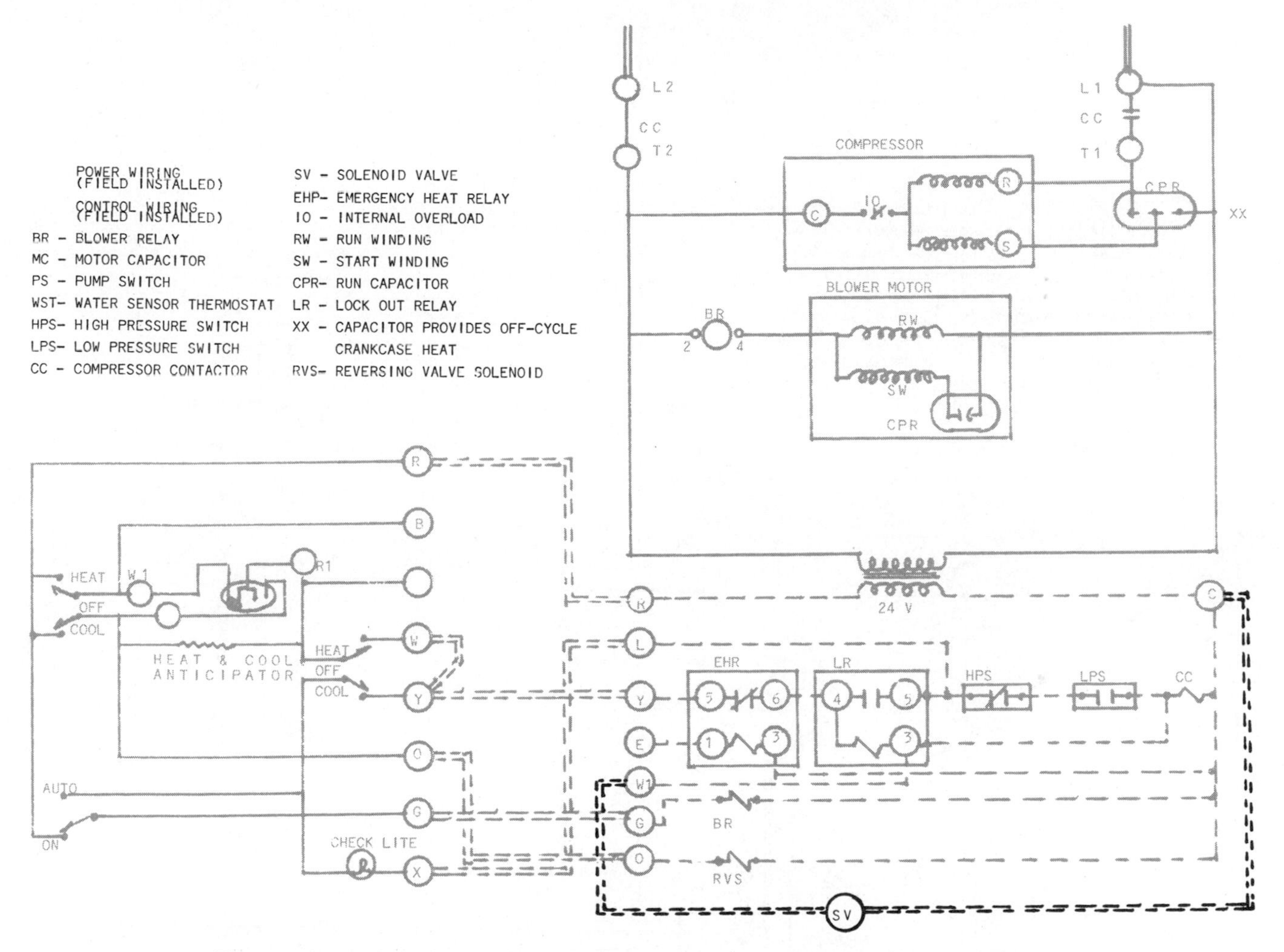

Figure 19–9 Auxiliary solenoid flow control (Courtesy Bard Manufacturing Co.)

All manufacturers include the electrical characteristics of their units in their specification literature. Supplied in table form, Figure 19–11 is a typical listing in the manufacturer's installation instructions. This table supplies the locked rotor amperes (LRA), the full load amperes (FLA), the minimum circuit ampacity (amperage capacity of the wire) based on 100' of wire, and the maximum size time-delay protection required.

The size of the wire is determined from the unit circuit ampacity requirement. For example, the ampacity of 12.2 for the HWPH22–1A unit will require 12 ga. wire.

Two things must be noted here:

1. Copper wire must be used. Aluminum wire is not approved without special connectors in power circuits because of the high locked rotor amperage involved.

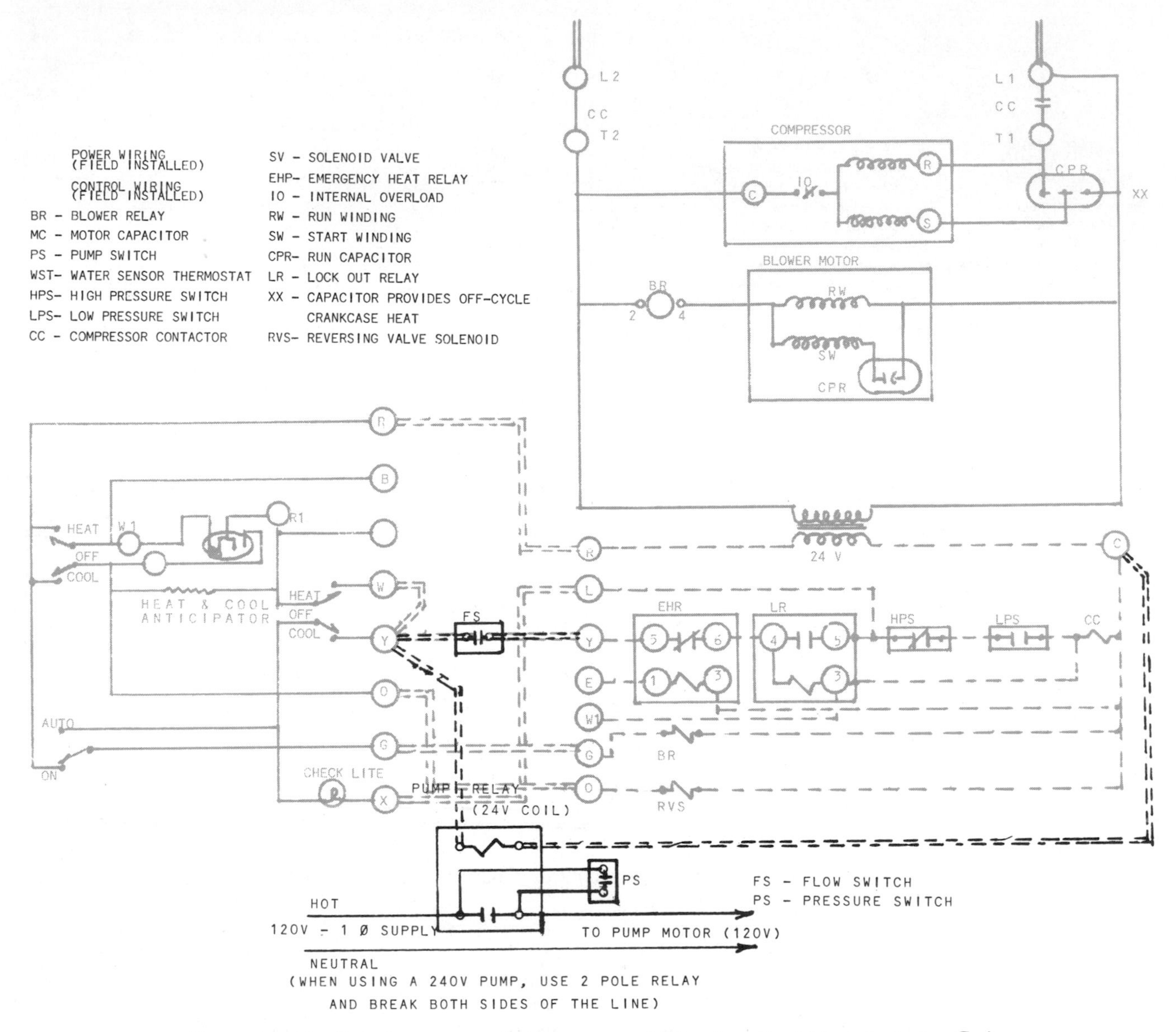

Figure 19–10 Supply pump or circulator control (Courtesy Bard Manufacturing Co.)

2. The fuse size is specified as "time delay type." When a motor, such as a compressor motor, starts, the inrush amperage (LRA) is many times higher than the running amperage (FLA).

For example, the HWPH22–1A unit selected for the Omaha, NB application has an LRA of 64 amperes and an FLA of 12.2 amperes. This unit will require No. 12 ga. wire. The maximum size time-delay fuse or disconnect will be the 20 amps. size for a supply voltage of 240 volts. If the supply voltage were 208 volts, the 35-ampere time-delay fuse would be used along with No. 12 ga. wire.

19–4 CONTROL WIRING

Residential and small commercial heat pump systems are controlled by low-voltage controls in practically all cases. The 18 ga. thermostat wire cable is used to supply control voltage to the contactors, relays, and solenoid valves. The recommended cable is the eight-conductor type, with each wire of a different colored insulation. Attempting to use multiple cables of like colors such as two three-wire cables having red, white, and blue colored wires only leads to confusion and possible equipment damage.

The use of telephone cable is strongly discouraged. This is 22 ga. wire and is too small to carry sufficient current to

MODEL NO.	HWPH22-1A	HWPH28-1A	HWPH34-1A
Standard Cooling Total	21,600	27,000	35,000
Capacity, BTUH (1) Sensible	16,945	18,630	25,000
Standard Heating Capacity, BTUH (1)	29,000	36,500	44,000
Electrical Characteristics, 60 Hz	208/230-1	208/230-1	208/230-1
Compressor Rated Load Amps.	8.3	10.2	14.9
Compressor Locked Rotor Amps.	53.0	64	75.0
Evaporated Blower Motor Hp.	⅕	⅕	½
Blower Mower Full Load Amps.	1.8	1.8	3.2
Operating Amps. (1) Cooling	9.7	12.0	15.2
Heating	9.9	13.2	15.5
EER	10.0	10.0	10.0
COP	3.7	3.7	3.5
Minimum Circuit Ampacity	12.2	14.6	19.2
Maximum Time Delay Fuse	20	20	30.0
Evaporator Face Area Sq. Ft.	1.67	2.08	2.67
Evaporator No. of Rows Deep	4	4	4
Rated Indoor Air Flow, CFM	860	970	1170
Evaporator Blower Type & Size	DD9-7	DD9-7	DD10x8
Evaporator Blower Motor Type	PSC	PSC	PSC
Evaporator Blower Motor RPM	1075	1075	910
Stand. Rated Conds. Water Flow, GPM	5.9	7	9.4
Water Valve Size (Pair) (2)	½"	½"	½"
Water Pressure Drop (PSI) Unit (3)	3.0	4.8	9.8
Valves (3)	3.5	5.0	8.2
Water Connection Size, FPT	¾"	¾"	¾"
Condensate Drain Size, FPT	¾"	¾"	¾"
Filter Size, 1"	20 x 20	20 x 20	20 x 20
Weight, Lbs. Operating	185	198	227
Shipping	205	218	247

Figure 19–11 Unit electrical characteristics (Courtesy Addison Products Co.)

properly operate the electrical devices in the unit. To be on the safe side and give the best operation and equipment life, use thermostat cable in all control circuits.

19–5 FREEZE UP PROTECTION

The electrical control system of a liquid-to-air heat pump is less complicated than the air-to-air heat pump. The major difference is the elimination of the defrost control system. The only protection against freeze up that is required is protection of the water coil in the unit.

To provide this freeze up protection, the unit is shut down whenever the refrigerant boiling point (suction pressure) reaches 36°F. Practically all manufacturers use a nonadjustable low-pressure control for this purpose. The control is nonadjustable to prevent change in the field that could result in unit damage.

When the liquid-to-air unit is applied to a ground loop system that requires an antifreeze solution for operation below 45°F, the nonadjustable control is removed from the control circuit and an adjustable one is substituted. *This substitution must only be done if an antifreeze solution is used in the unit coil circuit.*

After the adjustable control has been mounted in the unit, connected into the refrigeration system (usually by means of a Schraeder to male ¼" adapter on the Schraeder valve fitting provided in the unit), and wired into the control circuit, it must be set at the correct cutin and cutout points to properly protect the liquid-to-refrigerant coil.

Entering Fluid Temperature F.	Min. Leaving Fluid Tempurature F.	Antifreeze Protection Required Temperature F.	Pressure Control Setting (PSIG)	Pressure Control Open (PSIG)	Pressure Control Reset (PSIG)
25	15	5	38	28	38
30	20	10	43	33	43
35	25	16	49	39	49
40	30	22	55	45	55

Figure 19–12 Low-pressure control adjustments (R-22) (Courtesy Addison Products Co.)

The table in Figure 19–12 shows the pressure control settings for the adjustable low-pressure control. The setting must not be lower than allowable by the antifreeze solution freezing point. The table also gives the antifreeze protection required temperature for fluid temperatures from the earth loop, entering the unit, in the range from 25°F to 40°F.

For example, if the fluid temperature leaving the earth loop will have a minimum temperature of 30°F, the fluid leaving the unit can be expected to be 20°F and an antifreeze protection is required down to 10°F. Under these conditions, the low-pressure freeze protection control should be set for an R–22 refrigerant pressure of 33 psig open set point and 43 psig closing set point. This means that the control will open at 10°F and close at 20°F.

If the unit uses refrigerant other than R–22, the controls must be set for the corresponding pressures of 10°F and 20°F from the pressure/temperature chart for the particular refrigerant.

REVIEW QUESTIONS

1. The earth loop contains 20% glycol solution and the system uses R–22 refrigerant. What would be the cutin and cutout set points of the antifreeze protection control?
2. With liquid-to-air units, in the heating mode, what is the most common cause of the unit cutting out on the lockout protection?
3. With liquid-to-air units, in the cooling mode, what is the most common cause of the unit cutting off on the lockout protection?
4. An auxiliary flow control valve should always be used when water-regulating valves are used on the unit. True or False?
5. The device used to prevent compressor operation if the water supply fails is called a _____ _____.
6. When starting a liquid-to-air unit, using well water and operating in the cooling mode, the minimum temperatures are _____°F for water and _____°F for air.
7. The minimum air temperature for operating a liquid-to-air heat pump in the heating mode is _____°F.
8. When checking and tightening the electrical connections in a unit, the connections that are factory made should also be included. True or False?

20 STARTUP, CHECKOUT, AND ADJUSTMENT

20–1 GENERAL

To provide maximum heating and cooling capacity at the lowest energy cost, the system must be properly set up and adjusted after the installation is completed. The flow rate through the unit heat exchanger coil must be per design requirements. The CFM of air through the air coil must be correct and there must be proper suction and discharge pressure if water-regulating valves are used. The refrigerant charge in the unit must also be correct. The charge amount is supposed to be correct when the unit leaves the manufacturer, but this may not always be true.

To perform the task of setup, checkout, and adjustment, the technician performing the operation must have specific information. To maintain proper records on the performance and service history, certain information must be recorded. This is very important in the event of a disagreement over warranty questions.

The following items should be a part of the job record of the installation:

1. Owner's name, address, and telephone number. With most manufacturers, the address is the important item. Units are under warranty at the original installation address. The owner may change but the address does not. Practically all manufacturers will honor a change in ownership, but a change in address will cancel the warranty.
2. Unit identification. A record should be kept of the unit identification information
   a. Hi-side Model No. __________
   b. Hi-side Serial No. __________
   c. Air Handler Model No. __________
   d. Air Handler Serial No. __________
   e. Auxiliary Heat Model No. __________
   f. Auxiliary Heat Serial No. __________

   If the heat pump has been installed in conjunction with a fossil fuel unit, the inside coil information should be recorded:

   g. Add-on Coil Model No. __________
   h. Add-on Coil Serial No. __________
3. The type of pressure-reducing devices used on the air coil and water coil should be recorded. The operating characteristics of the system using T. X. valves are different than those using capillary tubes or restrictors. Therefore, this information is necessary for proper checkout and adjustment.
   a. Liquid-to-water heat exchanger
      ___ T. X. Valve
      Mfg. _____ Model No. _____
      _____ Capillary tubes or restrictors
   b. Air heat exchanger
      _____T. X. Valve
      Mfg. _____ Model No. _____
      _____Capillary tubes or restrictors.
4. Fuse protection. The size and type of fuses installed in the original installation must be recorded. Amateur electricians tend to change fuse sizes without thought to consequences. Installed circuit breakers will reduce this possibility as compared to cartridge-type fuses.
5. Power supply wire size. The size of the power supply wire as well as the length of the wire run for the power supply to the unit or to both sections of a split system should be recorded. If auxiliary hat is used, the wire size to this unit should also be recorded.
6. Inside air filter. The size and quantity of air filters in the unit should be recorded for future replacement. A check should be made to be sure air filters have been installed in the unit.
7. Type of heat energy supply. The type of heat supply system to the unit must be recorded. Pipe sizes and underground piping layout on a site plan must be included in the installer's file for future reference.

   A tracer system should have been installed with the underground pipe system, but a record of the layout should also be kept.

   _____Domestic well supply
   a. Submersible pump size _____
   b. Storage tank size _____
   c. Pressure control settings
      Cutin _____ psig
      Cutout _____ psig
      _____ Ground loop
      Circulator
      Mfg. _____ Model No. _____
      Water flow rate _____ GPM

When startup is performed on a liquid-to-air dual operation heat pump, the operating mode at startup will depend upon the season of the year and the end results required. The system has limits for each operating mode that cannot be exceeded if an accurate check is to be performed.

The heat pump will operate in the cooling mode down to water temperatures of 45°F and return air temperatures to the coil above 70°F. Water temperatures below 45°F or air temperatures below 70°F will not produce the necessary operating pressures to provide satisfactory operating characteristics. If the initial startup is done in the heating season, the final checkout on cooling should be made the following cooling season.

In the heating mode, the limitation on water temperature is 80°F if water-regulating valves are used.

Ground loop systems are not affected by the outdoor temperatures and in most areas, the system can be operated in the heating cycle and in the cooling mode long enough for testing. However, the air to the inside coil must not be over 75°F when the unit is operated in the heating mode. Excessive air temperatures will cause high head pressures and unit cutoff.

The unit must not be operated in the heating mode if the air temperature is below 65°F if the water supply is from a well and a constant flow valve is used for water flow rate control. With the air temperature below 65°F, the compressor discharge pressure can be low enough to cause the water coil evaporator pressure to drop below 32°F and cause freeze up of the coil.

If water-regulating valves are used, the cooling water control valve will open and increase the water flow. If the well supply is large enough, there will be no problem. However, if the well supply is border line, a freeze up could occur. This caution also applies to ground loop systems that do not contain antifreeze.

20–2 STARTUP CHECKLIST

When starting the system, three sets of information are needed, one before startup, one immediately after startup, and one after several minutes of operating time.

Prestarting Checklist

Several checks should be made before power is actually applied to the unit and the unit is put into operation.

1. Are all mechanical parts operating properly? The compressor has been released from the shipping hold-down means. The inside air blower is secure and properly aligned.
2. On split systems, are the refrigerant lines properly installed, securely fastened, and properly insulated?
3. Are evaporator drain pans properly sloped toward the drain outlet for proper condensate removal?
4. Is the system wired according to the wiring diagram that applies to this particular unit?
5. Are the different sections of the system properly installed?
6. Are all electrical connections tight, including those in the unit or units? Check and tighten *all* connections, including those done by the manufacturer.
7. Is the area thermostat level and correctly wired? Is it in a good location, away from drafts and heat sources such as lamps, TV sets, and so on? Is it mounted on a wall where a warm air supply duct is located in the wall?
8. Are clean air filters in place?
9. Is the duct work correctly installed and well insulated where necessary with vapor barriers where necessary and are all joints properly taped?
10. Are refrigerant couplings or joints tight and leakproof?
11. Have the necessary isolation shut-off valves been installed in the water supply system?

Some of these items may seem unnecessary, but they are items over which litigation has occurred.

Electrical Startup Procedure

The startup procedure begins with the main power in the "Off" position. The voltage at the distribution panel should be measured with the system off. This gives the no-load voltage.

When the system is started, two electrical checks should be made. Each section is checked—the low-side air handler first and then the high-side section. Any auxiliary heat package is checked separately.

To make each check, a voltmeter should be connected across the high-voltage supply to each section on the line side of the controlling contactor or relay. Place a clamp-type ammeter (preferably a digital ammeter with a high-amperage lock) around one of the hot leads to the contactor or relay.

The purpose of these applied voltage and amperage tests is to determine if the electrical supply system is of sufficient size to carry the load. When the contactor or relay closes, bringing on the electrical load, the voltage at the contactor or relay should not drop more than 5 volts. If it does, one or more of the following situations could be the problem:

1. The wire size could be too small for the length installed. Manufacturers are required to give the wire size in ampacity (amperage capacity) per wire length for each unit. If a specific length is not given, the National Electrical Code requires a size based on 100' of wire length.
2. The building distribution service is too small. The starting test should be repeated measuring the voltage at the circuit breaker in the distribution panel. If the starting voltage at the distribution panel drops less than 5 volts when the drop at the unit is more than 5 volts, the branch wires are too small. If the drop in starting voltage at the distribution panel is more than 5 volts, either the service drop to the building is too small or the distribution power transformer is too small.
3. The total connected load on the power distribution transformer may exceed the capacity of the transformer. To check this, turn on the heaviest electrical loads in the building. In a residence, this would be the electric range and oven. With both of these appliances in operation, the supply voltage to the building should not drop more than 2 volts.

At the same time that the voltage check is made, the amperage draw of the unit under test must be observed. Both starting and running amperages are measured and recorded.

The amperage draw must be observed during the startup to make sure the amperage drops from the inrush (LRA) to full load (FLA). If the amperage draw remains high, near the LRA, the unit must be disconnected immediately.

After the electrical problem is found and corrected, the test is repeated.

System Startup Procedure

To check the performance of the system, temperature and pressure readings have to be taken. The readings will depend upon if the unit is in the cooling or heating mode.

The basic rule for sizing heat pumps is "the heat pump has to be sized to handle the cooling load." Therefore, the most accurate testing and adjusting is done in the cooling mode. If the unit is started in the heating season, an approximate adjustment can be made, but the system must be rechecked and set during the next cooling season.

The actual startup and checkout will vary with the different types of liquid heat source arrangements. If the water flow rate is via a fixed flow rate valve, the flow rate is established according to the unit requirements before the unit is started. If the heat source is a ground loop system, the liquid flow rate is established according to the unit requirements before the unit is started. If the water to the unit is pressure controlled with suction and discharge pressure-regulating valves, the suction and discharge pressures are preset.

In all three types, a flow rate has to be determined to check unit capacity. Therefore, to aid in establishing the liquid flow rate, as well as measure the flow rate during operation, the inclusion of a flow meter in the original installation is highly recommended.

Fixed Flow Rate—Startup Procedure

Step 1. Be sure the main power to the unit is off.

Step 2. Set the thermostat season switch to "off" and the fan switch to "auto." This will keep the compressor and blower motors from operating when power is applied.

Step 3. Move the main power disconnect to "on." Power to the unit should be on for a time period of a minimum of four hours or one hour per pound of refrigerant in the system. The refrigerant quantity is given on the unit rating plate. This allows the crankcase heater to drive any liquid refrigerant out of the compressor pump. This procedure should be followed whenever the power has been off for a period of 12 hours or more. Except as required for safety when servicing the unit, *do not open the power disconnect to the unit.*

Step 4. Move the auto-on fan switch to the "on" position. The blower relay should pull in and the blower motor should operate. Be sure all registers and grills are open and air circulation has been established.

The proper CFM adjustment will be done after the unit has been operating. Move the fan switch to the "auto" position to stop the blower.

Step 5. Fully open the isolating valves in the water flow circuit.

Step 6. Check and establish the correct water or liquid flow rate by adjusting the outlet shut-off valve. For a unit with a well pump supply, manually open the solenoid valve controlling the water flow through the unit. Adjust the ball valve just ahead of the solenoid valve (on the outlet of the unit) to get the required GPM flow rate.

For a ground loop supply unit, start the loop circulator by closing the manual switch in the circulator relay control circuit. Set the flow rate by adjusting the ball valve on the outlet side of the unit.

The flow rate is adjusted with the unit off to prevent possible freeze up and coil damage in case the flow rate is not sufficient or established fast enough to prevent coil freeze up.

Step 7. Start the unit in the cooling mode by moving the season switch to "cool" and reducing the thermostat setting to 65°F.

Step 8. Check to see that the solenoid valve has opened or the circulator is operating and liquid flow has been established.

Step 9. Check and record the pressures, temperatures, and liquid flow rate to be able to test the system for capacity and operating efficiency.

This will be discussed in System Performance Check—Cooling.

Regulated Flow Rate—Startup Procedure

Step 1. Be sure the power to the unit is off.

Step 2. Set the room thermostat season switch to "off" and the fan switch to "auto."

Step 3. Move the main power disconnect to "on." This must be done for a period of time of a minimum of four hours or one hour per each pound of refrigerant in the system.

Step 4. Move the auto-on fan switch to the "on" position. The blower relay should pull in and the blower motor should operate. Be sure all the registers and grills are open and air circulation has been established. The proper CFM adjustment will be made after the unit has been operating. Move the fan switch to the "auto" position. The blower should shut off.

Step 5. Fully open the isolating valves in the supply and return circuits.

Step 6. Connect the gauge manifold to the unit to read suction and discharge pressures after they have been established. These pressures are needed to set the water-regulating valves.

Step 7. Start the unit in the cooling mode by moving the thermostat season switch to the "cool" position and reducing the thermostat setting to 65°F.

Step 8. Allow the unit to operate until water flow and the discharge pressure have been established.

Step 9. Read the discharge pressure and adjust the high-pressure regulating valve for the correct compressor discharge pressure according to the manufacturer's literature. If this information is not available, use the pressure equivalent of 105°F condensing temperature for the type of refrigerant in the system.

20–3 AIR SYSTEM CHECK

Because the heat pump system is selected to handle the cooling load, the amount of air over the air-to-refrigerant heat exchanger coil (the air coil), which is the evaporator in the cooling mode, must be correct in the cooling mode. With the correct amount of air through the air coil, the unit will produce the desired temperature drop in the air.

The inside design conditions are usually 75°F but ARI test conditions are 80°F DB and 50% RH. The objective of air adjustment through the air coil is to obtain the combination of temperature and relative humidity that will satisfy the greatest number of human occupants.

The first step to determine the correct ΔT°F of the air through the air coil is to measure the return air DB and WB temperatures at the return air grill. By measuring the temperatures at this point, the return air duct can be checked for inward air leaks.

Using the psychometric chart in Figure 20–1, the conditioned air relative humidity is determined. Using this information, the proper temperature drop through the air coil can be obtained from the load ratio chart in Figure 20–2.

The load ratio chart has been plotted for the various conditions of air within the operating range of the unit. Dry Bulb (DB) temperatures from 56°F to 92°F are given. Using the DB temperature and the %RH found in the air to the return air grill, we can determine the temperature drop of the air through the air coil that is needed to give the desired conditions over the widest range of outside conditions (the cooling load).

For example, assuming that the conditioned area has a measured dry bulb (DB) temperature of 80°F and a wet bulb (WB) temperature of 66°F, using the psychometric chart in Figure 20–1, the relative humidity would plot out at 50%. Using the DB of 80°F and the RH of 50%, these figures are located on the load ratio chart in Figure 20–2. Where the horizontal 80°F (DB) line and the 50% RH slope line cross, we find a vertical line that leads down to the bottom scale at 20ΔT°F. However, when other than these conditions of DB and RH are found, the 20ΔT°F would not give the desired results.

20–4 SYSTEM PERFORMANCE CHECK

General

All discussion has been in the cooling mode with only an occasional mention of the heating mode. This is because the cooling mode operation is the most critical. As has been discussed many times, the unit is sized to handle the cooling load and supplemental heat is used to make up for any additional heating capacity needed. Therefore, to check out the system, the first adjustments and tests are done in the cooling mode. The same adjustments are used for the tests for the heating mode with the exception of the air over the air coil. This is done in the cooling mode only.

With the following items checked out and/or properly adjusted, the system can be checked for performance capacity and operating efficiency:

1. Operating supply voltage is within 5 volts of the unit rating.
2. The condensing pressure has been set on systems using water-regulating valves.
3. The water flow rate has been set on units using constant flow valves and/or ground loop systems.
4. The temperature drop through the air coil has been set for the conditions of the return air in the conditioned area.

No reference is made to the flow rate in the heating mode because the cooling flow rate is the basic adjustment and the heating flow rate is the same.

System Performance Check—Cooling

The performance rating on a liquid-to-air heat pump in the cooling mode is the same as a water-cooled air conditioning unit. The most accurate way to calculate the net capacity of the unit is to determine the gross capacity and motor heat input. The gross capacity minus the motor heat input equals the net capacity. To find the gross capacity and motor heat input, the following readings have to be taken at the same time.

Remember, these readings depend upon the proper voltage being applied to the unit, the proper ΔT°F of the air over the air coil, and the proper GPM of liquid through the liquid coil. Before any further tests can be made, these three items must be correct.

To determine the gross capacity of the system in the cooling mode, the following items are determined:

1. Temperature of the liquid entering the unit liquid coil (EWT) ___°F.
2. Temperature of the liquid leaving the unit liquid coil (LWT) ___°F.
3. Temperature rise (ΔT°F) of the liquid through the liquid coil. Leaving temperature (LWT) minus the entering temperature (EWT) equals the ΔT°F.

$$\Delta T°F = LWT\ °F - EWT\ °F$$

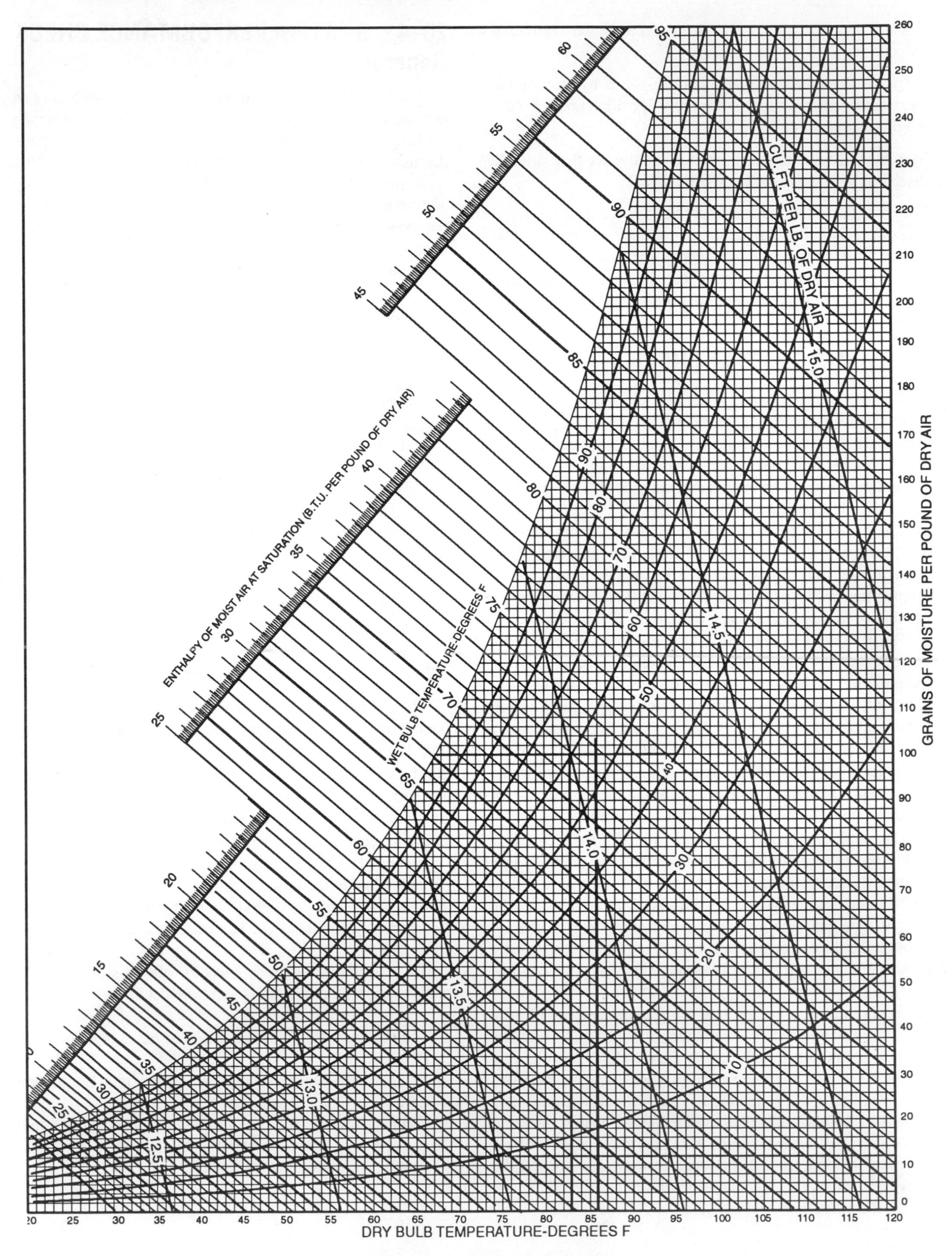

Figure 20–1 Psychrometric chart

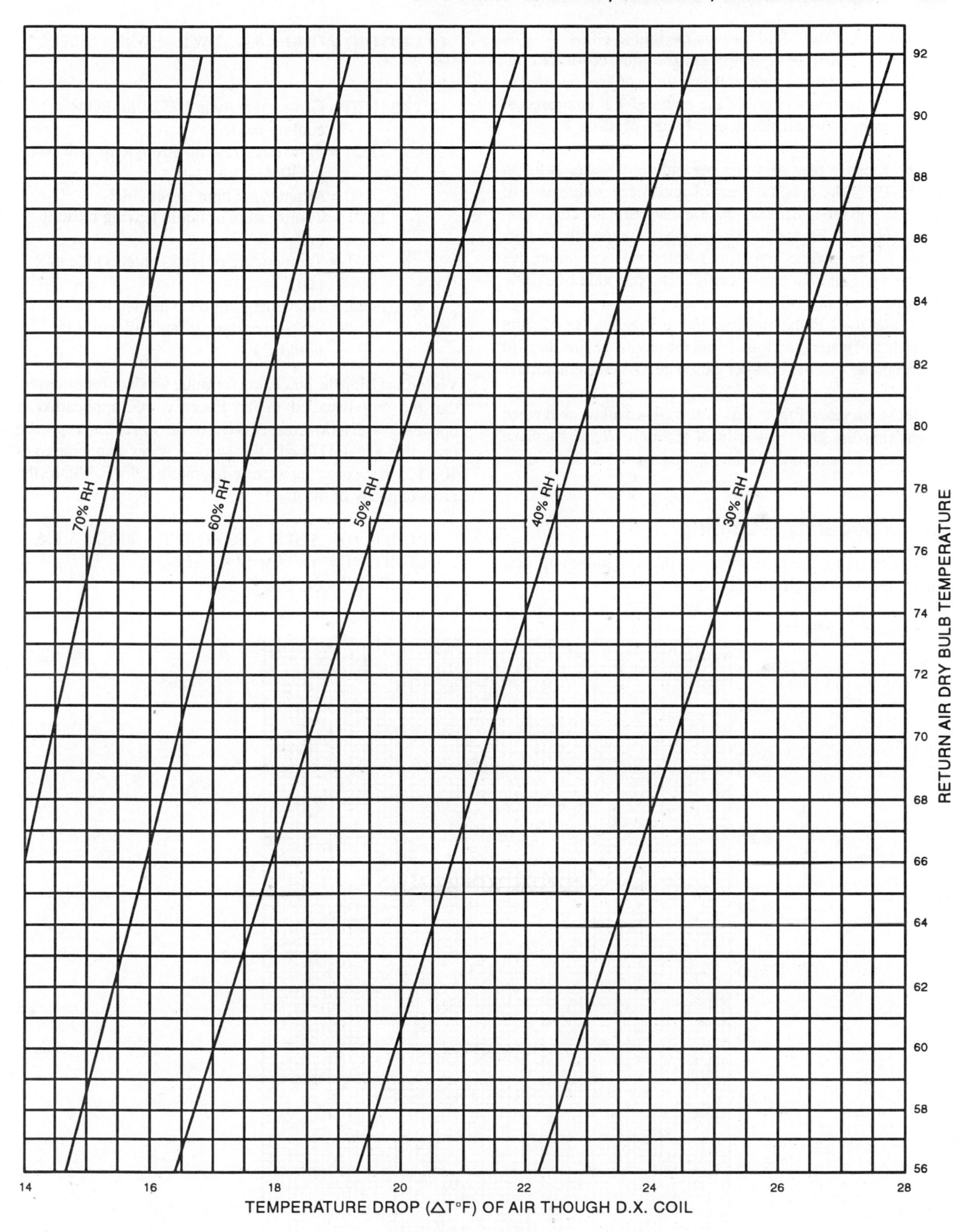

Figure 20–2 D.X. coil

4. Flow rate of the liquid through the unit in GPM.
5. Specific heat of the liquid through the liquid coil in the unit. Specific heat is the ratio of the quantity of heat required to change the temperature of the material 1°F compared to that required to change an equal mass of water 1°F.

The specific heat of water per gallon of water is 8.336 BTU/ΔT°F. The specific heat of antifreeze solutions will depend upon the antifreeze material used and the concentration percentage. Figure 20–3 is a graph of the specific heat per gallon for propylene glycol and water solution. From this graph, we find that the specific heat per gallon at 20% concentration is 8.058.

Figure 20–4 is the specific heat per gallon for calcium chloride and water solutions. From the graph, we find that, for example, the specific heat per gallon at 20% concentration is 7.35.

These curves in Figures 20–3 and 20–4 will be used when calculating the gross capacity of the unit when propylene glycol or calcium chloride antifreezes are used.

Gross Capacity—Cooling

To determine the gross capacity of the liquid-to-air heat pump, the following formula is used:

$$GCC(BTUH) = GPM \times 60 \times (LWT - EWT) \times SH/Gal.$$

where:

GCC(BTUH) = Gross capacity in BTUH/hr. in the cooling mode.
GPM = Flow rate of liquid through the unit coil in GPM.
60 = Minutes per hour to get GPH.
LWT = Temperature of liquid leaving the unit coil.
EWT = Temperature of liquid entering the unit coil.
SH/Gal. = Heat energy in BTU needed to change the temperature of one gallon of the liquid 1°F.

Water. Using the preceding formula, we find, for example, that the unit installed in the Phoenix, AZ application is operating with a flow rate of 6 GPM, the water has a specific heat of 8.336 BTU/Gal., the leaving water temperature is 102°F, and the entering water temperature is 90°F. What is the gross capacity of the unit?

$$GCC(BTUH) = 6\ GPM \times 60 \times (102°F - 90°F) \times 8.336$$
$$GCC(BTUH) = 6 \times 60 \times 12 \times 8.336$$
$$GCC(BTUH) = 36,011$$

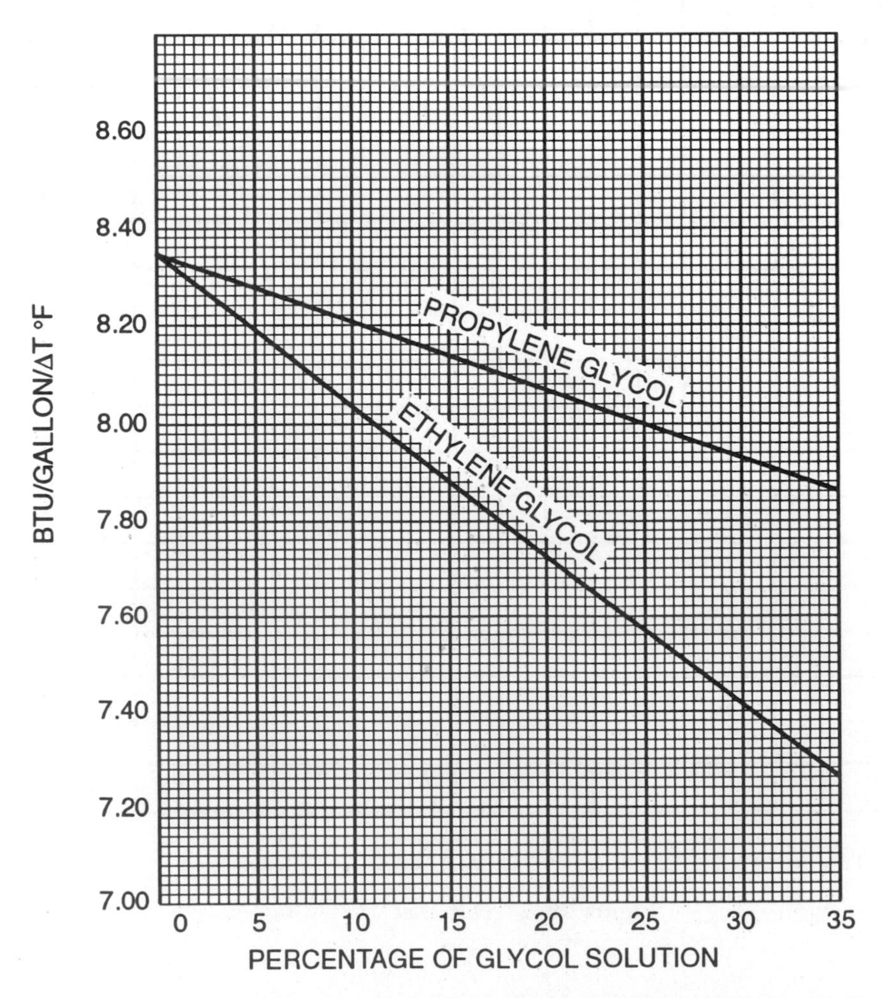

Figure 20–3 Specific heat per gallon for glycol and water solutions

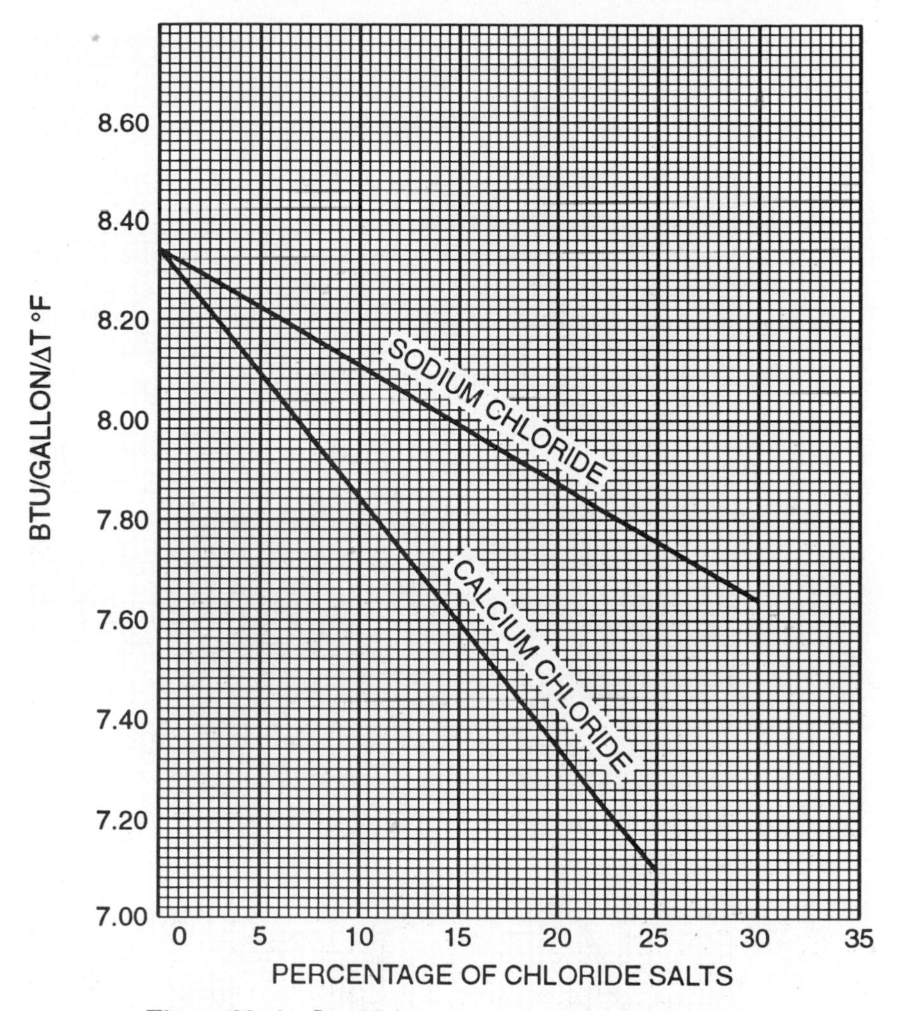

Figure 20–4 Specific heat per gallon for chloride salts

The readings calculate out to a gross cooling capacity of 36,011 BTUH. The rated capacity per the table in Figure 13–13 at a flow rate of 6 GPM and 90°F EWT is 35,500 BTUH. If the calculation results come within ±5% of the rating of the unit, the unit is operating satisfactorily. Field instrumentation, even though carefully handled and read, will only allow an accuracy of ±5%. In this application, a calculation result between 33,725 BTUH and 37,275 BTUH is satisfactory.

Antifreeze. To determine the specific heat per gallon of antifreeze mixtures, the percentage of concentration has to be determined. The easiest way to determine the percentage of concentration is to measure the specific gravity of the solution. How thick is the solution as compared to water, which has a specific gravity of 1.000?

A very inexpensive instrument for this test is a hydrometer and sample tube that can be purchased from any winemaking supply outlet. The hydrometer float measures specific gravity of liquids from 0.999 to 1.190. This range will handle the normal chloride and glyco mixtures down to 0°F freezing point, which is the lower limit of freeze protection needed for liquid-to-air heat pumps. The unit should not be subjected to liquid operating temperatures that require freeze protection below 0°F.

We will assume in our example that the application is a ground loop heat sink (in the cooling mode) in the Omaha, NB application, using a calcium chloride antifreeze solution. The test results show that the flow rate of the solution is 7.38 GPM with an EWT to the liquid coil of 64°F and an LWT from the liquid coil of 72°F. To find the specific heat per gallon, we must take a specific gravity sample of the liquid. Using the hygrometer in the sample tube filled with a sample of the liquid, we find the specific gravity to be 1.133. The graph is based on a liquid temperature of 60°F. Therefore, our sample should be as close to this temperature as possible.

Using the chart in Figure 20–5, we find the percentage of concentration is 18%. From the horizontal line at 1.133 to the slant comparison line, we find the cross point of these two lines is on the vertical line at 18%.

Applying the 18% concentration to the chart in Figure 20–4, we find that the vertical line at 18% concentration and the slant comparison line for calcium chloride cross on the horizontal line at 7.35 BTU/1°F/gallon.

Using the gross capacity formula:

$$GCC(BTUH) = GPM \times 60 \times (LWT - EWT) \times SH/Gal.$$
$$GCC(BTUH) = 7.38 \times 60 \times (72°F - 64°F) \times 7.35$$

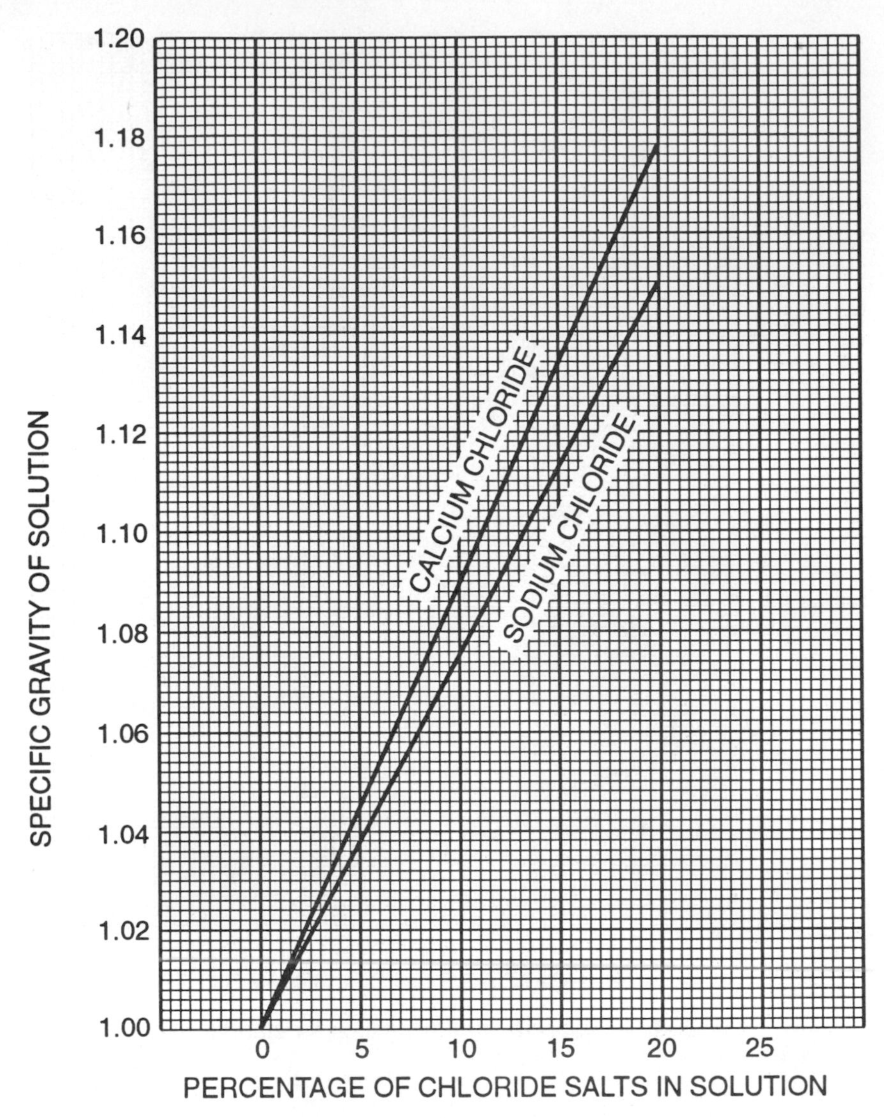

Figure 20–5 Percentage of solution/specific gravity for chloride salts

GCC(BTUH) = 7.38 × 60 × 8 × 7.35
GCC(BTUH) = 26,036

From the readings used in our example, we find the gross cooling capacity of the unit is 26,036 BTUH. If sodium chloride antifreeze solution is used, the same procedure and charts are used to figure the gross capacity.

Let us now assume that the ground loop in the Omaha, NB installation contains a propylene glycol antifreeze solution. The test results show that the flow rate is at the desired amount—6 GPM for water times the propylene glycol correction factor of 1.36 (Figure 16–6) or 8.16 GPM.

The specific gravity test shows from the graph in Figure 20–6 that the percentage of solution for propylene glycol is 24%. If the specific gravity reading were for ethylene glycol, the percentage of glycol solution would be 21%.

Using the 24% of propylene glycol solution found in Figure 20–6, we determine that the BTU/ΔT°F/gallon is 8.005.

The temperature readings of the liquid show an EWT of 65°F and an LWT of 72°F for a ΔT°F of 7°F.

Using the gross capacity formula:

GCC(BTUH) = GPM × 60 × (LWT – EWT) × SH/Gal/ΔT°F
GCC(BTUH) = 8.16 × 60 × (72°F – 65°F) × 8.005
GCC(BTUH) = 8.16 × 60 × 7 × 8.005
GCC(BTUH) = 27,435 BTUH

Motor Heat Input—Cooling

This does not mean that the BTUH total capacity is within the proper rating range. This will depend upon how much electrical energy the motor-compressor assembly needs to produce the calculated gross capacity.

Regardless of the type of material in the heat source, the calculation of the motor heat input is the same. The motor heat input is the BTU equivalent of the usable watts of electrical energy the compressor motor draws. To find the motor heat

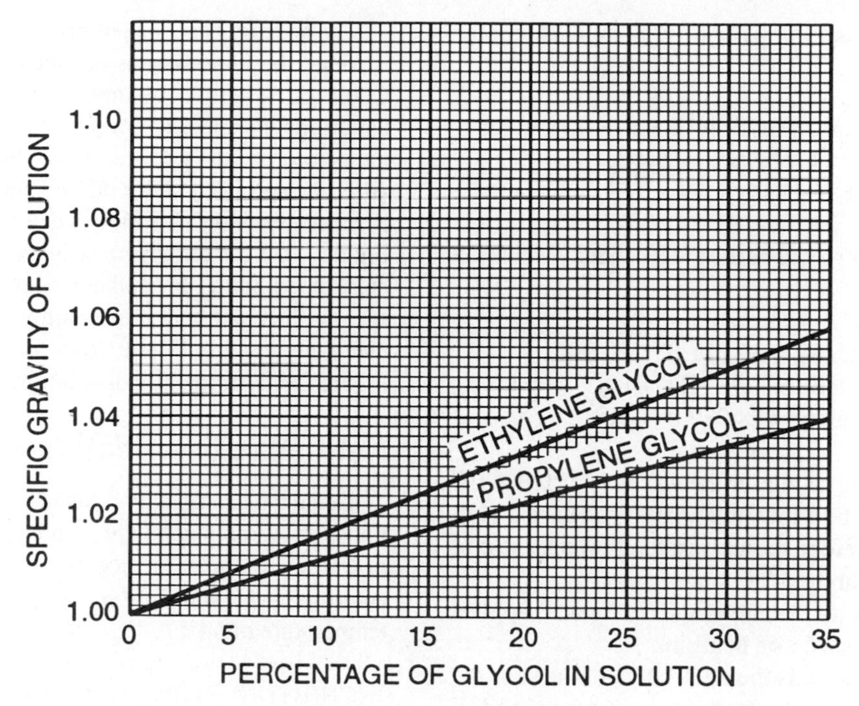

Figure 20–6 Percentage of solution/specific gravity for glycols

input, *at the same time that the liquid EWT and LWT are recorded, the applied voltage at the compressor contactor and amperage draw of the compressor motor are recorded.*

There must be as little time lapse as possible between recording the various test values. Changes in room air conditions as well as line voltage can make an appreciable difference.

To determine the motor heat input, the following formula is used:

MH(BTUH) = Volts × Amps. × Power Factor × 3.416 BTU/Watt

where:

MH(BTUH) = Heat energy equivalent of the usable electrical energy drawn by the motor.

Volts = Measured voltage across the load side of the compressor contactor while the compressor motor is operating.

For a single-phase motor, only one reading is needed. For a three-phase motor, the voltage across each of the three pairs of lead wires must be read and compared. The voltage readings between each of the pairs of leads must not be more than ±3% of the others. If an unbalance of more than 3% is allowed to continue, the motor life will be shortened considerably. For applications where an unbalance of more than 3% exists:

1. Check the electrical distribution panel for extra single-phase loads that may have been added because the panel is in a convenient location.
2. Contact the electric utility to provide balanced loads on their power distribution system.

To illustrate the calculation of motor heat input, we will assume that the unit in the Omaha, NB installation has 236 volts at the compressor contactor terminals while it is drawing 11 amperes.

Amps = The amperage draw of the compressor motor. In the case of an air-to-air heat pump, the amperage draw includes the amperage draw of both the compressor and fan motors. In the liquid-to-air heat pump, only the amperage draw of the compressor motor is included. There is no fan motor in the high-side section.

PF = The percentage of the measured watts that the motor is actually using. In capacitor run motors, the power factor conversion figure is very high, averaging 90%. Therefore, the power factor in this formula is 0.9.

3.416 BTU/Watt = The amount of heat energy equivalent of 1 watt of electrical energy.

MH(BTUH) = Volts × Amps. × PF × 3.416 BTU/Watt
MH(BTUH) = 236 × 11 × 0.9 × 3.416
MH(BTUH) = 7,981 BTUH

Net Capacity—Cooling

The gross cooling capacity minus the motor heat input equals the total amount of heat (sensible and latent) picked up by the air coil. The amount of heat picked up by the air coil is the net cooling capacity of the unit. The calculated net capacity should fall within a range of ±5% of the manufacturer's rating.

Water. The unit using water had a gross capacity of 36,011 BTUH. We will assume that the unit had a measured voltage of 236 volts and a measured amperage of 11 amps. The motor heat input would be:

MH(BTUH) = Volts × Amps. × PF × 3.416 BTU/Watt
MH(BTUH) = 236 × 11 × 0.9 × 3.416
MH(BTUH) = 7,981 BTUH

The unit net capacity would be the gross capacity minus the motor heat.

Gross Capacity = 36,011 BTUH
Motor Heat = 7,981 BTUH
Net Capacity = 28,030 BTUH

The standard cooling rating for the unit is 27,000 BTUH. The ±5% range would be a low of 25,650 BTUH and a high of 28,350 BTUH. Therefore, the test results show that the unit net capacity falls within the ±5% range.

The same test for motor heat input and net capacity applies regardless of the fluid used as a heat sink.

If the net capacity exceeds the ±5% range of the unit, the tests should be rerun to strive for more accuracy *after the ΔT°F of the air through the air coil is the correct amount.* This must be reestablished before any further testing is done.

If the net capacity calculates out below the rated net capacity range, the refrigeration system should be checked. This will be covered in Chapter 21, Troubleshooting

Gross Capacity—Heating

The total capacity of the unit is used for the heat energy supply to the conditioned area. The amount of heat picked up from the liquid through the unit coil (net capacity) plus the motor heat input equals the heat energy into the air through the air coil.

Net Capacity—Heating—General

The net capacity in the heating mode is calculated using the same formula and charts used to calculate the gross capacity in the cooling mode.

NCH(BTUH) = GPM × 60 × (EWT – LWT) × SH/Gal
NCH(BTUH) = The amount of heat picked up from the heat source, either well water or ground loop, using water or antifreeze solution.
GPM = The flow rate of the water or antifreeze solution through the unit water coil in gallons per minute.
60 = Minutes per hour.
EWT = Temperature of the water or antifreeze solution entering the unit liquid coil.
LWT = Temperature of the water or antifreeze solution leaving the unit liquid coil.
SH/Gallon = The specific heat of the liquid through the unit liquid coil in BTUs per gallon. The specific heat of a gallon of water is 8.336 BTU.

The specific heat of a gallon of chloride or glycol solution is found the same way as outlined under Gross Capacity—Cooling. A sample of the antifreeze solution is taken to determine the specific gravity. For accuracy of this test, the temperature of the solution should be allowed to rise to 60°F. After the specific gravity of the solution has been found, the percentage of concentration is determined using the graph in Figure 20–3 for the glycol solutions or Figure 20–4 for the chloride solutions. From the percentage of concentration, the specific heat per gallon of solution circulated is determined using the graph in Figure 20–5 for the chloride solutions or Figure 20–6 for the glycol solutions.

Net Capacity—Heating—Water

Using the net capacity—heating formula for the unit in the Phoenix, AZ application, we find that the flow rate of 6 GPM of water with a specific heat per gallon of 8.336 BTU has an entering water temperature of 62°F and a leaving water temperature of 54°F.

NCH(BTUH) = GPM × 60 × (EWT – LWT) × SH/Gallon
NCH(BTUH) = 6 × 60 × (62°F – 54°F) × 8.336
NCH(BTUH) = 6 × 60 × 8 × 8.336
NCH(BTUH) = 24,008 BTUH

Net Capacity—Heating—Chlorides

To determine the net capacity of the unit used in the Omaha, NB application using calcium chloride antifreeze solution, we find that operation in the heating mode produces an EWT of 43°F and an LWT of 36°F with a flow rate of 7.38 GPM and a specific gravity reading (at 60°F) of 1.035.

The specific gravity reading of 1.160, using the graph in Figure 20–5, shows a percentage of solution concentration of 18%. Using the graph in Figure 20–6, at 18%, we find a specific heat of 7.35 BTU/Gallon.

NCH(BTUH) = GPM × 60 × (EWT°F – LWT°F) × SH/Gallon
NCH(BTUH) = 7.38 × (43°F – 36°F) × 7.35
NCH(BTUH) = 7.38 × 60 × 7.0 × 7.35
NCH(BTUH) = 22,782

Net Capacity—Heating—Glycols

To determine the net capacity of the unit used in the Omaha, NB application using propylene glycol antifreeze solution, we find that operation in the heating mode produces an EWT of 49°F and an LWT of 37.2°F, with a flow rate of 8.16 GPM and a specific gravity reading (at 60°F) of 1.027.

The graph in Figure 20–3 shows a percentage of concentration of 24% for the specific gravity reading of 1.027. Using the graph in Figure 20–6, we find that the 24% concentration has a specific heat of 8.005 per gallon.

NCH(BTUH) = GPM × (EWT – LWT) × SH/Gallon
NCH(BTUH) = 8.16 × 60 × (43°F – 37.2°F) × 8.005

$$NCH(BTUH) = 8.16 \times 60 \times 5.8 \times 8.005$$
$$NCH(BTUH) = 22{,}731$$

Motor Heat Input—Heating

The calculation of the motor heat input is the same for the heating mode as for the cooling mode. Refer to Motor Heat Input—Cooling.

Gross Heating Capacity

By adding the net heating capacity and the motor heat input, the total amount of heat put into the air over the air coil is determined. This gross heating capacity should be within ±5% of the manufacturer's ratings. If the final capacity is more than the 5% tolerance, the test should be redone with extra effort for more accurate readings.

The CFM of air over the coil was set during the cooling mode performance testing. Therefore, the CFM of air over the air coil *should not be changed.* If the final capacity is less than the 5% tolerance, the refrigeration system should be checked.

REVIEW QUESTIONS

1. When a unit starts, the supply voltage at the unit should not drop more than _____ volts.
2. Because the unit is not located outdoors, the compressor does not require off-cycle heat. True or False?
3. When setting the CFM of air through the indoor coil, the heating mode is used. True or False?
4. The method for calculating the net capacity of a liquid-to-air heat pump in the cooling mode should not be confused with the method of calculating the net capacity of a water-cooled air conditioning unit. True or False?
5. The entering water temperature is 73°F and the leaving water temperature is 84°F. The flow rate is 6 GPM. What is the gross capacity of the unit operating in the cooling mode?
6. What would the gross capacity of the unit in Problem 5 be if the unit were operating on a closed loop containing a 20% propylene glycol and water solution?
7. What would the gross capacity of the unit in Problem 5 be if the unit were operating on a closed loop containing a 20% calcium chloride and water solution?
8. The density of a liquid as compared to water is called its _______ _______.

21 TROUBLESHOOTING

21–1 GENERAL

The majority of the problems that are encountered in liquid-to-air heat pump systems are the same as those that would be encountered in water-cooled refrigeration or air conditioning systems. In the cooling mode, the liquid-to-air heat pump operates no differently than the water-cooled "store cooler."

The difference between the water-cooled air conditioner and the liquid-to-air heat pump is that the heat pump operates in both the cooling and heating modes. This requires extra controls in the refrigeration circuit as well as the electrical controls.

In Chapter 2, the various testing equipment that is used on heat pumps are discussed. Many of the items listed have common usage when servicing heating and air conditioning systems as well as heat pumps. When used in conjunction with heat pumps, the basic heating and air conditioning test equipment will apply.

The discussion of troubleshooting in this text and the analysis of problems is confined to those problems that apply to liquid-to-air heat pump systems. Basic problems that apply to refrigeration and air conditioning systems in general, such as leak detection and repair, method of refrigerant charging, and so on, are covered in other texts on basic fundamentals. For basic information on air handling and air distribution, the best source is the Environmental Systems Library published by the Air Conditioning Contractors Association (ACCA).

To properly diagnose problems in a liquid-to-air heat pump system, four pressure readings, eight or more temperature readings, a liquid flow rate, and liquid specific gravity are required. Not all the readings apply to each type of system. To be able to cover each type of system without too much repetition, all the readings will be discussed. Separation by system will be done as each operating mode of each type of system is discussed.

21–2 PRESSURE READINGS

The six required pressure readings are:

1. Compressor suction pressure
2. Compressor discharge pressure
3. Liquid line pressure
4. Vapor line pressure
5. Liquid pressure entering liquid coil
6. Liquid pressure leaving liquid coil

Compressor Suction Pressure. The correct place to measure the compressor suction pressure is in the "suction line" between the reversing valve and compressor suction vapor connection. This is the only true "suction line" in the unit. Refrigerant vapor travels in one direction only—to the compressor—regardless of the operating mode.

A compound gauge is used to read this pressure.

Compressor Discharge Pressure. The correct place to measure the compressor discharge pressure is in the line between the compressor discharge connection and the reversing valve. This "hot gas line" carries vapor in one direction only—from the compressor to the reversing valve.

If a heat recovery unit is installed between the compressor and the reversing valve, the compressor discharge pressure must be measured in the hot gas line between the compressor and the heat recovery coil.

Liquid Line Pressure. The location to measure the pressure in the liquid line of a liquid-to-air heat pump depends upon the type of pressure-reducing device the unit uses. If the system is a SPCT (single-purpose capillary tube) or SPTXV (single-purpose T. X. valve), there is no need to measure the liquid line pressure as no check valves are involved.

The pressure in the liquid line in a liquid-to-air heating and cooling (dual-purpose) heat pump system using check valves is the liquid pressure in the line located between the pressure-reducing device/check valve combination (trombone) on each of the heat exchanger coils.

When the check valves in the system are operating properly, the liquid line pressure will be within 2 to 5 psig of the compressor discharge pressure. This is the normal flow resistance through the condenser coil in either of the operating modes. When the check valves malfunction, the results will be reflected in the line pressure.

Vapor Line Pressure. The pressure in the vapor line depends upon the operating mode of the unit. In the cooling mode, the vapor line pressure will be within 5 psig of the suction pressure. In the heating mode, the vapor line pressure will be within 5 psig of the compressor discharge pressure. Because the vapor line is subject to high compressor pressures in the heating mode, a 0 to 500 psig standard gauge must be used to measure the pressure.

Liquid Pressure Entering the Coil and Leaving the Coil. These pressures will give the flow pressure loss through the coil. These tests help to determine any unusual flow restrictions on startup as well as the effect of coil fouling from mineral deposits.

21–3 TEMPERATURE READINGS

A minimum of eight temperature readings are required to analyze problems in liquid-to-air heat pump systems:

1. Return air dry bulb
2. Return air wet bulb
3. Supply air dry bulb
4. Suction line temperature at evaporator coil outlet
5. Liquid line temperature
6. (EWT) liquid temperature entering the liquid coil
7. (LWT) liquid temperature leaving the liquid coil
8. Suction line temperature between the reversing valve and the compressor.

Return Air DB and WB Temperatures. A sling psychrometer and psychrometric chart (Figure 20–1) are used to determine the DB temperature and %RH in order to determine the total heat content of the return air and the heat load on the air coil (evaporator) in the cooling mode. This information is needed to determine the desired ΔT°F of the air through the coil.

Supply Air DB Temperature Leaving the Air Coil. This is necessary to determine the actual ΔT°F of the air through the coil.

Suction Line Temperature at the Evaporator Outlet. This is used in determining the evaporator operating superheat. The location of the thermometer on the system will change with the operating mode from cooling to heating and reverse.

In the cooling mode, the suction line temperature at the outlet of the air coil is measured. In the heating mode, the temperature is taken at the outlet of the liquid coil.

Liquid Line Temperature at the Outlet of the Condenser Coil. This is primarily a test when the unit is in the cooling mode. The temperature is then taken at the outlet of the liquid coil. The test is primarily for determining the amount of subcooling in the liquid refrigerant.

Liquid Temperature Entering the Liquid Coil (EWT) and Leaving the Liquid Coil (LWT). In the cooling mode, these temperatures (EWT and LWT) are used to determine the gross capacity of the system. In the heating mode, they are used to determine the net capacity of the system.

For troubleshooting, these two temperatures are important to test the heat transfer ability of the liquid-to-refrigerant heat exchanger coil. As the water coil in the unit fouls from mineral deposits, this causes a higher condensing temperature and pressure as well as reduces the temperature rise of the liquid through the coil. Whenever the temperature rise of the liquid through the coil is reduced more than 30%, the coil must be cleaned and demineralized.

Suction Line Temperature Between the Reversing Valve and the Compressor Suction Inlet Connection. This temperature reading is used to check for possible hot gas bypass through the reversing valve.

21–4 LIQUID CHARACTERISTICS

In addition to the operating pressure and temperature readings, the service technician must know the characteristics of the liquid through the liquid coil as well as the flow rate.

Specific Gravity. To be able to measure the heat transfer ability of the liquid through the liquid coil, the specific gravity of the liquid must be determined. Using the specific gravity float and sample tube, the specific gravity of the liquid is measured.

If the liquid is treated water, the specific gravity is 1.000. If it is an antifreeze solution, the specific gravity of the liquid will be higher as the percentage of concentration increases. From the specific gravity reading and using the charts in Figure 20–3 (Glycols) and Figure 20–5 (Chlorides), the specific heat in BTU/Gal. is determined.

Flow Rate. The flow rate of the liquid through the coil is measured. The fowling effect of the mineral in the water will add flow resistance to the liquid coil water circuit. This will result in the reduction in the flow rate. The measurement of the entering and leaving liquid pressures will help to verify if the flow rate reduction is from coil fouling or pump deterioration.

The specific heat per gallon and the flow rate are necessary in the procedure of determining the gross capacity in the cooling mode and the net capacity in the heating mode.

21–5 PROBLEM CATEGORIES

Liquid-to-air heat pump problems can be divided into four categories:

1. Air system problems
2. Refrigeration system problems
3. Electrical control problems
4. Liquid system problems

Air System Problems. If it is assumed that the CFM through the air coil was set properly in the cooling mode when the unit was installed, air problems will show up as an increase in the ΔT°F across the coil. It is not possible for the heat transfer capacity of the system to increase; the air quantity can only decrease. Therefore, the increase in the ΔT°F across the coil can only be due to a decrease in the CFM of air through the evaporator coil.

If the reduction in CFM becomes great enough, the reduced load on the coil will cause a reduction in the coil-operating temperature. This could cause a frosted or iced up coil. However, because a low-pressure switch is used in the unit, if the liquid coil heat source is water, the switch will open to prevent the iced up coil.

If the unit has been adapted to low-temperature operation to use antifreeze in the liquid circuit, the low-pressure switch setting will be low enough that this iced up coil protection will not exist.

In the heating mode, the reduction in CFM will cause a rise in compressor discharge pressure and an increase in the operating cost. If the reduction becomes great enough, the high-pressure switch will open. Opening of either the low-pressure switch or the high-pressure switch will cause the unit to cut out on the lockout relay system and light the emergency lite on the thermostat sub-base (if the thermostat sub-base has this feature). Immediate service is needed to prevent damage to the unit.

Air Filters. The heat pump system operates year-round, which necessitates more filter changes. To keep operating costs low and promote better air conditioning, the air filters should be changed at least every 60 days.

The best solution to air filter problems is the installation of an electrostatic air cleaner. They are easily cleaned and their operating flow resistance is very low. Coil surfaces operate much cleaner.

Blower Motor and Drive. The blower motor and drive should be inspected at least on an annual basis. Motor bearings and blower bearings should be lubricated once per year with not more than 10 drops of electric motor oil. *Do not use automobile oil.* Additives in automobile oil (detergents) will clog the bearing oil feed system and cause bearing burn out.

Duct Restrictions. In the cooling mode, closing off supply grills in unused rooms will have the effect of increasing the temperature drop across the evaporator coil and could result in ice formation on the air coil. In extreme cases, liquid refrigerant return to the compressor could cause damage.

In the heating mode, the reaction is more severe. Only a small reduction in CFM will have a large effect upon the compressor operation. High temperatures and pressures are very quickly reached. This situation is more hazardous in high water temperature areas.

21–6 REFRIGERATION SYSTEM PROBLEMS

As in refrigeration and air conditioning systems, refrigeration problems can be divided into two categories:

1. Refrigerant quantity
2. Refrigerant flow rate

Refrigerant Quantity. In a heat pump system, the refrigerant quantity will affect the system operation in both the operating modes. In the cooling mode, the reduction in cooling capacity will reduce the temperature drop across the air coil. In the heating mode, the temperature rise across the air coil will decrease.

In either case, the system has lost refrigerant via a leak. The leak must be found and repaired, and the system must be evacuated and recharged.

Refrigerant Charging Procedure. Because of the variety of liquid-to-air heat pumps that have been marketed, average readings of subcooling, compressor operating pressures, or amperage draws have not been reliable enough to even consider when charging the system with refrigerant. Therefore, the most accurate way to charge the system is to thoroughly evacuate the system and weigh in the correct refrigerant charge.

CAUTION: The dangers of evacuating a liquid-cooled refrigeration or air conditioning unit also apply to a liquid-to-air heat pump. *Before reclaiming any refrigerant from the system, liquid must flow through the liquid portion of the coil. Enough water must flow through the coil to keep the coil temperature above 32°F. If liquid refrigerant is allowed to boil off in the refrigerant portion of the coil without water flow, the coil could freeze and rupture.*

On units using regulating valves, the valve is forced open and wedged in the open position. On ground loop systems, the circulating pump must be operating *before* purging starts.

After the system has been evacuated, the refrigerant charge is weighed in. All manufacturers are required by UL standards to list the type and amount of refrigerant on the rating plate or portion of the system. In the case of a package unit, the entire system charge is given.

In the split system, usually the amount of normal operating charge is given for the sections marketed by the particular manufacturer. To determine the total charge, the amount of refrigerant in the liquid line and vapor line is calculated. Depending upon the type of refrigerant, each line will contain a certain weight per lineal foot.

For example, from the table in Figure 21–1, a 1/4" OD liquid line in a system using R22 will contain 0.39 ounces per foot. If the line were 25' long, the refrigerant charge for the liquid line would be 25' × 0.39 oz. per ft. or 9.75 oz. for the entire line.

If the vapor line were 1/2" OD and 25' long, the refrigerant charge for the vapor line would be 25' × 0.03 oz. per ft. or 0.5 oz. The total charge would be 9.75 oz. in the liquid line plus 0.5 oz. in the vapor line for a total of 10.25 oz. by measured weight.

This line refrigerant quantity, added to the refrigerant charge for the sections of the system is the total system charge needed for operation.

Refrigerant Line Size In. OD	Contents Per Liquid Line R-22	R-12	Foot in Ounces Suction Line R-22	R-12
¼"	0.39	0.54	0.01	0.01
⅜"	0.48	0.64	0.02	0.01
½"	1.12	1.28	0.03	0.02
⅝"	1.76	2.08	0.05	0.03
¾"	2.08	3.04	0.08	0.06
⅞"	3.84	4.16	0.10	0.08

Figure 21–1 Copper tube refrigerant volume

We assume that the reader has the step-by-step knowledge to be able to liquid charge the system.

Refrigerant Flow Rate. If the system has the proper quantity of refrigerant and the refrigerant flow rate is at the required amount, the system must perform as designed.

We have already discussed the handling of the refrigerant charge.

21–7 CHECK VALVES

The one item that does not apply to all four types of systems is the action of the check valves in the single-purpose capillary/check valve (SPCT/CV) and single-purpose T. X. valve/check valve (SPTXV/CV) systems. Problems with check valves sticking either in the closed or open position are possible. The position the check valve is stuck in will show up in only one of the operating modes, and not the other. This is why it is so important to operate the system in both modes to isolate the problem.

For example, if the check valve on the liquid coil were to stick closed, which operating mode would encounter the problem? The question really is in which mode is the check valve required to open and allow liquid refrigerant to bypass the pressure-reducing device?

The pressure-reducing device has to be bypassed when its companion coil is used as a condenser. In our example, the liquid coil is the condenser in the cooling mode.

If the system works in the heating mode but not in the cooling mode the check valve on the liquid coil is probably stuck closed. How would the stuck-closed check valve affect the operation of the system?

Capillary Tube System

The stuck-closed check valve forces the liquid refrigerant to flow through the capillary tubes on the condenser coil before entering the liquid line to travel to the capillary tubes on the evaporator. This causes the pressure drop of the liquid refrigerant to practically double before it enters the evaporator. Therefore:

1. Suction pressure will be extremely low. On liquid-to-air units with a standard low-pressure switch, the system will likely cut out on the lockout relay protection.

   On units using antifreeze solutions with the lower low-pressure switch setting, the suction pressure will be extremely low. It is possible to have a suction pressure indicating a completely frosted coil with very little if any frost on the coil.

   In the heating mode, the suction pressure regulating valve will open to allow heavy water flow with very little temperature drop in the water through the coil.
2. Evaporator superheat will be high. The refrigerant flow rate will be reduced because of the extra flow resistance.
3. In the heating mode, the compressor discharge pressure will be low. The amount of heat picked up in the evaporator will be low even though the water flow rate will be high. The system will reach a very low thermal balance and the condensing temperature (discharge pressure) will be low.

   In the cooling mode, in the system using water-regulating valves, discharge pressure will be approximately normal because it is maintained by the action of the water-regulating valve. The flow rate, however, will be considerably less due to the closing of the valve to try to maintain the head pressure setting.

   In systems using fixed water flow rate, the compressor discharge pressure will be low.

Before adding any refrigerant or recharging the system, operate the system in the opposite mode. Good performance in the opposite mode means the system is correct except for the device that was supposed to open in one mode and not the other, and failed to function properly.

The best indication of this problem is the pressure in the liquid line is considerably lower than 5 psig below the compressor discharge pressure measured at the hot gas line between the compressor and the reversing valve.

T. X. Valve System

If the check valve sticks closed in the system using T. X. valves, it is the same as closing a shut-off valve in the line. The T. X. valves used in the systems, even though they have special operating charges in the power element, will not allow refrigerant flow in the reverse direction. The high pressure under the diaphragm closes the valve and shuts off the refrigerant flow.

The indication of this problem is the system pumps down and shuts off on the lockout protection system in one of the operating modes but works fine in the opposite operating mode.

Check Valve Repair

Before replacing the check valve, which means reclaiming the system refrigerant and recharging the refrigerant after repair and evacuation, try using a magnet to free the valve. If the valve is stuck closed, start the magnet at the outlet end of the valve and move it to the inlet end. If this procedure is not successful, the valve must be replaced. To reduce the possibility of future problems, always replace with a ball-and-cage valve.

Before installing the valve, make sure the ball is loose in the guide cage. Shake the valve. If it rattles, use it. If there is no rattle, reject it.

When installing the new valve, be careful not to overheat and warp the ball cage. Use thermomastic material on the body of the valve when making solder connections.

21–8 REVERSING VALVES

As explained in Chapter 3, Component Parts, the reversing valve is a pilot valve operating a slide-action main valve. The entire valve operation depends upon a pressure difference between the compressor suction and discharge pressures of 75 psig or higher.

To expect proper operation of the reversing valve, the system must have a complete charge of refrigerant and operation in either mode to have at least the minimum pressure difference of 75 psig developed between the suction and discharge pressures. All testing must be done with the system operating.

Reversing Valves—Electrical Problems

Electrical problems are confined to the electrical solenoid coil. When voltage is applied to the coil, a click should be heard from the pilot valve plunger changing position. If no click is heard, check across the coil lead connections to make sure voltage has been applied to the coil circuit.

If power is supplied to the solenoid coil, the coil pull test should be done. Remove the nut and spacer washer holding the coil on the plunger housing. Pull the coil off the housing. A definite drag should be felt from the magnetic pull of the coil.

If the drag is felt, the coil is active and the pilot valve is stuck. Replace the coil and tighten it into place with the spacer washer and nut. Cycle the power to the coil several times to try to free the pilot valve. If this is not successful, replace the valve.

When the coil is removed and a magnetic pull is not felt, the coil is dead. Check the coil for continuity using an ohmmeter.

Check the coil connections and leads for opens. Coil replacement is required for an open coil. Lead connections can be replaced.

Reversing Valves—Mechanical Problems

Mechanical problems in the valve consist of failure of either the pilot valve or main valve to shift When power is applied to the coil and a pilot valve click is heard, *if the pressure difference is more than* 75 psig, the main slide valve should change position. If the valve does not shift after several on/off cycles of power to the coil, the valve should be replaced.

However, the slide valve may be in the mid-position with hot gas bypassing to the suction side directly through the valve. This can be detected by measuring the temperature of the suction line between the reversing valve and the compressor suction inlet. This line will rapidly rise in temperature as the compressor operates.

This mid-position stall can happen if the valve is energized immediately upon the start of unit operation. This means that the valve is free to slide whenever the pressure is high enough to cause movement. The slide action could be so slow that when it reaches the mid-position, the pressure loss due to the bypass of the vapor can cause the slide to remain in the mid-position.

To cause the slide to move off dead center, the system pressure has to be raised. The easiest way is to put the system in the cooling mode. This means that the liquid coil is the condenser. By closing off the liquid supply to the liquid coil, the compressor discharge pressure will rise very rapidly. Allow the unit to operate until the discharge pressure reaches the equivalent of 130°F condensing temperature.

Keep the unit operating and cycle the valve several times. This will usually cause the valve slide to move and put the valve into operation. If this procedure is not successful, the valve will have to be replaced

Replacing Reversing Valves

When replacing reversing valves, there are several rules to follow:

1. Always replace the valve with one of comparable size. A smaller valve would put too much vapor flow resistance into the system. A larger valve could cause sluggish valve operation.
2. The valve must always be mounted in a horizontal position with the pilot valve above the main body of the valve. This is to prevent oil from accumulating in the pilot valve and connecting lines. Oil accumulation in these parts can produce very sluggish valve operation and possible valve failure.
3. The valve body must never be subjected to temperatures above 250°F. The use of the thermomastic material on the valve body when soldering the tube connections is an absolute must. The surest way to require a second valve change is to overheat the first replacement valve.

21–9 PRESSURE-REDUCING DEVICE SYSTEMS

Several types of pressure-reducing device systems are used in liquid-to-air heat pump units. Basically they all have compressor, air coil, liquid coil, and reversing valves. The difference is the type of pressure-reducing device used. Following is a listing of the types of pressure-reducing device systems that are in common use:

1. Dual-purpose capillary tube or restrictor (DPCT)
2. Dual-purpose T. X. valve (DPTXV)
3. Single-purpose capillary tube or restrictor/check valve combination (trombone)—one for cooling and one for heating (SPCT/CV)
4. Single-purpose T. X. valve and check valve combination—one for cooling and one for heating (SPTXV/CV)

Dual Purpose Capillary Tube or Restrictor (DPCT). Referring to Figure 21–2, the refrigerant circuit shown uses a common set of capillary tubes as the pressure-reducing device. The same capillary tubes are used in both the heating and cooling modes.

When the unit has the proper water flow for each of the operating modes, the single capillary tube or restrictor operates satisfactorily. To operate with the common capillary tube or restrictor, the heating mode operation requires 105°F

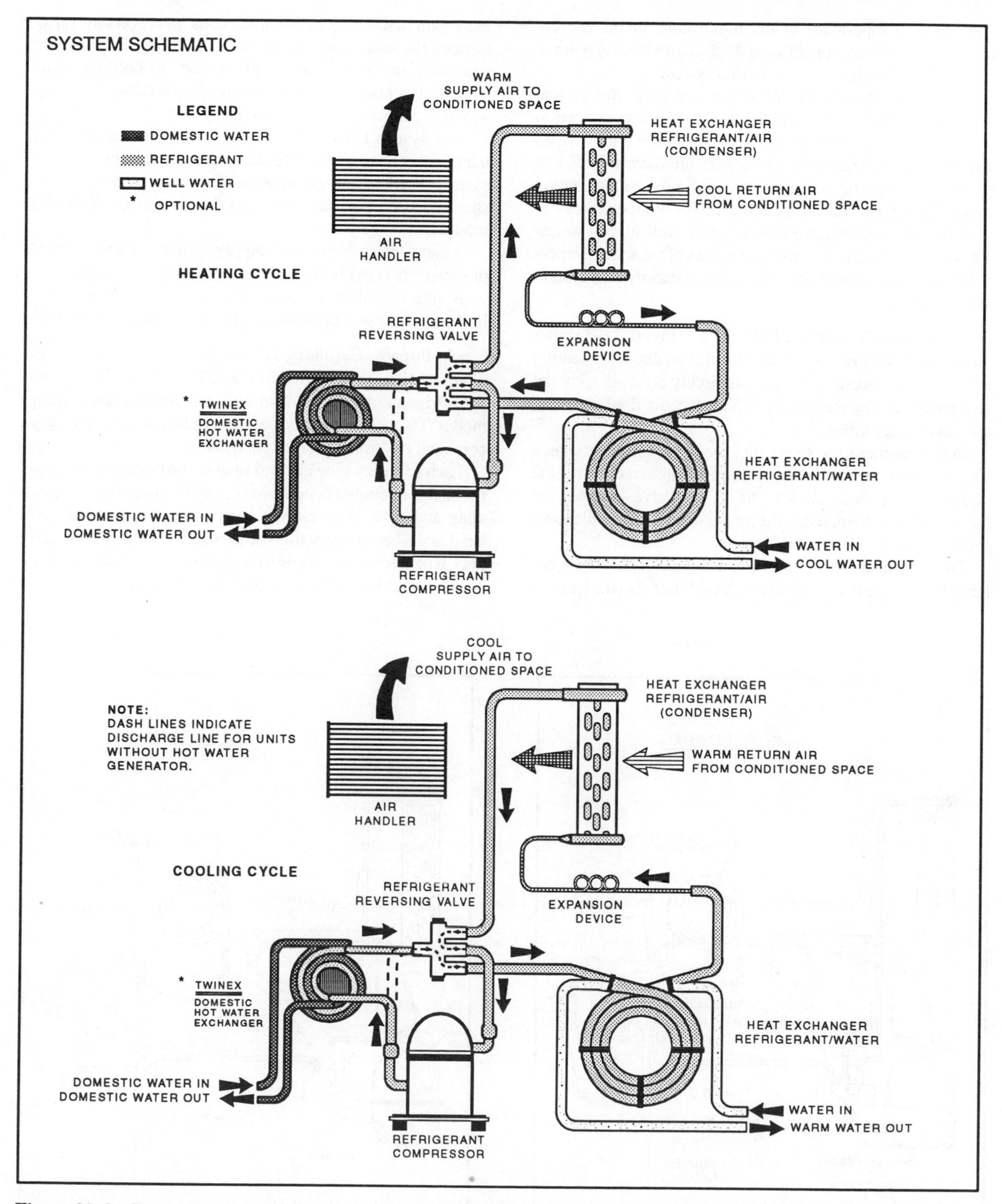

Figure 21–2 Dual-purpose capillary tube system (DPCT) (Courtesy Friedrich Air Conditioning & Refrigeration Co., Climate Master Division)

condensing temperature in the liquid coil. In the cooling mode, the liquid coil operates at 45°F liquid boiling point.

There are no check valves in this system.

For ground water temperatures above 70°F, this system does not require water-regulating valves if it is connected to a dedicated well water supply and the supply pump cycles with the unit. At ground water temperatures below 70°F, less water is used in the cooling mode and more in the heating mode. Therefore, water-regulating valves are required.

If the unit is connected to a domestic well supply, water-regulating valves are required regardless of the water temperature. This system will not operate satisfactorily on a ground loop system.

Dual Purpose T. X. Valve (DPTXV). Figure 21–3 shows the refrigerant flow circuit in both the cooling and heating modes. The system uses a multicircuit air coil with the refrigerant feed by means of a T. X. valve, distributor orifice, and four feeder tubes.

In the reverse flow or heating mode, the valve acts as a single-control orifice feeding the one refrigerant circuit in the liquid coil. In both modes, the T. X. valve controls the refrigerant flow to maintain the proper superheat in whichever coil is acting as the evaporator.

The superheat setting takes into account not only the vapor temperature rise in the coil (air or liquid) but also the temperature gain in the lines between the coils and reversing valve, through the valve and part of the suction line. Set for 10°F superheat, the T. X. valve will operate to hold the same superheat in both the heating and cooling modes.

No check valves are used in the system.

This system has more tolerance for differences in water temperatures as compared to the single capillary or restrictor system. For best overall operation, water-regulating valves should be used to control the water flow in both the operating modes.

A constant pressure water supply system, either dedicated or domestic, is the best type of heat source or heat sink.

Application of this type of system has not been satisfactory in areas where ground water temperatures drop below 60°F.

Single-Purpose Capillary Tube or Restrictor/Check Valve Combination (Trombone) (SPCT/CV). Figure 21–4 shows the refrigerant flow circuit in both the cooling and heating modes. The system uses a separate capillary tube/check valve combination (trombone) for each coil.

Each capillary tube is sized to give the design capacity at the design operating conditions of 105°F condensing temperature and 45°F evaporating temperature. Because separate capillary tubes or restrictors are used, the unit has a considerably wider tolerance for differences in liquid temperatures such as would be found in ground loop systems.

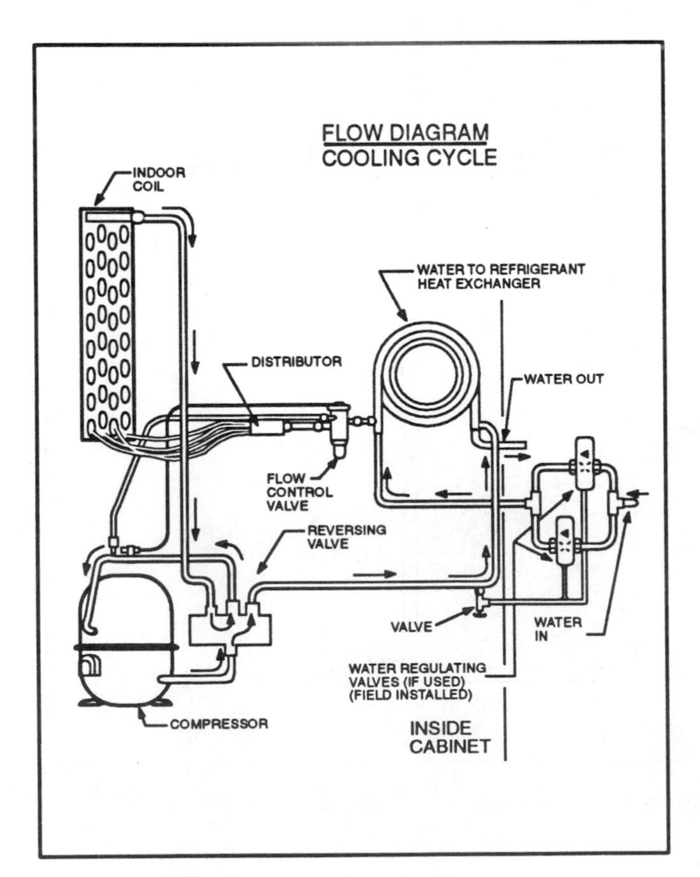

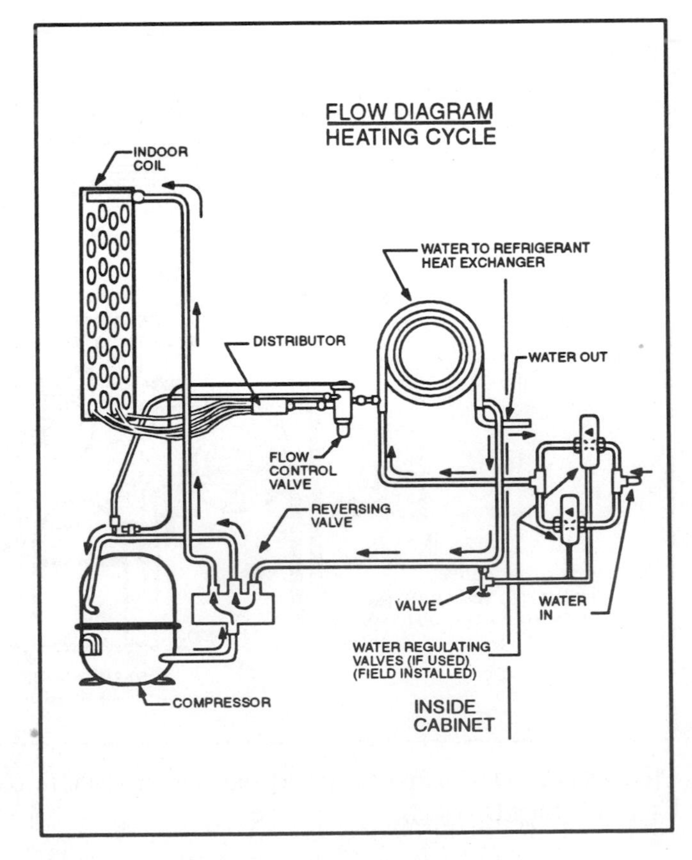

Figure 21–3 Dual-purpose T. X. check valve system (DPTXV) (Courtesy Addison Products Co.)

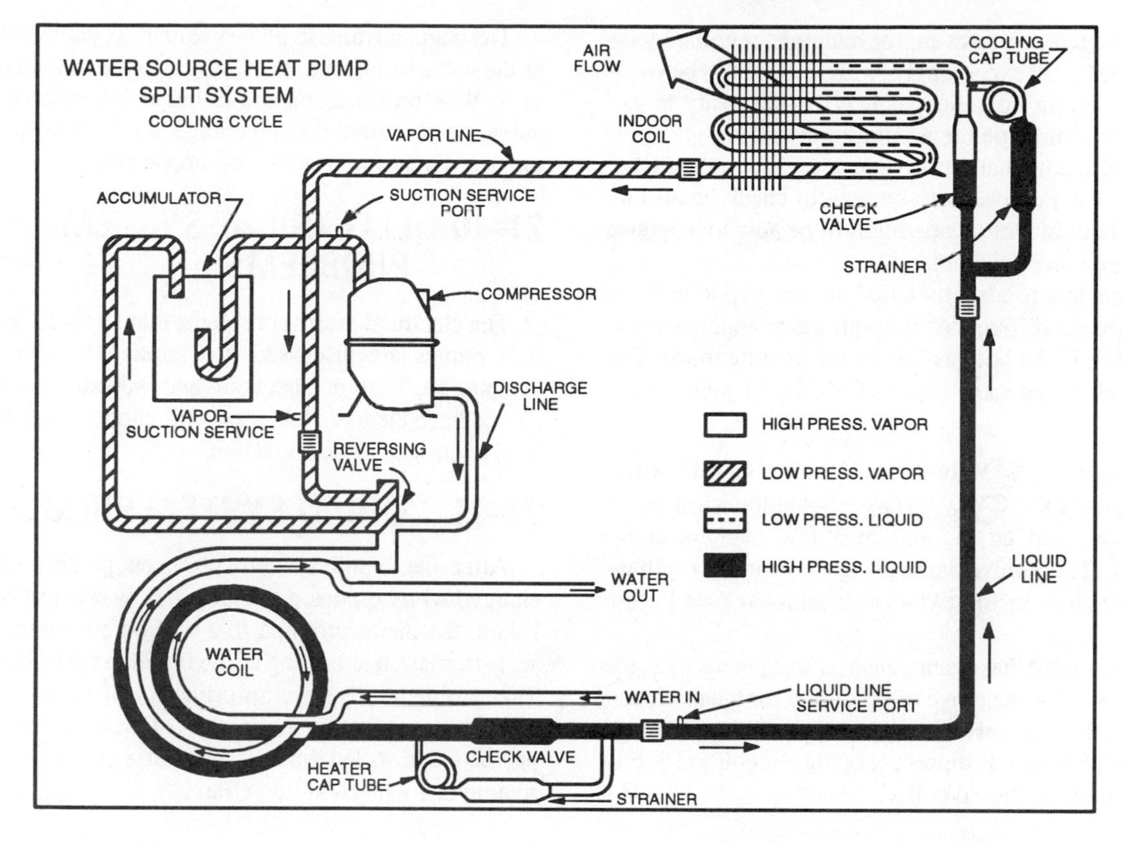

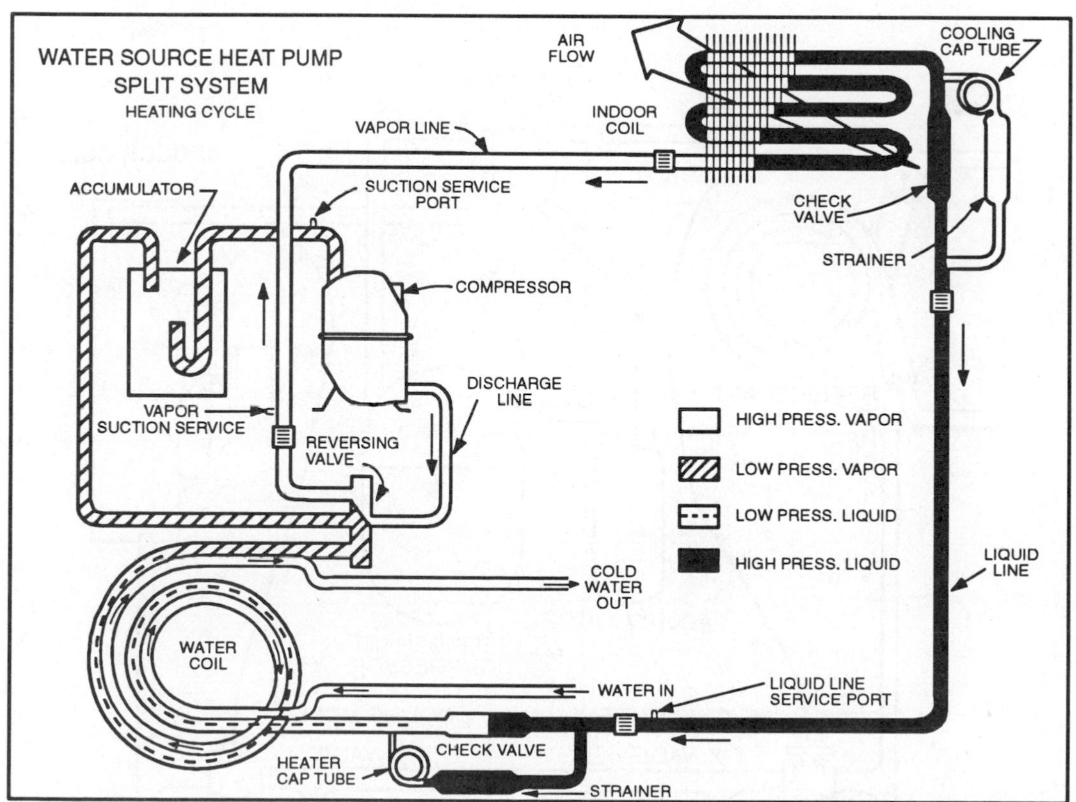

Figure 21–4 Single-purpose capillary check valve system (SPCT/CV) (Courtesy Bard Manufacturing Co.)

If the liquid temperatures in, for example, a ground loop application have an exceedingly wide difference between cooling and heating operation, it may be necessary to use multispeed circulators on the ground loop system.

You will also note that this system uses three pressure tap connections. The purpose is to be able to check liquid line pressures in both modes of operation to be able to diagnose possible check valve problems.

A pressure tap is also included in the vapor line for checking pressure differences through the evaporator coils. This vapor line is the hot gas line in the heating mode. The pressure should be checked using a 0–500 psig gauge.

Single-Purpose T. X. Valve/Check Valve Combination (Trombone) (SPTXV/CV). In larger liquid-to-air heat pump units, 7-1/2 hp and larger, and in a few manufacturers' smaller units, T. X. valves are substituted for the capillary tubes or restrictors in the two-coil trombones (see Figure 21–5).

Each T. X. valve has been selected to operate with the proper superheat for the range of operation for the particular type coil. Most T. X. valves in these applications are set to operate with a 5°F to 7°F superheat on the air coil and 3°F to 5°F superheat on the liquid coil.

The major advantage of the use of T. X. valves is the ability of the valve to adjust to the wide variation in load on the coil as well as pressures on the liquid refrigerant entering the valve. As a result of this advantage, the T. X. valve system is the best suited for ground loop operation.

21–10 ELECTRICAL SYSTEM PROBLEMS

The electrical system problems that apply to liquid-to-air heat pumps are discussed in Chapter 19, Electrical, and Chapter 20, Startup, Checkout, and Adjustment. The reader should have the basic knowledge of checking relays, contactors, control circuits, and so on.

21–11 LIQUID SYSTEM PROBLEMS

After the liquid system has been pressure tested and charged with liquid and the proper flow rate has been established, the major problem that will be encountered will be scale formation or fouling of the liquid in the refrigerant coil. This problem will develop primarily in well water supply systems as a continuous supply of new water is used. If purified or distilled water is used in a ground loop system, scaling or fouling should be rare.

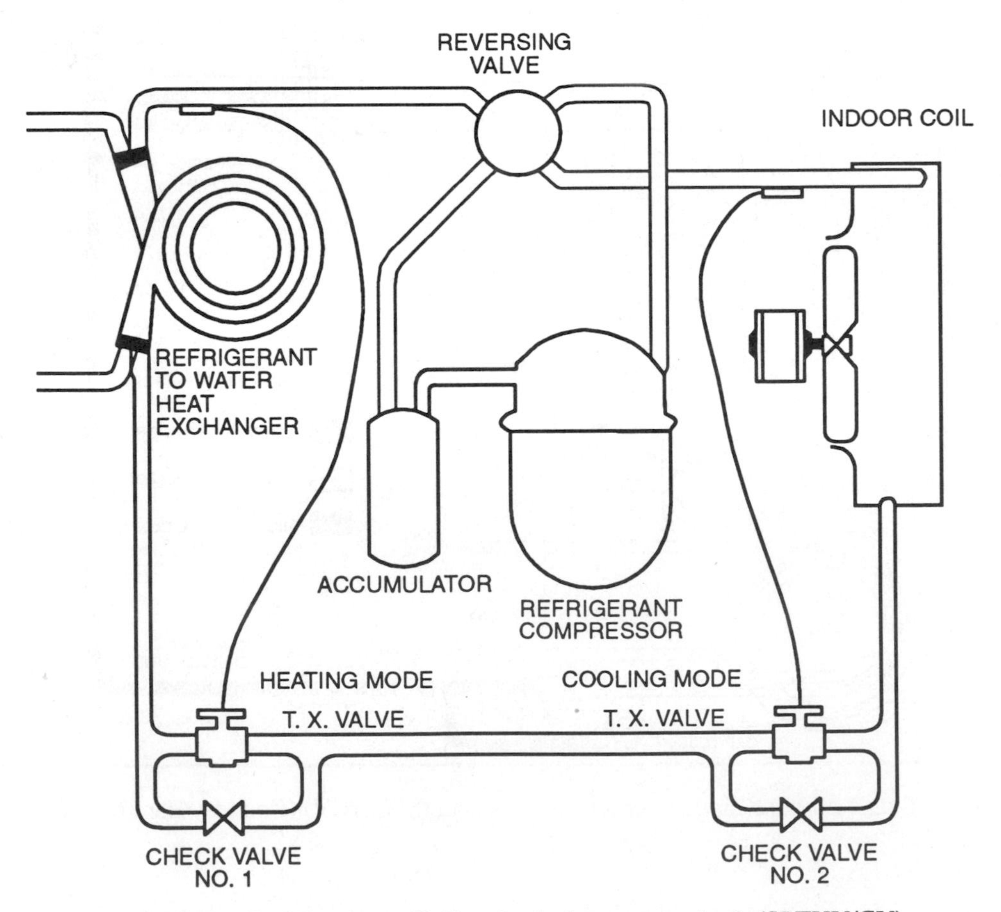

Figure 21–5 Single-purpose T. X. valve/valve system check (SPTXV/CV)

Regardless of the liquid system used, the fouling problem will be revealed in higher head pressure, higher amperage draw, and lower efficiency. This will be indicated to the owner by higher operating cost.

Figure 21–6 shows the piping arrangement for a system using water-regulating valves. Part of the arrangement is a coil bypass circuit incorporating two pet cock valves and a water pressure gauge. This arrangement should be a permanent part of the installation.

If it is not incorporated in the initial installation, shut-off cocks or P/T plugs must be installed upstream and downstream of the coil for measuring water pressure in the line.

When the unit was originally installed and checked out, the pressure drop through the coil was determined for the desired flow rate. This information should have been recorded on the checkout sheet for the installation.

Figure 21–7 shows a suggested checkout sheet that can be used for the initial information record. Items 3, 4, 5, and 6 record the information on the temperatures and pressures encountered when the water flow rate (item 7) is set.

As the coil is restricted due to scale, the temperature difference drops and the pressure difference increases. When the coil resistance increases to the point where a 30% increase in pressure drop occurs, the coil must be cleaned and descaled. The same equipment that was used for flushing and charging the earth loop can be used to descale the coil (see Figure 21–8).

There are several materials on the market that can be used for this purpose. As they are acid-type materials, care must be used to prevent damage to the soil. The cleaner manufacturer's instructions must be followed. After cleaning, the coil must be flushed thoroughly to prevent any of the cleaning material from entering the earth aquifer.

An annual check of the unit for coil fouling is very desirable.

21–12 QUICK REFERENCE TROUBLESHOOTING CHART

Figure 21–9 is a quick reference troubleshooting chart for liquid-to-air heat pumps that was published by Bard Manufacturing Co. of Bryan, Ohio. The common causes of problems are marked with a circle and occasional causes are marked with a triangle. The causes are divided into three categories:

1. Problems that occur in both heating and cooling modes
2. Problems that occur in the cooling mode
3. Problems that occur in the heating mode

The causes are also divided into four categories, three of which are subdivided:

1. Power supply
   a. Line voltage
   b. Control circuit
2. Water coil section (high side)
   a. Compressor
   b. Refrigerant system
   c. Reversing valve
   d. Check valve
   e. Water coil
3. Indoor section (low side)
   a. Indoor blower motor and coil
   b. Check valve
4. Auxiliary heat

This chart provides a quick reference to determine the probable causes of individual problems.

Not all causes apply to each type of system. For example, if the system uses a single dual-purpose capillary tube for the

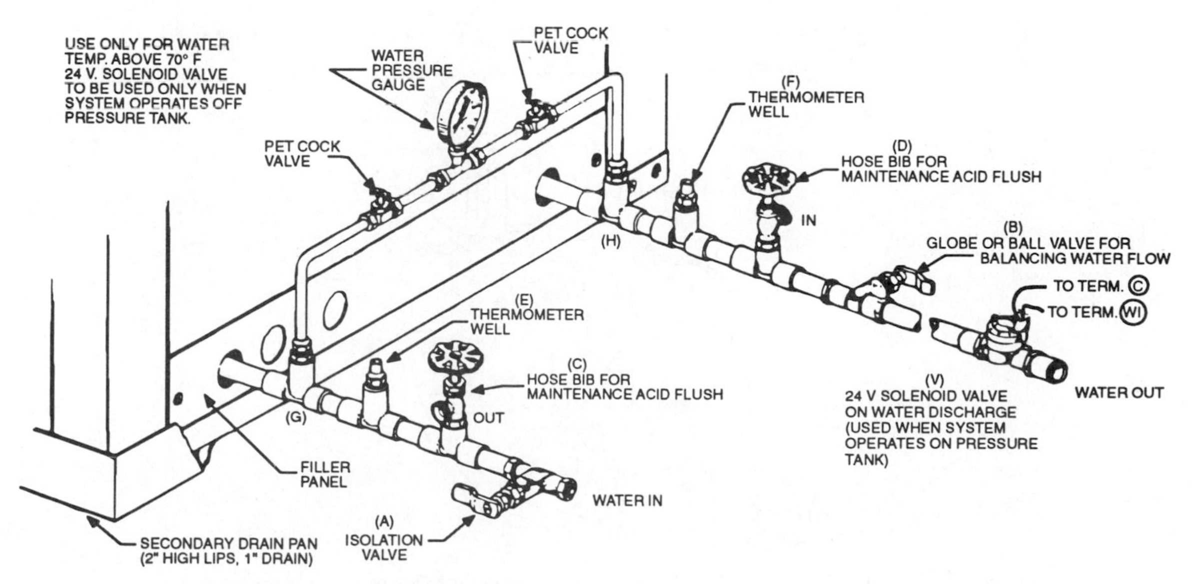

Figure 21–6 Typical piping arrangement (Courtesy Addison Products Co.)

PERFORMANCE CHECK
WATER SOURCE HEAT PUMPS

Installer Please Fill Out and Retain With Unit

Date of Installation ______________ Model No(s). ______________ Serial No(s). ______________

Item		Cooling	Heating
1. Head Pressure		_____	_____
2. Suction pressure		_____	_____
3. Water Temp. (In)		_____	_____
4. Water Temp. (Out)		_____	_____
5. Water Pressure (In)		_____	_____
6. Water Pressure (Out)		_____	_____
7. Water Flow (GPM)		_____	_____
8. Amperes (Blower)		_____	_____
9. Amperes (Compressor)		_____	_____
10. Line Voltage (Compressor Running)		_____	_____
11. Air Temp. (In)	D.B.	_____	_____
	W.B.	_____	_____
12. Air Temp. (Out)	D.B.	_____	_____
	W.B.	_____	_____
13. Desuperheater H_2O Temp. (In)		_____	_____
14. Desuperheater H_2O Temp. (Out)		_____	_____

Job Number

Name of Installer ______________

Name of Owner ______________

Address ______________

City ______________ State ______________

Field Comments: ______________

____

____

____

This PERFORMANCE CHECK SHEET should be filled out by installer and retained with unit.

Figure 21–7 Sample performance check list (Courtesy Bard Manufacturing Co.)

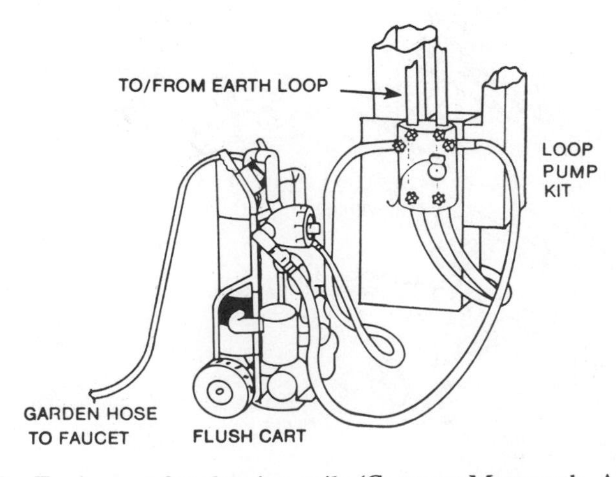

Figure 21–8 Equipment for cleaning coils (Courtesy Mammoth, A Nortek Co.)

QUICK REFERENCE TROUBLESHOOTING CHART FOR WATER-TO-AIR HEAT PUMP

DENOTES COMMON CAUSE ●

DENOTES OCCASIONAL CAUSE ▲

HEATING OR COOLING CYCLES

CYCLE

PROBLEM CONDITION / POSSIBLE CAUSE CONTROLS	POWER SUPPLY																		WATER COIL SECTION																										INDOOR SECTION								AUX.			
	LINE VOLTAGE										CONTROL CIRCUIT								COMPRESSOR					REFRIGERANT SYSTEM								WATER SOLENOID CIRCUL.			REV. VALVE		CHECK VALVE		INDOOR BLOWER MOTOR & COIL						WATERCOIL						CHECK VALVE		HEAT		GEN.	
	POWER FAILRE	BLOWN FUSE OR TRIPPED BREAKER	FAULTY WIRING	LOOSE TERMINALS	LOW VOLTAGE	DEFECTIVE CONTACTS IN CONTACTOR	COMPRESSOR OVERLOAD	POTENTIAL RELAY	RUN CAPACITOR	START CAPACITOR	FAULTY WIRING	LOOSE TERMINALS	CONTROL TRANSFORMER	LOW VOLTAGE	THERMOSTAT	CONTACTOR COIL	PRESSURE CONTROLS (HIGH OR LOW)	INDOOR BLOWER RELAY	DISCH. LINE HITTING INSIDE OF SHELL	BEARINGS DEFECTIVE	SEIZED	VALVE DEFECTIVE	MOTOR WINDINGS DEFECTIVE	REFRIGERANT CHARGE LOW	REFRIGERANT OVERCHARGE	HIGH HEAD PRESSURE	LOW HEAD PRESSURE	HIGH SUCTION PRESSURE	LOW SUCTION PRESSURE	NON-CONDENSABLES	UNEQUALIZED PRESSURE	SOLENOID VALVE STUCK CLOSED (HTG)	SOLENOID VALVE STUCK CLOSED (CLG)	SOLENOID VALVE STUCK OPEN (HTG or CLG)	LEAKING	DEFECTIVE VALVE OR COIL	STICKING CLOSED	LEAKING OR DEFECTIVE	PLUGGED OR DESTRICTED METERING DEVICE (Htg)	SCALLED OR PLUGGED COIL (HTG)	SCALLED OR PLUGGED COIL (CLG)	WATER VOLUME LOW (HTG)	WATER VOLUME LOW (CLG)	LOW WATER TEMPERATURE (HTG)	PLUGGED OR RESTRICTED METERING DEVICE (Clg)	FINS DIRTY OR PLUGGED	MOTOR WINDING DEFECTIVE	AIR VOLUME LOW	AIR FILTERS DIRTY	UNDERSIZED OR RESTRICTED DUCTWORK	STICKING CLOSED	LEAKING OR DEFECTIVE	AUX. HEAT UPSTREAM OF COIL			
COMPRESSOR WILL NOT RUN NO POWER AT CONTACTOR	●	●	●	●		▲	▲				●	●	●	▲	▲	●	●																																							
COMPRESSOR WILL NOT RUN POWER AT CONTACTOR			●	●	●		●	●	●	●										▲	●		●								●																									
COMPRESSOR "HUMS" BUT WILL NOT START			●	●	●	▲		●	●	●										●	●		●								●																									
COMPRESSOR CYCLES ON OVERLOAD			●	●	●	▲	●	●	●	●										▲		▲	▲	●	●	▲		▲		●					▲					▲	▲	▲	▲													
THERMOSTAT CHECK LIGHT LITE-LOCKOUT RELAY																	●																													▲	▲	▲	●	▲						
COMPRESSOR OFF ON HIGH PRESSURE CONTROL																	▲	●							●					●			●				●		▲		▲		●		▲	●	●	●	●	●	●		●			
COMPRESSOR OFF ON LOW PRESSURE CONTROL																		▲						●			●		●			●								▲		●		▲	●											
COMPRESSOR NOISY																			▲	▲		●																																		
HEAD PRESSURE TOO HIGH																									●					●			●			▲	▲		▲		●		●		▲	▲	▲	●	●	▲	▲		●			
HEAD PRESSURE TOO LOW																						●		●					●			▲			▲			▲		●		●		●								▲				
SUCTION PRESSURE TOO HIGH																						●			●	●									▲			▲														▲				
SUCTION PRESSURE TOO LOW																								●			▲					●					▲		▲	●		●		▲	▲	●	▲	●	●	▲	▲					
I.D. BLOWER WILL NOT START	●	●	●	●							●	●	●	▲	▲			●																																						
I.D. COIL FROSTING OR ICING																								●					●								▲								▲	●	●	●	●	▲						
HIGH COMPRESSOR AMPS					●				●											●	●		▲		●	●		●		●							▲		▲		●		●		▲	●	●	●	●	▲	▲		▲			
EXCESSIVE WATER USAGE																																		●						▲	▲															
COMPRESSOR RUNS CONTINUOUSLY — NO COOLING																						●		●					●												▲		▲		▲	●	●	●	●	▲						
LIQUID REFRIGERANT FLOODING BACK TO COMPRESSOR																						●			●	●																				●	●	●	●	▲		▲				
COMPRESSOR RUNS CONTINUOUSLY — NO HEATING																								●					●			●			▲				▲	●		●		▲												
REVERSING VALVE DOES NOT SHIFT			●	●																				▲			▲									●																				
LIQUID REFRIGERANT FLOODING BACK TO COMPRESSOR																									●	●						●			▲	▲				●		●		●												
AUX. HEAT ON I.D. BLOWER OFF			●	●							●	●			▲			●																													●									
EXCESSIVE OPERATING COSTS			▲	▲											▲									●	●							▲	▲		●	●				▲	▲	▲				▲	▲	▲								
ICE IN WATER COIL			▲	▲																				●								●				●				▲		●														

Figure 21–9 Quick reference troubleshooting chart (Courtesy Bard Manufacturing Co.)

pressure-reducing device, the "check valve" cause will not apply.

The "water valve" category also includes the circulator in the ground loop systems. Failure of the circulator will produce the same results as a stuck solenoid valve.

Use of this chart should be very helpful in analyzing system problems.

REVIEW QUESTIONS

1. When purging refrigerant from a unit, what is the easiest way to prevent freeze up of the liquid-to-refrigerant coil?
2. What would be the observation that would occur if the check valve on the water coil trombone were to stick closed with the unit in the cooling mode?
3. What would the observation be that would occur if the check valve on the water coil trombone were to stick closed in the heating cycle?
4. Before adding any refrigerant to the system, what must be done to the system?
5. List the four types of pressure-reducing devices used in liquid-to-air heat pump units.

Section FOUR
Air-to-Liquid Type Systems

AIR-TO-LIQUID 22

22–1 GENERAL

With the increased interest in using the heat pump for environmental control, several by-products of this industry have been developed. One of these by-products is the heating of domestic water by means of the refrigeration principle. The first contact this author had with such a product was in 1960 in Milwaukee, Wisconsin. A refrigerated water heater to be used in conjunction with a standard water heater was developed by a local refrigeration service company. This product, along with a refrigerated hair dryer for beauty shops was ahead of its time. Electric energy was inexpensive and savings via refrigeration systems were not enough to justify the investment. The heat pump industry was in the same situation.

With the rise in electrical energy cost, however, the heat pump industry grew rapidly, which helped promote the by-products. In the 1986 Consumer's Directory of Certified Water Heater Efficiency Ratings Published by the Gas Appliance Manufacturers Association, nine manufacturers are listed as marketing 86 brand names of "Heat Pump Water Heaters Without Tanks" and six manufacturers market 119 brands of "Heat Pump Water Heaters With Tanks."

Practically all of the manufacturers produce the heat pump water heater without tank to be applied to a tank of 52 to 58 gallon capacity. The COP ratings are in the 2.0 to 2.9 range.

The heat pump water heaters with tanks are practically all produced in four sizes from 50 to 120 gallon storage capacity.

This industry has not stopped at those sizes. Heat pump water heaters for commercial and industrial application as well as the heating of swimming pools and spas has become a large industry. In addition, many unique applications have developed, especially with the application to enclosed swimming pools.

The various products and their applications are discussed in further detail under each section heading.

22–2 HEAT PUMP WATER HEATER—REMOTE

The first heat pump water heaters were self-contained units designed to be connected to the present hot water supply tank. The system used two temperature controls—one to control the refrigeration unit and one to control the heat source of the hot water supply tank.

In these units the refrigeration unit becomes the main source of heat energy with the No. 1 thermostat set at the desired hot water temperature (usually 120°F to 130°F). The temperature control for the hot water supply tank energy source is set 10°F lower then the refrigeration unit temperature control. This keeps the hot water tank energy source as a reserve source in the event of a refrigeration unit malfunction or excessive load on the hot water supply.

Shown in Figure 22–1 is the style of unit that was marketed originally. This unit has a heating capacity of 13,700 BTUH at an air temperature of 80°F DB and 70°F WB (72.5% RH). Water is heated through the range of 60°F to 140°F for domestic hot water supply.

Figure 22–1 Hot water pump (Courtesy Borg-Warner Environmental Systems, Inc.)

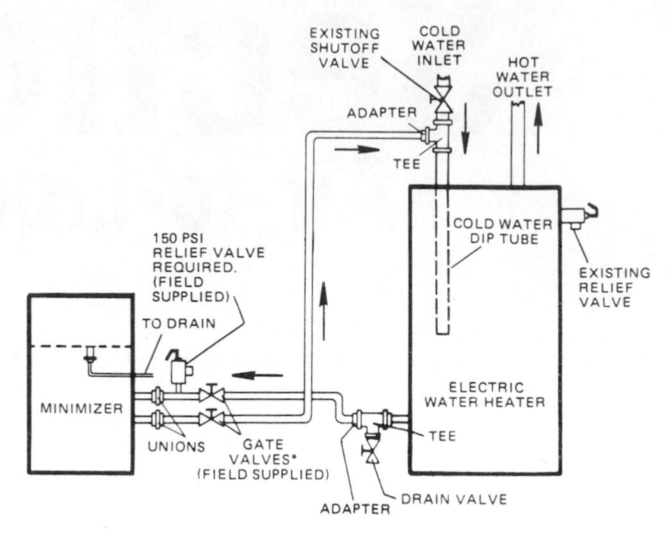

Figure 22–2 Piping diagram (Courtesy Borg-Warner Environmental Systems, Inc.)

Using the original electric water heater as the hot water storage means, the unit is connected to the hot water tank by means of tees in the cold water inlet and the tank drain outlet. This produces a circulation of water to the heat pump from the bottom of the tank. This reduces the dilution of hot water in the tank to maintain a more even supply water temperature.

Figure 22–2 shows the piping diagram of the unit connected to the hot water supply tank. Taking the water off the bottom of the tank, through the tee, the water flow is through the connecting pipe or hose (depending upon local code) to the unit. A shut-off valve (ball type for minimum flow resistance) is included in each of the connecting lines to permit removing the unit for service without draining the hot water system.

A drain connection is also required for the evaporator coil in the unit. The unit is also a very effective dehumidifier and requires a drain for disposing of the condensate water of the evaporator coil.

Figure 22–3 shows the internal construction of the unit. The active evaporator coil with the circulating fan is located above the compressor. Under the coil is the condensate pan with drain connection. The pressure-reducing device used is a T. X. valve. Both capillary tubes and restrictor valves are also used in these heat pumps.

To the left of the compressor is the refrigerant-to-liquid heat exchanger wrapped with insulation for higher efficiency of heat transfer. Above the heat exchanger is the circulating pump.

All controls and protection are via a solid state control module. This includes the maximum water temperature limit control as well as compressor overload protection. The thermostat in the water heater is the primary temperature control and is connected to the terminal block in the heat pump.

Figure 22–4 shows the wiring diagram for connecting the 230 volt, 1 phase, 60 hertz power to the unit and tank thermostat. Using the tank thermostat provides the operation of the heat pump as the primary source of heat energy and the water heater elements as an auxiliary source when needed.

The heat pump compressor and fan motor are 230 volt to match the power requirement of the electric water heater.

120 volt units are also marketed for connecting to gas- or oil-fired hot water heaters. These units have a separate control thermostat to provide the normal temperature control action.

Figure 22–5 shows a water heater heat pump of the wrap-around type with top fan discharge of the air through the evaporator. The unit resembles a round condensing unit. It contains the complete refrigeration system. The round evaporator is supplied refrigerant by capillary tubes. The compressor is in the center below the evaporator fan motor along with the refrigerant-to-liquid heat exchanger and water circulating pump.

Water temperature limit control and compressor overload protection are in the unit. The normal water temperature thermostat is part of the water connection adapter for connecting into the hot water heater.

This unit is applicable to any of the types of hot water heaters—gas, oil, or electric.

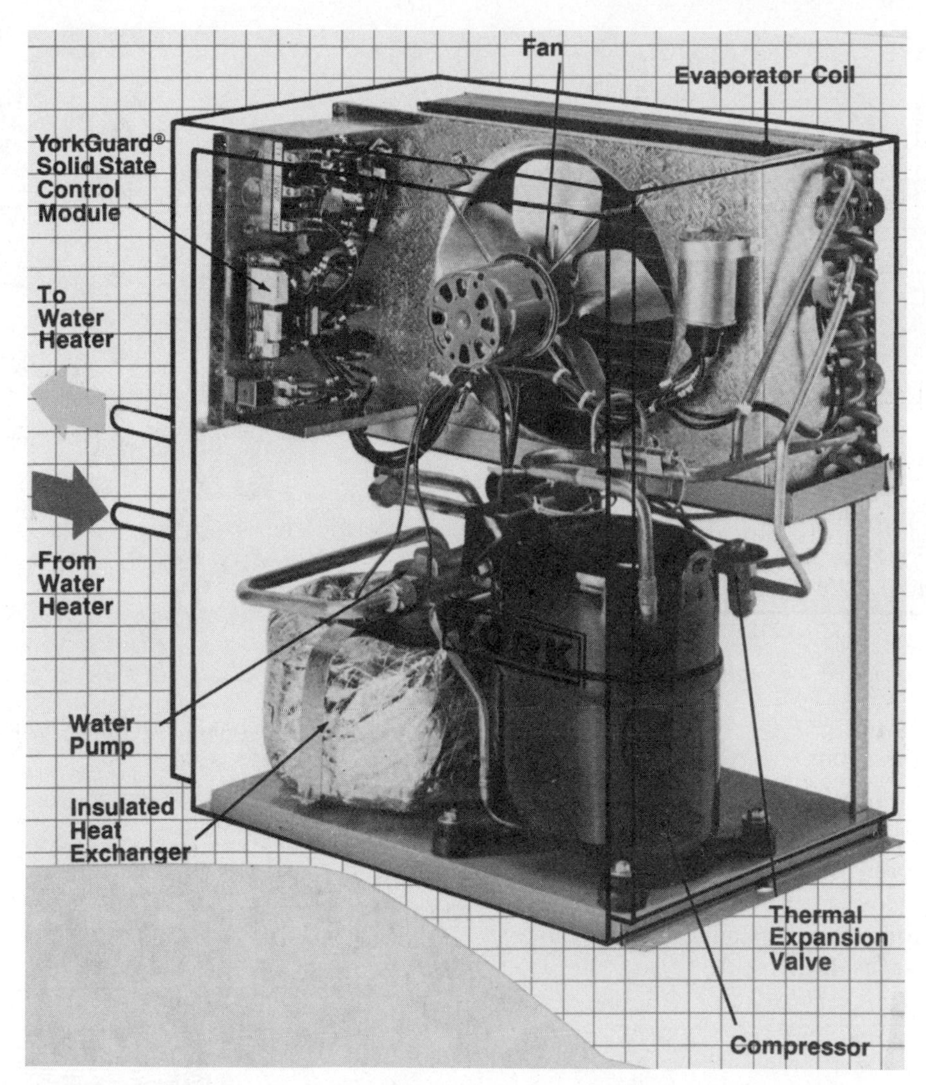

Figure 22–3 Interior view (Courtesy Borg-Warner Environmental Systems, Inc.)

Figure 22–6 shows the plumbing tree that is used to connect the unit to the hot water tank. The plumbing tree is actually a pipe within a pipe that includes a thermostat well. The union that fits the bottom tap of the tank (where the drain valve is normally located) has a side connection and shut-off valve to provide a flow of cold water out of the tank to the heat pump.

Heated water from the heat pump flows through the center pipe and through the extension pipe into the tank. The extension pipe forces an upward flow of the hot water supply to reduce the mixing with the cold water in the lower part of the tank. This helps to eliminate temperature layers in the tank and improves the operating efficiency.

A thermometer well is located in the hot water supply tube to provide a means of mounting the thermostat-sensing capillary tube without the necessity of entering the hot water system.

The built-in thermostat is mounted behind the control panel with an exposed temperature-setting knob. Included also are high-and low-pressure and temperature limit switches to protect the unit from damage due to extremes in both high and low ambient air temperatures, evaporator freeze up, loss of refrigerant, and severe restriction of air or water flow.

This particular unit has a nominal heat output to the condenser of 12,000 BTUH based on 75°F ambient air, 135°F tank water, and 55°F supply water temperature. At these conditions, the water heating rate is 16 GPH with a water temperature rise from 55°F to 135°F and 18 GPH with a water temperature rise from 70°F to 140°F.

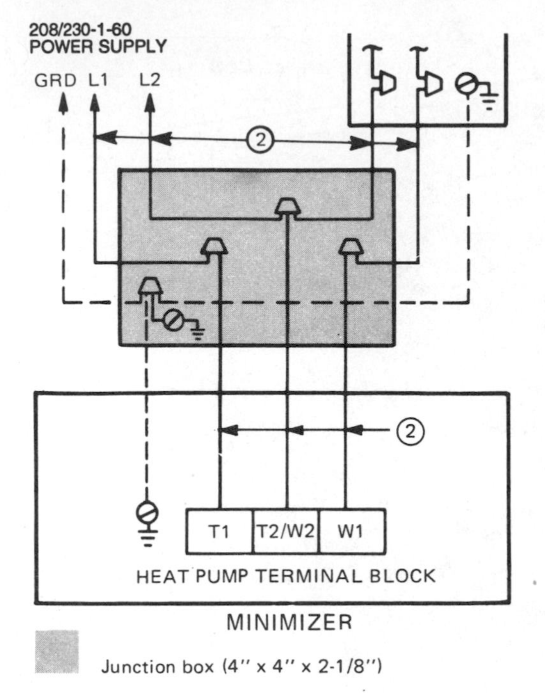

NOTES:
USE COPPER CONDUCTORS ONLY BETWEEN UNIT AND JUNCTION BOX. WIRE IN ACCORDANCE WITH ALL LOCAL AND NATIONAL ELECTRICAL CODES.

① Disconnect switch and overcurrent protection (30 amp max.) by others

② Select a wire size to handle the maximum wattage of the electric water heater.

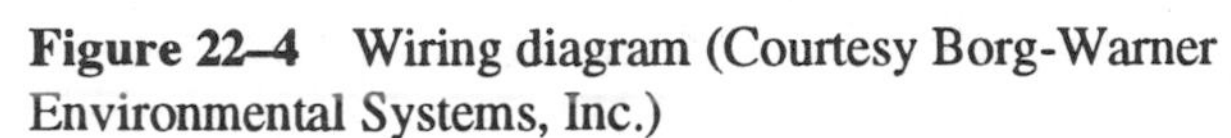

Figure 22–4 Wiring diagram (Courtesy Borg-Warner Environmental Systems, Inc.)

Figure 22–5 Round heat pump heater (Courtesy Energy Utilization Systems, Inc.)

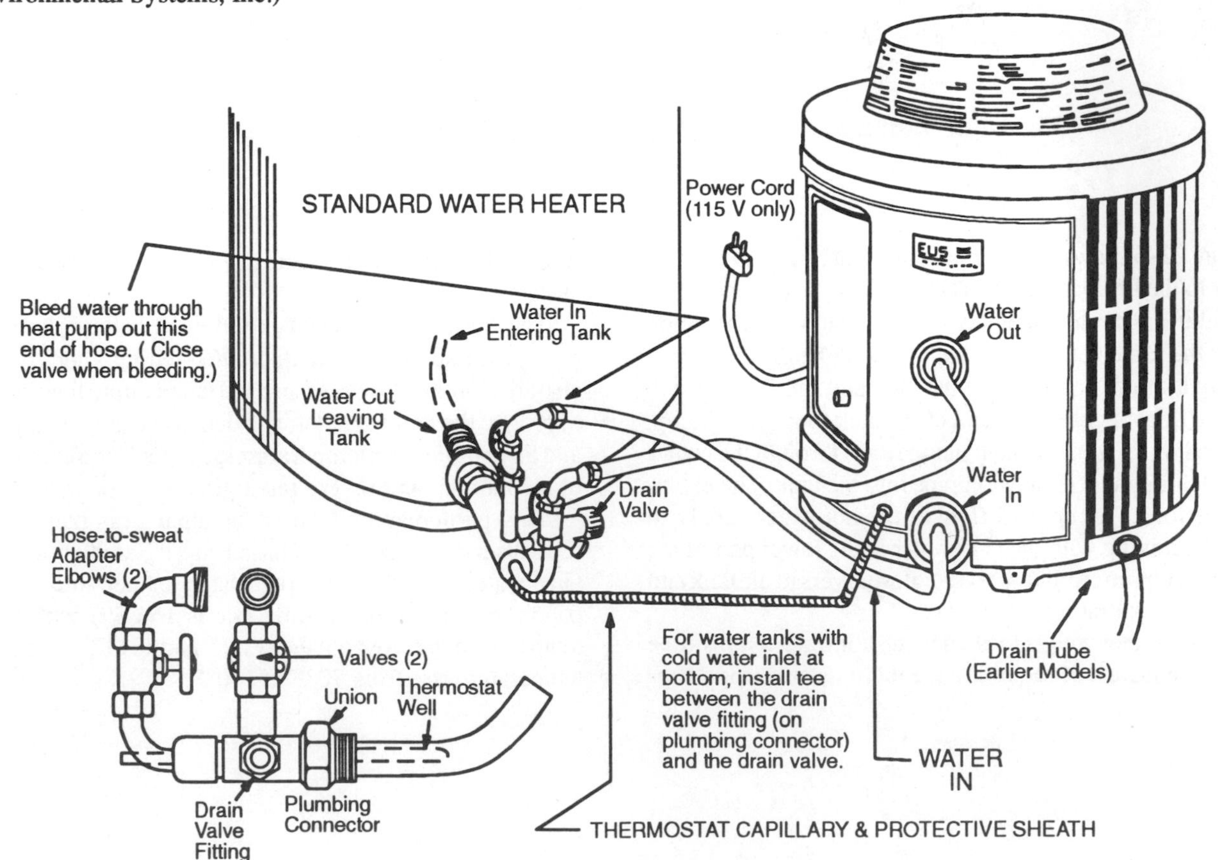

Figure 22–6 Piping diagram

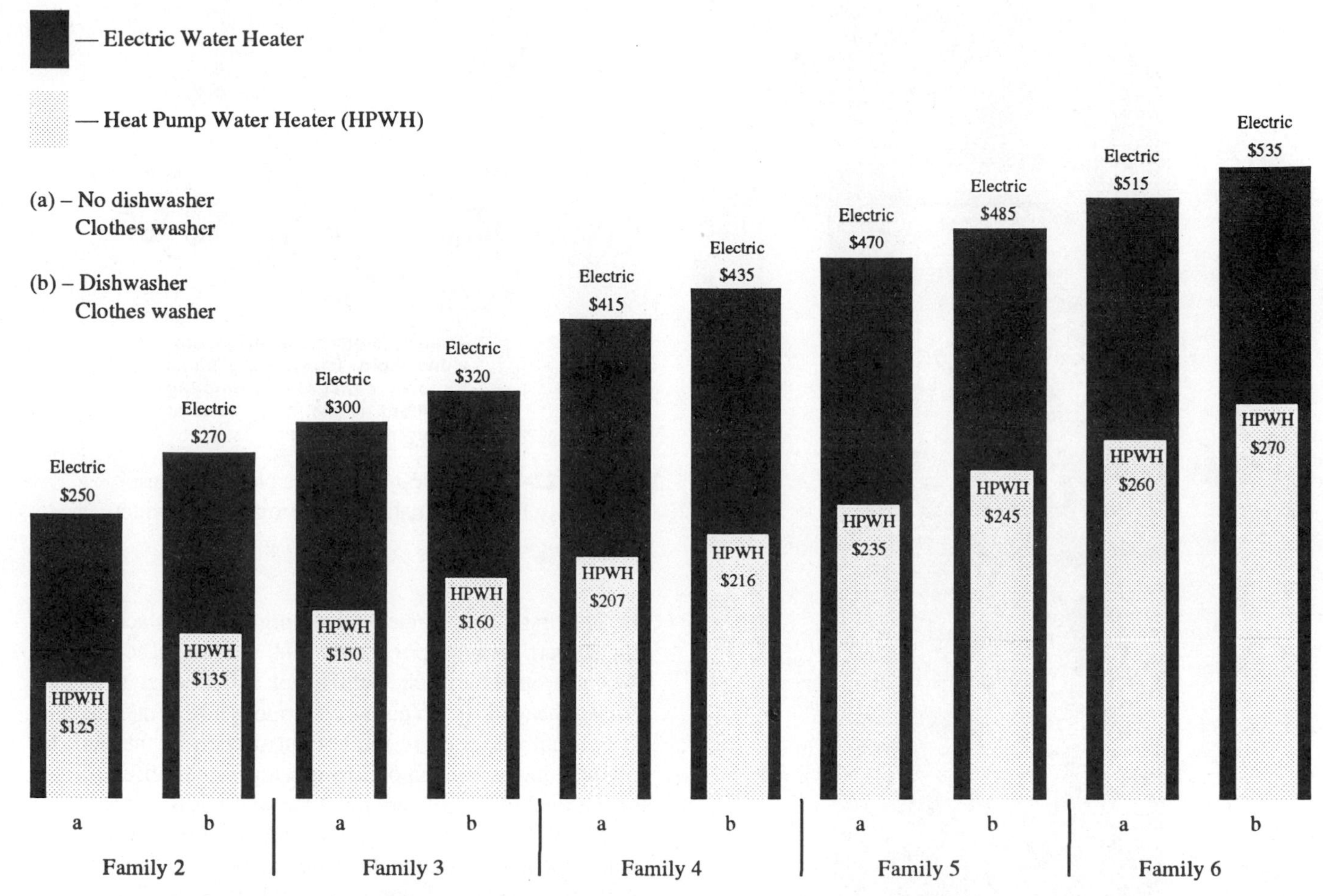

Figure 22–7 Bar graph comparison of the cost of heating water (Courtesy DEC International, Therma-Stor Products Group)

The rated watts drain of 1.25 kWh as compared to the equipment electric elements of 3.5 kWh (12,000 BTUH divided by 3,413 BTU/kW) produces a COP of 2.8. The unit will produce 2.8 times as much water heating energy per penny cost as compared to a resistance heating element.

Figure 22–7 is a bar graph comparison of the cost of heating water with a heat pump versus electric resistance heat based on an electric rate of $.05/kWh. Comparison is also made on average usage rates for five different size families. Using comparison costs such as these, the payback time can be calculated to determine the advisability of investment in this type of equipment.

22–3 HEAT PUMP WATER HEATER—SELF-CONTAINED

When the refrigeration system and storage tank are consolidated as a single unit, the self-contained water heater is created. These units usually have a single 2.5 kWh to 6.0 kWh element as a backup for hot water demand that exceeds the capacity of the refrigeration system. This element as well as the refrigeration system are thermostatically controlled. With the refrigeration system thermostat set at 135°F, the auxiliary thermostat is usually at 110°F. This provides a normal hot water temperature of 130°F with a minimum of 110°F.

Figure 22–8 shows a typical hot water heat pump of the self-contained type. All of the portions of the refrigeration system plus the controls, with the exception of the condenser coil and control thermostat, are located in this section. The condenser is an integral part of the water storage tank.

The specifications encountered in this unit are used by practically all the manufacturers in the industry:

1. The evaporator is aluminum fin—copper tube type.
2. Suction line accumulator.
3. Hermetic motor-compressor assembly.
4. Filter-drier assembly ahead of the pressure-reducing device.
5. System air conditioning capacity in the 2,800 BTUH to 12,200 BTUH range. This depends upon the hot water temperature. With lower water temperature, condensing temperatures are lower and capacity increases. The difference in capacity between 115°F water temperature and 135°F water temperature is around 5%.

Figure 22–8 Self-contained hot water pump (Courtesy DEC International, Therma-Stor Products Group)

6. Practically all manufacturers are now using urethane insulation of 1-1/2" to 2" thickness.
7. Backup heater element sizes are in the 2.5 to 6.0 kW size. The greater number use elements under 4.0 kW.
8. Hot water outlet connections are mostly 7/8" copper or 3/4" PVC.
9. Propeller fan and motor assembly. This unit uses a horizontal air movement with the fan mounted inside the cabinet for draw-through operation.
10. Insulated suction line to prevent condensate drip is standard.
11. Condenser coil. The condenser coil design varies among manufacturers. Two principles are involved:
    a. Direct contact between the hot refrigerant vapor and the tank water.
    b. Remote application with a means of moving the water from the tank, through the condenser coil, and back to the tank. This is the principle used for remote heat pump water heaters.

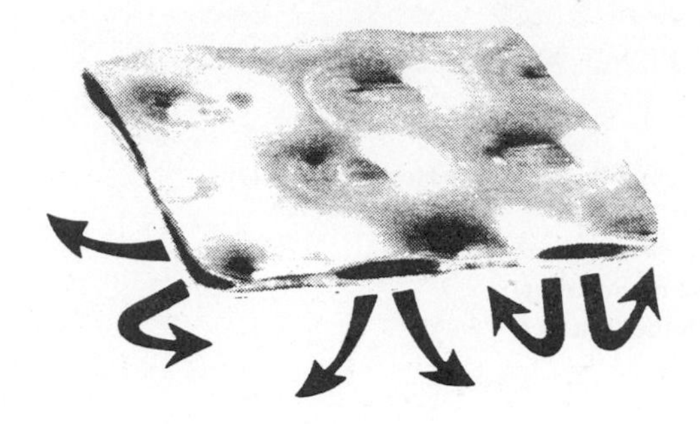

Therma-Stor® plate design, provides rapid, free-flowing paths for refrigerant gas — promoting excellent waste heat transfer throughout the tank.

Figure 22–9 Waffle-type double plate condenser (Courtesy DEC International, Therma-Stor Products Group)

The first self-contained units using direct contact between the hot refrigerant vapor and the tank water used copper tube wound around the outside of the tank and secured to the tank for best heat transfer. This is a continuous tube condenser coil. It is usually located in the bottom portion of the tank for maintaining as high an operating temperature difference between the refrigerant vapor and the water as possible.

Figure 22–9 shows a double wall construction of a condenser wrapped around the tank to provide full physical contact between the condenser and the tank. The spaces between the plates receive the hot vapor from the compressor at the top of the tank. The vapor travels down in the waffle-like spaces in a random manner with no set travel path. This allows for maximum heat flow rate between the vapor and the water but does not promote an even flow rate of refrigerant through the system.

This uneven refrigerant flow rate, however, is handled by the use of a suction accumulator in the suction line ahead of the compressor. The refrigerant charge is given as required on the rating plate. Even though an accumulator is included, the refrigerant charge must be held within ±1/2 oz. tolerance for systems using capillary tubes or restrictors.

The major advantage of this type construction is the fact that scale formations in the tank settle to the bottom of the tank and do not affect the condenser-to-water heat transfer ability. The disadvantage of this type unit is that when the unit needs to be serviced the hot water supply has to be shut off.

Figure 22–10 shows what appears to be a complete unit but is actually a field-assembled unit. The heat pump assembly is a complete refrigeration system containing all refrigeration and electrical components including a refrigerant to water heat exchanger and circulating pump.

It is designed to fit on the top of an electric hot water heater that has the standard 8" on center spacing between the hot and cold line connections.

This design makes it marketable in both the new construction as well as the retro-fit market.

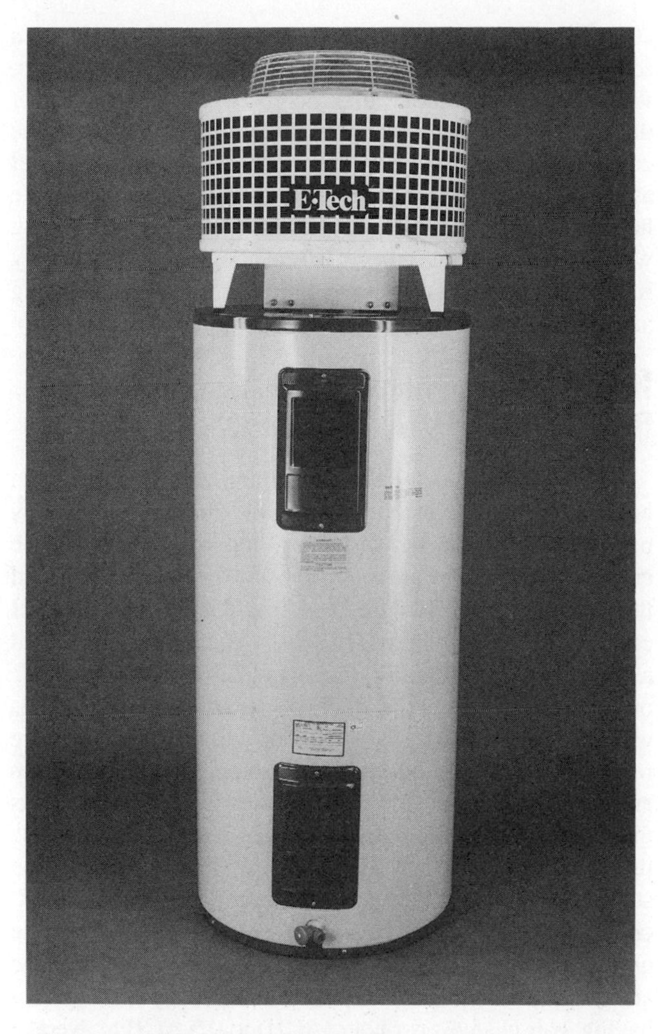

Figure 22–10 Tank-mounted heat pump water heater (Courtesy Crispaire Corporation)

The major advantage of this design is the fact that the heat pump assembly may be removed for service without interrupting the hot water supply. The disadvantage is the use of a circulating pump to carry water from the tank through the heat exchanger and back to the tank. The coil will require service to remove scale formation.

22–4 CONTROL

Practically all manufacturers of these products use a control system that controls the heat pump as the primary source of heat energy with a backup resistance element as a supplemental heat source. Figure 22–11 shows a typical control setup mounted on the front of the tank. Mounted in the location of the backup element, the control provides access to both the thermostats as well as the backup element.

In this particular control setup, the backup thermostat (A) is located above the backup element (B). The heat pump thermostat (C) is located below the backup element (B). This control setup is prewired by the factory. Therefore, only the two 240-volt hot leads and the ground lead need to be connected to the control setup.

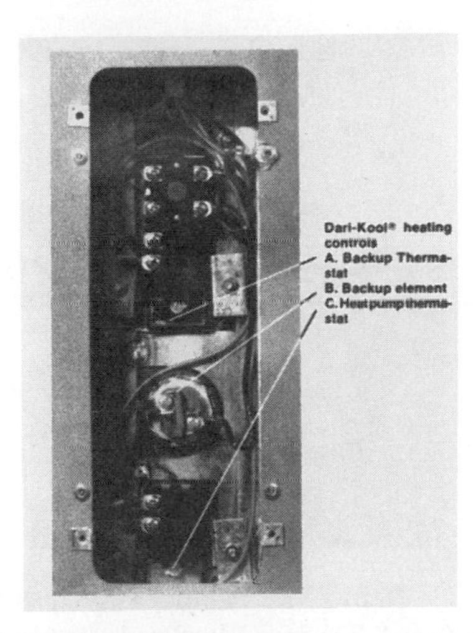

Figure 22–11 Thermostat control assembly (Courtesy DEC International Inc., Therma-Stor Products Group)

In the case of an add-on installation, the heat pump thermostat is added in the place of the lower element thermostat. The heat pump thermostat, set at 130°F, maintains the higher water temperature with operation of the heat pump unit.

If the demand for hot water exceeds the capacity of the heat pump and the water temperature drops to 110°F, the setting of the backup thermostat, the backup element is energized. The two heat sources then supply the full heating capacity of the unit.

In addition to the two thermostats, one manufacturer also provides a three-way selector switch that allows the choice of the heating mode—heat pump/off/electric resistance. In addition, to prevent operation of the heat pump below 50°F ambient air temperature, a control is included to switch the system from heat pump to resistance heat at 50°F. This allows freeze up protection for the heater.

22–5 HEAT RECOVERY WATER HEATER

With the development of the air-to-liquid water heater, a variety of multiple-use applications have been developed. A popular concept for commercial applications such as restaurants is the use of a remote air handler in the kitchen area to remove kitchen heat and use it to heat water. Suspending the air handler in the grill area can provide cooler, drier air for workers. Figure 22–12 show such an application.

As an aid to the air conditioning system in the restaurant proper, the system helps to provide more comfortable conditions in the kitchen area while the sensible and latent heat absorbed from the kitchen air is used to heat the hot water supply.

The air handler may also be used to help reduce the temperature of the refrigeration condensing unit room in supermarkets. This improves the operating efficiency of the general refrigeration systems as well as provides a waste heat source for heating the water.

Figure 22–12 Remote air handler application (Courtesy DEC International, Therma-Stor Products Group)

In larger 120-gallon storage capacity units a remote compressor assembly with multiple air handler evaporators can provide as much as 44,000 BTUH cooling and dehumidifying capacity. The 44,000 BTUH capacity unit will provide 88 gallons per hour of 145°F water or 102 gallons per hour of 120°F water for commercial use.

22–6 EHPWH (EXHAUST HEAT PUMP WATER HEATER)

Following is a reprint of a publication by Mr. Kenneth C. Gehring, V. P. of the Therma-Stor Products Group, Division of DEC International Inc., and reprinted with his kind permission. The section numbering system of the original publication has been changed to conform to the section numbering system in this text. The purpose of including this publication is to show an interesting application of this type of product to supply an answer to the problems encountered with ventilation of today's highly insulated and airtight homes.

Cost Effective Residential Ventilation and Water Heating (by Kenneth C. Gehring)

Therma-Stor Products Group,
Division of DEC International,Inc.

Abstract

Many new houses in the U.S. are air-tight enough that they experience excessive moisture and air quality problems. The ventilation devices commonly used in these homes are windows, bath fans, and kitchen hood fans. Wind, the "stack effect," internally-vented combustion devices, and air-handling systems can cause additional ventilation. Although a house may be adequately ventilated when winds are strong and the weather is very cold, until now no control strategy assured consistent, proper ventilation throughout the year. The ventilating water heater economically provides controlled ventilation according to the individual family's needs by using heat pump technology to recover waste heat from stale exhaust air. Controls allow ventilation to occur on a scheduled basis, and by strategically locating fresh air vents, whole house ventilation can be achieved with a minimum of ducting. While the initial investment is greater than the conventional equipment, a family of four can expect as short as a two year payback because of reduced water heating, space heating, and space cooling costs. In addition, the ventilating water heater can protect the structural integrity of the house by reducing the amount of warm moist air that penetrates ceiling and wall insulation.

Introduction: The Natural Air Change Rate of Today's New Houses

The factors that determine the natural air change rate of a house are the cracks in the structure, the difference in density between the outside and inside air, and wind velocity. The cracks in the structure are determined by the materials to build the house and the attention to sealing details at the time of construction. The difference in density occurs because heated air weighs less than cold air (see Figure 22–13). The buoyancy of the 68°F air inside the 2,000 sq. ft. house is 180 pounds when it is 0°F outside (like a hot air balloon). This pressure forces warm stale air out through the upper structural cracks and the cold fresh air in through the lower structural cracks. As the temperature of the outside air rises, the buoyancy of the inside air declines, which decreases the pressure on the structure. This decrease in pressure causes the natural air change rate to decline. This air change rate determines the quality of the air and moisture level inside the house. The airtightness of new homes being built today varies from having 2 air changes per hour (ACH) to 0.5 ACH at average winter conditions. Remember, as the outside temperature increases, the natural air change rate declines proportionately, At a 60°F outside temperature, very little natural ventilation takes place in either leaky or tight homes unless it is windy. Houses with an average natural 0.4 ACH will over ventilate during extreme winter conditions and under ventilate during mild conditions. Over-ventilation will require heat and humidification to be comfortable. 1,000 BTUs of heat are required to vaporize one pound of water plus the cost of the humidifier. Up to 5 gallons of water can be required to humidify a leaky house. This is 40,000 BTUs per day for humidification. Unfortunately the fresh air infiltrating the house must also be heated. This can account for an additional 380,000 BTUs per day. At today's energy prices, it pays to tighten up the house and eliminate excessive exfiltration.

The Recommended Air Change Rate

The stale air exfiltrating from a house contains the moisture and air pollutants generated by normal living activities. Most recommendations are that 0.4 ACH is necessary for moisture control and good air quality with typical family living habits. To maintain this air change rate in today's energy efficient house during all weather conditions, mechanical ventilation is necessary.

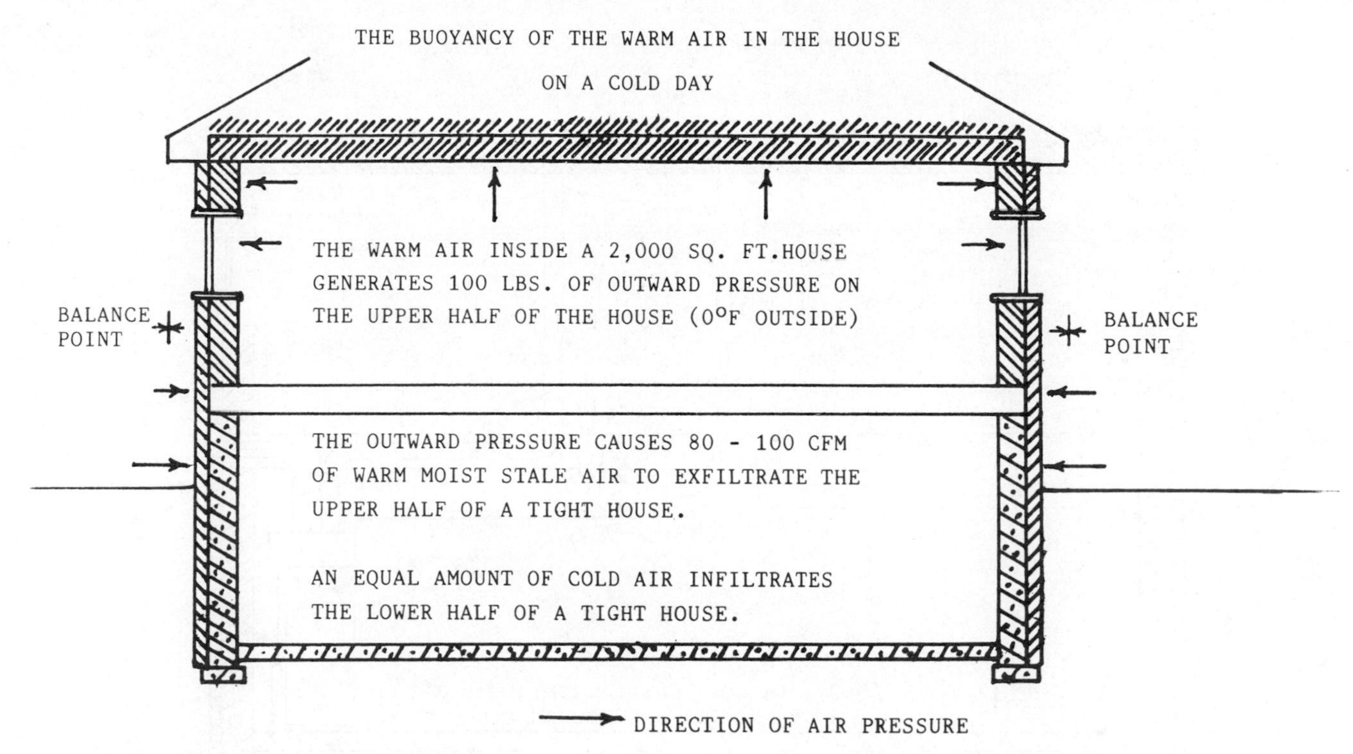

Figure 22–13 Cause of ventilation on a cold day (Courtesy DEC International, Therma-Stor Products Group)

The House and Mechanical Exhaust Ventilation Equipment

An energy efficient house should be tight enough to limit natural air change rate to 0.4 ACH during the coldest windy weather. At 0.4 ACH, 106 CFM of moist stale air is exfiltrating the upper cracks of the house. An equal amount of cold fresh air is infiltrating the lower cracks of the house. This house would demonstrate a 0.2–0.4 natural ACH during average winter conditions (or a blower door test). Now we have a house that naturally provides the ideal ventilation rate, 0.4 ACH, during the worst winter conditions. Instead of allowing the warm moist stale air exfiltrating through the cracks of the house, the air is exhausted mechanically. This exhaust rate is matched to the ideal ventilation rate (0.4 ACH) and continues whenever the house is occupied. By mechanically exhausting 106 CFM of warm moist stale air from a 2,000 sq. ft. house, ideal ventilation (0.4 ACH) is maintained during all weather conditions. The conclusion of a study at U. of California showed that this concept would provide the most uniform air change rate under varied weather conditions. In addition, by not allowing moist air to penetrate the structure, the potential for moisture condensing in the ceiling and wall insulation is eliminated.

Recovering Heat From the Warm, Moist, Stale Exhaust Air

The warm, moist, stale, exhaust air is an excellent source of heat for water heating and space heating. In Sweden, a high percentage of government financed housing uses the exhaust heat pump water heater (EHPWH) to recover the heat from the warm, moist, stale, exhaust air. The exhaust air is ducted through the evaporator of the EHPWH which removes the heat from the stale air (see Figure 22–14).

The cooled exhaust air temperature is 32°F to 35°F. This process produces 2.7 times more BTUs per kWh than electric resistance heat. In a super insulated house during moderate winter weather, this recovered heat will heat the water, the fresh cold infiltrating air, and the space.

A United States Version of EHPWH

Since 1976, Therma-Stor Products Group, Division of DEC International Inc. has manufactured energy efficient refrigerant-based water heating devices for the dairy and food processing industry. In 1980, the moist energy efficient water heater in the U.S. was introduced by Therma-Stor Products Group. From 1980 to 1983, an EHPWH was developed and tested.

The Product

The Main Components of the EHPWH and Their Function. This system is made up of the conventional insulated 80 gal. water heating tank and a small refrigeration system with its controls. This refrigeration system used a 0.6 hp refrigeration compressor, an evaporator to remove heat from the exhaust air, a condenser to heat water, a condenser to space heat, an exhaust blower to remove air from the house, and a space heating blower to recirculate inside air (see Figure 22–15).

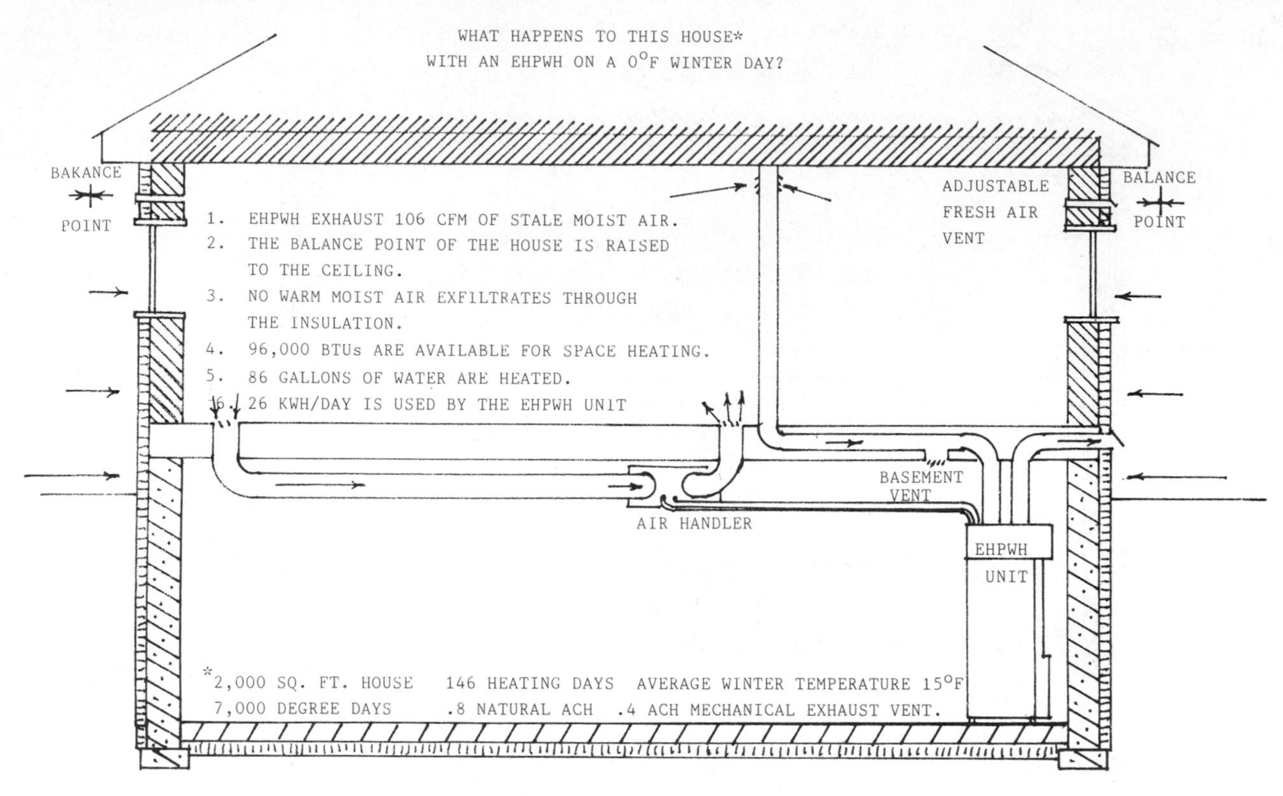

Figure 22–14 EWPWH—air flow and air pressure (Courtesy DEC International, Therma-Stor Products Group)

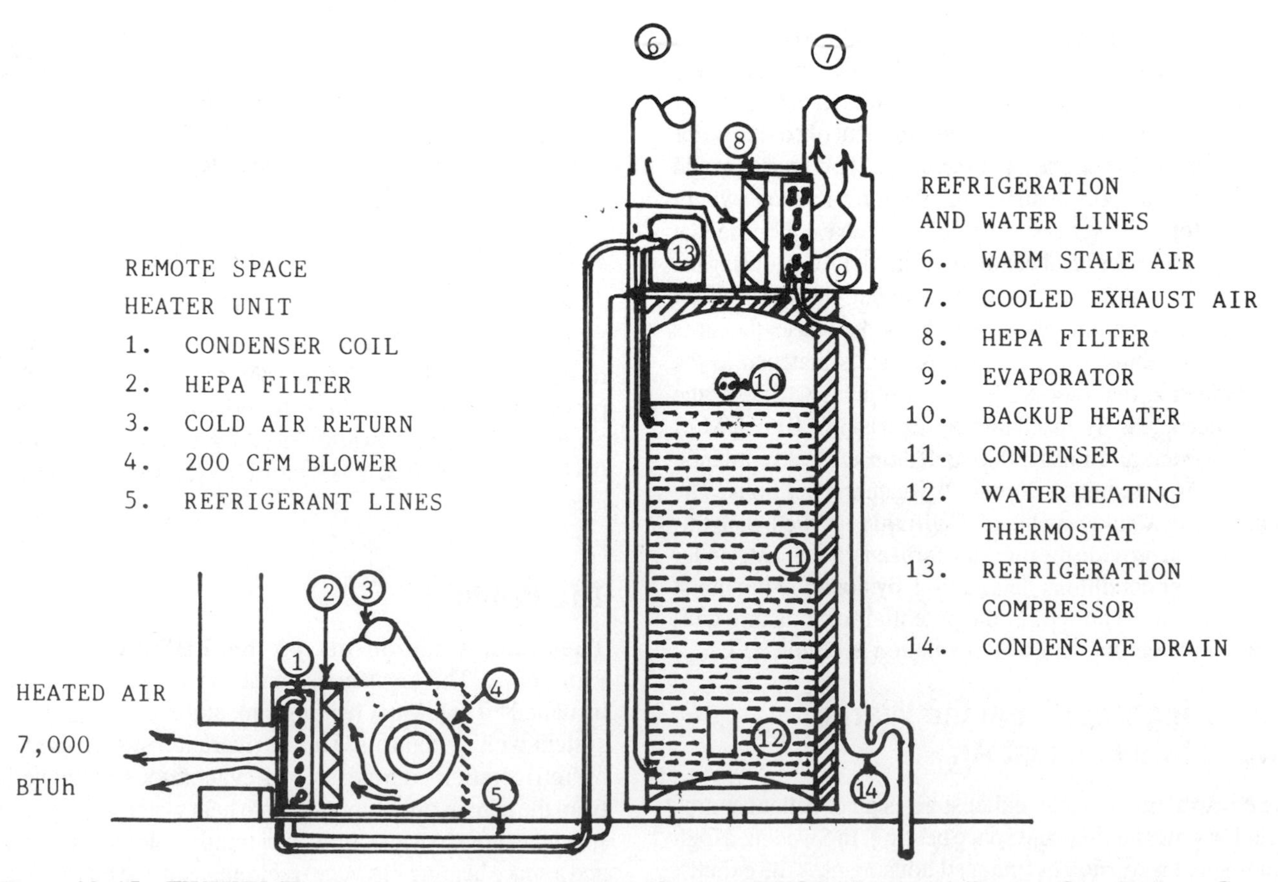

Figure 22–15 EWPWH (Exhaust heat pump water heater) (Courtesy DEC International, Therma-Stor Products Group)

Installation of the EHPWH in the House. The EHPWH is installed in the space allocated for the conventional water heater, which it replaces. Warm stale air is ducted from a central location (a high central wall between the kitchen and bath) to the EHPWH. The cooled stale air is ducted to the outside. This allows minimal ducting of the air to be heated. An insulated refrigerant copper line set connects the EHPWH to the space heating coil. The EHPWH space heating thermostat is located in the open area of the house. To assure minimal fresh air flow to the isolated occupied rooms of the house, individual fresh air inlets are recommended. These vents would be installed close to the ceiling in an outside wall. The vent features an adjustable opening, a filter, and a shut-off valve. They also balance the outside and inside pressure to prevent the whole house from being depressurized.

Operation of the EHPWH in Winter. The EHPWH operates in response to the water heating thermostat (primary), the space heating thermostat (secondary) or 24-hour timer control (exhaust fan only). The water temperature is adjustable from 110°F to 140°F. The space heating thermostat is adjustable from 55°F to 75°F. The exhaust timer should be set to operate during times of known occupancy in the event both water and space heating are satisfied. The quantity of exhaust air is adjusted at the time of installation to match the desired ventilation rate of the house (90 CFM to 200 CFM).

Operation of the EHPWH in Summer. As the cooling season approaches the internal temperature of the house rises. After the temperature of the house reaches an uncomfortable level, the air pivot valve on the top of the EHPWH is reversed. When water heating is needed, warm outside air is drawn through the EHPWH where it is dried and cooled. This air can be discharged into the basement for dehumidification or into the open part of the house for cooling. Water heating in summer will provide 10 hours of ventilation with cool dry air. Summer ventilation in an air conditioned house is more critical than winter because there is less difference between the inside and outside temperatures.

Performance

Water and Space Heating Capacity. The EHPWH heats 8 gallons of 80°F rise per hour or provides 6,900 BTUH of space heating. A backup 2.5-kW electric heating element is installed in the water tank, which heats an additional 12 gallons per hour. The space heat is ducted to the central open area of the house. A minimal efficiency space heating system is recommended because 65% of the total space heat can be supplied by the EHPWH. Occasional high water heating loads and extreme space heating require a normal space heating system.

Air Filtering. The air passes through the EHPWH evaporator and the space heater is filtered by a high efficiency particulate arresting (HEPA) filter. In winter, the recirculating air and in summer, the fresh air has the dust and pollen removed.

Control Strategy. The EHPWH water heating thermostat should be set at the minimum temperature needed for domestic hot water, which is usually 120°F. The EHPWH space heating thermostat should be set at the desired living space temperature. If individual room thermostats are used for the backup space heating system, set them at the minimum room setting. When a single central thermostat is used, it should be set a couple degrees below the EHPWH thermostat setting. A tight small super-insulated house can be heated by the EHPWH when the outside temperature is above 20°F.

Payback Using the EHPWH

Water Heating Savings Using the EHPWH. The cost of water heating is determined by cost of electricity, the quantity of hot water used, the temperature rise, and the efficiency of the water heater. The average cost of a kilowatt hour (kWh) in the U.S. is $.074. A family of four uses about 80 gals. of hot water per day. The average electric water heater is 80% efficient. The cost for hot water for a year with a conventional water heater with the above conditions is $551. With the identical conditions, the EHPWH produces hot water for $204. The savings are $347 per year. In addition, summer dehumidification or air conditioning (A/C) is provided.

Space Heating Savings Using the EHPWH. Whenever the EHPWH is not heating water, it is available for space heating and fresh cold infiltrating air. The days of space heating per year vary by region in the U.S. Seattle has 200 heating days. Although Madison, WI has 2,000 more heating degree days, it has only 150 heating days. It takes 10 hours a day to heat 80 gal. of water. This leaves 14 hours per day for space heating. The EHPWH produces 6,900 BTUs per hour for 14 hours consuming 0.65 kW of electricity. The total heat is 96,600 BTUs per day for 9.1 kWh per day. Electric resistance heat yields 3,414 BTUs per kWh. The 96,000 BTUs of space heat produced by EHPWH would require 28 kWh of resistance heat. This saves 18.9 kWh per day of 96,600 BTUs usable space heat. In Seattle, this would be 3,780 kWh or in Madison, it would be 2,835 kWh. The average savings in space heating in the northern U.S. could be $245.

Dehumidification or A/C by the EHPWH. The value of the dehumidification or air conditioning depends on how much hot water is needed. Using 80 gal. of hot water each day provides $17 per month worth of dehumidification or A/C. This is of value for 4–5 months per year. This is an additional $68 per year. As the EHPWH is applied further south in the U.S., the space heating value decreases and the A/C benefit increases.

Additional Investment of EHPWH Versus Conventional Equipment. EHPWH replaces the water heater, which would install for $600. A $250 dehumidifier is avoided. A properly designed house should have a minimum of an exhaust air handling system, which costs $600. An air-to-air heat exchanger is an alternative which would cost $1,000 installed. The EHPWH with space heater costs $2,300 installed. The increased investment in EHPWH instead of the water heater, the dehumidifier, and simple mechanical exhaust ventilation is less than $1,000. The combined savings of using the EHPWH water heating, supplemental space heating, and dehumidification (or A/C) is $800 per year. This assumes electric water and space heating at $.075 per kWh, 80 gallons of hot water per day, ventilation 24 hours per day for 175 days of space heating. If compared to natural gas fuel, savings would be reduced by 50% to 70% (depending on equipment used and the cost of fuel). The cost of the alternative fuels available on the site ultimately determines the choice of equipment used. If electricity was $.15 per kWh and natural gas was $.50 per therm, there would be no payback using the EHPWH. Usually paybacks range from less than 2 years to more than 5 years.

Conclusion

The House. Build the house to have less than 0.4 ACH infiltration under coldest windy winter conditions. This usually requires the house to have a natural 0.1 to 0.2 ACH. This will prevent the house from overventilating (exceeding 0.4 ACH) during the coldest winter condition. This paper does not address level of optimum wall and ceiling insulation. We have equipment installed in R–19 walls and R–30 ceilings houses all the way up to R–30 walls and R–50 ceilings. The better a house is insulated, the less critical the efficiency of the heating system. In super insulated houses, the annual energy loads could be as follows:

1. Water heating is 29 M BTUs.
2. Heating ventilation makeup air is 24 M BTUs (0.4 ACH).
3. Heat loss through the structure is 10 M BTUs.
4. The cooling load is 7 M BTUs.

Water heating and heating fresh air are clearly the largest energy loads in an energy efficient house.

Mechanical Systems Replaced by EHPWH. The EHPWH replaces the ventilation equipment, the water heater, and the dehumidifier (or reduces the size of the air conditioner). It is possible for the EHPWH to provide a constant 0.4 ACH during the heating season, 80 gallons of hot water per day, provide 99,600 BTUs of space heat per day, and provide 15 pints of dehumidification per day (or marginal A/C) for 4,040 kWh at $.074/kWh or $299 per year. The payback could be as low as two years.

EHPWH Limitations. The EHPWH cannot be the total heating system for the house. The house requires a backup system for extreme cold conditions. The higher the insulation and tightness level, the smaller the backup heating system can be. In winter, the EHPWH will provide 80 gallons of hot water and 96,000 BTUs of space heating for 19 kWh using 100 CFM of warm, moist, stale air. In summer, the amount of dehumidification, A/C, is limited by the quantity of hot water needed.

Utilization by the Home Owner. A Washington State University study of 117 homes with independent mechanical ventilation systems showed 70% of the systems off or seldom used for various reasons. The EHPWH must operate to have hot water. This assures that there is minimal ventilation. On 100 installations of ventilating water heating systems, all malfunctions were reported because of lack of hot water. The need for hot water will guarantee ventilation.

End of quote from publication from Therma-Stor Products Group, Division of DEC International, Inc.

22–7 SPA HEAT PUMP WATER HEATERS

A growing market for air-to-liquid heat pumps is the heating of swimming pools and/or spas (hot tubs). With the introduction of the heat pump water heater, the reduction in energy cost of maintaining the pool or spa temperature has been the largest factor in the promotion of these products. Heat pump water heaters for both open and enclosed swimming pools are discussed in later sections of this chapter. In this section, the application of the heat pump water heater to the control of temperature in a spa or hot tub is discussed.

The heat pump for heating the spa or hot tub is, in reality, a water-cooled air conditioning unit using the water from the spa or hot tub to remove the heat from the condenser coil. Figure 22–16 shows the basic water and refrigeration circuit. Basically, the unit removes sensible and latent heat from the air surrounding the unit and transfers the heat into the water in the spa.

During the process, the air surrounding the unit is cooled and dehumidified. The most desirable result is the dehumidifying of the air. If the spa is in an enclosed area, the 105°F water in the tub will have an increased evaporating rate, which raises the ambient relative humidity. Latent heat removed from the air heats the water in the spa.

The humidity increase, however, can be kept to a minimum by keeping the pool covered with an insulating cover when it is not in use. The recommended cover should be one that is equal to 2" of foam. Urethane foam sheets can be used. Usually, covers are available for the spa or hot tub from the installing dealer as an optional item.

Operating Cost Comparison

We will assume that the spa heater in Figure 22–5 is the unit to be installed with the spa. This unit from the manufacturer's literature is the Model SE11B—1 HP—115 volt. The capacity with 70°F ambient air temperature and 100°F water temperature is 14,000 BTUH. These conditions will require a draw of 1.25 kWh with a COP rating of 3.3. The nominal water flow through the unit condenser is 25 GPM with a pressure drop of 1.5 psig. The maximum operating water temperature is 105°F. The normal control thermostat setting is 105°F.

To make the operating cost comparison, we will assume the following conditions:

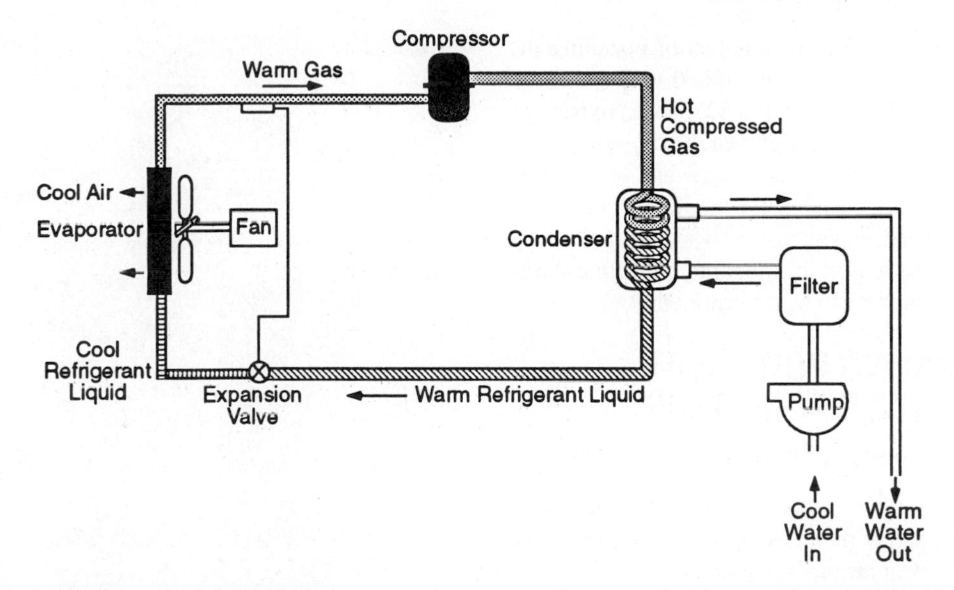

Figure 22–16 Basic circuit for a spa heat pump water heater (Courtesy EUS, Energy Utilization Systems, Inc.)

1. The spa (hot tub) is a 5' × 5' (25 sq. ft.) unit.
2. The water temperature will be maintained at 105°F.
3. The ambient air temperature of the unit is 70°F.
4. The spa will be used for 1 hour each day.
5. The unit will operate 24 hrs. per day on the control of the thermostat.
6. The unit will be covered when not in use.
7. The spa is an above ground installation.

These conditions will result in a 9.7°F temperature loss from the water in the spa per day. To maintain the 105°F water temperature, a 2.0 kWh electric resistance heater will be required to operate 23.53 hours per day.

Under these conditions, the heat pump water heater will operate 10.5 hours per day. These conditions will result in the annual need of 15,713 kWh for the electric heater and 4,790 kWh for the heat pump.

If we assume an average annual electric rate of $.08 per kWh, the operating cost per year of the electric water heater would be $1,257.04. For the heat pump unit, the annual cost would be $383.20. This results in an annual cost savings of $873.84. Assuming that the increase in installation cost of the heat pump system is $700.00 over the cost of the electric heat system, the difference in cost is recovered in 0.8 years (292 days). The entire cost of the installation would be recovered in less than 5 years as the 5 year savings would be $4,369.20.

Operating Cost with Tub Cover

The largest heat loss from the spa is from the surface of the water to the air via both sensible and latent heat loss. If a cover with an insulating value equal to 2" of urethane foam (R–12) is used, this surface loss is reduced approximately 83%.

This would reduce the annual operating cost for the electric heater to $213.70 and $65.14 for the heat pump unit. The savings would not be as high as without the cover. Therefore, it would take approximately 4.7 years to recover the difference in cost between the two systems. This payback period still represents a 19% return on the initial investment.

Inground Spa (Hot Tub)

Inground hot tubs installed outdoors will have heat loss to earth as well as ambient air. Outdoor spas are usually larger in size and have longer usage. Units installed in housing communities can have a usage time as high as 6 hours per day. The average time is usually 2 hours.

We will assume that the inground spa is a 50 sq. ft. (8' diameter) spa and is operating under the following conditions:

1. Maintained at 105°F
2. 70°F ambient air temperature
3. Usage 2 hours per day
4. Uncovered 24 hours per day
5. Used 12 months (365 days) per year

The heat loss of the unit at the 70°F air temperature and 105°F water temperature is 12,842 BTUH. The SE11B unit delivers 14,000 BTUH. This means that the unit will operate 92% of the day to handle the load. The unit operating continuously will handle the heating load requirement down to 64°F ambient air temperature. If the spa will be used below these temperatures, auxiliary heat will be used to supply the additional requirement.

Under the stated conditions, the annual operating cost for electrical resistance heat will be approximately $2,627; for natural gas ($.87/therm), $1,225; for L.P. gas ($1.42/gallon), $2,172; and for the heat pump, $969. This means an annual savings of operating cost of $1,658 as compared to electric resistance heat, $256 as compared to natural gas, and $1,203 as compared to L.P. gas.

If we assume that the difference in cost of installing the heat pump over a 5 kWh electric heater is $650, over a 125,000 BTUH natural gas or L.P. gas boiler is $275, the payback or time required to recover the difference in the equipment investment would be 0.4 years (4.8 months) for electric resistance, 1.1 years (13.2 months) for natural gas, and 0.23 years (2.76 months) for L.P. gas.

From the figures, the reason for the rapid rise of the use of heat pumps for spa heating can be determined.

22–8 SWIMMING POOL HEAT PUMP WATER HEATERS

Swimming Pool—Outdoor Installation

The heating of an outdoor pool is the same arrangement as heating the inground spa with the exception of the size of the unit as well as the piping arrangement needed.

Equipment

The equipment for outdoor swimming pools is of a larger range of sizes than used for a spa or hot tub. Pool heaters, at 50°F outdoor ambient temperature, vary from 24,600 BTUH to as high as 360,000 BTUH, and from a unit using one 2-hp motor-compressor to one using three 10-hp motor-compressor assemblies.

This means a water flow rate of pool water through the unit of 10 GPM for the 2-hp unit to 80 GPM for the 30-hp unit. Units up to 10 hp are usually single compressor systems that control the pool water temperature by cycling on and off.

From 12 hp to 20 hp, two systems are usually used. From 25 hp up, three systems are usually used.

This design of multiple systems in the larger units helps to keep the operating cost to a minimum by operating the unit in stages according to the load demand. It also limits the peak demand of the starting amperage by having a built-in delay of the starting time of each system.

Swimming Pool Heaters—Small Size. Figure 22–17 shows a typical unit in the 2-hp to 7.5-hp range. Units in this size range usually use a three-row coil with a propeller fan for moving the outdoor air through the coil. Designed to be in the outdoor atmosphere, no duct work is required. A top discharge propeller fan is very satisfactory from a noise as well as performance standpoint. All the other components—compressor, pressure-reducing device, and piping—are standard air conditioning system type.

The refrigerant-to-liquid condenser is usually cupronickel for longer performance life. One manufacturer, TempMaster Enterprises, Ltd. has marketed a unit using a refrigerant-to-liquid heat exchanger using a cupronickel tube for the inner refrigerant tube surrounded by a nonelectrolytic virgin plastic outer tube.

The smooth surface tube heat exchanger uses 4.5 to 6 times the tube length for the same size unit. The smooth surface reduces the buildup of lime and sediment, which reduces the need for acid treatment.

Figure 22–17 Typical small size swimming pool heat pump (Courtesy Drake Industries Inc.)

The smooth plastic/cupronickel tube expands and contracts with the wide temperature change from the off operating mode to the on operating mode. This expanding of the plastic tube tends to remove scale buildup, which is washed away by the water flowing through the tube.

Because of the ability of the plastic to expand and contract, the lime buildup can be removed by actually causing the heat exchanger to freeze and then thaw out. The freezing causes the lime to leave the plastic surface because of the expansion of the plastic tube.

Upon thawing out and contracting, the scale deposits are flushed out of the tube by the water flow. As a result, this type of heat exchanger does not require acid washing.

Swimming Pool Heaters—Large Size. Figure 22–18 shows the typical configuration of a large size swimming pool heater. Units in this size range usually use a 5- or 6-row coil with a centrifugal blower (usually belt driven) sized large enough to supply the required CFM through the deeper coil. The result of using the deeper coil is an increase in the latent heat removal capacity of the unit.

With the centrifugal blower, a duct distribution system can be added to the unit. This allows a wide range of application. The unit does not have to be in the open atmosphere. A

Figure 22–18 Typical large size swimming pool heat pump (Courtesy TempMaster Enterprises, Inc.)

machine room application can be done with entering and leaving air ducts.

In the smaller sizes of this size range, 7-1/2 hp and 10 hp, the systems usually have one refrigeration circuit (one motor-compressor assembly) and are controlled by on/off thermostat operation. In the mid-size range, 12 hp to 25 hp, two refrigeration circuits are generally used. This allows step control operation of the unit operation for closer control of pool water temperature as well as allowing for reserve heating capacity.

Sizes above the 25 hp unit are usually three refrigeration circuits. These applications require a greater variety of control to provide a wider range of operating characteristics.

Size Calculation

Heat is lost from a swimming pool via four different ways:

1. Heat conducted to ground
2. Heat conducted to air
3. Heat radiated to air
4. Evaporation

The greatest heat loss, approximately 75%, is by evaporation. The next greatest loss, approximately 14%, is by conduction to the air above the pool. The third greatest heat loss, approximately 10%, is via heat radiation to the air above the pool. The fourth greatest heat loss is heat conducted into the earth, approximately 1%. From these figures, it is obvious that the greatest loss is from evaporation.

With air above the pool at 30% relative humidity (RH) and a 5-mph wind over the surface of the pool, from a pool of 1,000 sq. ft. of water surface, the heat loss by evaporation can be as high as 240,000 BTUH. A 1,000 sq. ft. pool that is uncovered for 24 hours per day can lose as much as 5,760,000 BTU/day.

By using a pool cover when the pool is not in use, this loss can be reduced considerably. If the pool is in use 5 hours per day and covered for 19 hours, the heat loss can be reduced as much as 70%.

In areas where swimming pools are used 12 months per year and night time temperatures drop below 70°F, a pool cover is an absolute requirement. A cover is recommended, however, on all pool installations to reduce the heat loss as well as help to reduce filtration and chemical costs.

To determine the swimming pool heater size, several factors must be determined:

1. Square feet of the pool water surface
2. Quantity of water in the pool
3. Desired pool temperature
4. Average daily temperature
5. Average relative humidity
6. Average wind speed at the pool water surface
7. Hours per day of pool use

Square Feet of Pool Water Surface. The square feet of the pool water surface is needed to estimate the surface heat loss as well as water quantity.

Quantity of Water in the Pool. The total square feet of the pool surface times the average depth of the pool will give the quantity of water in the pool in cubic feet. This figure times 62.425 pounds per cubic foot will give the number of pounds of water in the pool. This total is also the BTUs needed to change the pool temperature 1°F.

For example: assume a 25' × 50' pool as an average depth of 6'. The pool will contain 7,500 cu. ft. of water or 468,188 pounds of water. This means that for each 1°F change in water temperature, 468,188 BTUs of heat energy are required.

Desired Pool Temperature. The desired pool temperature will depend upon the usage. Normally, 75°F to 82°F temperatures are maintained for most pool users.

If the pool is for very active use such as competitive swimming and/or physical fitness programs, 65°F to 68°F temperatures are usually maintained. If the pool is used primarily for the senior citizen age group, an 84°F to 88°F water temperature is usually desired.

Average Daily Temperature. This information can be obtained from the United States Weather Bureau. An excellent source of this information is the Installation Standards for Heating, Air Conditioning, and Solar Systems published by the Sheet Metal and Air Conditioning Contractors National Association (SMACNA).

For example, if the pool is used year-round in the Phoenix, AZ area, the greatest heating load on the pool heater will be in January when the average daily temperature is 51.8°F. The pool heat loss will be the difference between the average daily temperature (52°F) and the maintained pool temperature (82°F) or a ΔT°F of 30°F.

Average Relative Humidity and Average Wind Speed at the Pool Water Surface. The relative humidity in the air will have a direct effect upon the evaporation rate of the water from the surface of the pool. This, along with the average wind speed at the pool surface, will determine the amount of water evaporated from the pool.

From a 1,000 sq. ft. uncovered pool with a 5-mph wind on the pool surface and 30% relative humidity, the evaporation rate is approximately 250 pounds of water or 30 gallons of water per hour. This is a latent heat loss from the pool of 242,500 BTUH with average wind speed at the pool water surface.

This information is obtainable from the United States Weather Bureau. To keep this wind effect to an absolute minimum and for maximum safety, the pool should be surrounded by a wind resistant fence at least 6' high.

Hours of Pool Use. The pool should be covered when not in use. For pools of general public use, the covers must be in place during the pool closed hours. The heating capacity of the heat pump, however, must be calculated for the hours that the pool is uncovered and in use.

For example, we will assume that the swimming pool is a 25' × 50' surface area with an average depth of 6' (8' deep for a distance of 20' and a slanting up to a depth of 3' at the shallow end). This pool will have 1,250 sq. ft. of surface area. With an average depth of 6', the pool will contain 7,500 cu. ft., 67,762 gallons, or 468,188 pounds of water.

During the open or uncovered hours, the average heat loss per hour from the pool can be approximately 0.5°F for every 15°F difference in temperature between the pool temperature and the average air temperature.

In the example for the Phoenix, AZ area, with a 30°FΔT, the pool loss per hour will be approximately 1°F with a BTUH loss of 466,188 pounds of water × 1°F or 466,188 BTUH.

We will assume pool open hours from 6:00 AM to 11:00 PM or 17 hours per day. The pool will be covered from 11:00 PM to 6:00 AM or 7 hours per day.

Covering the pool will reduce the heat loss by approximately 70%. Therefore, the heat loss during this time period is not included when calculating the equipment heat load. The uncovered pool hours (17) will result in a heat loss of 466,188 BTU × 17 hours for a total of 7,925,196 BTUs.

The planned running time for a pool heater should be 10 to 15 hours. Using a 15-hour running time, the pool heater will need a heating capacity of 528,346 BTUH. With a capacity requirement of 528,346 BTUH, at the 52°F average temperature in January, the unit size required will be one 20-hp unit and one 25-hp unit.

Figure 22–19 is a typical rating table giving unit capacities from 7.5 hp to 30 hp with ambient air temperatures from 48°F to 95°F. Using this table, we find that the model PA–20–W unit with 48°F ambient air is rated at 240,000 BTUH and the PA–25–W unit is rated at 300,000 BTUH. The total of 540,000 BTUH capacity will handle the pool heat loss.

The use of two units on this installation, with independent parallel flow through each unit, will result in step control of the pool water temperature. This will produce more economical operation as one unit will carry the heating requirement the majority of the season. The second unit will operate only when extreme heat loss exists.

In our example of the 25' × 50' pool in the Phoenix, AZ area, the 20-hp unit will require a 54 GPM flow rate to maintain a maximum temperature rise in the water circuit of 10°F. The 25-hp unit will require a 65 GPM to maintain the same temperature rise of 10°F. The maximum of 10°F rise is desirable to keep the water entering the pool within safe limits for anyone occupying the pool.

The total GPM flow for the two units will be 119 GPM. If a single pool filter and filter pump is used it must have a capacity of 130 GPM or more. Usually, in double unit installations, a pool filter and pump combination is used with each heat pump. A GPM capacity of 10% minimum over the GPM required for the unit is used to size the pool filter and pump.

This double setup is also more economical as a single filter will handle the load most of the time. The second filter can be used when the load is heavy or when unusually dirty atmospheric conditions exist.

Piping distances must be kept to a minimum to keep pump size down. If the flow resistance through the heat pump bypass circuit is higher than the flow resistance through the balance valve circuit, a booster pump may be required to ensure adequate flow through the heat pump coil.

Swimming Pool—Enclosed Installation

Enclosing the swimming pool reduces the heating requirements of the pool. Wind effect and direct sunshine effect are eliminated. The heat loss of the building and evaporation from the pool have to be taken into account. In order to provide a satisfactory pool water temperature as well as building temperature and relative humidity, several items have to be seriously considered:

1. Pool users usually desire pool water in the 75°F to 82°F range. Senior citizen users normally require temperatures 4°F higher—79°F to 86°F. This means that the building temperature has to be maintained a minimum of 2°F above the water temperature.
2. Indoor swimming pool buildings require more heat energy per cubic foot than practically all other occupied areas.
3. The heating of pool water increases the evaporation rate. This creates a need to reduce the relative humidity of the area to reduce the discomfort as well as damage to the structure from condensation.

Model		PA-7, 5-W	PA–10–W	PA-12-W	PA-15-W	PA-20-W	PA-25-W	PA-30-W
RATING (Heating Cap.)								
At 80° Water and								
60° Ambient	(BTU/HR)	90,000	120,000	144,000	180,000	240,000	300,000	360,000
C.O.P.	3.2 to 4.6							
AIR HEAT EXCHANGER				High Efficiency Air Coil With Copper Tubing and Aluminum Pins				
Air Flow CFM		3,600	6,000	7,200	8,200	12,000	14,000	18,000
Min. Air Inlet Temp.	°F			48				
Max. Air Inlet Temp.	°F			95				
POOL WATER HEAT EXCHANGER				Co-Axial Tube in Tube (Polybutylene and Cupronickel 90/10)				
Water Flow at to 10°F	GPM	18	27	33	36	54	65	80
Pressure Drop	PSI				Maximum 15			
Refrigerant				R-22 With Expansion Valve				
	LBS.	13.0	20.0	2 × 10.0	2 × 13.0	2 × 20.0	2 × 21.0	3 × 20.0
Blower Size		15-15	18-18	(2) 15-15	(2) 15-15	(2) 18-18	(2) 18-18	(2) 18-18
Motor	HP	14	2	(2) 14	(2) 14	(2) 2	(2) 3	(3) 2
	Voltage-PH-HZ			208/230-3-60				
	FLA	5.3	6.0	2 × 5.3-10.6	2 × 5.3-10.6	2 × 6.0-12.0	2 × 8.8-17.6	3 × 4.0-18.0
	LRA	26.5	30.0	2 × 26.5-53.0	2 × 26.5-53.0	2 × 30.0-60.0	2 × 44.0-88.0	3 × 30.0-90.0
COMPRESSOR								
	Voltage-PH-HZ			208/200-3-60				
	RLA*	20.2	29.0	2 × 18.0-1-36.0	2 × 20.2–40.4	2. × 29.0–58.0	2 × 36.0–72.0	3 × 29.0–87.0
	LRA	135.0	183.0	2 × 126.0–252.0	2 × 135.0–270.0	2 × 183.0–366.0	2 × 229.0–458.0	3 × 183.0–549.0
Total	RLA	25.5	35.0	46.6	51.0	70.0	89.6	105.0
Main Circuit Breaker (Time Relay Fuse)	AMPS	40	50	60	70	100	125	150
Physical Demensions W × D × H	IN.	44 1/2 × 34 1/2 × 70	60 × 34 1/2 × 80	73 1/2 × 37 × 59	73 1/2 × 37 1/2 × 70	86 × 45 1/2 × 93	86 × 45 1/2 × 99	127 1/2 × 45 1/2 × 93
Water Connections	IN MPT.	1 1/2	1 1/2	1 1/2	1 1/2	2	2	2 1/2
Weight	LBS.	740	810	1180	1220	1520	17800	2280

*Compressor Manufacturer's Rating
Specification to change without notice

Figure 22–19 Typical specifications for an air-to-liquid pool heater (Courtesy TempMaster Enterprises, Inc.)

4. Ventilation or dehumidification normally requires a great amount of energy cost. If ventilation is used, the cost of heating make-up air in cold climates or dehumidification in hot humid climates is a major part of the energy cost.
5. Showers and locker rooms, hot tubs, and sauna baths are usually associated with enclosed swimming pools. These add to the building sensible and latent load problems.
6. Because of the evaporation of pool water, design of the building must include vaporproof wall, ceiling, and other surfaces and high enough insulation R-values to prevent condensation on the inside surfaces at outside wintertime design conditions.

Therefore, three requirements are involved in indoor pool areas:

1. Room temperature control. Any wide variation in area temperature will cause occupant discomfort and complaint. This also tends to increase the heat loss of the pool by evaporation as well as promote higher relative humidity in the area.
2. Relative humidity control. Wide variation in relative humidity causes structural damage when it is excessive, as well as occupant discomfort. Low relative humidity causes an increase in the pool evaporation rate and heating cost, as well as occupant discomfort.
3. Pool temperature control. Control is absolutely necessary to reduce discomfort as well as operating cost.

Most experts recommend an 80°F pool water temperature with the air conditions of 82°F DB and a relative humidity (RH) between 55% and 65%. In most indoor pools, a combination of separate systems is used to produce the desired results. In many of these installations, the balance of equipment is difficult and operating costs are high.

The conditioning of indoor swimming pool rooms, including dehumidifying, heating, and cooling, involves equipment with high energy use. Pool water temperature is usually provided by and under the control of a separate heating unit. Thus a piece of equipment uses energy to provide heat while another removes heat and moisture.

One method of dehumidifying the occupied area is to use a fresh air system combined with an exhaust system. This a balanced pressure ventilating system. This method exhausts the high temperature and humid air by providing low temperature and humidity into the area. The energy needed to heat the fresh air into the building is exhausted after absorbing the moisture from the pool water. Thus, the heat energy supplied by the pool heater is carried to the outside atmosphere.

Because of the amount of energy consumed by the equipment used to accomplish temperature and humidity control, different dehumidifying equipment has been installed with the make-up system. Usually the dehumidifying equipment is an air-to-air heat pump using the evaporator to cool and dehumidify the air and the condenser to reheat the air. The air travels first through the evaporator coil and then through the condenser coil before reentering the conditioned area.

Used only for dehumidifying, it is not sized to handle the building heat loss or gain. A separate system is used to accomplish this. Pool water temperature is usually under the control of a third heater system.

In 1969, a heat pump designed strictly for indoor swimming pools was introduced by TempMaster Industries Ltd. of Orlando, FL. This unit is a two-stage liquid-to-air heat pump designed strictly for the enclosed swimming pool market.

ECP—Environmental Control Package

The Environmental Control Package (ECP) was originally designed for residential enclosed swimming pools where the pool water evaporation rate is not as great as a commercial or public enclosed pool. The ECP unit can provide the entire heating for the enclosure, the total dehumidification capacity to maintain between 55% and 65% RH, and 55% to 85% of the pool heating requirements, depending on the size of the ECP.

Using the pool water as a heat energy source as well as a heat energy sink, the unit performs as follows:

1. When maintaining room temperature, the ECP unit uses heat from the pool water to heat the air. It can provide 100% of the air heating load.
2. When maintaining the relative humidity in the air, the ECP acts as a dehumidifier and absorbs heat from the air and transfers it back into the pool water. The ECP is capable of dehumidifying and then heating the air at the same time.
3. The ECP is capable of maintaining the temperature of the pool water up to 85% of average winter pool water heating requirements, depending upon the unit size in relation to the pool heating requirements. Some additional heat is required to maintain the pool water temperature.

The performance ability of this type unit will reduce the cost of operation of an indoor swimming pool by 25% to 75% as compared to systems using separate units for each energy load.

ECP—Equipment

The ECP unit consists of two separate refrigeration circuits, each equipped with reversing valves for dual operation. Operation is in two stages. The first stage is the primary control stage and the air passes through the stage 1 coil last before leaving the discharge opening of the unit blower (see Figure 22–20). The air enters the unit through the second stage coil.

Each system has its individual refrigerant-to-liquid heat exchanger with a common circulating pump serving both circuits.

ECP Control

This type of unit operates in one of three different modes (see Figure 22–21):

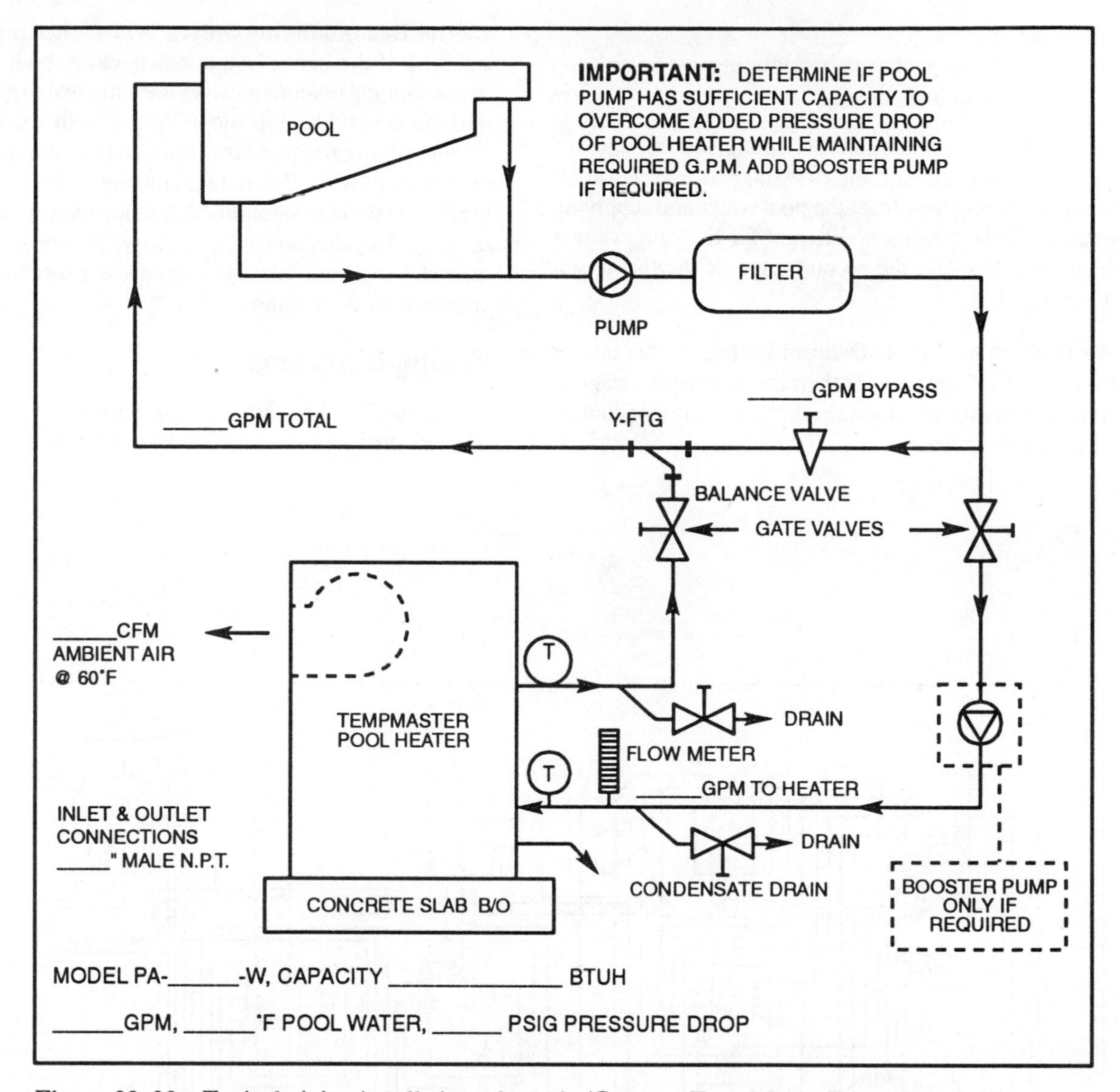

Figure 22–20 Typical piping installation schematic (Courtesy TempMaster Enterprises, Inc.)

FUNCTION MODE	STAGE 1	STAGE 2
Heating Demand Always Override	First On	Last Off
Mode 1 Call for Heating	↑ On	↑ On
Mode 2 Call for Heating St. 1 and Dehumidifying St. 2	↑ On	↱ Reverse
Mode 3 Heating Satisfied Call for Dehumidifying	↱ Reverse	↱ Reverse
Heating & Dehumid. Satisfied	Last Off	First On

Figure 22–21 ECP control functions (Courtesy TempMaster Enterprises, Inc.)

1. Call for heating
2. Call for heating as well as dehumidifying
3. Call for dehumidifying only

Call for Heating. Upon a call for heating, the first stage of the thermostat will bring on the first stage as a liquid-to-air heat pump removing heat from the pool water and supplying heat to the air in the conditioned area. If the first stage cannot handle the requirement, the second stage of the thermostat activates stage 2.

Call for Heating as Well as Dehumidifying. This call for heating by the first stage of the thermostat activates stage 1. The humidistat contacts activate both the compressor contactor and reversing valve.

Call for Dehumidifying Only. With the heating thermostat satisfied, if the humidistat contacts close, both compressor contactors and reversing valves are activated to put both units into the dehumidifying mode. When both the heating and dehumidifying requirements are satisfied, the stage 2 cycles off before stage 1. This reduces the unit capacity from 100% to 50%. As the heating and/or dehumidifying demand increases, stage 1 cycles on (50% capacity) before stage 2 (100% capacity). This provides more even space conditions at lower amperage peak demand.

Wiring Diagrams

Figure 22–22 is the wiring diagram for the ECP. The diagram shows the field wiring of 208/230-volt, single-phase,

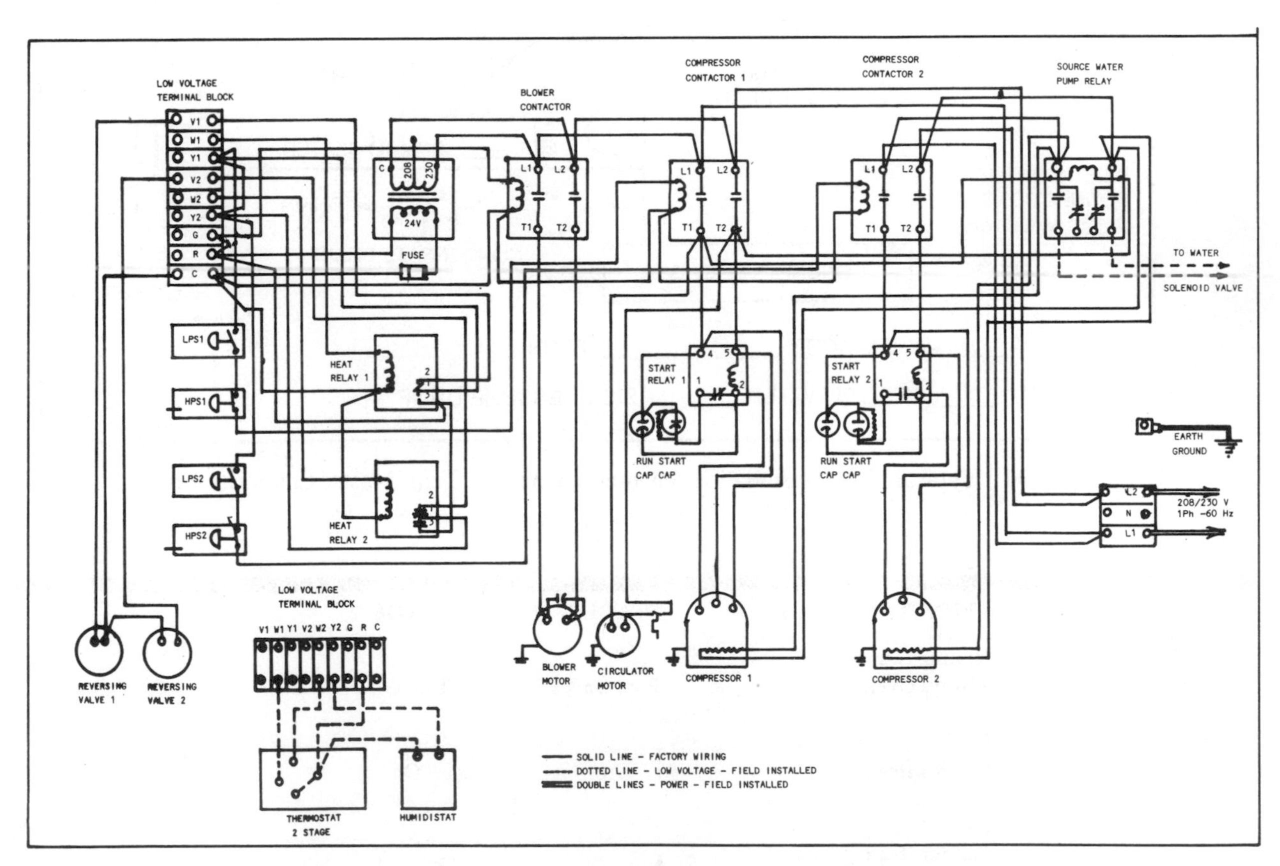

Figure 22–22 ECP wiring diagram (Courtesy TempMaster Enterprises, Inc.)

60-hertz power supply to the unit, control circuit to the bypass control solenoid valve (and booster pump if used), and the thermostat and humidistat control combination.

Wiring Diagram—Power On—No Heating or Dehumidifying Required. Examining the emphasized circuit in Figure 22–23 shows there is no control in the low-voltage power circuit to the blower contactor. The purpose of this is to ensure continuous air circulation (CAC) in the occupied area. Cycling blower operation produces very poor control. CAC is an absolute must.

Examination of the reversing valve coil circuits shows that the reversing valves are in the heating position when the operating coils are de-energized. The reversing valves are "normally heating" when the power is off.

Wiring Diagram—Mode 1—Heating Only. With the room temperature lower than the setting of the thermostat, the thermostat contacts between R and W1 and between R and W2 close (see Figure 22–23).

When heating relay No. 2 is energized, it opens the normally closed contact (1 to 2), which prevents power to the stage 1 reversing valve. The thermostat stage 1 contact (R to W1) is closed and the circuit to the heating relay No. 1 coil is powered. This causes the relay to pull in, opening the NC contact (1 to 2), which prevents power to the stage 1 reversing

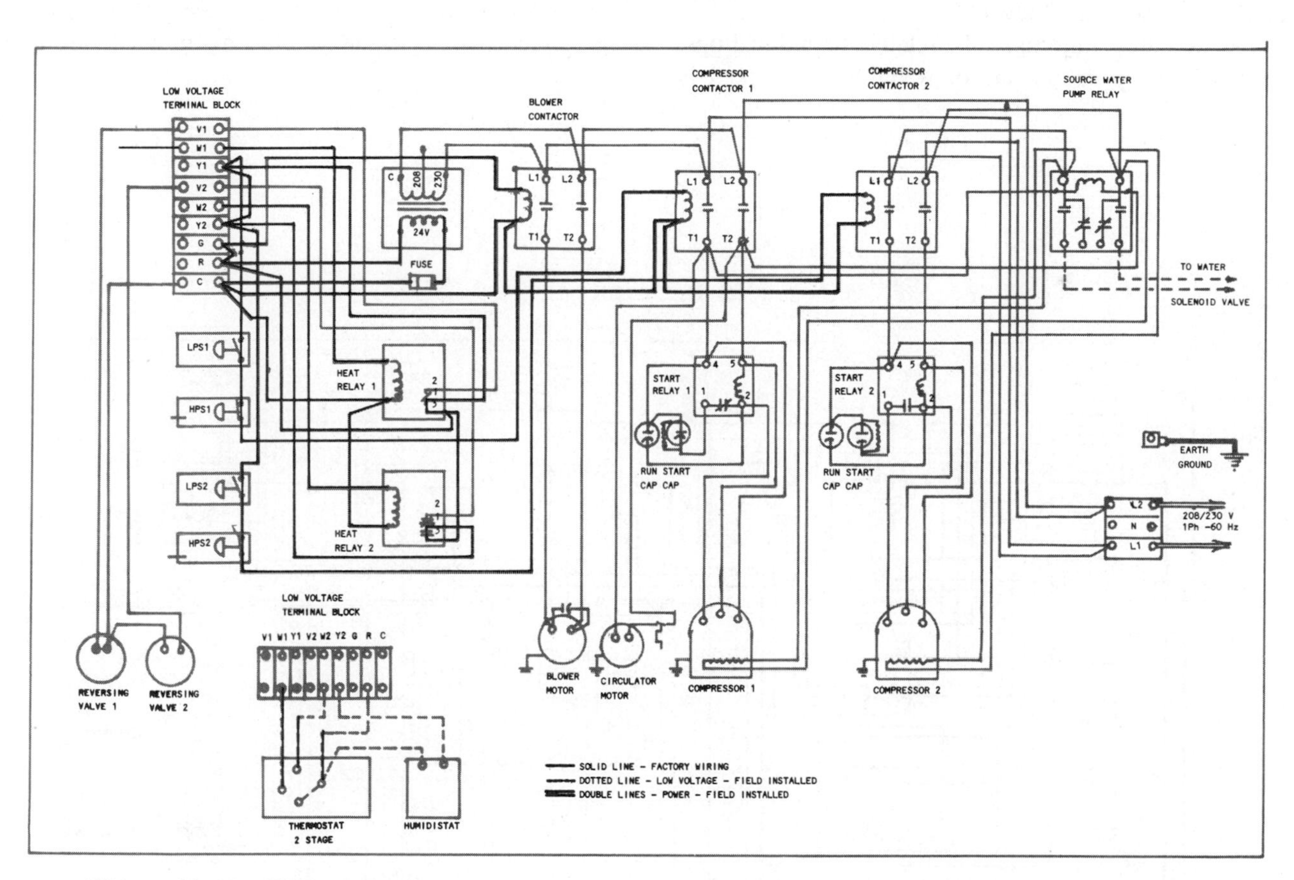

Figure 22–23 ECP wiring diagram—mode 1—heating only (Courtesy TempMaster Enterprises, Inc.)

valve. The NO contact (1 to 3) closes and completes the circuit to terminal "Y" on the low-voltage terminal block, through the stage 1 low- and high-pressure controls to the low-voltage coil in compressor contactor No. 1, and back to the 24-volt power source.

Contactor No. 1 pulls in and supplies power to the stage 1 compressor and starting components. At the same time, power is supplied to stage 2 via a jumper wire between Y1 and Y2 on the low-voltage terminal block, which supplies power through the stage 2 low- and high-pressure controls, through the low-voltage coil in the stage 2 compressor contactor coil. This causes the contactor to pull in and energize the stage 2 compressor and starting components.

Both compressors start as a unit. If the electric utility requires a delay start between the stages to reduce the amount of inrush or starting amperage, a time-delay relay of 6-second delay can be connected in the circuit between Y2 and the stage 2 low-pressure control.

Both systems operate until the heating requirement is satisfied. With CAC, air is supplied at low velocity to increase temperature to the desired level.

As the air temperature rises, the relative humidity drops. However, the evaporation rate of water from the pool is reduced to a minimum as the total heat input to the air is extracted by the unit from the pool water. No auxiliary heat to the water is needed.

Wiring Diagram—Mode 2—Heating and Dehumidification. As the temperature rises and reaches a temperature 2°F below the setting of the thermostat, the second stage contact in the thermostat (R to W2) opens and the system automatically switches to Mode 2 (see Figure 22–24). This mode 2 means that stage 2 is cooling and removing moisture from the air while stage 1 is heating the air.

Stage 2 is removing latent heat from the air and transferring the heat to the pool water and raising the temperature of the water. At the same time, stage 1 is removing heat from and lowering the temperature of the pool water.

The control action is the deenergizing of heat relay No. 2 by opening the second stage thermostat contact (R to W2). When the stage 2 heat relay is deenergized, the NO contacts open and the NC contacts close. This completes a power supply from Y2, through the NC contact (1 to 2) in the relay, to the V2 terminal on the low-voltage terminal block, to reversing valve No. 2, and back to the common (C) side of the power source.

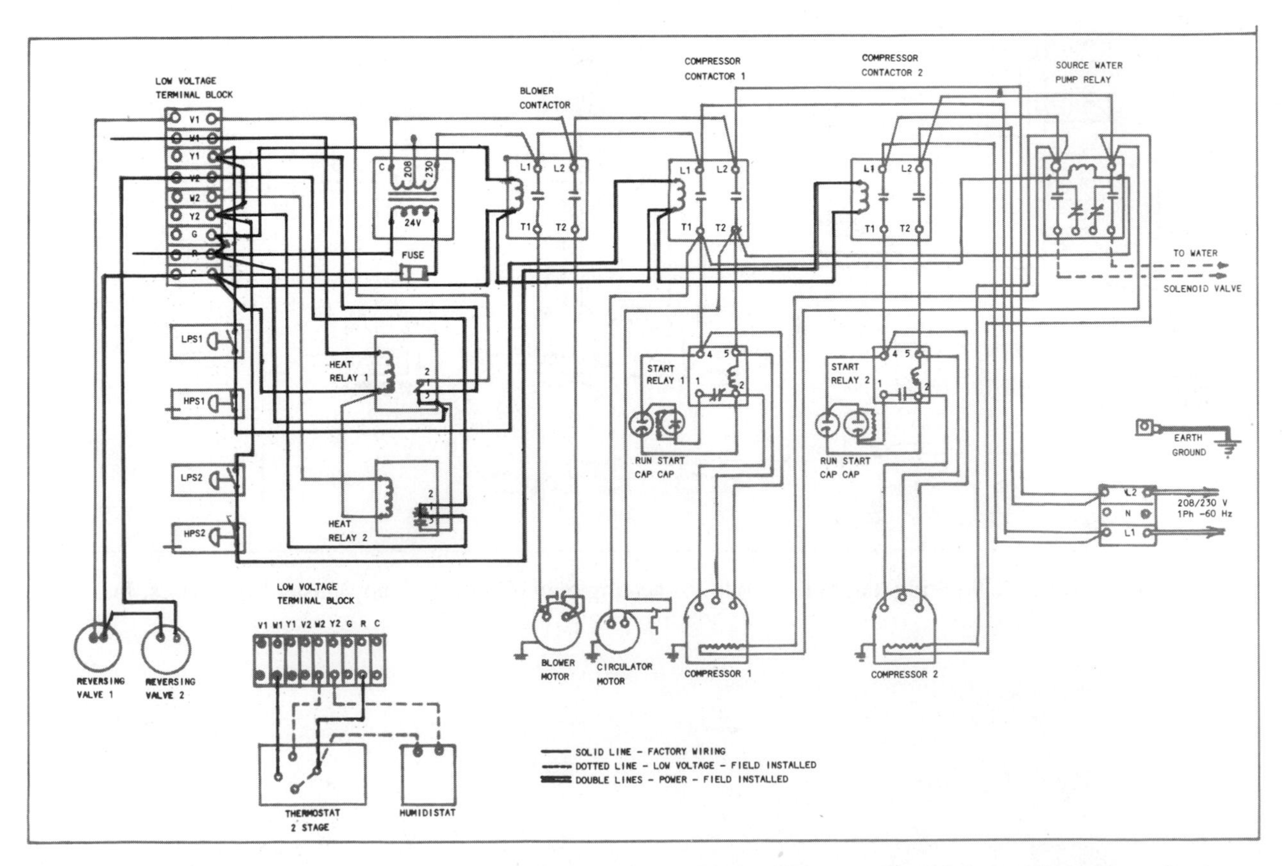

Figure 22–24 ECP wiring diagram—mode 2—dehumidifying (Courtesy TempMaster Enterprises, Inc.)

The stage 2 system continues to operate but now operates in the dehumidifying (cooling) mode. The air through the unit is cooled to remove moisture and then reheated.

The heating capacity of the system is reduced, which slows down the temperature rise of the room. If the room temperature drops below the setting of the second stage of the thermostat (2°F below the thermostat dial setting), the unit returns to the two-stage heating capacity.

Stage 2 will oscillate between the heating and dehumidifying operation modes to provide the desired air temperatures while reducing the relative humidity. Although there is only a small amount of heat added to the pool water during the mode 2 operation, the pool water temperature is maintained.

When moisture is condensed, heat is given up. Heat that would be wasted by condensation on ceilings, roofs, walls, structures, glass, floors, and furnishing is carried directly to the pool. This almost balances the amount lost by the pool because of the lower humidity in the air.

Wiring Diagram—Mode 3—Dehumidification Only—No Heating. With the thermostat satisfied, if the relative humidity exceeds the setting of the humidistat, the humidistat contact closes (see Figure 22–25). This makes a circuit between R (power) terminal of the thermostat to terminal Y2 of the low-voltage terminal block.

Connected directly to the Y2 terminal is the control circuit of stage 2. Connected through the jumper from Y2 to Y1 is the control circuit of stage 1. Both compressors operate.

Power is also carried from terminal Y2 to the common (C) terminal of heating relay 2 and from Y1 to the common (C) terminal of heating relay 1. The heating relays have not been energized by the thermostat contacts. Therefore, in both circuits, the reversing valves are energized and both systems are dehumidifying (cooling) the air through the unit.

The latent heat from the air is put into the pool water. As the unit cools the air, when the air temperature drops to the setting of the thermostat, the first stage contact in the thermostat closes. This puts the unit into mode 2 operation. The unit will alternate between the mode 3 and mode 2 operation until the humidistat is satisfied. By careful sizing and proper installation practice, the fuel-wasting short cycling and control overriding is kept to a minimum.

To maintain the proper conditions of comfort both in the air and the water, continuous operation of the blower is required. Also, the system must be under the control of the thermostat and humidistat at all times.

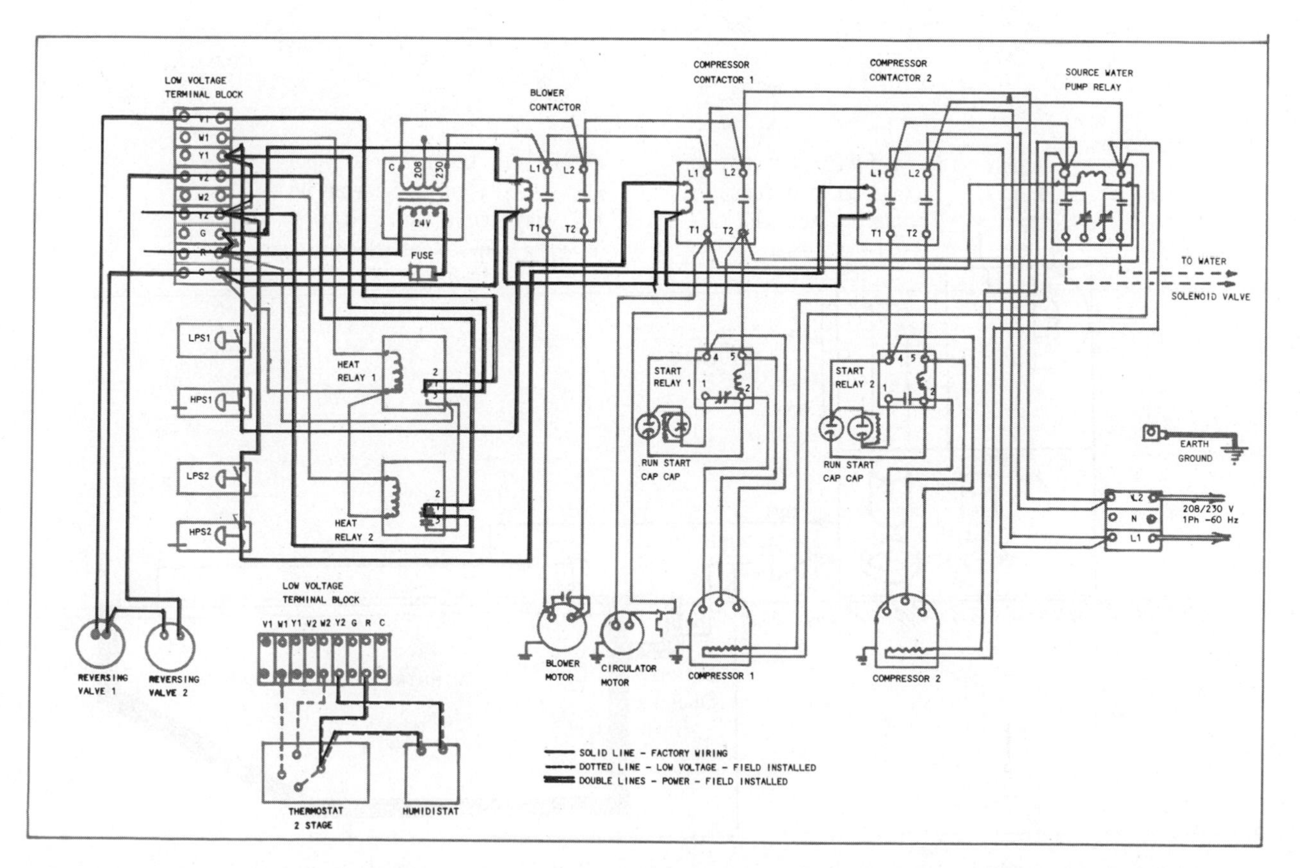

Figure 22–25 ECP wiring diagram—mode 3—dehumidifying only (Courtesy TempMaster Enterprises, Inc.)

Piping Arrangement

Figure 22–26 shows a typical piping schematic connecting an air-to-liquid heat pump heating a swimming pool. The piping arrangement is the same for outdoor and indoor pools. The water is taken from the bottom of the pool and the skimmer inlets, and passes through the pump, which forces it through the main filter back to the outlets in the side walls of the pool. Leaving the main filter, the flow is in two parallel paths, one through the heat pump and one bypassing the heat pump.

The total amount of water flow through the main filter required to maintain pool cleanliness is higher than the required flow rate through the heater. Therefore, a bypass circuit through the heater is required as part of the system.

This bypass circuit has ball valves for isolation in case of service requirement as well as a drain valve on the inlet and outlet of the unit coil. These provide access to the coil for use when a deliming procedure is needed.

In addition, a flow meter such as the one pictured in Figure 2–14, is required to establish and check the flow rate through the unit coil. P/T fittings are required to provide temperature and pressure checks to check unit performance.

To regulate the water flow through the unit, a balance valve is installed in the main line between the bypass line take off and return. The closing down of the balance valve forces more water through the unit.

Several items of information are located on the diagram that are needed to determine pipe size as well as pump and fitting sizes

1. The total GPM through the system will be determined by the quantity of water in the pool and the necessary flow rate through the filter to maintain the desired condition of the water.
2. Inlet and outlet connections on the heat pump unit will determine the "minimum" pipe size to be used.
3. The CFM of ambient air that must be supplied to the unit. The unit in the piping installation schematic has a centrifugal blower. This unit will allow the installation of duct work on the supply and/or return up to 0.15" WC of the 60°F ambient air. If the unit uses a propeller fan for air movement through the evaporator coil, no duct work can be added to the unit.

Installation Practices

There are certain standard practices regarding the installation of enclosed swimming pool heaters that will help to lengthen the life of the equipment as well as promote better operating results.

Supply Air Duct System. Perimeter supply systems with the supply grills in the floor at the outer walls will provide the best air distribution. The drier supply air along the outer walls

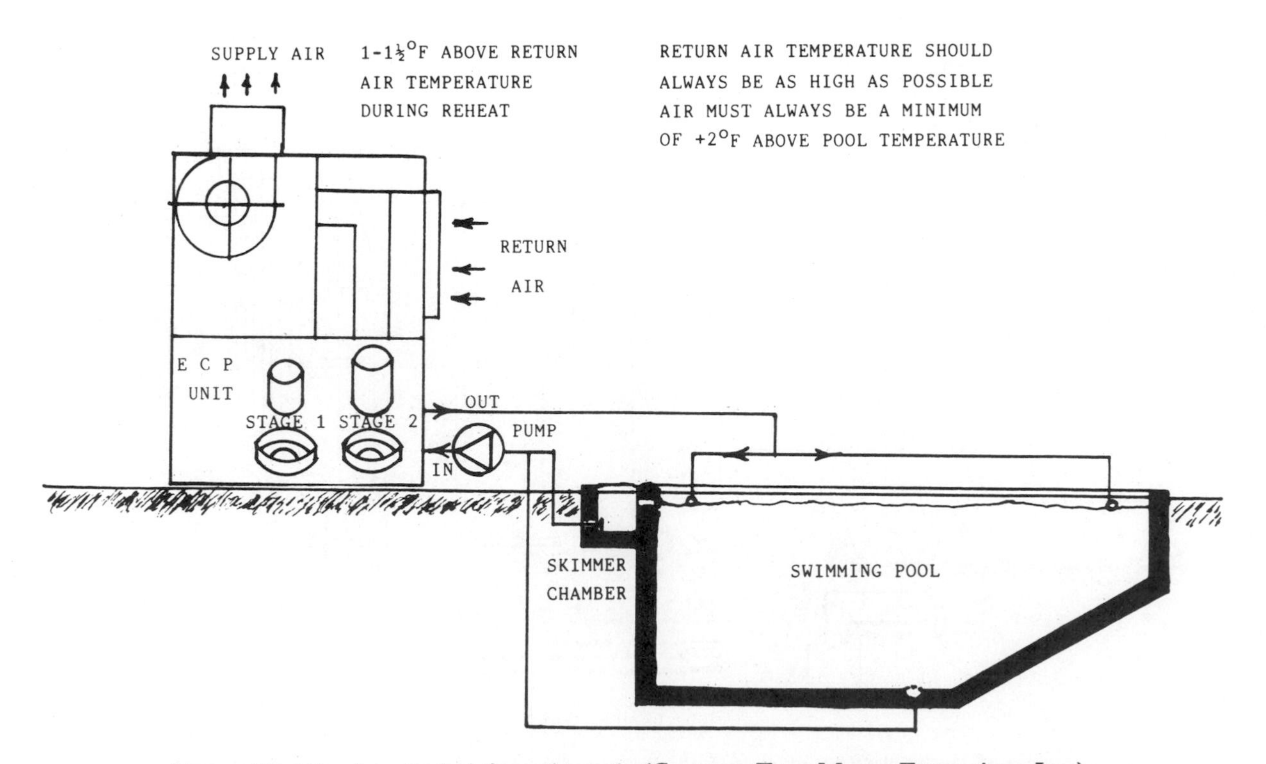

Figure 22–26 A typical piping schematic (Courtesy TempMaster Enterprises, Inc.)

will help to reduce condensation on the outer walls.

All supply ducts are sized for a pressure drop of 0.15" WC per 100 equivalent feet for the CFM of air required. All supply registers should be of extruded aluminum or marine epoxy paint or equal to reduce rusting or corrosion. All under-the-floor ducts should be insulated per Figure 22–27.

The temperature rise of the air through the unit is from 80°F return air to 100°F supply air for a 20°FΔT. A 5°F loss in air temperature as it travels through the duct system is a 25% loss in unit heating capacity.

The installation in Figure 22–27 (A) may be the most expensive over the long haul. Figure 22–27 (B) is the minimum amount of insulation that should be used. However, for best performance and lowest operating cost, the 2" Styrofoam insulation as shown in Figure 22–27 (C) is required.

Return Air Duct System. All return air should be from the high side wall or ceiling return air grills. Ceiling grills are preferred. Sizing of return air duct systems should be based on 0.10" WC per 100 equivalent feet of duct. All return air grills should be of extruded aluminum or marine epoxy paint or equal.

Moisture Barrier. All areas of ceilings, walls, and floors should be covered with polyvinyl sheet. The recommended thickness is 0.004 mil poly, though 0.002 mil poly is barely acceptable.

All walls and ceilings should be painted or coated with moisture-resistant sealing paint.

Insulation. All areas of walls, ceilings, and foundations must be insulated. The minimum R-numbers are:

- Walls—Minimum R–16, with R–21 recommended. Fiberglass insulation plus flexible outside vapor barrier sheathing.
- Ceilings—Minimum R–30, with R–40 recommended. Blown fiberglass batts installed in a criss-cross method to reduce air passage ways.
- Foundation— Minimum R–6, with R–12 recommended. Styrofoam sheet should extend at least 24" down from the floor surface.
- Glass—All windows, sliding glass doors, and sky lights should at least be double glass. Triple glass is recommended.
- Doors—All outside doors should be insulated with a storm door. On commercial installations, an entry vestibule, with inner and outer doors is highly recommended to keep traffic load to a minimum. All adjoining rooms must have self-closing doors.

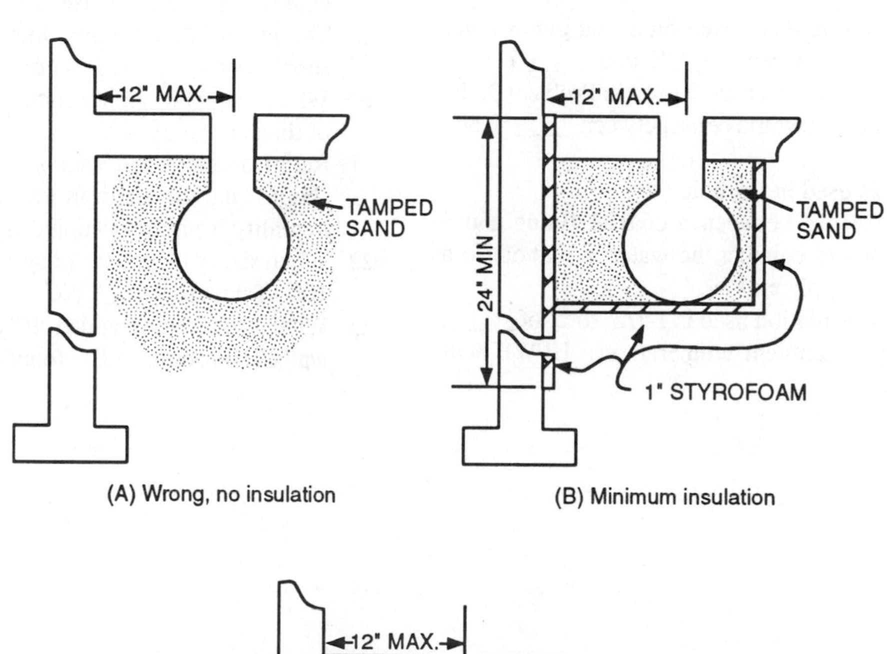

(A) Wrong, no insulation

(B) Minimum insulation

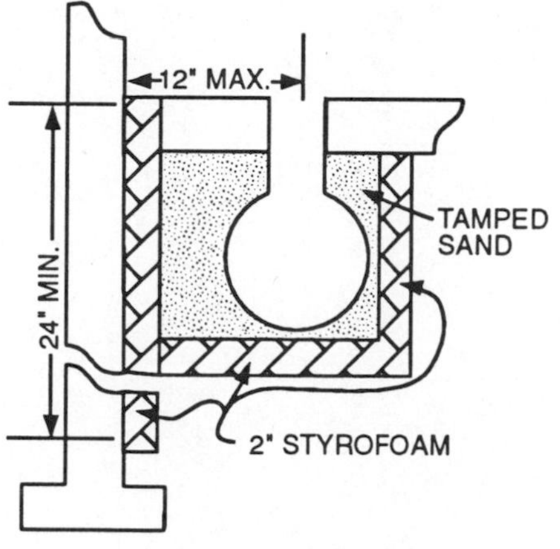

(C) Recommended insulation

Figure 22–27 Supply duct work under the floor (Courtesy TempMaster Enterprises, Inc.)

Exhaust Air. For odor control and for function in summer, an exhaust air system is recommended. A balanced pressure system is the best for this purpose. Forced entry of fresh air to balance forced exit of exhaust air. Fresh air supply systems should have a tempering unit to prevent cold air striking the pool occupants. All installations must meet local, state, and national codes.

22–9 SPECIAL APPLICATIONS

There are units marketed for special conditions of extremely high humidity, commercial applications as well as extra environmental control. Discussion in this chapter has been limited to the basic design of air-to-liquid units for water heating. For more advanced information and for assistance in sizing units for specific applications, the manufacturer of the equipment proposed for the installation should be contacted.

REVIEW QUESTIONS

1. Heat pump water heaters (HPWH) are made in two configurations. Name them.
2. The COP range of heat pump water heaters is between _____ and _____.
3. The water temperature thermostat on a heat pump water heater is usually set between _____°F and _____°F.
4. If a separate hot water heater were used in Problem 3, the water heater thermostat would be set between _____°F and _____°F.
5. List three controls used in the unit.
6. What is the difference between a coaxial fitting and a plumbing tree used to connect the water in and out to a single port on the supply tank?
7. The most popular insulation used is 1-1/2" to 2" of _____.
8. What is the lowest ambient temperature a HPWH will safely operate in?
9. The maximum number of air changes per hour (ACH) for proper ventilation in a residence is _____ACH.
10. What is the recommended maximum operating temperature for a spa or hot tub?
11. The largest heat loss from a spa or hot tub is:
    a. Through the sides of the unit to the air.
    b. Through the sides and bottom of the unit to the earth.
    c. From the surface of the water to the air.
12. List four ways that heat is lost from a swimming pool.
13. The greatest heat loss from a swimming pool is by _____.
14. The normal range of general-purpose swimming pool water temperatures is _____°F to _____°F.
15. If the swimming pool is located in a senior citizen project, the normal range of pool water temperatures is _____°F to _____°F.
16. To reduce the wind effect and for maximum safety, the pool should be surrounded by a windproof fence at least ___' high.
17. For an enclosed swimming pool, what are the recommended pool water temperature, building air temperature, and building relative humidity in the enclosed area?
18. In enclosed pool temperature control systems, the air circulating blower only operates with the unit to prevent cold drafts. True or False?
19. The preferred air supply duct system will have the air supply outlets located where?
20. When sizing the supply ducts, the pressure drop per 100' of duct is _____" WC.
21. Return air systems in enclosed swimming pool buildings should take the air from low wall grills to pick up the humidity from the occupied area. True or False?
22. When sizing the return air system, the pressure drop per 100' of duct is _____" WC.
23. What is the recommended "R" factor for insulation in the walls? The ceiling? The foundation?

Section FIVE
Liquid-to-Liquid Type Systems

LIQUID-TO-LIQUID 23

23–1 GENERAL

Another by-product of the rise in interest in the heat pump is the heating of domestic and/or commercial hot water supply by means of a refrigeration system using a liquid heat source. Where sufficient water is available, the water becomes an excellent source of heat energy for heating potable (consumable) water. Such a source could be brackisk groundwater such as in southern Florida, reclaimed water from sewerage treatment, or even hot water from a restaurant kitchen or a commercial laundry. Some units are designed to use sea (salt) water. Anywhere heat is being carried away via liquid, the heat can be removed very economically by means of a liquid-to-liquid heat pump.

Designed to operate in conjunction with a hot water storage tank, the units are intended to take the place of an electric or fossil fuel heater. The heat pump can be connected to operate with an electric hot water heater with the heat pump as the primary source of heat energy. The electric elements in the hot water are used as an emergency source of heat in case of heat pump malfunction.

The liquid-to-liquid heat pump can roughly be classified into three categories: domestic, commercial, and marine. There are units available for special applications. These are classified in the commercial category.

23–2 HEAT PUMP WATER HEATER—DOMESTIC

Unit Characteristics

Figure 23–1 shows the exterior of a domestic hot water heat pump. This heat pump unit is intended as a replacement for a conventional electric water heater. Using a refrigeration system to extract heat from a liquid heat source, it can provide an operating cost savings of up to 70% over the resistance-type electric hot water heater.

The system is a nonreversing type, intended for heating water only by extracting heat from a liquid source and expelling the heat through a condenser coil into potable water.

Figure 23–2 shows the interior of the unit. Two refrigerant-to-liquid heat exchangers are stacked in the front of the unit. The top coil is the evaporator. This coil extracts heat from the liquid source. The coil is not insulated. This allows the coil to absorb heat from the compartment and keep the cabinet temperature low.

The lower coil is the condenser. It transfers heat from the hot refrigerant vapor off the compressor to the domestic water. This coil is insulated to promote higher system efficiency.

The compressor, with service gauge taps above it, is seen in the upper portion of the unit.

A third pressure tap is provided in the liquid circuit downstream from the evaporator coil. This pressure tap is used to determine the pressure drop through the liquid portion of the evaporator coil to check for liming and/or fouling conditions.

The unit also contains the necessary pressure-reducing device (usually a capillary tube or restrictor), water-circulating pump, and controls. The water-circulating pump is in the domestic water circuit to the hot water storage tank.

In addition to the compressor operating controls and the thermostat mounted on the hot water storage tank, a solenoid valve is also included. This valve, cycling with the compressor, is used to prevent reverse gravity flow of the hot water in the "off" mode of the operating cycle. This reduces the heat loss from the hot water and reduces the operating cost.

Figure 23–1 Domestic hot water heat pump (front)

Figure 23–2 Interior view of hot water heat pump (rear)

Flow-regulating valves are also used to maintain the design evaporator temperatures regardless of the source water temperature and pressure. Using the source pressure for supply to the evaporator, no pump for this purpose is supplied in the unit. To include a source pump would be difficult because of the wide possibility of the types of water source.

Performance Rating

Figure 23–3 shows the performance rating of one particular manufacturer's unit. The unit is rated at three different final tank temperatures of 115°F, 125°F, and 135°F. Each final tank temperature rating is also based on four different source water temperatures of 50°F, 60°F, 70°F, and 80°F.

Final Tank Temperature 115°

Source Water Temp. (°F)	Average Gal/Hr.	Water Heating Average BTU/Hr.	Average kW	Average COP
80	29.3	14600	4.28	3.43
70	26.3	13100	3.84	3.22
60	23.1	11500	3.37	3.00
50	19.9	9900	2.90	2.80

Final Tank Temperature 125°

Source Water Temp. (°F)	Average Gal/Hr.	Water Heating Average BTU/Hr.	Average kW	Average COP
80	24.6	14300	4.19	3.29
70	22.0	12800	3.75	3.08
60	19.5	11300	3.31	2.88
50	16.7	9700	2.84	2.68

Final Tank Temperature 135°

Source Water Temp. (°F)	Average Gal/Hr.	Water Heating Average BTU/Hr.	Average kW	Average COP
80	21.1	14000	4.10	3.14
70	18.9	12500	3.66	2.94
60	16.6	11000	3.22	2.74
50	14.3	9500	2.78	2.56

Figure 23–3 Performance ratings

This unit uses a 1-hp compressor with a double-wall copper-to-copper heat exchanger for the domestic water heat exchanger and a cupronickel heat exchanger for the water source side. Cupronickel is used to protect against possible corrosive water source conditions.

A magnetically coupled potable water pump is used to eliminate the need for a shaft seal and the possibility of air and contaminates in the water supply (see Figure 16–1). This pump provides a nominal flow rate of 2 GPM of domestic water through the unit.

Using a flow-regulating valve, the source water flow is controlled to maintain a constant evaporator pressure in the refrigeration system. With 45°F water source temperature, the source water flow rate is approximately 2 GPM. With 90°F water source temperature, the flow rate is approximately 0.6 GPM. To supply the necessary water flow rate, the source water pressure must not drop below 15 psig or exceed 125 psig.

If a cycling water source is used, a pressure-type storage tank is required to maintain water source pressure above 15 psig.

If the water source pressure can exceed 125 psig at any time, a regulator ahead of the unit is required. If a water pressure regulator is used, the recommended output setting is 60 psig.

The unit control system contains the standard compressor starting and running controls. In addition, high- and low-pressure safety controls are included. These are to prevent system damage in the event of domestic water pressure loss (high-pressure cutout) as well as coil freeze up in the event of water source pressure loss (low-pressure cutout).

From the table in Figure 23–3, the average COP (Coefficience of Performance) ranges from 3.43 with a tank temperature of 115°F and 80°F water source temperature down to 2.56 with 135°F tank temperature and 50°F water source temperature.

In all cases, the amount of heat energy per penny operating cost for heating domestic hot water is more than 2.5 times the amount of heat energy available per penny from an electric resistance heat source. This difference in operating cost is the main factor for the rise in popularity of the heat pump water heater.

Piping

Piping the unit to the hot water tank is the same as used for the air-to-liquid water heating heat pump. Figure 23–4 shows the recommended piping arrangement for such an installation. The unit is connected to take the cold water from the bottom of the electric water heater or storage tank, whichever is used. A tee is provided for a drain valve out the branch of the tee at the bottom for service draining. Connected to the unit with removable unions (for major service), valves are required in both the cold and hot water lines. These valves should be ball type for minimum flow resistance. Globe-type valves have a high flow resistance due to the two 90° turns in the valve, and their use is not recommended.

The hot water supply line must be connected to the system between the main water service line and the tank. This is to prevent the shut off of water flow through the unit while the unit is capable of operating. Possible damage to the unit could result.

A relief valve is required by most local codes. Even if not required by code, a relief valve should be installed to prevent damage to the system from excessive hydrostatic pressure.

Special adapter assemblies are supplied by some manufacturers to permit easy installation of the system to a storage tank. Figure 23–5 shows such an assembly marketed by Friedrich Climate Master, Inc. of Utica, NY. This assembly, called a coaxial thermostat assembly by the manufacturer, allows the unit to be connected to the tank through an existing tank drain connection without disturbing any other plumbing.

The actual pipe connection (1) is a 3/4" pipe and tee assembly connected to the tank drain opening. The cold water inlet to the heat pump (outlet from the tank) is off the top of the tee to prevent air lock. For illustration simplicity, a standard hose bib valve is shown. This allows the use of flexible hose between the heater and the tank.

On the end of the 1/2" supply pipe (4), another valve is used to provide a service shut off on the hot water side. This also is

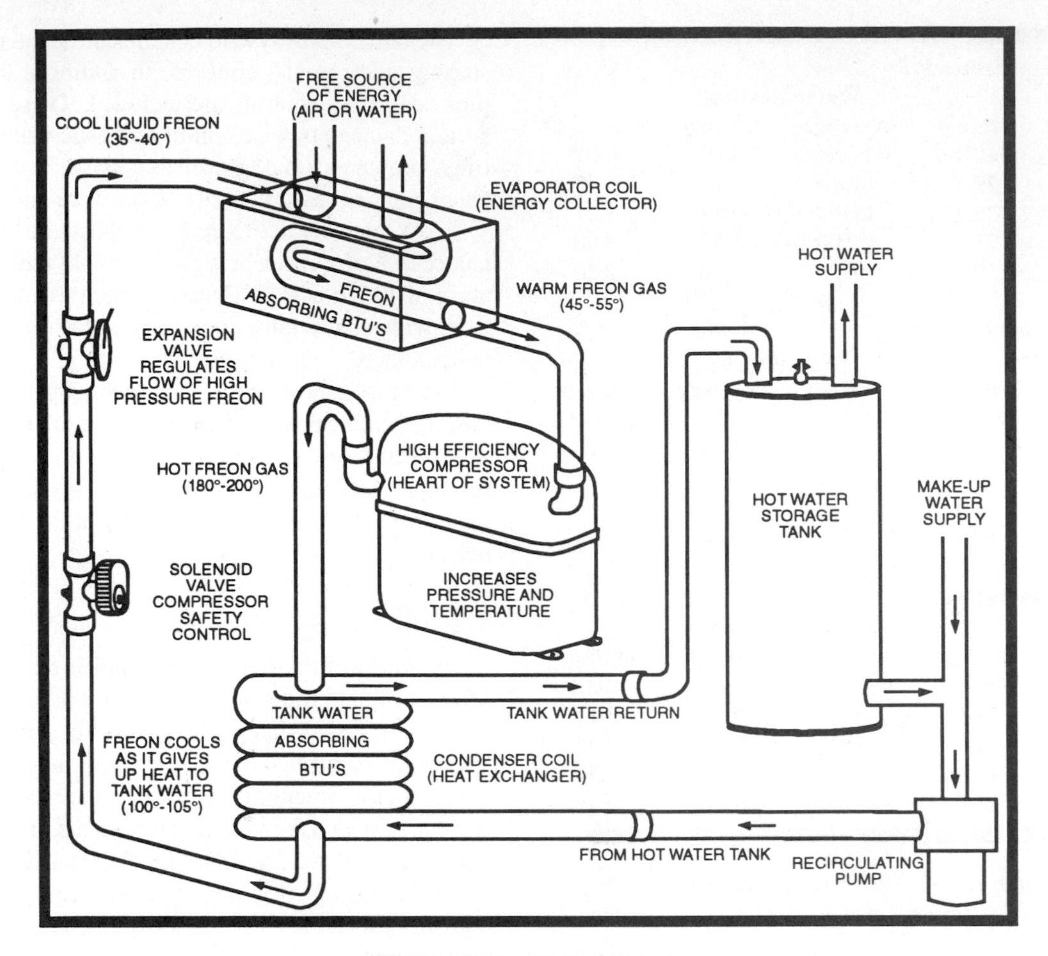

Figure 23–4 Piping diagram

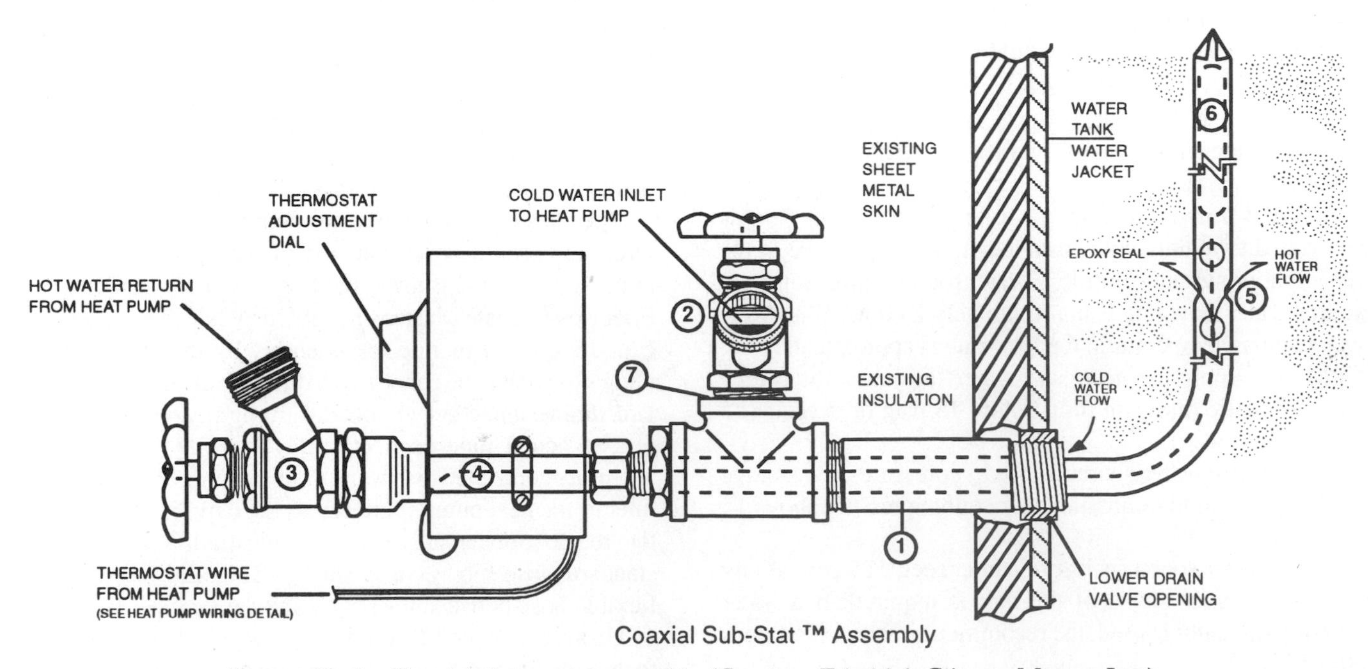

Figure 23–5 Coaxial thermostat assembly (Courtesy Friedrich Climate Master, Inc.)

pictured as a hose bib valve to permit the use of flexible hose. If copper pipe or PVC pipe is used, these valves should be gate or ball type.

The 1/2" pipe (4) fits through the pipe nozzle and tee assembly (1) and into the tank and has a 90° bend to a vertical up position. The purpose of this bend is to promote the flow of water to the top of the storage tank to reduce the stratification of the water in the tank (5).

The end of the pipe contains the feeler bulb of the thermostat. The bulb is sealed in the end of the pipe to reduce the corrosive effect of water on the bulb material.

Connected by the capillary tube between the bulb and the thermostat (4), the water temperature is monitored at the end of the flow tube.

This assembly provides both connection means and temperature control with the minimum of installation effort. The thermostat is 24 volts to reduce the installation cost and potential hazard from electrical circuits in a water situation. All phases of the installation must conform to local, state, and national codes.

23–3 HEAT PUMP WATER HEATER—COMMERCIAL

The commercial application of liquid-to-liquid heat pump units can be from hydronic heating systems in a residence to high-volume water heating to make-up water heating in conjunction with a master water heater such as used in swimming pool applications.

The heat source may be city water supply, well water supply, or possibly a large storage tank, which may act as a heat source and/or heat sink for multiple units.

Domestic Hydronic System

Figure 23–6 shows the packaged heat pump water heater marketed by Drake Industries of Port St. Lucie, FL. With an average heating rating of 40,000 BTUH using water at 55°F and a water flow rate of 6 GPM, the unit operates at an average COP of 4.6. The unit will provide the same heating capacity as a 13.7 kWh electric heating unit at approximately 75% of the operating cost.

By connecting the unit into the existing electric or fossil fuel hydronic system, the heat pump becomes the primary source of the heating energy. The more expensive heat energy source is then used as an emergency backup in situations of extreme weather conditions.

Commercial Water Heating

Figure 23–7 shows a package-type water source water heater that is used in commercial applications. Available in a range of sizes from 40,000 BTUH to 565,500 BTUH, the unit is used to supply heated water for a wide range of uses.

The refrigeration circuit includes a foam-insulated condenser, steel jacketed with a copper outer tube and cupronickel water tubes. The refrigerant-to-liquid evaporators are also copper refrigerant tube and cupronickel water tubes.

The safety controls in this type of unit usually consist of:

1. Water flow switch. A safety device to prevent compressor operation in the event that the water flow rate in either or both coils decreases to a point that may cause the unit to cut out on high head pressure or freeze up the evaporator.
2. Lockout relay system. U.L. requires a manual reset high head pressure safety cutout system. The most popular type is the use of the lockout relay system. This provides manual reset by power interruption and does not require access into the unit itself.
3. Evaporator freeze protection. This may be done by means of a thermostat with the control feeler bulb clamped to the evaporator coil liquid outlet or by means of a low-pressure control.

Figure 23–6 Packaged water source heat pump (Courtesy Drake Inductries, Inc.)

Figure 23–7 Commercial-size heat pump water heater (Courtesy Drake Industries, Inc.)

4. Cross ambient thermostat. Where precise water temperatures must be maintained, such as a pool or spa water temperature, a cross ambient thermostat calibrated with a differential of 1°F is included.

   For commercial hot water applications, an ordinary hot water heater thermostat mounted on the hot water tank is used.
5. Motor-compressor assembly. Hermetic compressors with start and run capacitors and capacitor relay are used.

Multistage Units

These units may have as many as three separate water heating circuits. Figure 23–8 shows a three refrigeration system capable of providing 360,000 BTUH water heating capacity. The top section contains the three motor-compressor assemblies and accumulators. The middle section is the refrigerant-to-liquid condensers with PVC piping to provide liquid flow via parallel circuits through the condensers. Each condenser has a manual control valve to be able to set the desired flow rate through the individual condensers.

The lower section is the evaporator section. This section contains the liquid-to-refrigerant evaporators. In this unit, T. X. valves are used as the pressure-reducing devices. In larger units of 7-1/2 hp or more, the use of T. X. valves is the predominant method of refrigerant control.

Figure 23–8 Three-circuit water source heat pump (Courtesy Drake Industries, Inc.)

Included in the refrigerant liquid line are normally closed liquid line solenoid valves. These solenoid valves operate with the motor-compressor assembly in the particular circuit. The purpose of using the positive shut off of the liquid flow during the off cycle of the compressor is to prevent refrigerant migration from the condenser to the evaporator. This can occur because the water flow through the coils continues as long as one system operates. The water temperature difference between the evaporator circuit and condenser circuit will cause the temperature and pressure difference to cause the refrigerant to flow to the evaporator of the systems in the off mode.

23–4 HEAT PUMP WATER HEATER—MARINE

Figure 23–9 shows a small-size single circuit unit designed for marine applications. Built into the hull or machinery room of a ship, there is no need for a cabinet or other cosmetic items. This reduces the unit weight.

The refrigerant and water circuits are the same as for the domestic units. The evaporator, located on the far side of the unit, which will remove the heat energy from salt (sea) water is made of 90/10 cupronickel. A T. X. valve is used for the pressure-reducing device and a liquid line solenoid is used for refrigerant flow control on the evaporator coil.

The condenser coil is a copper tube refrigerant-to-water heat exchanger, which is foam insulated and housed in the steel cabinet.

Water-circulating pumps are not included in marine units because they vary according to the type of ship equipment. They usually, however, contain the pump controls to work in conjunction with the compressor controls.

Large-size Marine Unit

Figure 23–10 shows a larger two-circuit marine heat pump water heater. The copper tube condensers are shown as

Figure 23–9 Small-size marine unit (Courtesy Drake Industries, Inc.)

Figure 23–10 Large-size marine unit (Courtesy Drake Industries, Inc.)

Figure 23–11 Multiple-unit application (Courtesy Drake Industries, Inc.)

wrapped coils on the upper portion at each end of the unit. The motor-compressor assemblies and accumulators are located between the condensers. On the front of the upper section is the control panel for both systems. The lower cabinet contains the cupronickel evaporators with T. X. valves and liquid line solenoid valves for refrigerant flow control. The unit does not have a finished cabinet to save space and weight.

23–5 CONTROL

The control of liquid-to-liquid units is predominately by 24-volt controls. The master control is usually a return fluid type thermostat that is set to provide a maximum water temperature to the domestic water storage system. These units must be used with a storage system. Their water heating capacity is not sufficient to be used as "flash type" or instant water heaters.

Usually the thermostat mounted on the water storage tank is the master control. For small single-circuit units, a single thermostat is used. Multiple systems, such as shown in Figure 23–8, use either three single-stage thermostats or a sequencing type multistage thermostat. In the multistage setup, usually a 2°F differential is used between the stages of the thermostats.

To prevent freeze up of the evaporator coil, flow switches must be included in the evaporator water circuit to prevent compressor operation if the liquid flow rate falls more than 2 GPM below the required operation rate. A low-pressure control is also included in the evaporator refrigerant circuit as extra protection.

The heated water circuit (condenser liquid circuit) should also have a flow switch included along with a manual reset high-pressure safety cutout system for the protection of the motor-compressor assembly.

The electrical power supply to marine units is from on-board generators or may be from the main engines. To protect against low-voltage power as well as power surges, electrical current "surge" and "short cycle" time-delay protection is standard on these units.

The basic control circuit for the motor-compressor assembly is the standard wiring system using start and run capacitors and a start capacitor relay for starting and running the system against unbalanced pressures. Using T. X. valves and liquid line solenoids, the system pressures will usually not balance out. This requires the high starting torque components.

23–6 APPLICATION

The application of liquid-to-liquid heat pumps and hot water heaters is limited to situations where a sufficient heat source/sink water supply is available. Given the adequate source of water for the heat source/sink, the unit will supply heated or chilled water for whatever application requires it.

A typical application is shown in Figure 23–11. This application shows six liquid-to-liquid heat pumps used to control the temperature of the water in the various ponds in a fish hatchery. Holding the water between 50°F and 55°F regardless of the outdoor weather and building conditions is much easier using a single-circuit ground water supply loop as the heat source/sink than with an air source.

Using units that reverse the heat flow, each pond is kept at the desired temperature independently of the rest. Each system is complete control wise for the independent operation. The unit is controlled by a single-stage heat, two-stage cool remote-bulb-type thermostat, which controls the temperature of the return water from the individual pond. The automatic changeover from heating to cooling and reverse is done by the first-stage cooling thermostat contact controlling the reversing valve. The minimum water temperature differential is 5°F.

Figure 23–12 shows three 7-1/2-hp units used to supply temperature-regulated water for a motel application. Operating in conjunction with a 3,000 gallon buried hot water supply tank, the units supply the heating capacity for the hot water needs of the motel guest rooms and restaurant kitchen.

Located in Florida where brackish groundwater is available, this arrangement supplies hot water at a considerably lower cost than using electric resistance elements. Using three units rather than a single larger unit increases the operating

Figure 23–12 Multiple large-unit application (Courtesy Drake Industries, Inc.)

efficiency by operating in stages according to the hot water demand.

The most popular use for liquid-to-liquid heat pumps is the control of swimming pool water temperatures. Figure 23–13 shows a typical swimming pool piping layout using a liquid-to-liquid heat pump to control the temperature of the water in the swimming pool with a unit using a liquid heat source/sink. The illustration shows the application to a well water system with well return. An optional method would be the use of a ground loop.

Because the sizing and servicing of well supply and ground loop systems is covered in Section 3, Liquid-to-Air, it is not repeated here. The handling of swimming pool applications is covered in Chapter 22, Air-to-Liquid.

23–7 SERVICE

The major item of service is the maintenance of the liquid circuits through the liquid-to-refrigerant coils. For a review of this subject, refer to Section 3, Liquid-to-Air Type Systems.

REVIEW QUESTIONS

1. Liquid-to-liquid heat pumps can roughly be classified into three categories. Name them.
2. The evaporator coil in a heat pump water heater is heavily insulated to improve the operating efficiency. True or False?
3. A slow-acting solenoid valve in the water circuits is a necessary item. True or False?
4. The water source pressure should not drop below _____ psig.
5. The water source pressure should not exceed_____psig.
6. If a water-pressure regulator is used, the recommended outlet pressure is _____ psig.
7. The use of gate or ball valves is recommended for control of water flow in the heated water circuit. Why?
8. Liquid line solenoid valves are used in liquid-to-liquid heat pumps to provide automatic pump down of the refrigerant from the evaporator. True or False?
9. One of the optional items used on liquid-to-liquid heat pumps is liquid flow switches in the water circuits. True or False?
10. Flow switches are set to cut off compressor operation if the flow rate drops more than 2 GPM below the design rate. True or False?

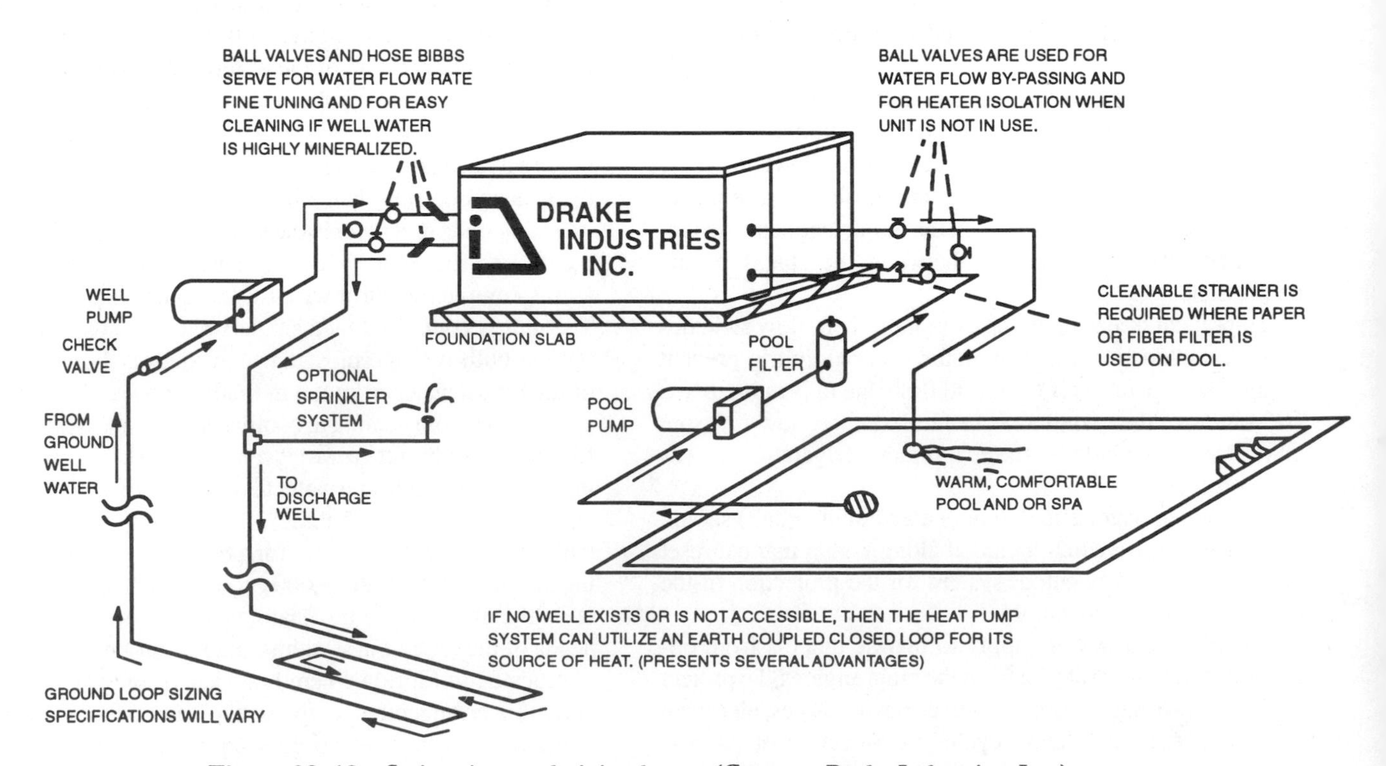

Figure 23–13 Swimming pool piping layout (Courtesy Drake Industries, Inc.)

WORKSHEETS

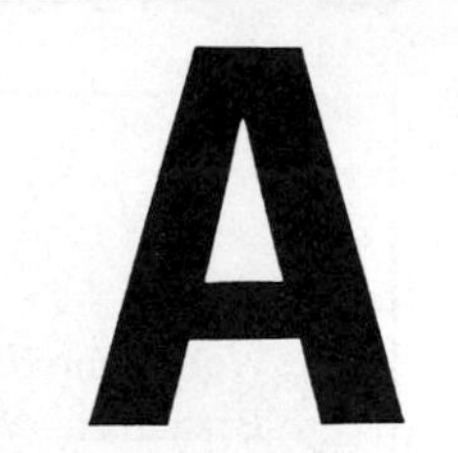

WATER SYSTEM WORKSHEET SE3WKS-1
(Method applicable to submersible pumps only)

A. Well Pump Sizing

Branch A—Well Pump—Piping from pump to pressure tank.
Branch B—Heat Pump Water Supply—Piping from tank to heat pump to drain.
Branch C—Domestic Water Supply—Piping from tank to house fixtures.

1. Determine the household needs from Figure 12-2. Enter here.	Branch C. ________	GPM
2. Heat pump GPM from unit specifications.	Branch B. ________	GPM
3. Add lines 1 and 2 for total water flow.	Branch A. ________	GPM

Note: If piping layout has more branches (C1, C2, C3, etc.), determine the flow rate for these from Figure 12-2 and include in the total.

B. Determine Water Pressure Requirements Pipe Sizing For Each Branch—Household plumbing, Branch B, may be assumed to have a total pressure requirement of 30 psig (69 FTHD).

	Br. A	Br. B	
4. Tentatively select a pipe size from Figure 12-5 and enter here.	______	______	ID
5. Using Figure 12-6, determine the equivalent length of all the fittings and shut-off valves and enter here (equiv. ft.)	______	______	EF
6. Determine the total lineal feet of pipe in each branch. (physical ft.)	______	______	ft.
7. Add lines 5 and 6 and enter here.	______	______	
8. From Figure 12-5, for pipe size and GPM needed for each branch, determine each branch friction loss. (FTHD/100')	______	______	
9. Multiply line 8 by line 7; divide by 100 to determine the total branch friction loss. (FTHD/100')	______	______	
10. From the heat pump specifications, enter the unit pressure drop. (FTHD)	______	______	
11. Pressure drop through the controls, from Figures 12-9, 12-11, and 12-12. (FTHD)	______	______	
12. Calculate the total pressure drop. Branch A pressure drop is the same as line 9. Enter Branch A here. (FTHD)	______		
Branch B, by adding lines 9, 10, and 11 (FTHD)	______	______	
Branch C.	______	______	
13. Multiply line 12 by 0.433 to convert to psig. Enter here.	______	______	

14. From the piping layout, determine parallel flow among the branches. Beginning at the well pump, add the friction loss in psig for the well pump branch (Branch A) to the branch having the higher pressure, Branch B or C.
Note: If more than three branches are required by the piping layout, select that branch that has the highest pressure drop and add this pressure drop to Branch A. Enter on line 14 the number obtained as total pressure loss due to pipe friction. ______ (psig)

15. Add 20 psig to line 14 to obtain the pressure control cutout set point. ______ (psig)

16. Multiply line 15 by 2.31 to convert to FTHD. Enter here. ______ (FTHD)

17. Pump requirement.

________ GPM at ______ (FTHD)

Vertical distance to water in well ______ ft. lift

TOTAL PRESSURE ______ (FTHD)

C. Pressure Tank Sizing (Bladder or diaphragm types only)

18. Enter desired minimum off time of the well pump in minutes. (Never less than two minutes.) ______ minutes
19. Enter the pressure control cutin set point. ______ psig
20. Enter pressure cutout set point from line 15. ______ psig
21. Multipy line 3 by line 18 to determine the minimum drawdown. ______ GPM
22. Refer to Figure 12-3 or the pressure tank specifications for a specific model of tank(s) to select the nominal capacity of tank needed, using information from lines 19, 20, and 21.

Pipe Length Calculation Worksheet SE3WKS-2
Horizontal Earth Coils
Data Collection

Job Name ______________________ Pipe Length—Cooling __________ Ft.
Location ______________________ Pipe Length—Heating __________ Ft.

A. Pipe Information
1. Pipe Material ______________________
2. Pipe Size __________
3. Pipe Resistance (Rp) ______________ (Figure 14–6)

B. Soil Information
4. Soil Type ______________
5. No. of pipes in trench ________
6. Horizontal spacing between pipes ________
7. Soil Resistance (Rs) ________ (Figure 14–5)

C. Location Information
8. Mean earth temperature (Figure 14–1) Tm ________ °F
9. Soil Surface temperature (Figure 14–1) As ________ °F
10. Soil temperature variation (maximum 15°F) STV ________ °F
(If multiple pipes, average the temperatures at each pipe depth.)
11. Horizontal earth coil depth (Figure 14–3) D ________ Ft.
(Average depth for multiple coils.)
12. High soil temperature (Tm + STV) Th ________ °F
13. Low soil temperature (Tm – STV) T_L ________ °F

D. Heat Pump Information
14. Highest EWTc temperature (Tmax) (Figure 12–1) ________ °F
15. Lowest EWTh temperature (Tmin) (no antifreeze) (Figure 12–1) ________ °F
16. Lowest EWTh temperature (Tmin) (20% antifreeze)(Figure 14–15) ________ °F
17. Unit cooling capacity (UC) at ________ GPM at Tmax ________ BTUH (Figure 12–1)
18. Unit heating capacity (UH) (no antifreeze) at ________ GPM at Tmin ________ BTUH (Figure 14–15)
19. Unit heating capacity (UH) (20% antifreeze) at ________ GPM at Tmin = ________ BTUH
20. Unit cooling COPc at Tmax ________ °F at ________ GPM ________ (Figure 12–1)
21. Unit heating COPh (no antifreeze) at Tmin ________ °F at ________ GPM ________ (Figure 12–1)
22. Unit heating COPh (20% antifreeze) at Tmin ________ °F at ________ GPM ________ (Figure 14–15)

E. Unit Run Factor
23. Cooling outside design temperature ________ °F
24. Cooling design heat gain ________ BTUH
25. Heating design outside temperature ________ °F
26. Heating design heat loss ________ BTUH
27. Cooling run factor (Fc) (Figure 14–20) ________
28. Heating run factor (Fh) (Figure 14–20) ________

F. Ground Loop Heat Exchanger Length (This does not include supply and return pipes or pipe headers.)
Cooling
Pipe Length/12,000 BTUH Cooling Capacity

29. $$\text{Pipe Length} = \frac{12{,}000\ \text{BTUH} \times \dfrac{\text{COP}+1}{\text{COP}} \times [\text{Rp} + (\text{Rs} \times \text{Fc})]}{(\text{Tmax} - \text{Th})}$$

(Insert values from information section)

30. $PLc = \dfrac{12{,}000\ BTUH \times \frac{\quad +1}{\quad} \times [\quad + (\quad \times \quad)]}{(\quad - \quad)}$

31. Solve $(A) = \dfrac{COP+1}{COP} = \dfrac{\quad +1}{\quad} = \dfrac{\quad}{\quad} = \underline{\qquad}$

32. $PLc = \dfrac{12{,}000\ BTUH \times (A) \quad \times [\quad + (\quad \times \quad)]}{(\quad - \quad)}$

33. Solve (B) = (Rs × Fc) = (×) = ________

34. $PLc = \dfrac{12{,}000\ BTUH \times (A) \quad \times [\quad + (B) \quad]}{(\quad - \quad)}$

35. Solve (C) = [Rp + (B)] = (+) = ________

36. $PLc = \dfrac{12{,}000\ BTUH \times (A) \quad \times (C)}{(\quad - \quad)}$

37. Solve (D) = (Tmax – Th) = (–) = ________ °F

38. $PLc = \dfrac{12{,}000\ BTUH \times (A) \quad \times (C)}{(D)} = \underline{\qquad}$ Ft.

39. $Total\ PLc = \dfrac{UC\ 22{,}300\ BTUH \times \quad Ft./12{,}000\ BTUH}{12{,}000\ BTUH} = \underline{\qquad}$ Ft.

40. Cooling heat exchanger pipe length = ________ Ft.

Heating (No antifreeze)

41. $Pipe\ Length = \dfrac{12{,}000\ BTUH \times \frac{COP-1}{COP} \times [Rp + (Rs \times Fh)]}{(T_L - Tmin)}$

42. $PLh = \dfrac{12{,}000\ BTUH \times \frac{\quad -1}{\quad} \times [\quad + (\quad \times \quad)]}{(\quad - \quad)}$

43. Solve $(A) = \dfrac{COP-1}{COP} = \dfrac{\quad -1}{\quad} = \dfrac{\quad}{\quad} = \underline{\qquad}$

44. $PLh = \dfrac{12{,}000\ BTUH \times (A) \quad \times [\quad (\quad \times \quad)]}{(\quad - \quad)}$

45. Solve (B) = (Rs × Fh) = (×) = ________

46. $PLh = \dfrac{12{,}000\ BTUH \times (A) \quad \times [\quad + (B)]}{(\quad - \quad)}$

47. Solve (C) = [Rp +(B)] = [+] = ________

48. $PLh = \dfrac{12{,}000\ BTUH \times (A) \quad \times (C)}{(\quad - \quad)}$

49. Solve (D) = (T_L – Tmin) = (–) = ________

50. $PLh = \dfrac{12{,}000\ BTUH \times (A) \quad \times (C)}{(D)} =$ ______ Ft.

51. $Total\ PLh = \dfrac{UH \quad BTUH \times \quad Ft./12{,}000\ BTUH}{12{,}000\ BTUH} =$ ______ Ft.

Heating (20% antifreeze)

52. $Pipe\ Length = \dfrac{12{,}000\ BTUH \times \dfrac{COP - 1}{COP} \times [Rp + (Rs \times Fh)]}{(T_L - Tmin)}$

53. $PLh = \dfrac{12{,}000\ BTUH \times \dfrac{\quad -1}{\quad} \times [\quad (\quad \times \quad)]}{(\quad - \quad)}$

54. $Solve\ (A) = \dfrac{COP - 1}{COP} = \dfrac{\quad -1}{\quad} = \dfrac{\quad}{\quad} = \dfrac{\quad}{\quad}$

55. $PLh = \dfrac{12{,}000\ BTUH \times (A) \quad \times [\quad + (\quad \times \quad)]}{(\quad - \quad)}$

56. Solve (B) = (Rs × Fh) = (×) = ________

57. $PLh = \dfrac{12{,}000\ BTUH \times (A) \quad \times [\quad + (B) \quad]}{(\quad - \quad)}$

58. Solve (C) = [Rp + (B)] = [+] = ________

59. $PLh = \dfrac{12{,}000\ BTUH \times (A) \quad \times (C)}{(\quad - \quad)}$

60. Solve (D) = (T_L – Tmin) = (–) = ________

61. $PLh = \dfrac{12{,}000\ BTUH \times (A) \quad \times (C)}{(D)} =$ ______ Ft./12,000 BTUH

62. $TPLh = \dfrac{UH\ 13{,}500\ BTUH \times \quad Ft./12{,}000\ BTUH}{12{,}000\ BTUH} =$ ______ Ft.

Heating heat exchanger pipe length is ________ Ft.

Run Factor Worksheet SE3WKS-3

Name ______________________________

Location ____________________________

Temperature Bin	(A) Weather Data (HR/YR)	(B) Loss/Gain (MBTUH PER HR)	(C) Total Heat Loss/Gain A × B (MBTUH)	(D) Calculations
–35 to –30				(D) Heating
–30 to –25				Total Heat Loss
–25 to –20				(in 1,000)
–20 to –15				
–15 to –10				(E) Total Hours
–10 to –5				Per Season
–5 to 0				
0 to +5				(F) Avg. BTU/HR
+5 to +10				(in 1,000)
+10 to +15				
+15 to +20				(G) Avg. BTU/HR
+20 to +25				
+25 to +30				(H) Unit Heating
+30 to +35				Capacity (BTUH)
+35 to +40				
+40 to +45				(I) Fh =
+45 to +50				(J) Cooling
+50 to +55				Total heat Gain
+55 to +60				(in 1,000)
+60 to +65				
+65 to +70				(K) Total Hours
+70 to +75				Per Season
+75 to +80				
+80 to +85				(L) Avg. BTU/HR
+85 to +90				(in 1,000)
+90 to +95				
+95 to +100				(M) Avg. BTU/HR
+100 to +105				
+105 to +110				(N) Unit Cooling
+110 to +115				Capacity (BTUH)
+115 to +120				
+120 to +125				(O) Fc

Pipe Length Calculation Worksheet SE3WKS–4
Vertical Earth Coils
Data Collection

Job Name ______________________________ Pipe Length—Cooling ____________ Ft.
Location ______________________________ Pipe Length—Heating ____________ Ft.
Number of boreholes ________ Size ________" Depth ________ Ft.

A. Pipe Information
1. Pipe Material ______________________________
2. Pipe Size __________
3. Pipe Resistance (Rpe) ________________________ (Figure 14–6)

B. Soil Information
4. Soil Type ____________________________
5. Soil Resistance (Rs) ________ (Figure 14–5)

C. Location Information
6. Mean earth temperature (50' to 150' depth) (Figure 14–1) Tm ________ °F
7. Disturbed earth variation DEV ________ °F
8. High earth temperature (Tm + DEVc) Th _______ °F
9. Low earth temperature (Tm – DEVh) T_L _______ °F

D. Heat Pump Information
10. Highest EWTc temperature (Tmax) (Figure 12–1) ________ °F
11. Lowest EWTh temperature (Tmin) (no antifreeze) (Figure 12–1) ________ °F
12. Lowest EWTh temperature (Tmin) (20% antifreeze) (Figure 14–14) ________ °F
13. Unit cooling capacity (UC) at ________ GPM at Tmax ________ BTUH
14. Unit heating capacity (UH) (no antifreeze) at ________ GPM at Tmin ________ BTUH (Figure 14–14)
15. Unit heating capacity (UH) (20% antifreeze) at ________ GPM at Tmin ________ BTUH
16. Unit cooling COPc at Tmax ________ °F at ________ GPM ________
17. Unit heating COPh (no antifreeze) at Tmin ________ °F at ________ GPM ________
18. Unit heating COPh (20% antifreeze) at Tmin ________ °F at ________ GPM ________

E. Unit Run Factor
19. Cooling outside design temperature ________ °F
20. Cooling design heat gain ________ BTUH
21. Heating design outside temperature ________ °F
22. Heating design heat loss ________ BTUH
23. Cooling run factor (Fc) (Figure 14–14) ________
24. Heating run factor (Fh) (Figure 14–14) ________

F. Vertical Ground Loop Heat Exchanger Length (This does not include supply and return pipes or pipe headers.)
Cooling
Pipe Length/12,000 BTUH Cooling Capacity

25. $$\text{Pipe Length} = \frac{12{,}000 \text{ BTUH} \times \frac{\text{COP}+1}{\text{COP}} \times [\text{Rp} + (\text{Rs} \times \text{Fc})]}{(\text{Tmax} - \text{Th})}$$

(Insert values from information section.)

26. $$\text{PLc} = \frac{12{,}000 \text{ BTUH} \times \frac{\quad +1}{\quad} \times [\quad + (\quad \times \quad)]}{(\quad - \quad)}$$

27. $$\text{Solve (A)} = \frac{\text{COP}+1}{\text{COP}} = \frac{\quad +1}{\quad} = \frac{\quad}{\quad} = \underline{\quad}$$

28. $PLc = \dfrac{12{,}000\ BTUH \times (A) \quad \times[\quad + \quad (\quad \times \quad)]}{(\quad - \quad)}$

29. Solve (B) = (Rs × Fc) = (×   ) = ________

30. $PLc = \dfrac{12{,}000\ BTUH \times (A) \quad \times[\quad +(B) \quad]}{(\quad - \quad)}$

31. Solve (C) = [Rp + (B)] = [+   ] = ________

32. $PLc = \dfrac{12{,}000\ BTUH \times (A) \quad \times (C)}{(\quad - \quad)}$

33. Solve (D) = (Tmax – Th) = (–   ) = ________

34. $PLc = \dfrac{12{,}000\ BTUH \times (A) \quad \times (C)}{(D)} =$ ______ Ft.

35. $Total\ PLc = \dfrac{UC \quad BTUH \times \quad Ft./12{,}000\ BTUH}{12{,}000\ BTUH} =$ ______ Ft.

36. Cooling heat exchanger pipe length = ________ Ft.

Heating (No Antifreeze)

37. $Pipe\ Length = \dfrac{12{,}000\ BTUH \times \dfrac{COP-1}{COP} \times [Rp + (Rs \times Fh)]}{(T_L - Tmin)}$

38. $PLh = \dfrac{12{,}000\ BTUH \times \dfrac{\quad -1}{\quad} \times [\quad +(\quad \times \quad)]}{(\quad - \quad)}$

39. $Solve\ (A) = \dfrac{COP-1}{COP} = \dfrac{\quad -1}{\quad} = \dfrac{\quad}{\quad} = \underline{\quad}$

40. $PLh = \dfrac{12{,}000\ BTUH \times (A) \quad \times[\quad +(\quad \times \quad)]}{(\quad - \quad)}$

41. Solve (B) = (Rs × Fh) = (×   ) = ________

42. $PLh = \dfrac{12{,}000\ BTUH \times (A) \quad \times[\quad +(B) \quad]}{(\quad - \quad)}$

43. Solve (C) = [Rp +(B)] = [+   ] = ________

44. $PLh = \dfrac{12{,}000\ BTUH \times (A) \quad \times (C)}{(\quad - \quad)}$

45. Solve (D) = (T_L – Tmin) = [–   ] = ________

46. $PLh = \dfrac{12{,}000\ BTUH \times (A) \quad \times (C)}{(D)} =$ ______ Ft.

47. $Total\ PLh = \dfrac{UH \quad BTUH \times \quad Ft./12{,}000\ BTUH}{12{,}000\ BTUH} =$ ______ Ft.

48. Total bored hole length ________ Ft.

Circulating Pump Worksheet SE3WKS-5
Data Collection

Job Name ______________________ Pipe Length—Cooling __________ Ft.
Location ______________________ Pipe Length—Heating __________ Ft.
Type of ground loop system ______________________

A. Unit Information
1. Model No. ______________
2. Flow Rate Required ________ GPM
3. Operating water temperature: Heating ________ °F, Cooling ________ °F

Pipe and Fitting Information

B. Pipe (Figure 16-8)
Material ______________ Size ________ Length ________ Ft. Flow Loss ________ FTHD/100'

$\frac{\text{Pipe Length} \quad \text{Ft.} \times \quad \text{FTHD/100'}}{100} =$ FTHD

Material ______________ Size ________ Length ________ Ft. Flow Loss ________ FTHD/100'

$\frac{\text{Pipe Length} \quad \text{Ft.} \times \quad \text{FTHD/100'}}{100} =$ FTHD

Material ______________ Size ________ Length ________ Ft. Flow Loss ________ FTHD/100'

$\frac{\text{Pipe Length} \quad \text{Ft.} \times \quad \text{FTHD/100'}}{100} =$ FTHD

Material ______________ Size ________ Length ________ Ft. Flow Loss ________ FTHD/100'

$\frac{\text{Pipe Length} \quad \text{Ft.} \times \quad \text{FTHD/100'}}{100} =$ FTHD

Material ______________ Size ________ Length ________ Ft. Flow Loss ________ FTHD/100'

$\frac{\text{Pipe Length} \quad \text{Ft.} \times \quad \text{FTHD/100'}}{100} =$ FTHD

4. Total fitting flow loss ________ FTHD

C. Fittings (Figure 16-9)
Type ______________ Size ________ Quantity ________ Equivalent Length ________ Ft.

$\frac{\text{Quan.} \quad \times \text{Equiv. Length} \quad \times \quad \text{FTHD/100'}}{100} =$ FTHD

Type ______________ Size ________ Quantity ________ Equivalent Length ________ Ft.

$\frac{\text{Quan.} \quad \times \text{Equiv. Length} \quad \times \quad \text{FTHD/100'}}{100} =$ FTHD

Type ______________ Size ________ Quantity ________ Equivalent Length ________ Ft.

$\frac{\text{Quan.} \quad \times \text{Equiv. Length} \quad \times \quad \text{FTHD/100'}}{100} =$ FTHD

Type ______________________ Size ________ Quantity ________ Flow Loss ________ FTHD

$\frac{\text{Quan.} \quad \times \text{Equiv. Length} \quad \times \quad \text{FTHD/100'}}{100} =$ FTHD

Type ______________________ Size ________ Quantity ________ Flow Loss ________ FTHD

$\frac{\text{Quan.} \quad \times \text{Equiv. Length} \quad \times \quad \text{FTHD/100'}}{100} =$ FTHD

Type ______________________ Size ________ Quantity ________ Flow Loss ________ FTHD

$\frac{\text{Quan.} \quad \times \text{Equiv. Length} \quad \times \quad \text{FTHD/100'}}{100} =$ FTHD

5. Total fitting flow loss ________ FTHD

D. Controls (Figures 12-7 and 12-24)

Type ______________________ Size ________ Quantity ________ Equivalent Length ________ Ft.

$\frac{\text{Quan.} \quad \times \text{Equiv. Length} \quad \times \quad \text{FTHD/100'}}{100} =$ FTHD

Type ______________________ Size ________ Quantity ________ Equivalent Length ________ Ft.

$\frac{\text{Quan.} \quad \times \text{Equiv. Length} \quad \times \quad \text{FTHD/100'}}{100} =$ FTHD

Type ______________________ Size ________ Quantity ________ Equivalent Length ________ Ft.

$\frac{\text{Quan.} \quad \times \text{Equiv. Length} \quad \times \quad \text{FTHD/100'}}{100} =$ FTHD

Type ______________________ Size ________ Quantity ________ Equivalent Length ________ Ft.

$\frac{\text{Quan.} \quad \times \text{Equiv. Length} \quad \times \quad \text{FTHD/100'}}{100} =$ FTHD

6. Total control flow loss ________ FTHD

E. Pump Size Calculation.

7. Unit water coil flow loss (From manufacturer's specification literature) (Figures 13–4 and 13–13) ________ psig × 2.31 FTHD/1 psig = ________ FTHD
8. Total pipe flow loss ________ FTHD
9. Total fitting flow loss ________ FTHD
10. Total control flow loss ________ FTHD
11. Total flow loss ________ FTHD
12. Antifreeze solution ______________________
13. Antifreeze conversion factor ________ (Figure 16-6)
14. Unit flow Rate ________ GPM × antifreeze conversion factor ________ = Adjusted Flow Rate ________ GPM

F. Pump Selection

15. Circulating pump Mfg. ________ and Model No. ________ (Figures 16-3 and 16-4)
16. Number of pumps required ________

If multiple pumps in series are required, all pumps must be of the same make and model number to have the same performance characteristics.

B BIN WEATHER DATA

Bin Weather Data

OUTDOOR TEMPERATURE (F)	-35 to -30	-30 to -25	-25 to -20	-20 to -15	-15 to -10	-10 to -5	-5 to 0	0 to 5	5 to 10	10 to 15	15 to 20	20 to 25	25 to 30	30 to 35	35 to 40	40 to 45	45 to 50	50 to 55	55 to 60	60 to 65	65 to 70	70 to 75	75 to 80	80 to 85	85 to 90	90 to 95	95 to 100	100 to 105	105 to 110	110 to 115
Akron, Ohio					•	4	10	17	44	127	214	422	654	819	706	680	605	684	736	762	801	639	454	261	112	30	6	•		
Albany, N. Y.				1	2	2	20	62	90	174	254	376	552	814	751	677	694	694	720	769	741	559	387	230	115	31	10			
Albuquerque, N.M.							1	2	4	19	76	142	319	523	666	714	710	712	678	783	838	780	629	506	364	228	66	3		
Allentown, Pa.						1	3	7	23	61	129	248	501	851	848	712	661	702	792	773	812	657	471	305	153	43	12	•		
Amarillo, Tex.					1	2	1	5	34	52	115	191	315	499	595	612	684	714	671	752	824	804	607	475	381	288	129	24	1	
Anchorage, Alaska			5	28	46	94	124	211	285	376	560	662	795	831	713	637	741	995	930	499	180	50	8							
Atlanta, Ga.								2	•	6	14	20	82	255	455	627	651	700	817	908	978	1188	880	620	361	172	23	2		
Augusta, Ga.										2	9	35	94	210	368	497	570	666	785	875	957	1167	959	652	531	274	88	17	1	
Austin, Tex.									•	2	15	13	51	78	220	355	472	567	717	768	1032	1105	1251	819	600	438	215	46	1	
Bakersfield, California													7	77	247	541	746	908	977	966	898	631	742	613	474	371	254	100	18	•
Baltimore, Md.							•	2	7	43	89	184	328	642	755	770	673	683	696	729	794	883	655	438	263	109	23	2		
Baton Rouge, La.										1	5	6	41	103	199	336	482	598	664	828	992	1346	1307	826	622	376	50	2		
Billings, Mont.		1	3	30	51	71	81	89	121	163	236	372	522	730	785	781	721	741	749	686	564	447	340	242	144	76	17	3		
Binghamton, New York				1	1	5	23	67	156	263	350	515	568	873	666	598	595	673	694	808	781	548	318	145	31	2				
Birmingham, Ala.									3	4	13	48	111	273	423	489	581	621	767	875	1044	1166	915	631	471	256	57	8	1	
Bismarck, N.D.	7	21	41	71	102	161	214	278	304	349	358	436	558	702	575	482	501	549	612	632	560	442	346	233	139	68	15	2	•	
Boise, Idaho							4	2	18	47	142	272	498	854	881	865	809	790	748	663	609	484	376	307	214	135	39	6		
Boston, Mass.					•	1	9	4	35	74	151	256	429	674	848	828	757	766	781	804	819	676	433	245	127	93	10	•		
Brownsville, Tex.												1	13	9	25	73	187	270	390	564	963	1419	1901	1579	862	489	20			
Buffalo, N.Y.						•	2	19	66	125	230	427	605	821	796	684	643	727	752	771	755	621	396	232	76	15	1			
Burbank, Calif.													•	10	83	292	661	1186	1562	1562	1163	808	565	431	257	129	48	9	3	
Burlington, Ia.					3	13	29	55	108	161	233	356	542	797	708	572	485	528	613	714	753	743	584	377	199	95	31	1		
Burlington, Vermont		1	2	5	17	39	81	135	216	272	332	491	561	752	716	637	603	655	694	703	670	573	362	189	53	9	1			
Calgary, Alta.			12	35	75	109	171	194	224	303	335	436	619	820	838	781	797	839	698	558	594	273	173	89	28	5				
Casper, Wyoming			2	3	15	30	45	73	116	200	324	495	683	806	831	782	670	606	642	592	532	423	347	283	201	66	3			
Charleston, S.C.											3	14	58	150	308	433	546	633	792	1006	1134	1311	1156	736	436	148	34	1		
Charleston, W. V.							1	7	22	73	135	252	356	630	633	607	667	661	689	767	949	912	606	471	270	57	3			
Charlotte, N. C.									2	5	23	64	166	360	515	634	684	730	752	839	908	1115	747	567	397	203	52	5		
Chattanooga, Tenn.								1	4	10	32	87	215	423	538	631	681	696	746	776	930	986	713	535	388	260	87	16	1	
Cheyenne, Wyo.			1	5	4	13	34	55	92	177	279	499	608	797	810	883	765	769	771	637	514	409	301	214	93	32	3	•		
Chicago, Ill.					4	11	18	63	79	117	207	337	517	842	826	616	561	617	581	676	767	697	531	355	209	96	32	4		
Cincinnati, Ohio					1	2	6	10	36	60	101	218	475	709	692	642	619	661	692	738	855	822	613	431	250	105	24	2		
Cleveland, Ohio						1	23	9	49	109	198	346	581	809	760	614	615	633	645	733	831	721	562	311	155	56	8	1		
Cold Bay, Alaska							5	39	39	132	303	476	673	1245	1755	1327	1484	1042	211	31	6	2	•							
Colorado Springs, Col.				1	1	7	18	43	75	136	262	448	626	755	673	678	725	760	805	784	600	477	397	276	176	46	1			
Columbia, S.C.										1	11	36	115	286	394	496	564	661	778	838	941	1126	964	655	502	279	93	17	2	
Columbus, Ohio					4	5	6	24	25	71	136	177	537	739	740	661	645	655	701	737	814	741	558	392	212	83	18	1		
Corpus Christi, Texas											•	3	9	27	83	180	302	444	551	748	1041	1175	1538	1408	785	436	36	1		
Dallas, Tex.									1	4	17	34	91	231	371	504	576	629	656	693	795	831	942	880	659	493	273	79	9	•
Dayton, Ohio				•	1	5	16	23	55	99	182	309	558	786	698	601	576	580	627	717	832	817	607	402	202	65	8	1		
Denver, Colorado			1	•	1	6	22	36	78	119	216	359	553	721	717	692	704	678	731	783	684	549	437	332	236	103	10	1		
Des Moines, Ia.					9	26	56	81	131	195	280	409	583	763	630	492	515	521	597	694	783	709	562	378	223	94	30	5	•	
Detroit, Michigan						1	4	17	61	131	248	377	618	884	808	595	566	592	633	695	783	721	516	314	148	47	9	•		
Duluth, Minn.	2	9	20	50	142	131	190	229	284	373	499	617	638	766	585	528	605	699	733	633	458	323	194	86	23	2	•			
Edmondton, Alta.	4	10	19	64	105	154	196	281	333	338	404	426	571	707	644	515	680	814	750	597	423	283	182	78	23	4				
El Paso, Tex.									•	3	8	35	90	205	342	514	590	712	759	839	893	970	865	740	586	406	204	42	2	
Evansville, Ind.			•	1	•	1	4	9	20	47	81	146	380	654	683	658	643	678	658	670	735	811	721	530	346	199	83	12		
Fairbanks, Alaska	121	126	159	206	270	332	401	401	379	401	447	457	455	495	409	429	513	637	658	515	361	202	118	54	14	1				
Fargo, N.D.	2	19	33	70	124	182	237	255	274	360	385	439	578	657	569	445	449	513	641	680	626	505	362	216	108	30	6			
Flint, Michigan			•	1	2	11	34	75	142	208	347	487	707	863	675	565	597	574	638	746	745	588	411	253	88	11	•			
Fort Wayne, Ind.				•	2	5	15	32	61	101	183	366	615	890	754	639	586	613	664	682	762	667	502	353	180	80	15	1		
Fort Worth, Texas									1	3	12	44	132	294	422	538	591	648	622	689	774	889	982	788	596	440	246	54	3	
Fresno, California												•	34	168	426	673	952	1036	1006	921	803	709	607	490	392	297	192	56	5	
Galveston, Tex.											1	5	10	27	47	115	279	486	706	1195	1236	885	1129	1738	848	56	1			
Grand Rapids, Mich.				1	1	2	6	28	67	157	256	454	701	983	794	589	564	589	671	729	718	594	415	280	122	38	6	•		

*Less than one hour

AVERAGE NUMBER OF HOURS EACH TEMPERATURE SHOWN OCCURS IN A YEAR

Source of data: Lennox Ind.

(Continued)

OUTDOOR TEMPERATURE (F)	-35 to -30	-30 to -25	-25 to -20	-20 to -15	-15 to -10	-10 to -5	-5 to 0	0 to 5	5 to 10	10 to 15	15 to 20	20 to 25	25 to 30	30 to 35	35 to 40	40 to 45	45 to 50	50 to 55	55 to 60	60 to 65	65 to 70	70 to 75	75 to 80	80 to 85	85 to 90	90 to 95	95 to 100	100 to 105	105 to 110	110 to 115	OUTDOOR TEMPERATURE (F)
Great Falls, Mont.	1	5	15	55	49	58	97	92	112	154	231	383	553	712	809	822	847	884	805	635	501	374	270	160	86	36	5	1			Great Falls, Mont.
Green Bay, Wisconsin				3	19	42	95	160	231	321	373	515	689	820	649	542	522	598	720	758	658	474	331	176	6	9					Green Bay, Wisconsin
Greensboro, N.C.									*	15	36	92	261	466	614	621	686	721	817	866	963	955	659	504	318	135	33	2			Greensboro, N.C.
Halifax, N.S.						1	12	44	81	174	245	353	528	959	998	799	809	884	941	956	547	251	105	29	2						Halifax, N.S.
Harrisburg, Pa.								3	9	33	87	170	393	766	901	758	682	685	757	753	823	790	545	345	189	61	14	1			Harrisburg, Pa.
Hartford, Connecticut				*	2	3	11	33	77	153	233	370	552	825	807	683	575	649	752	751	755	617	419	274	118	29	3				Hartford, Connecticut
Helena, Mont.	1	10	28	61	69	73	99	86	153	211	286	389	645	809	900	826	776	761	690	560	465	344	234	164	92	29	5				Helena, Mont.
Hilo, Hawaii																			11	361	2272	3001	2126	975	24						Hilo, Hawaii
Honolulu, Hawaii																			6	122	722	2789	3458	1552	117	1					Honolulu, Hawaii
Houston, Tex.											4	9	25	50	100	220	396	553	684	796	1069	1263	1621	922	676	323	53	1			Houston, Tex.
Huron, South Dakota		1	7	21	56	83	145	208	262	305	419	476	571	652	574	502	488	513	569	614	624	554	443	318	205	103	47	9	*		Huron, South Dakota
Indianapolis, Ind.				1	2	3	7	28	53	74	135	279	543	752	751	653	614	620	601	708	763	760	604	420	238	102	23	2			Indianapolis, Ind.
Jackson, Miss.								1	3	3	4	17	64	176	325	485	557	605	669	810	989	1196	1051	719	546	383	143	16			Jackson, Miss.
Jacksonville, Fla.												1	10	52	133	262	350	468	688	895	979	1329	1658	949	607	316	70	2			Jacksonville, Fla.
Kansas City, Mo.						*	7	23	51	93	147	251	389	600	643	660	581	598	617	593	740	767	687	569	367	221	107	36	8	1	Kansas City, Mo.
King Salmon, Alaska	4	22	43	87	144	221	222	250	237	259	323	435	628	994	1047	872	923	984	595	289	131	43	13	3							King Salmon, Alaska
Knoxville, Tenn.									5	15	26	60	209	442	565	639	698	676	709	847	935	1042	793	578	335	156	32	3			Knoxville, Tenn.
LaCrosse, Wis.	1	1	8	19	29	52	97	140	188	272	305	489	600	784	654	518	481	520	655	738	754	638	458	225	103	32	2				LaCrosse, Wis.
LaGuardia, N.Y.								1	6	22	69	139	262	507	804	825	814	788	765	771	842	882	659	368	170	51	14	2			LaGuardia, N.Y.
Lake Charles, La.										1	5	6	14	58	126	303	450	622	717	838	1038	1302	1444	834	640	327	40	2			Lake Charles, La.
Lansing, Mich.					*	2	12	33	76	158	299	451	734	920	741	575	592	628	696	706	725	596	409	256	121	30	3				Lansing, Mich.
Laredo, Texas											*	1	8	27	82	192	343	471	563	687	838	958	1346	1161	801	626	469	189	6		Laredo, Texas
Las Vegas, Nevada											1	7	44	194	396	591	716	769	786	699	644	651	669	678	602	474	431	301	101	16	Las Vegas, Nevada
Lexington, Kentucky						*	7	16	35	80	144	238	441	654	627	629	611	644	656	710	898	957	630	464	263	62	2				Lexington, Kentucky
Little Rock, Ark.							1	1	2	6	23	41	136	306	469	603	702	652	704	782	847	996	964	691	497	288	100	11			Little Rock, Ark.
Los Angeles, Calif.														1	15	151	532	1183	2130	2331	1458	670	202	68	18	4	1	*			Los Angeles, Calif.
Louisville, Ky.				*	*	2	2	8	13	37	66	137	324	631	703	646	663	650	700	702	757	831	717	538	350	209	64	14			Louisville, Ky.
Lubbock, Texas						*	1	5	7	33	86	180	346	490	546	620	618	642	700	688	829	833	708	544	447	322	109	14	2		Lubbock, Texas
Macon, Georgia										3	11	39	109	210	362	487	545	668	755	800	905	1239	995	665	538	354	85	3			Macon, Georgia
Madison, Wis.	1	1	1	6	13	31	60	93	171	215	307	459	659	896	690	539	528	565	636	724	680	586	455	264	131	42	5				Madison, Wis.
Medford, Ore.									*	1	7	48	224	660	984	1143	1086	1000	901	715	566	443	329	274	207	124	55	6			Medford, Ore.
Memphis, Tenn.						1	*	1	4	7	12	54	157	345	518	595	645	659	716	788	855	926	902	672	471	297	120	17	1		Memphis, Tenn.
Miami, Fla.															4	26	71	147	277	452	810	1708	2463	1795	888	125	2				Miami, Fla.
Midland, Texas									1	5	23	80	163	337	451	602	669	631	678	720	793	914	865	673	532	426	177	28	1		Midland, Texas
Milwaukee, Wis.			1	2	4	18	47	83	116	176	285	421	459	913	774	611	591	585	634	749	753	597	390	226	96	32	6	*			Milwaukee, Wis.
Minneapolis, Minn.		3	8	50	46	72	124	177	219	288	357	526	598	666	623	504	468	522	609	735	713	620	432	268	128	33	2				Minneapolis, Minn.
Missoula, Mont.				2	9	19	38	43	89	170	284	465	736	1016	952	848	812	797	666	553	418	320	223	159	97	41	7	*			Missoula, Mont.
Mobile, Ala.										1	4	5	23	74	180	370	488	567	657	898	1135	1486	1338	785	523	220	38	2			Mobile, Ala.
Moline, Ill.				*	8	13	39	66	111	175	247	376	551	827	714	551	537	551	591	684	744	694	574	387	209	88	19				Moline, Ill.
Montgomery, Ala.										3	5	17	63	159	329	455	567	663	708	799	965	1235	1121	751	500	332	89	7			Montgomery, Ala.
Montreal, Que.			3	4	25	79	107	180	280	318	355	437	523	719	706	555	548	589	666	693	740	593	388	103	43	3					Montreal, Que.
Nashville, Tenn.						2	2	2	5	22	43	89	241	444	562	643	601	644	729	836	876	919	804	583	404	220	76	13	2		Nashville, Tenn.
Newark, New Jersey								2	11	38	109	202	360	637	846	784	701	692	697	755	814	819	615	383	200	79	21	2	*		Newark, New Jersey
New Orleans, La.											1	4	7	18	94	143	448	556	719	844	1047	1259	1848	966	606	259	16	1			New Orleans, La.
New York, New York								1	10	26	2	188	330	603	858	838	796	722	745	754	877	926	604	263	96	28	5	*			New York, New York
Norfolk, Va.											6	38	108	312	543	737	782	681	795	858	995	1113	909	481	273	113	20	1			Norfolk, Va.
Oklahoma City, Okla.							*	2	19	41	66	135	232	401	498	609	684	683	725	802	769	932	847	606	421	240	89	20	1		Oklahoma City, Okla.
Oakland, California													*	20	155	509	971	1858	2431	1498	756	339	135	61	26	8	2	*	*		Oakland, California
Omaha, Neb.				*	6	21	54	88	127	180	261	374	521	683	672	584	567	527	544	637	743	703	579	435	263	134	45	14	1		Omaha, Neb.
Orlando, Florida													4	30	84	156	245	377	564	830	1123	1680	1732	1004	666						Orlando, Florida
Philadelphia, Pa.									9	32	100	189	335	654	818	758	701	663	710	735	808	864	655	420	225	74	17	1			Philadelphia, Pa.
Phoenix, Arizona													8	57	182	391	540	659	769	767	776	762	774	798	698	611	507	324	129	15	Phoenix, Arizona
Pittsburgh, Penn.						1	7	30	60	159	233	360	569	774	688	631	587	637	678	799	910	722	503	311	98	12					Pittsburgh, Penn.
Pocatello, Idaho					4	10	34	58	94	151	250	462	678	891	835	743	703	688	638	585	536	467	359	283	189	88	17	*			Pocatello, Idaho
Portland, Me.					1	9	15	64	117	190	271	384	573	750	841	805	810	773	857	808	642	387	263	129	61	14					Portland, Me.
OUTDOOR TEMPERATURE (F)	-35 to -30	-30 to -25	-25 to -20	-20 to -15	-15 to -10	-10 to -5	-5 to 0	0 to 5	5 to 10	10 to 15	15 to 20	20 to 25	25 to 30	30 to 35	35 to 40	40 to 45	45 to 50	50 to 55	55 to 60	60 to 65	65 to 70	70 to 75	75 to 80	80 to 85	85 to 90	90 to 95	95 to 100	100 to 105	105 to 110	110 to 115	OUTDOOR TEMPERATURE (F)

*Less than one hour

AVERAGE NUMBER OF HOURS EACH TEMPERATURE SHOWN OCCURS IN A YEAR

Source of data: Lennox Ind.

(Continued)

OUTDOOR TEMPERATURE (F)	-35 to -30	-30 to -25	-25 to -20	-20 to -15	-15 to -10	-10 to -5	-5 to 0	0 to 5	5 to 10	10 to 15	15 to 20	20 to 25	25 to 30	30 to 35	35 to 40	40 to 45	45 to 50	50 to 55	55 to 60	60 to 65	65 to 70	70 to 75	75 to 80	80 to 85	85 to 90	90 to 95	95 to 100	100 to 105	105 to 110	110 to 115	OUTDOOR TEMPERATURE (F)
Portland, Ore.							1	2	4	15	31	52	97	301	702	1291	1313	1312	1359	959	546	367	208	128	53	21	3				Portland, Ore.
Providence, R.I.						1	12	12	34	66	134	257	418	709	801	690	822	824	827	844	803	664	419	228	73	22	2				Providence, R.I.
Pueblo, Colo.	*	1	1	1	1	8	26	39	52	117	203	365	496	570	556	664	674	677	708	769	714	587	484	383	296	210	68	10			Pueblo, Colo.
Raleigh, N.C.										7	23	72	197	383	520	589	668	728	813	915	968	1049	726	518	356	174	49	5			Raleigh, N.C.
Rapid City, S. D.			1	5	10	46	79	128	194	246	273	420	594	742	694	675	621	632	666	655	590	488	370	275	198	108	52	10	1		Rapid City, S. D.
Regina Sask.	12	27	51	99	168	221	238	281	330	349	441	533	563	648	559	499	530	616	693	533	481	386	282	168	92	47	10				Regina Sask.
Reno, Nevada						1	4	15	37	101	227	387	530	733	829	890	909	845	690	572	477	418	371	333	243	120	35	2			Reno, Nevada
Richmond, Va.										13	58	108	269	497	617	688	695	710	759	812	854	914	726	483	330	163	59	12			Richmond, Va.
Roanoke, Virginia								1	11	28	86	186	307	533	702	710	712	695	704	758	924	928	620	459	305	91	8				Roanoke, Virginia
Rochester, N.Y.					*	1	6	30	69	135	233	403	606	823	816	695	661	691	746	750	712	590	378	241	125	43	9				Rochester, N.Y.
Sacramento, Calif.													8	93	355	701	1049	1298	1329	1071	773	630	486	375	276	192	93	34	5	*	Sacramento, Calif.
Salem, Oregon									*	3	20	46	172	435	707	1100	1398	1352	1163	833	540	405	261	169	100	47	14	5	*		Salem, Oregon
Salt Lake City, Utah							1	16	46	90	150	308	572	842	841	725	652	690	661	632	641	588	461	363	272	150	43	3			Salt Lake City, Utah
San Antonio, Tex.									1	1	8	15	36	81	161	340	407	529	675	853	1019	1180	1310	863	623	427	215	22			San Antonio, Tex.
San Diego, Calif.															11	88	393	972	1863	2389	1821	832	270	99	22	4	*				San Diego, Calif.
San Francisco, Calif.														15	121	517	1319	2469	2278	1174	571	192	65	31	10	2					San Francisco, Calif.
San Juan, P.R.																				*	158	1450	3780	2521	837	19					San Juan, P.R.
Savannah, Ga.											2	9	35	106	235	367	496	586	736	935	1054	1313	1279	820	488	206	66	8			Savannah, Ga.
Scranton, Pennsylvania						2	9	31	78	178	264	392	575	848	805	628	592	629	719	784	804	666	413	254	83	14					Scranton, Pennsylvania
Seattle, Wash.										5	21	42	133	536	1121	1605	1337	1295	1132	699	406	233	105	36	17	4	1				Seattle, Wash.
Shreveport, La.								1	1	3	8	25	43	161	306	464	584	658	695	811	934	1065	1112	767	583	394	133	18	*		Shreveport, La.
Sioux City, Ia.			*	2	14	38	73	97	152	247	361	468	595	732	642	550	523	527	575	642	701	618	507	365	207	97	26	4	1		Sioux City, Ia.
Sioux Falls, S.D.		*	2	16	43	59	102	152	208	293	448	520	585	712	625	501	498	522	605	669	684	566	443	277	155	67	7	1			Sioux Falls, S.D.
South Bend, Indiana				*	1	7	26	47	81	166	250	449	661	870	694	564	526	567	608	728	806	698	507	318	150	41	3				South Bend, Indiana
Spokane, Wash.				*	3	8	16	29	52	91	153	302	625	1060	974	853	805	786	715	633	525	414	294	212	136	66	16	1			Spokane, Wash.
Springfield, Ill.					3	3	20	34	68	88	150	273	531	832	688	623	588	569	573	651	757	725	645	486	257	152	39	7	1	*	Springfield, Ill.
Springfield, Missouri						1	4	14	29	68	139	235	437	588	621	616	621	602	620	759	846	876	647	475	331	169	60	6	2	*	Springfield, Missouri
St. John, N.B.				2	14	42	97	128	201	264	335	427	669	986	784	692	796	941	1022	887	390	207	79	11	1						St. John, N.B.
Syracuse, N. Y.				1	1	7	10	59	79	141	241	370	564	800	752	708	669	709	753	742	752	604	406	255	109	28	4				Syracuse, N. Y.
Tallahassee, Florida												9	57	126	219	331	428	568	760	895	1061	1618	1275	739	530	149	4				Tallahassee, Florida
Tampa, Fla.													1	10	48	137	216	345	570	877	1187	1387	1910	1126	752	195	6				Tampa, Fla.
Toledo, Ohio						7	12	28	50	113	214	337	597	920	811	640	567	611	661	716	782	655	491	306	169	61	18	*			Toledo, Ohio
Topeka, Kan.						2	10	28	58	105	173	280	500	681	654	601	581	590	707	625	729	723	649	490	334	214	110	34	7		Topeka, Kan.
Toronto, Ont.					2	14	45	98	195	291	339	413	600	872	745	616	577	581	700	691	701	577	387	224	84	14					Toronto, Ont.
Tucson, Arizona												2	25	98	248	417	598	716	800	763	781	870	959	777	656	520	357	152	31	*	Tucson, Arizona
Tulsa, Oklahoma							1	2	12	29	75	159	265	438	535	611	637	622	636	671	752	838	816	649	481	333	149	43	10	1	Tulsa, Oklahoma
Vancouver, B.C.										5	13	21	110	386	753	1364	1618	1321	1217	936	559	310	108	18	3	1					Vancouver, B.C.
Waco, Texas										1	3	24	84	216	354	501	558	651	622	701	830	909	1101	829	612	482	249	42	1		Waco, Texas
Wake Island, Pacific																					5	621	3336	3944	863	1					Wake Island, Pacific
Washington D.C.										11	48	104	213	532	768	798	715	690	723	745	823	947	735	496	284	101	31	1			Washington D.C.
West Palm Beach, Fla.				-										2	18	57	115	202	291	455	835	1672	2413	1664	860	183	3				West Palm Beach, Fla.
Wichita Falls, Texas								*	4	10	27	114	228	388	497	581	636	627	606	677	714	784	825	741	560	406	260	83	2		Wichita Falls, Texas
Wichita, Kan.						1	3	22	47	78	140	236	349	549	614	618	658	669	670	669	757	768	673	503	341	213	105	47	11	2	Wichita, Kan.
Wilmington, Del.							*	1	9	39	107	200	369	682	816	752	708	668	731	721	804	849	630	390	209	66	16	2			Wilmington, Del.
Winnipeg, Man.	8	17	60	134	187	248	278	303	333	361	406	454	505	667	532	451	476	552	637	632	702	424	290	152	65	17	2				Winnipeg, Man.
Winston Salem, N.C.								1	4	18	43	116	245	456	622	708	704	710	694	801	907	1086	703	518	323	105	7				Winston Salem, N.C.
Yakima, Wash.				1	1	1	6	7	22	50	118	304	629	904	807	841	876	823	752	694	582	450	342	271	165	86	31	2			Yakima, Wash.
Youngstown, Ohio						3	10	31	63	158	242	420	645	840	681	645	620	664	698	773	821	641	421	266	102	21	2				Youngstown, Ohio
OUTDOOR TEMPERATURE (F)	-35 to -30	-30 to -25	-25 to -20	-20 to -15	-15 to -10	-10 to -5	-5 to 0	0 to 5	5 to 10	10 to 15	15 to 20	20 to 25	25 to 30	30 to 35	35 to 40	40 to 45	45 to 50	50 to 55	55 to 60	60 to 65	65 to 70	70 to 75	75 to 80	80 to 85	85 to 90	90 to 95	95 to 100	100 to 105	105 to 110	110 to 115	OUTDOOR TEMPERATURE (F)

*Less than one hour

AVERAGE NUMBER OF HOURS EACH TEMPERATURE SHOWN OCCURS IN A YEAR

Source of data: Lennox Ind.

ANSWERS TO PROBLEMS C

CHAPTER 1

1. False, the heat pump operates on the same principal as the air conditioning system.
2. False, the basic principle is the same as the air conditioning system.
3. True, the heat picked up from the outside atmosphere is used to heat the conditioned area.
4. False, heat picked up plus the motor input equals the total capacity.
5. False, the COP is the total heat rejected compared to the heat energy equivalent of the electrical energy used.
6. In four categories: air-to-air, liquid-to-air, air-to-liquid, and liquid-to-liquid.
7. False, some units, such as the air-to-liquid unit, are made to produce heating only; for example, a heat pump water heater.
8. False, subcooling is needed to improve the system COP.
9. False, the change in flow direction is by means of the reversing valve.
10. To close and force the refrigerant through the pressure reducing device when the matching coil is used as the evaporator.
11. False, the accumulator is used to protect the compressor when the defrost operating mode is terminated.
12. False, any liquid may be used as the heat source as long as the heat energy available is capable of satisfying the unit's requirements and the liquid is not detrimental to the system components.
13. Inside: 70°F DB; Outside: +45°F and +17°F.
14. Inside: 80°F DB and 66.5°F WB; Outside: 95°F.
15. True.
16. False, standard design is down to 45°F liquid; special units are built for operation below 45°F liquid.

CHAPTER 2

1. True.
2. a. Dry bulb temperature, b. Wet bulb temperature.
3. Psychrometric.
4. a. Liquid line, b. Suction line, c. Vapor line.
5. a. Standard, b. Compound.
6. a. Standard, b. Compound.
7. Standard.
8. False, AC only.
9. True
10. False, used to measure air quantity from supply and/or return air grills in the air distribution system.
11. Inclined manometer.
12. Hydrometer.
13. True.
14. 1.000
15. False, each type antifreeze solution has a different correction factor.
16. pH meter.
17. c. 7.
18. 1–o, 2–n, 3–i, 4–s, 5–m, 6–f, 7–a, 8–g, 9–p, 10–e, 11–j, 12–i, 13–d, k and q, 14–h, 15–l, 16–b, 17–c.

CHAPTER 3

1. Air to refrigerant and liquid to refrigerant.
2. False, will transfer heat in either direction with equal efficiency.
3. Flat, "U," "A," and "H" shaped.
4. False, no duct work can be used with a propeller fan.
5. False, the main purpose is to insure complete removal of ice and frost from the bottom of the coil.
6. False, the increase in subcooling as well as condenser capacity can have an effect of increasing the capacity of the unit up to 20%.
7. The use of a T. X. valve allows a wide range capability to produce better operation of the coil during extreme frost and load conditions.
8. False, the liquid feed is always into the bottom of the coil in the cooling mode to have the vapor feed into the top in the heating mode. The flow of refrigerant must be downward in the heating mode.
9. False, the liquid and refrigerant flow in opposite directions.
10. False, the same type coil works equally well in either direction.
11. False, T.X. valves for heat pumps have a special type power element and are not interchangeable with A/C type.
12. False, capillary tubes are selected to fit each particular model unit by each individual manufacturer.
13. False, the valve should always be mounted with the pilot valve above the main valve body.

14. To the small single connection on the valve body opposite the three connection cluster.
15. To the center connection on the three connection cluster side of the main valve body.
16. 75 Psig.
17. d. The reversing valve electrical solenoid coil.
18. b. 250°F.
19. c. Vertical up refrigerant flow.
20. c. Between the reversing valve and the compressor.
21. a. Operate the reversing valve to put the system into the cooling mode.
    b. Cut off the outdoor fan.
    c. Keep the system in the defrost mode until the coil is clear of frost and ice.
    d. Operate the auxiliary heat to reduce room chill.
22. Not used.
23. False, it must be normally closed to operate the fan when the relay is de-energized.
24. False, depends if the reversing valve is energized or de-energized when the unit is operating in the cooling mode.
25. False, circuitry is different in the heat pump thermostat.
26. False, the changeover relay is connected to the "Y" terminal.
27. False, manual reset is required.
28. d. Temperature of the liquid refrigerant leaving the outdoor coil.
29. b. Hot gas.
30. a. Temperature initiation/temperature termination.
    b. Time initiation/temperature termination.
    c. Static pressure initiation/temperature termination.
    d. Static pressure-time initiation/temperature termination.
31. c. Insufficient air over the indoor coil.
32. d. The electrical resistance value.
33. False, power is taken off the load side of the compressor contactor to cycle with the compressor.
34. False, power off terminates the defrost cycle.
35. c. Excessive amount of reheat causing the room thermostat to cut the compressor operation off.
36. d. Too much reheat.
37. False, the coil must get cold enough to actuate the defrost termination thermostat.
38. d. Action of the holding contact in the defrost relay.
39. a. Coil air flow resistance must increase to the setting of the D20 pressure switch.
    b. The coil temperature must be below 26°F.
40. c. 15 min.
41. a. 5 min.
42. b. 10 KWh.
43. c. Outdoor ambient thermostat, e. Hold back thermostat.
44. False, should not exceed 80% of the total capacity of the unit in the cooling mode.
45. False, controls the maximum temperature of the air off the fossil fuel unit before it enters the heat pump coil.

CHAPTER 4

1. False, always size to the cooling load.
2. False, the amount of auxiliary heat capacity must equal or exceed the design heat loss.
3. False, in the 80°F to 105°F range.
4. True.
5. d. The heat gain of the conditioned area.
6. False, accuracy is required; the cubic content method can result in an error of ±20%.
7. Decreases.
8. Increases.
9. a–I, b–O, c–O, e–I, f–I and O, i–0.
10. Gross heating capacity.
11. True.
12. 4.814 KWh times 3,413 Btu/KW equals 16,430 Btuh. 46,000 Btuh divided by 16,430 Btuh equals 2.75 COP.
13. False, keep the compressor operating to prevent liquid migration.
14. False, the cooling capacity is the total amount of heat (sensible and latent) picked up by the inside coil.
15. e. 0% to +10%.
16. 2. 32,765 Btuh divided by 3,413 Btu/KW equals 9.6 KWh. Two 5 KWh elements are required.
17. 3. 5.0 KWh (5,000 watts) divided by 240 volts equals 20.83 amperes. 240 volts divided by 20.83 amperes equals 0.0868 ohms resistance per element. 0.0868 ohms times 208 volts equals 3,744 watts per element times 3.413 Btu/watt equals 12,788 Btuh/KWh element. 32,765 Btuh heating load divided by 12,778 Btuh/5 KWh element equals 2.56 elements. Three 5 KWh elements are required.
18. 42,000 Btuh minus 15% (6,300 Btuh) equals 35,700 Btuh.

CHAPTER 5

1. False, to reduce traffic and snow problems.
2. False, only one system can operate at one time.
3. c. Initial Balance Point.
4. a. 65°F.
5. e. 2nd Balance Point.
6. Four.
7. e. Length of the connecting lines.
8. False, aluminum wire is not recommended.
9. 18 Gauge.
10. False, telephone cable is 22 Gauge and is too light.
11. False, duct sizes are larger for heat pump systems.
12. False, mounted in the blower discharge so as to get maximum air velocity over the elements.
13. False, a noncombustible base is required for electric heat as well.
14. False, each pan must have a separate drain system.

15. a. Outdoor Ambient Thermostat, b. Second stage of the area thermostat using a changeover relay.
16. 2°F.
17. The oil burner operating controller has its own power source, which places two power sources in the control circuit. An isolation relay is required.
18. False, to prevent excessive air temperatures over the inside coil during the defrost mode.
19. c. Vapor Line.
20. True.
21. False, line must be in a horizontal loop position.
22. a and c.
23. False, outdoor air should never be mixed with return air.
24. True.
25. d. 1 in. in 10 ft.
26. True.
27. c. 115°F.
28. c. 70°F.
29. True.

CHAPTER 6

1. 5.
2. Cooling mode.
3. False, the actual temperature drop required also depends on the amount of moisture in the air.
4. False, heat gain and loss in the duct can result in very inaccurate readings.
5. 9°F Superheat. 68 psig + 2 psig loss in the suction line equals 70 psig equals 41°F boiling point. 50°F coil outlet temperature minus 41°F boiling point equals 9°F superheat.
6. 10°F. 243 psig discharge pressure equals 115°F condensing temperature. 115°F minus 105°F liquid temperature equals 10°F subcooling.
7. b. Undercharged.
8. a. Charge to correct superheat, b. Charge to correct subcooling, c. Complete evacuation and weighing in the correct refrigerant charge.
9. c. Complete evacuation and weighing in the correct refrigerant charge.
10. 28,190 Btuh – (110°F – 98°F) × 2,580 × 1.1 = 35,244 Btuh. 232 volts × 14 amperes × .9 PF = 2,923.2 watts × 3.413 = 7,054 Btuh. 35,244 (gross capacity) – 7,054 Btuh (motor input) = 28,190 Btuh.
11. Yes, within 10% of factory rating.
12. 1,166.67 CFM 8 in. × 20 in. duct = 160 sq. in. divided by 144 = 1.111 sq. ft. × 1,050 ft./min. velocity = 1,166.67 CFM.
13. 1,693 CFM – 62.5 amperes × 240 volts × 3.413 Btu/watt = 51,195 Btuh divided by 28°F divided by 1.08 = 1,693 CFM.
14. 38,397 Btuh – 1,693 CFM × 21°F × 1.08 = 38,397 Btuh.
15. Yes, the unit is only producing 80% of the rated capacity.
16. The problem is in the test procedure; the results show the unit operating at 128% of the rated capacity.
17. 1,026 CFM—19 seconds on the meter means 95 cu. ft./hr. of 1,050 Btu/cu. ft. natural gas or 99,750 Btuh input. This input × .80 = 79,800 Btuh output divided by 72°F divided by 1.08 = 1026 CFM.
18. 1,111 CFM .80 GPH × 144,000 Btu/gal. = 115,200 Btuh input. This input × .75 efficiency = 86,400 Btuh output. 86,400 Btuh output divided by 72°F divided by 1.08 = 1,111 CFM.

CHAPTER 7

1. b. Compressor discharge pressure.
2. 62 psig—57 psig plus 5 psig tolerance = 62 psig.
3. a. Air system problems, b. Refrigeration system problems, c. Electrical control problems.
4. Dirty air filters.
5. Check valve on the inside coil stuck closed or screen in the valve plugged.
6. 75 psig.
7. 250°F.
8. e. 65°F.
9. c. 35°F.
10. True.
11. False, remains the same.
12. a. Reduction in both operating modes;
    b. Increase in both operating modes;
    c. Increase in one operating mode, none in the other;
    d. Reduction in one operating mode, none in the other;
    e. Increase in one operating mode, none in the other;
    f. Reduction in both operating modes;
    g. Reduction in both operating modes.

CHAPTER 8

1. Shut off air to evaporator to see if it frosts up.
2. "Y" on the outdoor unit (compressor contactor circuit) connected to "Y" on the thermostat subbase instead of "W1."
3. To prevent short cycling of the system between the heating and cooling modes.
4. 71°F.
5. 1/2 ohm.
6. Change from a mechanical-type to an electronic-type thermostat that has outdoor temperature compensation.
7. a. Will not initiate the defrost function,
   b. Will not terminate the defrost function,
   c. Unnecessary or nuisance defrost cycles,
   d. Does not complete the defrost function.
8. Too much reheat causing defrost termination by the room thermostat.
9. Excessive operating condensing pressure.
10. Insufficient air through the inside coil.

11. Loose termination thermostat or feeler bulb not obtaining the proper temperature.
12. a. Coil temperature must drop below the control point of the termination thermostat,
    b. Air flow resistance through the outdoor coil must be high enough to close the pressure switch.
13. c. 70% to 80%.
14. Loose termination thermostat not receiving proper coil temperature.
15. The holding contact in the defrost relay.
16. False, should not exceed the cooling sensible capacity of the heat pump.
17. a. Turn off the outdoor air supply,
    b. Put the reversing valve in the cooling mode,
    c. Bring on the reheat,
    d. Sustain the defrost mode until the defrost function is complete.

CHAPTER 9

1. True.
2. a. Well water source—open circuit,
   b. Well water source—closed circuit,
   c. Water reservoir,
   d. Buried closed loop.
3. a. Unit must be sized to handle the cooling load.
   b. Enough auxiliary heat must be included to handle the design heat loss.
   c. Air distribution system must be free of drafts handling 105°F to 110°F supply air.
   d. Installation, evacuation, and charging the system must be in accordance with best industry practices.
   e. Maintenance programs are a must.
4. True.
5. True.
6. False, no defrost controls are required in liquid-to-air heat pumps.
7. False, the de-superheater size must be limited to prevent refrigerant condensation in the de-superheater.
8. False, counterflow produces the highest rate of heat transfer overall through the coil.
9. False, the liquid lines are exposed to possible freezing.
10. The requirement of having the highside section in an area above 32°F caused the package type to be the major portion of the market.
11. False, most manufacturers produce units of multiposition versatility of the inlet and outlet duct connections.
12. b. 65°F.
13. False, groundwater temperature varies very little.
14. To compensate for wind effect above 7-1/2 MPH, 2°F to 4°F is added.

CHAPTER 10

1. a. Stored water quantities;
   b. Drilled well, surface return;
   c. Drilled well, drilled well return;
   d. Horizontal pipe, ground loop;
   e. Vertical pipe ground loop;
   f. Solar storage.
2. a. Quantity of water required in GPM,
   b. Temperature of water supply during both the heating and cooling seasons,
   c. Quantity of water available.
3. False, cooling mode only.
4. a. 3 GPM,
   b. 3.5 GPM
5. 10°F.
6. 34°F
7. Returning the water directly back to the source of supply.
8. a. Maintain the water pressure,
   b. Limit the temperature rise.
9. Outlet side.
10. c. 20°F.
11. a. Cupro-nickel liquid to refrigerant heat exchangers,
    b. PVC or Polybutylene pipe and fittings.
12. Galvanic action.
13. Incrustation.
14. By means of a dry well located next to the pond or lake and filled with gravel.
15. 1 acre.
16. 5 ft.
17. 15 ft.
18. False, only centrifugal type.
19. True.
20. False, all water lines must be below the level of the water source to insure the lines are full at all times.

CHAPTER 11

1. Aquifer.
2. Consolidated.
3. Unconsolidated.
4. Water table.
5. Static water head.
6. Pumping water head.
7. Draw down.
8. Cone of depression.
9. Cone of impression.
10. False, the most common is the drilled well.
11. False, actual usage is the one time function requirement.
12. False, peak demand is the probable use time and single use quantity used to figure the total water needed at one time.
13. Percolation test.
14. c. 100 ft.
15. Return—2 to 1 ratio.
16. To at least 4 ft. below the static water level.
17. When the supply water is high in sand and other particles.
18. To wash the return well screen to remove the collected sand and particles from the system.
19. Geothermal.

20. a. Dedicated geothermal well,
    b. Domestic geothermal well, sand geothermal well.
21. a. Dedicated geothermal well,
    b. Domestic geothermal well.
22. False, no well casing is used through solid consolidated material.
23. Domestic geothermal well.
24. a. Horizontal,
    b. Vertical.
25. False, they must be deep enough to be in moist earth.
26. a. Series flow,
    b. Parallel flow.
27. a. Close,
    b. Reverse return.
28. False, the major concern is upward slope of the pipe in the direction of liquid flow to promote removal of air from the pipe.
29. 5.
30. Reservoir system.

CHAPTER 12

1. False, the first step is water quantity needed to handle the potential load.
2. Pump head.
3. a. Lift,
   b. Friction,
   c. Pressure from the pressure tank.
4. c. 2.31.
5. Lift.
6. Back pressure.
7. e. 30 psig to 50 psig.
8. 2 min.
9. Draw down capacity.
10. The use of these materials promotes electrolysis action as well as scale formation.
11. a. PVC,
    b. Polybutylene,
    c. Rubber.
12. a. Size of pipe,
    b. Flow rate in GPM.
13. True.
14. Direct acting.
15. 105°F.
16. Reverse acting.
17. 45°F.
18. False, it should be installed ahead of the tank to also keep the tank clean.
19. 23.1 FT HD.
20. Slow acting water shut-off valve.

CHAPTER 13

1. 7 FT HD.
2. False, the vertical lift is from the pump up to the top tee fitting in the supply pipe.
3. a. Highest GPM flow rate,
   b. Total system flow resistance in FT HD,
   c. Vertical distance to the water in the well.
4. Geothermal temperature gradient.
5. 3°F.
6. Thermal conductivity.
7. False, the more dense the material, the higher the heat transfer rate (K factor).
8. The portion below the static water level.
9. 38 ft. × 0.433 psig/FT HD + 10 psig = 26.454 psig.
10. Cut in—23 psig + 8 psig = 30 psig; Cut out—30 psig + 20 psig differential = 50 psig.
11. 9 GPM × 2 min. = 18 GPM draw down.

CHAPTER 14

1. Low as 25°F and high as 100°F.
2. To take advantage of the sun effect recharging the earth thermal mass.
3. a. To prevent damage from freeze up of the liquid in the pipe,
   b. To take advantage of the latent heat of freezing of the surrounding earth.
4. b—300 ft.
5. False, they remain the same.
6. Sunshine.
7. True.
8. Thermal diffusivity.
9. Heat of rejection.
10. Heat of extraction.
11. Disturbed earth effect.
12. Resistance to heat flow through the earth material.
13. Light dry soil.
14. False, the Rp of plastic pipe is different for vertical position than for horizontal position.
15. a. 10 ft.
16. b. 15 Ft.
17. b. 15 ft.
18. a. An adjustable low pressure control must be installed,
    b. An antifreeze solution must be used in the pipe loop.
19. False, the flow rate must increase according to the antifreeze material used.
20. Run factor.
21. a. Single continuous horizontal loop,
    b. Single double back horizontal loop,
    c. Multiple horizontal loop,
    d. Series vertical loop,
    e. Parallel vertical loop.
22. 82%.
23. 1,000.
24. True.
25. c. 80 ft.
26. False, the casing is pulled to increase the heat transfer efficiency.

CHAPTER 15

1. True.
2. Socket fused, butt fused.
3. False, minimum pressure is 50 psig.
4. Just ahead of the unit. The pump discharges into the unit.
5. c. 50 psig.
6. 10.
7. Tracer system.
8. 18.

CHAPTER 16

1. False, circulators are piped for series operation.
2. False, circulators must be the same size. This application would require two circulators, each capable of delivering against 10 FT HD at the GPM requirement.
3. False, only one loop resistance is needed. A properly designed loop system has the same flow resistance in all the loops.
4. 5.6 GPM. Minimum flow rate per loop of 1.4 GPM × 4 loops = 5.6 GPM.
5. 13.2 Minimum flow rate of 3.3 per loop × 4 loops = 13.2 GPM.

CHAPTER 17

1. 2 feet per second.
2. False, the coldest expected temperature leaving the unit coil determines the required freezing point of the liquid.
3. False, the stronger the solution, the lower the heat transfer rate.
4. Chlorides, Glycols, Alcohols.
5. Glycols.
6. 7.
7. Decreases.
8. Zinc.
9. 60°F.
10. 10 gallons. 50 gallons of water × 0.2 = 10 gallons.
11. b. 75°F.

CHAPTER 18

1. False, the amount of heat recovered in the short cooling season may not be enough to justify the additional investment.
2. False, a special three tube coil must be used to provide a two tube barrier between the refrigerant and the domestic water.
3. 125°F to 130°F.
4. 16.07 gallons.
5. Co-axial fitting.
6. 150°F.
7. 130°F.

CHAPTER 19

1. a. Cut in—28°F or 52.4 psig,
   b. Cut out—18°F or 40.9 psig.
2. Dirty air filters.
3. Low water supply.
4. True.
5. Flow switch.
6. 7 GPM.
7. Water, 45°F; Air, 75°F
8. True.

CHAPTER 20

1. 5 volts.
2. False, a temperature difference with the "indoor" coil at a temperature higher than the compressor temperature will cause refrigerant migration.
3. False, only the cooling mode is used.
4. False, the methods are the same.
5. 32,987 BTUh. (84°F – 73°F) × 6 GPM × 60 min. × 8.33 #/gallon = 32,987 BTUh.
6. 31,910 BTUh. (84°F – 73°F) × 6 GPM × 60 min. × 8.058 Sg. = 31,910 BTUh.
7. 29,106 BTUh. (84°F – 73°F) × 6 GPM × 60 min. × 7.35 Sg. = 29,106 BTUh.
8. Specific gravity.

CHAPTER 21

1. Keep liquid flowing through the coil by forcing open the water regulating valves or keep the circulator operating.
2. Heavy water flow with very little cooling capacity.
3. Normal operation. Check valve must be closed in the heating mode.
4. Operate in both the heating and cooling modes to verify if additional refrigerant is needed.
5. a. Dual purpose capillary tube or restrictor (DPCT),
   b. Dual purpose T.X. valve (DPTXV),
   c. Single purpose capillary tube or restrictor/check valve combination (SPCT/CV),
   d. Single purpose T.X. valve and check valve combination (SPTXV/CV).

CHAPTER 21

1. a. Remote, b. Self-contained.
2. 2.0 and 2.9.
3. 120°F and 130°F.
4. 110°F and 120°F.
5. a. Water temperature control thermostat,
   b. Compressor protection overload cut out,
   c. Water temperature limit control.

6. The "Plumbing Tree" also contains the water temperature control thermostat.
7. Urethane Foam.
8. 50°F.
9. 0.4 ACH.
10. 105°F.
11. c. From the surface of the water to the air.
12. a. Heat conducted to the earth,
    b. Heat conducted to the air,
    c. Heat radiated to the air,
    d. Evaporation.
13. Evaporation.
14. 75°F to 82°F.
15. 84°F to 88°F.
16. 6 ft.
17. a. 80°F, b. 80°F, c. 55% to 65%.
18. False, continuous air circulation (CAC) is an absolute requirement.
19. Floor-type outlets at the outer walls.
20. 0.15 in. wc.
21. False, return air should be taken off the ceiling.
22. 0.10 in. wc.
23. a. R–21, b. R–40, c. R–12.

CHAPTER 23

1. a. Domestic,
   b. Commercial,
   c. Marine.
2. False, no insulation is used to help keep the unit cabinet temperature low.
3. True.
4. 15 psig.
5. 125 psig.
6. 60 psig.
7. The straight through flow results in minimum flow resistance.
8. False, to prevent refrigerant migration from filling the evaporator.
9. False, liquid flow switches are a necessary safety protection item.
10. True.

ABBREVIATIONS D

A Ampere. Unit of electrical quantity.
AC Alternating current. Variable reversing voltage.
ACCA Air Conditioning Contractors of America.
ACH Air change per hour.
ADR Auxiliary defrost relay.
ARI Air Conditioning and Refrigeration Institute.
AS Annual soil surface temperature swing.
AUTO Fan or blower operation by cycling with the motor-compressor assembly.

BFL Branch friction loss.
BP Boiling point. The temperature at which a liquid boils at a given pressure.
BR Blower relay.
BTU British Thermal Unit. The amount of heat energy added or removed to change the temperature of one pound of water one degree Fahrenheit.
BTUh The amount of heat transferred in British Thermal Units per Hour.

CAC Continuous air circulation.
CC Compressor contactor coil.
CFM Cubic feet of air per minute.
CLG-1 First stage cooling contact in the thermostat.
CLG-2 Second stage cooling contact in the thermostat.
CO Carbon Monoxide.
CO_2 Carbon Dioxide.
COPc Coefficience of Performance, cooling.
COPh Coefficience of Performance, heating.
CR Cooling Relay.
CR1 Cooling relay contact set number.
CV Check Valve.

D Average depth of horizontal earth coil.
DB Dry bulb air temperature.
DC Direct current. Voltage is constant.
DEF Disturbed earth factor.
°F Degree of temperature on the Fahrenheit scale.
°C Degree of temperature on the Centigrade scale.
ΔT Difference in temperature. Δ is the Greek symbol for "Delta," meaning difference.
DP Dew Point. Wet bulb temperature of saturated air.
DPCT Dual Purpose Capillary Tube or Restrictor.
DPDT Two sets of single pole-double throw contacts mechanically linked together to both operate by means of a single operating device.
DPST Double set of single pole contacts mechanically linked together to both operate by means of a single operating device.
DPTXV Dual Purpose Thermostatic Expansion Valve.

E Voltage
ECP Environmental control package.
EER Energy Efficiency Ratio.
EF Equivalent Feet. A means of comparing the flow resistance of a liquid or air carrying device to the length of straight pipe or duct.
EHPWH Exhaust Heat Pump Water Heater.
ESP External Static Pressure. Pressure in an air duct distribution system external to the blower or air handling unit. Total of both the positive pressure in the supply and negative pressure in the return added together as though both pressures are positive.
EWTc Entering Liquid Temperature, cooling.
EWTh Entering Liquid Temperature, heating.

FBP Final Balance Point.
FC Unit Cooling Run Factor.
FH Unit Heating Run Factor.
FLA Full Load Amperes.
FPM Speed of travel in feet per minute.
FR Fan Relay Coil.
FT HD Feet of Head. Flow resistance against pump performance consisting of height of lift, pipe flow resistance, and system operating pressure.
FT H_2O Feet of head for water.
FT^3/HR Flow rate of gas or vapor in cubic feet per hour.

Ga Gauge. Used in sizing wire. Based on the diameter of the metal in the wire.
GCC Gross Capacity, Cooling.
GCH Gross Capacity, Heating.

GPH Flow rate of a liquid in gallons per hour.
GPM Flow rate of a liquid in gallons per minute.
Gs Specific Gravity. Weight of a quantity of a liquid as compared to an equal quantity of water (Gs of water) is 1.000.

HCl Hydrochloric Acid.
HEPA High Efficiency Particulate Arresting Filter.
HPA High Pressure Cut Out Control.
HPS High Pressure Switch.
HPWH Heat Pump Water Heater.
HR/YR Hours per Year.
HT Holdback thermostat.
Hs Specific Heat. Heat quantity needed to change a quantity of a liquid one degree Fahrenheit as compared to an equal quantity of water.
HTG-1 First stage heating contact in the thermostat.
HTG-2 Second stage heating contact in the thermostat.

I Amperes.
IBP Initial Balance Point.
ID Inside diameter of pipe. Used in the plumbing industry.

K Thermal conductivity of a material or a combination of materials.
KWh Kilowatts (1,000 watts) of electrical energy used per hour.

L' Lineal Feet. Actual measured length of a material in feet.
Lc Length of pipe needed per 12,000 BTUh cooling capacity.
Lh Length of pipe needed per 12,000 BTUh heating capacity.
LPG Liquified Petroleum Gas.
LPS Low Pressure Switch.
LR Lockout Relay.
LRA Locked Rotor Amperage.
LWT Leaving Liquid Temperature.

MBTUh British Thermal Units per Hour expressed in thousands.
MG Manufactured Gas.
MH Motor Heat. Heat energy (in BTUh) equivalent of the usable electrical energy drawn by an electric motor.

NC Normally Closed. Contact closed when the device is de-energized or at standard conditions: 70°F temperature, 0 psig pressure.
NCC Net Capacity, Cooling
NCH Net Capacity, Heating.
NFPA National Fire Protection Association.
NG Natural Gas.
NO Normally Open. Contact open when the device is de-energized or at standard conditions: 70°F temperature, 0 psig pressure.

OAT Outdoor Ambient Thermostat.
OD Outside diameter of pipe; refrigeration industry.

PE Polyethylene Plastic.
PF Power Factor. The percentage of electrical energy recorded on a wattmeter compared to the total electrical energy used.
PH Pump Head. Total flow resistance against the pump operation.
pH Acid/Alkaline ratio of a liquid. Neutral is No. 7 on a scale of 1 to 10; below 7 is acid, above 7 is alkaline.
PLc Pipe length needed for each 12,000 BTUh cooling capacity.
PLh Pipe length needed for each 12,000 BTUh heating capacity.
psig Pressure in pounds per square inch as measured on a standard atmospheric-type gauge.
P/T Pressure/Temperature test fitting.
PVC Polyvinylchloride plastic.

R Resistance in ohms.
RH Relative Humidity. Percentage of moisture in the air as compared to saturated air.
Rp Pipe heat transfer resistance, horizontal pipe.
Rpe Pipe heat transfer resistance, vertical pipe.
RS Soil heat transfer resistance.
Rv Reversing Valve.

SBP Second Balance Point.
SEER Seasonal Energy Efficiency Ratio.
SFS Safety Fan Switch.
SH Superheat. Heat energy added to a vapor to raise the temperature of the vapor above the boiling point.
SPCT Single Purpose Capillary Tube or Restrictor.
SPCT/CV Single purpose capillary tube or restrictor and check valve combination (trombone).
SPDT Two sets of contacts mechanically connected to one action device; arranged to have only one of the contacts closed at one time.
SPST One set of electrical contacts closing in one direction of mechanical operation.
SPTXV Single purpose thermostatic expansion valve.
SPTXV/CV Single purpose thermostatic expansion valve and check valve combination (trombone).
STV Soil Temperature Variation.
SV Solenoid Valve.

TBFL Total Branch Friction Loss.
TDR Time Delay Relay (Sequencer).

TEF	Total Equivalent Feet
Th	High soil temperature at peak day of year.
TI/TT	Temperature Initiation/Temperature Termination.
T_L	Low soil temperature at low point day of the year.
TM	Timer Motor.
T_M	Mean annual earth temperature.
T_{max}	Maximum earth or liquid temperature needed.
T_{min}	Minimum earth or liquid temperature needed.
TPLc	Total pipe length required for cooling.
TPLh	Total pipe length required for heating.
TXV	Thermostatic Expansion Valve.
Uc	Unit net capacity, cooling.
Uh	Unit gross capacity, heating.
UL	Underwriters Laboratories.
V	Volts. Unit of electrical pressure.
WB	Wet Bulb Temperature; air saturation temperature.
WC	Inch Water Column. Measurement of air pressure in inches on a water column.

GLOSSARY

Accumulator: A storage vessel located in the suction line ahead of the compressor. Used to limit liquid refrigerant return to the compressor and store excess refrigerant in the heating mode.

Air conditioner: A piece of equipment used to provide control of temperature, humidity, air cleanliness, and air distribution in a conditioned area.

Air conditioning: The simultaneous control of temperature, humidity, cleanliness, and air distribution for human comfort.

Air changes: A method of expressing the amount of air leakage into or out of a building or room interior or the number of building or room volumes exchanged.

Air-cooled condenser: Heat picked up by the refrigeration system is transferred by the air-cooled condenser into the surrounding atmosphere by either an active or passive method.

Air cooler: A device used to lower the temperature of air passing through it.

Air cushion tank: See Expansion tank.

Air diffuser: An air distribution outlet designed to direct air flow into desired directions.

Air handler: A device used to move air. Contains a blower assembly for active air movement.

Air movement (active): Air moved by means of a mechanical device such as a fan or blower.

Air movement (passive): Air moved by the difference in the weight of the air due to temperature difference of the air.

Air recirculation: The movement of air directly from a unit discharge opening to the unit intake opening. This seriously affects the operation of the system by creating adverse temperatures.

Air return: Air returned from a conditioned space.

Air, saturated: A sample of air that contains the maximum amount of water vapor. The percentage of relative humidity in the air is 100%.

Air, specific heat of: The amount of heat needed to change the temperature of one pound of the air (at the existing temperature) one degree Fahrenheit as compared to the amount for an equal amount of standard air.

Air, standard: Air with a density of 0.075 pounds per ft^3 at 70°F and 29.92 in. Hg atmospheric pressure.

Air vent: Device that automatically vents air out of a liquid system when required.

Air vent riser: A vertical length of pipe at the highest point in the system to gather and retain air in the system as the liquid flows past the riser. The air vent is located at the top of the riser.

Alcohol brine: A mixture of alcohol and water that remains a liquid below 32°F.

Algae: One-cell plants found in water of ponds and swamps. Includes sea weed, scum, etc.

Alternating current (AC): Electrical current that reverses polarity or direction. As the voltage or electrical pressure varies—rising to maximum, falling to zero, reverses direction, rises to maximum, and falls to zero—the current in the circuit follows the voltage change. This 360° reversal is called a Hertz. Sixty 360° reversals per second is called 60 Hertz frequency.

Ambient temperature: Temperature of the air that surrounds an object.

Ammeter: An electric meter used to measure electric current (quantity) in units called amperes.

Ampacity: A term used to designate the amperage carrying capacity of wire or conductors.

Ampere: A unit of electrical energy quantity.

Anemometer: An instrument used to measure the flow rate of air.

Antifreeze: Materials—such as salt, alcohol, or Glycol products—that are added to water to lower the fusion point.

Aquastat: Water temperature control thermostat.

Armature: The moving or rotating portion in a motor, relay, or solenoid valve or other electromechanical devices.

Automatic defrost: A control system used to control the frequency and duration of system operation for removal of frost and/or ice from the evaporator coil.

Auxiliary drain pan: A complete drain pan large enough to contain the entire unit. Used to prevent water damage to surfaces under the unit in the event of blockage of the coil condensate drain.

Available head: The difference in pressure that can be used to circulate water in a system. The difference is the pressure used to overcome the flow resistance in the system at the desired flow rate.

Azeotropic mixture: A mixture of two or more substances that do not combine chemically but produce different characteristics than any of the individual substances.

Backhoe: A mobile piece of equipment that uses a controlled shovel/bucket to dig trenches.

Back pressure: Pressure in the low side of the refrigeration system. Also called low side pressure or suction pressure.

Balance fitting: A pipe fitting or valve designed so that its resistance to flow may be varied. This type of fitting is used to obtain the desired flow rate through parallel circuits.

Balance point: The outdoor temperature at which the heating capacity of a heat pump in a particular installation is equal to the heat loss of the conditioned area.

Balance point, initial: The outdoor ambient temperature at which the heating capacity of the heat pump only balances the heat loss of the conditioned area.

Balance point, second: The outdoor temperature at which the heating capacity of the heat pump plus the auxiliary heat capacity balances the heat loss of the conditioned area.

Balanced pressure ventilating system: Mechanical fresh air supply is balanced against mechanical exhaust to maintain a constant air pressure in the conditioned area.

Ball check valve: A check valve that uses a ball against a seat as a shut off means.

Balometer: The trade name of a balance meter marketed by the Alnor Instrument Co.

Barometer: An instrument for measuring atmospheric (barometric) pressure. The predominate calibration is in inches of Mercury (Hg) in a vertical column.

Bentonite: A clay material used in well drilling to seal the walls of the bore hole to reduce the loss of drilling mud into the earth.

Bimetal strip: Used in temperature controls and indicators. Composed of two metals with different expansion rates with temperature change, welded together to produce a bending motion with temperature change.

Bin weather data: Tables published by the Air Conditioning Contractors Association showing the number of hours per year the local ambient temperature will be in temperature categories of 5°F spread.

Blower: A device using a centrifugal fan that provides pressure to move air (active action).

Boiling point: Temperature at which a liquid boils from a liquid to a vapor at a given pressure.

Bore holes: Holes made with a rotary drill rig and bit without a well casing. Bore holes are used for vertical ground loops.

Branch: That portion of the piping system that supplies a portion of the load off the main portion of the system.

Break: Electrical discontinuance of current flow by opening the circuit.

Bridging: The arch over of dirt that allows an air space under the earth's surface. Bridging will reduce the capacity of the buried loop.

Brine: A solution of salt and water to prevent the freezing of the water below 32°F.

Brine charging valve: A valve in the piping system, usually at the low point of the unit, used to insert antifreeze (brine) solution into the system.

British Thermal Unit: The quantity of heat energy needed to change the temperature of one pound of water one degree Fahrenheit.

Cable: A combination of insulated wires in a common sheath.

Calculator, gas service: A slide rule calculator used to determine the orifice size for a vapor burning heating unit.

Cantilevered mount: The unit is mounted on a platform fastened to a vertical surface and braced by angle pieces.

Capacitor: Two electrode plates separated by insulating material. The application of voltage across the plates builds an electrical energy quantity in the device. The time needed to produce the buildup results in a delay in the current buildup in the circuit of any connected load.

Capacity, net cooling: The cooling capacity of an air conditioning unit or heat pump in the cooling mode. This is made up of the sensible and latent heat picked up by the evaporator.

Capillary tube: A fixed restriction-type pressure reducing device. Usually consists of lengths of small inside diameter tubing. The flow restriction produces the necessary reduction in pressure and boiling point of the liquid refrigerant before entering the evaporator.

Celsius: The German language word for Centigrade used in the metric system temperature scale.

Celsius (Centigrade) scale: Temperature scale used in the metric system. Water freezes at 0°C and boils at 100°C.

Chain trencher: A piece of portable equipment that digs a trench in the earth by means of a series of shovel/bucket combinations mounted on a continuous travel chain.

Charge: The amount of refrigerant in the system.

Charging: The process of putting the refrigerant charge into the system.

Charging the loop: Filling the earth loop with the correct mixture and purging all the air from the loop.

Check lite: The light on the thermostat that indicates the system has been shut down by the lock-out control.

Check valve: A flow control valve that permits flow in one direction only.

Circuit: A tubing, piping, or electrical wire installation that permits flow from an energy source to a potential usage (load) and return to the energy source.

Circulator: A motor driven device used to circulate liquid in a closed loop system. Also called a pump or circulating pump.

Close header: See Header, close.

Co-axial fitting: A fitting that connects to a single port of a hot water storage tank. A tube in a fitting. The hot water through the tube into the tank. The cold water out of the fitting surrounding the tube.

Co-axial thermostat assembly: A co-axial fitting that also contains a remote bulb thermostat as an integral part of the assembly.

Code installation: A heat pump installation that conforms to the local, state, and national codes for a safe installation.

Coefficient of conductivity: The measure of the rate at which different materials conduct heat energy as compared to a base standard material (copper).

Coil, de-ice subcooler: A section in the outdoor evaporator coil used to increase the liquid subcooling during the cooling mode as well as act as an extra defrost surface during the defrost mode.

Coil, inside: The coil located in the inside portion of the heat pump system. Performs as the evaporator in the cooling mode and the condenser in the heating mode.

Coil, outside: The coil located in the outside portion of an air-to-air heat pump system. Performs as an evaporator in the heating mode and as the condenser in the cooling mode.

Comfort chart: Chart used in the heat pump cooling mode to show the dry bulb temperature and humidity conditions for human conditions.

Comfort zone: The area on a psychrometric chart that shows the conditions of temperature and humidity in which most people are comfortable.

Compound gauge: See Gauge, compound.

Compressor: The pump of a refrigeration system that raises the pressure and condensing temperature of the refrigerant vapor to a required pressure and condensing temperature.

Compressor, hermetic: A motor-compressor assembly in which the motor and compressor are one assembly inside the welded together housing.

Compression gauge: See Gauge, high pressure.

Compression tank: See Expansion tank.

Condensate: Water that condenses out of air passing over an evaporator coil.

Condensate drain trap: A pipe arrangement that provides a water seal in the drain line to prevent air flow through the drain line.

Condensate pan: A pan located under an evaporator so as to catch condensate off the coil and carry it to the drain line.

Condensation: Liquid that forms when a vapor is cooled below its condensing temperature or dew point.

Condense: The action of changing a saturated vapor to a saturated liquid.

Condenser: The part of the refrigeration system that receives the high temperature, high pressure vapor from the compressor and extracts the heat in the vapor reducing it to a high pressure, medium temperature liquid.

Condenser, air-cooled: A heat exchanger that transfers heat energy from the refrigerant vapor to air.

Condenser fan: A device used to force air through an air-cooled condenser.

Condenser, liquid-cooled: A heat exchanger that transfers heat energy from the refrigerant vapor to the liquid heat sink.

Condenser water pump: A device used to force the liquid through a liquid-cooled condenser.

Condensing temperature: The temperature at which a vapor changes to a liquid at a given pressure.

Condensing unit: The portion of the refrigeration system that converts low pressure, low temperature vapor to high pressure, medium temperature liquid. Commonly called the "high side."

Conductance, surface film: The time rate of heat flow per unit of surface area (per sq. ft.) per degree temperature difference between the ambient air or liquid and the material surface.

Conductance, thermal: The process of transferring heat energy through a material by affecting the activity of the molecules in the material without affecting the material itself.

Cone of depression: The drop in water level that surrounds a well casing caused by the well drawing water from the aquifer. The point of the cone is the pumping water level.

Cone of impression: The mounding of the water surrounding the well casing when water is forced into the aquifer.

Connected load: The total of the load or loads connected to a unit or power source.

Cooling coil: See Evaporator.

Cooling mode: The operating phase of a system that removes heat and/or moisture from a conditioned area.

Cooling run factor: The percentage of time the unit is expected to operate to handle the cooling load during the warmest month of the year (August).

Consolidated formation: The materials below the surface of the earth that contain little water, such as granite, limestone, sandstone, shale, etc.

Contactor: A device with a 20 ampere or higher rating that controls an electrical circuit by means of electromagnetically operated contacts. No overload cut-outs are included.

Control: Automatic or manual device used to start, stop, and/or regulate the flow of liquid, vapor, and/or electricity.

Control, compressor: See Motor control.

Control, defrost: A control used to automatically defrost the evaporator when needed and to keep the system in the defrost mode until the defrost process is completed.

Control, low pressure: A pressure operated control connected to the suction side of the refrigeration system to prevent unit operation below a set pressure or coil operating boiling point.

Control, motor: A temperature or pressure operated device to control the operation of a motor.

Control, refrigerant: A device used to provide the necessary pressure reduction of the liquid refrigerant to obtain the proper boiling point of the refrigerant in the evaporator.

Control, temperature: A device that uses changes in temperature to operate contacts in an electrical circuit.

Convection: Transfer of heat energy by means of a flow of liquid or vapor.

Convection, active: Transfer of heat energy by forced movement of a liquid or vapor.

Convection, passive: Circulation of a liquid or vapor due to differences in the weight of the liquid or vapor due to differences in temperature.

Corrosion: The destruction of metals by the compounds dissolved in water.

Counterflow: Two liquids and/or vapors flowing in a direction opposite of each other.

Crankcase heater: A heating device fastened to the "crankcase" or lower portion of the compressor housing intended to keep the oil in the compressor at a higher temperature than the rest of the system to reduce the migration of the refrigerant.

Cut-in: Switch or contact action to close and cause current flow through a circuit.

Cut-in control: Temperature or pressure actuated device that closes a control circuit at a predetermined set point.

Cut-out: Switch or contact action to open and stop current flow through a circuit.

Cut-out control: Temperature or pressure actuated device that opens a control circuit at a predetermined set point.

Cylinder, refrigerant: Cylinder in which refrigerant is purchased and from which refrigerant is dispensed. Color coded according to the type of refrigerant contained.

Daily range: The range of outdoor ambient temperatures encountered in an area over a 24-hour period. The temperatures are taken at the design high and average low temperature for the area.

Daily range, high: A range of outdoor ambient design temperatures both high and low that exceeds 25°F. A low humidity area.

Daily range, low: A range of outdoor ambient design temperatures both high and low that is less than 15°F. A high humidity area.

Daily range, medium: A range of outdoor ambient design temperatures both high and low that is in the 15°F to 25°F range. A medium humidity area.

Dedicated geothermal well: Water is drawn from the top of the well and is returned to the bottom of the well. The well supplies only the heat pump.

Defrost control: A control system used to detect frost and/or ice build up on the outside coil of an air-to-air heat pump during the heating mode and causes the system to reverse to remove the frost and/or ice from the coil.

Defrost mode: The portion of the operation of the system in which the evaporator frost and/or ice is removed.

Defrost timer: A device connected into the electrical control system that controls the frequency and duration of the defrost operating mode.

Dehydrate: To remove water in all forms from a material or system.

Dehydrator: See Drier.

Dehumidify: To remove water vapor from the atmosphere.

Density: Closeness of texture or consistency.

Desicant: Material used to collect and hold moisture in refrigeration systems. A drying agent such as Silica Gel or Activated Aluminum.

Design temperature difference: The difference between the design indoor and outdoor temperatures.

De-superheater: A heat exchanger in the hot gas line between the compressor discharge and reversing valve. Used to heat domestic water by removing the superheat from the hot refrigerant vapor.

Design water temperature difference: The difference between the temperature of the water leaving the unit and entering the unit when the system is operating at design conditions.

Detector, leak: A test instrument used to detect and locate refrigerant leaks.

Dew point: Temperature at which water vapor begins to condense out of air. Air at the dew point—100% relative humidity—is referred to as "Saturated Vapor."

Diaphragm: A flexible membrane usually made of a thin metal, plastic, or rubber. Pressure tanks in heat pump liquid systems must contain a rubber diaphragm.

Dichlorodifluoromethane (R–12): A refrigerant having the designation of R–12. The chemical formulae is CCL_2F_2. Cylinder color code is white. Boiling point at 29.97 in. Hg. Standard atmospheric pressure is –21.6°F.

Differential: As applied to controls, the differential between the cut-in and cut-out temperature or pressure set points of a control.

Diffusivity: The ability of earth material to take on or give off heat energy depending on its density and moisture content.

Dill Fitting: See Schroeder valve.

Direct expansion evaporator: An evaporator coil using a pressure reducing device to supply liquid refrigerant at the correct pressure and boiling point to obtain the desired rate of heat absorption into the liquid refrigerant.

Disturbed earth effect: Denotes the heat content change that affects the earth temperature by the addition to or removal from of heat energy by the heat pump system.

Domestic geothermal well: Water is drawn from the top of the well. Only the water from the heat pump is returned to the bottom of the well. The difference is used for building requirements.

Draft: A current of air. Usually refers to the pressure difference that causes movement of air or gases through a flue, chimney, vent, or space.

Draft gauge: An instrument used to measure pressure differences that produce drafts.

Draft indicator: An instrument used to measure draft in vents or chimneys. Draft is measured in 0.01 in. water column.

Drain cock: A valve installed in the piping system at the lowest point of the tank or unit to promote draining the tank.

Draw down: The difference between the static water level and the pumping water level in a well.

Draw down, pressure tank: The amount of liquid the tank will deliver per minute within a specified pressure range before the pump starts.

Drier: A substance or device used to hold moisture from circulating through a refrigeration system.

Drilled supply well: Drilled wells are small diameter holes bored into the earth by a screw-type shaft attached to a drilling rig, using a stream of water or drilling mud to flush out the material loosened by the drill. The hole is bored to find water up to 500 ft. if necessary.

Drip pan: See Condensate pan.

Dry bulb temperature: The actual (physical) temperature of a substance. Usually refers to air.

Dry bulb thermometer: An instrument with dry sensing element that measures the physical temperature of a substance.

Dry well disposal: Disposing of return water into a gravel-filled large hole in the ground to return the water directly to the aquifer.

Duct: Round or rectangular sheet metal or fiberglass pipe that carries the air between the conditioning unit and the conditioned area.

Earth aquifer: The water quantity below the surface of the earth.

Earth, disturbed effect: See Disturbed earth effect.

Earth temperature, maximum (T_{max}): The highest temperature the earth will reach in the summer season. Will depend on location and weather conditions.

Earth temperature, minimum (T_{min}): The lowest temperature the earth will become in the winter season. Will vary with location and weather conditions.

Effective area: The actual opening in a grille or register through which air can pass. The effective area is the gross (overall) area of the grill face minus the area of the deflector vanes or bars.

Effective temperature: The combination of the overall effect on a human being by the temperature, humidity, and movement of air.

Effective temperature difference: The difference between the room air temperature and the supply air temperature as it enters the room.

Electric circuit: The complete path for electric current flow from a power source to a connected load through any control devices and return to the power source.

Electric heating: A heating system that uses electric resistance elements as the heat energy source.

Electric heating element: A unit consisting of resistance wire, insulated supports, and connection terminals for connecting the resistance wire to a source of electrical power.

Electric water valve: An electrically operated valve used to control the flow of a liquid.

Electromagnet: A coil of wire that creates a magnetic field when carrying an electrical current.

Electronic leak detector: An electronic instrument that measures the changes in electron flow across an electrically charged gap that indicates the presence of refrigerant vapor molecules.

Electronics: Field of science dealing with electronic devices and their applications.

Encrustation: The buildup of slimy orange-brown deposits on the water side of pipes caused by iron bacteria in the water.

Energy efficiency ratio (EER): The comparison of the heat transfer ability of a refrigeration system and the electrical energy used as expressed in Btuh per watt.

Evaporation: A term used to describe the changing of a liquid to a vapor by the addition of heat energy.

Evaporative cooling tower: A device that dissipates heat energy by the evaporation of water. The water is recirculated to cool refrigeration condensers.

Evaporator coil: A piece of equipment made of tubing (liquid) or tubing and fin (air) used to transfer heat energy from the material into refrigerant.

Evacuate: See Dehydrate.

Exfiltration: Outward flow of air from an occupied area through openings in the building structure.

Exhaust opening: An opening through which air is removed from an area.

Expansion tank, air cushion: A closed tank connected into a water circulating system in such a manner that when the system is filled with liquid, air is trapped above the liquid in the tank. Air and liquid are in direct contact. Because the liquid will absorb the air when under pressure, this type of tank is not recommended for use in closed loop systems.

Expansion tank, bladder-type: A bladder-type expansion tank has a rubber bladder that separates the air from the liquid in the tank. No absorption of the air into the liquid. This is the only expansion tank that should be used in closed loop or supply-type, liquid-type heat source/sink systems.

External equalizer: A pressure connection from the area below the diaphragm of a T.X. valve and the suction outlet of an evaporator.

Fahrenheit scale: On a Fahrenheit thermometer, under standard atmospheric pressure of 29.97 in. Hg., the boiling point of water is +212°F and the fusion point of water is +32°F.

False defrost: The system goes into the defrost mode even if the outdoor has no frost and/or ice build up.

Fan: A radial or axial flow device using blades in a venture for moving air.

Fan, centrifugal: A fan rotor or wheel within a blower scroll that includes either a direct drive motor or a pulley and belt combination.

Fan, propeller: A blade-type wheel within a mounting ring (venture) with a direct drive motor or pulley and belt combination.

Feet of heat (FT HD): The term used to designate the flow resistance the pump must overcome to deliver the required amount of liquid. 2.307 FT HD = 1 psig.

Feet of water (FT H_2O): See Feet of heat (FT HD).

Field tile disposal: Disposal of return water through a grid of perforated tile buried in the earth.

Filter: A device for removing foreign particles from vapor or liquid.

Fire efficiency finder/stack loss rule: A slide rule that uses the CO_2 percentage and temperature of a vapor to determine the efficiency of a fossil fuel heating unit.

Flash gas: Vapor that is produced by the boiling of liquid refrigerant to reduce the temperature of the liquid refrigerant from the condensing temperature minus the subcooling to the boiling temperature of the evaporator.

Flow hood: A direct reading instrument that measures air quantity (CFM) out of or into a duct system and directing the air over a sensing element.

Flow meter: Instrument used to measure the volume of a liquid flowing through a pipe or tube in quantity per time (gallons per minute, pounds per hour, liters per minute, etc.).

Flow switch: See Switch, flow.

Fluid: Substance in a liquid or vapor state that moves and changes position without separation of the material.

Fluid flow: The movement of a fluid by a mechanically created pressure difference or a difference in fluid density created by a temperature difference.

Flushing the loop: Forcing liquid through the loop at sufficient velocity to remove all forms of foreign material and air.

Foot of water: A measure of pressure. One foot of water is the pressure created by a column of water 1 ft. high. This is equal to 0.433 psig.

Forced convection: Movement of a liquid or vapor by a mechanical means.

Free area: The total area of the opening in a grill or register through which air passes.

Freeze up: Frost and/or ice formation on an evaporator that restricts air or liquid flow through the evaporator.

Freezing: Change in state of a substance from a liquid to a solid.

Freezing point: The temperature at which a liquid becomes a solid upon removal of latent heat. At standard atmospheric conditions (29.97 in. Hg.) water freezes at +32°F/0°C.

Freon: Trade name for refrigerants manufactured by the DuPont Co.

Friction head: In a liquid flow system, the friction head is the loss in pressure due to the flow resistance created by the liquid moving through the pipe.

Friction loss in pipe: See Pipe, friction loss.

Frost back: A condition in which liquid refrigerant boils in the suction line and produces a frosted surface on the line.

Frost control, automatic: A control that automatically reverses a refrigeration system to remove frost/ice from the evaporator.

Frost line: The depth to which the ground freezes over a winter season. The "frost line" is usually set at 12 in. below the deepest recorded frost penetration.

Fuse: An element designed to melt at a predetermined current flow value.

Fuse, instant break: A thermal fuse with an element that melts and opens the circuit immediately if the current flow through the fuse exceeds the rating.

Fuse, time delay: A thermal fuse that has two elements. One will open immediately on a high surge of current due to a short circuit. The other element will carry up to 300% of the fuse rating for 5 seconds. This delay is enough to start a motor and still have full load ampere protection.

Fusing: The joining of plastic by heating both pipe and/or fitting ends until pliable and then pressing together to make a tight joint.

Fusion, butt-type: Pipe ends are heated to the proper melting point and then joined together until cool. No couplings are used.

Fusion, socket-type: The pipe is inserted into the fitting. Prior to insertion, the pipe and fitting are heated to the proper temperature to the plastic stage. The pipe is inserted and allowed to cool. Fittings are used on all joints.

Fyrite indicator, CO_2: An instrument used to measure the percentage of CO_2 in a vapor sample.

Fyrite indicator, O_2: An instrument used to determine the percentage of Oxygen (O_2) in a vapor sample.

Galvanic action: The destruction of one metal by the generation of electrical energy when two different metals are joined.

Gas: A substance in the vapor state.

Gas supply meter: A device that measures and records on indicator dials the amount of gas in cubic feet that flows through the meter.

Gas valve: A device that controls the flow of gas or vapor.

Gate valve: A valve designed to have the valve control the flow of liquid by means of a wedge or gate between two sealing surfaces. The flow of the liquid is straight through the valve body.

Gauge, compound: A pressure indicating instrument that measures pressures both above and below atmospheric pressure.

Gauge, draft: A vacuum gauge used to determine the negative pressure of a heating unit vent.

Gauge, gas manifold pressure: A direct reading gauge for measuring the gas manifold pressure of a vapor burning heating unit. A more convenient substitute for a "U" tube manometer. Measures in increments of inches of water.

Gauge, high pressure: An instrument used to measure pressures above atmospheric pressure in the 0 to 500 psig range.

Gauge, low pressure: An instrument used to measure pressures above atmospheric up to 50 psig.

Gauge, manifold: A device that contains a combination of gauges and control valves to control the flow of vapors through the device.

Gauge, standard: An instrument designed to measure pressures above atmospheric.

Gauge, vacuum: An instrument used to measure pressures below atmospheric.

Geothermal well: A drilled well in which the return water is returned to the same well from which the supply water is taken.

Grade level: The top surface of the earth.

Gravity, specific: The specific gravity of a solid or liquid is the ratio of the mass of the material to the mass of an equal volume of water at standard temperature (65°F). The specific gravity of a vapor is usually expressed in terms of dry air at the same temperature of the vapor.

Grill: An ornamental or louvered opening through which air leaves a duct system or area.

Gross capacity, heating: The gross capacity of a heat pump in the heating mode is the total amount of heat energy transferred to the inside air or material. This is made up of the heat energy picked up from a heat source (air or liquid) plus the heat energy equivalent of the electrical energy required to operate the heat pump.

Ground coil (earth coil): A heat exchanger coil buried in the earth that is used to either remove heat energy from the earth or add heat energy to the earth.

Ground loop; double "U" bend, vertical: A multiple pipe system in vertical holes, connected to divide the liquid flow among the loops.

Ground loop; four layer, horizontal: Two two-layer pipe loops located in the same trench, one above the other.

Ground loop; single layer, horizontal: One continuous length of pipe laid horizontally in the earth.

Ground loop; single "U" bend, vertical: A single run of pipe in one or more vertical holes, connected in series where all the liquid travels through all portions of the loop.

Ground loop; two layer, horizontal: A single loop that doubles back on itself to reduce the earth area needed.

Halide refrigerants: Refrigerants that contain Halogen chemicals in their molecular structure (chlorine, fluorine).

Halide torch: A device that produces a reaction in a flame when it detects Halogen chemicals in the combustion air.

Head: The term *head* refers to a pressure difference across a pump. See pressure head, pump head, available head.

Head pressure: The pressure in the condensing or high side of the refrigeration system.

Head, pressure control: A pressure operated control that opens an electrical circuit if the pressure exceeds the cut-out point of the control.

Head, static: Pressure of a liquid or vapor in terms of the height of a column of a liquid or mercury.

Head, velocity: In a flowing liquid or vapor, the height of a liquid column, either water or mercury, equal to the velocity pressure of the liquid or vapor.

Header: A piping arrangement for interconnecting two or more supply or return lines into a common line.

Header, close: Supply and return headers for multiple loop vertical systems are spaced close together with the multiple loops arranged in a circle around the headers.

Header, reverse return: With the vertical loops arranged in a straight line, the headers between are connected to supply and return from opposite ends. Supply is to the closest loop, return is from the distant vertical loop.

Heat: The energy that affects the molecular activity of a substance. This is reflected in the temperature of the substance. Addition of heat energy increases the molecular activity and temperature. Removal of heat energy lowers the molecular activity and temperature.

Heat, latent: Heat energy used to change the state of a substance; solid to liquid and liquid to vapor by addition, or vapor to liquid and liquid to solid by removal without a change in the temperature of the material at the time of change.

Heat, sensible: The heat energy used to change the temperature of a material—solid, liquid, or vapor—without a change in state.

Heat exchanger: A device used to transfer heat energy from a higher temperature to a lower temperature. Evaporators, condensers, and earth coils are examples.

Heating coil: A heat transfer device designed to add heat to liquid or vapor.

Heating control: A device that controls the operation of a heating unit to maintain a set temperature in a material or vapor.

Heating mode: The operating phase of a system that is adding heat energy to an occupied area.

Heating run factor: The percentage of the time the unit is expected to operate during the coldest month of the year (January).

Heating unit (electrical): A device containing one or more electrical resistance elements, electrical connections, safety and control devices in a frame or casing.

Heat load: The amount of heat energy, measured in Btuh, that is required to maintain a given temperature in a conditioned area at design conditions both inside and outside the conditioned area.

Heat motor: An electrical device that produces motion by means of the temperature change of bimetal elements. Used to produce time delay in the operation of the device.

Heat of fusion: The heat energy needed to accomplish the change in state of a material between a liquid and solid; addition for solid to liquid, removal for liquid to solid.

Heat pump: A name given to an air conditioning system that is capable of either heating or cooling an area on demand.

Heat pump, air-to-air: A device that transfers heat between two different air quantities in either direction on demand.

Heat pump, air-to-liquid: A device that transfers heat from an air source to a liquid by means of a refrigeration system. Units in this category are nonreversible.

Heat pump, liquid-to-air: A device that transfers heat between liquid and air in either direction on demand.

Heat pump, liquid-to-liquid: A device that transfers heat between two different liquids in either direction on demand.

Heat pump, water heater: An air-to-liquid refrigeration system used to heat domestic water with air as the heat source.

Heat pump; water heater, remote: The refrigeration system is a complete unit designed to be connected to an existing water tank.

Heat pump; water heater, self-contained: The component parts of the refrigeration system (condenser) is an integral part of the hot water tank.

Heat recovery unit: See De-superheater.

Heat sink: A place or material into which heat energy is placed.

Heat source: A place or material from which heat energy is obtained.

Heat, specific: The heat required to change the temperature of a quantity of a material. Heat quantity per pound of material per 1°F or heat quantity per gram per 1°C.

Heat, total: The sum of both the sensible and latent heat energy in air. Expressed as Btu/pound of air.

Heat transfer: Movement of heat energy from one body or substance to another. Heat may be transferred by any combination of or all of the three methods (radiation, conduction, and convection).

Heating value: The amount of heat energy released by the burning of a fuel or operation of an electric element. It is usually expressed in Btu/Kw for electricity, Btu/cu. ft. for vapor and Btu/pound for solid fuels.

Hermetic motor: A motor completely sealed in a welded case.

Hermetic system: A refrigeration system that uses the hermetic-type motor-compressor assembly in a totally welded together system.

High pressure cutout: An electrical pressure control switch operated by the high side pressure set to stop compressor action operation at maximum pressure safety limits.

High side: See Condensing unit.

High vacuum pump: A pump designed to create a vacuum on the intake side of less than 1,000 micron pressure.

Hole, wetted: The portion of the well casing or bored hole below the static water level in the aquifer.

Horsepower: A unit of power equal to 33,000 foot pounds of work. One electrical horsepower equals 746 watts.

Hot gas defrost: A defrosting system in which hot refrigerant gas from the compressor is directed through the evaporator to remove frost and/or ice from the evaporator.

Hot gas line: The refrigerant tube that carries the high pressure, high temperature refrigerant vapor from the compressor to the condenser.

Hot tub: An oversize container of water not over 105°F for the pleasure or treatment of people. Average size hot tub will hold four people. Also called a spa.

Humidity: Moisture in air.

Humidistat: A control that is affected by changes in the relative humidity in air. Used to control the operation of a humidifier.

Hydrometer: A floating instrument used to measure the specific gravity of a liquid.

Hygrometer: An instrument used to measure the percentage of moisture (relative humidity) in air.

Ice ring: An accumulation of ice at the bottom of the outdoor coil due to incomplete defrost operation.

Impeller: The rotating part of a centrifugal-type liquid pump.

Inches of mercury column: A unit of pressure measurement. One inch of mercury column is equal to a pressure of 0.491 psig.

Inches of water column: A unit of pressure measurement. One inch of water column is equal to a pressure of 0.578 ounces per sq. in. or 0.04817 psig.

Ions: Electrically charged atoms suspended in water that can cause electrolysis action and corrosion of metal.

Joints, brazed: A solder-type pipe connection or joint obtained by the joining of the metal parts with metallic mixtures or alloys that have a melting temperature above 1,000°F up to 1,500°F.

Joint, soldered: A solder-type connection or joint obtained by the joining of the metal parts with metallic mixtures or alloys that have a melting temperature below 1,000°F.

Joint, welded: A solder-type pipe connection of joint obtained by the joining of the metal parts with metallic mixtures or alloys that have a melting temperature above 1,500°F.

Junction: A place in an electrical circuit where two or more wires or conductors are joined together.

Junction box: A metal or plastic box into which wires and/or cables are inserted and joined together. Required for safety and protection.

Kilowatt: A unit of electrical energy equal to 1,000 watts.

Leak detector: A device or instrument used to detect and locate leaks of gases or vapors.

Lift: The work the pump must do to overcome the gravity pull on the water to push the water uphill to the level of the pressure tank. The weight of the water.

Limit control: A device used to open or close electrical circuits if temperature or pressure reaches preset limits.

Limit line: The refrigerant pipe or tube that carries high pressure, medium temperature refrigerant liquid from the condenser or receiver outlet to the pressure reducing device inlet.

Litigation: The act or process of carrying on a law suit.

Load: The amount of heat per hour that the refrigeration system is required to supply at design conditions. The amount of electrical energy expected to be connected to an electrical power supply.

Low pressure control: A device used to start or stop the system operation when the low pressure (low side pressure, suction pressure) drops to preset limits.

Low side: The portion of the refrigeration system that operates at the evaporator pressure. Consists of the pressure reducing device, evaporator, and suction line.

Low side pressure: See Back pressure.

Main: The pipe used to carry liquid between the unit and buried earth loop headers.

Makeup air (fresh air): The air supplied to a building to replace air exhausted from the building.

Makeup water line: The line connected to the loop system for filling or adding liquid as necessary.

Manifold: A piece of pipe or tube that has two or more branch lines connected.

Manifold, service: See Gauge manifold.

Manometer: An instrument used to measure the pressure of gases or vapors. The gas pressure is balanced against a column of liquid, such as water or mercury, in a U-shaped tube open to atmospheric pressure.

Manometer, inclined: A manometer on which the liquid tube is inclined from horizontal to produce wider, more accurate readings over a smaller range of readings.

Mean annual earth temperature: The mean average temperature of the earth as it changes throughout the year. The largest factor for earth temperature change is sunshine.

Melting point: The temperature, at a given pressure, at which a solid becomes a liquid.

Meter: A metric unit of length or distance equal to 39.37 inches. An instrument used to measure.

Meter, flow: An instrument used to indicate the rate of a liquid flow in the system. The instrument is calibrated for the specific application.

Metric system: A decimal system of measures and weights based on the meter and gram in decimals.

Micron: A metric unit. For length it is one millionth (1/1,000,000) of a meter. For pressure it is one 254,000th of an inch of mercury (1/254,000 of 1 in. Hg.)

Micron gauge: An instrument used to measure pressure below atmospheric (vacuum) close to perfect vacuum.

Milli-: A term used to denote one one-thousandth (1/1,000) of a unit. For example, a milli-ampere is 1/1,000 of an ampere.

Module: A combination of components in a circuit that are combined in a replaceable package. Replacement is by the module in its entirety.

Monel: A trademark name for a metal alloy consisting primarily of copper and nickel.

Monochlorodifluoromethane: A refrigerant designated by the R–12 symbol. The chemical formulae is $CHCLF_2$. The cylinder color is white. It will be a discontinued refrigerant because of its effect on the earth's ozone layer.

Monoxor: An instrument used to measure the concentration in parts per million of carbon monoxide (CO) in a vapor sample.

Motor burnout: A condition in a motor where the insulation has deteriorated due to overheating.

Motor, capacitor-type: A single phase induction motor that uses both running and starting windings in the running mode. The starting winding is connected in series with a capacitor to change the electrical characteristics of the winding.

Motor control: A control to start and stop a motor at preset pressures or temperatures.

Motor, two pole: A motor with a synchronous speed of 3,600 RPM. Full load speed of 3,450 RPM.

Motor, four pole: A motor with a synchronous speed of 1,800 RPM. Full load speed of 1,725 RPM.

Motor, six pole: A motor with a synchronous speed of 1,200 RPM. Full load speed of 1,150 RPM.

Motor starter: A high capacity electrical contact operated by an electromagnetic coil and containing properly sized overload cutouts.

Natural air change rate: Is determined by the amount of cracks and openings in the building, the temperature difference between the building and the outside ambient, and the wind velocity.

Natural convection (passive): Movement of a vapor or liquid caused by differences in weight of the material as a result of the difference in temperatures of the material.

Neoprene: A synthetic rubber that is resistant to oils and vapors.

Nitrogen (N_2): An inert gas used in portions of tubing and/or parts that are subject to high heat during the assembly process. The nitrogen replaces the oxygen in the air in the tube or part and reduces the formation of scale (copper oxide).

Nominal size tubing: Tube measurement that has an outside diameter the same as the inside diameter of iron pipe of the same given size.

Noncombustible base: A base made of metal with adequate ventilation to reduce heat transfer from the unit to combustible floor materials.

Null: Zero.

Occupied area: The area that is conditioned by the heat pump, air conditioning, or heating system.

Off mode (cycle): That part of the operating mode when the system has been shut down by the controls.

Ohm's law: A fundamental factor of electrical circuits that states that "the current flow (amperage) through a circuit is equal to the electrical pressure (voltage) divided by the flow resistance (ohms) of the circuit."

Open circuit: An interrupted electrical circuit that prevents electrical flow.

Orangeburg: A trade name for polyethylene plastic pipe manufactured by Orangeburg Industries, Inc. of Ashville, NC.

Orifice: An accurately sized opening that controls the flow of vapor or liquid at given pressures across the opening.

Orifice, oil metering: A small hole in the pick up tube in the accumulator used to insure oil return to the compressor in small easily handled quantities.

Outside air: Atmosphere exterior to the conditioned area.

Overload: Load greater than the design load of the device.

Overload protector: A device—either electrical, temperature, or pressure operated—that will open and stop operation if preset conditions are exceeded.

Oxidizing: The deterioration of a metal due to oxygen penetration into the metal. With iron as a material, the process is called "rusting."

Oxy-Acetylene: A type of torch that produces a flame by the combustion of Acytelene and Oxygen.

Pilot valve: See Valve, pilot.

Pinch-off tool: A tool used to press the walls of tubing together to stop the flow of material through the tube.

Pipe, friction loss: The flow resistance of liquid through the pipe. Dependent on pipe size and flow quantity. Expressed in feet of head (FT HD) per 100 ft. of pipe.

Pitch: Pipe or tube slope in the direction of flow that causes gravity to enhance the flow of material through the pipe or tube.

Pilot tube: A tube, used in conjunction with a manometer, to measure air velocities and pressures.

Plenum: A cube-shaped duct or box on the supply or return side of an air handling unit to connect the supply or return duct system to the unit.

Point of vaporization: The location of the position in the evaporator coils where the last bit of liquid refrigerant is vaporized.

Pole: The part of a magnet where magnetic flux is concentrated or where they leave the magnet.

Polybutylene: A plastic used for pipe material that has excellent creep resistance as well as high resistance to stress cracking. Recommended for earth loops.

Polyethylene: A plastic used for tubing for ground loop systems, cold water lines, and heat pump piping.

Polyphase motor: An electric motor constructed to be used with three-phase electrical power.

Potable water: Water that is suitable for human consumption. See Domestic water.

Potential: The amount of voltage or electrical pressure between two points of an electrical circuit.

Power: Time rate at which work is done or energy consumed. Source or means of supplying energy.

Power element: Temperature sensitive element that is used to operate circuit control devices.

Pressure: An energy impact on a unit area. Force or thrust against a surface.

Pressure, back: The pressure the pump must overcome to be able to force the liquid into the pressure tank. The pressure in the tank.

Pressure drop: The pressure difference between two locations in a circuit. The difference being a result of flow resistance in the circuit.

Pressure head: See head.

Pressure reducing device: The device used to reduce the pressure on the liquid refrigerant and thus the boiling point before entering the evaporator.

Pressure, static: The pressure in the system when the pump is idle.

Pressure, suction: The pressure in the low pressure or evaporator section of the refrigeration system.

Pressure tank draw down: See Draw down, pressure tank.

Pressure, water valve: Device used to control water flow according to the connected pressure.

Primary voltage: The voltage of the power circuit supplying power to the input side of a transformer.

Protector, circuit: An electrical device that will open an electrical circuit if the amperage flow exceeds the rating of the device. See Circuit breaker.

Psychrometer: An instrument used to measure the dry bulb and wet bulb temperatures of air.

Psychrometric chart: A chart that graphs the relationship of the temperature, pressure, and moisture content of air.

Psychrometric measurement: Using a psychrometer to measure the dry bulb and wet bulb temperatures of air and a psychrometric chart to determine the characteristics of the air.

P/T plugs: Pressure/temperature plugs that allow entrance into the system by the gauge stems of thermometers and pressure gauges without draining or removing the pressure of the system.

Pull down: An expression indicating the action of removing refrigerant from any or all parts of a refrigeration system.

Pull-in voltage: The voltage or electrical pressure that creates enough current flow to produce the amount of magnetic force needed to cause the relay armature to seat on the pole face.

Pump: See Circulator.

Pump head: The system resistance against which the pump must deliver liquid.

Pumping water level: The level to which the water in a well drops when the pump is removing the full load water quantity.

Pump, jet: A pump system that uses high pressure water from the pump through a jet venturi to pick up additional water for the supply system. The pump handles three to five times as much water as is supplied by the system.

Pump, short cycling: The capacity of the pump is too large for the system, causing very short on and off periods.

Pump, submersible: A pump that is located below the level of water in the well. Uses multistage impellers to produce the pressure to lift water up the pipe. The motor is cooled by the flow of water through the pump.

Pump, suction: A pump that depends on negative pressure in the inlet to lift water from the earth. Actually, the pressure difference between the atmospheric pressure and the pump inlet pressure forces water up the well casing. The maximum practical operating height is 15 ft.

Purging: Releasing compressed gas into the atmosphere through some ports or parts of a refrigeration system to remove excess refrigerant or contaminants.

Quick connect coupling: A device that permits an easy fast means of connecting two fluid lines or fittings together without the use of solders.

R–12, Dichlorodifluoromethane: A synthetic chemical refrigerant suspected of causing breakdown of the earth's ozone layer. Manufacture of the refrigerant is expected to stop.

R–22, Monochlorodifluoromethane: A synthetic chemical refrigerant. At standard atmospheric pressure, the boiling point is –41°F.

R–502: An aziotropic mixture of R–22 and R–115. At standard atmospheric pressure, the boiling point is –50°F.

Radiation: Transfer of heat energy by heat rays.

Range: Temperature or pressure settings of a control. Limits of operating pressure or temperature.

Reclaim: To reprocess refrigerant to new product specifications by means that may include distillation.

Recovery: To remove refrigerant in any condition from a system and store it in an external container without testing or processing it in any way.

Recycling: To clean refrigerant for reuse by oil separation and single or multiple passes through devices that reduce moisture, acidity, and particulate matter.

Refrigerant: A liquid material used in a refrigeration system to absorb heat in the vaporization (liquid to vapor) process and reject heat in the condensing (vapor to liquid) process.

Refrigerant charge: The designed quantity of refrigerant in the system.

Refrigerant migration: The transfer of refrigerant from a high temperature location in the system to a low temperature location in the system by the pressure difference created by the temperature difference.

Refrigerating effect: The amount of heat energy in Btuh or cal./hr. the system is capable of transferring under the given circumstances.

Refrigeration: The process of transferring heat energy from one place to another by the change in state of a liquid refrigerant.

Refrigeration system: A system composed of parts necessary to produce the boiling point and condensing temperatures necessary to result in heat transfer by the change in state of a liquid.

Register: A device that combines a grill and damper assembly to control quantity and direction of air flow.

Reheat: The addition of heat energy into air after it has been cooled.

Reinjecting: Returning the water from a liquid heat source heat pump directly back into the earth aquifer.

Relative humidity: The percentage of moisture in air as compared to the amount of moisture in saturated air at the same pressure and temperature conditions.

Relay: Electrical mechanism that uses an electromagnetic coil of wire at a given voltage and/or current requirement to control an electrical circuit of a different voltage, a higher current flow requirement, or multiple circuits that cannot be connected together.

Relay, lock out: A relay used in conjunction with an automatic reset high pressure control to cause manual reset interruption of the unit operation in the event the head pressure exceeds the control cut-out set point. Reset is by power interruption.

Remote power element control: A device that has the sensing element located a distance away from the operating portion of the control.

Remote (split) system: A refrigeration system that has the high side and low side in different locations, connected together by refrigerant lines.

Reservoir system: A closed geothermal loop system using a reservoir storage system (swimming pool, storage tank, pond, lake, etc.) for the heat energy source.

Return branch: The piping used to return liquid from a loop to the return header or manifold.

Return main: The piping used to return liquid from the ground loop manifold to the unit.

Return tapping: The connection on the unit where the pipe used to return liquid from the loop is connected.

Reverse cycle defrost: A method of defrosting the evaporator by means of flow valve(s) to move hot vapor from the compressor into the evaporator.

Reverse cycle refrigeration system: Commonly called a "heat pump." A refrigeration system capable of reversing its operation and direction of heat transfer.

Reverse return header: See Header, reverse return.

Reversing valve: A device used to change the direction of refrigerant vapor flow between the evaporator to compressor and compressor to condenser depending on the heating or cooling effect desired.

Roof mounted: The unit is mounted on a platform designed to distribute the weight of the unit over as wide an area of the roof as possible.

Run factor: The percentage of the time the unit can be expected to operate to handle the heating load during the coldest month (January) and the cooling load during the warmest month (August).

Run winding: The main electrical winding of an electric motor that provides the major portion of the torque (rotary power) available from the motor.

Safety control: Device that will stop the operation of a device if preset conditions of the control are not maintained.

Sampling assembly: A hand-held rubber squeeze pump used to force samples of flue products through various testing devices.

Sand geothermal well: Water is drawn from the bottom of the well casing and returned to the top of the well casing. Limited in application to where the sand and/or gravel layers holding the water are 75 ft. or more in depth.

Saturation: A condition existing when a substance contains the maximum of another substance for that temperature and pressure. Moisture in air, antifreeze in water, etc.

Scaling: The formation of lime and other deposits on the water side surfaces of heat exchangers.

Schroeder valve: A fitting that has a spring-loaded shut-off core that permits fluid flow when the core is depressed.

Sealed unit: See Hermetic system.

Secondary voltage: The voltage produced by a transformer in the output or secondary winding.

Sensible heat: Heat energy added to or removed from a material that causes a change in temperature of the material without a change in state of the material.

Sensor: A material or device that changes characteristics (electrical or mechanical) with a change in temperature or pressure.

Sequencer: A control device used to control electrical circuits that uses a heat motor for time delay of the operation of the device.

Serviceability: An installation is classified as serviceable when sufficient area is provided for access to control panels, electrical connections, and refrigerant lines. The better the access, the higher the serviceability of the system.

Service conductors: The portion of the electrical service system that is the conductors that carry electrical energy from the utility transformers to the entrance pipe or conduit on the building.

Service drop: That portion of the overhead or buried conductors that carry the electrical energy from the power transformers to the first point of attachment to the building.

Service (electrical): The wires (conductors) and equipment for delivering electrical energy from the power distribution system of the electrical utility to the building using the electrical energy.

Service equipment: The necessary equipment, usually consisting of a circuit breaker or fuse panel and cabinet, located near the point of entrance of the service drop and is the main control and means of electrical supply cut off for the building.

Shell and tube heat exchanger: A device used to transfer heat energy between a liquid and refrigerant by means of a coil of tubing in a cylindrical shell.

Short circuit: A high current flow connection (usually accidental) between two paths of an electrical circuit.

Short cycling: A device or system that cycles on and off more than normal.

Shroud: A housing over an evaporator or condenser. Used to concentrate and improve the performance of propeller-type fans.

Silver brazing (silver soldering): A brazing or soldering process using silver bearing alloy for the fastening process in the operation.

Single circuit system: An earth loop that consists of one continuous pipe.

Site plan: A layout of the building to be conditioned and its location on the ground, as well as all buried utility lines, cables, pipes, and/or conduits.

Skimmer inlets: Openings at the top water level that carry water from a swimming pool to the filter system. Designed to trap any foreign material that may be floating on the surface of the water.

Slab mounted: The unit sets on a level flat slab located high enough above the surface of the earth (grade level) to eliminate snow problems.

Slide valve (heat pumps): See Valve, slide.

Sling psychrometer: An instrument used to measure the dry bulb and wet bulb temperatures of air by producing air movement over the sensing bulbs of the thermometers by rotation of the instrument in the air.

Soil swing curve: The effect of air and solar changes on the soil temperatures from the surface to a depth of 12 ft.

Soil temperature variation: An assumed figure that is a balance between the allowable earth temperature and the depth of the pipe loop location.

Solar heat: Heat from the invisible and visible energy waves from the sun.

Soldering: Joining two metals by the use of a third metal or alloy adhering to the metals to be joined.

Solenoid valve: A flow control valve using a shut-off needle and seat controlled by a moving core under the influence of an electromagnetic coil.

Solenoid valve, slow acting: A solenoid valve whose needle is activated by a heat motor that decreases the speed of the plunger to give slow opening and closing operation.

Spa: See Hot tub.

Specific capacity: Specific capacity of a well is the quantity of water flow in gallons per minute for each foot of draw down.

Specific gravity: The weight of a quantity of a liquid at a given temperature and pressure as compared to a like quantity of water at the same temperature and pressure. The specific gravity of water is 1.000.

Specific heat: The amount of heat energy needed to change the temperature of a material at a given pressure as compared to an equal quantity of water or air at the same pressure.

Specific volume: The volume of a given weight of a substance at a given temperature.

Split: The difference in temperature between the temperature of the air or liquid entering a coil and the coil operating temperature.

Split system: See Remote system.

Spot smoke tester: An instrument used to measure the free carbon content of the flue products from an oil-fired heating unit. The result is obtained by comparison to a standard using spots of carbon or black color of varying density.

Squirrel-cage fan: A fan assembly that has blades equal distance around a parallel axle and moves parallel to the axle.

Stack effect: The difference in pressure in a building due to the difference between the building temperature and the outside ambient temperature.

Standard atmospheric conditions: Used as a basis for calculations and testing. Air is at 14.7 psia (absolute) (29.97 in. Hg.) and at 70°F.

Starting relay: A control device that senses the amperage draw or voltage across the start winding to connect or disconnect the starting function of the motor.

Starting winding: The winding in an electrical motor used to aid in the starting of the motor. Is also used as the auxiliary winding (second winding) in capacitor-type motors.

Static pressure: The pressure exerted against the inside surfaces of a container or duct. Is sometimes defined as "bursting pressure."

Static water level: The level to which water rises in a well casing when no water is being removed from the well.

Stator, motor: The stationary winding assembly in an electric motor.

Steam: Water in a vapor state.

Strainer: A device, such as a screen or filter, used to retain any solid material while liquid or vapor flows through.

Stratification of air: Condition in which the air movement is less than 50 ft./min.

Subcooling: The reduction of the temperature of a liquid below its condensing temperature.

Suction line: Tubing or pipe used to carry refrigerant vapor from the evaporator to the compressor in single function systems or from the reversing valve to the compressor in dual function systems.

Suction pressure: See Back pressure.

Superheat: The heat added to a vapor to raise the sensible temperature of the vapor above its boiling point.

Superheater: A device used to transfer heat energy from a high temperature liquid to a low temperature vapor to raise the temperature of the vapor.

Supply branch: The section of piping used to supply liquid to a branch circuit of the earth loop.

Supply main: The portion of the earth loop system that carries liquid from the unit to the supply header or manifold.

Supply tapping: The opening on the unit to which the supply main is connected.

Supply time, well water: The supply time for a well is considered to be a period of 20 of 24 hours of a day up to 10 days at a time.

Surface water disposal: Disposing of return water into an open area such as a lake, pond, or stream.

Sweating: This term has two definitions in air conditioning or heat pump work: (1) Formation of moisture on the outside of cold pipes or ducts; and (2) joining of two metals by the adhesive action of a third metal; see Soldering.

Sweet water: A term used to describe water fit for human consumption.

Switch, flow: A control installed in the liquid circuit. Used to prevent operation of the unit when the liquid flow is not up to the required rate.

Switch, season selector: The three-position selector switch on the area thermostat that selects the type of operation of the system (cooling, off, heating). Sometimes a fourth position (auto) is included.

Synthetic rubber, neoprene: Soft resilient material made from petroleum-based chemical compounds.

Tap-a-line: A device used to puncture or tap a tube where there is no service valve in order to measure pressures in the line. Sometimes called a needle valve.

Tee: A pipe fitting designed to connect three pipes together. Two of the connections are in line (main connections) and one is at a 90° angle (branch connection).

Temperature: The degree of heat energy in a material as measured by a thermometer.

Temperature/humidity index: The temperature and relative humidity of a sample of air as compared to standard conditions.

Terminal: A point of connection for electrical conductors.

Test dial, gas supply meter: One of the recording dials that indicates a quantity of gas per revolution of the dial.

Thermal balance: The heat absorbing ability and the heat rejecting ability of the refrigeration system are stabilized and pressures and temperatures are constant.

Thermal cut-out: An overcurrent protection device that contains a heater element that affects a bimetal element designed to open a circuit in the event of electrical current flow above the rated amount of the device.

Thermal conductivity, earth: The rate at which heat energy flows through an earth material. Expressed in Btu per sq. ft. of the material surface times the temperature difference per thickness in feet of the material.

Thermistor: An electrical device that changes electrical resistance with a change in temperature of the device.

Thermocouple: A device that develops an electrical pressure (voltage) due to temperature differences across hot and cold junctions of two dissimilar metals welded together.

Thermocouple thermometer: Electrical instrument that uses the amount of voltage generated by a thermocouple to indicate temperature.

Thermo-mastic: A type of material that has a high heat absorbing ability. Used to prevent overheating of a device when making solder-type connections.

Thermometer: An instrument used to measure the temperature of materials.

Thermometer, digital: A thermometer that uses solid state circuitry and direct reading digital numbers for temperature read out. Easier to read and more accurate than ordinary column type.

Thermometer, stack temperature: A thermometer with a range of 200°F to 1,000°F used to measure the temperature of the flue products from fossil fuel heating units.

Thermostat: A control device that responds to surrounding air temperatures.

Thermostat, outdoor ambient: A control used to limit the amount of auxiliary electric heat according to the outdoor ambient temperature to reduce electrical surge and cost.

Thermostat, termination: A thermostat mounted on the outdoor coil that interrupts the defrost mode when the temperature of the coil reaches the cut-out set point of the control.

Thermostatic control: A device that controls the operation of equipment according to the temperature of the air surrounding the control.

Thermostatic expansion valve: A pressure reducing device that varies the refrigerant flow into the evaporator to maintain a constant superheat in the vapor off the coil. The pressure in the coil connected under the diaphragm of the valve and the temperature of the feeler bulb of the valve fastened to the outlet of the coil are the controlling factors.

Timers: Mechanism used to control the time cycling of an electrical circuit.

Timer, defrost: A timer that operates at the same time as the refrigeration system. After a set period of operating time, the timer trips to initiate a defrost operation if the termination thermostat is closed.

Ton of refrigeration: Refrigerating effect is equal to the melting of one ton (2,000 pounds) of ice in 24 hours: 288,000 Btu/24 hrs. = 12,000 Btu/hr.

Total pressure: The sum of the static pressure and the velocity pressure at the point of measurement.

Tracer system: A steel wire that is buried 18 in. in the earth directly above the ground loop. By means of a portable metal detector, the location of the ground loop can be determined. For vertical loops, steel rods are inserted upright above the loops.

Transformer: An electromagnetic device with two or more coils of wire linked by magnetic lines of force used to increase or decrease voltage in direct proportion between the number of turns of wire in each coil. The watts of power out is equal to the watts of power in minus the small loss in efficiency.

Trombone: The portion of a heat pump refrigeration system consisting of the pressure reducing device and check valve in parallel.

Tube-in-a-tube (co-axial): A heat exchanger constructed of a tube inside a tube sealed off from each other. Usually liquid through the inner tube and refrigerant through the outer tube.

Tube-in-a-tube-in-a-tube (double co-axial): Three tubes, where one is inside a second, which is inside the third. The innermost tube carries refrigerant and the outer tube carries domestic water. The space between the first and second tube is open to atmosphere. This is to prevent contamination of the water in case of a leak in the refrigerant tube that would be vented to atmosphere. This tube construction is required in domestic water heating applications.

Tubing, precharged: A refrigerant line that is fitted with quick couple-type sealed fittings on each end. The line has been factory evacuated and filled with refrigerant the same as is used in the system.

Two well reversible system: A system using two submersible pumps and a supply and return line in each well to supply a screen flushing means.

Unconsolidated formation: A mixture of loose or granulated material—such as sand, gravel, soft clay, soil, etc.—that contains large quantities of easily obtainable water.

Unit foundation: The horizontal surface on which the unit rests.

Vacuum: Reduction in pressure below atmospheric.

Vacuum pump: A high efficiency vapor pump used for creating deep vacuum in refrigeration systems for testing and/or drying purposes.

Valve: A device for controlling fluid flow.

Valve, check: A valve that will permit fluid flow in only one direction. Sometimes called a one-way valve.

Valve, check, spring loaded: A valve at the end of the return pipe in the well to maintain a minimum pressure in the system when the supply pump shuts down or the unit cuts off.

Valve, pilot: A small capacity pressure control valve used to control the action of a larger capacity control valve.

Valve, reversing: A valve used to change the direction of refrigerant flow in a heat pump system. Because there are four pipe connections, it is also called a four-way valve.

Valve, service: A device used by service technicians to connect pressure gauges into the refrigeration system.

Valve, slide: The slide valve portion of the reversing valve that shifts the refrigerant flow.

Valve, solenoid: A flow control valve controlled by an electromagnetic coil actuating a plunger off a seat.

Valve, T. X., bi-flow: A thermostatic expansion valve that is designed to provide pressure reduction and refrigerant flow control in either direction.

Valve, water: Valves used to control the flow of water. May be electrically operated with electromagnetic coils or pressure operated by a pressure bellows.

Vapor: A fluid in gaseous form.

Vapor barrier: A thin sheet of plastic or metal used to prevent moisture penetration into insulation.

Vapor line: Found only in dual action heat pumps. It is the suction line in the cooling mode and the hot gas line in the heating mode.

Vapor, saturated: A vapor whose temperature has been reduced to the point of condensation but condensation has not started.

Velocimeter: An instrument used to measure air velocities on a direct reading air speed indicating dial.

Ventilation: The introduction of outdoor air into a building by mechanical (active) means.

Vibratory plow: A piece of equipment used to bury ground loop pipes by wedging a path through the earth by means of a vibrating blade, followed by a pipe inserted into the slit made in the earth.

Viscosity: A term used to describe the resistance to flow of liquids.

Voltmeter: An instrument used to measure electrical pressure (voltage) in an electrical circuit.

Volume, specific: The volume of a substance per unit mass at a given temperature. The opposite of specific density. Vapor is measured in cubic feet per pound. Liquid is measured in cubic feet per gallon or pound.

Water-cooled condenser: A heat exchanger that uses water to remove the heat from the high temperature compressor discharge vapor.

Water-cooled condensing unit: A condensing unit (high side) that is cooled by the use of water.

Water, domestic: Pure water intended for human consumption.

Water hammer: The excessive vibration caused by heavy pressure buildup against the shut-off means when rapid cut-off of liquid flow occurs.

Water, return: Water off the liquid source heat pump that is returned to the source of supply.

Water source, buried closed loop: Water is circulated through a closed loop pipe system to use heat transfer between the loop and the earth as a heat energy source or sink.

Water source, reservoir: A body of water, such as a large lake or stream, is used as a heat energy source or sink.

Water table: The top surface of the water in the water bearing unconsolidated formation. The top surface of the aquifer.

Well casing: A perforated, noncorrosive pipe (PVC or steel) inserted into a well to keep the walls from collapsing. Allows both the insertion of a submersible pump and free flow of water into the well.

Well casing screen: A section of well casing made of stainless steel screen located below the pumping level of the well to allow water entry into the casing.

Well water source, closed loop: A system that removes water from the earth by means of a drilled well and returns it to the earth by means of a separate drilled well.

Well water source, open loop: A system that removes water from the earth by means of a drilled or bored well and returns the water to the earth through a separate disposal system.

Wet bulb: A thermometer that uses a wetted sac on the testing tip or bulb to measure the evaporation rate of the air sample. Evaporation of the moisture lowers the temperature of the wet bulb thermometer as compared to the dry bulb thermometer. The two readings and a psychrometric chart are used to determine the characteristics of the air.

Wetted hole: The portion of the well below the static water level in the well.

Wind effect: The increase in evaporation rate due to air travel over a water surface.

INDEX

Note: Page numbers in **bold type** reference non-text materials.